5th International Symposium on Material, Mechatronics and Energy (ISMME 2018)

IOP Conference Series: Materials Science and Engineering
Volume 619

Gowa, Indonesia
6 - 8 November 2018

Editors:

Rafiuddin Syam
Keigo Watanabe
Adi Maimun bin Abdul Malik
Azwar Hayat
Andi Amijoyo Mochtar
Muh. Zubair Alie Muis

ISBN: 978-1-5108-9826-4
ISSN: 1757-8981

Printed from e-media with permission by:

Curran Associates, Inc.
57 Morehouse Lane
Red Hook, NY 12571

proceedings.com

Some format issues inherent in the e-media version may also appear in this print version.

Institute of Physics
Dirac House, Temple Back
Bristol BS1 6BE UK

Phone: 44 1 17 929 7481
Fax: 44 1 17 920 0979

techtracking@iop.org

Additional copies of this publication are available from:

Curran Associates, Inc.
57 Morehouse Lane
Red Hook, NY 12571 USA
Phone: 845-758-0400
Fax: 845-758-2633
Email: curran@proceedings.com
Web: www.proceedings.com

TABLE OF CONTENTS

TRANSACTIONS OF REAL-TIME MEDICAL SERVICES USING SMART CARD TO SUPPORT THE ROLE OF SMART CAMPUS 1
Dewiani, Elyas Palantei, Misa Amira Azwar and Sitti Wahyuni Artini

EFFECT OF CURRENT AND WIRE SPEED ON SURFACE ROUGHNESS IN THE MANUFACTURING OF STRAIGHT GEAR USING WIRE-CUT EDM PROCESS 10
Rusdi Nur, M. Muas, Apollo and Syah Risal

MODELING OF VERTICAL DYNAMIC VIBRATION CHARACTERISTICS ON VEHICLES SUSPENSION SYSTEM 16
Simon Ka'ka, Syukri Himran, Ilyas Renreng and Onny Sutresman

DEVELOPMENT OF QUADCOPTER FOR TRACKING OBJECT USING IMAGE PROCESSING 24
Muhammad Ifan Nugraha, Ardiansyah Tri Utomo, Akhmad Taufik, Remigius Tandioga and Rafiuddin Syam

THE EFFECT OF POLE DENSITY ON PERMEABLE GROIN DUE TO CONTROLLING LONGSHORE CURRENT 33
Hasdinar Umar, Sabaruddin Rahman, Chairul Paotonan and A.Y. Baeda

AN ANALYSIS OF THE USE OF LOCAL BAMBOO AS AN ALTERNATIVE ENERGY SOURCE 39
Sallolo Suluh, Petrus Sampelawang and Normalia Sirande

IMPACTS OF GRID FREQUENCY VARIATION ON DYNAMIC PERFORMANCE OF DFIG BASED WIND TURBINE 43
A. M. Shiddiq Yunus

PROGRESSIVE COLLAPSE BEHAVIOUR OF VLCC UNDER LONGITUDINAL BENDING 51
Tsumiratin Rizkiani, Muhammad Iqra Ramadhan and Muhammad Zubair Muis Alie

PROTOTYPE DEVELOPMENT OF PORTABLE HEAD FEED HARVESTING ROBOT 57
Muh Rifai Anugrah, Jeky, Abdul Kadir Muhammad, Akhmad Taufik and Rafiuddin Syam

ULTIMATE STRENGTH INVESTIGATION OF RO-RO SHIP 63
Muhammad Zubair Muis Alie, Juswan, Taufiqur Rachman and Chairul Paotonan

SODIUM HYDROXIDE AND POTASSIUM PERMANGANATE TREATMENT ON MECHANICAL PROPERTIES OF COCONUT FIBERS 69
Muhammad Arsyad

THERMAL PERFORMANCE OF SHALLOW SPIRAL-TUBE GROUND HEAT EXCHANGER FOR GROUND-SOURCE COOLING SYSTEM 76
Jalaluddin, Akio Miyara, Rustan Tarakka, Andi Amijoyo Mochtar and IR Muhammad Anis

ENVIROMENTAL GEOCHEMISTRY OF BAWAKARAENG MOUNTAIN SOIL: IMPLICATION FOR ANTHROPOGENIC IMPACT GOWA SOUTH SULAWESI INDONESIA 82
Adi Tonggiroh and Rafiuddin Syam

COMBUSTION REACTIVITY OF LOW RANK COAL BY THE MIXTURE OF CANDLENUTS SHELL 94
Daud Patabang, Effendy Arif, Jalaluddin and Nasruddin Azis

THE INFLUENCE OF TOPOGRAPHY TO THE DISTRIBUTION OF NI-LATERITE DEPOSITS OF MANGGURUH AREA, SEBUKU ISLAND, SOUTH KALIMANTAN 105
U R Irfan, I Alimuddin and I B Pasalli

CALCAREOUS NANNOFOSSIL ASSEMBLAGES OF TONASA FORMATION PALAKKA AREA, SOUTH SULAWESI: IMPLICATION OF PALEOENVIRONMENTAL APPLICATION 115
M Farida, A Jaya and T Sato

BENEFICIATION OF LATERITIC IRON ORE FROM MALILI AREA, SOUTH SULAWESI, INDONESIA USING MAGNETIC SEPARATOR 122
Sufriadin, S Widodo and M Trianto

ON SOME EXAMPLES OF BEM SOLUTION TO ELASTICITY PROBLEMS OF ISOTROPIC FUNCTIONALLY GRADED MATERIALS 129
S. Hamzah, M. I. Azis and E. Syamsuddin

GARBAGE COMPACTION MECHANISM OF THREE-WHEELS ELECTRICAL MOTORCYCLE 142
Alfaroq, Andi Cahya Adi Putera, Abdul Rahman, Abdul Kadir Muhammad and Rafiuddin Syam

CONTROL DESIGN FOR PREVENTING DENGUE INFECTIONS USING INPUT-OUTPUT LINEARIZATION METHOD 150
Kasbawati, Firman and Waode Ainun Apriyani Amrin

CAR DRIVER DROWSINESS RECOGNITION ANDROID-BASED SYSTEM 164
Indrabayu, Siti Khumaerah Mufti and Intan Sari Areni

NUMERICAL EXAMINATION OF METAMATERIAL CONCENTRATOR DESIGN FOR HIGH EFFICIENT THERMAL ENERGY HARVESTER 176
Zaenab Muslimin, Rapha Nichita Kaikatui, Elyas Palantei, Intan Sari Areni, Syafruddin Syarif, Ansar Suyuti, Indrabayu and Merandy Pala'biran

EVALUATION OF FINITE CONTROL SET - MODEL PREDICTIVE CONTROL FOR DYNAMIC RESOURCES MANAGEMENT IN A SHARED-RESOURCES SYSTEM 187
D Aryani, L Wang and N A Hidayatullah

ANALYSIS OF NATURAL LIGHT DISTRIBUTION IN THE BUILDING 196
Nurul Jamala, Ramli Rahim and Rosady Mulyadi

A CLASS OF VARIABLE COEFFICIENT ELLIPTIC EQUATIONS SOLVED USING BEM 208
N. La Nafie, N. Ilyas, M. I. Azis and A. K. Amir

AUDIAL COMFORT OPTIMALIZATION OF THE LANRAKI BIRINGKANAYA MAKASSAR CHURCH ROOM 221
Muhammad Taufik Ishak, Asniawaty and Dwi Retno Sulistyowati

EXPERIMENTS ON MOTION AND OSCILLATION CONTROLS OF A GANTRY CRANE SYSTEM USING PARALLEL PROPORTIONAL CONTROLLERS 234
A K Muhammad, R Tandioga, S Ka'ka, L Lewi, A Rahman and A Taufik

OPTIMIZATION OF STIRRING PARAMETERS ON THE RHEOCAST MICROSTRUCTURE AND MECHANICAL PROPERTIES OF ALUMINUM ALLOY ADC12 239
Syaharuddin Rasyid, Ilyas Renreng, Effendy Arif, Hairul Arsyad and Muhammad Syahid

PERFORMANCE ANALYSIS OF RACK TYPE SOLAR DRYERS WITH MASS VARIATIONS OF DRIED MATERIAL AND TYPES OF FINS 248
Jamal Jamal, Sri Suwasti and Sukma Abadi

MJO ANOMALIES RELATIONSHIP WITH VOLCANIC ERUPTION IN INDONESIAN ARCHIPELAGO 255
Achmad Yasir Baeda, Dharmawan Abdullah and Chairul Pao'tonan

EXPERIMENTAL STUDY ON THE INFLUENCE OF DRAFT RELATIVE DUE TO WAVE TRANSMISSION ENERGY COEFFICIENT ON HANGING SHEET PILE BREAKWATER 261
Chairul Paotonan, Sabaruddin Rahman, Hasdinar Umar, Taufiqur Rachman, Achmad Yasir Baeda, M Agustinus and Alif Hidayat

IDENTIFICATION OF GROUND–FAULT CURRENT IN VARIOUS TRANSFORMER CONNECTION AND GENERATOR GROUNDING USING SYMMETRICAL COMPONENTS METHOD 269
A.R. Sultan, M.W. Mustafa, M. Saini and A. Gaffar

APPLICATION OF HYBRID SOLAR AND WIND ENERGY GENERATION FOR PADDLE WHEEL AERATOR 278
Musrady Mulyadi and A. M. Shiddiq Yunus

FUEL PRODUCTION AND MATERIALS SYNTHESIS BY IN-LIQUID PLASMA 284
Shinfuku Nomura

PRODUCTION OF CARBON DIOXIDE, METHANE AND NITROUS OXIDE GAS EMISSIONS IN BIODEGRADATION OF WASTE OIL SLUDGE 292
Fahruddin

EXPERIMENTAL STUDY FOR MEASURING DENSITY OF SLAG MELTS BY AN ESTABLISHED ARCHIMEDEAN DOUBLE-BOB METHOD 298
Arman and Hiromichi Takebe

ANALYSIS OF IGNITION DELAY AND BURNING RATE OF DAMMAR RESIN 307
Jamal Jamal

DESIGN AND DEVELOPMENT OF STIRRING AND COOKING MACHINES FOR PALM SUGAR 313
Anwar Mazmur and Syarifuddin

THE DEVELOPMENT MECHANISM OF MICROCONTROLLER-BASED WATER LEVEL AND TEMPERATURE CONTROL 319
S Ka'ka, R Tandioga, Lewi, A K Muhammad, A Rahman and A Taufik

INVESTIGATION OF CREEP BEHAVIOUR OF AL-9ZN-4MG AND AL-9ZN- 4MG-5CU ALLOYS 330
Muhammad Syahid and Bondan T. Sofyan

EXPERIMENTAL STUDY ON MACHINABILITY IN SUB-MILLIMETER RANGE OF THE ALUMINUM ALLOY USING FACE TURNING METHOD 335
Hairul Arsyad, Lukmanul Hakim and Nurul Annisa

PERFORMANCE OF CANDLENUT SHELL (ALLEURETUS MOLUCCANA) BASED SUPERCAPACITOR ELECTRODE WITH ACID ELECTROLYTES AND THEIR SALTS 342
M. Zakir, H. Kasim, I. Raya, Y. Lamba, Nuratisah and A. Belen Jorge

PRELIMINARY STUDY ON ACOUSTICAL CHARACTERISTICS OF SAGO MIDRIB COMPOSITE 350
Asniawaty Kusno, Sherly Asriani, Kusno Kamil and Nurul Jamala
ANALYSIS OF DAYLIGHT DISTRIBUTION ON BUILDING HEIGHT AND SPACE DEPTH. 358
Nurul Jamala, Annajma Nurul Wika and Asniawaty Kusno
CONTROL OF A TETHERED QUADROTOR USING A QUATERNION FEEDBACK 367
Ryuta Takeuchi, Keigo Watanabe and Isaku Nagai
THRUST ANALYSIS OF PROPULSION MECHANISMS WITH PECTORAL FINS IN A MANTA ROBOT 376
Tomoya Watanabe, Keigo Watanabe and Isaku Nagai
DEVELOPMENT OF A FOLDABLE PERSONAL MOBILITY VEHICLE WITH ONE DRIVE WHEEL 382
Takanori Sakata, Isaku Nagai and Keigo Watanabe
INTERACTING PROPELLER THRUST FOR NAVIGATING OMNI-DIRECTIONAL REMOTELY OPERATED VEHICLE 390
Rahimuddin, Haryanti Rivai and Hasnawiyah Hasan
CLITORIA TERNATEA FLOWER AS NATURAL DYES FOR DYE-SENSITIZED SOLAR CELLS 400
Azwar Hayat, A. Erwin E. Putra, Novriany Amaliyah and Shyam. S. Pandey
FINITE ELEMENT ANALYSIS ON VIBRATION OF A FLEXIBLE SINGLE-LINK MANIPULATOR MOVED TRANSLATIONALLY 406
D Dermawan, H Abbas, R Syam, Z Djafar and A K Muhammad
COMPUTATIONAL SIMULATION ON FACIAL EXPRESSIONS USING FINITE ELEMENT ANALYSIS 414
Andi Amijoyo Mochtar and Rafiuddin Syam
PERFORMANCE OF A 90° -BENT BLADED VERTICAL AXIS WIND TURBINE MODEL WITH VARIOUS NUMBERS OF BLADES 421
Luther Sule, Andi Amijoyo Mochtar and Jalaluddin
EXPERIMENTAL ANALYSIS OF SMART GREEN HOUSE WITH ROTARY GARDEN 432
Onny S. Sutresman, Rafiuddin Syam, Abdul Hay Muchsin, Zulkifli Djafar, Syahrier Arief, Andi Amijoyo Mochtar, Azwar Hayat, Onny S Sutresman, Amiruddin and Abdul Mufti Radja
KINEMATIC MOTION CONTROL FOR ROBOT MOBILE MANIPULATORS AS FIRE FIGHTERS 440
Rafiuddin Syam, Onny S. Sutresman, Fahruddin, Adi Tonggiroh, Amiruddin and Abdul Mufti Radja
SYNTHESIS AND CHARACTERIZATION OF NI-DOPED ZNO THIN FILMS GROWN BY SOL-GEL SPIN COATING 450
P.L. Gareso, Monika Pakabu, N. Rauf, E. Juarlin and Didik Aryanto
OPTIMIZATION OF ROTATION SPEED PARAMETERS AND NUMBER OF GRINDING WHEELS ON THE QUALITY AND PRODUCTION CAPACITY OF CHICKEN FEED PELLETS 457
Syaharuddin Rasyid, Muas Muchtar and Tri Agus Susanto
CODES DEVELOPMENT OF INDUSTRIAL MECHATRONIC SYSTEM 466
Sri Hasnawati Ulandari, Irdyah Inayah, Zulfhis Shara Firstiawaty, Lewi, Simon Ka'ka and Rafiuddin Syam
NUMERICAL SOLUTIONS TO BVPS GOVERNED BY THE ANISOTROPIC MODIFIED HELMHOLTZ EQUATION FOR TRIGONOMETRICALLY GRADED MEDIA 480
N. La Nafie, M. I. Azis and Fahruddin
BEM SOLUTIONS TO A CLASS OF ELLIPTIC BVPS FOR ANISOTROPIC TRIGONOMETRICALLY GRADED MEDIA 489
St. N. Jabir, M. I. Azis, Z. Djafar and B. Nurwahyu
NUMERICAL SOLUTIONS TO ANISOTROPIC BVPS FOR QUADRATICALLY GRADED MEDIA GOVERNED BY A HELMHOLTZ EQUATION 499
S. Hamzah, M. I. Azis and A. K. Amir
NUMERICAL SOLUTIONS TO ANISOTROPIC FGM BVPS GOVERNED BY THE MODIFIED HELMHOLTZ TYPE EQUATION 508
R. Syam, Fahruddin, M. I. Azis and A. Hayat
Author Index

IOP Conf. Series: Materials Science and Engineering **619** (2019) 011001 doi:10.1088/1757-899X/619/1/011001

The 5th International Symposium on Material, Mechatronics and Energy The 5th ISMME 2018

Editors:
Dr Rafiuddin Syam Hasanuddin University, Indonesia
Professor Keigo Watanabe Okayama University, Japan
Professor Adi Maimun bin Abdul Malik UTM-Malaysia
Azwar Hayat Hasanuddin University, Indonesia
Dr Andi Amijoyo Mochtar Hasanuddin University, Indonesia
Dr Muh. Zubair Alie Muis Hasanuddin University, Indonesia

Engineering Faculty of Hasanuddin University
Graduate School of Hasanuddin University
Research and Community Services Institute of Hasanuddin University

November 2018

IOP Conf. Series: Materials Science and Engineering **619** (2019) 011001 doi:10.1088/1757-899X/619/1/011001

Foreword

First, we would like to thank all researcher who are already send the results of scientific research papers and participated in the 5th International Symposium on Material, Mechatronics and Energy 2018. All papers in this volume has presented at ISMME 2018 by oral presentation. The papers have been peer reviewed through processes administered by the proceedings Editors. Reviews were conducted by expert referees to the professional and scientific standards expected of a proceedings journal published by IOP Publishing.

Our theme is Challenges and Opportunities of Materials Engineering, Mechatronics and Energy towards independence of independent and sustainable technology products. Themes have been given an important role of Indonesian Development of Industrial Manufacture strategic plan, where the Indonesian people are still in desperate need of technology in these areas, material, mechatronics and energy.

Today Issues is still on Industry 4.0, they are five items should be considered:
1. Scalability; The automation principle of Industry 4.0 could help to facilitate improved scalability among companies in the manufacturing sector.
2. Security; One of the foremost concerns about Industry 4.0 among manufacturers is the possibility of mishaps due to glitches in cognitive computing.
3. Control and Visibility; As manufacturing networks globalize, it is crucial to make digital processes visible to all points of a system. When fully implemented, the principles of Industry 4.0 support responsiveness by making information available worldwide within a fraction of a second.
4. Customer Satisfaction; The process will be fully transparent along all stops on the manufacturing chain, from the moment someone places an order or submits a design until the moment when shipments arrive. Industry 4.0 will facilitate co-creation capabilities between manufacturers and related entities on a global scale.
5. Customization; Industry 4.0 could take customization to new levels with the use of commercial 3-D printers, which there are 23,000 of in use worldwide.

We hope many researchers play on such conditions. Finally, thanks to all of my college in Faculty of Engineering Hasanuddin University, Okayama University, Graduate School of Unhas, Research and Community Services Institute of Unhas and Polytechnik State of Ujung Pandang.

Makassar-Gowa, November , 2018
Yours

Dr. Ir. Muhammad Arsyad Thaha, MT
Dean of Engineering Faculty of Hasanuddin University

IOP Conf. Series: Materials Science and Engineering **619** (2019) 011001 doi:10.1088/1757-899X/619/1/011001

Peer Review Statement

All papers published in this volume of Journal of Physics: Conference Series have been peer reviewed through processes administered by the proceedings Editors. Reviews were conducted by expert referees to the professional and scientific standards expected of a proceedings journal published by IOP Publishing.

IOP Conf. Series: Materials Science and Engineering **619** (2019) 011001 doi:10.1088/1757-899X/619/1/011001

Editors

- Rafiuddin Syam, PhD Hasanuddin University, Indonesia
  Email: rafiuddin@unhas.ac.id

- Professor Keigo Watanabe Okayama University, Japan
  Email: watanabe@sys.okayama-u.ac.jp

- Professor Adi Maimun bin Abdul Malik UTM-Malaysia
  Email: adi@utm.my

- Dr Azwar Hayat Hasanuddin University, Indonesia
  Email: azwar.hayat@gmail.com

- Dr Andi Amijoyo Mochtar Hasanuddin University, Indonesia
  Email: andijoyo@yahoo.com

- Dr Muh. Zubair Alie Muis Hasanuddin University, Indonesia
  Email: bair_offshore@yahoo.co.id

IOP Conf. Series: Materials Science and Engineering **619** (2019) 011001 doi:10.1088/1757-899X/619/1/011001

Steering Committee

Rafiuddin Syam, PhD - Unhas- Indonesia (Chairman)
Prof. Keigo Watanabe-Okayama University-Japan
Prof. Shinfuku Nomura- Ehime University-Japan
Prof. Dr. Adi Maimun bin Abdul Malik-UTM-Malaysia
Prof. Satoru Shibata-Ehime University-Japan
Prof. Mitsuhiro Okayasu-Okayama University-Japan
Prof. Mitsu Okamura-Ehime University-Japan
Prof. Kiyotaka Izumi—Saga University-Japan
Prof. Jackrit Suthakorn, PhD- Mahidol University, Thailand
Prof. Sherwin Guirnaldo, PhD – Mindanau State University-Philipine
Prof. Lanka Udawatta, PhD – Moratuwa University - Srilanka
Prof. Samy F. M. Assal, PhD, Alexandria University- Egypt
Prof. Bondan T. Sofyan-UI-Indonesia
Dr. Indrawanto-ITB-Bandung-Indonesia
Heru Santoso Budi Raharjo-Ir.M.Eng.Dr-UGM-Yogyakarta-Indonesia
Dr. Gesang Nugroho, ST, MT-UGM-Yogyakarta-Indonesia
Dr. Kusmono–UGM-Yogyakarta-Indonesia
Dr.Ir. Yudi Satria Gondokaryono ITB-Bandung-Indonesia
Dr. Ir. Samiadji Herdjunanto, M. Sc – UGM-Yogyakarta -Indonesia
Ir. Oyas Wahyunggoro, M.T., Ph.D–UGM-Yogyakarta-Indonesia
Hendro Nurhadi, Dipl. Ing., Ph.D–ITS-Surabaya-Indonesia
Dr. Syaeful Bakhri -BATAN-Indonesia
Dr. Abdul Muis, UI-Jakarta-Indonesia
Prof. Dr.-Ing Nandy Setiadi Djaya Putra-UI-Indonesia
Prof. Dr.H.Hammada Abbas – Hasanuddin University-Indonesia
Prof. Effendi Arief– Hasanuddin University-Indonesia
Prof.Dr. Syamsul Arifin– Hasanuddin University-Indonesia
Prof. Dr. Ir. Salama Manjang— Hasanuddin University-Indonesia
Dr. Zulfajri Basri Hasanuddin – Hasanuddin University-Indonesia
Dr.-Ing Wahyu H Piarah— Hasanuddin University-Indonesia
Dr. Johannes Leonard – Hasanuddin University-Indonesia
Dr. Ir. Arsyad Taha, MT – Hasanuddin University-Indonesia

IOP Conf. Series: Materials Science and Engineering **619** (2019) 011001 doi:10.1088/1757-899X/619/1/011001

Organized Committee

Azwar Hayat, ST, M.Sc., Ph.D
Dr. Kimiko Matonaka, MEng
Dr. Eng Armin Lawi, MT
Dr. Jalaluddin, ST,MT
Dr. Andi Amijoyo, MT
Dr. Muh. Syahid ST,MT
Dr. Faizal Arya Samman
Rahimuddin, ST,MT, PhD
Dr. Eng Armin Lawi, MT
Dr. Abdul Kadir Muhammad
Dr. Meuthia Farida, ST, MT
Muhammad Zubair Muis Alie, ST,MT, PhD
Dr. Wiwik Wahida Usman, MT
Dr. Asniawati Kusno, MT
Dr. Chaerul Paotonan
Dr. Adi Tonggiroh, MT
Dr. H. Ilyas Renreng, MT
Dr. Abd. Mufti Radja
Dr. Muh. Syahid, ST, MT
Dr. Haerul Arsyad, ST, MT
Dr. Indrabayu
Dr. Intan Sari Areni, MT
Dr. Andi Haris Muhammad
Dr. Ir. Ulfa Ria Irfan, MT
Dr. Dewiani, MT
Dr. Hasdinar Umar, ST,MT
Dr. Sufriadin, ST, MT
Dr. Nurul Jamala
Arham Hamid, SE , MSi
Amiruddin, ST, MT

Peer review statement

All papers published in this volume of *IOP Conference Series: Materials Science and Engineering* have been peer reviewed through processes administered by the proceedings Editors. Reviews were conducted by expert referees to the professional and scientific standards expected of a proceedings journal published by IOP Publishing.

Transactions of Real-Time Medical Services Using Smart Card to Support the Role of Smart Campus

Dewiani[1], Elyas Palantei[2] , Misa Amira Azwar[3], Sitti Wahyuni Artini[4]
[1,2,3,4]Electrical Engineering Department, University of Hasanuddin, Indonesia

[1]dewiani@unhas.ac.id [2]e_palantei@yahoo.com [3]misaazwar13@gmail.com
[4]artiniwa@gmail.com

Abstract. To improve health service, we can do by using new technology. One example of technology development in health is telemedicine. In this case, the patient data management integrated with electronic checks payments with one card. For that, in this final project is made using smart card system integrated with Unhas Clinic Website. Smart cards that are used by researchers are based on the Consortium Smart Card Indonesia iRFC v 1.0, while the PHP programming language is used to design the website. The creation of MySQL databases and Apache web server responsible for the request-response HTTP are combined in an application called XAMPP. This forms a system that uses a smart card which can save patients' medical records and perform electronic payment transactions (e-payment). There are 10 people who have been registered as patients on the Website System of Unhas Clinic where 2 of them are insurance patients and have successfully completed the examination. In addition, the examination process at the Website of Unhas Clinic can be done on short distance (0 Km) and long distance (20,3 Km) by using Wireless Fidelity (Wi-fi) network operating system (uplink: 2,94 Mbps; downlink: 20.37 Mbps) and cellular networks such as Provider 1 and Provider 2 (average uplink: 7.1 Mbps; downlink: 5.15 Mbps) with fast connectivity.

Keywords. Smart card, Website, Medical Record, e-Payment.

1. Introduction

According to the World Health Organization (WHO), health is defined as a state of physical, mental, and social well-being and not just the absence of disease or weakness. Excellent health can support humans to engage in productive activities. To provide early detection of the disease, one must undergo a medical examination at the hospital. However, some patients have difficulty. For example, the hospital and the patient's residence are far enough away.

Limitations of health services can be overcome with technology, one of which is Information and Communication Technology (ICT) based on wireless technology, and has become a necessity in various aspects of life. With this ICT, information can be presented in real time, fast and certainly very accessible to interested parties [1]. In the field of health, ICT is needed to monitor the health condition of patients so that the combination of ICT and medical science is what creates biomedical techniques.

With the development of increasingly sophisticated biomedical techniques today, many studies design long-term treatment. For example, medical personnel can monitor vital signs of a patient's health condition without requiring direct interaction in the same room so that medical personnel problems can be minimized. Indoor interactions differ from medical personnel and patients known as telemedicine.

In the implementation of science and technology development, especially in healthcare overview with advanced vision and prioritization of convenience, many researchers develop Health Service Applications based on smart cards [2].

Research [3]; [4]; and [5] have developed the process of recording medical data using Java Card cards and access cards Unified Medical Record Access and Analytics (UMRAA) so that clinics or hospitals no longer need to store patient data in paper form. The methods used are different, some use the Integrated Electronic System Recorder System (IEHRS) as a patient medical record storage database and some use the US Health Care System database associated with the UMRAA card. In this regard, it would be more sophisticated if a smart card containing data or information of patients who can be input online can also make electronic payment transactions so that the work of hospital staff or clinics can be easier in providing services to patients. Therefore, this research designs an integrated web-based payment service system using one smart card, so students can check their health at campus clinic and can use it in canteen and library. In other words a smart card can support the role of smart campus.

2. System design

As result of development of information technology, various approaches have made to improve health services by integrating both telemedicine and electronic patient data management. As for the benefits to be obtained from this system are if in an emergency, the patient does not need to come to the hospital for treatment, medical record is no longer written manually, and patients can interact online with the doctor. In this regard, the main concentrations of this research are: first, the telemedical system that has been designed consists a website which is a storage of patient medical records database, various types of medical examination and communication links between doctors and patients. Second, in facilitating patients to seek treatment, the web server is integrated with a smart card that is useful for storing patient data and facilitate in making payments electronically by top up the card balance first. Third, the mechanism of long-distance communication relationship between doctor and patient is implemented through chatting mode. This system is systematically designed as shown below:

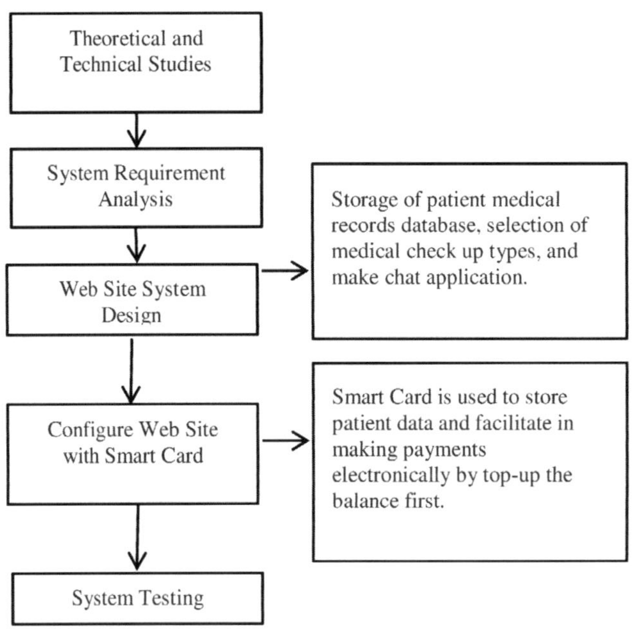

Figure 1. Stages of Medical Service Transaction System Design

The initial stage of the design process begins with a theoretical and practical review to be tailored to the needs of the technical specifications of the desired Unhas Clinic Website System. In order to meet the first stage well, more intensive and extensive literature review activities should be undertaken to obtain the formulation of the Medical Service Transactions System that are different from previous studies and perform more robust. The next step is to analyze the needs of Medical Service Transaction System. At this stage the researchers perform analysis of the system, what are the needs of the system to be designed, such as software, hardware, or operating systems and so on. Based on the results of the two stages, then the next process information system design can be more effective and accuratThe next stage is the design of the Website System where in the website the patient will conduct health checked transactions such as the selection page of the types of examination. There are types of examinations that patients can choose such as body temperature, blood tension, general blood check-up, urine, headache, stomachache, asthma, cough, vomiting, diarrhea, chest pain, allergies, and emergency checks. In this type of emergency check, it can be selected if the patient is in an emergency condition that requires immediate relief. Patients can pay for medical expenses after they are handled by medical staffs or doctors. In addition, it can store patient medical records in database that can only be filled by admins and doctors and visible for the patient but cannot be changed by the patient. Furthermore, to facilitate the patient to communicate with the doctor then the researchers made a chat application so that patients can consult a doctor easily without having to go to the clinic especially if there is an emergency situation. This stage is very important to establish the desired medical service transaction system by writing the code of the program (coding) and system testing. In this research, program code is built using PHP programming language with MySQL database creation and Apache web server which is responsible for HTTP request-response where to make it easier for researchers to use XAMPP application. To link the clinic website that has been created, then it will be integrated with the smart card in accordance with the Consortium of Smart Card Indonesia Version 1.0. The use of the Smart Card itself is to store patient data and facilitate in making payments electronically (by top-up the card balance, first). While the functional testing of medical service transaction system conducted intensively to determine the overall system performance whether the resulting output is in accordance with the input data. In general, a schematic diagram of a medical ser vice transaction system that has been designed can be seen on Figure 2.

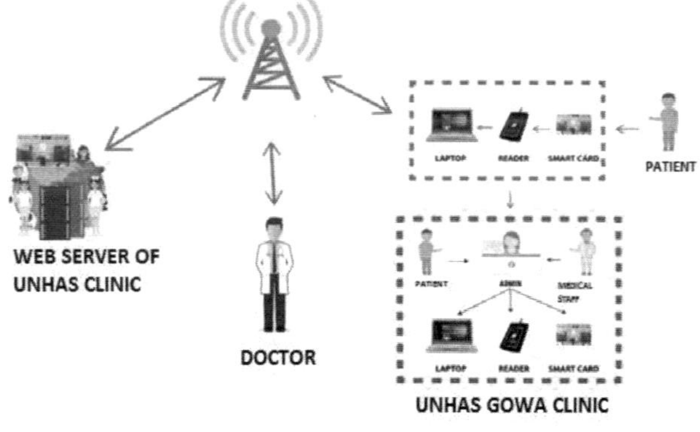

Figure 2. Overview of Medical Service Transaction System

3. Trial test of medical transaction system

Examination system at Unhas Clinic that located on Gowa Engineering Campus can be seen in figure 2. When the patients come to check their health, they must bring smart card to the clinic. The Smart Card is inserted into the reader to access patient information, perform the types of check-ups and to facilitate health payment transactions.

After the information about the patient has been read, then the patient is assisted by the medical staff to check or choose what examinations are desired, then the patient pays the examinations using the smart card. If the patient does not have sufficient balance to payment transaction, then the patient must top up the smart card balance first by paying cash to the medical staff, then officers will top up through the system.

When the payment process is complete, then the patient will be checked by a medical staff at the clinic according to selected examination before. Then the medical staff will fill the examination parameters and examination conclusions are inputted through the Unhas Clinic Website System so that doctors that deal with the disease can see the results of examination from the medical staff and provide diagnosis even though the doctor is not in the Unhas Clinic. The medical staff or admin can print all patient examination results and receipt of payment after receiving the diagnosis result from a specialist doctor, then the examinations process is done. The flowcharts of patient examination flow system at Unhas Gowa Clinic can be seen below.

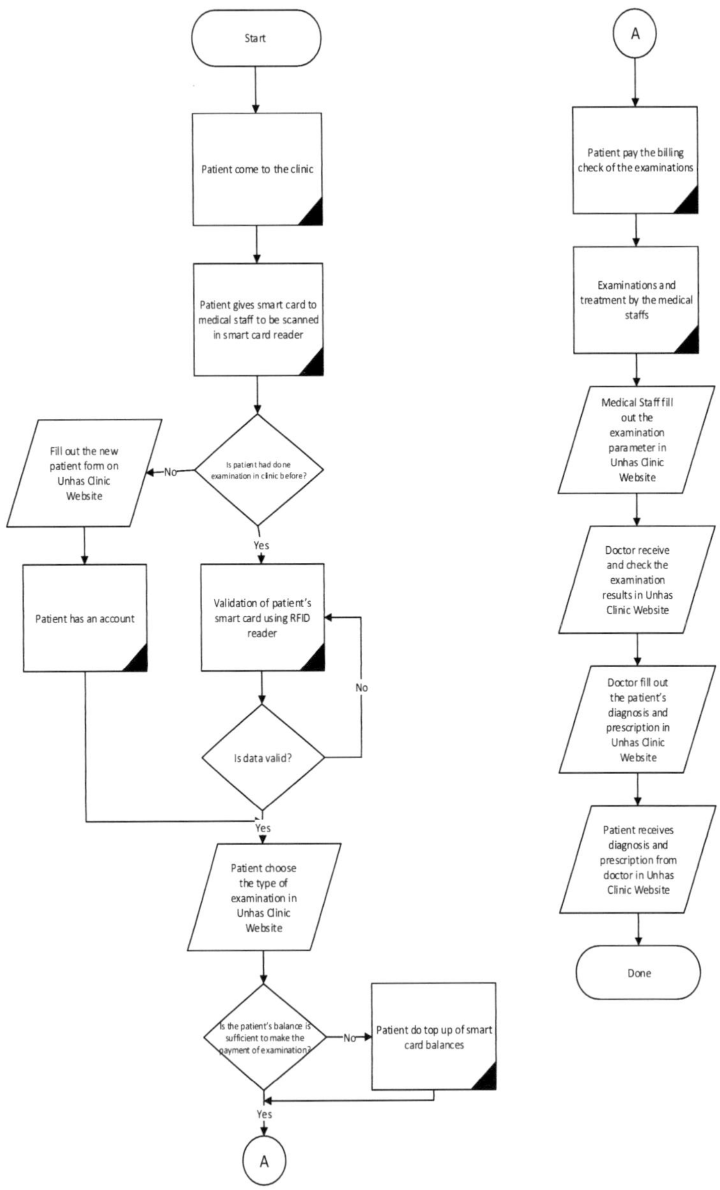

Figure 3. Flowchart of System Flow

Figure 4. The Main Page of Unhas Clinic Website

Figure 4 shows the front page of the Unhas Clinic website that can be accessed on the web browser by typing the web page www.klinik-unhas.com. There are 10 patients have been registered on the Unhas Clinic website where the example of patients' medical records is shown in figure 5. The web page view of the medical record website is intended to see the patient's examinations. In the web page view, it contains the check code, name, check-up date, status, and detail. Status column serves to see the status of the examination of the patient whether it is paid off, has not been paid or is finished. While the detail column is to continue the examination.

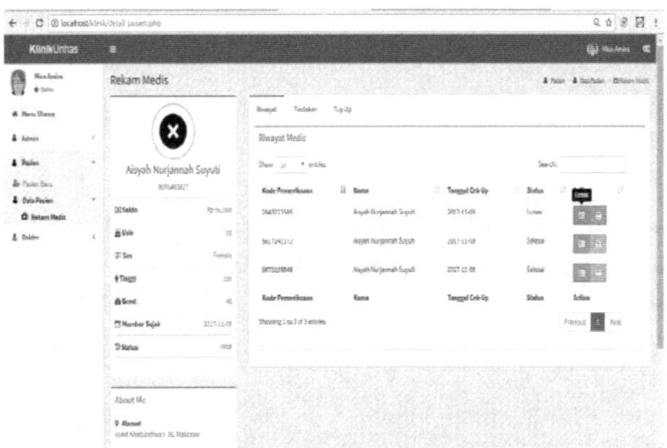

Figure 5. Patient Medical Record Page

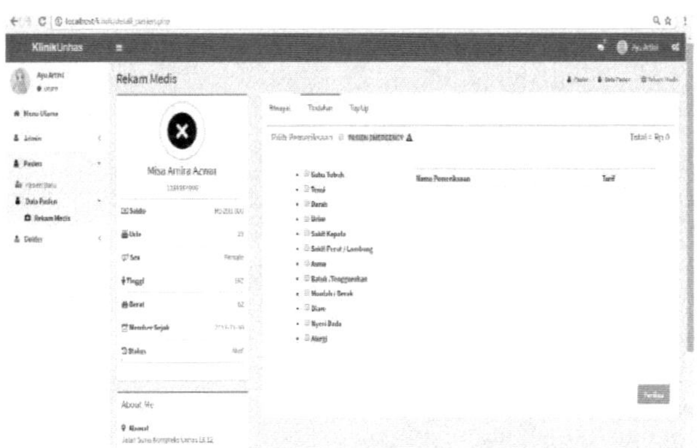

Figure 6. The Types of Examinations Page

Figure 6 shows the action page. There are types of examinations that patients can choose such as body temperature, blood pressure, general blood check-up, urine examination, headache, stomachache, asthma, cough, sore throat, vomiting, diarrhea, chest pain, allergies, and emergency checks. In this type of emergency check, it can be selected if the patient is in an emergency situation requiring immediate relief. And the patient can pay for medical treatment after they handled by medical staffs or doctor.

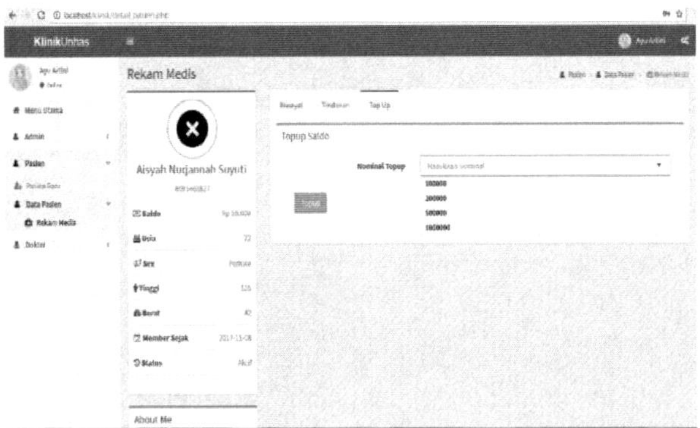

Figure 7. Top-Up Page

Figure 7 shows a Top Up page that is ready to add balance to the card as desired by the smart card holder. For now, if the patient wants to top up the balance of the smart card then they must pa y using cash to the medical staff or admin at the Unhas Gowa Clinic. After making a payment it will open an Invoice page like Figure 8 which is a Payment Check of the type of examination that has been done. This invoice can be printing out or not, depend on the patient needs.

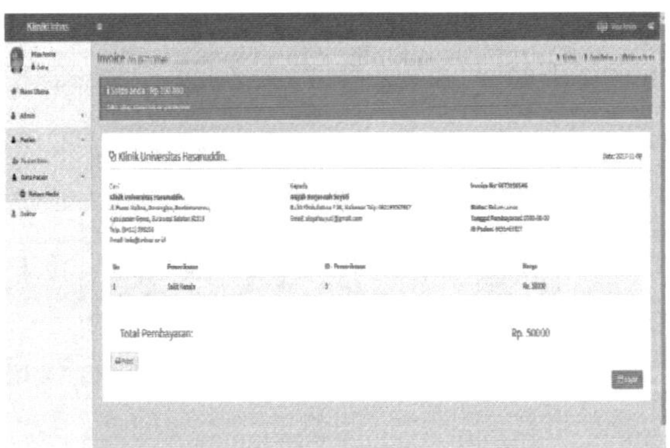

Figure 8. Transaction Invoice Page

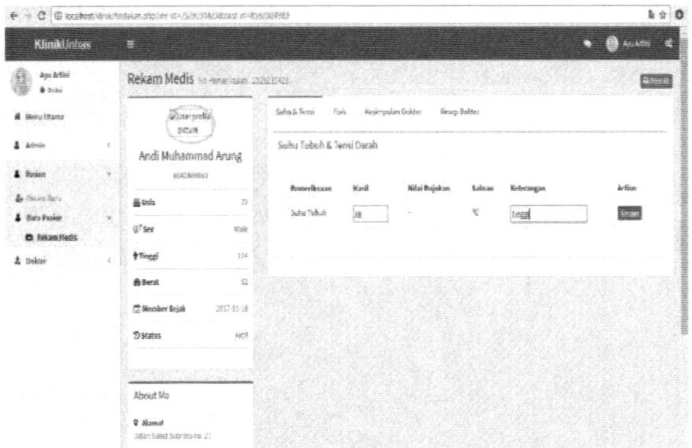

Figure 9. Examinations Parameter and Checkup Results on Admin Side

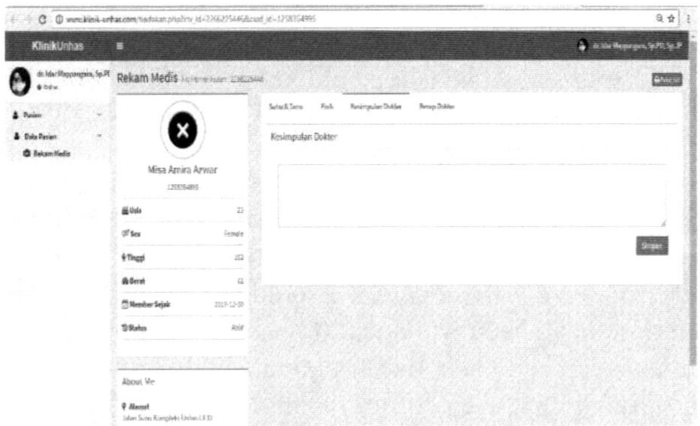

Figure 10. Examination Parameters and Examination Results on the Doctor Side

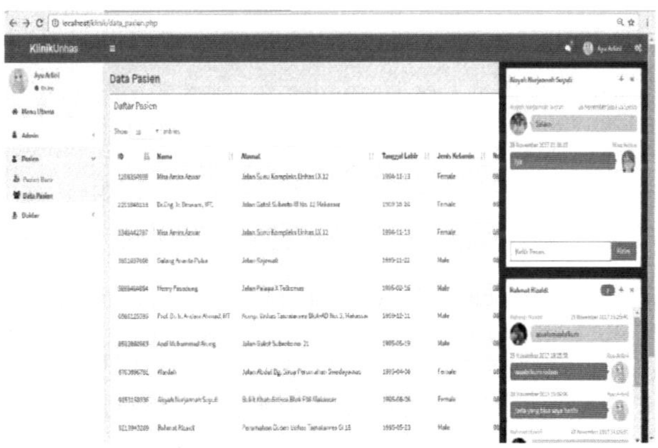

Figure 11. Chat Feature on Admin or Medical Staff Side

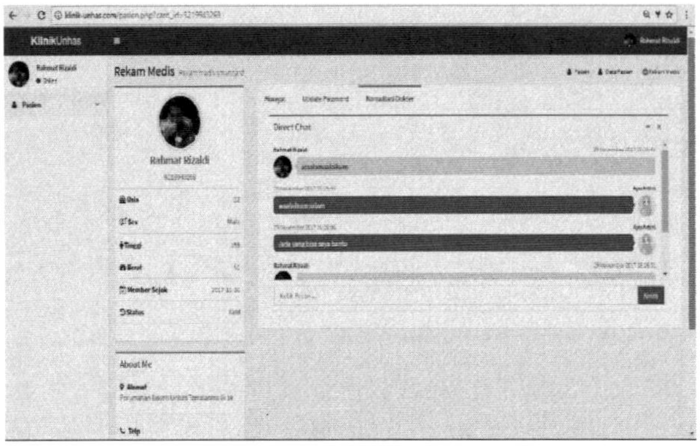

Figure 12. Chat Feature on Patient Side

Figure 9 is the display of parameters and examination results that have been filled by medical staffs who are in Unhas Gowa Clinic. This page displays several options of medical check-up such as body temperature and blood pressure, general blood check-up, urine examination, physical examination, conclusions of the examinations and prescriptions which aims to fill out the examination results along with its parameters. While for Figure 10, it contains a conclusion column where the doctor will fill out the diagnosis result after reading the examinations result from Unhas Gowa Clinic and fill the doctor's prescription column if patient needs medicine for the illness. To fill the conclusion and prescription, all the doctors should be careful because after the results of examinations and prescriptions have been typed and stored it cannot be changed again.
Figure 11 and 12 illustrates consultations between patients and medical staffs or admins who are at the Unhas Gowa Clinic through chat. In addition, it also works if the patient is in an emergency situation in which case the patient is unable to perform a medical examination at the clinic, then can send an emergency message to this chat page.

4. Conclusion

By integrating smart card with Unhas Clinic Website it can facilitate the patient to do the examinations without having to queue for waiting for medical record to be found which mostly still written using paper because in smart card there is patient information and can make payment transaction easily. In addition, health monitoring can be done by looking at the parameters of patient examination results that its directly inputted by the admins or medical staffs in the Unhas Clinic Website System. Doctors can also provide diagnosis and prescribe medication through the system according to the parameters and examination results of the patient. Consultation between the medical staffs or the admins in the clinic with the patient can be done through the chat feature of the Unhas Clinic so if the patient has an emergency that is not possible to visit the clinic, they can send emergency messages to this chat feature.

5. References

[1] Santi, "Database System Optimization On Mobile Computing In The Interactive Telemedical Network" *Thesis*, Universitas Hasanuddin Makassar, 2014.
[2] V. Mustafa Sarinanto, Afrias Sarotama, "Development of Smart Card Program at Hospital", *Proceedings Komputer and Sistem Inteligen (KOMMIT)*, pp. 78-86, 2002
[3] Riri Fitri Sari, Donny, "Smart Health Application Implementation on UI Javacard based Smart Card," *Asian Journal of Information Technology*, 7(3): 117-125, 2008, ISSN:1682-3915
[4] V. R. Kumar, S. Dayanidhi, and K. Vignesh, "Health Care Recording System Using Smart Card Technology," *IJARSET*, vol. 3, no. 3, pp. 1697–1699, 2016.
[5] S. Hussain, T. Ariyachandra, and M. Frolick, "Patient Healthcare Smart Card System : A Unified

Medical Record for Access and Analytics," *Proc. Conf. Inf. Syst. Appl. Res.*, pp. 1–14, 2016.
[6] B. T. Joko, "Benefits of Telemedicine for Society," 2016.
[7] A. Pusparini, "Comparison Of Wireless And Satellite Transmission Media," *Final Project*, no. 9061002004, 2009.
[8] E. Nurmiati, U. Islam, N. Syarif, and H. Jakarta, "Analysis and Design Web Server Handphone," *Jurnal Sistem Informasi*, vol. 5, no. 2, pp. 1–17, 2012.
[9] D. Setiawan, R. Setiawan, R. Karunia, and I. W. S. Wicaksana, "Comparing Web Browser Performance," *Web Brows.*, pp. 1–6, 2009.
[10] A. Solichin, "Web Programming with PHP and MySQL," Jakarta, 2005.
[11] R. Sovia and J. Febio, "Building E-Library Applications Using Html, Php Script, And Mysql Database," *Jurnal Ilmiah Sistem Informasi, teknologi dan Sistem Komputer*, vol. 6, no. 2, pp. 86–101, 2017.
[12] M. S. Al-ani, "Billing System Design Based on Internet Environment," *International Journal of Advanced Computer Science and Applications*, vol. 3, no. 9, pp. 224–230, 2012.
[13] "iRFC1-Data Specification of Smart Card Based Interoperability Mapping For Non-Financial Functions Inter-Institutions In Indonesia." Konsorsium Smart Card Indonesia.

Effect of Current and Wire Speed on Surface Roughness in the manufacturing of Straight Gear using Wire-cut EDM Process

Rusdi Nur[1,2,*], Muas M.[1], Apollo[1], and Syah Risal[3]

[1] Mechanical Engineering Department, Politeknik Negeri Ujung Pandang Jl. Perintis Kemerdekaan Km. 10 Makassar 90245 South Sulawesi Indonesia
[2] Center for Materials and Manufacturing, Politeknik Negeri Ujung Pandang Jl. Perintis Kemerdekaan Km. 10 Makassar 90245 South Sulawesi Indonesia
[3] Engineering Department, Politeknik Ilmu Pelayaran Jl. Tentara Pelajar No.173, Makassar 90165 South Sulawesi Indonesia

Corresponding author: rusdinur@poliupg.ac.id

Abstract. The gears are an important part of a mechanism. Making gears requires high precision to avoid backlash. The machining process that can produce high precision is the process of Wire cut EDM. This study aims to determine the roughness of the workpiece surface and investigate the effect of current and wire speed on the roughness of the straight gear surface through wire cut machine EMS 45. The wire cut machining parameters used are 3, 5, and 7 amperes for current, and wire speed of 8, 10, and 12 mm/min. While the Wire material is a type of brass with 0.25 mm in diameter. The results presented that the increased use of current and wire speed will result in lower surface roughness, and vice versa. Therefore, it can be concluded that use low machine parameters to produce a smooth workpiece surface.

Keywords. Current, Wire Speed, Surface Roughness, Wire-cut EDM, Straight Gear.

1. Introduction

Wire-cut Electric Discharge Machining (EDM) is one of the greatest innovations in the tooling and machining industry. This process has brought about dramatic improvements in accuracy, quality, productivity, and earnings. Before wire EDM, costly processes were often used to produce finished parts. Now with the aid of computer and wire EDM machines, extremely complicated shapes can be cut automatically, precisely and economically even in materials as hard as carbide. As more design engineers incorporate new designs into the drawings, therefore it becomes important for contract shops to understand wire EDM as today's drawings are calling for tighter tolerances and shapes that can be efficiently machined only with wire EDM. Hence WEDM plays a significant role in the industries to attain better surface finish of the components. Wire cut machine is one of the CNC machines that can work automatically through a computer control system. The wire cut machine has the ability to produce high precision products with varying levels of desired roughness [1]. Surface roughness, especially on some high precision products, is a parameter to note. Products produced with a certain surface roughness value will determine the functionality of the product. The accuracy of the surface roughness value of a product will determine whether the product is working properly or not.

Gears are precision products that are commonly found in a variety of propulsion mechanism systems, which in practice sometimes require low surface roughness (smooth). The straight gear is a type of gear that has a simple shape. Consistent gear contours result in a straight gear machining process relatively easier to perform than any other type of gear machining process [2].

A wire cut machine is a conventional machine that is often used in various research objects. Research on setting cutting cut parameters on straight gears has been done before. Tosun et al. [3] were conducted a study on the effect of cutting parameters on surface roughness on wire cut with steel workpiece SAE 4140. From this research concluded that pulse duration, open circuit voltage, and wire speed greatly affect the surface roughness. The increase of dielectric fluid pressure can be obtained by a finer cutting surface. Sarkar et al. [4] were optimized for the parameter machining on cutting speed, surface finish and dimensional deviation when wire-cut EDM of titanium aluminide alloy. Sivaraman et al. [5] also were investigated the control parameters of wire-cut EDM for titanium on material removal rate and surface finish using the Taguchi approach. While Mahapatra and Patnaik [6] have also conducted research on the optimization of process parameters on the wire cut by using Taguchi method. The result of this research stated that the amount of current, frequency, and velocity of dielectric fluid flow and its interaction give a significant influence on cutting in increasing MRR, minimizing roughness of cutting result, and narrow cut width. This research also produces setting parameters of cutting wire cut process.

Based on the above study, this paper investigates the effect of cutting parameters on surface roughness through the wire cut process of EMS 45 steel to manufacture of straight gear.

2. Wire Cut Process

The wire cut process can only cut a workpiece that is conductive, whose base uses electricity. The hot wire is caused by an electrical pulse of DC generated between the wire and the workpiece. This is similar to other EDM processes, the wire becomes a negative pole and the workpiece becomes a positive pole so it can lead to spark jumps. Wire cut is a cutting process that uses a wire cutter that moves through the workpiece.

In the wire-cut process, there is deionized water called a dielectric. The deionization process will make pure water that serves as an insulator. in addition, the fixed water contains minerals, so it makes the wire very conductive. Meanwhile, to set the conductivity of water, then made the process of water circulation on the wire cut system. Schematic of the wire-cut process was shown in the Fig. 1.

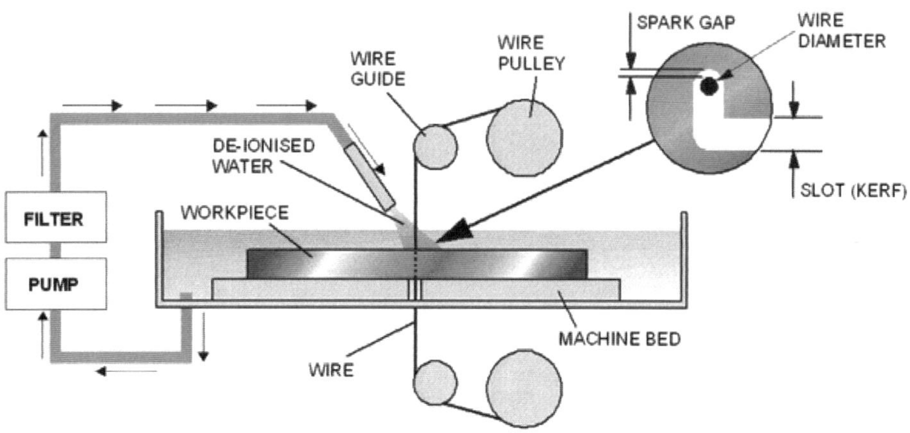

Figure 1. Schematic of wire cut process

Water will be deionized when the system is electrically powered. Furthermore, there is an electrical spark jump between the wire and the workpiece and erode a small part of the workpiece. Electric pulses occur repeatedly thousands of times per second. While pressurized dielectric fluid is flowed to assist the process of cooling the workpiece and cleaning up the scraping of wire and workpiece.

3. Experimental Setup

3.1. Materials
The material specification used in this experiment is EMS 45 steel. Meanwhile, the material used for wire cut electrode is wire brass type (brass) with a thickness of 0.25 mm.

3.2. Cutting Parameters
In this study, the experimental setup is using the cutting parameters as follow:

Table 1. Parameters for wire cut process

Parameters	**Range**
Current (Ampere)	3, 5, and 7
Wire Speed (mm/m)	8, 10, and 12

Surface roughness testing was performed using surface roughness tester SURFTEST SJ-310. Data on Surface roughness was measured three times for each position.

4. Results and Discussion
The results of surface roughness on the straight gear making are shown in Table 2.

Table 2. The result of surface roughness

No	Current (Ampere)	Wire Speed (mm/min)	Surface Roughness (μm)			
			Ra_1	Ra_2	Ra_3	Average
1	3	8	2.766	2.929	2.871	2.855
2	3	10	2.729	2.820	2.660	2.736
3	3	12	2.693	2.781	2.954	2.809
4	5	8	2.828	2.700	2.755	2.761
5	5	10	2.872	2.823	2.709	2.801
6	5	12	2.695	2.649	2.553	2.632
7	7	8	2.656	2.567	2.496	2.573
8	7	10	2.296	2.457	2.450	2.401
9	7	12	2.767	2.577	2.615	2.653

In order to gain a clear understanding, the test results data can be illustrated in Figs 2 and 3.

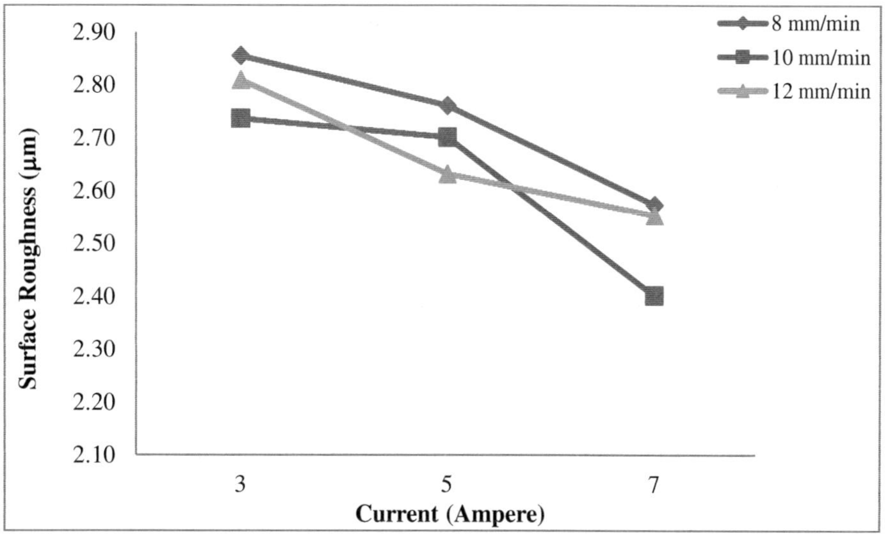

Figure 2. Data of surface roughness for different of current.

Fig. 2 shows that surface roughness is strongly influenced by currents. it is very clear that the surface roughness tends to be smoother if the current value is greater. If the current is minimized then the surface roughness will be rougher.

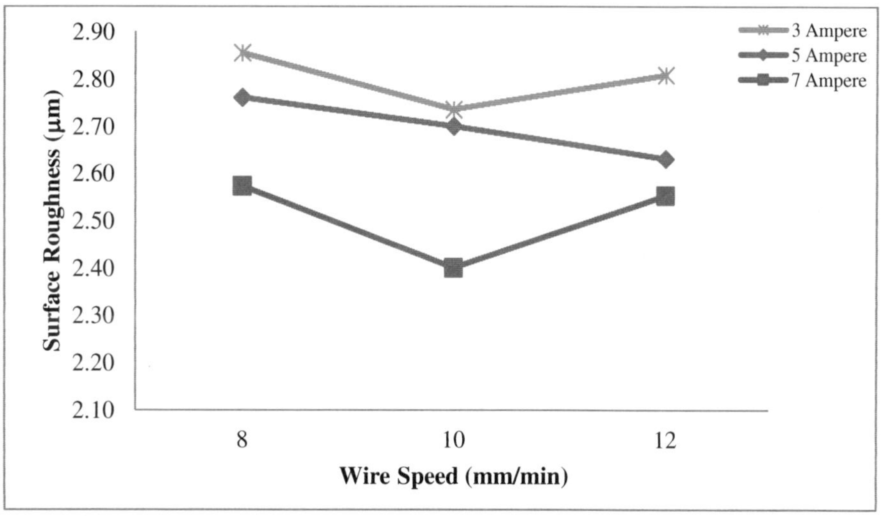

Figure 3. Data of surface roughness for different of wire speed.

Fig. 3 presents that surface roughness is influenced by wire speed, where the lower wire speed the surface roughness value will be rougher. Likewise, if the wire speed is improved then the surface roughness value will be smoother, but if wire speed is added then it will make the workpiece surface rough again. It had been proven by Hasan et. al. [7], it was found that an excellent machined finish can be obtained by setting the machine parameters at low pulsed current and small pulse-on duration. This can be attributed to the fact that as the pulsed current decreases, discharges strike the surface of the sample less intensely, and the resulting better erosion effect leads to smoother surface. Furthermore, as the pulse-on duration decreases, the amount of heat energy transferred to the sample surface decreases, and so less material melts. The fact that the surface roughness decreases with decreasing discharge energy has been described in the literature [8, 9, 10].

Displaying the results of gear made with the wire cut process can be seen in Fig. 4.

Figure 4. Display the results of making gear straight

The wire cut process using two variables (i.e. current and wire speed) has been running well. The result of surface roughness and cutting time is significantly influenced by strong variables of current and wire cut. This is in accordance with the facts proposed by Tosun et al. [3] was stated that wire speed is very influential on the roughness of the cutting surface. This phenomenon is also reported by a study conducted by Mahapatra and Patnaik [6] who examined the optimization of the parameters of wire cut EDM machining process by using the Taguchi method with the result that the magnitude of the current and the interaction gave a significant effect on cutting in increasing the MMR, minimizing roughness the result of cutting, and able to narrow the width of cutting [6].

As implied in Figs. 2 and 3, the surface roughness is dominated by variations of the two given variables, i.e. current strength and wire speed. The slower wire speed used then the surface roughness value will be rougher, and if the wire speed is increased when the surface roughness will be smoother. Likewise with the current variable, if the current value used is low then the surface roughness value will be rougher, and if the current value is added gradually will make the surface roughness value smoother. Ravi Kumar et al [11] find out on this work is aimed to optimize the parameters of Wire-cut EDM process by considering the effect of input parameters viz. Time On, Time Off, Wire Speed & Wire Feed. Experiments have been conducted with these parameters in three different levels of data related to process responses viz. Metal removal rate, surface roughness (Ra) have been measured for each of the experimental run. These data have been utilized to fit a quadratic mathematical model (RSM) for each of the responses, which can be represented as a function of the process parameters.

The experimental work by Selvam and Kumar [12], it reveals the following conclusions on Wire EDM operations on Hastelloy C–276 workpiece material. Main objective of this work is to develop the empirical model using GA. The genetic algorithm methodology is one of the best techniques to identify the effects of machining parameter on Wire EDM process. The voltage and pulse off time have the significant effect on machining time, the higher level of current produce lower machining time. The voltage, Current, Pulse on time and pulse off time are have significant effect on surface roughness. The higher level of current, voltage, pulse on time and pulse off time are produce poor surface finish.

5. Conclusion

Based on the investigation done in the process of wire cut on the straight gear making, it can be concluded that the current variable gives a significant influence on the roughness of the straight gear surface. The greater the current use, the roughness of the straight gear surface is lower, the smaller the current is used, the higher the roughness of the straight gear surface.

References

[1] Juneja, B.L., 2003. *Fundamentals of metal cutting and machine tools*. New Age International..

[2] Kalpakjian, S., Vijai Sekar, K.S. and Schmid, S.R., 2014. *Manufacturing engineering and technology*. Pearson.

[3] Tosun, N., Cogun, C. and Inan, A., 2003. The effect of cutting parameters on workpiece surface roughness in wire EDM. *Machining Science and Technology*, **7**(2), 209-219.

[4] Sarkar, S., Mitra, S. and Bhattacharyya, B., 2005. Parametric analysis and optimization of wire electrical discharge machining of γ-titanium aluminide alloy. *Journal of materials processing technology*, **159**(3), pp.286-294.

[5] Sivaraman, B., Eswaramoorthy, C. and Shanmugham, E.P., 2015. Optimal control parameters of machining in CNC Wire-Cut EDM for Titanium. *International Journal of Applied Sciences and Engineering Research*. **4**(1) 102-121.

[6] Mahapatra, S.S. and Patnaik, A., 2007. Optimization of wire electrical discharge machining (WEDM) process parameters using Taguchi method. *The International Journal of Advanced Manufacturing Technology*, **34**(9-10) 911-925.

[7] Hassan, M.A., Mehat, N.S., Sharif, S., Daud, R., Tomadi, S.H. and Reza, M.S., 2009. Study of the surface integrity of AISI 4140 steel in wire electrical discharge machining. In *Proceedings of the International MultiConference of Engineers and Computer Scientists* (Vol. 2).

[8] Ramasawmy, H., Blunt, L. and Rajurkar, K.P., 2005. Investigation of the relationship between the white layer thickness and 3D surface texture parameters in the die sinking EDM process. *Precision Engineering*, **29**(4) 479-490.

[9] Guu, Y.H., Hocheng, H., Chou, C.Y. and Deng, C.S., 2003. Effect of electrical discharge machining on surface characteristics and machining damage of AISI D2 tool steel. *Materials Science and Engineering: A*, **358**(1-2) 37-43.

[10] Keskin, Y., Halkacı, H.S. and Kizil, M., 2006. An experimental study for determination of the effects of machining parameters on surface roughness in electrical discharge machining (EDM). *The international journal of advanced manufacturing technology*, **28**(11-12) 1118-1121.

[11] Kumar, K. and Ravikumar, R., 2013. Modeling and optimization of wire EDM process. *International Journal of Modern Engineering Research*, **3**(3) 1645-1648.

[12] Selvam, M.P. and Kumar, P.R., 2017. Optimization Kerf width and surface roughness in wire cut electrical discharge machining using brass wire. *Mechanics and Mechanical Engineering*, **21**(1) 37-55.

Modeling of Vertical Dynamic Vibration Characteristics on Vehicles Suspension System

Simon Ka'ka,[1] Syukri Himran,[2] Ilyas Renreng,[3] Onny Sutresman[4]

[1]Mechanical Engineering Department, State Polytechnic of Ujung Pandang, South Sulawesi, Indonesia

[2,3,4]Mechanical Engineering Department, Engineering Faculty, Hasanuddin University, South Sulawesi, Indonesia

The corresponding E-mail address: simon_kaka@poliupg.ac.id

Abstract. Vibrations often interfere with driving comfort for riders and passengers. Variations in loads, especially for four-wheeled vehicles, bumps, hollow road surfaces, and other forms of road damage will greatly affect the vehicle suspension work system. This research aims to (1) further test the effect of vertical dynamic load of the vehicle and change the dimension of resistance on the road surface, (2) the contribution of spring and shock absorber to the load fluctuation of the vehicle. Experimentally these load fluctuations are replaced by pneumatic actuator forces of varying magnitude based on the regulatory pressure of the regulator. The deviations generated by the varying load work are measured by placing a proximity sensor along the spring movement. The vertical dynamic load transformation up to the road surface is measured using a "Load cell" mounted under the wheels of the vehicle. Characteristics of vertical dynamic vibration occurring due to several dimensional barriers, U (cm) obtained using mathematical modeling method with 2 DOF suspension system transfer function. The results showed a condition on the body and wheels of vehicles experienced a brief overshot for 0.14 seconds with deviation of 0.178 m. From the graph shows that the rate of deviation that occurs is large enough that Y2d = 1.03 m / s caused by a sudden shock that occurred on the wheels of the vehicle. This condition does not last long that is only duration t = 0.22 s, because the contribution of springs and shock absorbers that can absorb vibration is 25% to the vibrations caused by the vertical load of the body and the axis of the vehicle.
Keywords; Vibration, suspension, pneumatic actuator, deviation, vertical load.

1 Introduction

The increasing frequency of recurrent vertical loads of four-wheel vehicles passing through the road will decrease the ability of the road to receive the load. Vibrations originating from vertical dynamic loads of vehicles often fluctuate due to unstable drivers and passengers. Such conditions will further weaken the ability of the road structure to accept the fluctuating load. The spring and shock absorber mounted on each of the wheels of the vehicle is expected to be able to overcome and reduce the vertical dynamic load of vehicles overloading the road structure. The vehicle suspension system according to [1-3] is composed of a spring and a shock absorber arranged in parallel. The main function of the suspension system is to support the weight of the vehicle, to provide comfort for the rider, to keep traction of the wheel on the road surface condition, and to maintain the alignment of the front wheel and rear wheel [4-6]. The shape and mechanism of the suspension system is shown in

Fig.1 (a), with sprung mass (m_2), un-sprung mass (m_1), suspension spring (k_2), shock absorbers (c) and tire elastic constants on wheels (k_1), [7-9].

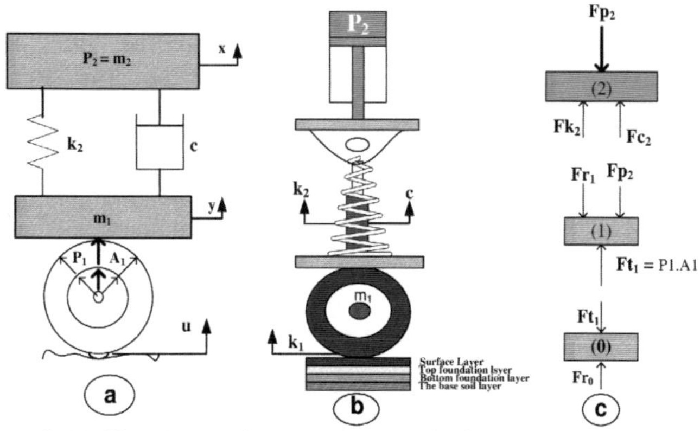

Figure 1 (a) The shape of suspension mechanism, (b) Pneumatic system experimental loading mechanism, (c) Free body diagram

The magnitude of the pneumatic cylinder thrust force generated by the air pressure P_2 (bar) according to [7, 10] is formulated as;

$$F_k = \frac{\pi}{4} D^2 P_2 (\mathrm{N}) \tag{1}$$

If the dimensions of the cylinder, D = 100 mm (0.100 m) are substituted to equation (1) then the compression force of the suspension mechanism is obtained:

$$F_k = 785\ x\ P_2\ and\ F_{p2} = 706.5\ P_2 = 707\ P_2\ (\mathrm{N}). \tag{2}$$

Based on the free-body diagram of Figure 1 (c), the equation of the wheel loading the road structure is;

$$\left.\begin{aligned} F_{t1} &= F_{p2} + F_{r1} \\ F_{t1} &= (k_2 + c)x_2 + (m_1 . g) \end{aligned}\right\} (\mathrm{N}) \tag{3}$$

If in equation (3), the spring constant k_2 (N / mm), the coefficient of damping c (Ns / mm), the weight of the wheel m_1 (kg) and the gravitational acceleration g (m / s^2) are obtained;

$$F_{t1} = F_{ro} = 707\ P_2 + (m_1 . g)\ (\mathrm{N}) \tag{4}$$

The mass sprung of m_2 supported by the spring and the shock absorber with the spring stiffness of k_2 and the damping coefficient c will overload the vehicle wheel as the mass unsprung of m_1 further gives the action force to the contour of the road surface. The reaction force of the road contour to the tire will be distributed in the direction of u through the tire elasticity constant of k1 [9]. Thus the transfer function of a quarter vehicle suspension system for a front wheel with two degrees of freedom (2DOF) is obtained by comparing the output and input as follows:

$$\frac{Y(s)}{U(s)} = \frac{k_1(cs + k_2)}{m_1 m_2 s^4 + (m_1 + m_2)cs^3 + [k_1 m_2 + (m_1 + m_2)k_2]s^2 + k_1 cs + k_1 k_2} \tag{5}$$

Calculation of spring constant value, k_2 by [11, 12] respectively in the book "Machinery's Handbook, 29th Edition and the journal Mechanism and Machine Theory are formulated as follows:

$$k_2 = \frac{Gd^4}{8n_a D^3} \tag{6}$$

G, is the modulus of stiffness (N / cm^2), *d* is the diameter of the wire (cm), n_a is the number of active coils and *D* is the mean coil diameter (cm) obtained from the difference between the outer diameter of the coil and the diameter of the wire. The number of active coils, n_a is usually less than the total coil, n whereas the spring compress area is between the minimum compression of 20% and the maximum compression of 80%. If the number of active coils, n_a = 80% x n, and the mean coil diameter, D = Do - d then the value of the spring constant, k_2 can be obtained by using equation (6).

2 Research Method

Experimental tests conducted on the mechanical work of suspension on the wheels of the vehicle is by adjusting the working pressure P_2 (bar) on the regulator ranging from 1 to 8 bar [7]. Furthermore, the loading of the spring and the shock absorber by the pneumatic actuator begins to occur when the 5/2 directional control valve is operated. The change in the position of compression motion on the spring and the shock absorber is measured by placing the proximity sensor along the spring movement. The amount of deviation detected by the sensor can be transferred to the LCD in digital form. It should be noted that the change in the size of the readings on the LCD screen is correlated with the regulatory pressure settings. By placing a Load cell gauge just below the wheel of the vehicle, the measurement of the vertical dynamic load of the vehicle successfully transformed to the road surface can be obtained through a direct reading on the LCD screen.

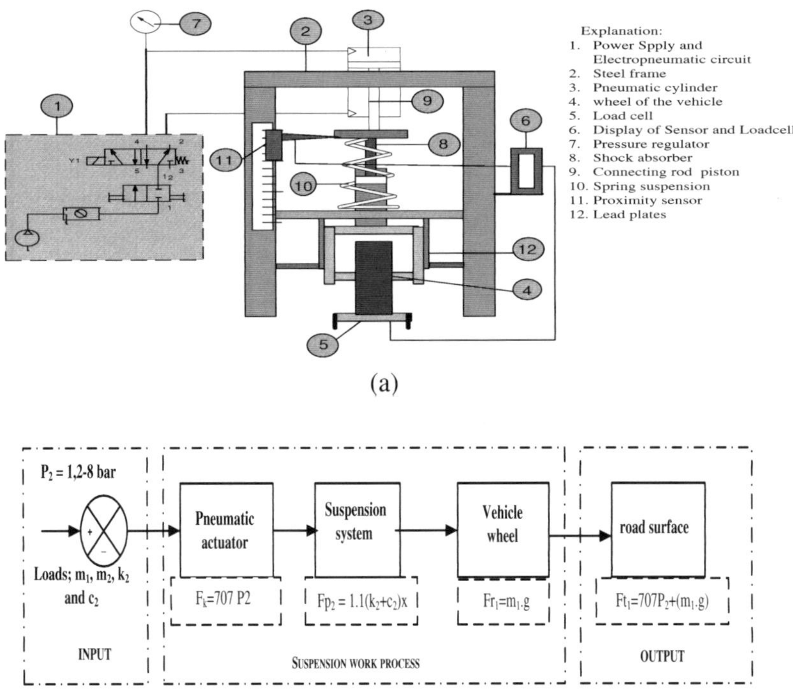

(a)

(b)

Figure 2 (a) Loading experiments with pneumatic actuators (b) Block diagram of vehicle suspension loading process

Optimization of physical models, using pneumatic cylinders as dynamic vertical drive test simulators on wheel drive suspension work has been discussed by [7].

The type of pneumatic cylinder selected is, DNU-100-300-PVA and sourced from [13] with specifications; Diameter of piston D = 100 mm, stroke L = 300 mm, Weight piston W = 3.864 kg = 38.64 N, and Weight piston rod / 10 mm w = 0.090 kg . The forward and reverse thrust force of cylinder piston at a working pressure of 6 bar are respectively F_k = 4496 N and F_m = 4221 N.

Mathematical model optimization is done by making the weight data of wheel axis, m_1 (kg) and body weight, m_2 (kg) as INPUT as shown in Figure 2 (b). Similarly, the constant values of k_1, k_2, and c_2 are the input data to create the program in MatLab.

The test spring used in the suspension system mechanism is a kind of "Helical" with the following specifications: inside and outside coil diameters are Di = 12.985 cm and Do = 15.815 cm, diameter of spring wire, d = 1.415 cm, and number of coils, n = 5 pieces. The spring material used is from Chrome Vanadium, ASTM A231 with Stiffness Modulus, G = 7.929E + 10 Pa = 7.929 x 10^6 N / cm^2.

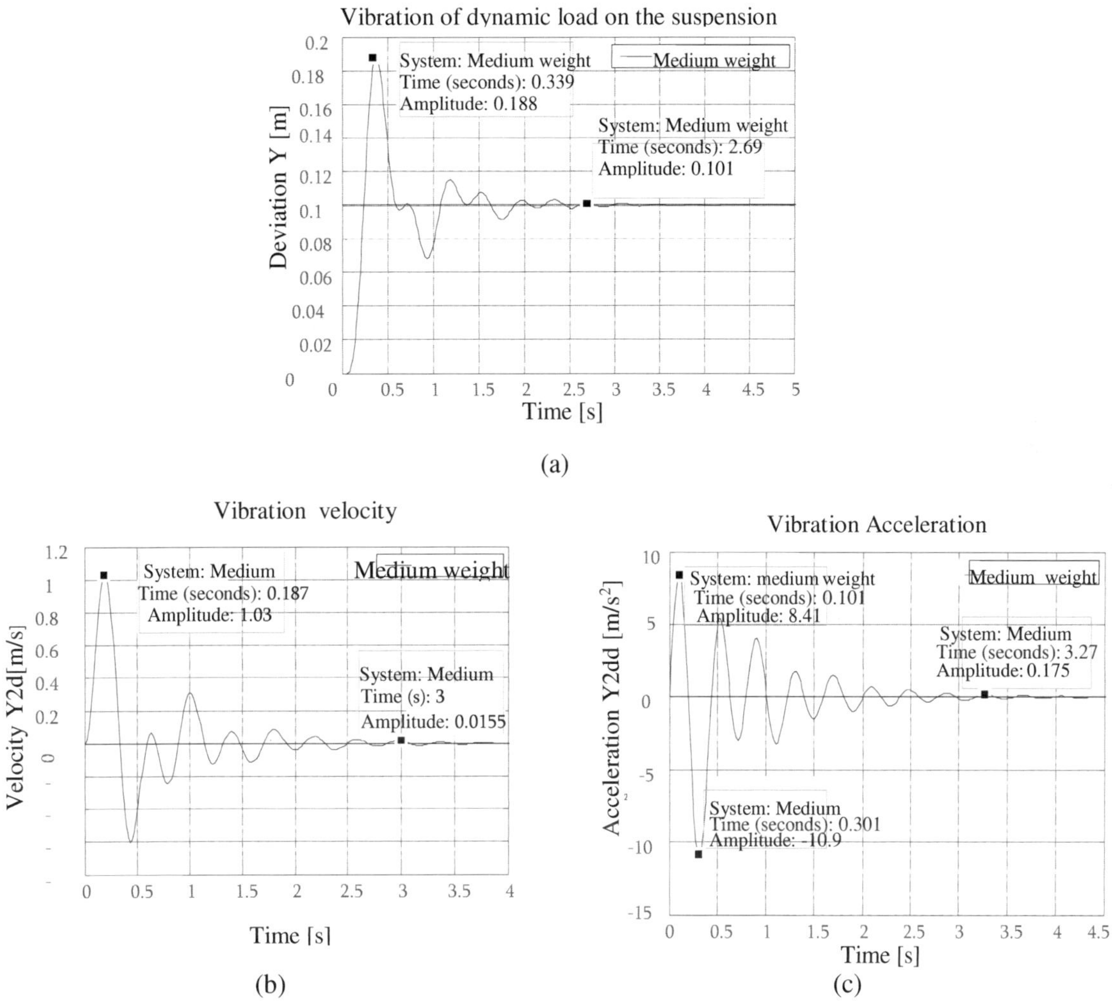

Figure 3 (a) Characteristics of vibration on the body and wheels with deviation Y (m), (b)Th characteristics of vibration velocity on the body and wheels, V (m/s), (c) The characteristics o vibration acceleration on the body and wheels, a (m / s^2).

3 Results and Discussion

Programming with Mat Lab software referring to equation (5) will produce the suspension vibration characteristics with small and large deviations as shown in Figure 3(a), (b) and (c). The execution result of the program has obtained the characteristic graph of vibration on the suspension as the impact of vertical dynamic load. In Figure 3(a) if the INPUT resistance is U = 0.1 m, then the suspension on the wheels will experience vibration with maximum deviation, Y = 0.188 m with a vibration time of 0.339 s. then two seconds later, the minimum deviation occurs at Y = 0.101 m.

Proportionally author [14] has obtained 0.12 m deviation on the measured vehicle wheel as a strain using the WIM sensor that placed below the tire.

The characteristic shape of the vibration in the wheel suspension as in Figure 3(a) is caused by a mound as high as U = 0.1 (m) on the road surface. Similarly, the reaction to the total dynamic load of the vehicle can result in a vertical overshot deviation of Y = 0.188 (m) with a duration of 0.1 seconds. Furthermore, within a time interval of 2.34 seconds later, the vibrations begin to shrink and stabilize at 0.66 seconds.

Table 1, Characteristics of Experimental Results and Theoretical Deviation

Pressure	**Experimental**		**Theoretical**	
P_1(bar)	t (s)	Y (m)	t (s)	Y (m)
1	0.5	0.059	0.5	0.11
2	1.0	0.067	1	0.087
3	1.6	0.078	1.5	0.11
4	2.1	0.086	2	0.1
5	2.7	0.088	2.5	0.098
6	3.3	0.092	3	0.1
7	3.9	0.096	3.5	0.1
8	4.6	0.098	4	0.099

The result of the research in Figure 3(b) shows that the largest deviation rate Y2d = 5.51 m / s occurs at Amplitude, A= 1.03 m and the vibration time, t = 0.187 s, while the lowest deviation rate is Y2d = 0.0051 m / s when the vibration time t = 3 s. From the graph shows that the rate of deviation that occurs largely enough caused by a sudden shock that occurs on the wheel. Such conditions can not last long because of the resistance of the spring and shock absorber. The change in the drift velocity shown in Figure 3(c) has a significant slowdown of amplitude, A = (- 10.9 + 8.41) m = - 2.59 m, at the vibration time, t = 0.2 s. The lowest acceleration Y2dd = 0.014 m / s^2 occurs at the vibration time, t=0.175 s.

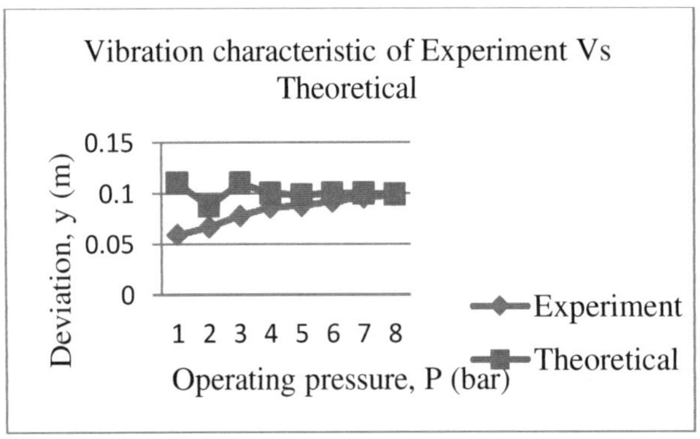

Figure 4. Relationship of vibration characteristics between experimental and theoretical studies.

It should be noted that the perceived comforts of the vehicle are often disrupted by the uneven, hollow, bumpy surface of the road surface and are filled with mound-shaped materials. Comfort can also be felt by passengers and drivers when the mechanism of shock absorber damper is able to reduce vertical dynamic load of the vehicle well.

Given the circulation of liquid or gaseous fluid determined by the regulation of the flow control valve in the cylinder chamber, the performance of the absorbent shock is getting better. The amount of

deviation experienced by the body and the vehicle's axle while experiencing vertical dynamic loads of experimental results and modeling using Mat Lab is shown in Figure 3 above.
The comparison of deviation values between the experimental results and the programming with Mat-Lab is shown in Table 1.

The change in deviation occurring according to the experimental results shows a significant increase with increasing work pressure P_1 (bar) and vibration time t (s). In contrast to the results of deviations obtained through theoretical studies, whereby each increase in vibration time of t = 0.5 (s) then the change of deviation Y (m) occurs, does not occur regularly.

The experimental vibration characteristics and theoretical studies show a trend that starting at the working pressure of 4 bars vibration deviation is increasingly blessing each other and finally at the working pressure of 8 bars the deviation is at Y = 0.1 m.
The mean deviation of the experimental results is Y = 0.083 m, with a mean duration of 2.5 s, while the average deviation of the programming results is Y = 0.1 (m) in the mean duration, t = 4.5 s.

The rate of deviation and the acceleration of movement of the vehicle body that weighs 300 kg by [4] are 0.03 m and 1.8 m/s^2 respectively and stabilized at t = 3.5 (s).
The shape of the deviation graph shows that at the time of vibration, t = 0.5 s there is a deviation gap of Y = 0.051 m, when pressure P = 1 bar, but after reaching the vibration time t = 2.1 s when pressure P = 4 bar, then the deviation gap starts to decrease to Y = 0.014 m.

The magnitude of the vehicle's actual vertical load that overlies the road structure is obtained by multiplying the 80% dynamic factor with the force according to equation (4). If the weight of the vertical load, after getting suspense resistance on the wheels of the vehicle is Fr1 = 22197 (N) then the load of the vehicle that weighs the road is Ft1 = 0.8 x 22197 (N) = 17758 (N). The magnitude of the reduced vertical load by the spring and the shock absorber is 4439 (N) or equivalent to 25%.

4 Conclusion

Based on the results of experimental studies and analyzes that have been done on the vehicle suspension work performance, it can be concluded as follows:

1 Effect of suspension working mechanism on body and wheel of vehicle for medium-weight vehicle type if experiencing disturbance U = 0,1 m that happened the biggest deviation Y = 0,18 m with duration 0,34 second. In the span of 2.34 seconds later, the vibrations begin to shrink and eventually stabilize at 0.66 seconds. The highest suspension vehicle vibration rate occurs at Y2d = 5.51 m / s and the lowest, Y2dd = 0.0051 m / s occurs at, t = 3 s. At almost the same time, the lowest acceleration is Y2dd = 0.054 m / s^2. The occurrence of change in deviation speed at the work of the suspension will reduce the driving comfort felt directly by the passengers and drivers.

2 The contribution of the spring and the shock absorbers to the fluctuating loads on the vehicle is to reduce the vertical load by 25%.

NOMENCLATURE

Symbol	Unit	Meaning
A, A_o	Cm^2	The area of the tire contact on the road
A_1	m^2	The inside area of the tire
c_2, c	Ns/mm	Coefficient of shock absorber damping
C	-	Coefficient of vehicle distribution
COG	Cm,cm	Center of gravity
D	mm	Piston diameter
Di	cm,	The inside diameter of the coil
DLC	mm	Dynamic Load Coefficient
Do	-	The outer diameter of the coil
DOF	cm,	Degree Of Freedom

d	mm	Diameter of spring wire
di	-	Diameter of the air channel
E	-	The vehicle's axis equivalent number
Ee	Pa	Elastic Modulus of the road
F_{Ao}	N	Road stability force
FBD	-	Free Body Diagram
Fc	N	Damping force of shock absorber
F_{Dv}	N	Vertical dynamic vehicle load force
F_{ef}	N	Effective force of piston forward
Fk	N	Theoretical force of piston cylinder
Fp_2	N	Effective force of piston forward
Fr_1	N	Force of the wheels
Ft_1	N	Force of the road surface
Fg	N	Friction force of tires
Fh	N	Horizontal force of tire
fg	-	Coefficient of the road surface
f	1/s	The frequency of vibration
g	m/s^2	Acceleration of gravity
G	Pa	Spring stiffness modulus
k	-	The standard axis correction factor
k_1, k_e	*N/m*	Tire elastic constants
k_2, k_s	*N/m*	The suspension spring constant
Ls	Ton	Axis loads of vehicles
L	cm	The length of the road surface contact area
m_1	Kg	Unsprung mass of vehicle
m_2	Kg	Sprung mass of vehicle
n	pieces	The number of spring coils
na	pieces	Number of active coils
P_1,p_1	bar	Air pressure in the tire
P_2	bar	Working air pressure
t	s	Vibration time
u	mm	Deviation shock absorber
W	N	Total vehicle weight
y	mm	deviation on the suspension spring.

5 References

[1] M. F.-D. J. Brembeck, "LPV Control of Full-Vehicle Vertical Dynamics using Semi-Active Dampers," *Science Direct,* vol. 49-11, p. 8, 2016.

[2] M. A. A. Mohd Nizam Sudin1, Shamsul Anuar Shamsuddin1, Faiz Redza Ramli1, Musthafah Mohd Tahir1, "Review of Research on Vehicles Aerodynamic Drag Reduction Methods," *Review of Research on Vehicles Aerodynamic Drag Reduction Methods,* vol. Vol.14, p. 35, 2014.

[3] S. W. B. G.R. Chavhan*, Dr.L.P.Dhole, "Analysis Of Shock Absorber Using Different Material Of Spring," *International Journal of Advanced Engineering Technology,* vol. V, pp. 19-21, 2014.

[4] M. [a,] TongyiXu [a,] ChuanLi [b,*,] ShuaiYang [a], "Design and analysis of a shock absorber with variable moment of inertia for passive vehicle suspensions," *Journal of Sound and Vibration,* vol. 355, p. 20, 2015.
[5] E. A. Finney, "Dynamic Aspect Of Vehicle Size and Weight," p. 23, 2010.
[6] R. B. Łukasz Konieczny1, "Modern suspension systems for automotive vehicles and their test methods," *JVE International LTD. Vibroengineering Procedia,* vol. 14, p. 5, 2017.
[7] Simon Ka'ka[1,] Syukri Himran[2], Ilyas Renreng[2] and Onny Sutresman[2], "Pneumatic Actuator As Vertical Dynamic Load Simulator On The Suspension Mechanism Of A Quarter Vehicle Wheels," *ARPN Journal of Engineering and Applied Sciences,* vol. 12, 2017.
[8] X. J. Y. Z. W. Ma, "Nonlinear dynamic characteristics of a micro-vibration fluid viscous damper," *Springer Science+Business Media, B.V.,* vol. 92, p. 10, 2018.
[9] V. V. N. I.V. Ryabova, A.V. Pozdeev a,*, "Efficiency of Shock Absorber in Vehicle Suspension," *International Conference on Industrial Engineering, ICIE 2016,* vol. 150, pp. 354 – 362, 2016.
[10] Simon Ka'ka[1,] S.Himran[2].,Ilyas Renreng[2] and Onny Sutresman[2], "The Pneumatic Actuators As Vertical Dynamic Load Simulators On Medium Weighted Wheel Suspension Mechanism," *Journal of Physics,* vol. 962, p. 10, 2018.
[11] E. Edge, "Helical Compression Spring Design Equation and Calculator," *Became An Engineers Edge Contributor,* 2017.
[12] P.-A. Y. Hassen Trabelsi , *, Jamel Louati, Mohamed Haddar "Interval computation and constraint propagation for the optimal design of a compression spring for a linear vehicle suspension system," *Mechanism and Machine Theory , journal homepage: www.elsevier.com/locate/mechmt,* vol. 48, pp. 67–89, 2015.
[13] K. Festo, "Automation With Pneumatic," in *Festo Didactic, KG* vol. NR.23099-01/1, ed. Esslingen: Festo Pneumatic, 2015.
[14] Wenbin Zhang 1, Chunguang Suo 2 and Qi Wang 1, "A Novel Sensor System for Measuring Wheel Loads of Vehicles on Highways," *Sensors* www.mdpi.com/journal/sensors, vol. 8, p. 19, 2008.

ACKNOWLEDGEMENT
The author would like to thank to Kemenristekdikti, team supervising commission, Mechatronics Study Program of State Polytechnics of Ujung Pandang, and the Department of Mechanical Engineering of Hasanuddin University have much support and facilitate the authors in the form of grants, guidance, and support of laboratory equipment and workshops.

IOP Conf. Series: Materials Science and Engineering **619** (2019) 012004 doi:10.1088/1757-899X/619/1/012004

Development of Quadcopter for Tracking Object Using Image Processing

Muhammad Ifan Nugraha, Ardiansyah Tri Utomo, Akhmad Taufik*, Remigius Tandioga, Rafiuddin Syam**

Center for Mechatronics and Control System· Departement of Mechatronics Engineering, State Polytechnic of Ujung Pandang Makassar, Indonesia
**Mechanical Engineering Dept., Hasanuddin University, Indonesia
*Corresponding Authors E-mail:** fike76@gmail.com

Abstract. Quadcopter is an unmanned aircraft that could be used as an image recorder at various view angle that can't be placed by a human. This research is done to planning, building and implement a quadcopter that is equipped with a camera. Then, we apply for a tracking object program in the quadcopter which is called image processing. Therefore, this research begins by designing the mechanical and electrical of the quadcopter. While the designing of the software is done by making image processing program using C++. Data collection is done by testing the device in each section. Based on the result of research and testing, the result shows that quadcopter could follow a straight moving human, then it turns 90° to right as far as 10 meters. But, there are any movements that we did not expect before, which is in the altitude of the quadcopter, it because of the instability of the loiter flight mode.

1. Introduction

A quadcopter is an unarmed aircraft that has 4 motors equipped with propellers on each motor. Its function is to produce an up force and for quadcopter movement. Moreover these main tools, some quadcopter has Global Positioning System (GPS) that use for quadcopter navigation, Inertial Measurement Unit (IMU) sensor for measuring the acceleration and movement direction orientation, an ultrasonic sensor to detect the presence of any things and other sensors that support the quadcopter performance [1][9].

Among those components, there only a few quadcopters have components that can make the quadcopter follow the movement of an object automatically. The support system that has the capability to follow the movement of an object is image processing.

It has designed a quadcopter but it can't detect and follow an object yet, because there are no camera and image processing program on that quadcopter. [2]Based on that problem, so we develop a quadcopter that can detect and follow a moving object using image processing.

2. Quadcopter and Image Processing

2.1. Quadcopter

Quadcopter is an Unmanned Aerial Vehicle (UAV) devices that are currently being researched because of its ability to take-off and landing vertically and move horizontally using only four propellers. [3]

2.2. Brushless Motor

Motor dc brushless is built by three phase with Y as the midpoint of its configuration. Y is a neutral point on the motor and a permanent magnet is positioned in front of the rotor by facing between the north and south poles. Rotation control of Brushless motor used Electronic Speed Control (ESC) which drives each phase alternately using Pulse Width Modulation (PWM). When the current through stator there will be an interaction between the electric charge and the rotor magnetic field so that the motor rotates. [4]

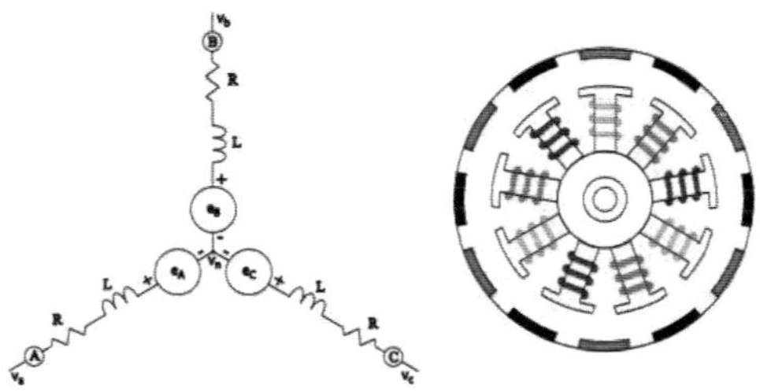

Figure 1. Brushless Motor

2.3. Electronic Speed Controller (ESC)

Electronic Speed Controller (ESC) is an electronic circuit that functions as a regulator of motor rotation speed on a quadcopter. This tool can also reverse the direction of motor rotation and break dynamically.

Basically, ESC *Brushless* creating three phase ac power, like in inverter (*Variable Speed Drive*) to run the brushless motor. Computer programmable speed control is generally possible with a low voltage cut-off limit, time, acceleration, braking and direction of rotation. Reversing the direction of the motor can also be done by sending three of the three leads from ESC to the motor. [5]

Figure 2. Electronic Speed Control

2.4. Flight Controller

Flight Controller (FC) is a set of several components and sensors that are used to keep the quadcopter balanced and manageable. At this time all quadcopter flight controllers have a balance sensor in the form of a gyro and an accelerometer. Some of the latest FC even have an air pressure sensor (barometer), compass (magnetometer) and GPS. The barometer sensor functions to maintain the height of the quadcopter at a certain height then the magnetometer and GPS are used to maintain orientation, autopilot, and failsave features. [6]

2.5. Wireless Audio/Video Transmitter and Receiver

Wireless audio/video transmitter and receiver are modules used to send and receive video/audio signals that sent by the transmitter and received by the receiver wirelessly. This tool uses 2.4 GHz radio waves with many channel choices. [7]

Figure 3. Flight controller

Figure 4. Video Transmitter and Receiver

2.6. Wireless Audio/Video Transmitter and Receiver
Image Processing is a technique to process an image or bitmap which transform input images to other images for a better output quality than the input quality.[8] Analog images are divided into N rows and M columns so that those become discrete images. Crosses between certain rows and columns are called pixels. For example is an image/discrete point on line N and column M is called pixel.

3. Research Methods
The design diagram as a guide in this process of developing followers object quadcopter based image processing is to obtain a good tool in terms of quality and still consider the economic aspects.

3.1. Wireless Audio/Video Transmitter and Receiver
Design of following object quadcopter using image processing can be seen in Figure 7.

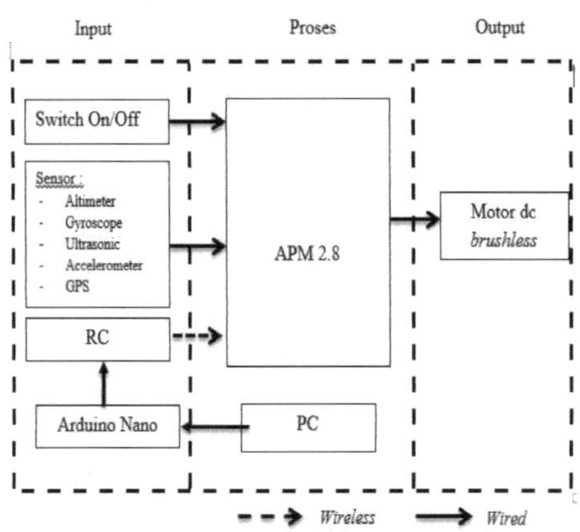

Figure 5. Quadcopter Design

3.2. Wireless Audio/Video Transmitter and Receiver

The implementation and testing of this research are as follows:

a. Testing of each part
b. Combining the system form several parts to be the whole system.
c. Tools response testing tosensor parameter
d. Testing of the whole circuit.
e. Testing quadcopter to follow an object as in figure 8.

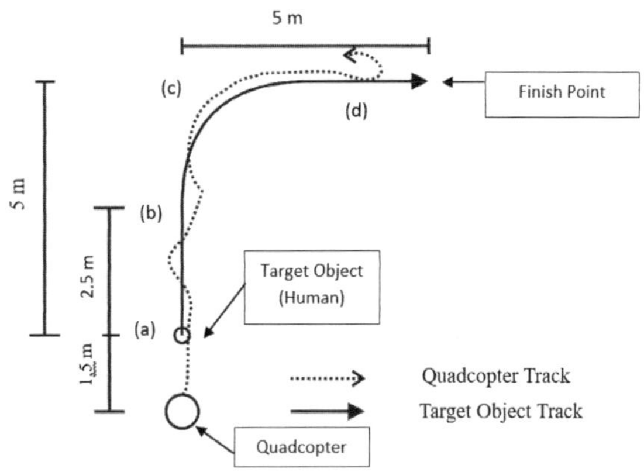

Figure 6. The path of the Target Object for Experiment

4. Results and Disussion

4.1. Hardware Results

The path of voltage and data distribution can be seen in Figure 9. While the final results of the installation of quadcopter components can be seen in Figure 10.

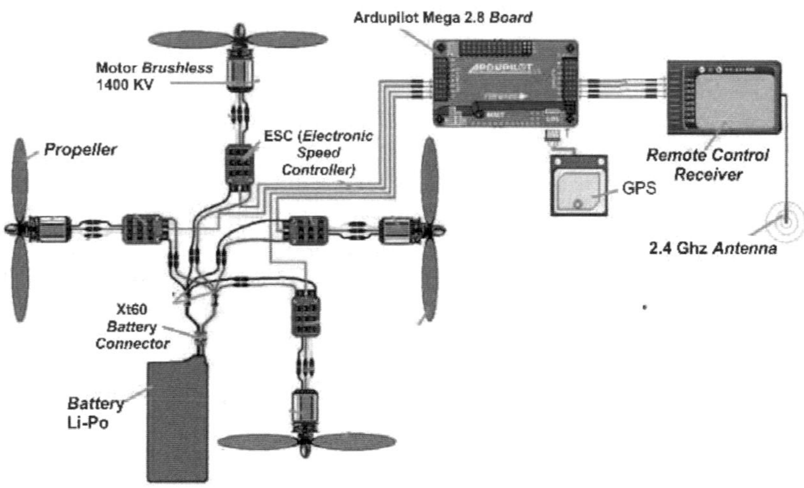

Figure 7. Circuit of Quadcopter Components

IOP Conf. Series: Materials Science and Engineering **619** (2019) 012004 doi:10.1088/1757-899X/619/1/012004

4.2. Calculation Result of Lifting Force Quadcopter

This calculation is done by taking the lift force data of thrust brushless motor1400-kv on the specific throttle. As for other quadcopter data as follows:

1. Thrust 40 % = 245 g, 50 % = 358g , 60 % = 461g, 70 % = 535g, 80 % = 628 g, 100% = 696 g
2. Motor quatity = 4 units, Gravity (g) = 9,8 m/s^2 and Mass (m)= 1,6 kg

By using those data above, the lift force is obtained as in table 1.

Figure 8. Circuit of Quadcopter Components Results

Table 1 Calculation Data Results of Quadcopter Lift

Thrust	**Lift Force (N)**	**Explenation**
40%	-6,076	*Quadcopter* can't fly
50%	-1,646	*Quadcopter* can't fly
60%	2,391	*Quadcopter* can fly
70%	5,292	*Quadcopter* can fly
80%	5,409	*Quadcopter* can fly
100%	11,603	*Quadcopter* can fly

4.3. Test Result of Arduino Nano PWM

This test aims to determine the modification of the quadcopter motion command signal. By Applying the PWD Arduino Nano Output which is connected to the Volage output pin on the Potentiometer. The results of this test can be seen in Table 2.

Table2Test Result Data of Arduino PWM

DAC Out Value of Arduino (analogWrite)	**RC Signal Value Output in YAW Channel (pwm)**	**Quadcopter Movement**
140	1506	Stay in Position (*Mid*)
100	1403	Turn to Left (*Left*)
180	1607	Turn to Right (*Right*)

4.4. Image Processing Test Result

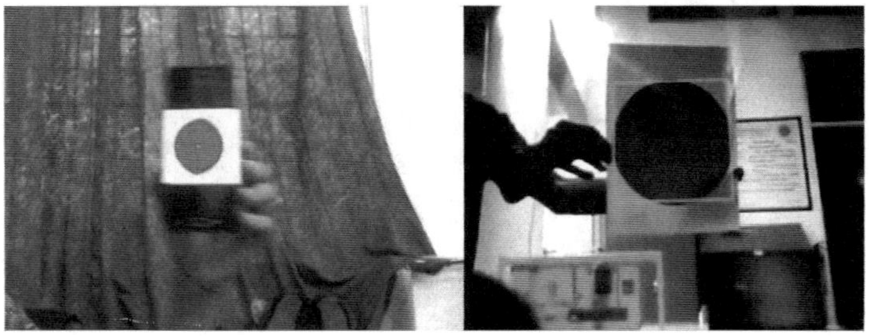

(a) (b)

Figure 9 Image Processing Test Result. (a) Tracking Program Using Color in circle Shape. (b) tracking Program Using in any Shape

Testing of area-based image processing program, the author did to get the mapping coordinates for the tracking system, this program is suitable for all objects, contours, and movements and colors because this program locks the shape, color, and movement of objects accurately and sends coordinates mapping results to Visual Studio which is then manipulated to move the quadcopter.

This program is better than the previous program that the authors tested because in addition to tracking objects, this program also maps the coordinates of object movements in real time, unlike the previous program which can only be tracking round objects and contrasting colors.

4.5. Quadcopter Flying Test Results Using Image Processing in Following Objects

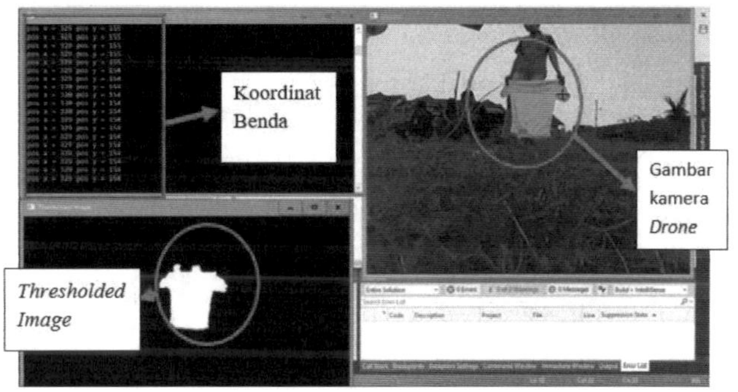

Figure 10. Display of Visual Studio Image Processing before TheQuadcopter Takeoff

Initially, HSV (Hue Saturation Value) of quadcopter camera set up so that it can detect a single clump of color as can be seen from Threshold Image, as the trial is yellow, then the X and Y coordinates appear as the image coordinates of the object.

When the program detects an object located on the right side of the screen, Arduino sends a signal of 180 analogWrite which is the same with the command to the left on Channel YAW RC. Then the quadcopter rotates on its axis to the left, but there is an additional vertical downward and horizontal downward movement which is a form of quadcopter instability when image processing is executed.

Figure 11. Display of Visual Studio Image Processing After The Quadcopter Takeoff

4.6. Quadcopter Flying Test Results Using Image Processing in Following Human

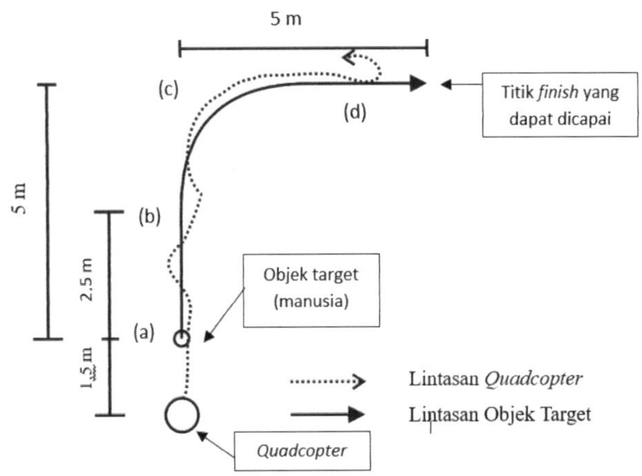

Figure12 Ilustration of Quadcopter Following Human Movement

At this point, the object being followed is human and the color sample detected is the clothes worn by humans. Tests are carried out to find out how far the quadcopter can move when following humans with movements as shown in Figure 11.

When the quadcopter is at a height of 1.5 m, the tracking switch on the RC is changed from the manual to the automatic tracking position. Then the quadcopter rotates in place looks for the middle position of humans based on the camera. When humans move forward the quadcopter moves forward (pitch) slowly also and moves to rotate (yaw) left and right. Based on the author's observation, the quadcopter can follow human movements as in Figure 18 as far as 10 meters, after that the quadcopter maneuvered around humans because of the lack of stability in the quadcopter.

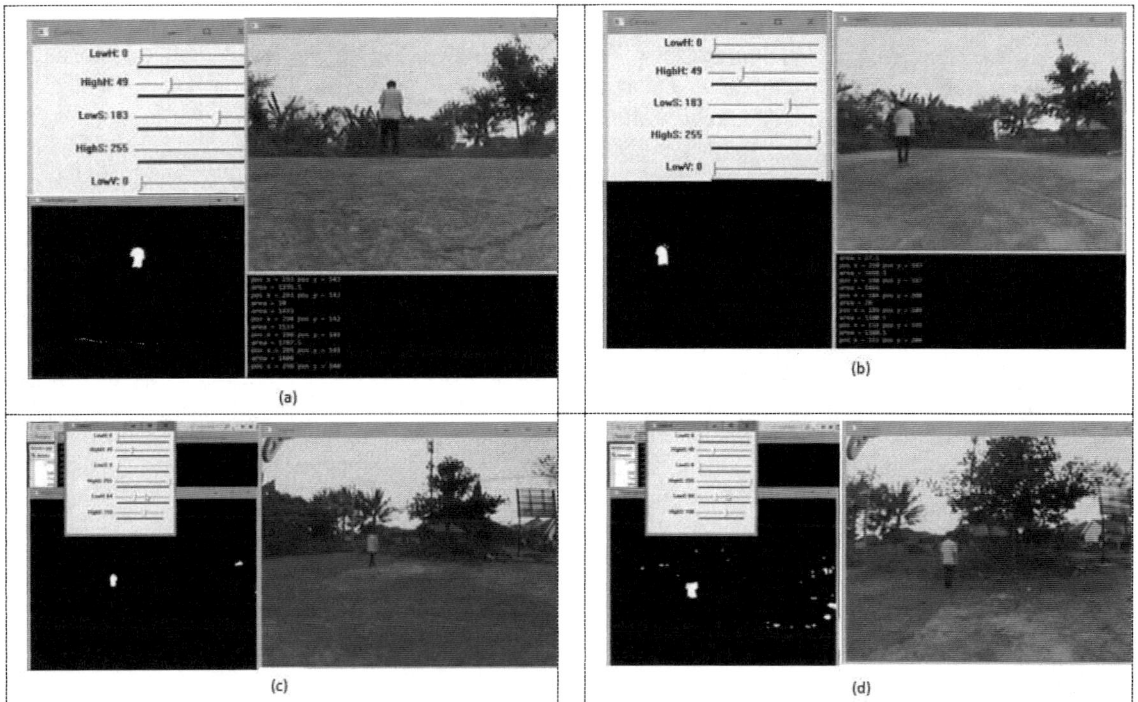

Figure 13(a) Display of*Image Processing* Visual Studio before *Quadcopter* Takeoff and Trackinghuman, (b) *Quadcopter*starts movingfollows the Object (c) *Quadcopter* Turn 90° to Left (d) *Quadcopter* Move Forward

5. Conclusions

Based on the results and discussion, some conclusions can be drawn as follows:
Modifications and quadcopter assembly have been carried out. The cuadcopter has been equipped with an altimeter sensor, accelerometer, GPS and camera sensor. Image processing testing to detect and follow objects using quadcopter has been carried out. The test results show that the quadcopter can follow a straight moving object and then turn around 90 ° as far as 10 meters. However, there are some unwanted quadcopter movements in the altitude or height of a quadcopter caused by quadcopter instability when flying in loiter mode.

References

[1] Anonyms, Droneindonesia.com. 2015. Assemble Mini Drone Quadcopter with Minimum Cost (Original in Bahasa: Merakit Mini Drone Quadcopter Dengan Biaya Minimal). http://www.droneindonesia.com/2015/02/merakit-mini-drone-quadcopter-dengan-biaya-minimal/.Acsessed on 6th February 2018.

[2] LisuallodanMuh. Sirmawan Suwardi. 2017. Designing Image Processing Object Based Drone Followers (Original in Bahasa: Rancang Bangun Drone Pengikut Obyek Berbasis Pengolahan Citra). Makassar: State Polytechnic of Ujung Pandang.

[3] Anshori, Syaifuddin. 2016. Designing Quadcopter for Optimum Route Search on Peatland Fires Using Particel Swarm Optimization Method (Original in Bahasa: Rancang Bangun Quadcopter untuk Pencarian Rute Optimum pada Kebakaran Lahan Gambut Menggunakan Metoda Particel Swarm Optimization). Universitas Islam Negeri Maulana Malik Ibrahim Malang, Indonesia.

[4] Saputra, Roni,et all.2011. Design and Initial Testing of Brushless Dc Motor Control for Independent 4-Wheel Drive Platform Robot Rev-11 (Original ini Bahasa: Perancangan dan Pengujian Awal Kendali Motor Dc Brushless untuk Independent 4-Wheel Drive Platform Robot Rev-11) *Journal of Mechatronics, Electrical Powers, and Vahicular Technology Vol. 02 No 2, pp 85-94,* Bandung:LIPI.

[5] Tefay,B.,et all.(2011). *Design of An Integrated ElectronicSpeed Controller for Compact Robotic Vehicle. Proceedings of Australasian Conference on Robotics and Automation* .Australia: Monash University.

[6] Anugerah, Risha. 2016. Flight Controller on Quadcopter Systems Using IMU (Inertial Measurement Unit) Based on ATmega 2560 Microcontroller (Original in Bahasa: Flight Controller pada Sistem QuadcopterMenggunakan IMU (Inertial Measurement Unit) Berbasis Mikrokontroler ATmega 2560). Yogyakarta: University of Sanata Dharma.

[7] Zilog, Inc. 2008. *Electric Bike BLDC Hub Motor Control* (PDF). Zilog, Inc.

[8] Hendy et all. 2011. Identifying and Tracking Objects Based on Real-Time Image Processing (Identifikasi dan Tracking Objek Berbasis Image Processing secara Real Time). Surabaya: Institut Teknologi Sepuluh Nevember.

[9] Keigo Watanabe, Kouki Tanaka, Kiyotaka Izumi, Kensaku Okamura, Rafiuddin Syam, Discontinuous Control and Backstepping Method for the Underactuated Control of VTOL Aerial Robots with Four Rotors, Intelligent Unmanned Systems: Theory and Applications Studies in Computational Intelligence Volume 192, 2009, pp 83-100

IOP Conf. Series: Materials Science and Engineering **619** (2019) 012005 doi:10.1088/1757-899X/619/1/012005

The Effect of Pole Density on Permeable Groin Due to Controlling Longshore Current

Hasdinar Umar, Sabaruddin Rahman, Chairul Paotonan, A.Y. Baeda

Department of Ocean Engineering, Faculty of Engineering, Universitas Hasanuddin

Email: hasdinar.umar@gmail.com, hasdinarumar@unhas.ac.id

Abstract. Erosion and sedimentation problems that occurs, usually solved firstly by finding the cause of erosion and sedimentation literally, due to determining the action to overcome them. In coastal engineering, there are two approaches that can be used to solve such kind of problems; which are the soft approach, i.e. beach nourishment, vegetation, sand by passing, coastal boundary policy; while the hard approach or structural approach are building breakwater, jetty, groin, sea wall and so on. In structural approach, the uses of permeable groin nowadays is one of the new ways to overcome the erosion and sedimentation problems in beaches. To find out how efficient the permeable groin can control the flow along the coast, so called longshore current, a research needed to be conducted focusing on the effect of pole density at permeable groin itself. The ability of controlling longshore current will be the main analysis, in such various variations of pole densities at the permeable groin. It found out that the pole densities of permeable groin can significantly controlled the longshore current; which is if the pole density become higher than the speed of longshore current can be reduced. It also found out that by implementing 40% of pole density, the longshore current can be reduce up 50% respectively.

1. Introduction

Groin is a coastal protective structure made perpendicular to the shoreline and used to control the movement of sedimentary materials mainly to control the amount of current along the coast and sediment transport along the coast [1]. In addition, groins can also be used to withstand the entry of sediment transport along the coast to the port or river mouth. The groin structure has been widely used as a shoreline structure, for example at the port of Richards Bay, which uses four impermeable groin structures mounted in series to protect the parrots along the Richards Bay harbor. The groin structure successfully protects the shoreline from the storm without causing damage to the structure and shoreline. In addition to the coastline of Richards Bay Harbor channel, the groin structure has also been used on Westhampton, Long Island, New York [2] and along the Northwestern Coast [3]. At Northwestern Beach, a coastline overview was observed prior to the existence of a groin structure and after a groin structure, the results of the analysis showed that the series groin structure provides larger current and sediment transport reductions than a single structure. In Indonesia the use of groin structures as shoreline protectors has also been widely used, for example in Pasauran Beach, Serang District, Banten Province [4]. On the coast of Pasauran sediment transport amounted to 7,514.88 m^3/year, sediment transport is causing the retreat of the coastline which has been a residential and tourist area.

The use of the groin structure described above is an example of an impermeable groin structure. A high impermeable groin and its length covering the width of the surf zone resulted in sediment transport along the retained coast in the up drift groin section and on the down drift groin of sediment transport continues to occur so that erosion occurs in the down drift groin due to the absence of sediment supply to the down drift. Therefore there is a considerable difference in coastline changes in the up drift and down drift groins. To overcome this, groin permeable can be used. The permeable groin has an aperture structure so that there is still a current passing through the structure so that sediment transport to the down drift groin can still occur. The amount of current and sediment transport along the coast in the permeable groin is controlled by the small gap density in the groin structure. The permeable groin structure has also been widely used in particular the permeable groin structure of the mast. One example of the use of permeable pole groins is discussed in [5], the permeable groin on the Baltic Sea coast. The permeable structure of the poles can reduce the flow and sediment transport along the coast but still provide the sediment supply to the down drift groin.

2. Groin Permeable Surface Current Flow through Groin Permeable Obstacles

The longshore current equation based on [6] is the coastal drift caused by differences in radiation stress, where they have studied the theory of radiation stress i.e. the excess flow of momentum generated by waves. The stress radiation theory can be used to describe the various phenomena occurring in the surf zone, for example explaining the setup, parallel currents, rip currents and others. In [7] the current equation along the coast through the permeable groin barriers can be developed from the Longuet-Higgins equation. The development of the current velocity equation along the coast through the permeable groin of the pole is derived from the y-direction momentum equation (parallel to the shoreline) after the obstacles of the groin structure, hence then obtained by the following equation [7],

$$\langle v\rangle_{groin} = \frac{\frac{5}{8}H_b\sqrt{\frac{g}{h_b}}(\tan\beta\ \sin\alpha_b)}{\left(\frac{2}{\pi}C_f+\frac{4C_d p h_r}{2\pi d_t}\right)} \tag{1}$$

Through the linear wave theory approach for shallow water in [8] the current velocity equations along the coast after permeable groins can also be written as follows [7],

$$\langle v\rangle = \left(\frac{h}{h_b}\right)\times\begin{cases} v_b & h < h_b \\ 0 & h > h_b \end{cases} \tag{2}$$

With

$$v_b = \frac{\frac{5}{8}H_b\sqrt{\frac{g}{h_b}}(\tan\beta\ \sin\alpha_b)}{\left(\frac{2}{\pi}C_f+\frac{4C_d p h_r}{2\pi d_t}\right)} \tag{3}$$

with groin density (p) can be formulated as follows [7]

$$p = \frac{N\pi {d_t}^2}{4L_g d_t}\times 100\% = \frac{N\pi d_t}{4L_g}\times 100\% \tag{4}$$

3. Experimental Setup

The effect of parameters of groin permeable density on the control of coastal streams are two things to be observed, i.e. the phenomenon of coastal drift in the permeable groin structure and the permeable groin structure parameters that affect the current along the coast after the permeable groin structure. Theoretically, the parameters of breaking wave height (Hb) and breaking wave angle (αb) are required to determine the amount of current along the coast (v). So in this research phase will be observed changes that occur to the parameters after the permeable groin structure. In addition to current parameters, this stage of the research will also be observed parameters of permeable groin structure i.e. pole diameter (*dt*), pole spacing (a), and distance between groin structure (*B*) which influence to the amount of current along the coast.

In the early stages of this study, a model design of the permeable groin structure will be used in accordance with the model design in Table 1. Furthermore, the permeable groin model and measuring instrument are placed in accordance with the drawings of the research design. After the groin model and the measuring instrument are placed in a predetermined position then observations and measurements of influential parameters are taken.

Table 1. Permeable Groin Model Variation Design

Model	Diameter (d_t) (cm)	Pole Spacing (a) (cm)
M2	1,5	0,7
M3	1,5	0,7
M4	1,25	0,7
M5	1,25	0,7
M6	1,5	1,0
M7	1,5	1,0
M8	1,25	1,0
M9	1,25	1,0

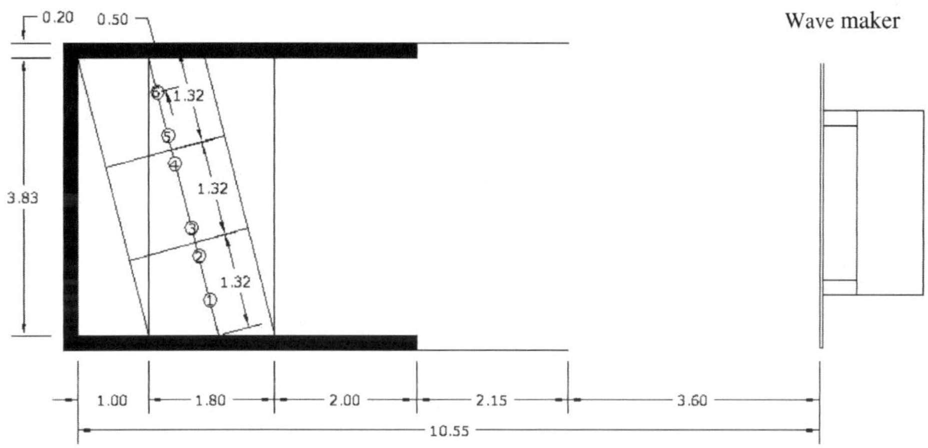

Figure 1. Plan view of fixed bed beach design model and current measurement point

4. Experimental Result

Measurements of coastal currents were carried out on 6 variations of the measurement condition i.e. condition without groin structure, condition with impermeable groin and condition with groin permeable 54%, 51%, 47% and 43%. If an impermeable or groin permeable structure is used to control the current along the shore, the flow along the mean coast (v) will be reduced due to the obstacles of the groin structure. The reduction of coastal current velocity is shown in Tables 2, 3, 4, 5 and 6. While the ratio of current velocity along the coast to the condition without groin, impermeable groin and permeable groins are shown in Figure 2 and 3.

Table 2. Reduction of current velocity along the coast in impermeable groin conditions

Breaking Wave Height, H_b (cm)	Average Velocity without Groin (*buoys*) (cm/s)	Average Velocity after *Impermeable Groin* (cm/s)	Reductions (%)
5.16	27.94	5.94	80.08
3.52	21.47	4.43	82.49
3.08	18.40	1.75	91.2

Table 3. Average velocity reduction with permeable groin of 54%
(d_t = 1.5 cm; a = 0.7 cm)

Breaking Wave Height, H_b (cm)	Velocity without Groin (cm/s)	Velocity in Groin permeability of 54% (cm/s)	Reductions (%)
5.16	27.94	12.95	53.65
3.52	21.47	9.55	55.50
3.08	18.40	7.84	57.40

Table 4. Average velocity reduction with permeable Groin of 47%
(d_t = 1.5 cm; a = 1.0 cm)

Breaking Wave Height, H_b (cm)	Velocity without Groin (cm/s)	Velocity in Groin permeability of 47% (cm/s)	Reductions (%)
5.16	27.94	13.88	50.32
3.52	21.47	10.90	49.20
3.08	18.40	8.94	51.42

Table 5. Average velocity reduction with permeable groin of 51%
(d_t = 1.25 cm; a = 0.7 cm)

Breaking Wave Height, H_b (cm)	Velocity without Groin (cm/s)	Velocity in Groin permeability of 66.67% (cm/s)	Reductions (%)
5.16	27.94	11.27	59.66
3.52	21.47	9.13	57.49
3.08	18.40	8.53	53.66

Table 6. Average velocity reduction with permeable groin of 43%
(d_t = 1.25 cm; a = 1.0 cm)

Breaking Wave Height, H_b (cm)	Velocity without Groin (cm/s)	Velocity in Groin permeability of 80% (cm/s)	Reductions (%)
5.16	27.94	12.3	56.02
3.52	21.47	10.44	51.36
3.08	18.40	9.52	48.25

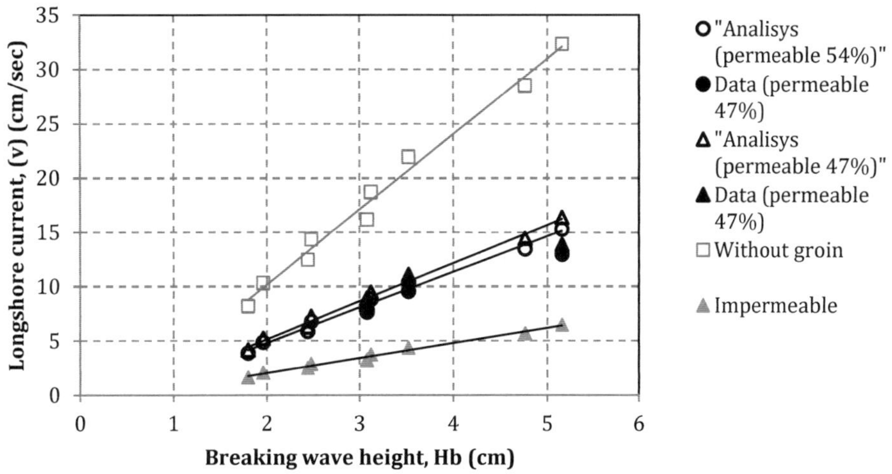

Figure 2. Relationship between Breaking Wave Height (H_b) with Longshore Current Velocity (v) from groin with permeability of 54% and 47% at d_t = 1.5 cm.

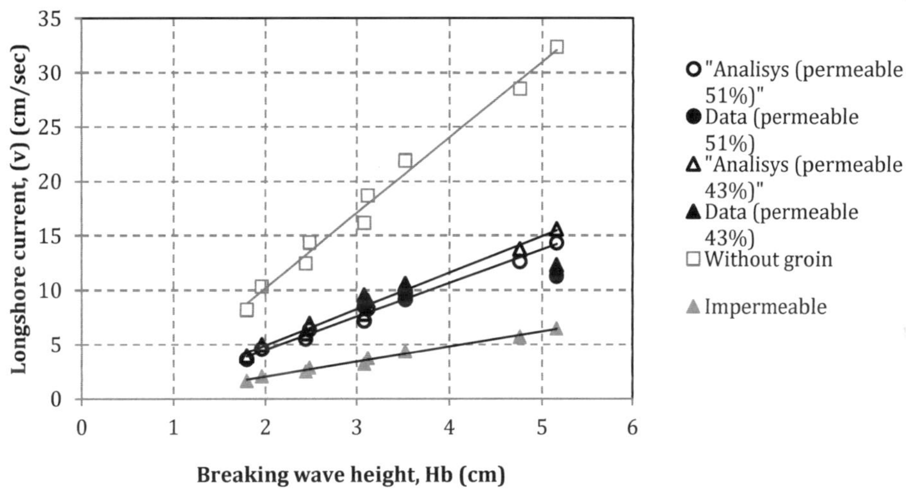

Figure 3. Relationship between Breaking Wave Height (H_b) with Longshore Current Velocity (v) from groin with permeability of 51% and 43% at d_t = 1.25 cm.

Figures 2 and 3 shows that the longshore current velocity occurring after the permeable groin structure is greater than the speed after the impermeable groin. This indicates that the permeable groin current is not completely dissipated by the groin structure. The longshore current happens to be controlled by the size of the groin density; which is the greater percentage of the groin density, the ability of the groin structure in passing the stream will decrease respectively.

5. Conclusion

Groin density (p) is a permeable groin parameter that affects the current along the coast; which the smaller the percentage of the groin density than the greater the current that can pass through the groin structure. So it can be concluded that the longshore current can be controlled by the groin structure through the large percentage of its density. The results showed that 54% of the groins density was able to reduce the current by 53.65% and 47% of the groin density was able to reduce the current by 50.32%. To determine the amount of current to be reduced can be done by adjusting the size of the groin density. This behavior can be expected to be on the same pattern on sediment transport along the coast, respectively.

6. References

[1] Yuwono N, 1992, *Dasar-Dasar Perencanaan Bangunan Pantai*, Pusat Antar Universitas, Pusat Studi Ilmu Teknik, Universitas Gadjah Mada, Yogyakarta.

[2] USACE, 2002, *Coastal Engineering Manual*, Washington, D.C.

[3] Abdellah and Balah, 2001, *Application of Permeable Groins On Tourist Shore Protection*, 4th Inter. Symp. On ocean wave measurement and analysis, ASCE, San Francisco, California, USA, Sep. 3-5.

[4] Dedi Junarsa, 2006, *Study of erosion and Solving system at Pasauran beach, Kabupaten Serang – Propinsi Banten*. Magister Tesis ITB, Bandung.

[5] Raudkivi, 1996, *Permeable Pile Groin*, Journal of Waterway, Port, Coastal, and Ocean Engineering, ASCE.

[6] Longuet-Higgins, 1970, *Longshore Currents Generated by Obliquely Incident Sea Waves,1*, Journal of Geophysical Research, Vol.75, No. 33, 20 November 1970.

[7] Hasdinar Umar, Nur Yuwono, Radianta Triatmadja, Nizam, 2013, *Theoretical Approach of Longshore Current Reduction Coefficient Through Permeable Groin*, Prociding of 4th International Seminar of HATHI, 6-8 September 2013, Yogyakarta.

[8] Longuet-Higgins, 1970, *Longshore Currents Generated by Obliquely Incident Sea Waves,2*, Journal of Geophysical Research, Vol.75, No. 33, 20 November 1970.

IOP Conf. Series: Materials Science and Engineering **619** (2019) 012006 doi:10.1088/1757-899X/619/1/012006

An Analysis of the Use of Local Bamboo as an Alternative Energy Source

Sallolo Suluh[1], Petrus Sampelawang[2], Normalia Sirande[3]

[1]Departement of Mechanical Engineering, Faculty of Engineering, Christian University of Indonesia Toraja,Tana Toraja, South Sulawesi, Indonesia
[2]Departement of Mechanical Engineering, Faculty of Engineering, Christian University of Indonesia Toraja,Tana Toraja, South Sulawesi, Indonesia

E-mail: sallolonel@gmail.com, petrussampelawang145@gmail.com, lhiyar1@yahoo.com

Abstract. The use of local bamboo biomass bricket is very effective and eficient to be used as an alternative fuel. This reasearch aimed to (1) evaluate the proximation and calor value of three types of local bamboo based on the variasi jenis bambu, (2) to determine therma efficiency of the bamboo charcoal bricket. This research applied experimental method by using three types of material through efficiency evaluation on the bricket stove. The result of experiment on the comparison of calor value was found that bricket type B1 (*Petung Bamboo)* has the highest calor value as 5176,333 cal/gram, B2 (*Wulung Bamboo*) has the value calor as 4873 cal/gram, and B3 (*Apus Bamboo*) has value of the calor as 5025 cal/gram. Meamwhile, the result of burning experiment on the stove showed that B1 (*Petung Bamboo*) is the best in terms of the ability to boil water for five (5) times by thermal efficiency 1 in the amount of 56,91%.
***Key words*:** charcoal bricket, petung bamboo, wulung bamboo, apus bamboo, calor value, thermal efficiency

1. Introduction

By the development of this increasingly advanced era, fuel energy consumption is increasing and focused only on the use of petroleum fuel which is limited and the price is increased. Therefore, it needs to do a variety of breakthroughs to get alternative energy sources, in addition to the use of oil and gas fuel.

Beside petroleum, there are three kinds of hydrocarbon sources namely coal, natural gas and biomass. Of the three sources of energy, only biomass that has the character can be renewed. Biomass is generally known as the dry matter of organic material or the material remaining after the water contain in a plant or organic material is being removed.

Biomass is very easy to be found in agricultural activities, plantations, livestock, fisheries and other wastes. Biomass waste becomes one of alternative energy sources. An example of the utilization of biomass energy derived from the forestry activities of bamboo. Bamboo is a plant of grass that grows a lot in our country. These plants can grow in both hot and cold climates. Most of the rural areas of bamboo plants are allowed to grow wild, but although not getting treatment, bamboo can grow well. Bamboo grows clustered to form clumps, shoots easily out of the rhizomes and form new plants. This new plant will grow together with its predecessor plant and will eventually form a clump with many bamboo reeds. Single-leaved bamboo arranged alternately at the tip of the reed or twigs. The root of bamboo is very strong because the rhizome branches and have a strong bond that is difficult to separate. Therefore, bamboo is widely planted in sloping areas or riverbanks to prevent erosion or landslides.

Today, many people make handicrafts from bamboo. There are also musical instruments made from bamboo such as flute and angklung. The bamboo stems are round, hollow, fully segmented, can be split from vertical and flexible. The presence of the segment will increase the strength of the bamboo rod. Specific gravity of bamboo is about 0.6 to 0.9 (dry air) lighter than water. Although bamboo grows fast but remains at its maximum properties, i.e for the tropics 6 months after the buds arise. The use of bamboo itself, especially in South Sulawesi, especially North Toraja, provides more value not only used for handicraft, but also used in the making of temporary houses if there is an event of Rambu Solo' (Funeral Ceremony) and Rambu Tuka' (Thanksgiving) as the place to sit for the guests. After the event, the bamboo is wasted away. Therefore, I will utilize it in another form so it can be worth again for renewable energy.

The description above encourages authors to examine the calorific value of bamboo under the title: "Local Bamboo Experimental Studies As Alternative Energy Sources". This study aims to determine the calorific value of bamboo in terms of variations of its type, and to determine the burning efficiency value of bamboo in the review of the variation of its type.

2. Basic Theory

Bamboo is a plant of grass that grows a lot in our country. These plants can grow in both hot and cold climates. Most of the rural areas of bamboo plants are allowed to grow wild, but although not getting treatment, bamboo can grow well.

Bamboo grows clustered to form clumps, shoots easily out of the rhizomes and form new plants. This new plant will grow together with its predecessor plant and will eventually form a clump with many bamboo reeds. Single-leaved bamboo arranged alternately at the tip of the reed or twigs. Root bamboo is very strong, because the rhizome branches and have a strong bond that is difficult to be separated. Therefore, bamboo is widely planted in sloping areas or riverbanks to prevent erosion or landslides.

Today many people make handicrafts from bamboo. There are also musical instruments made from bamboo such as flute and angklung. The bamboo stems are round, hollow, fully segmented, can be split from vertical and flexible. The presence of the segment will increase the strength of the bamboo rod. Specific gravity of bamboo is about 0.6 to 0.9 (dry air) which is lighter than water. Although bamboo grows fast but remains at its maximum properties, i.e for the tropics 6 months after the buds arise. This is the characterictic that distinguishes bamboo from ordinary trees. Bamboo felled after 4 years, the number of bamboo per hectare between 100 to 500 clumps or about 2000 to 14000 stems, depending on the type and fertility of the soil, while the diameter between 2 to 10cm.

The types of bamboo that are found in Indonesia are: rope bamboo (*apus*), petung bamboo, ater bamboo, spotted bamboo, wulung bamboo and others.

2.1. **Binder/Adhesive material**

To bind the particles of substances in the raw materials in the process of making briquettes, the are some necessary binding materials to produce a compact briquettes, they are:

2.1.1 Clay

Clay is a silicate base fractional element of silicates less than 4 micrometers in diameter. Clay contains a fine silica and/or aluminum fused. These elements, silicon, oxygen, and aluminum are the elements that make up most of the Earth's crust. Clays are formed from the process of weathering silica rocks by carbonic acid and partly generated from geothermal activity.

2.2 Adhesives/Sago Flour

Sago flour is flour obtained from sago stalk. Sago has adhesive properties, so it is very good to be made as an adhesive briquette in this study.

The testing consisted of two parts : briquette burning/water boiling and calculations efficiency (η_{th}) :

$$\eta_{th} = \frac{Q_w + Q_p}{LHV \times M_{bb}} \tag{1}$$

$$\eta_{th} = \frac{(Ma \times Cp \times (T_a - T_b)) + \left(M_p \times Cp_{al} \times (T_c - T_b)\right) + (M_u \times H_L)}{LHV + M_{bb}} \tag{2}$$

where :

η_{th} : thermal efficiency of briquette burning (%).
M_a : initial water mass (kg),
M_{bb} : remaining briquette mass in the stove (kg).
M_u : mass of water vapor (kg).
H_L : vapor latent heat (kJ/kg).
Cp_{air} : water specific heat 4.1769(kJ/kg ^{0}C).
Cp_{al} : aluminum/pot material specific heat (kJ/kg ^{0}C).
LHV : briquette lower heating value (kJ/kg).
T_b : water's ambient temperature
T_a : water vapor temperature (100 ^{0}C)
T_c : Pot temperature (^{0}C)

Figure 1. Petung bamboo (*Dendrocalamus asper)* Grows up in the Toraja area

Figure 2. Wulung bamboo (*Gigantochloa atroviolacea* Grows up in the Toraja area

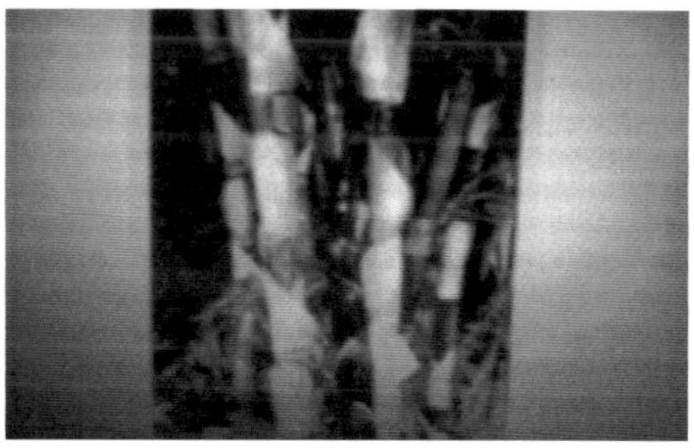

Figure 3. Apuss bamboo (*Gigantochloa apus* Kurz*)* Grows up in the Toraja area

3. Research Materials and Methods

The research method used was experimental method by utilizing local bamboo as rope bamboo (*apus*), briquettes as alternative fuel on the stove tested. The form of charcoal briquettes used is the form of the wasp nest because based on some previous researches that the form of the wasp nest has a larger surface of the flame field. The variation of bamboo charcoal species with a combination of raw materials can be that B1 briqutesss (850 gram fuel petung bamboo+75 gram clay+75 gram sago flour); B2 Briquettes

(850 gram fuel wulung bamboo+75 gram clay+75 gram sago flour); B3 briquettes (850 gram fuel apus bamboo+75 gram clay+75 gram sago flour). The next is testing proksimate test, pysical test, the calorific value and burning efficiency of the stove.

4. Results and Discussion

The content of heating value obtained in briquettes was B1 5176 cal/gram, B2 4873 cal/gram and B3 5025 cal/gram. The average of all three types of briquettes based on the type of bamboo obtained was 5033,777 cal/gram. The result of calculation of burning efficiency for the three types of bamboo charcoal briquette, can be seen in Figure 2. In the graph above, it can be seen that the results of the calculation of thermal efficiency shows that B1 has the highest efficiency of 57.93%, followed by B2 of 49.96% and the lowest efficiency is in B3 that is equal to 35.48%. B1 type briquettes have a very high thermal efficiency because they have a very high calorific value of 5176 cal / gram, with the ability to boil water 5 times with a mass of boiling water of 4000 ml which consumes 0.28 kg of briquettes.

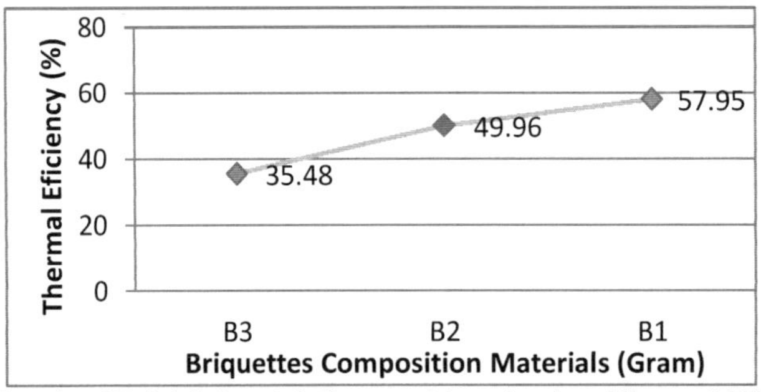

Figure 4. Graph of Brick Material Composition Against Thermal Efficiency

5. Conclusions

By the results of this study, it can be concluded that:The highest calorific value was petung bamboo briquettes (B1) of 5176 cal/gram, followed by apus bamboo briquettes (B3) of 5025 cal/gram, and the lowest is wulung bambo briquettes (B2) of 4873 cal/gram. The efficiency value of burning in briquette stove shows that bamboo bamboo briquettes have the highest efficiency of 57.93%, followed by bulu bulul bulse (B2) of 49.96% and the lowest efficiency value is bamboo briquette briquette (B3)) of 35.48%.

6. Reference

[1] Arif, E dan sallolo suluh., 2014 Study of Performance Improvement Possibility of Various Stoves with Waste BiomassFuel Briquettes. The 1st International Symposium On Smart Material And Mechatronics, Graduate School Of Mechanical Engineering University of Hasanuddin, . 2014.

[2] Pilimon Unukly dkk. 2016. The quality of charcoal briquettes as an alternative fuel made from corn cobs and bamboo waste.(Original in Bahasa: Kualitas briket arang sebagai bahan bakar alternative berbahan baku limbah tongkol jagung dan bambu) http://jurnalee.files.wordpress.com., Acessed April 20 2017.

[3] Taufik Iskandar, Hesti Poerwanto. 2015. Identification of calorific value and time of flame resulting from a combination of particle size and strength in the bamboo bio-briquettes (Original in Bahasa : identifikasi nilai kalor dan waktu nyala hasil kombinasi ukuran partikel dan kuat takan pada bio-briket bambu), http://Jurtek.akprind.ac.id/sites/default/files/116_123_putra.pdf.com, Acessed 10 Juli 2017.

[4] Hijra Purnama Putra., Melrdania Mokodompit. 2013. The study of the characteristics of briquettes made from bamboo waste using rice adhesive (Orginil in Bahasa : Study karakteristik briket berbahan dasar limbah bambu dengan menggunakan perekat nasi., http://Jurtek.akprind.ac.id/sites/default/files/116_123_putra.pdf.com., Acessed 10 Juli 2017

Impacts of Grid Frequency Variation on Dynamic Performance of DFIG Based Wind Turbine

A. M. Shiddiq Yunus

Energy Conversion Study Program, Mechanical Engineering Department, Politeknik Negeri Ujung Pandang, Indonesia

shiddiq@poliupg.ac.id

Abstract. Wind energy is one of popular sources of renewable energy that massively used to generate electric power nowadays. Among wind generator technologies, Doubly Fed Induction Generator (DFIG) is placed as the first dominant wind generator type that is installed worldwide since 2004. Its popularity is due to the advantages of its capability to extract more energy compare to the fixed speed type. Moreover, it uses only one-third converter size which reduced the total construction cost compared to its closest rival, Full Converter Wind Turbine Generator Type. Although DFIG widely installed worldwide, it is in fact, very sensitive to the grid disturbances such as frequency variation. Based on IEEE standard for 60 Hz system, a power system should not be allowed to work in less than 59.3 Hz or more than 60.5 Hz. In this paper, DFIG-grid connected is extensively simulated in the condition of frequency variation that is carried out using Matlab/Simulink. The results show that some important parameters such as power output of the generator, voltage at point of common coupling, voltage at DC link and generator speed are experiencing significant rise/drop which in turn might damage the wind turbine generator.

1. Introduction

It is inevitable for every country all around the world to intensively explore their renewable energy sources to mitigate the detrimental impacts of conventional energy sources to environment. Moreover, the deposit sources of these conventional energy sources decreases from time to time which may lead to global economic instability. Therefore, many countries, particularly for some groups of developed countries have installed large number of renewable based power plants such as hydro, wind and solar [1, 2]. For the last decades, wind turbine generators have significantly increased in installation number all around the world by about 540 GW until February 2018 [3] as can be seen in Fig. 1.

Among other types of wind turbine generators (WTGs), Doubly Fed Induction Generator (DFIG) becomes the most popular since last decade [4]. Its popularity is due to several advantages including capability to extract more energy compare with the fixed speed type [5], and its benefit in reducing the construction total cost by having only one-third capacity of converters [6] compare with its closest rival, Full Converter Wind Turbine Generator. The general configuration of DFIG is shown in Fig. 2.

As can be seen in Fig. 2, a DFIG consists of two converters; Rotor Side Converter (RSC) a converter that connects to the rotor of the induction generator and Grid Side Converter (GSC), a converter that connects with the grid side. These two converters is linked by a capacitor or so-called DC link capacitor. These one-third sizes of converters play an important role to allow absorb and deliver some amount of reactive power from/to the grid. Its 30% converters size could reduce the construction cost significantly compared to its closer rival Full Converter Wind Turbine Generator [7]-[9].

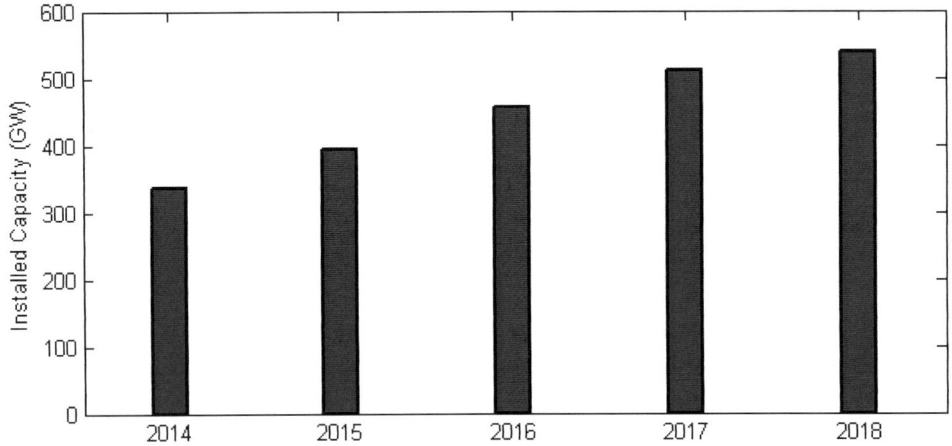

Figure 1. Number of wind turbine generator installation worldwide until February 2018 [3].

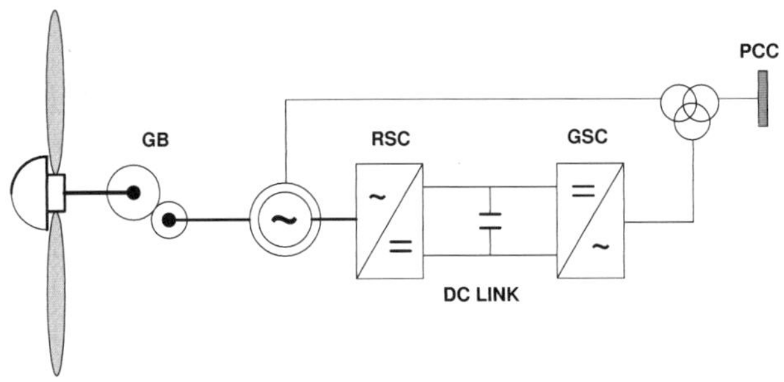

Figure 2. Typical configuration of a DFIG

As large scale type of WTG, DFIGs normally connected to the grid and capable to provide limited reactive power when required by the grid. However, DFIGs are sensitive to the grid faults that may lead to the disconnecting it from the grid. Disconnecting a large MWs DFIG from the grid means large economic loses to the WTGs' owner. Some papers discusses the impact of DFIGs' dynamic performance during grid faults such as voltage sags [10] and swell [11] as well as internal faults event that might occur at the converters of DFIG [12, 13]. However, not much attention is given to the case of grid frequency variation impacts on the dynamic performance of DFIG.

Based on IEEE standard [14] for 60 Hz system, a power system should not be allowed to work in less than 59.3 Hz or more than 60.5 Hz. However in the real cases, many frequency variations occur even more severe than the limit of IEEE Standard as can be seen in several places all around the world such as summarize in table 1.

In this paper, two cases of grid frequency variation are applied to investigate the dynamic responses of the DFIG. All system are carried out and extensively simulated in Matlab/Simulink environment.

IOP Conf. Series: Materials Science and Engineering **619** (2019) 012007 doi:10.1088/1757-899X/619/1/012007

Table 1. Major Frequency Deviation Cases from 1970s-2000s

Places	Frequency Deviation	Causes	Year	Ref.
East Ontario, Canada	58.7 Hz-62.6 Hz	Islanding Electrical Grid	1972	[15]
New York City, USA	< 47.5 Hz	Islanding Electrical Grid	1977	[16]
West and South Coast of England	47.3 Hz	Islanding Electrical Grid	1981	[17]
France	49.6 Hz	Cascading Fault	1985	[17]
Italy	NA	Cascading Transmission Line	1994	[15]
Perth, Australia	<3.5 Hz/s	Cascading Outage	1994	[17]
Brazil	55.25 Hz-58.0 Hz	Cascading Fault	1996	[15]
Malaysia	49.1 Hz	Cascading Fault	1996	[15]
Italy	47.0 Hz	Lack of Generation	2003	[18]
Europe	NA	Cascading Outage	2006	[19]

2. System under Study

The system under study consist of six 1.5 MW DFIG, 60 Hz that are connected to the grid via 30 km distribution line as depicted in Fig. 3. The system parameters are provided in Table 2-3.

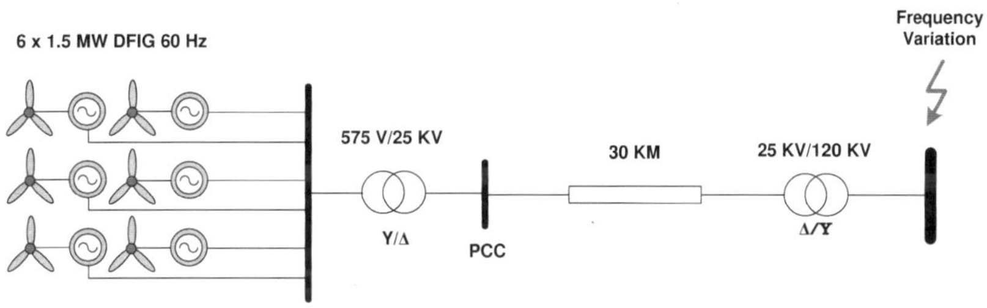

Figure 3. System under Study

Table 2. Parameters of DFIG

Parameters	DFIG
Rated Power (MW)	6 x 1.5 MW
R_s (p.u.)	0.023
H(s) (p.u.)	0.685
V_{dc} (V)	1150

Table 3. Parameters of Distribution Line and Grid

Parameters	Distribution line
R_1 (ohms/km)	0.1153
R_0 (ohms/km)	0.413
L_1 (H/km)	1.05e-3
L_0 (H/km)	3.32e-3
C_1 (F/km)	11.33e-9
C_0 (F/km)	5.01e-9

3. Simulation Results and Discussion

In this study, there are two cases are investigated. Where the two cases are applied to the system under study for about 8 cycles from 0.5s-2.0s for the case I and II.

3.1. Case I

In this case, a frequency drop of -3.5Hz/s is applied and lasting for 1.5s. Some of the most important parameters of dynamic performance of DFIG during the fault are shown in Fig. 3(a)-3(d).

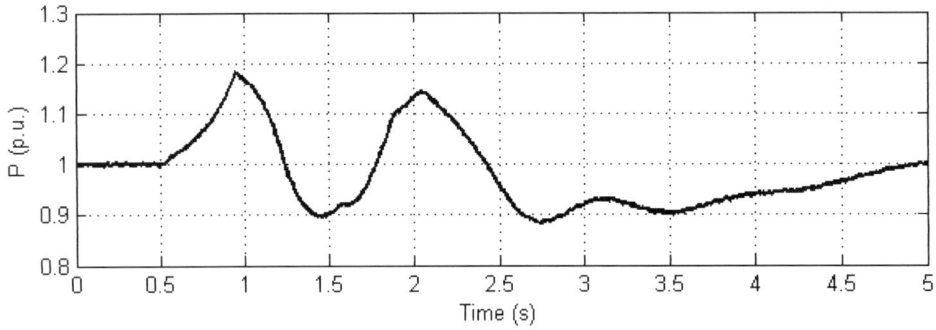

(a)

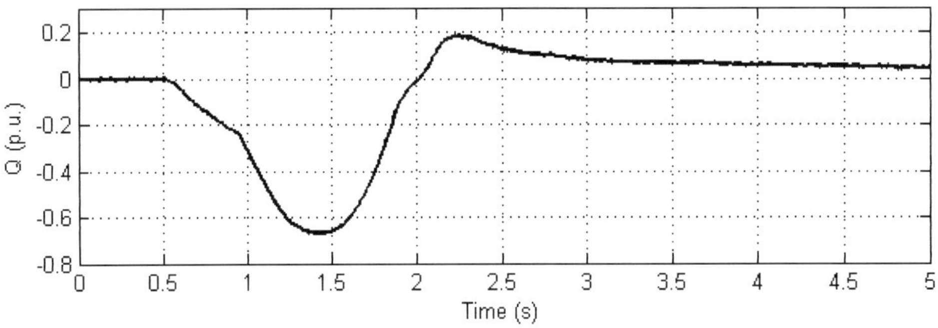

(b)

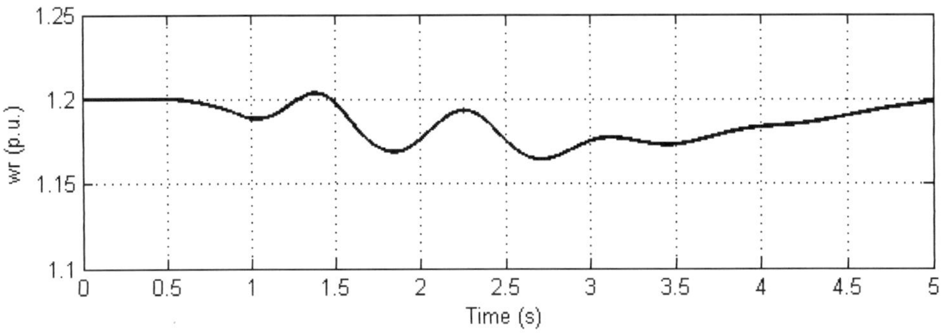

(c)

IOP Conf. Series: Materials Science and Engineering **619** (2019) 012007 doi:10.1088/1757-899X/619/1/012007

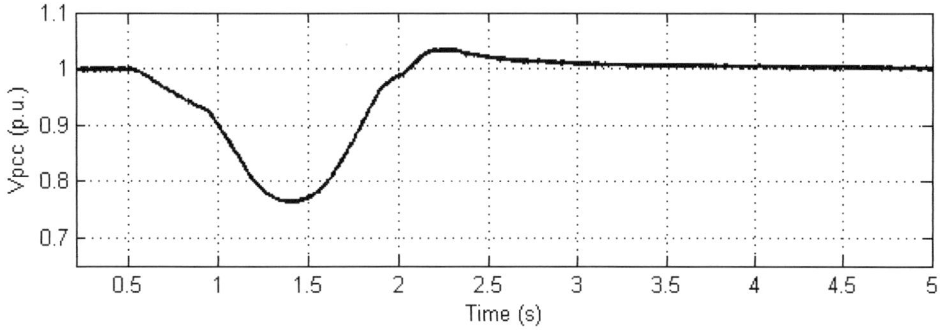

(d)

Figure 4. Dynamic responses of DFIG during frequency drop for -3.5 Hz/s; (a) P output response, (b) Reactive Power output response, (c) Generator Speed (ω_r) response, and (d) Voltage Output profile at PCC.

Fig. 3(a) shows that there is an oscillation on the generated power when grid frequency drop started at 0.5s. Moreover, although the drop of grid frequency is returned to normal in 2.0s, the trend of generated power to return to normal is achieved after 5.0s. This condition might influence the critical load that connected to the grid and rely on the power supply from the DFIGs.

As can be seen in Fig. 3(b), the frequency drop of -3.5 Hz/s from 0.5s to 2.0s causes the drop of reactive power output of DFIG by about 0.7 p.u. meaning that there is a momentary reactive power supply of DFIGs to the grid. However, the control mechanism of the DFIG could return the generated reactive power to normal condition quickly after the frequency drop return to normal. Fig. 3(c) shows the similar trend with the Fig. 3(a) as the generated power is related to the rotor speed of the induction generator. Meanwhile, Fig. 3(d) tends to give the same response with the reactive power profile shown in Fig. 3(b) as they influence each other. The voltage drop at PCC (Fig. 3(d)), although dropped about 0.25 p.u., is still within the safety margin of almost international grid codes.

3.2. Case II

The extensive simulation for Case II, is aimed to investigate the impacts of increased frequency change about +3.5Hz/s. The simulation results are shown in Fig. 4(a)-4(b).

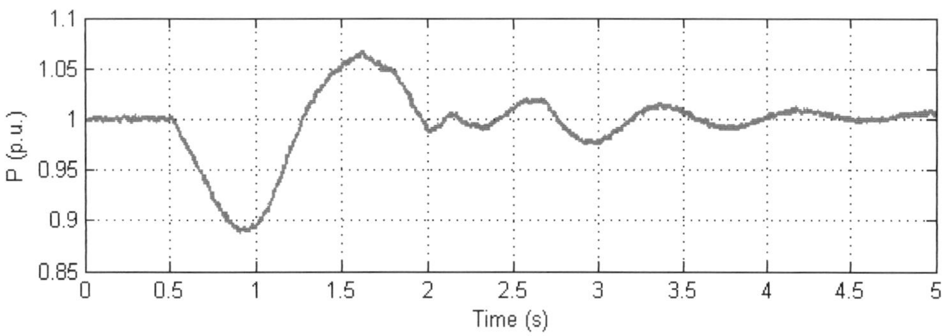

(a)

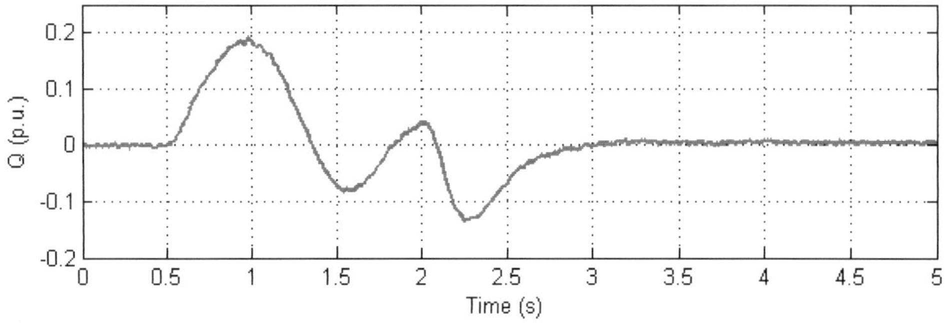

(b)

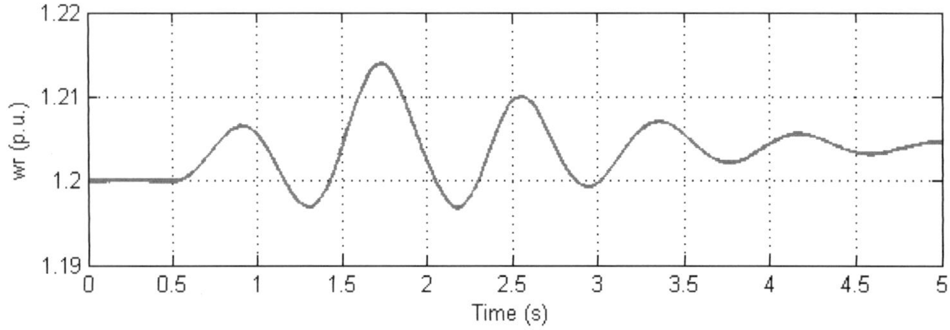

(c)

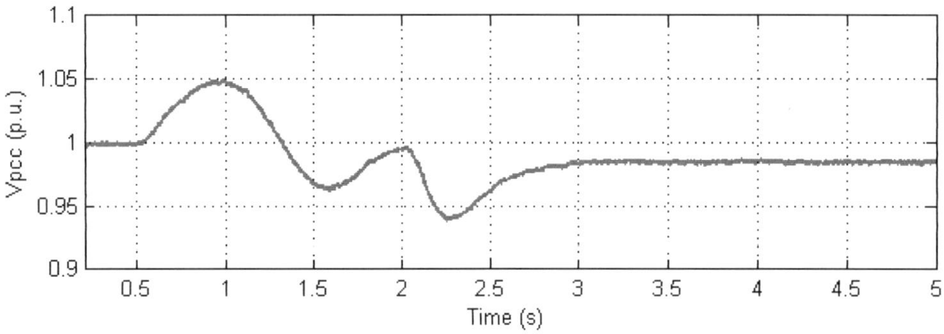

(d)

Figure 5. Dynamic responses of DFIG during frequency ramp for +3.5 Hz/s; (a) P output response, (b) Reactive Power output response, (c) Generator Speed (ω_r) response, and (d) Voltage Output profile at PCC.

Fig. 5(a)-(d) shows the dynamic response of DFIGs during frequency ramp of +3.5 Hz/s that is started from 0.5s to 2.0s. In this case, oscillation of generated power is lesser than the first case and reaches faster steady state as shown in Fig. 4(a). This condition might not risk the critical loads that are connected to the grid. Reactive power, Q, is oscillating right after frequency ramp is started at 0.5s but short after fault cleared out at 2.0s, generated Q returned to its defined state at 3.0s (Fig. 4(b)).

Generator speed (ω_r) is swing right after the frequency fault started and tends to reach stable condition after 5.0s as plotted at Fig. 4(c). Meanwhile, low amplitude oscillation of voltage profile at PCC (as shown in Fig. 4(d)) occurs due to the grid frequency fault and might reach its defined value longer than the first case. However, its level does not violate the most of common international grid codes as normal voltage level must not lower or higher than ± 5%.

IOP Conf. Series: Materials Science and Engineering **619** (2019) 012007 doi:10.1088/1757-899X/619/1/012007

4. Conclusion

In this paper, dynamic responses of DFIGs-Grid connected are investigated during grid frequency change. The dynamic responses of DFIG during grid frequency drop of -3.5 Hz/s (Case I) are worse than the grid frequency ramp of +3.5 Hz/s (Case II) for the studied systems. The oscillation reaches higher amplitude and longer steady state time for both P and Q in the first Case. Moreover, voltage profile at PCC experiencing larger drop in Case I than Case II. This study might be continued for the sustained frequency drop/ramp as future work and the results might be considered for the related authorities in designing and developing the WTGs grid connected.

References

[1] Yunus, A.M. S., Saini, M., 2016, "Overview of SMES units application on smart grid systems", 2016 International seminar on intelligent technology and Its Application, ISITIA 2016. pp. 465-470

[2] Firman, Yunus, A.M.S., Yunus, M. Y., 2017, "Sediment characteristic on hydropower plant Bakaru, South Sulawesi", AIP Conference Proceedings.

[3] http://www.wwindea.org/2017-statistics/ (acessed: 01 July 2018)

[4] Shiddiq Yunus, A.M., Abu-Siada, A., Masoum, M.A.S., 2012, "Improving dynamic performance of wind energy conversion systems using fuzzy-based hysteresis current-controlled superconducting magnetic energy storage", IET Power Electronics, 5(8), pp.1305-1314.

[5] Shiddiq Yunus, A.M., Abu-Siada, A., Masoum, M.A.S., 2011, "Effect of SMES unit on the performance of type-4 wind turbine generator during voltage sag", IET Conference Publications 2011 (579CP), pp.94.

[6] Mitra, A. and Chatterjee, D., 2015, "Active Power Control of DFIG-Based Wind Farm for Improvement of Transient Stability of Power Systems", IEEE Trans. On Power Syst., Vol. PP, Issue: 99, pp.1-12.

[7] www.windpowermonthly.com (accessed: 03 July 2018)

[8] Yunus, A.M.S, Masoum, M.A.S., Abu-Siada, A., 2012, " Impact of intermittent misfire and fire-through on the performance of full converter based WECS", 22nd Australasian Universities Power Engineering Conference: "Green Smart Grid Systems", AUPEC 2012, 6360194.

[9] Yunus, A.M.S., Abu-Siada, A., Masoum, M.A.S., 2011, "Improvement of LVRT capability of variable speed wind turbine generators using SMES Unit", Innovative Smart Grid Technologies Asia (ISGT), IEEE PES, pp. 1-7.

[10] Khamaira, M.Y., Shiddiq Yunus, A.M., Abu-Siada, A., 2013, " Improvement of DFIG-based WECS performance using SMES unit", Australasian Universities Power Engineering Conference, AUPEC 2013, 6725380.

[11] Alharby, Y.M., Shiddiq Yunus, A.M., Abu-Siada, A., 2012, " Application of STATCOM to improve the high-voltage-ride-through capability of wind turbine generator", 22nd Australasian Universities Power Engineering Conference: "Green Smart Grid Systems", AUPEC 2012, 63630253.

[12] Yunus, A.M.S., Abu-Siada, A., Masoum, M.A.S., 2012, "Impact of DC-link fault on the dynamic performance of DFIG", IEEE Power and Energy Society General Meeting, 6345278.

[13] Yunus, A.M.S., Abu-Siada, A., Masoum, M.A.S, 2011, "Effects of SMES on Dynamic Behaviors of Type D-Wind Turbine Generator-Grid Connected during Short Circuit". IEEE Power and Energy Society General Meeting, 6039276.

[14] IEEE, 2003, "IEEE standard for interconnecting distributed resources with electric power systems", IEEE Std. 1547TM.

[15] N. Janssens, P. Baratella, D. Bickers, D. Bones, A. Cooke, E. Gaglioti, E. Grebe, R. Hartman, M. Houry, T. Inoue, J. Jardim, K. Karoui, P. Kundur, M. O"Malley, F. Thoden, R. Schulz, D. Tong, P. Scarpellini and F. Casamatta,, 1998, "Analysis and Modeling Needs of power Systems under Major Frequency Disturbances", IEEE /CIGRE Task Force 38.02.14 Rep.

[16] B. Weedy, B. Corny, N. Jenkins, J. Ekanayake and G. Strbac, 2012, "Electric Power System", 5th ed. John Wiley & Sons Ltd, p. 152.

[17] M. Altin, 2012, "Dynamic Frequency Response of Wind Power Plants", Ph.D, Aalborg University, Denmark.

[18] S. Crosi and C. Sabelli, 2004, "General Blackout in Italy Sunday September 28, 2003, h. 03:28:00", IEEE, Power Engineering Society General Meeting, pp. 1691-1702.

[19] C. Li, S. Yuanzhang and X. Chen, 2007, "Analysis of the Blackout in Europe in November 4, 2006", Power Engineering Conference, 2007. IPEC 2007. International, Singapore, pp. 939-944.

IOP Conf. Series: Materials Science and Engineering **619** (2019) 012008 doi:10.1088/1757-899X/619/1/012008

Progressive Collapse Behaviour of VLCC under Longitudinal Bending

Tsumiratin Rizkiani*, Muhammad Iqra Ramadhan, Muhammad Zubair Muis Alie

Department of Ocean Engineering, Hasanuddin University, Indonesia

Email: *tsumiratinr@gmail.com

Abstract. The aim of this study is to analyse the ultimate strength of VLCC subjected to longitudinal bending moments under hogging and sagging conditions. One-frame space in the mid-ship section of VLCC is taken to be assessed. The Nonlinear Finite-Element Analysis method (NLFEA) is adopted in this study. The boundary and loading conditions are applied on both sides of the ship cross section. The Multi Point Constraint (MPC) is imposed to the cross section based on the neutral axis references. The element type is SHELL 181 is used on the model. The meshing size is 500 mm**,** and the arc length method is applied to solved the load and shortening curve. The initial imperfection, cracks and welding residual stress are ignored. The result obtained by NLFEA method for ultimate strength in hogging and sagging are $10{,}83 \times 10^{12}$ Nmm and $-10{,}34 \times 10^{12}$ Nmm, respectively. It is shown that the ultimate strength obtained by NLFEA is larger than Smith's method, and the ratios of the ultimate strength comparison are 7,71% and 11,27% for hogging and sagging conditions, respectively.

1. Introduction

Many tankers were operated using single hull on 1960-1990. In that year, there were many ship accidents results pollution and impact on the destruction of the marine environment. In this regard, due to structural degradation, the ship lost its stiffness because of damage then collapse occur. To avoid this accident, the ultimate strength of ship structure need to be investigated before and after damage takes place. Some tools of the calculation may be used to assess and/or evaluate the ultimate strength including their progressive collapse. One of them is The Non-Linear Finite Element Analysis (NLFEA) method that is very accurately and recommended by the classification and many researchers to calculate the ultimate strength of complex structures.

Many researches have been done regarding to the ultimate strength analysis of ships. Van Vu [8] performed finite element analysis to predict the ultimate longitudinal strength of intact ship using RISK method that was modified in ABAQUS software. The reliability of this approach was estimated on the comparison between results of experiment models and finite element models. Muis Alie, M.Z *et al* [4] assessed the ultimate hull girder strength of Ro-Ro ship after damage. The cross section of Ro-Ro ship was considered to be analyzed. Muis Alie, M.Z *et al* [3] analyzed the hull girder ultimate strength of asymmetrically damaged ships using Finite Element Method. The collision damage was modelled by removing the plate and stiffened plate elements. Muis Alie, M.Z *et al* [2] analyzed the residual strength of ship hull girder with bottom damage. The nonlinear finite element method was used and the fully cross section was considered in the calculation. Soares *et al* [7] evaluated the ability of structural analysis of simple method based on Smith's formulation to predict the power limit of hull girder in detriment.

The incremental-iterative approach in progressive collapse analysis of various hull girder structures, Andric, et.al [1], compares the hull girder bending moment values using 14 different methods to five different ship types, one of which is the single hull VLCC tanker. Based on the research conducted, the strength value of bending moment limit on hogging condition is greater than sagging.

The previous description has also been presented by Paik [5], for the double hull ship type. The boundary strength value is greater in the hogging condition because the bottom has a structural configuration that can massively withstand the compression force and tensile force better than the structure configuration on the deck. Configuration of double-hull and double-bottom structure cross-section models has greater border-strength than single hull and single bottom structures with the IACS design rules.

Paik et al [6], analyzed the accuracy of the ALPS / HULL calculations in the progressive collapse analysis of hull girder by comparing with the analysis method of element as well as nonlinear. The moment of inertia in the cross section and the limit strength of the girder hull is not the goal of optimization. The purpose of the optimization is on the weight of the structure and strength of the local boundary girder. This indicates that optimizing the local strength of the plate can increase the strength of the global boundary of the hull girder.

2. Method of Analysis

In the present study, the non linerar finite element analysis is used to investigate the progressive collapse behavior of VLCC under vertical bending moment in hogging-sagging condition. The breadth and depth of VLCC are 42 m and 20.3 m, respectively. The cross section of VLCC is shown in Figure 1 as follow.

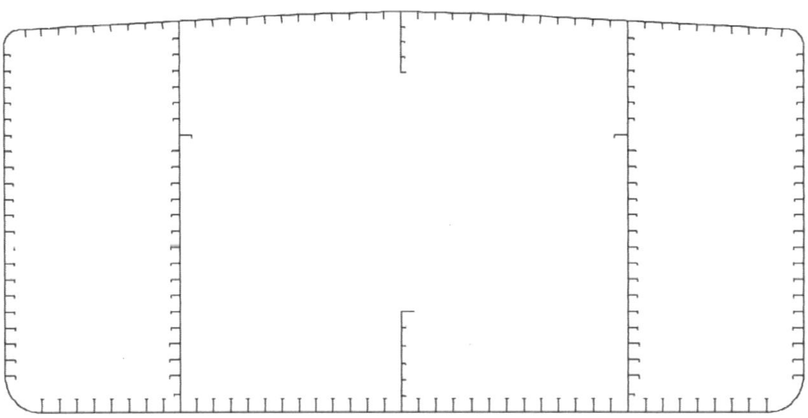

Figure 1. Cross section of VLCC.

The 3D of the finite element model is created by using shell element type 181 with configuration and material specifications for each element as shown in Table 1. The mesh size is 500 mm and it is applied for whole model. The material properties of VLCC using AH 27 as shown in Table 1.

In the present study, the ultimate strength is analysed under moment control process in order to know the progressive collapse of VLCC in intact condition. The Multi Point Constraint (MPC) is applied at both sides of the cross section as reference point. It should be noted that the MPC is placed where the neutral axis position is located. The verrtical bending moment 1.6 x 10^{13} N-mm is given on the MPC point. Then, all nodes at the fore-end are rigidly linked to the neutral axis reference point. The illustration of boundary condition and loading as shown in Figure 2.

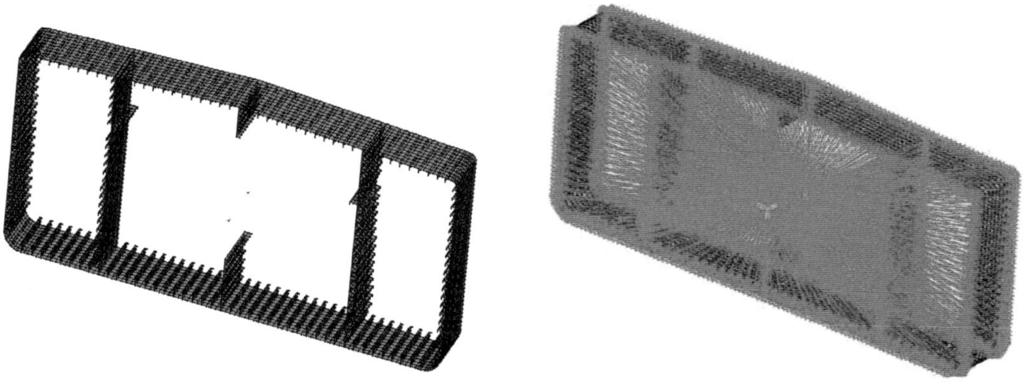

Figure 2. Modelling and boundary condition of VLCC by FE method

Table 1. Material Properties of VLCC

Density (kg/m^3)	7850
Modulus Young (N/mm^2)	210000
Poisson's Ratio	0,3
Yield Strength (N/mm^2)	290
Tangent Modulus (N/mm^2)	625

3. Results and Discussion

Table 2 shows result of the ultimate bending moment strength calculated at each percentage of intact condition in both sagging and hogging conditions using NLFEA. In intact condition, the ultimate strength of VLCC obtained as 10.83x10^{12} Nmm and -10.34x10^{12} Nmm under hogging and sagging conditions, respectively. Figure 3 shows the moment-curvature relationship of VLCC under hogging and sagging condition.

Table 2. The Ultimate Strength of VLCC (in Nmm).

Condition	NLFEA Method Nmm
$M_{u\,Hogging}$ x 10^{12}	10,83
$M_{u\,Sagging}$ x 10^{12}	-10,34

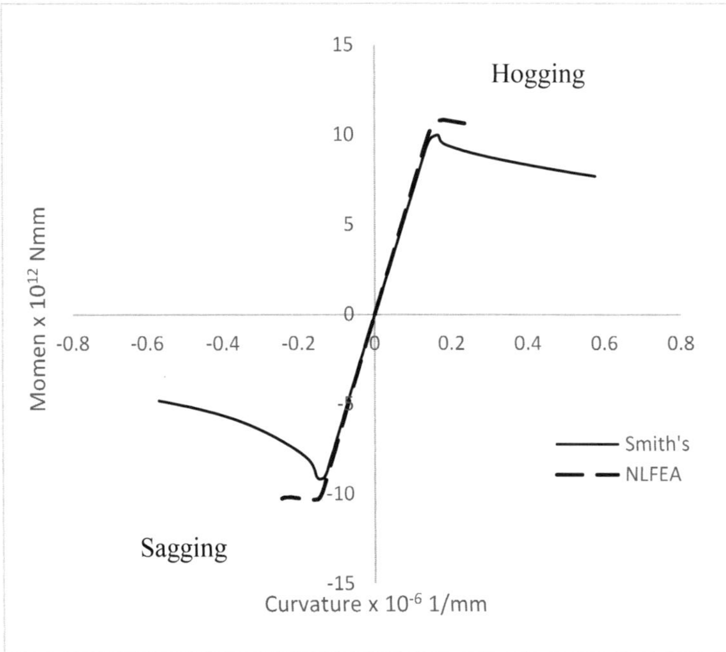

Figure 3. Moment-curvature relationship in intact condition.

Table 3. The Ultimate Strength of VLCC

Condition	Smith's Method x 10^{12} Nmm	NLFEA x 10^{12} Nmm	Ratio
$M_{u\ Hogging}$	9,986	10,82	1,084
$M_{u\ Sagging}$	-9,175	10,34	1,127

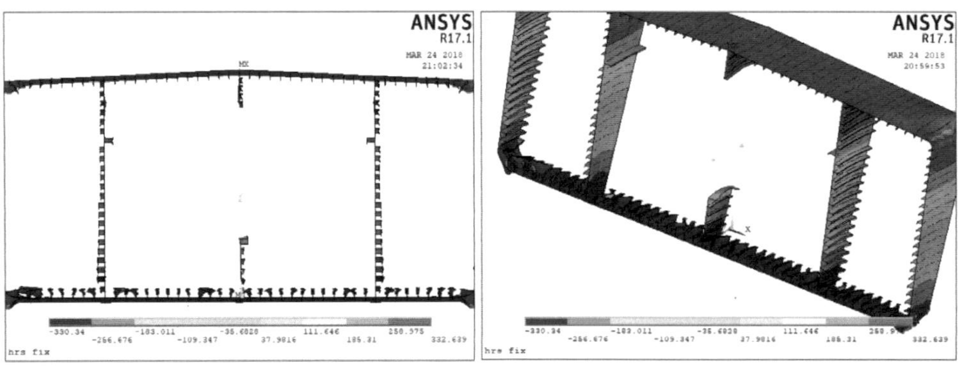

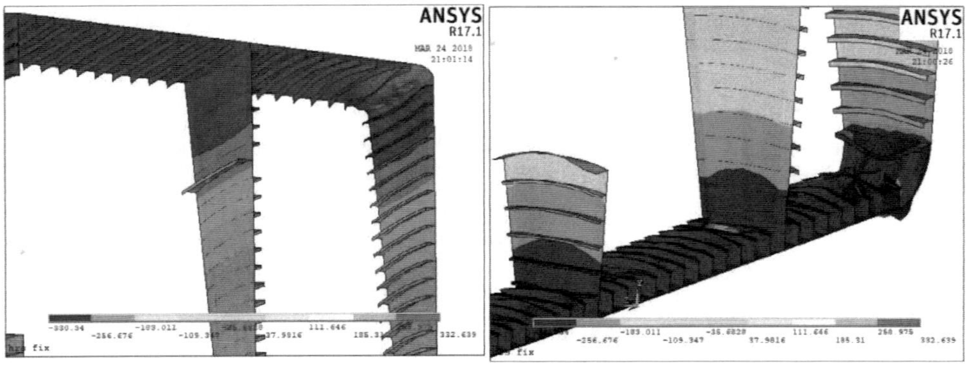

Figure 4. Stress distribution of VLCC at the ultimate strength in hogging condition.

Based on the result of the analysis using NLFEA method, the working stress on the VLCC tanker at hogging is -330.34 N/mm^2 at the bottom and 332,639 N/mm^2 on the deck as shown in Figure 4. At the hogging condition, the longitudinal section of the deck will experience the pull and the bottom will be pressed, so that the amount of stress value will vary. This is due to the process of control moment. The deformation that occurs in the bottom encounters a change in shape, where in the stiffeners and the plates on the side are bending.

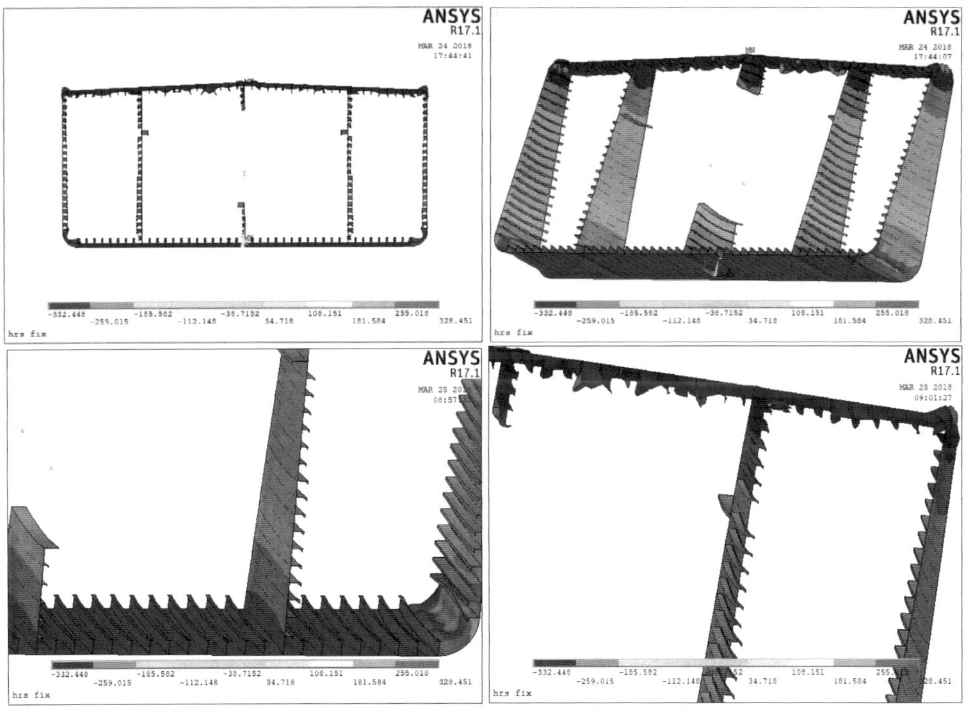

Figure 5. Stress distribution of of VLCC at the ultimate strength in sagging condition.

Based on the result of the analysis using NLFEA method, the working stress on the VLCC tanker at sagging is -332,448 N/m^2 on the deck and 328,451 N/mm^2 on the bottom as shown in Figure 5. At the sagging condition, longitudinal section of the bottom will experience the drag and the deck will be pressed. So that the amount of stress value will vary. This is due to the process of control moment. The deformation that occurs in the deck encounters a change in shape, wherein stiffeners and plate on the side are bending.

The use of the ANSYS application applying the method of NLFEA on this research has shown the ratio of the values resulted from the NLFEA method that is greater than the SMITH method. The NLFEA method can estimate the load redistribution and interaction between complex local and global failures that the accuracy of estimation of the NLFEA method is greater than the SMITH method.

4. Conclusions

The Nonlinear Finite Element Analysis (NLFEA) method to investigate the ultimate strength of VLCC under vertical bending moment in hogging and sagging condition may be summerized as follow; the comparison of the ultimate strength obtained by NLFEA is grather than Smith's method. The results obtained by the Smith method, for the hogging condition, the ultimate strength is 9,986 x 10^{12} Nmm, while in sagging condition, the ultimate strength is -9,175 x 10^{12} Nmm. The ultimate strength ratio between Smith's method and NLFEA in hogging and sagging conditions are 7.71% and 11.21%, respectively. This is because the NLFEA method can calculate the load redistribution of elements for local and global failures.

References

[1] Andrić J., et al. 2014. IACS Incremental-Iterative Method In Progressive Collapse Analysis of Various Hull Girder Structures

[2] Muis Alie M Z 2016 *Residual Strength Analysis of Asymmetrically Damaged Ship Hull Girder using Beam Finite Element Method* Makara J. Technol. 20/1 p 7-12

[3] Muis Alie M Z 2016 *Finite Element Analysis On The Hull Girder Ultimate Strength Of Asymmetrically Damages Ships* Proc. of the ASME 2016 35th International Conference on Ocean and Artic Engineering, OMAE.

[4] Muis Alie, M.Z et al. 2016 a. The Influence of Superstructure on the Longitudinal Ultimate Strength of a RO-RO Ship. Proceedings of the Twenty-sixth International Society of Offshore and Polar Engineers (ISOPE)

[5] Paik J. K. 2007. Ultimate limit state performance of oil tanker structures designed by IACS common structural rules. *Journal Thin-Walled Structures*, 45(12):1022-1034

[6] Paik, J.K.; Kim, B.J.; Seo, J.K. (2008), Methods for ultimate limit state assessment of ships and ship shaped offshore structures Part III: Hull girders, Ocean Engineering 35, pp.281-286

[7] Soares, C. G. et al. 2008. Benchmark Study on the Use of Simplified Structural Codes to Predict the Ultimate Strength of a Damaged Ship Hull. Journal International Shipbuilding Progress, 55:87-107.

[8] Van-Vu, Hunh. 2005. *Prediction The Ultimate Strength of intact Ship by Finite Element Method.* International Journal of Mechanical Engineering and Aplications (IJMEA)

Prototype Development of Portable Head Feed Harvesting Robot

Muh Rifai Anugrah[1)], Jeky[1)], Abdul Kadir Muhammad[1)], Akhmad Taufik[1)], Rafiuddin Syam[2)]
[1]Center For Mechatronics And Control System, Mechatronic Engineering, Polytechnic Of Ujung Pandang.

E-mail: rifaianugrah1@gmail.com, kadir.muhammad@poliupg.ac.id, afike76@gmail.com, rafiuddinsyam@gmail.com

Abstract. This study aims to develop a prototype of portable head feed harvesting robot that can be installed in a tractor so that the users of tractors that initially only function during the farming season can also function during the harvesting season. This machine is divided into 3 main parts. In the first part there is a cutting mechanism that functions to cut rice using a DC electric motor with a crank mechanism connection to the cutting shaft. The second part is a rice lifting mechanism which functions to lift the rice cut into the conveyor belt position. The third part is a connection that connects the rice cutting machine to the used tractor using a bolt connection. The machine has been tested on rice field and work successfully.

1. Introduction

The application of the science of mechatronics in agriculture, agro-mechatronics, increasingly sophisticated and modern, it can be found in various regions in Indonesia. Some of the technology in the field of agriculture are tractors used by farmers in the planting and harvesting machines are used in the harvesting process. The machine has several drawbacks one of which is the price that is relatively expensive and have only one purpose[5-7].

In this study, researchers conducted a design prototype cutting machine rice that are portable and can be mounted on tractors owned by farmers so that tractors were initially work only when the process of planting after pairing with a machine that researchers designed, the tractor can function well during the process of harvesting the rice.

2. Literature Review

2.1. Rice

Rice is a plant that is very easy to find, especially we who live in rural areas. Expanse of rice fields are filled with rice. Most make rice as a staple food source[2].

2.2. Harvester Machine / Cutting Type Rice Combine Harvester

This type rice harvester machine is a machine that can melaukan process of cutting rice paddy harvesting, threshing rice and the latter was able to enter the results of threshing rice into a sack and is ready to be marketed.

IOP Conf. Series: Materials Science and Engineering **619** (2019) 012009 doi:10.1088/1757-899X/619/1/012009

Figure 1. Rice

Figure 2. Machine Harvesters / Cutting Rice *Combine Harvester*

2.3. DC Electric Motor

The motor of the regulator rotates clockwise or opposite direction and is equipped with a transmission gear box. DC motors using DC electric energy and magnetic energy to produce mechanical energy. Motor operation depends on the interaction of two magnetic fields. In simple terms it is said that the electric motor works on the principle that two magnetic fields can be made to interact to produce movement or torque [1].

Figure 3. Motor DC

Figure 4. Driver Motor

2.4. Driver Motor

Driver used is current driver with 12 VDC inputs, functioning as inverting the direction of rotation of DC motors as well as set the rotation speed of the DC motor[3].

2.5. Eye Knife Rice

The blade used in cutting machine rice generally use the same rotating blade to blade mower is moving in rotation.

IOP Conf. Series: Materials Science and Engineering **619** (2019) 012009 doi:10.1088/1757-899X/619/1/012009

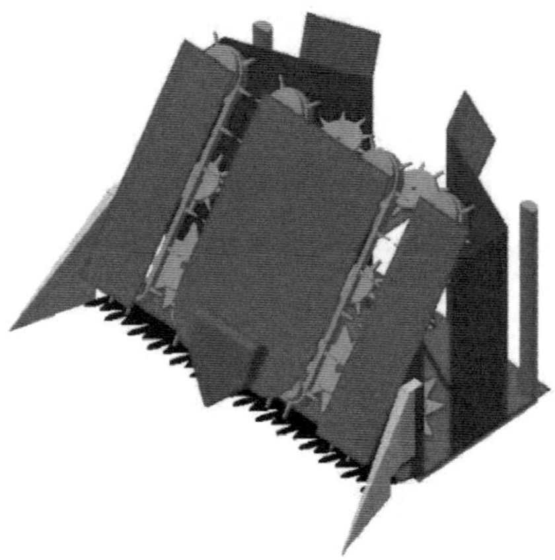

Figure 5. Design of Harvesting Machine

3. Research Methods

To obtain a good tool in terms of quality as well as considering the economic aspect, the design steps taken to make Portable Cutting Machine Rice is found below:

3.1. Design

The design of the cutting machine rice portabe like in the picture below

3.2. Procedure Machine Working System

The design of this portable rice cutting tools necessary preliminary picture of how systems work on the tool created. Broadly speaking, it can be seen in the Figure 8, below

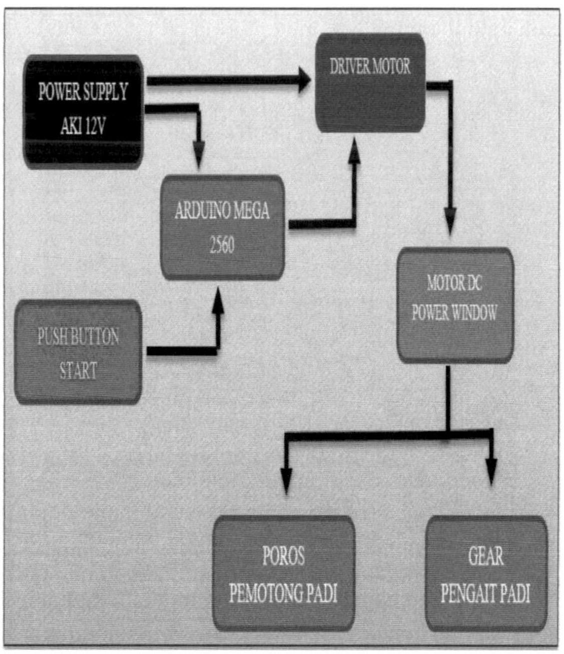

Figure 6. Procedure Machine Operating System

4. Result and Discussion

4.1. Mechanical Design Result
After doing some research, the researchers got a prototype portable cutting machine rice as shown below:

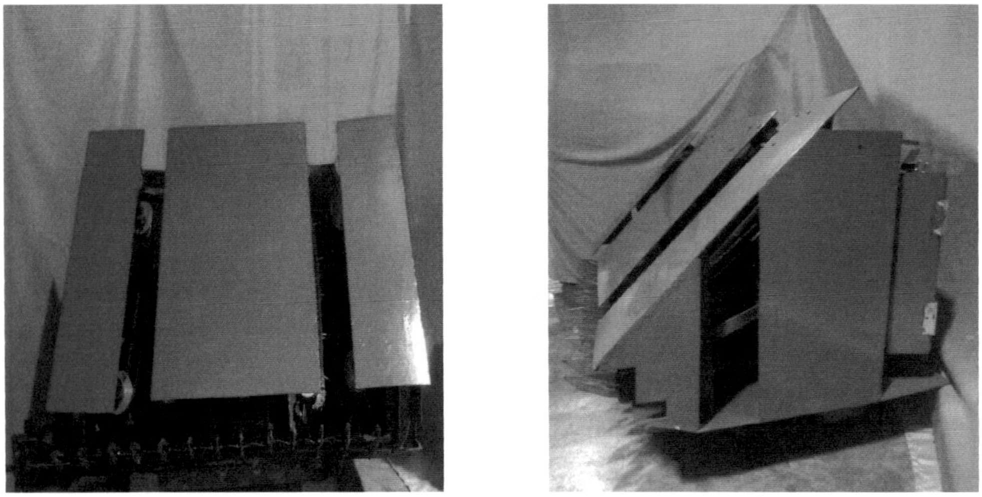

(a) (b)

Figure 7. Design of Harvesting System

4.2. Electronic Design Result
In designing the electronics in these machines is quite simple because it uses open control system, this machine uses only the input of rotary switch that serves to regulate the entire motor speed on the machine and two switches, one of which serves to activate and deactivate the microcontroller and the other function to activate and deactivate the motor[5].

Figure 8. The circuit design Results Electronics

4.3. *Result of Program Design*

For programming is done on the Arduino software that governs the workings of all the equipment that has been made. The type of program movement imposed on the microcontroller arduino as follows.

1) Setup settings input output pins on the Arduino.
2) The main program of motor speed regulation with input from the rotary switch.
3) Sub program to activate the motor cutter rice.
4) Sub program to activate the motor hook rice.
5) Sub program to activate the motor lifter rice.
6) Sub program to disable the entire motorcycle.

4.4. *Testing Results Rice Portable Cutting Machine*

In this test machine test machine installed in the tractor hands of farmers and examine the cutting mechanism of rice.

(a) (b)

Figure 9. (a) Rice Cutting Machine Assembly (b) Rice Cutting Machine Connection Mounted on tractor farmer

4.5. *Testing Analysis*

In the results of the tests the researchers have conducted in Julupamai village and Bontoala village, the palangga sub-district of the stranded Gowa district there are 2 test analyzes, namely the testing of the installation on the tractor and testing of the rice cutting mechanism without the installation of a tractor.

4.5.1. Analysis of Machine Testing with Instalations on Farmer's Tractor. Result is testing the installation of rice cutting machine portable type feed head in the farmer's hand tractor, the cutting machine is installed properly but due to the excessive cutting machine load makes the hand tractor difficult to control, so testing by installing the device in the farmer's hand tractor is not optimal.

4.5.2. Analysis of Cutting Mechanism Testing Without Installation in Farmer's Tractor. Result of testing the rice cutting mechanism without installing the machine into the tractor, it runs well on speed testing, cutting the initial stage of rice can be cut well, but it takes 7-8 cutting movements to be able to cut 1 rice clump perfectly. While at the speed of cutting the second stage, rice can be cut well but it takes

4-5 cutting movements to be able to cut 1 rice clump perfectly. While at the speed of cutting the third stage, rice can be cut well but it takes 1-2 cutting movements to be able to cut 1 rice clump perfectly.

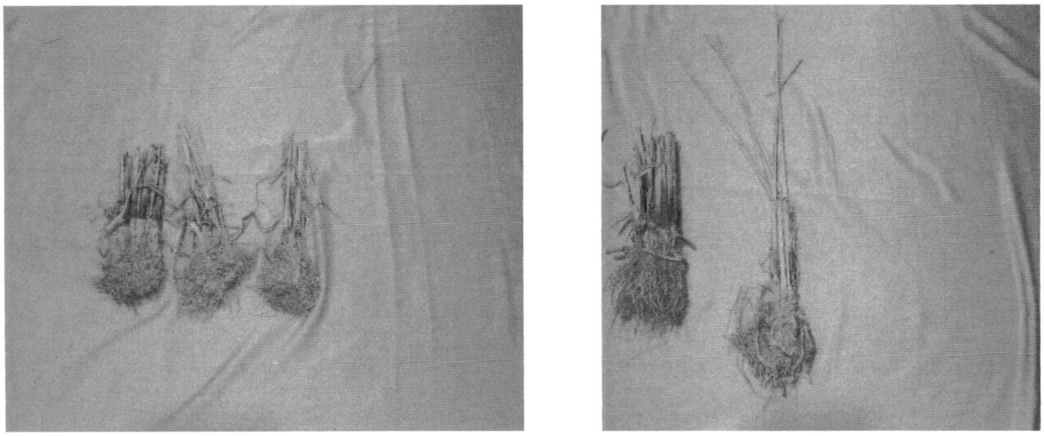

(a) (b)

Figure 10. Results of Cutting Rice

5. Conclusion

From the result and discussion of this study, some conlusions can be drawn as follows:

- Design of prototype rice cutting machine portable type head feed has been carried out. Testing of the machine without a tractor has also been carried out, where the test results show that the rice can be cut well.Testing Analysis.
- Testing prototypes of rice cutting machines for portable head feed types on farmers' hand tractors has also been carried out. However, the test results have not been maximized because the excess tractor load causes the tractor to be difficult to control.

References

[1] R Syam, K Watanabe, K Izumi 2005 Adaptive actor-critic learning for the control of mobile robots by applyıng predictive models Soft Computing 9 (11), 835-845.

[2] Anonym, Arduino 2012. *Arduino Mega*. (https://www.arduino.cc/en/Main/arduinoBoardMega).

[3] Efendy, Syahrul. 2016. The Meaning of Rice Plant (Original inBahasa:*Pengertian Tanaman Padi*). (https://petaniindomodern.wordpress.com/2016/02/10/pengertian-tanaman-padi-3/).

[4] Kho, Dickson. 2016. Understanding Relay and Function (Original in Bahasa: *Pengertian Relay dan Fungsi Relay*).(http://teknikelektronika.com/pengertian-relay-fungsi-relay/).

[5] Muslikin, Chusnan and Nanda Kusuma Arum. 2017. Papers Know the Tool and Machine Paddy Harvesters (Original In Bahasa: *Makalah Mengenal Alat dan Mesin Pemanen Padi*). Gresik. Agrotechnology Study Program Faculty of Agriculture, University of Muhammadiyah Gresik. Indonesia.

[6] Riadi, Muchlisin. 2012. Push Button (Push botton) (Original in Bahasa: *Tombol Tekan*). http://www.kajianpustaka.com/2012/10/ tombol -tekan-push-botton.html).

[7] Sulistiaji, K., 2007. Book Tool and Engineering (Alsin) Harvest and Threshing Rice in Indonesia (Original in Bahasa: *Buku Alat dan Mesin (Alsin) Panen dan Perontokan Padi di Indonesia*). Serpong. Agricultural Engineering Research Center for Agricultural Research and Development Agency Indonesia.

Ultimate Strength Investigation of Ro-Ro Ship

Muhammad Zubair Muis Alie, Juswan, Taufiqur Rachman and Chairul Paotonan,

Department of Ocean Engineering, Hasanuddin University, Indonesia
zubair.m@eng.unhas.ac.id

Abstract. Ro-Ro ship has special characteristic for cargo handling because the cargo is transferred in horizontal direction. For example, cars can move by using ramp door which is located in front and/or aft of the ship. In conjunction with this, car passenger and other decks such as top including bottom part need to be investigated for local and global of Ro-Ro ship structure. In the present study, the ultimate strength investigation of Ro-Ro ship is conducted. There are three types of Ro-Ro ships are taken as the ship's subject. The cross section of Ro-Ro is considered in the analysis and it is modelled having plate and stiffened plate. The one-frame is represented in longitudinal direction. The simply supported of boundary condition is applied and the cross section is assumed to be remained plane. The calculation of the ultimate strength using in-house integrated program. The ultimate strength of Ro-Ro ship is investigated under hogging and sagging conditions. The vertical bending moment is imposed to the cross section to obtain the ultimate strength including their progressive collapse. The result obtained by in-house program for three types of Ro-Ro ships are summarized and the collapse behaviour for local and global of the decks to the ultimate strength of Ro-Ro ship are also presented in this study.

1. Introduction

Generally, a Ro-Ro ship functioned to move cargo in horizontal direction. The ship has passenger and car decks. One of the deck on Ro-Ro ship is car deck where it is very important to be assessed when the ship is under voyage. The ship's hull resists applied loads such as internal and external. Loading and unloading process may have an impact to structural damage of the ship. Therefore, the ultimate strength of ship hull is one of the most important aspects in the ship design criteria and must be taken into account.

Many studies have been conducted to assess and/or evaluate the ultimate strength of ship structure. The researcher [2] assessed the ultimate hull girder strength of Ro-Ro ship after damage. The cross section of Ro-Ro ship is considered to be analysed. Another researcher [1] focused on the ultimate strength of typical bottom structure in container ships under both longitudinal and transversal loads in corrosive environment. Muis Alie, M.Z et al [3] analysed the hull girder ultimate strength of asymmetrically damaged ships using Finite Element Method. The collision damage was modelled by removing the plate and stiffened plate elements. Muis Alie, M.Z [4] analysed the residual strength of ship hull girder with bottom damage. The nonlinear finite element method was used and the fully cross section was considered in the calculation. Shu, Z [5] investigated the ultimate strength of a capsize bulk carrier hull girder under combined global and local load in the hogging and alternate hold loading (AHL) condition using nonlinear finite element (FE) analysis with ABAQUS. Amlashi, H.K.K [6] contributed to establish rational ultimate longitudinal strength criteria for the hull girder under combined loading. An important issue was the significant double bottom bending in empty holds in AHL due to combined global hull girder bending moment and local loads. Paik, J.K [7] investigated the ultimate strength

IOP Conf. Series: Materials Science and Engineering **619** (2019) 012010 doi:10.1088/1757-899X/619/1/012010

characteristics of ship hulls with large hatch opening under torsion. Axial (warping) as well as shear stress is normally developed for thin-walled beams with open cross sections subjected to torsion.

In this study, the investigation of Ro-Ro ship to the ultimate strength is conducted. There are three types of Ro-Ro ships are analysed. The cross section for each type of Ro-Ro ship is considered. The vertical bending moment is given both left and right sides of the cross section. For the simple calculation, one-frame space represents the length of ship is taken. The ultimate strength calculation is done by considering hogging and sagging condition.

2. Method of Analysis

The ultimate strength of ship hull is performed using the Smith's method considering the cross sectional approach. The Smith's method is adopted and implemented in-house program developed by [8] is used with some modification in the process. The moment-curvature relationship for the ultimate hull girder strength calculation is expressed by,

$$\begin{Bmatrix} \Delta P \\ \Delta M_H \\ \Delta M_V \end{Bmatrix} = \begin{bmatrix} \bar{D}_{AA} & \bar{D}_{AV} & \bar{D}_{AH} \\ \bar{D}_{HA} & \bar{D}_{HH} & \bar{D}_{HV} \\ \bar{D}_{VA} & \bar{D}_{VH} & \bar{D}_{VV} \end{bmatrix} \begin{Bmatrix} \Delta \varepsilon_0 \\ \Delta \varphi_H \\ \Delta \varphi_V \end{Bmatrix} \quad (1)$$

Where

$$\begin{aligned} \bar{D}_{AA} &= \sum_{i=1}^{N} D_i A_i \\ \bar{D}_{HH} &= \sum_{i=1}^{N} D_i y_i^2 A_i \\ \bar{D}_{VV} &= \sum_{i=1}^{N} D_i z_i^2 A_i \\ \bar{D}_{AH} = \bar{D}_{HA} &= \sum_{i=1}^{N} D_i y_i A_i \\ \bar{D}_{AV} - \bar{D}_{VA} &= \sum_{i=1}^{N} D_i z_i A_i \\ \bar{D}_{HV} = \bar{D}_{VH} &= \sum_{i=1}^{N} D_i y_i z_i A_i \end{aligned} \quad (2)$$

To obtain the ultimate hull girder moment capacity, the above expression can be simply expressed by,

$$\begin{Bmatrix} 0 \\ \Delta M_H \\ \Delta M_V \end{Bmatrix} = \begin{bmatrix} \bar{D}_{AA} & 0 & 0 \\ 0 & \bar{D}_{HH} & \bar{D}_{HV} \\ 0 & \bar{D}_{VH} & \bar{D}_{VV} \end{bmatrix} \begin{Bmatrix} \Delta \varepsilon_0 \\ \Delta \phi_H \\ \Delta \phi_V \end{Bmatrix} \quad (3)$$

Where

$$\begin{aligned}\overline{D}_{AA} &= \sum_{i=1}^{N} D_i A_i \\ \overline{D}_{HH} &= \sum_{i=1}^{N} D_i \left(y_i - y_G\right)^2 A_i \\ \overline{D}_{HV} = \overline{D}_{VH} &= \sum_{i=1}^{N} D_i \left(y_i - y_G\right)\left(z_i - z_G\right) A_i \\ \overline{D}_{VV} &= \sum_{i=1}^{N} D_i \left(z_i - z_G\right)^2 A_i\end{aligned} \tag{4}$$

There are three types of Ro-Ro ships as shown in Figure 1. The investigation of Ro-Ro ship to the ultimate strength is conducted by considering cross section. Welding residual stress, crack and initial deflection are not included in the calculation.

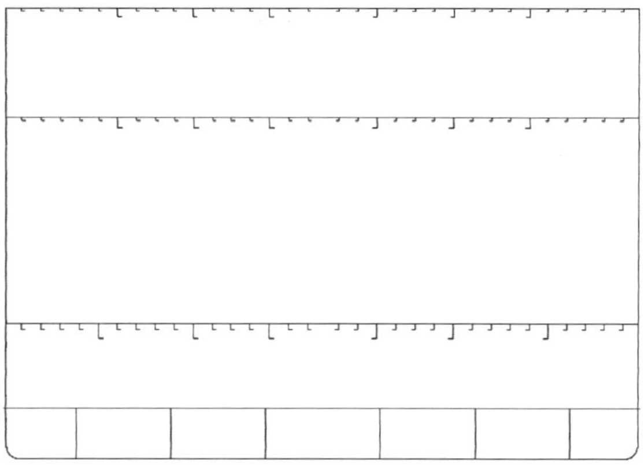

(a) Type-1

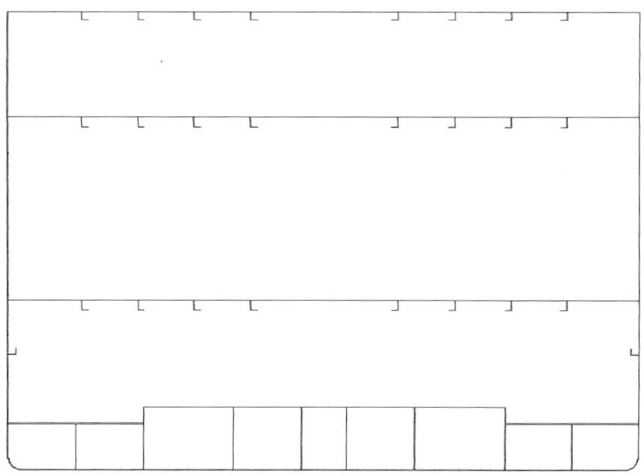

(b) Type-2

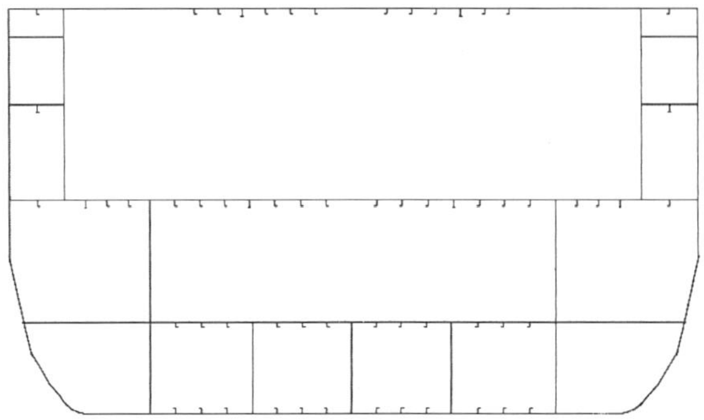

(c) Type-3

Figure 1 Cross sections of Ro-Ro types

The ultimate strength calculation is done under vertical bending moment in hogging and sagging conditions. All the types of Ro-Ro ships are calculated for intact only. Damages such as collision, grounding and so on are not considered in the analysis. The vertical bending moment is given on the cross section in right and left sides. The one-frame space represents the length of ship is taken as a fundamental calculation. The calculation of the ultimate strength of Ro-Ro ship considers two frame-spaces under hogging and sagging conditions.

3. Results and Discussions

The bending moment-curvature relationships for three types of Ro-Ro ship are shown in Figures 2-5 for two frame-spaces under hogging and sagging conditions, respectively. The larger ultimate strength of Ro-Ro ship is found on type-3 than followed by type-1 and type-2. These differences may be caused by dimension of scantling, configuration shapes, section properties of the hull girders and so on of the Ro-Ro ships.

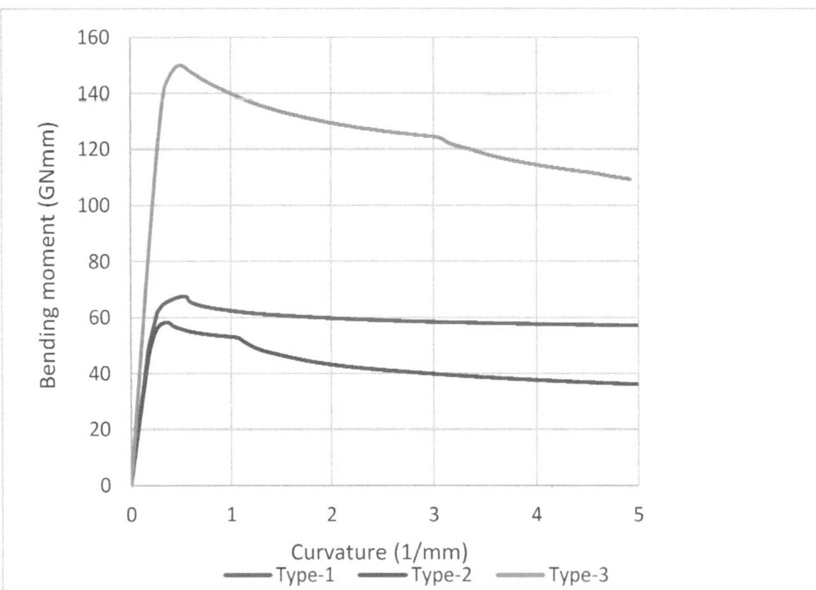

Figure 2 Bending moment-curvature relationship of 550 mm frame-space in hogging

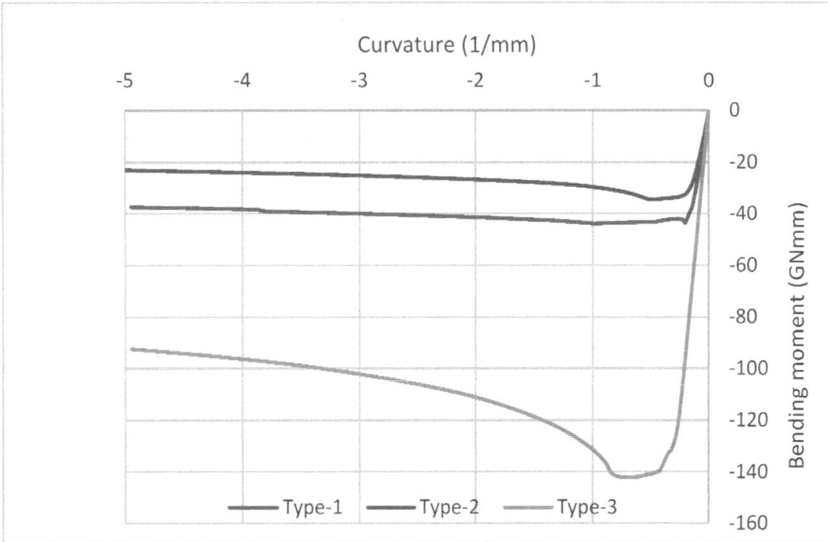

Figure 3 Bending moment-curvature relationship of 550 mm frame-space in sagging

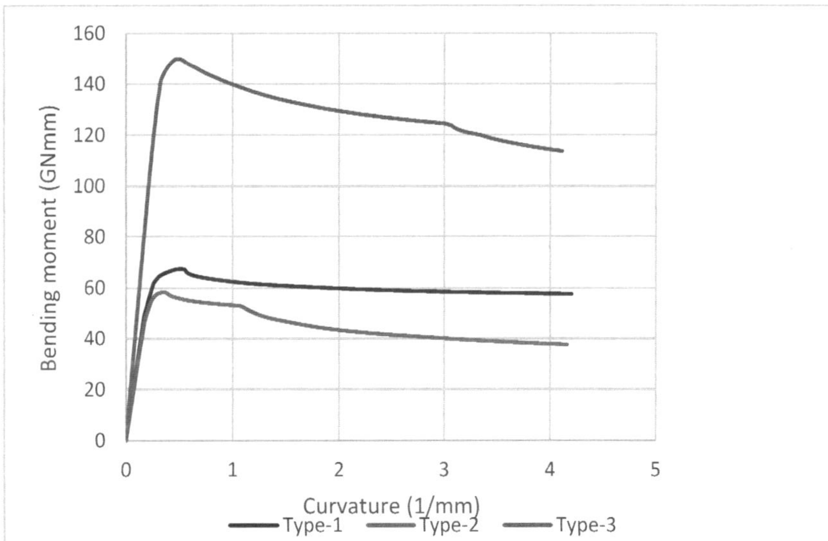

Figure 4 Bending moment-curvature relationship of 650 mm frame-space in hogging

It is also observed that the bending stiffness for type-3 is different with type-1 and type-2 for hogging and sagging. The tendency of the bending stiffness for type-3 tends to coincide to the vertical axis. This behavior may be influenced by the shape of bilge. It should be noted that the cross section of Ro-Ro ship for type-1 and type-2 is almost identic for the case of bilge. Perhaps, it is also affected by the shape of side shell for type-3, because there is a little different for the bottom part near to the bilge shell. This part is generally contributed to the effect of the ultimate strength including their tendency of the stiffness.

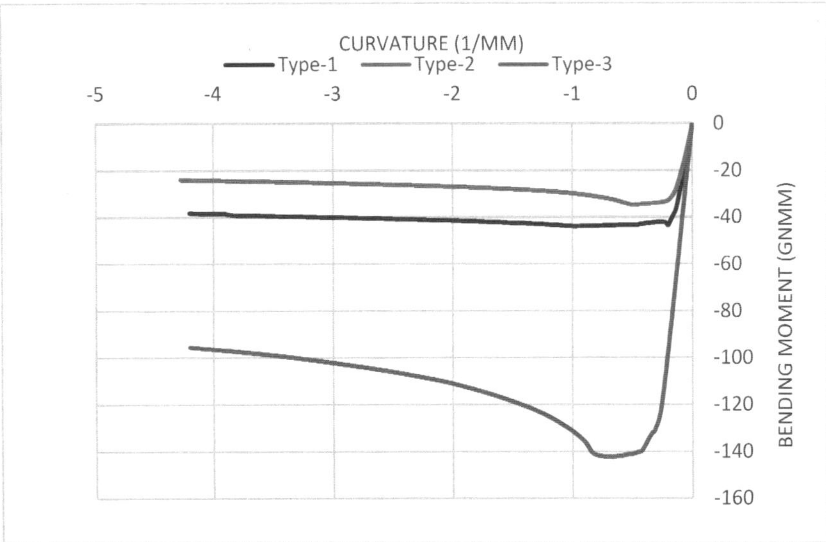

Figure 5 Bending moment-curvature relationship of 650 mm frame-space in sagging

4. Conclusions

The investigation of the ultimate strength on Ro-Ro ships has been done using Smith's method. The following conclusions may be summarized as; the ultimate strength of Ro-Ro ship for type-3 gives significant influence toward type-1 and type-2. This effect cause by the shape of cross sectional including their dimensions and hull girders. For the case of the influence of the frame-space, the effect is not so significant. The percentage of the biggest one is only about 0.5%.

References

[1] Cui, J. Wang, D. Ma, N. A Study of Ultimate Strength for Container Ship Bottom Structures under Bi-axial Loads Considering Corrosion Effects. Proc. 27th Int Offshore and Polar Eng Conf, ISOPE, 990–997, 2016.

[2] Muis Alie, M.Z, Sitepu G, and Latumahina, S.I, Proc. The Assessment of the Ultimate Hull Girder Strength of Ro-Ro Ship after Damages. Proc. 27th Int Offshore and Polar Eng Conf, ISOPE, 913–919, 2017.

[3] Muis Alie, M.Z, Sitepu G, Juswan, Wahyuddin, Nugraha, A.M and Alamsyah. Finite Element Analysis on the Hull Girder Ultimate Strength of Asymmetrically Damaged Ships. Proc. 35th Int Conf. on Ocean, Offshore and Arctic Engineering, OMAE, 2016.

[4] Muis Alie, M.Z. Finite Element Analysis on the Hull Girder under Longitudinal Bending with Bottom Damage. Proc. 3rd Int Conf. on Ship and Offshore Technology, ICSOT, 1-4, 2014.

[5] Shu, Z and Moan, T, J. Mar Sci. Technol, 94-113, 2012

[6] Amlashi H, K.K and Moan, T. Ultimate Strength Analysis of a Bulk Carrier Hull Girder uder Alternate Hold Loading Condition – A Case Study : Part 1 : Nonlinear Finite Element Modelling and Ultimate Hull Girder Capacity. Marine Structures, 327-352, 2008.

[7] Paik, J.K, Thayamballi, A.K, Pedersen, P.T and Park, Y.K. Ultimate Strength of Ship Hulls under Torsion. Ocean Engineering, 1097-1133, 2001.

[8] Yao, T and Nikolov, P.I. Progressive Collapse Analysis of a Ship's Hull Girder under Longitudinal Bending (2nd Report). J. Soc. Naval Arch of Japan, 437-446, 1992.

Sodium Hydroxide and Potassium Permanganate Treatment on Mechanical Properties of Coconut Fibers

Muhammad Arsyad

Mechanical Engineering Department, Politeknik Negeri Ujung Pandang, Jalan Perintis Kemerdekaan KM. 10 Makassar 90245 South Sulawesi Indonesia

Corresponding author: arsyadhabe@poliupg.ac.id

Abstract. The purpose of this research is to determine the tensile stress and interfacial shear stress of coconut fiber that has been treated Sodium Hydroxide and Potassium Permanganate. This research is divided into three stages: soaking for 3 hours, testing, and data analysis. Coconut fiber is soaked in a solution of 5%, 10%, 15%, and 20% sodium hydroxide solution. Then, soaking in a solution of Potassium Permanganate with concentrations of 0.25%, 0.50%, 0.75%, and 1.00%. After that, the coconut fiber is dried in the furnace at 90°C for 5 hours. Then, coconut fiber was tested for tensile stress and interfacial shear stress. Based on the results of the test can be concluded that the treatment of Sodium Hydroxide and Potassium Permanganate affect tensile strength, and the shear strength of coconut fiber. The largest tensile stress was obtained in the treatment of Sodium Hydroxide with a concentration of 20% ie 289.94 N/mm^2, while the highest interfacial shear stress was obtained in the treatment also on Sodium Hydroxide with a concentration of 20% ie 3.09 N/mm^2.

Keywords. treatment, tensile, shear, stress, coir fiber.

1. Introduction

Lignocellulose is the name used for materials containing lignin, cellulose, and hemicellulose. Lignocellulose is a constituent component of plant cell wall especially on the stem section. Hemicellulose and cellulose are polysaccharides that can be decomposed into monosaccharides. The availability of abundant lignocellulosic materials on earth has resulted in the study of the use of lignocellulosic materials to be very attractive[1].

Natural fibers containing hemicellulose, cellulose, and lignin are hydrophilic which is a feature that likes to water. It is this property that causes natural fibers to hardly unite or bind with a hydrophobic matrix of a dislike of water. Therefore, there is a need to improve the properties of the natural fibers in order to establish a good bond with the matrix. Treatment of lignocellulosic materials can be done in several ways as chemically, physically, and microbiologically. Each method has advantages and disadvantages. The efficiency and effectiveness of its use may vary, depending on the source of the material and the purpose of the process. Chemical treatment is the most commonly used method because it is easier, more effective, faster and does not use too high energy. The bond between fiber and matrix has an effect on the mechanical properties of fiber-reinforced composites. Specifically, the composite tensile strength is affected by the efficiency of the load transfer from the matrix to the fiber through shear at the interface. Therefore, a number of mechanical tests have been developed to measure the interface capacity to transfer the stress from the matrix to the fibers in the composite. Several test methods have been used to evaluate the bond capability between fibers and matrices such as: (1) pull outs, (2) microtension, (3) microcompression, (4) fragmentation. The interconnection of the fiber with the matrix consists of several bond models: (a) chemical bonds, (b) ionic electrostatic bonds, (c) molecular reaction interdiffusion, and (d) mechanical bonds (*interlocking*) [2].

The composite mechanical properties are strongly influenced by the bond between the matrix and the fibers. Treatment with sodium hydroxide has the highest effect on tensile strength and tensile modulus, resulting in composites with the best tensile properties. The strength of the composites treated with sodium hydroxide increased significantly by about 53% compared to composites made from untreated fibers and 33% compared to the non-fiber composite [3]. The alkali treatment of the fibers will give two effects to the fibers: (1) increasing the surface roughness of the fibers so as to produce better interlocking, (2) increasing the release of cellulose [4]. The reaction between fiber and sodium hydroxide are [1] [5]:

$$\text{Serat-OH} + \text{NaOH} \rightarrow \text{Serat-O-Na} + H_2O$$

Permanganate is a compound containing MnO4- permanganate groups. Permanganate strengthening is used on cellulose fibers that turn into radicals through the formation of permanganate ions. Then the highly reactive Mn3 + ion as the initiation of the graph copolymerization [6] [8].

Most permanganate strengthening is carried out by using a solution of potassium permanganate (KMnO4) or in acetone with different concentrations within 1 to 3 hours after alkaline strengthening, strengthening the hemp fiber in permangant solution at a concentration of 0.033: 0.0626: and 0.125% in acetone for one minute. As a result of permanganate strengthening is a reduction in the fiber, therefore the water absorption of fiber-reinforced composites increases. The hydrophilic properties of the fibers will decrease with the increase of KMNO4 concentration, but at 1% KMnO4 concentration there is cellulose degradation which will produce polar groups between the fibers and the matrix [6].

2. Material and Methods

The materials used in this research are: coconut fiber, polyester resin, metyl etyl keton peroksida catalyst, sodium hydroxide (NaOH), potassium permanganate (KMnO4), and aquades. The equipments used in this study are digital scales, immersion media, manila cartons, glue, scissors, tensile test equipment. Coconut fiber is obtained from coconut husk with manually method. Coconut fiber is soaked in sodium hydroxide solution for 3 hours with concentrations of 5%, 10%, 15%, and 20%. Then soaked for also for 3 hours in solution 0.25%, 0.5%, 0.7.5%, and 1.00%. Subsequently, coconut fiber was dried in a furnace at 90°C for 5 hours.

The stress occurring at a maximum limit indicates a load capable of being retained by a material called the maximum stress. The maximum stress held by the specimen before break is called the maximum tensile stress that is the ratio between the maximum load and the cross-sectional area of the material. Single fiber tensile specimens were prepared according to ASTM 3379-02 Standard as shown in Fig. 1, whereas the shearing mechanism between fiber and matrix is shown in Fig. 2. The maximum tensile stress value is calculated by using equation 1, while the shear stress value between the fibers and the matrix is calculated using equation 2. The single tensile fiber and pull out test is performed by using the LR10K Plus 10 kN Universal Materials Testing Machine.

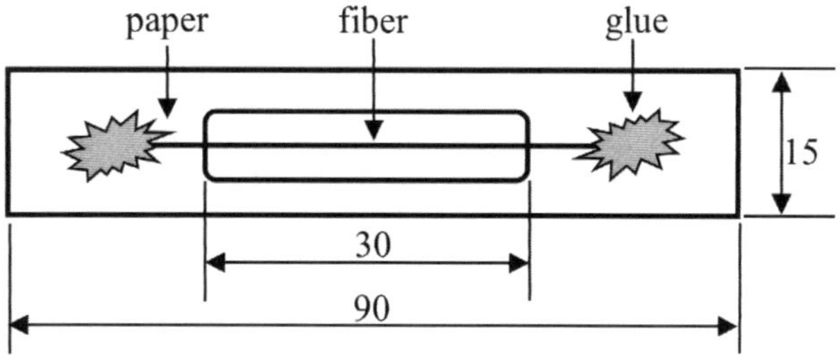

Figura 1. Speciment of single fiber test ASTM 3379-02

$$\sigma = \frac{P}{A} \tag{1}$$

with σ = Maximum tensile stress (MPa), P = Maximum load (N), A = Sectional area (m^2).

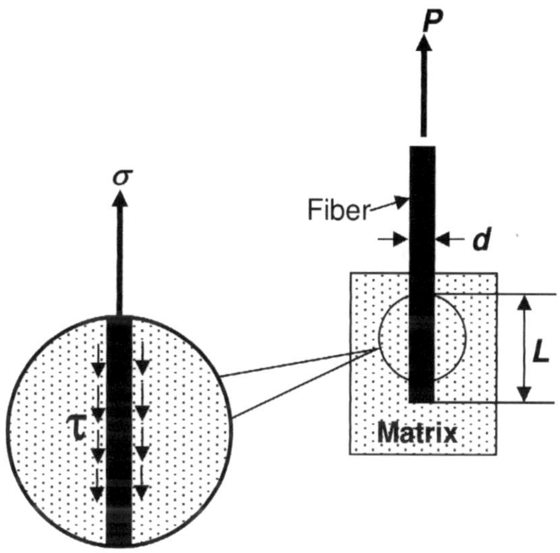

Figure 2. *Single Fiber Pull-Out test* mechanism (Arsyad, 2017)

$$\tau = \frac{P}{\pi d L} \tag{2}$$

with: τ = Interfacial shear stress (MPa), P = Maximum load (N), d = Fiber diameter (mm), L = Embedded fiber length (mm)

In order for the research to be done well, each research variable is given different notation for each treatment as in Table 1.

Table 1. Nomenclature used for the different coconut fibre treatments

No	Notation	Description
1	WT	Without treatment
2	N05	Soaking in alkali 5%
3	N10	Soaking in alkali 10%
4	N15	Soaking in alkali 15%
5	N20	Soaking in alkali 20%
6	P025	Soaking in Potasium 0,25% after N05
7	P050	Soaking in Potasium 0,50% after N10
8	P075	Soaking in Potasium 0,75% after N15
9	P100	Soaking in Potasium 1,00% after N20

3. Results and Discussion

The results of tensile test are shown in Table 2 and Figure 3.

Table 2. The tensile stress of coconut fiber

No	Treatment	Tensile Stress (MPa)	Strain (%)
1	WT	186,42	28,33
2	N05	144	50
3	N10	113,09	29,17
4	N15	52,65	11,67
5	N20	280,94	11,25
6	P025	138,37	31,25
7	P050	82,7	16,25
8	P075	195,37	16,67
9	P100	102,87	24,58

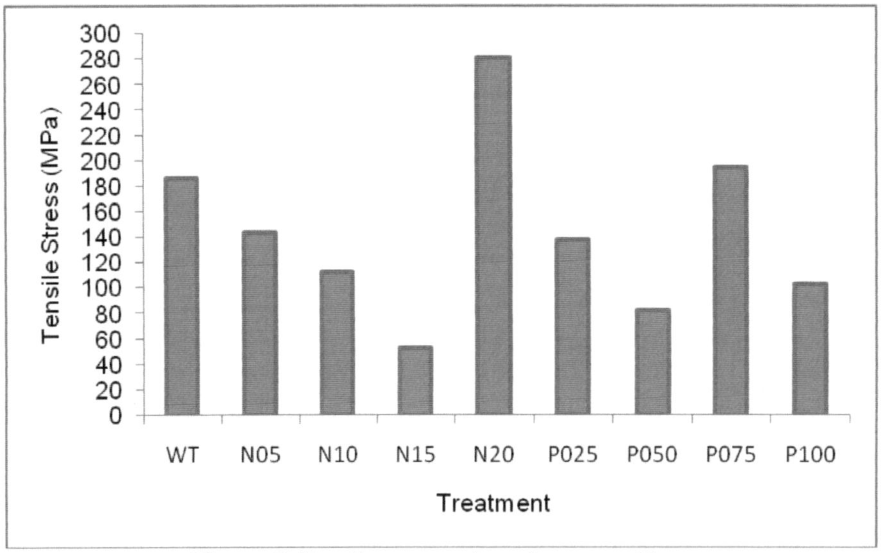

Figure 3. Tensile stress value of coconut fiber

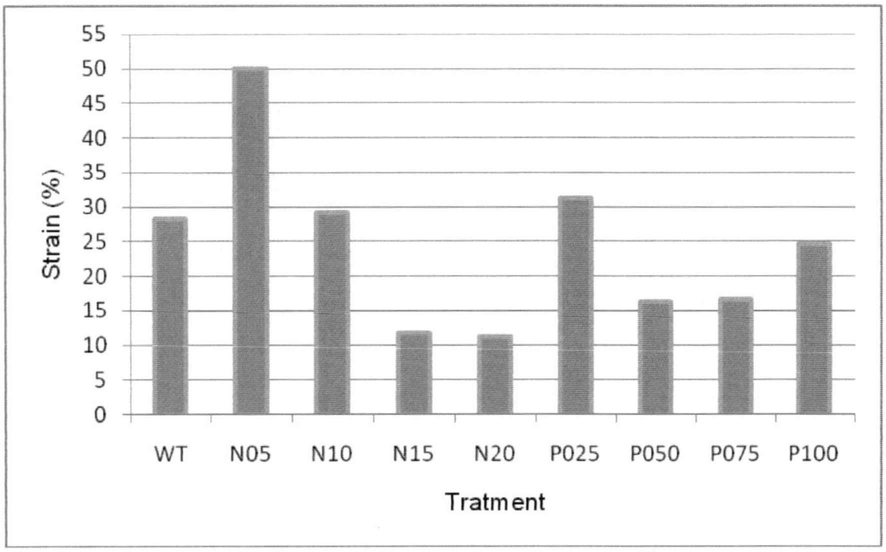

Figure 4. Strain value of coconut fiber

Table 2 and Fig. 3 show that the highest tensile stress is obtained on coconut fiber treatment with 20% alkali which is 280,94 MPa, and this value is higher than coco fiber without treatment ie 186,42 MPa. This indicate that the alkali treatment of coconut fiber can increase the tensile strength of coconut fiber [3][4] [6]. In this treatment coconut fiber contains the lowest lignin that is only 6.1%. This value is lower than the lignin content of coconut fiber without treatment ie 33.5% as shown in table 4 and fig 6. This represent that the alkali treatment of coconut fiber results in lignin degradation so that influence the tensile strength of coconut fiber [7]. While the highest strain of coconut fiber is obtained at 5% alkali treatment which is 50% as shown in fig 4. In this treatment, coconut fiber contains the lowest hemisellulose that is 11% lower than coconut fiber without treatment ie 15.5% . This shows that alkaline treatment degrades the number of coconut fiber hemisellulose [6].

The results of pull out test are shown in Table 3 and figre 5.

Table 3. The interfacial shear stress of coconut fiber

NO	Perlakuan	D	F	τ
		Mm	N	MPa
1	WT	0,31	3,7	1,85
2	N05	0,39	5,43	2,32
3	N10	0,29	3,03	1,72
4	N15	0,33	3,4	1,57
5	N20	0,35	7,38	3,09
6	P025	0,3	4,85	2,42
7	P050	0,35	6,1	2,82
8	P075	0,29	3,4	1,92
9	P100	0,31	2,36	1,19

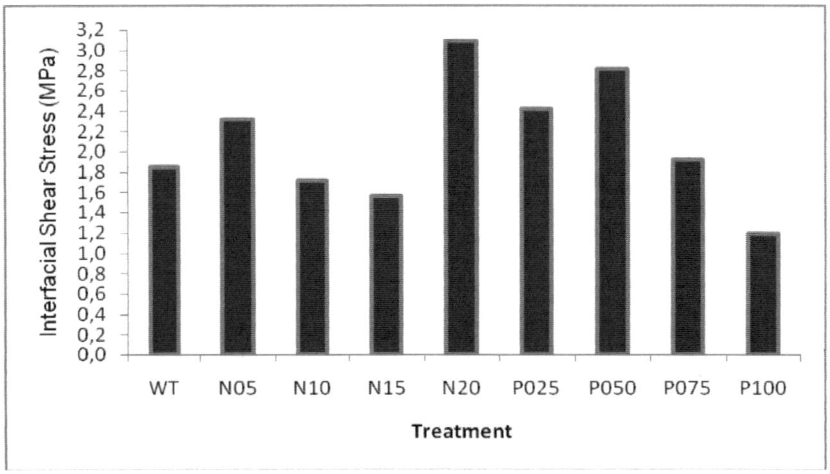

Figure 5. Interfacial shear stress value of coconut fiber

Table 3 and Fig 5 shows the shear stress of coconut fiber, both of which have been treated or untreated. The highest shear stress is found in coconut fiber treated with alkali 20% that is 3.09 MPa. The shear stress value is higher than the value of the shear stress without treatment coconut fiber is 1, 85 MPa. This indicate that the alkali treatment of coconut fiber has compatibility to polyester matrix. While on the other treatment obtained at the treatment of alkali 10% and potassium permanganate 0.50% that is equal to 2.82 MPa. Overall, potassium permanganate treatment after alkali treatment increased the shear stress of coco fiber with a polyester matrix except at 1% potassium permanganate treatment after 20% alkali treatment, the shear stress decreased from 2.09 MPa to 1.19 MPa. The treatment of potassium permanganate caused a reduction in the amount of cellulose, the highest amount of cellulose obtained in coconut fiber without treatment ie 37.9% while the lowest amount of cellulose was obtained in potassium permanganate treatment with 0.75% concentration of 21.6% [6,9].

Table 4. Content of hemicellulose, cellulose, lignin of caconut fiber

No.	Treatment	Content (% weigth)			
		Hemicellulose (%)	Cellulose (%)	Lignin (%)	Others (%)
1	WT	15,5	37,9	33,5	13,1
2	N05	11	37	29	23
3	N10	11,3	36,8	27,3	24,6
4	N15	12,7	22,2	37,5	27,6
5	N20	40,9	22	6,1	31
6	P025	24,5	37,7	13,2	24,6
7	P050	20,4	29,2	24,8	25,6
8	P075	18,7	21,6	26,6	33,1
9	P100	8,6	35	17,1	39,3

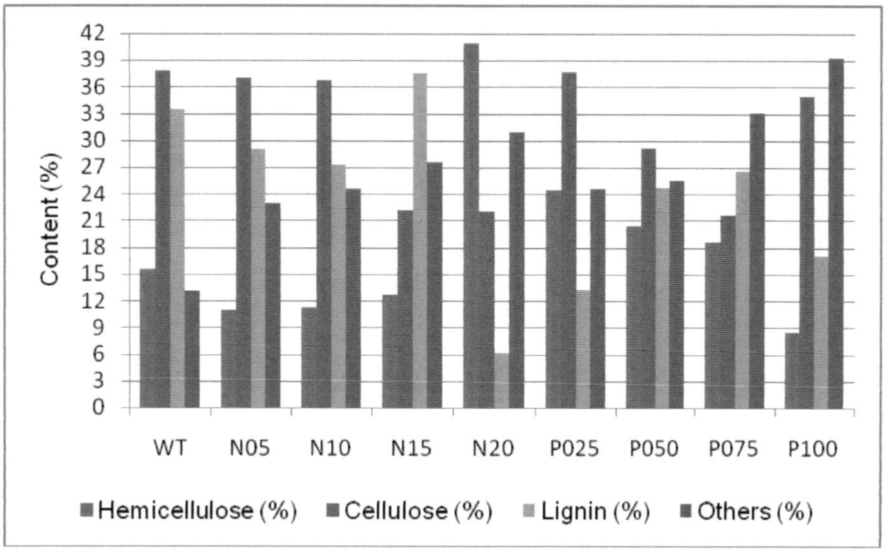

Figure 6. Content of hemicellulose, cellulose, lignin of caconut fiber

4. Conclusion

Based on the testing and discussion that has been done then concluded:

a. The tensile stress of coconut fiber is increased by sodium hydroxide treatment but decreases in potassium permanganate treatment.

b. The treatment of sodium hydroxide, and potassium permanganate increases the compatibility between coconut fiber and polyester matrix.

References

[1] Arsyad, M. 2017. Effect of alkali treatment on the coconut fiber surface. ARPN Journal of Engineering and Applied Sciences. 12(6):1870-1875.

[2] Joshy, M.K., 2007. Isora fibre reinforced polyester and epoxy composites. Disertation. India : Cochin University of Science and Technology.

[3] Rosaa, M.F., Chiou, B., Medeiros, E.S., *et.al.*, 2009. Effect of fiber treatments on tensile and thermal properties of starch/ethylene vinyl alcohol copolymers/coir biocomposites. *J. Bioresource Technology* 100:5196–5202.

[4] Mohanty, A.K., Misra,M., Drzal,L.T., 2005. Natural Fibers, Biopolymers, And,Biocomposites. New York : CRC Press Taylor & Francis Group.

[5] Carvalho,K.C.C., Mulinari,D.R., Voorwald,H.J.C., Cioffi, M.O.H. 2010. Chemical Modification Effect on The Mechanical Properties of HIPS/Coconut Fiber Composites. Bioresources 5(2): 1143-1155.

[6] Li, Xue., Tabil, L.G., Panigrahi, S. 2007. Chemical Treatment of Natural Fiber for Use in Natural Fiber Reinforced Composite: A Review. *J.Polym Environ* 15:25-33

[7] Daulay, L.R., 2009. Adhesion Strengthening Pulp Powder Empty Fruit Bunch Oil is esterified with a Polyethylene Matrix Composites : Disertation. Indonesia : Universitas of North Sumatera.

[8] Dixit S, Verma P. 2012. The Effect of Surface Modification on The Water Absorption Behavior of Coir Fibers. Advances in Applied Science Research 3(3):1463-1465.

[9] Arsyad M, Wardana ING, Pratikto, Irawan YS. 2015. Bonding Ability of Coconut Fiber with Polyester Matrix As A Result of Chemical Treatment. *International Journal of Applied Engineering Research*. **10**.04:9561-9570.

Thermal performance of shallow spiral-tube ground heat exchanger for ground-source cooling system

Jalaluddin [1,*], ***Akio* Miyara**[2], ***Rustan* Tarakka**[1], ***Andi Amijoyo* Mochtar**[1], ***Muhammad* Anis IR**[1]

[1] Department of Mechanical Engineering, Hasanuddin University, Tamalanrea, Makassar, 90245, Indonesia
[2] Department of Mechanical Engineering, Saga University, 1 Honjomachi, Saga-shi, 840-8502, Japan

[*] Corresponding authors Email: jalaluddin_had@yahoo.com

Abstract. This study present an investigation of thermal performance of spiral-tube ground heat exchanger (GHE) buried in shallow depth of borehole. The spiral-tube GHE consisting of spiral pipe installed in the borehole provides a better performance in application of ground-source cooling system. Experimental study was carried-out by circulating water through the spiral-tube GHE which is buried in the ground of 3 m depth. Temperatures of inlet and outlet water in the GHE were measured and recorded periodically. The thermal performance of the GHE was calculated with different inlet water temperatures. Heat exchange rate for different inlet water temperature of the spiral-tube GHEs are 44 W/m (35 ^oC), 126 W/m (40 ^oC) and 110 W/m (45 ^oC). The results show that the utilization of shallow borehole of spiral-tube GHE is appropriated for application in ground-source cooling system especially for the hot climate like Indonesia.
Key words: Ground heat exchanger, spiral-tube GHE, ground-source cooling system, thermal performance

1 Introduction

The ground source cooling system shows potential technology in engineering application for air conditioning system in the building. Ground heat exchanger (GHE) is used in this system for exchanging heat with the ground. Thermal performances of a number type of GHEs installed vertically and horizontally have been investigated. The performances of U-tube, double-tube and multi-tube types show that the heat exchange rate of the double-tube has the highest [1]. Operation modes of the GHEs including short-time period, discontinuous and continuous operations affected their heat exchange rates [2, 3]. Pipe thermal interferences, inlet water temperature and borehole depth also affected the heat exchange rate of the GHEs [4]. Recently, a spiral tube GHE is gaining interest due to its better thermal performance. In the spiral tube GHE, a spiral pipe is installed in the borehole or building foundation pile. Some studies have been carried-out to investigate the thermal performance of types of GHE. Analytical solutions for spiral coil type of GHE have been developed by Man et al. [5], Cui et al. [6], Man et al. [7] and Li and Lai [8]. In addition, comparison study of helical GHE with double U-tube and triple U-tube models have been presented by Zarella et al. [9, 10]. It is found that the helical GHE performance is better than others. The performance and pressure drop along the pipes of spiral tube GHE is a significant parameter in the GHE design [11]. Large investment cost is required to install a deep borehole of spiral-tube GHE. Several parameters should be considered in the design of spiral-tube GHE such as pumping power due to pressure drop and ineffective of outlet pipe due to thermal interference.

In order to provide the possibility for reducing a borehole depth, a shallow spiral-tube GHE is taking interest. Dehghan et al. studied the performance and distance between shallow spiral-tube GHEs [12].

The heat transfer rate of shallow spiral-tube GHE becomes an important issue in application. However, there is a limited number on performance investigation of shallow spiral-tube GHEs. This study presents an experimental investigation of thermal performance of shallow spiral-tube GHE in order to study the possibility of application. The thermal performance of the GHE was calculated with different inlet water temperatures.

2 Experimental study

2.1 Spiral-tube ground heat exchanger

Spiral-tube GHE consists of a spiral pipe used as inlet tube and a straight pipe used as outlet pipe is shown in figure 1. Inlet and outlet pipes of the spiral-tube GHE are PEX-AL-PEX which is a multi-layered composite tubing consisting of an interior aluminum tubing lined with inner and outer layers of crosslinked polyethylene tubing with an inner diameter of 12 mm. The geometric parameter and material thermal properties of the spiral-tube GHE is shown in Table 1.

Fig. 1. Spiral-tube GHE.

Table 1. The geometric parameter and material thermal properties of the spiral-tube GHE.

Parameters	Value	Unit
Outer diameter, d_o	0.016	m
Inner diameter, d_i	0.012	m
Thermal conductivy, k_{pipe}	0.45	W/m K
Spiral diameter, D	0,25	m
Pitch (Spiral distance), p	0.2	m

2.2 Experimental set-up

The spiral-tube GHE is installed in the borehole of 3 m depth. The schematic diagram of experimental set-up is shown in figure 2. The spiral-tube GHE is placed 1 m depth from the ground level to protect from the effect of ambient climate. The experiment was carried-out by circulating water through the spiral-tube GHE with different inlet water temperatures. Inlet and outlet temperatures of circulated water and ambient air temperature were periodically recorded. Water was circulated into the spiral-tube GHE. Different inlet temperatures were applied of 35, 40 and 45 °C in the experiments. The flow rate of circulated water was set to be constant of 3.6 L/min.

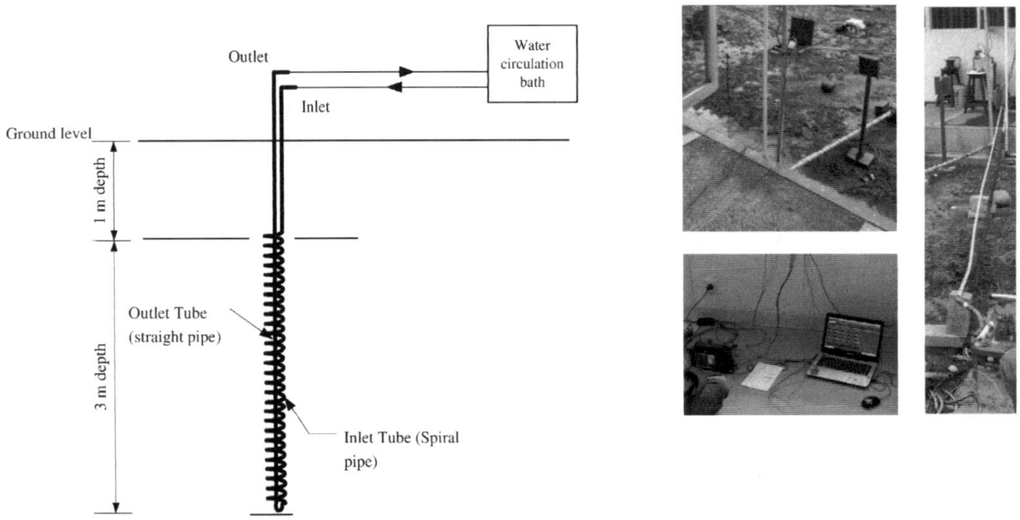

Fig. 2. The Schematic Diagram of Experimental Set-up.

3 Results and discussion

3.1 Temperature distributions

The temperatures of inlet and outlet water of the spiral-tube GHE, ground and ambient air were measured periodically as shown in figure 3. Local ground temperature at Hasanuddin University Gowa campus (119° 30' 06.1" E and 05° 13' 52.4" S) was measured at 3 m depth. This ground temperature is also shown in figure 3. The initial ground temperature surrounding the borehole is 27-28 °C. Water was circulated through the spiral-tube GHE. The temperature of water decreased gradually along the spiral pipe. The average temperature differences of inlet and outlet water are 0.5 °C for inlet water temperature of 35 °C; 1.3 °C for inlet water temperature of 40 °C and 1.1°C for inlet water temperature of 45 °C.

3.2 Heat exchange rate

The thermal performance of the spiral-tube GHE is evaluated by calculating its heat exchange rate. The heat exchange rate, Q, is calculated by the following equation:

$$Q = \dot{m} c_p \Delta T \,, \quad (1)$$

where $\dot{m}$ is flow rate, c_p is specific heat, and ΔT is the temperature difference of inlet and outlet water.

For simplicity, the heat exchange rate per meter of borehole depth, $\overline{Q}$, is defined as

$$\overline{Q} = Q / L \,, \quad (2)$$

where L is borehole depth of spiral-tube GHE.

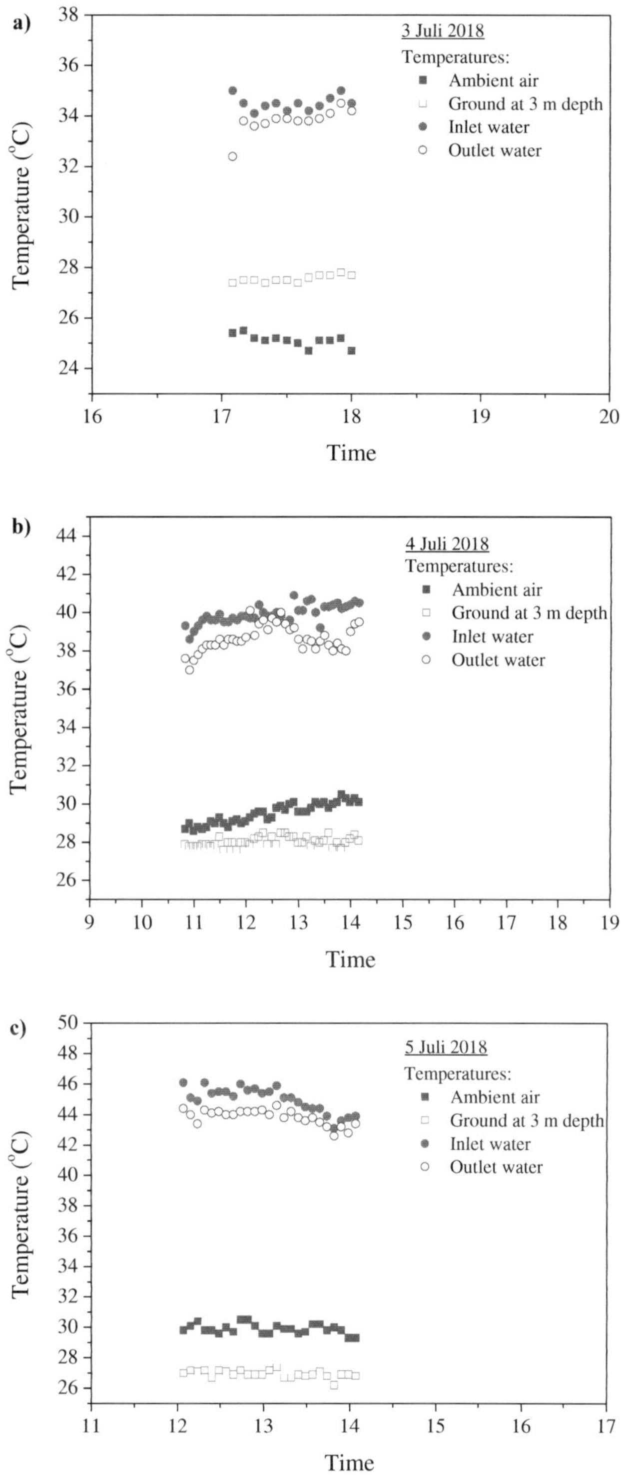

Fig. 3. Temperatures Distribution of Spiral-tube GHE.

The heat exchange rates of the spiral-tube GHE with different inlet temperatures are shown in figure 4. Heat is rejected to the ground surrounding the borehole through water flowing in the spiral-tube GHE.

The high rejected heat to the ground will increase the performance of the GHE. The thermal performance of the GHE was calculated with the different inlet water temperature. Heat exchange rate for different inlet water temperature of the spiral tube GHEs are 44 W/m (35 °C), 126 W/m (40 °C) and 110 W/m (45 °C). The high heat exchange rate of the shallow spiral-tube GHE indicates that this GHE is appropriated for application in ground-source cooling system especially for the hot climate like Indonesia.

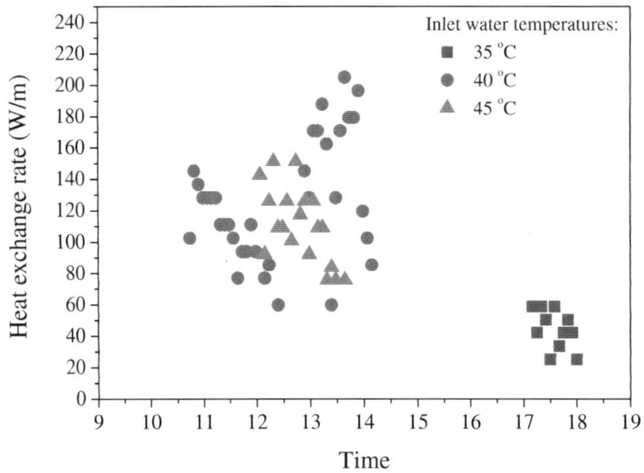

Fig. 4. Heat Exchange Rate of the Spiral-tube GHE.

4 Conclusions

The experimental study of shallow spiral-tube GHE which was buried in the ground of 3 m depth has been carried-out. The thermal performance of the GHE was evaluated by calculating its heat exchange rate. Based on the results of this study, the following conclusions are drawn:

1) The average temperature differences of inlet and outlet water are 0.5 °C for inlet water temperature of 35 °C; 1.3 °C for inlet water temperature of 40 °C and 1.1°C for inlet water temperature of 45 °C.
2) Heat exchange rate for different inlet water temperature of the spiral tube GHEs are 44 W/m (35 °C), 126 W/m (40 °C) and 110 W/m (45 °C).
3) The high heat exchange rate of the shallow spiral-tube GHE shows that the utilization of this type of GHE can be applied in ground-source cooling system especially for the hot climate like Indonesia.

References

[1] Jalaluddin, A. Miyara, K. Tsubaki, S. Inoue, K. Yoshida, Experimental study of several types of ground heat exchanger using a steel pile foundation, Renewable Energy, 36:764-771 (2011).

[2] Jalaluddin, A. Miyara, K. Tsubaki, K. Yoshida, *Thermal performance of three types of ground heat exchangers in short-time period of operation*, Proceedings of the 13th International Refrigeration and Air Conditioning Conference at Purdue, West Lafayette, USA (2010).

[3] Jalaluddin and A. Miyara, 2012,Thermal performance investigation of several types of vertical ground heat exchangers with different operation mode, Applied Thermal Engineering, 33-34:167–174.

[4] Jalaluddin and A. Miyara, 2016, Thermal performances of vertical ground heat exchangers in different conditions, Journal of Engineering Science and Technology, 11 (12): 1771-1783.

IOP Conf. Series: Materials Science and Engineering **619** (2019) 012012 doi:10.1088/1757-899X/619/1/012012

[5] Y. Man, H. Yang, N. Diao, J. Liu, Z. Fang, 2010, A new model and analytical solutions for borehole and pile ground heat exchangers, Int. J. Heat Mass Transfer, 53:2593–2601.
[6] P. Cui, X. Li, Y. Man, Z. Fang, 2011, Heat transfer analysis of pile geothermal heat exchangers with spiral coils, Applied Energy, 88:4113–4119.
[7] Y. Man, H. Yang, N. Diao, P. Cui, L. Lu, Z. Fang, 2011, Development of spiral heat source model for novel pile ground heat exchangers, HVAC&R Research, 17 (6):1075–1088.
[8] M. Li and A.C.K. Lai, 2012, Heat-source solutions to heat conduction in anisotropic media with application to pile and borehole ground heat exchangers, Applied Energy, 96:451–458.
[9] A.Zarrella, A. Capozza, M. De Carli, 2013, Analysis of short helical and double U-tube borehole heat exchangers: A simulation-based comparison, Applied Energy, 112:358–370.
[10] A. Zarrella, M. De Carli, A. Galgaro, 2013, Thermal performance of two types of energy foundation pile: Helical pipe and triple U-tube, Applied Thermal Engineering, 61:301-310.
[11] Jalaluddin and A. Miyara, 2015, Thermal performance and pressure drop of spiral-tube ground heat exchangers for ground-source heat pump, Applied Thermal Engineering, 90:630-637.
[12] B. Dehghan, A. Sisman, M. Aydin, 2016, Parametric investigation of helical ground heat exchangers for heat pump applications, Energy and Buildings, 127:999–1007.

Acknowledgment
This study was supported by LP2M Hasanuddin University and financed by grant of DIKTI (Directorate General of Higher Education of Indonesia).

IOP Conf. Series: Materials Science and Engineering **619** (2019) 012013 doi:10.1088/1757-899X/619/1/012013

Enviromental Geochemistry of Bawakaraeng Mountain Soil: Implication for Anthropogenic Impact Gowa South Sulawesi Indonesia

Adi Tonggiroh [1)] **Rafiuddin Syam** [2)]

[1] Engineering Chemistry Lab,LBE Environment,Geochemistry, Engineering Faculty of Hasanuddin University, Indonesia
[2] Mechanical Dept. Engineering Faculty of Hasanuddin University, Indonesia

the corresponding author's e-mail: adi_unhas@yahoo.com

Abstract. Determination of heavy metal are done on Mount Bawakaraeng soil to find out the impact of anthropogenic, using range proximity matrix independent variables on camps, fired, garbage. There are two concentrated heavy metal zone and two particular zone (sample point 29, zone I; sample point 3 and 5, zone II) as concentrated heavy metal on rocks fractures, lavene statistic probability show the source of heavy metal Cr are different. Surface soil samples are taken on mapping variable, forest fires, especially for top and bottom of soil only garbage variable are taken. Furthermore, using Inductively Coupled Plasma Oscilloscope Emission Spectrometer (ICP-OES), X-Ray Fluorescence (XRF). Rock Samples analysis are done using thin section method on polarized microscope to find out the composition of volcanic minerals. EF range are higher and lower than 2 are showing the enrichment of heavy metal are naturally formed and EF relatively moderate by human activities. (Igeo) camp are Cr (0.606), Cu (0,529), Mn (0.383). Zn (0.590) Cd (-0.779). Pb (0.440); fired are Cr (0.586), Cu (0.514), Mn (0.383), Zn (0.590), Cd (-0.770), Pb (0.447); garbage is Cr (0.584), Cu (0.461), Mn (0.343), Zn (0.579), Cd (-1.110), Pb (0.458). the difference of *EF* and *Igeo* value are affected by sample point 3;5;29 and indexes value 1.5 on volcanic rocks soil. Friedman test are showing relatively small differences between human activities and nature factor, which indicating that Bawakaraeng Mount morphology are easily affected by exogenic and endogenic factors.
Key Words : Bawakaraeng mountain, anthropogenic, heavy metals, proximity matrix.

1. Introduction

Heavy metal contamination are happened caused by anthropogenic interaction, precipitation of surface water [5], which accumulated on soil [12]. These factors are important on research location which composed by soil from parent materials [3] [14] [31]. Vulcanic having permeability properties, same porosity and different fractures quantity. The difference on fractures are caused by resistances of trachyte, andesite and volcanic breccia along with fault zone.

On volcanic rocks environment, heavy metal transported on surface following soil particles [2] [4] [9] [19] [24] [29] [32] [22], permeabilities, rock fragment size and concentrated. The metal content in soil is a sum of metals originating from natural processes and human activity. It is estimated that the contribution of metals from anthropogenic sources in soil is higher than the contribution from natural ones [13].

Study area located in south part of Sulawesi island on S 5°16'30", E119°55'48"; S18°18'7.2", E119°57'21.6"; S5°19'15.6", E119°56'56.4"; S5°19'15.6", E119°56'2.4"; S5°17'9.6", E119°54'57.6" (see figure 1).

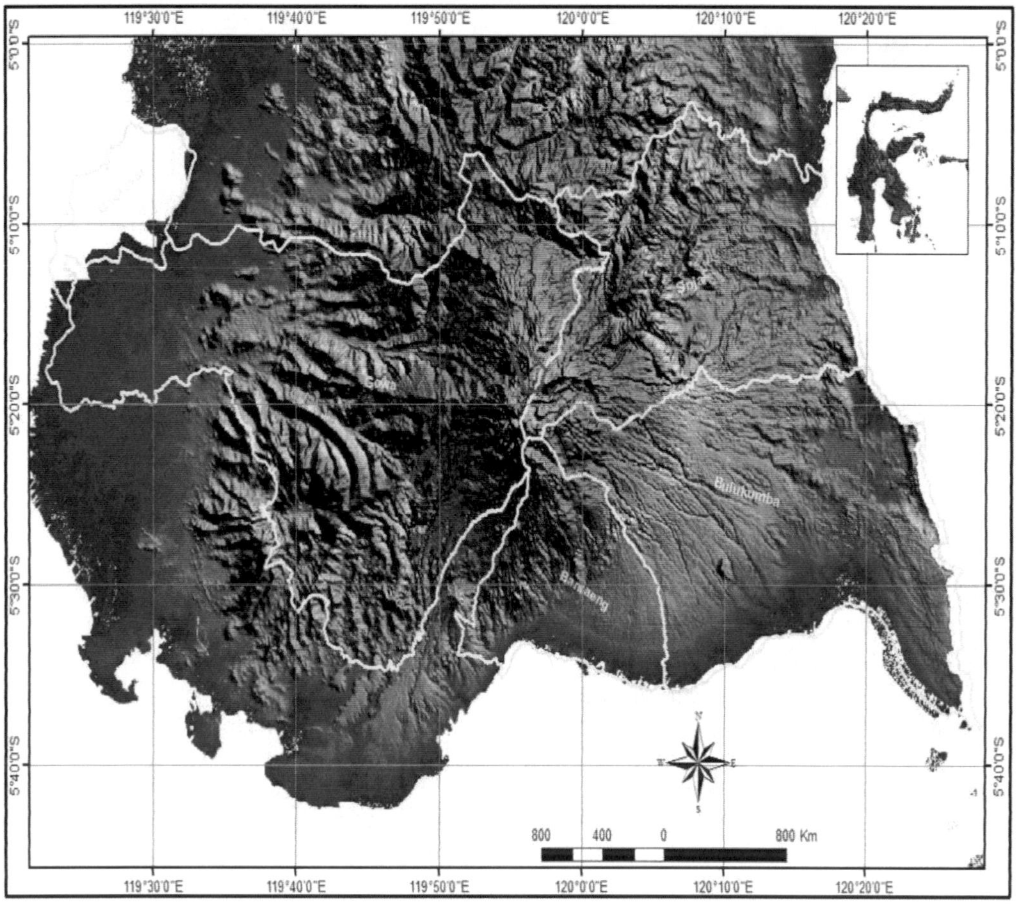

Figure 1. Area of Research

Mount Bawakaraeng are dominated by volcanic rocks, fault structures, fractures, and the mount becomes important as Jeneberang River upstream (current area of Jeneberang river). As tropical wet forest are having yearly average rainfall (1991 to 2000) 380.85 mm/yr, (2001 to 2010) 267.66 mm/yr and 270.80 mm/yr (2011 to 2017) (calculation of raw data from Meteorological Climatology and Geophysical Agency Pagentungan Unit Gowa Regency, South Sulawesi Indonesia), temperature range from 2°C to 6°C (before period of 2000 year) and 8°C to 14°C (after period of 2000 year) (fig 2).

Note from Tribunnews (2015), human activities are increase to thousands people per year, this causing the land clearing for camping, losses of soils layer, garbage and forest fires (figure 3).

This research is funded by Research and Community Services of Hasanuddin University. The samples are taken on March to May 2017 using methods of geochemist mapping, enrichment factors, ige and multivariate statistic as heavy metal concentrate determination on soils.

2. Regional Geology

Mount Bawakaraeng are part of south regional geology of mandala West Sulawesi, two mountain alignment which elongated relatively north to south, Mount Bone and Mount Lompobattang landscape. The two mountains are separated by Walanae Fault (Jaya, 2013) and Mount Bawakaraeng (2284 m asl) as Pleistocene volcanic mountain with tis cone called Mount Lompobattang (2874 m asl). In western side laid Mount Baturape with its height 1124 m asl and in northern lain Mount Cindako with its height reaching 1500 m asl.

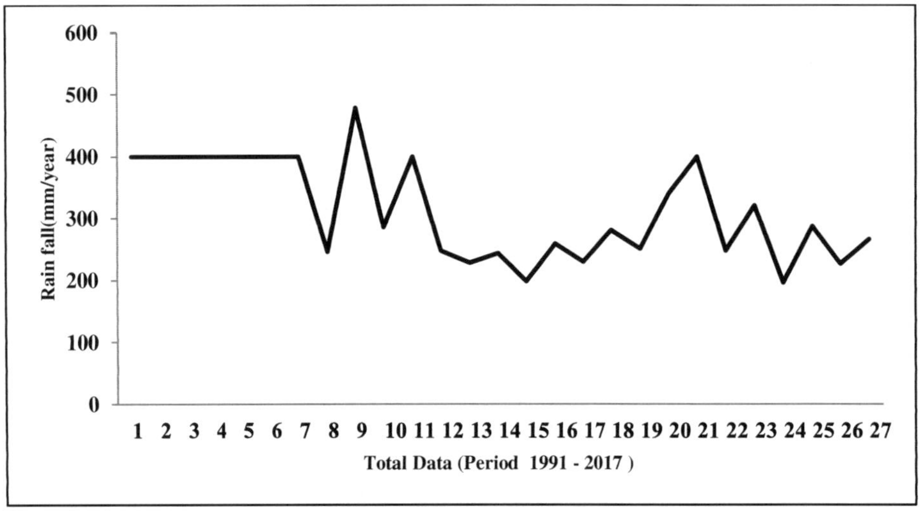

Figure 2. Rainfall data the Periods from 1991 to 2017

3. Material and Methods

Using 31 soil samples which taken from camp sites, waste disposal and fires area, there are difficulties to determine distribution boundary location because the changes of soil surfaces by human activities. Sampling are done vertically on top layer and bottom layer 10 cm to 40 cm. Soil samples are dried and sieved to separate size < 0.062 mm using sieved machine in chemical engineering laboratory of Unhas Engineering Faculty, afterward packed in sample plastic bags then analyzed using ICP-OES, XRF.

The rocks which going to be analyzed are sliced and polished to 0.03 mm and 0.05 mm thick, placed on top of preparation glass to observe under polarized microscope in mineral optic laboratory of Geology department, Faculty Engineer, Hasanuddin University

Geochemical data is consisting of major and minor element processed using hierarchy cluster proximity matrix statistic method and factor analyze on software SPSS IB v.22, Statistica v.10.

4. Geology Local

4.1. Geomorphology and drainage pattern

Geomorphology are grouped into 2 unit (fig 4).

- **Very Steeply Mountain Morphology**

This morphology having percentage of slope 62,66% - 104,44% Scattering in southern and north north-eastern of Bulu Bawakaraeng, Moncong Lompobattang, Bulu Porong, Bulu Baria, Bulu Ganting and Bulu Asumpolong.

- **Steep Mountain Morphology**

This morphology having percentage of slope 25,86% - 38%. Scattering in northern of Jeneberang river area. Tanralili Lake and its Surrounding and on the western of Bulu Porong alongated to south.

The slope are affecting rocks resistance, the increases of rainfall on layer of soil surface and human feet trace as early forming of rill erosion, gully erosion which irregularly formed on volcanic rocks morphology.

Figure 3. (a) human activities (b) camp (c) garbage soil sampling (d) fires traces

4.2. Lithology

Vulcanic breccia. Lithology of volcanic breccia fragment are consisting of phenocryst (plagioclase, peroxin, hornblende, biotite, opaque minerals) and base mass in form of microcrystalline plagioclase, **andesite porphyry.** Lithology of volcanic breccia matrix are consisting of (rock fragment, plagioclase, orthoclase, muscovite, biotite, volcanic glass, opaque mineral), **Crystal tuff.** Petrography of thin section (pyroxene, plagioclase, opaque mineral, crystallite plagioclase, and base mass glass), **Basalt Porphyry** (Travis, 1955). Petrography analysis on thin section (pyroxene, biotite, hornblende, orthoclase, plagioclase, opaque mineral, crystallite plagioclase and base mass glass, **Andesite Porphyry** and petrography analysis on thin section (plagioclase, pyroxene, orthoclase, biotite, opaque mineral, crystallite plagioclase and base mass glass, **Trachyte** (fig. 3).

Weathering processes modify the mineralogy, petrography (microfabric) and geochemistry of rock materials [26] (fig 5). Rocks fragment microscopic identification on volcanic breccia showing phenocryst plagioclase minerals, pyroxene, hornblende, biotite ace equigranularity (A,B,C,E), Plagioclase crystallite (E) pyroxene crystallite (F), bad sortation on tuff (D) are showing this kind of rocks are easily affected by weathering. Increases of weathering degree are influenced by altitude [26]. Autochthonous of tuff type are easily weathered by climate changes from hot to cold, groundwater circulation and effect of some cynabacterial [17], the weathering residues forming silt and clay as association with Pb, Zn, Cr and Cu [1].

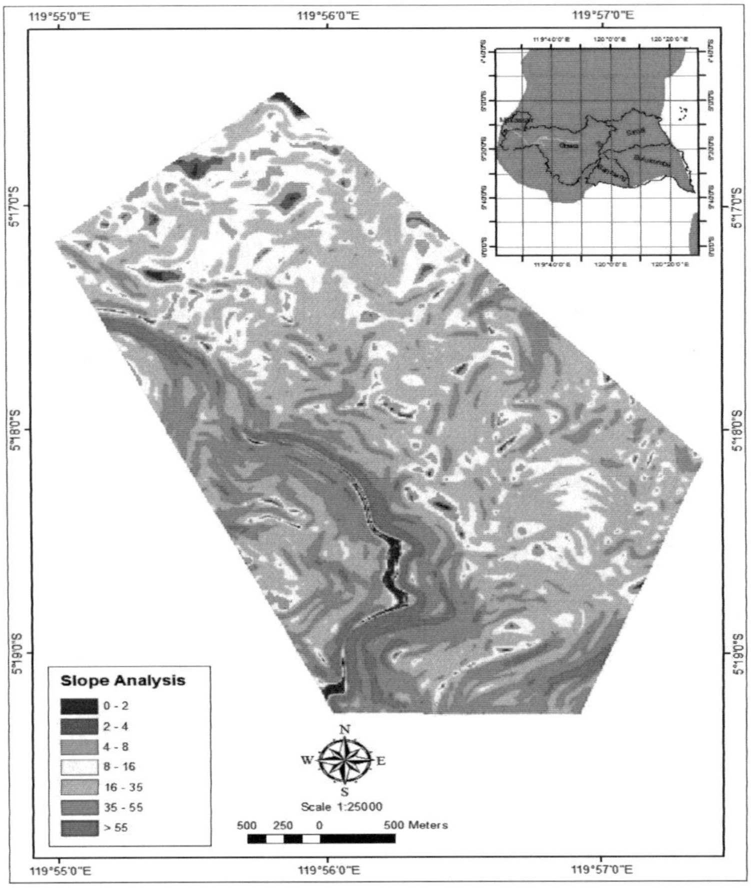

Figure 4. Slope Pattern

4.3 Structure Geology

Systematic joint in lithology of trachyte are analyzed using fan diagram which resulting greatest principal stress (σ_1) are relatively west north west – east southeast (N285°E) and least principal stress (σ_3) relatively north northeast – south southwest (N 15° E) (fig. 6) which is impact of Walanae Fault to south [10].

Bawakaraeng Strike Slip Fault. The fault structure relatively elongated from Northeast to Southwest and cut across by Lompobattang Strike Slip Fault, which formed by pressure on volcanic rocks which relatively west northwest – east southeast with shifting relatively righting (dextral). **Lompobattang Strike Slip Fault.** The fault structure relatively elongated from northwest to southeast. The fault structure and volcanic rocks fractures directly correlated on slope forming and landslide. Indirect correlation as tracking route, decreases of vegetation and surface erosion.

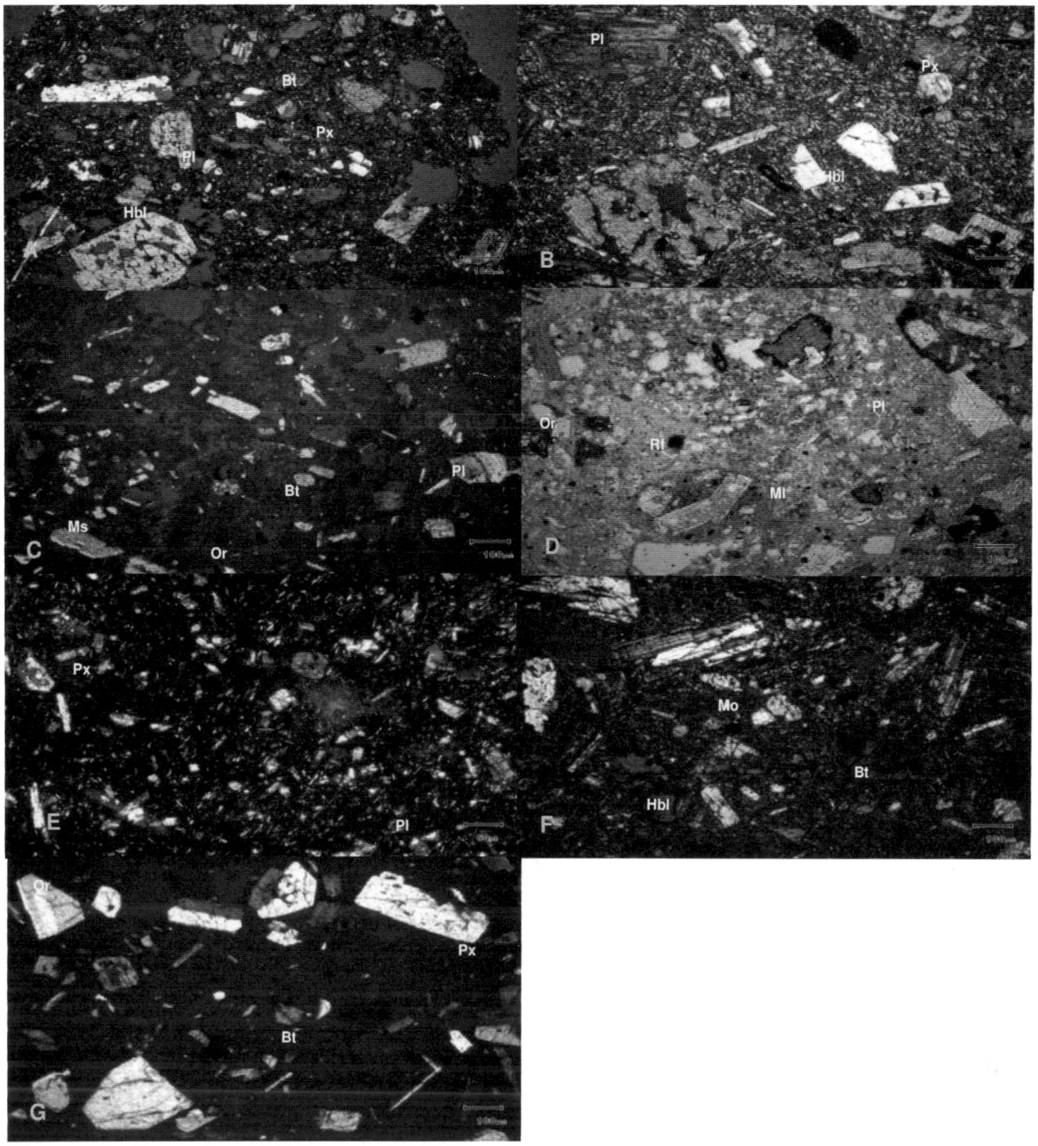

Figure 5. Micrograph photo of thinsection magnified 50x: (A) porhyry fragmen (B) porphyry basalt (C) trachyte porphyry fragmen (D) vulcanic breccia matrix (E) porphyry basalt (F) porphyry andesite (G) trachyte

5. Geochemistry

5.1 Index Geoaccumulation

Effect of metal concentration and contamination soil level are using Boxplot statistic and Geoaccumulation index and (Igeo) was originally stated by [16]. I geo is calculated as follows:

I geo = log2 [$Cn/1.5Bn$]

Where: Cn: Element concentration from soil samples (0.04 mm to 0.06mm), *Bn* : background value 1.5 (Muller, 1969).Classification: Class 0 ($I_{geo} \leq 0$) Practical Uncontamined; Class 1 ($0 < I_{geo} \leq 1$) Uncontamined to Moderately Contamined; Class 2 ($1 < I_{geo} \leq 2$); Class 3 ($2 < I_{geo} \leq 3$) Moderately to Heavily Contamined; Class 4 ($3 < I_{geo} \leq 4$) Heavily Contamined; Class 5 ($4 < I_{geo} \leq 5$) Heavily to Extremely Contamined; Class 6 ($I_{geo} > 5$) Extremely Contamined.

Geo-accumulation Index (I_{geo}) Value for camp sample point are Cr (0.606), Cu (0,529), Mn (0.383). Zn (0.590) Cd (-0.779). Pb (0.440); fired are Cr (0.586), Cu (0.514), Mn (0.383), Zn (0.590), Cd (-0.770), Pb (0.447); garbage are Cr (0.584), Cu (0.461), Mn (0.343), Zn (0.579), Cd (-1.110), Pb (0.458).

Negaive value results from Cd (-0.779; (-0.770); (-1.110) indicated there are some soils yet Uncontaminated to Contaminated by heavy metal

5.2 Enrichment Factor

Simpilication to find out environment contamination, using Enrichment Factor (EF), such as are used to review contamination on various environment media [6] [8] [18] [33] [20].

$$EF = [C_x / C_{ref}]_{sample} / [Cx / C_{ref}]_{background}$$

Where Cx is the concentration total of trace element, Cref is element reference contamination (Cr,Cu,Mn,Zn,Cd,Pb) from soil variable on sample point of garbage, fired, and camp site without anthropogenic references (Table 1).

Table 1. Enrichment Factor for heavy metals in three variables

Variable	Cr(ppm)	Cu(ppm)	Mn(ppm)	Zn(ppm)	Cd(ppm)	Pb(ppm)
Camp	0.377±1.2655	0.340±1.275	0.700±1.143	0.810±1.134	0.394±1.197	0.731±1.308
SD	19.54	17.87	17.119	9.374	0.176	6.394
Skewness	-1.294	-1.651	0.134	-0.261	-2.052	0.282
Curtosis	1.897	3.370	1.157	-1.774	5.060	2.569
Fired	0.214±1.933	0.144±1.677	0.338±1.897	0.492±1.337	0.342±1.514	0.765±1.297
SD	6.522	10.96	5.157	14.284	0.326	7.760
Skewness	0.04	-0.292	0.204	-0.680	-0.590	-0.171
Curtosis	-1.682	-1.959	-1.244	-1.317	-1.592	0.631
Garbage	0.375±2.660	0.299±2.657	0.622±1.779	0.869±1.195	0.521±2.065	0.709±1.384
SD	29.108	22.659	21.767	16.900	0.256	12.665
Skewness	1.572	1.446	0.420	0.729	1.324	0.937
Curtosis	1.603	1.702	-1.205	1.826	0.684	0.297

There are average enrichment value on all variable which dominated by garbage>fired>camp where EF>2 states deficiency to minimal enrichment and EF 2-5 moderate enrichment.

5.3 Ranking

Chain model from 112 respondent on exogenous factor (human activities) and endogenous (weathering, water, climate, rock, slope, geology structure, vegetation). That square factor (R^2) exogeneous value 92.1% and endogenous is 7.9% (Tabel 2).

Table 2. Variable effect on metal sources, Model Summary[b]

Model	R	R Square	Adjusted R Square	Std. Error of the Estimate	Durbin-Watson
1	.959[a]	.921	.905	18.973	1.474

a. Predictors: (Constant), water infiltration, properties structure, slope quantity, weathering intensity, resistence of rock; b. Dependent Variable: influence human

Table 3. Hypotesis on human activities

Ranks 3a

Variable	Mean Rank
Human Activities	0.377±1.2655
Properties Rock	19.54
Structure Geology	-1.294
Slope Degree	1.897
Climate Change	0.214±1.933

Statistics Test

N	25
Chi-Square	65.787
df	4
Asymp.Sig.	.000

5.4 Distribution

Zona I. The lowest range proximity matrix (0,131) and the highest (0,967) are heavy metal concentration on the peak of Mount Bawakaraeng distributed to north of Bulu Sarongan. Particular sample 3 and 5, does not have effect on proximity matrix as distribution indication and heavy metal concentration on rocks fractures. Zone II (Fig.6).

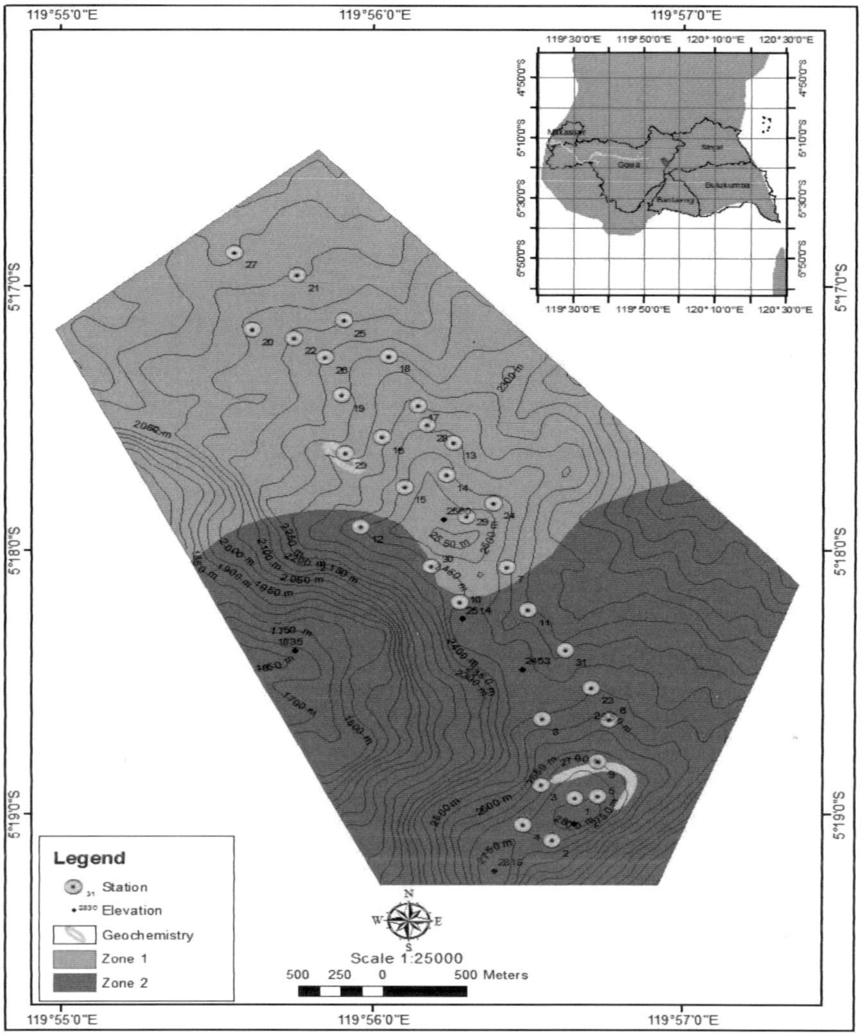

Figure 6. Zone of heavy metals distribution

The lowest range proximity matrix (0,191) and the highest (0,886) indicated on sample 21, 19, 27 heavy metal concentration distributed to north of Bulu Sarongan, Bulu Sarobaia (Fig. 7).

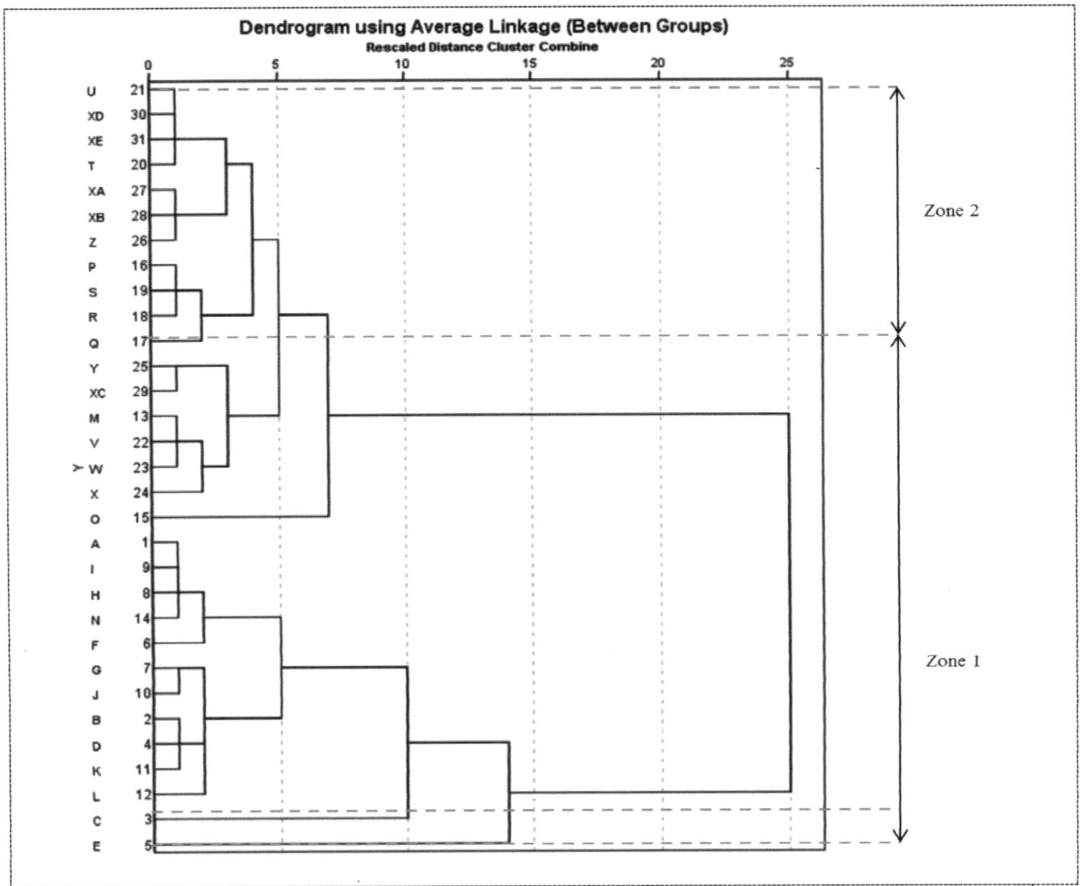

Figure 7. Zonation distribution of heavy metal

Analysis One Way Anova are showing the average of heavy metal increases: Mn (629.42 ppm)>Cr (106.35 ppm)>Zn (74.13 ppm)> Cu (49.23) > Pb (49.80) > Cd (0.65 ppm). Minimum rate of Mn (213.0 m); Cr (22.0 ppm); Zn (38.0 ppm); Cu (7.0 ppm); Pb (33.0 ppm); Cd (0.19 ppm), and maximum are Mn (1210.0 ppm); Cr (198.0 ppm); Zn (103.0 ppm); Cu (86.0 ppm); Pb (80.0 ppm); Cd (1.06 ppm). Lavene Statistic probabilities value (0.00 < 0.05) are considered the heavy metal variants aren't same and F value (1.34) are showing the source of heavy metal are significantly different especially Cr and Cd (Table 4).

Table 4. One Way Anova Analysis

Element	**Cr Ppm**	**Cu ppm**	**Mn ppm**	**Zn ppm**	**Cd ppm**	**Pb ppm**
Average	106.35	49.22	629.41	74.12	0.64	49.80
Max	198	86	1210	103	1.06	80
Min	122	7	213	38	0.19	33
Mean	106.35	49.23	629.42	74.13	0.65	49.80
	Test of Homogeneity of Variances					
Lavene Statistic	df1	df2	Sig.			
75.346	5	180	0.00			
	Multiple Comparison					
Cr	Cd		0.007			
Cu	Mn		0.000			
Cr	Cu		0.398			
Zn	Cr		0.889			

Squared Euclidean Distance garbage variable there are similarity on heavy metal distribution on sample 1, 6 and 8 with distance average (0.437) and on sample 2 (3.602). Fig 8a, show the forming of three cluster which are: (1, 2, 3, 7, 9, 12, 14); II (18.25) and III (5). On camp variable heavy metal distribution are showed on sample 22, 31, 29 (1.820) on sample 14 (2.470) with four cluster, which are: I (9); II (14, 18, 23, 24, 25); III (22, 29, 31) and IV (30), figure 8b. Fired variable are indicated by increasing of distance average value on sample 17, 20, 21 (2.280) and sample 4, 6, 17 (6.250) with four cluster, which are: I (4, 11, 12); II (6, 16); III (17, 20) and IV (2, 27, 28) on fig 8c.

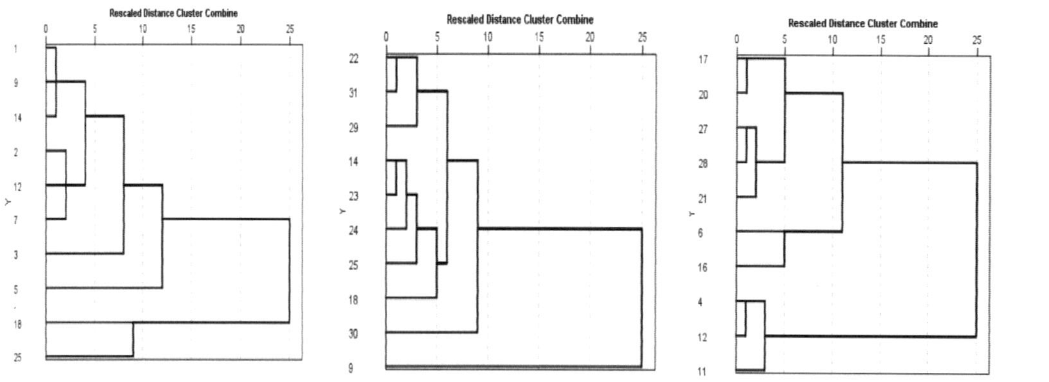

Figure 8. Heavy metal distribution on (a) garbage (b) camp (c) fired

6. Conclusion

Increases of heavy metal Mn>Cr>Zn>Cu>Pb>Cd, that heavy metal source actually heterogenous enough on soil mainly heavy metal Cr (lavene probabilistic = $0.00 < 0.05$), this is could be caused by Cr having relatively constant mobilities. The difference value of range proximity matrix is showing that heavy metal distribution heterogeneously divided into two zone, indicated by changes of homogeneity to two zone heterogeneities that almost all camp, fired and garbage variable as soil sampling point are contaminated by human activities. The differences on EF and Igeo value, could be affected by homogeneity of sample point 3 and 5 (zone I), sample point 29 (zone II), and uses of background value on soils from weathered volcanic rocks. Friedman Test showing there are small differences between human activities and natural factor, showing that geomorphology of Bawakaraeng Mountain unstable enough towards the changes from exogeneous and endogenous.

References

[1] Adamo P.,Denaix L.,Terribile F, Zampell M.,2003.Characterization of Heavy Metas in Contamined Volcanic Soils of the Solofrana river valley (southern Italy).Geoderma 117.346-366

[2] Adams, M., Zhao, F., McGrath, S., Nicholson, F., Chambers, B., 2004. Predicting cadmium concentrations in wheat and barley grain using soil properties. J. Environ. Qual. 33,532–541

[3] Alloway, B.J., 2012. Heavy Metals in Soils: Trace Metals and Metalloids in Soils and Their Bioavailability Springer

[4] Baize, D., Bellanger, L., Tomassone, R., 2009. Relationships between concentrations of trace metals in wheat grains and soil. Argon. Sustain. Dev. 29, 297–312

[5] Botte SE.,Freije,RH.,Marcovecchio,J.E.,2007.Disssolved Heavy Metal (Cd,Pb,Cr,Ni) Concentration in Surface Water and Pore Water From Bahia Blanca Estuary Tidal Flats.Bull.Environ.Contam.Toxicol.79,415-421

[6] Chester, R., Stoner, J.H., 1973. Pb in particulates from the lower atmosphere of the eastern Atlantic. Nature 245, 27–28.
[7] Flores V, L.M., Ducaroir, J., Jaunet, A.M., Robert, M., 1996. Study of the distribution of copper in an acid sandy vineyard soil by three different methods. Eur. J. Soil Sci.47
[8] Hernandez, L., Probst, A., Probst, J.L., Ulrich, E., 2003. Heavy metal distribution in some French forest soils: evidence for atmospheric contamination. Science of the Total Environment 312, 195–219.
[9] Huang, M., Zhou, S., Sun, B., Zhao, Q., 2008. Heavy metals in wheat grain: assessment of potential health risk for inhabitants in Kunshan, China. Sci. Total Environ. 405, 54–61
[10] Jaya A,Nishikawa.2013.Paleostress reconstruction from calcite twin and fault-slip data using the multiple inverse method in the East Walanae fault zone:Implication for the Neogene contraction in South Sulawesi,Indonesia.Journal of Structural Geology.55.p.34-49
[11] Jimenez MS.,Osuna FP.,Fernandez ACR.,2003.Geochemical evidences of the Anthropogenic alteration of trace metal composition of the sediments of Chiricahueto marsh (SE Gulf of California), Environmental Pollution 125. 423–432
[12] Karimi A.,Haghnia GH.,Safari T.,Hadadian,2017.Lithogenic and anthropogenic pollution assessment of Ni, Zn and Pb in surface soils of Mashhad plain, northeastern Iran. Catena 157.151–162
[13] Loska K, Wiechula, Irena Korus,2004, Metal contamination of farming soils affected by industry, Elsevier
[14] Luo, L., Ma, Y., Zhang, S., Wei, D., Zhu, Y.-G., 2009. An inventory of trace element inputs to agricultural soils in China. J. Environ. Manag. 90, 2524–2530.
[15] Meteorological Climatological and Geophysical Agency Pagentungan Unit Gowa Regency South Sulawesi,Data of Rain Fall Period of 1991-2017.
[16] Muller, G.,1969. Index of geoaccumulation in sediments of the Rhine River. Geo. J., 2, (3): 108-118.
[17] Nash DJ.,Laren SJ.,2007. Geochemical Sediments & Landscape,Blacwell Publ.1.173-199
[18] N'guessan, Y.M., Probst, J.L., Bur, T., Probst, A., 2009. Trace elements in stream bed sediments from agricultural catchments (Gascogne region, S-W France): where do they come from? Science of the Total Environment 407, 2939–2952.
[19] Oliver, D.P., Hannam, R., Tiller, K.G., Wilhelm, N.S., Merry, R.H., Cozens, G.D., 1994. The effects of zinc fertilization on cadmium concentration in wheat grain. J. Environ. Qual. 23, 705–711.
[20] Preston W, Silva Y J.A.B, Nascimento C WA, Cunha Karina PV,
[21] Silva DJ, Ferreira HA,2016, Soil contamination by heavy metals in vineyard of a semiarid region:An approach using multivariate analysis, Elsevier
[22] Ran J.,Wang D.,Wang C.,Zhang G.,Zhang H.,,2016 Heavy metal contents, distribution, and prediction in a regional soil–wheat system, Science of Total Environment 544,422-431
[23] Salomons, W., Fo¨ rstner, U., 1984. Metals in the Hydro cycle. Springer Verlag, Berlin
[24] Shi, G.L., Zhu, S., Bai, S.N., Xia, Y., Lou, L.Q., Cai, Q.S., 2015. The transportation and accumulation of arsenic, cadmium, and phosphorus in 12 wheat cultivars and their relationships with each other. J. Hazard. Mater. 299, 94–102
[25] Szefer, P., Skwarsec, B., 1988. Distribution and possible sources of some elements in the sediment cores of the southern Baltic. Marine Chemistry 23, 109–129.
[26] Taboada T.,Lado LR, Vázquez CF., Stoops G., Cortizas AM.,2016. Chemical weathering in the volcanic soils of Isla Santa Cruz (Galápagos Islands, Ecuador).Geoderma 261.p.160-168
[27] Tonggiroh.A , 2013.Application of Principal Analysis Component and Mobility of Heavy Metals in Mine Settling Ponds Nickel Laterite , Konawe North, Southeast Sulawesi. Advances in Environmental Biology, 8(22) , p. 461-467
[28] Tribunnews.com,17 Agustus 2015.
[29] Wang, C., Ji, J., Yang, Z., Chen, L., Browne, P., Yu, R., 2012. Effects of soil properties on the transfer of cadmium from soil to wheat in the Yangtze River Delta region, China—a typical industry–agriculture transition area. Biol. Trace Elem. Res. 148, 264–274.

[30] Wang, C., Wang, J., Yang, Z., Mao, C., Ji, J., 2013. Characteristics of lead geochemistry and the mobility of Pb isotopes in the system of pedogenic rock–pedosphere–irrigated riverwater–cereal–atmosphere from the Yangtze River delta region, China. Chemosphere 93, 1927–1935.
[31] Wei, B., Yang, L., 2010. A review of heavy metal contaminations in urban soils, urban road dusts and agricultural soils from China. Microchem. J. 94, 99–107.
[32] Wu, J., Norvell, W., Hopkins, D., Welch, R., 2002. Spatial variability of grain cadmium and soil characteristics in a durum wheat field. Soil Sci. Soc. Am. J. 66, 268–275.
[33] Redon PO., Bur Thomas., Guiresse M., Probst JL., Toiser A ., Revel JC ., Jolivet C., Probst .,A. 2013. Modelling trace metal background to evaluate anthropogenic contamination in arable soils of south-western France Geoderma, vol. 206 . pp. 112-122

Acknowledgments
We would like to appreciate any support by NVJT Foundation,Kosmik Unhas, BEM Economic Unhas, Mapala-PNUP-Kompas Sospol-Equlibrium-Panditas-Nobel-PAL;KPA-Reichas-Celebes-Kriyaw-Lontara-Wirpala, Primpungan Ana'Gowa.

Combustion reactivity of low rank coal by the mixture of candlenuts shell

Daud Patabang[1], Effendy Arif[2], Jalaluddin[2,*], Nasruddin Azis[2]

[1] Departement of Mechanical Engineering of Tadulako University, Palu, 94118, Indonesia
[2] Departement of Mechanical Engineering,Hasanuddin University, Bontomarannu, Gowa, 92171, Indonesia
[1] Departement of Mechanical Engineering of Tadulako University, Palu, 94118, Indonesia

the corresponding author's e-mail: jalaluddin_had@yahoo.com

Abstract. Mixture of Low-Rank Coal (LRC) with Candlenuts Shell (CNS) has been carried-out to improve the quality of combustion. The quality of combustion is affected by combustion reactivity, thermal efficiency and combustion emission. This study investigated the effect of the mixture of CNS and LRC to find-out its combustion reactivity. An experimental study has been done to investigate the combustion reactivity of mixing LRC and CNS with the addition of CNS to LRC of 10%, 30%, 50%, 70% and 90% respectively. Testing methods including proximate and ultimate analysis following ASTM Standard and sample combustion in electric furnaces were applied with a combustion temperature of 500 ^{O}C, 625°C, 750 ^{O}C and 875°C respectively. Oxygen was used as an oxidizer in this experiment. The result of the experimental test shows that the addition of CNS into LRC generally increased its combustion reactivity. This fact indicates that adding CNS into LRC improves significantly the combustion quality of CNS and LRC mixture. The optimum addition of CNS into LRC to increase combustion reactivity is 10%.

Keywords. Low-rank coal, Candlenuts shell, Proximate Analysis, Ultimate Analysis, combustion reactivity

1. Introduction

Combustion is a rapid reaction of fuel and air as an oxidizer. Co-combustion of Coal mixed biomass is due to improving combustion reactivity, thermal efficiency and decreasing global warming effect. Indonesia has the potential of coal resources reaching 120 billion tons and reserves of 31 billion tons. Most of them (60%) are low-rank coal (LRC), Coal which has a calorific value of <5,000-6,000 kcal/kg, moisture content (30-40%) [1] and low ignition temperature [2].The use of coal as fuel in power plants and industries is the largest contributor to the global warming effect [3].In order to improve combustion reactivity of LRC, it is necessary to mix LRC with biomass that containing high volatile matters (VM), high oxygen (O_2) and low moisture (M) content.

Recently, previous studies investigated the effect of biomass mixed with coal in order to improve the quality of coal combustion, among others: Co-firing of coal with peat showed that the ash particles of the combustion product were smoother. This fact indicates that the reactivity of combustion and combustion is almost complete with the addition of peat into LRC [4]. The study of the effects of biomass addition of wheat and wood straws mixed into coal shows the decrease of NO and CO emissions and also reduces unburned carbon (UBC) [5], NO emissions from biomass combustion mixed with LRC coal using the oxy-fuel atmosphere obtained NO emission reduction [6]. The addition of biomass (corn stalk) into the coal increases the combustion rate of the fuel mixture [7]. The investigation on co-firing of coal with biomass through Thermo gravimetric and mass spectrometric (TG-MS) testing at temperature range 150-750 ^{0}C concluded that by increasing the addition of biomass into coal resulted in increased mass reduction in combustion and this suggests that there is an increase in reactivity combustion of a mixture of biomass with coal [8]. Performance evaluation of co-firing various kinds of biomass such as wood pellet, empty fruit bunch pellet, palm kernel shell, walnut shell with LRC in a 500 MWe coal-fired power plant is obtained maximum plant efficiency this is due to the torrified so as to increase grind ability of biomass before mixed with LRC [9]. The addition of biomass peanut husk and rise husk into Changzhi coal has implications for the decrease of ash fusion temperature [10].

Based on the description it is clear that actually biomass greatly improve the quality of coal combustion. Indonesia has the potential energy from biomass waste of 32,000 MW and which has been utilized only around 1740.4 MW [1]. One of the least optimal biomass used as a fuel in improving the quality of LRC is the Candlenuts Shell (CNS). According to data from the Indonesian Bureau of Statistics 2015, Indonesia's national candlenut production is 100,700 tons[11] and from this potential is 65,700 tons of candlenut shell waste. The potential of this CNS is large enough to be used as one of the fuels that can be co-fired with LRC.

The aim of this study is to obtain the characteristic of combustion reactivity of LRC mixture with CNS on various mixed compositions and combustion temperatures performed in a conditioned electric furnace.

2. Material and Method

2.1. Material

LRC derived from Sangatta, Coal mining in East Kalimantan, Indonesia and CNS obtained from the plantation of the people in Tinambung, West Sulawesi, Indonesia. The raw materials of LRC and CNS are as shown in figure 1, and figure 2.

Figure 1. Low-Rank Coal (LRC)

Figure 2. Chandelnut Shell (CNS)

2.2. Method

The CNS sample was firstly cleansed from the inherent impurities and then dried with dry torrefaction method at 250-300 ^{O}C for 30 minutes in electric furnace [12], each of LRC and CNS sample was milled by separate crusher with particle size 40-60 mesh [13], both of them CNS samples and the LRC is mixed in the mixing machine for 30 minutes, this is intended to allow the sample mixture to be homogeneous. The composition of CNS addition to LRC was 10%, 30%, 50%, 70% and 90%, respectively by mass basis as shown in figure 3(a-g).

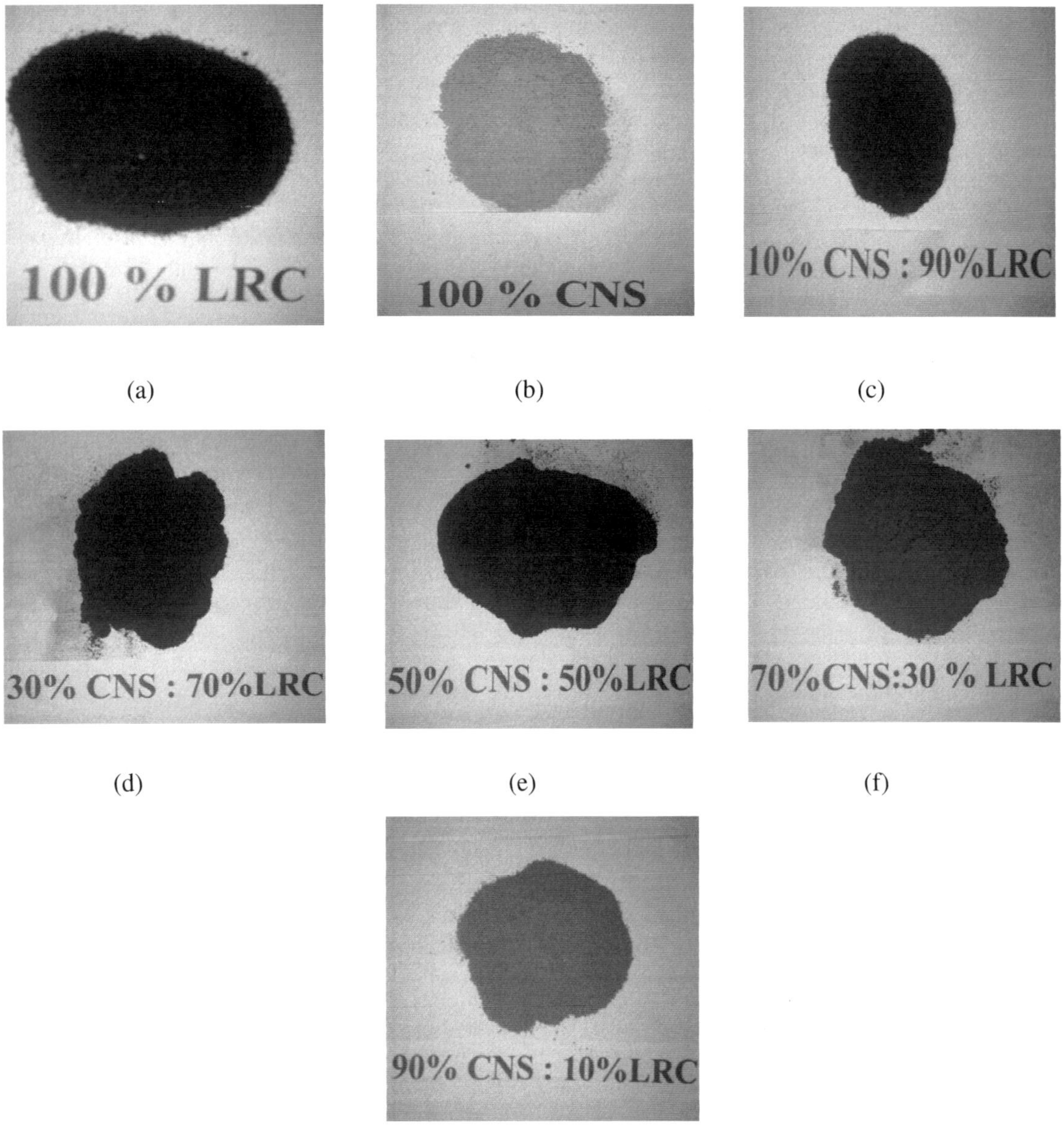

(g)

Figure 3(a-g), The sample of LRC, CNS and their blended.

The research method is done by experiment with the characterization of the sample consisting of Proximate Analysis to find out of thermal characteristic and Ultimate Analysis to obtain chemical characteristics, both of the measurement methods are based on the ASTM standard [14]. A 5-gram sample is placed in a cup sample suspended on a digital balance and then burned in an electric furnace by using oxygen as an oxidizer. The oxygen used in this experiment is taken from an oxygen tube by adjusting the oxygen flow rate in the oxygen gas regulator. The amount of oxygen used in combustion is 0.5 l/min. The burning of each sample is carried out at temperatures: 500 ^{O}C, 625 ^{O}C, 750 ^{O}C, and

875 ^{o}C [15]. The temperature setting is done by adjusting the flow of electricity with the electrical regulator. The test data taken decreases the sample mass to the constant mass of samples recorded on the digital balance and the flame temperature of the samples shown on the thermocouple panel. The experiment set-up of combustion as shown in figure 4.

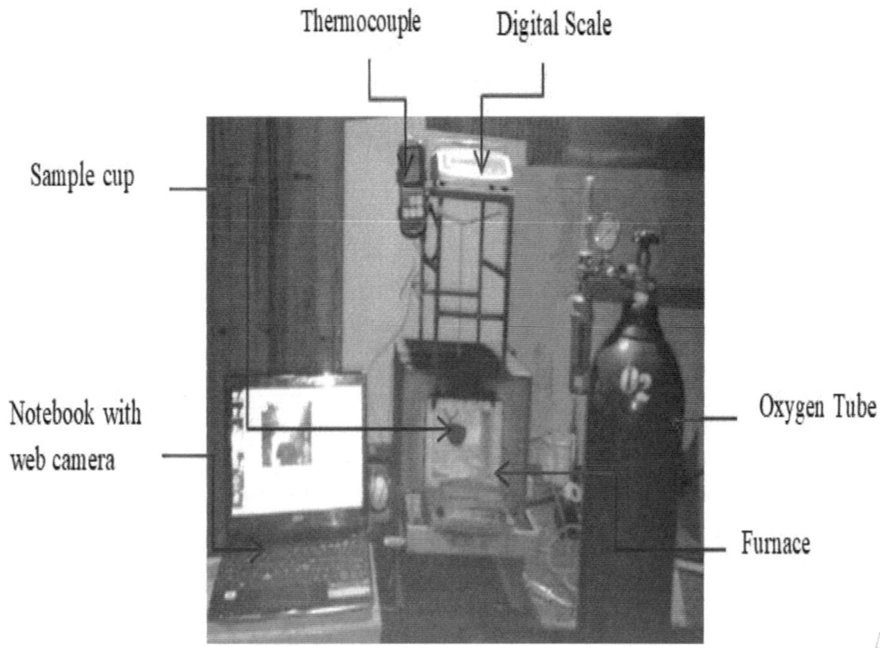

Figure 4, The experiment set-up

3. Results and discussion

3.1 The proximate and ultimate analysis

The proximate analysis is carried out to obtain thermal characteristics consisting of High Heating Value (HHV), Volatile Matters (VM), Moisture (M) Ash (A) and Fixed Carbon (FC) .The Ultimate analysis to obtain chemical characteristics consisting of Carbon C), Hydrogen (H_2), Nitrogen (N_2), Sulfur (S), and Oxygen content (O_2). Both of these tests are done following ASTM Standard. The test results are as shown in table 1.

Table 1, Proximate and Ultimate analysis

Proximate Analysis(%)adb	CNS 100%	LRC 100%	(LRC/ CNS): (90:10)%	(LRC/ CNS): (70:30)%	(LRC/C NS): (50:50)%	(LRC/ CNS): (70:30)%	(LRC/ CNS): (90:10)%	ASTM Standard
Volatile Matters (VM)	63,00	39,80	42,25	47,20	51,51	56,99	62,03	D 3175
Moisture (M)	6,72	15,15	14,10	12,15	10,56	8,51	6,93	D3173
Ash(A)	0,47	2,38	1,95	1,55	1,34	1,08	0,53	D3174
Fixed Carbon(FC)	28,69	42,67	41,70	39,10	36,59	33,42	30,31	by difference
HHV(kcal/kg	4936	5406	5329	5216	5187	5104	5049	D5865-12

Ultimate Analysis (%) adb								
Carbon(C)	35,53	19,47	32,11	28,98	26,31	22,79	22,56	D5373-02
Nitrogen (N)	1,05	0,33	0,81	0,75	0,57	0,46	0,39	D5373-02
Hidrogen (H2)	9,72	15,49	10,36	11,97	13,11	14,43	14,68	D5373-02
Sulfur (S)	0,148	0,032	0,133	0,110	0,085	0,061	0,038	D 3177
Oxygen (O2)	51,17	64,21	54,64	56,64	58,59	61,18	61,80	by difference

3.2 Mass change analysis

3.2.1 The effect of composition of co-fuel on combustion behaviour
The sample mass changes are shown in figure 5(a-d).

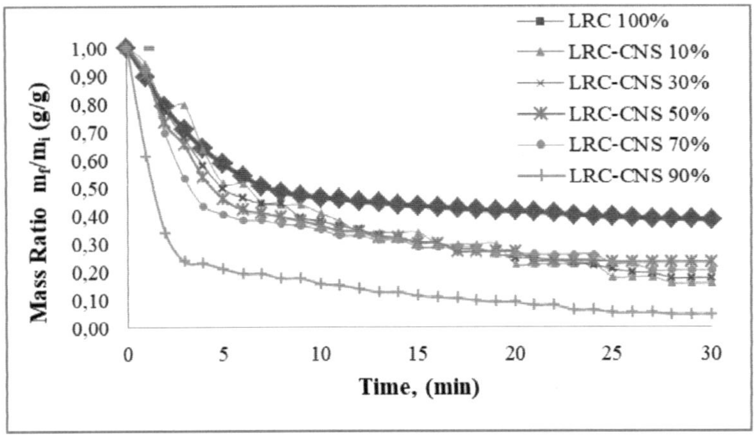

Figure 5(a), Combustion history of LRC mixed CNS at combustion temperature of 500 ^{O}C.

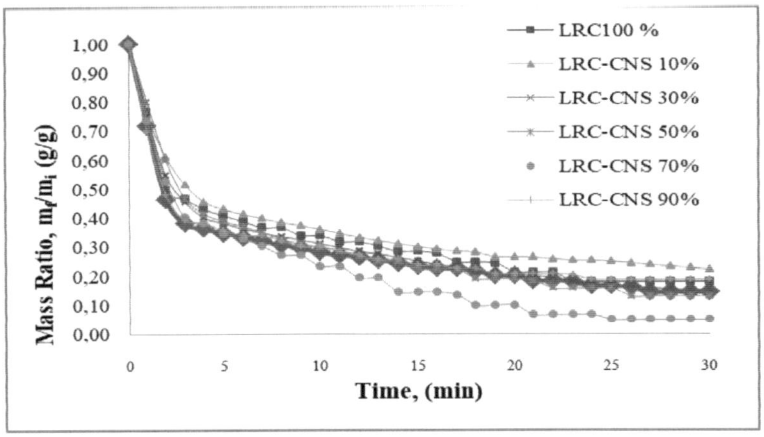

Figure 5(b), Combustion history of LRC mixed CNS at combustion temperature of 625 ^{O}C

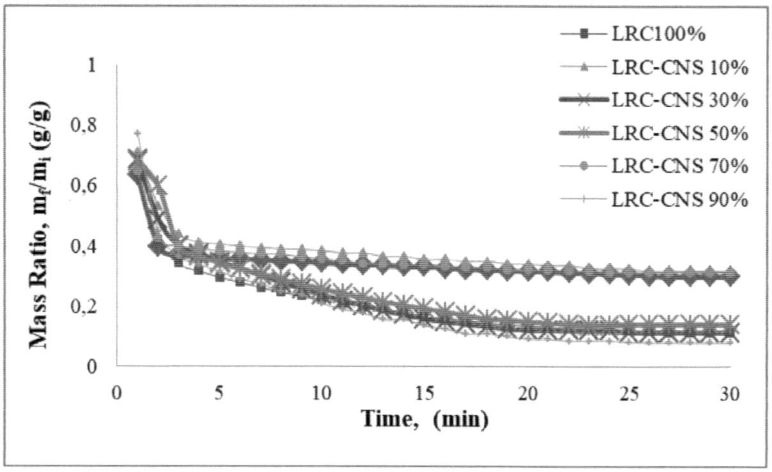

Figure 5(c), Combustion history of LRC mixed CNS at combustion temperature of 750 ^{O}C.

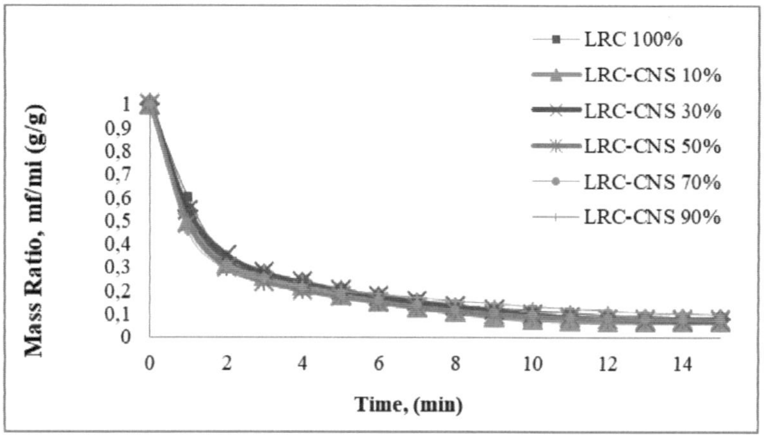

Figure 5(d), Combustion history of LRC mixed CNS at combustion temperature of 875 ^{O}C.

Based on the combustion history of LRC mixed CNS in Figure 5 (a-d) it can be seen that basically, combustion consists of two zones. They are evaporation of moisture and the devolatilization zone occurs at the beginning of combustion and continues in about 4 minutes and the charcoal burning zone that occurs thereafter. The evaporation zone and devolatization are shorter in time, compared to longer charcoal zones, depending on the Fixed Carbon content in the fuel. The addition of CNS to the LRC increases the combustion rate, but due to the smaller content of Fixed Carbon from CNS than the LRC, it has implications for the shortness of the charcoal combustion.

In general, the addition of CNS into the LRC shows a larger decrease in the ratio of mass reduction of the sample. This suggests that the implications of adding CNS to the LRC increase the reactivity of combustion. The greater the addition of CNS into the LRC results in increased reactivity of combustion. This indicator is indicated by the increasing ratio of mass reduction of the sample. Some studies of coal combustion with biomass results increasing combustion reactivity [14-19]. This fact shows that as a result of the VM, H_2, and O_2 content on the CNS is higher than in the LRC, and the M content on the CNS is lower than the LRC, based on proximate and ultimate analysis as presented in Table 1.

3.2.2 Effect of Temperature on combustion behaviour

In order to find out of the impact of combustion temperature on the decrease of sample mass, it is plotted correlation between the behaviour of decreasing sample mass to the combustion of combustion temperature inside the furnace, as in figure 6(a-e).

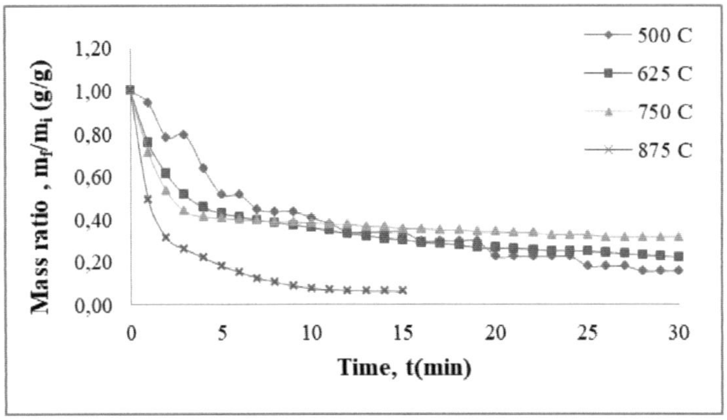

Figure 6(a), Combustion history on the addition of CNS 10%

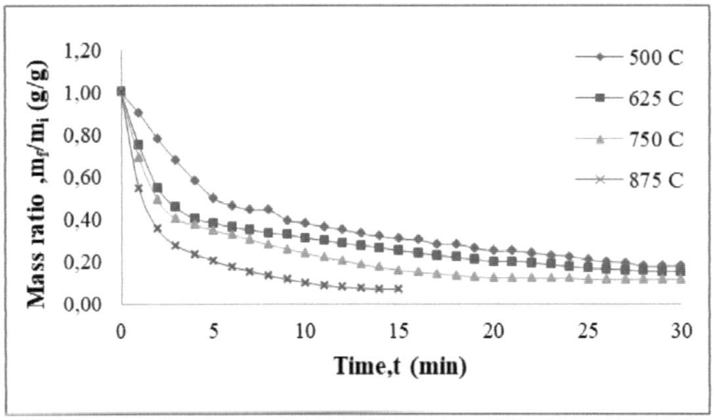

Figure 6(b), Combustion history on the addition of CNS 30%

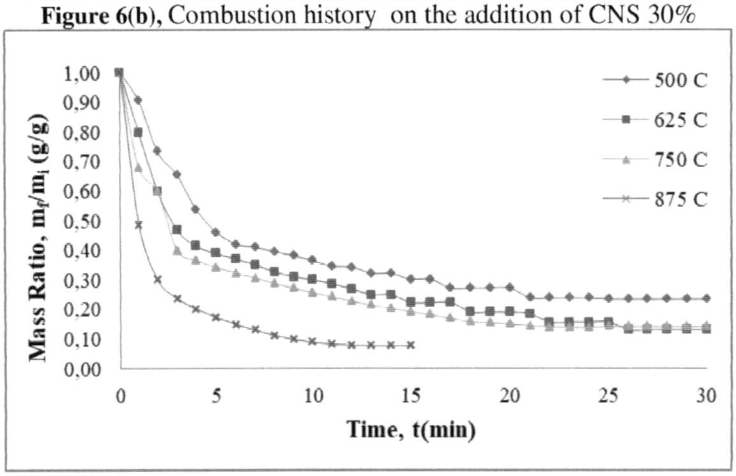

Figure 6(c), Combustion history on the addition of CNS 50%

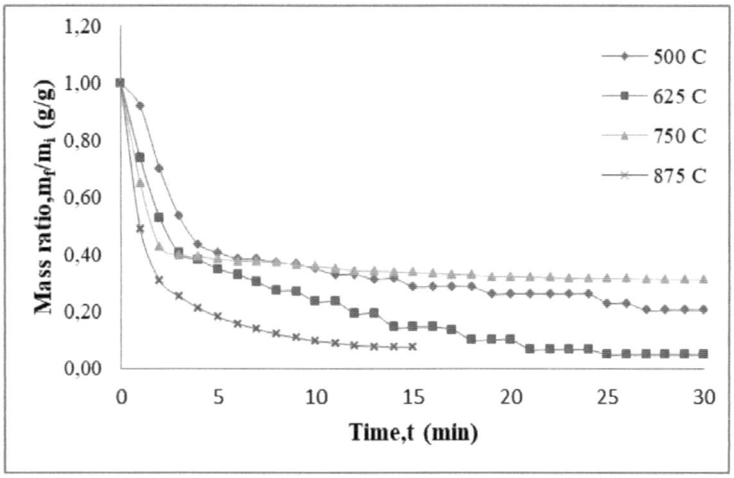

Figure 6(d), Combustion history on the addition of CNS 70%

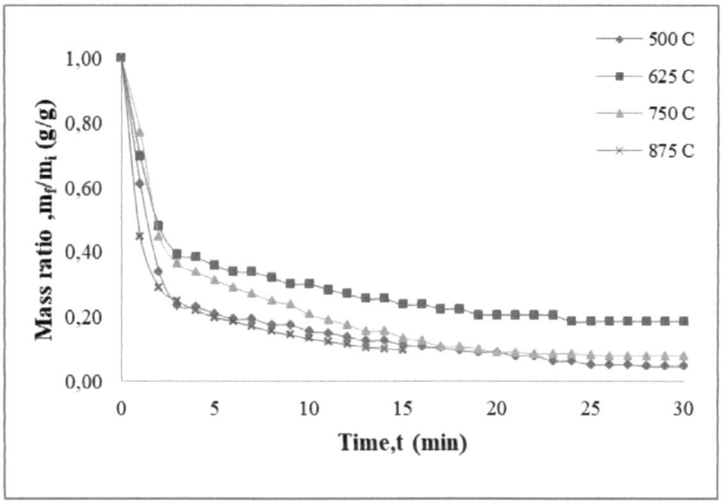

Figure 6(e), Combustion history on the addition of CNS 90%

Based on the behaviour of the mass decrease to the combustion temperature inside the furnace then the following is obtained. The higher sample combustion temperature resulted in a larger decrease in sample mass ratio indicated the higher the reactivity of the combustion. As a comparison of the effects of biomass type i.e. Japanese cedar, rice straw, and seaweed, the performance of LRC has a significantly impact on co-gasification, either at high combustion temperatures even at low combustion temperatures [18].

3.2.3 Reactivity of combustion

The reactivity of combustion based on maximum average of loss mass and the initial weight of sample according to the formula below [19]

$$R = \left(\frac{-1}{W_o}\right)(dW/dt) \quad (1)$$

Where (dW/dt) the maximum average of mass loss, and W_o is the initial mass of sample (g).

The results of the reactivity calculation of sample combustion are presented in figure 7(a-d).

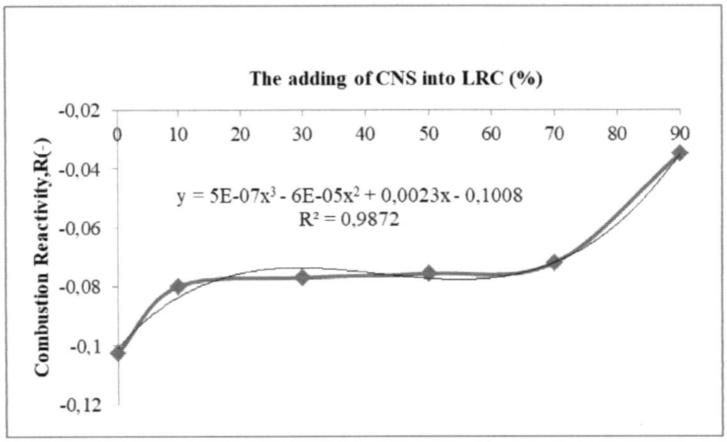

Figure 7(a), The combustion reactivity of the sample at combustion temperature of 500°C

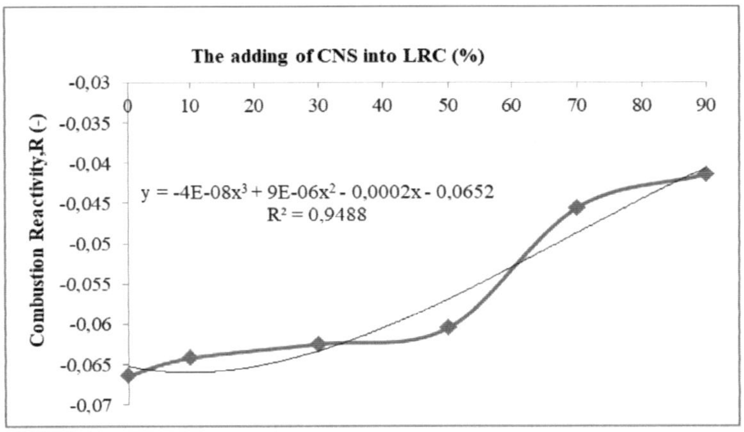

Figure 7(b), The combustion reactivity of the sample at combustion temperature of 625°C

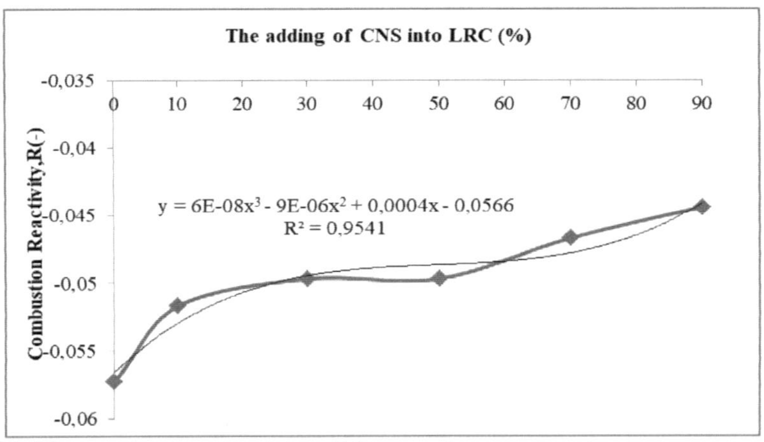

Figure 7(c), The combustion reactivity of the sample at combustion temperature of 750 ^{O}C

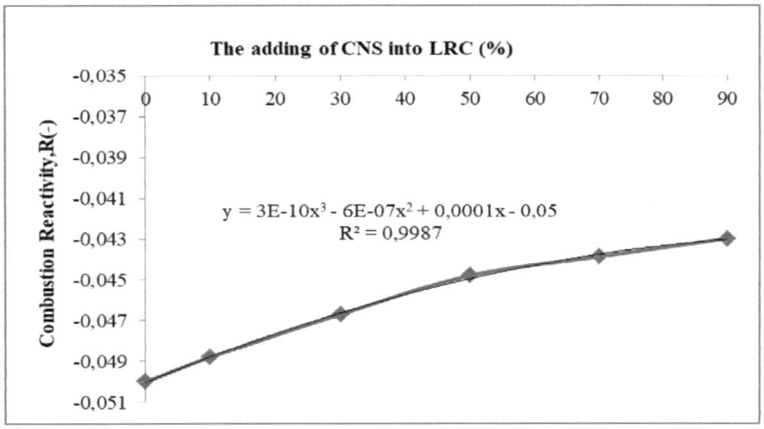

Figure 7(d), The combustion reactivity of the sample at combustion temperature of 875 ^{O}C

Based on the results of the calculation of combustion reactivity presented in Figure 7 (a-d), the increasing of CNS added into the LRC results in increased combustion reactivity. It can be presented based on the correlation equation between the addition of CNS into the LRC to the combustion reactivity number following the equation of the power three polynomials, as presented in figure 7 (a-d) From the results of combustion reactivity, a significant increase in combustion reactivity is obtained by adding 10% CNS.

4. Conclusions

Based on the results of the combustion experiment as an impact of the co-firing of LRC with CNS then it can be concluded as follows:

1. Addition of CNS into the LRC shows an increase in the rate of loss sample mass, indicating that there is a significant increase in reactivity of combustion.
2. Increased sample combustions temperatures have implications for increased reactivity of combustion
3. The optimum of CNS added into LRC is obtained at 10%.

5. References

[1] Renstra KSDEM RI 2015-2019, *Renstra Kementrian Sumber Daya Energi dan Mineral Republik Indonesia*. (in Bahasa) (2015-2019).
[2] Khatami R and Levendis Y. A,(2016) *Combust. Flame*, **164** :22–34.
[3] Zuwała J and Lasek J, (2017) *Co-combustion of low-rank coals with biomass*. **6.**Elsevier Ltd.
[4] Hutrindo E, Hardianto T. and Suwono A,(2006) *Characterization of The Ash Residues from The Co-combustion of Coal Mixed with Various Biomass / Peat*, no. Table **1**:1–7.
[5] Wang G*et al.*(2016)., *Energy Convers. Manag.*, **124**:414–426.
[6] Riaza J, Álvarez L, Gil M V, Pevida C, JPis J J, and Rubiera F,(2013)., *Ignition and NO emissions of coal and biomass blends under different oxy-fuel atmospheres*, **37**:1405–1412
[7] Liu Y *et al* (2018)., *Appl. Energy*, **215**:523–531
[8] Jayaraman K, Versan M, and Gokalp I,(2017)., *Renew. Energy*, **101**:293–300 (2017).
[9] Mun T, Zelalem T, Lee U, and Yang W,(2016)., *Performance evaluation of co- fi ring various kinds of biomass with low rank coals in a 500 MWe coal- fi red power plant*, **115**: 954–962 .
[10] Xiu-wei M A, Feng-hai L I, Ming-jie M A, and Yi-tian F,.(2018) *Fuel Chem. Technol.*, **46**(2):129–137

[11] Badan Pusat Statistik Indonesia (2015), *Badan Pusat Statistik Indonesia* (in bahasa)
[12] Ullah H, Liu G, Yousaf B, Ubaid M, and Abbas Q (2017)., *Bioresour. Technol.*, **245**:73–80
[13] Neminda B, Gan S, Eastwick C, and Kiat H,(2017)., *Fuel Process. Technol.*, **159**:287–305
[14] Adekunle J, Ibrahim J, and Kucha E, (2015) *Br. J. Appl. Sci. Technol.*, **7**(1):114–123
[15 Collard F,and Blin J, (2014).,*Renew. Sustain. Energy Rev.*, **38**:594–608
[16 Bhuiyan A A, Blicblau A S, A. Islam K M S, and Naser, J.(2016)., *Energy Inst.*, **91**(1):1-18
[17] S. V. Vassilev, C. G. Vassileva, and V. S. Vassilev, Fuel, 158:330–350 (2015).
[18] Rizkiana J *et al.*, (2014)., *FUEL*, 6–11.
[19] Qing W, Hao X, Hongpeng L, Chunxia J, Jingru B, (2011)., *Oil Shale*, **28**(2):284-295

IOP Conf. Series: Materials Science and Engineering **619** (2019) 012015 doi:10.1088/1757-899X/619/1/012015

The Influence of Topography to the distribution of Ni-laterite deposits of Mangguruh Area, Sebuku Island, South Kalimantan

U R Irfan[1], I Alimuddin[2] and I B Pasalli[3]

[1]Mineral Geochemistry Laboratory, Department of Geological Engineering, Faculty of Engineering, Hasanuddin University, Gowa 92119, Indonesia
E-mail: ulvairfan@unhas.ac.id
[2]Geocomputation Laboratory, Department of Geological Engineering, Faculty of Engineering, Hasanuddin University, Gowa 92119, Indonesia
E-mail: ialimuddin@hotmail.com
[3]PT. Sebuku Iron Laterite, South Kalimantan, Indonesia.
E-mail: irfanpasalli@gmail.com

Abstract. One important factor that influences the optimization of the formation of nickel laterite deposits is the slope of a region's topography. The Mangguruh area in Sebuku Island in South Kalimantan is composed of ultramafic rocks on a gently flat topography but forms laterite deposits of varying thickness. Based on the exploration data from 33 drilling points and field data acquisition, this study aims to: 1) create topographic slope groups with varying ranges to determine the range most suited to real conditions in the field, 2) determine the laterite thickness, 3) analyze the concentration of Ni and Fe in limonite and saprolite layers, 4) analyze the effect of topographic slope to the laterite thickness. The method used is the measurement of topographic slope in the field with compass and using ArcGIS software applications in preparing the maps, analyzing laterite profiles from the drilling data and sampling rocks and ores. The study performed a laboratory analysis consisting of petrographic analysis and X-ray Fluorescence testing of the elements. The result revealed that the appropriate topographic slope groups in the study area were 0-5%, 5-10%, 10-15% and ≥ 15%. The laterite thickness of > 10m lied on the slope of 0-5% and 5-10% and contains a moderate-high nickel content of 1.2% -> 1.4%. This indicates that laterite nickel enrichment is influenced by topography. While the distribution of iron ore with medium and high levels of concentration (45-50% and > 50%) has commonly occurred in each topographic slope group indicating that topography is less influential on the distribution of iron ore. Earlier research has indicated that effect of topography on different morphologies is clear and many of the studies were conducted on a regional scale, but interestingly in Sebuku area, the topography is gently flat but the thickness and nickel content is varied. So the results of this study can be applied to slightly flat slope areas and therefore needs more detailed mapping.

1. Introduction

Nickel laterite deposits are formed due to further weathering of ultramafic rocks carrying Ni-Silicates, generally in the sub-tropical to tropical regions of the equator. The geological conditions and tectonic patterns of Eastern Indonesia conduct to the formation of potential laterite deposits in eastern Sulawesi

(Sorowako, Pomalaa, Bahodopi), eastern Halmahera (Gebe, Tanjung Buli, Sangaji, Pakal Island), and northern Irian Jaya (Waigeo, Gag , Sentani). On Sebuku Island, South Kalimantan found laterite nickel deposits that have different characteristics from other regions. The laterite deposits have a low nickel content compared to iron content [1]. Laterite iron ore reserves reached 426.5 million tons, which is the largest iron ore reserves in Indonesia

In addition to protolith and climate, the topography is also an essential factor in the formation of laterite nickel deposits. Laterite nickel can be formed in areas that have moderate reliefs controlled by structure and fracture density [2]. In areas with variate topographic slopes will form laterite deposits with different thicknesses. The thickness of the limonite zone is inversely proportional to the topographic slope which indicates that there is a correlation between the percentage of slope and the thickness of laterite deposits [3] [4]. Sebuku Island has a maximum topographic elevation of 125 m which elongate in the east. The research area is located in the western part, the topography is included in the geomorphology of the terrain with an uneven slope between 2 - 25 meters. On the west coast, there are freshwater swamps which are bordered by mangrove zones up to 500 m.

This research was conducted based on exploration drilling data which was confirmed by field observation data obtained from the Mangguruh Island of Sebuku Island, Kota Baru Regency, South Kalimantan Province. The area is part of the employment contract of PT. Sebuku Iron Lateritic Ores. The main focus of this study was to determine the optimal topographic slope group as a location for the formation of laterite nickel deposits, especially in relatively flat areas.

2. Materials and Method

2.1 Geology of Sebuku Island

Regional Geology of Sebuku Island is in Kotabaru Sheet, Kalimantan, published by the Geological Survey, Indonesia in 2006 [5]. The oldest rock that is Jura-aged is ultramafic consisting of harzburgite, dunite, serpentinite, gabbro, basalt, and serpentinized pyroxenite. In the Late Cretaceous, Pitap Formation is deposited in shallow marine environments, consisting of conglomerate, sandstone, and siltstone interfering, limestone insertion, breccia, claystone, conglomerate, and basalt. In Eocene, the Tanjung Formation was deposited which consisted of a tangle of conglomerates, sandstones, and claystone with shale, coal and limestone inserts. The Tanjung Formation overlaps the Pitap Formation and the Haruyan Formation. The youngest sediment in the study area is Alluvium (Qa) sediment, which is the result of erosion from older rocks whose sedimentation process still continues to the present. The lithology consists of gravel, sand, silt, clay, and mud. These deposits are found as swamp deposits, rivers and beaches. Its spread mainly occupies the plains around the Sebuku Strait.

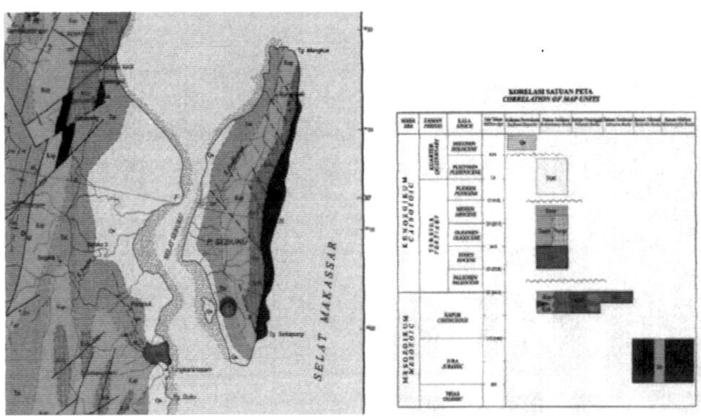

Figure 1. Regional Gology of Sebuku Island [5].

2.2 Method

The study was conducted in the Mangguruh Region, Sebuku Island, South Kalimantan, is a sub-district located to the east of Laut Island at 3 ° 24 ′ 23 ″ North Latitude and 116 ° 24 ′ 21 Timur East Longitude. The research location can be reached for 10 hours by road from Banjarmasin to Batulicin Ferry Port, crossing to Pulau Laut by ferry for 2 hours to arrive at Tanjung Serdang Port, followed by a trip on the Sea Island to the Pier to then go to Gosong Bay. From Gosong bay use a speedboat crossing to Tanjung Nusantara pier on Sebuku Island for 2 hours (Fig. 2).

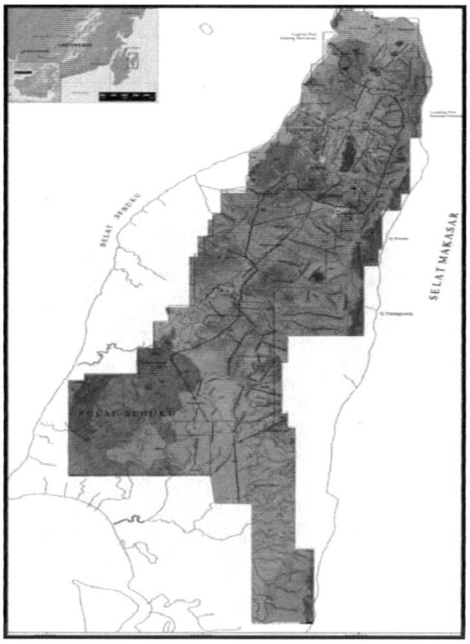

Figure 2. Map of the research area

The research method consists of field observations, slope value analysis, analysis of geochemical data from drilling, and modifying the classification of rays based on geochemical data. Observations in the field are surveys of surface geological mapping through predetermined trajectories, including morphological observations, exposed rock observations, observations of alteration zones, and rock sampling. Measuring the slope using a geological compass, and determining the position on the map by field orientation and using GPS. Samples from the selected field were then prepared in the form of the thin section of rock for petrographic analysis using a polarizing microscope in the Department of Geological Engineering, Hasanuddin University, which aims to determine rock types based on mineral composition and texture and microstructure in rocks.

Physical analysis of rocks is done on 33 core drilling directly at the time of drilling and on the results of the previous drilling. Physical characteristics analysis consists of color, mineral composition, grain size and minerals which aim to determine the profile of laterite deposits consisting of layers of limonite, saprolite, and bedrock. Geochemical analysis using X-ray Flourosense (XRF) instrument, to analyze oxide compounds contained in nickel laterite deposits oxide compounds using spectrometry methods such as SiO_2, MgO, CaO, Al_2O_3, Ni, Fe and Co.

Slopes can be defined in the form of percentages, comparing height difference to horizontal difference. The topographic analysis used ArcMap software to create slope maps in the study area. Furthermore, to find the slope classification at the surface that corresponds to the thickness of the deposits from the results of the analysis on the borehole, three slope classifications are made, namely: topography with slope 0 - 2°, 2° - 7°, 7° - 15° and ≥15°; 0 - 5°, 5° - 10°, 10° - 15° and ≥ 15°; 0 - 7°, 7° - 14°, 14° - 21° and ≥ 21°.

3. Result and Discussion

3.1 Geomorphogy and Topography

Morphogenesis analysis is the process of natural results of the results of weathering, erosion, and material movement processes. As a result of the denudational process that works on ultramafic rocks continuously, the land surface in the area decreases in height and forms almost flat surface. Based on the morphometric approach, the geomorphology unit in the study area is on low hills with slope forms generally convex to straight, and roughness is quite smooth and uniform. Based on the classification of reliefs can be divided into three units, namely flat topography, surging with sloping and irregular slopes with sloping hills. Flat topography is characterized by a slope of 0 - 2% and a difference in altitude <5 meters. The geomorphological unit is wavy in areas with slopes of 3 - 7% and differences in altitude of 5 - 50 m. While areas with slopes of 8 - 15% include undulating geomorphology units.

Figure 3. Plain and denudation hill of morphology unit

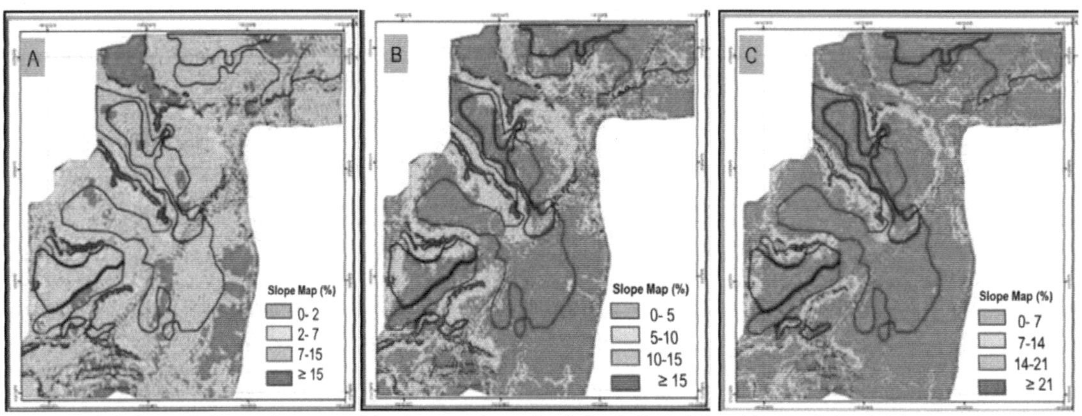

Figure 4. Map of slope degree with slope value: A. 0-2, 2-7, 7-15, ≥15; B. 0-5, 5-10, 10-15, ≥15, C. 0-7, 7-14, 14-21, >21

Based on the slope distribution which is measured directly in the study area and compared with topographic maps, three slope percentage distributions can be grouped. The slope classification considers the largest to the smallest slope angle and the density of the slope percentage value. The three classifications with slope percentage values as follows A. 0 – 2%, 2 – 7%, 7 – 15% and >15%; B. 0 – 5%, 5 – 10%, 10 – 15% and >15%, C. 0 – 7%, 7 – 14%, 14 – 21% and > 21% (Fig. 4). The percentage slope of classification A shows that there is an overlap in the range 0 - 2% with 2 - 7%, as

IOP Conf. Series: Materials Science and Engineering **619** (2019) 012015 doi:10.1088/1757-899X/619/1/012015

well as 2 - 7% with 7-15%. In the B classification overlap between 0 – 5% and 5 -10% can be minimized as well as percentage 5 – 10% with 10 – 15%, while 10 - 15% and > 15% the more specific the distribution area. Classification of percentage slope C shows the distribution of 0 – 7% is very broad compared to the range 7 – 14%, percentage slope 14 – 21% and >21% the smaller the distribution. Based on the classification of the slope, the classification corresponding to the study area is topographic classification B, because it can describe the topographic state in accordance with the study area, supported by the slope data measured directly.

3.2 Research Study Area

Based on lithostratigraphy, the research area is composed of two non-formal rock units namely the Peridotite Unit and Gabbro Unit (Fig. 5).

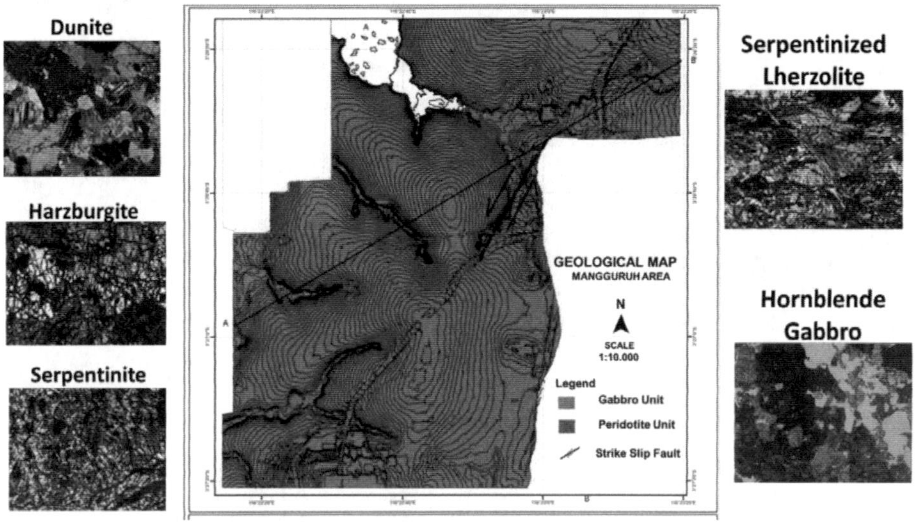

Figure 5. Geological map of Mangguruh area

3.2.1 Peridotite Unit

Peridotite units occupy 80% of the research area spread from north to south. The exposed rock can be grouped into four based on the level of serpentinization. The first group is rocks that do not have serpentinization; the second group of rocks that experienced serpentinization was 10-40%; the third group is peridotite rock whose serpentine level is 41-70% and the fourth group is serpentinite rock.

The exposed peridotite rocks are characterized by greenish gray, massive textured, phaneritic, composed of olivine and pyroxene minerals. Peridotite under a microscope is called dunite which is composed of olivine (50 - 60%), pyroxene (20 - 27%), plagioclase (8 - 10%) and magnetite minerals (2 - 3%). The rocks have altered by serpentine mineral by 8%. Olivine mineral in the form of anhedral measuring 0.2 - 0.9 mm, there are cracks in olivine individuals forming a small portion, and partially replaced by serpentine minerals. Orthopyroxene is brownish green, subhedral in shape; two-cleavage form a 90° and some have been changed by serpentine minerals. This first group rock was found in an amount of at least about 10% in peridotite units.

The second group of peridotite rocks is exposed to gray, hypocrystalline, composed of olivine and pyroxene minerals. Observations on a thin section of rock are called harzburgite which is composed of augite minerals (35 - 45%), olivine (20-25%), hypersthene (10-20%), plagioclase (5-20%) and opaque minerals (5-10%). Serpentine minerals alter the pyroxene and olivine minerals in the fracture, the space between minerals and some individual minerals is completely changed.

The third group is peridotite which is exposed to yellowish gray, hypocrystalline which is composed of pyroxene and olivine minerals and serpentine minerals and quartz veins. Analysis of rocks under a microscope is called lherzolite which is composed of minerals augite (40-50%), olivine (20-25%), and hypersthene (10-25%). The space between minerals is changed by chrysotile serpentine minerals characterized by light brown color, fibrous forming mesh texture, and parallel extension.

The fourth group of peridotite rocks is greenish gray, foliated textured and glassy due to serpentine mineral content. Petrographic analysis showed greenish-brown color, low relief, fibrous shape, partly formed mesh texture, and partly filled veins. There is a subhedral to an anhedral distribution of magnetite minerals and is 0.05 - 0.1 mm in size (Fig. 6).

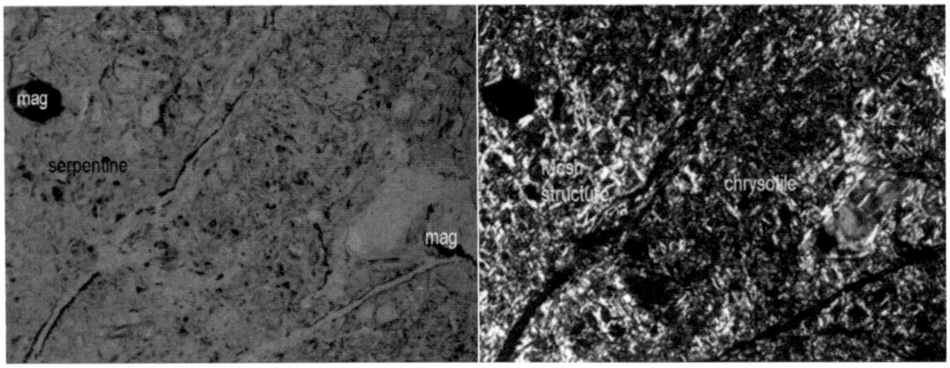

Figure 6. Serpentinite under microscope with mesh texture

3.2.2 Gabbro Unit

The exposed gabbro unit has experienced high weathering, and shows reddish color changes. Gabbro rocks at the borehole show a blackish gray to brown. Holocrystalline texture, composed of clinopyroxene minerals (30-50%), labradorite (20-30%) olivine (10-25%) and honblede (10-25%). Minerals form are subhedral to anhedral and 0.5 - 1.6 mm in size.

3.3 Laterite Profile

Data for determining the laterite profile from observations in the field and results of drill hole analysis. Differences in layers of laterite deposits are found in the physical characteristics of color, grain size, and minerals found in each layer, in accordance with the standardization of PT. SILO (Sebuku Iron Lateritic Ores). Observation in the field in outcrops found limonite, saprolite and bedrock layers. The limonite layer is brownish red to yellowish, measuring grains of sand to clay, layer thickness of 1.5 - 2.0 meters. Found minerals goethite, magnetite and secondary quartz. Saprolite layer is characterized by greenish brown color deposits, rock fragments of origin are still visible between grains of sand to clay. There are textured boxwork veins, serpentine veins and chalcedony. Saprolite layer thickness between 1.0 - 2.0 m. The bedrock layer is yellowish gray, characterized by massive rocks and partly in the form of large fragments (Fig. 7).

Differences in physical characteristics and grades of Ni and Fe in a borehole are used to determine the layers of nickel laterite deposits. The limonite layer is reddish brown, measuring silt to clay and not solid, there are goethite, hematite and magnetite minerals. Saprolite layer is yellowish brown in color, the size of sand grains to glacial, and there are origin rock fragments of 5.0-10 cm in size, there are garnierite and serpentine minerals. The bedrocks layer is composed of gray peridotite, massive texture and several parts in the form of fragments measuring 5.0-30 cm (Fig. 8).

Figure 7. Laterite profile at observation station 11, which shows limonite, saprolite and bedrock layers.

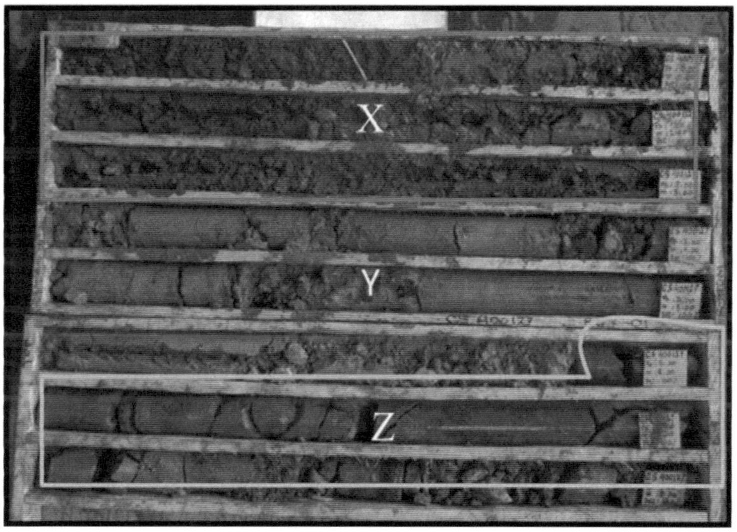

Figure 8. The appearance of the laterite profile on drilling results No. 7, which shows the limonite (X), saprolite (Y) and bedrock (Z) layers.

3.4 The thickness of laterite deposits

The determination of laterite thickness based on the total thickness of the limonite and saprolite layers which are distinguished based on the grade of Ni and Fe. Geochemical analysis is very important in determining the laterite profile, because each layer has a different chemical element. Elemental analysis is carried out qualitatively and quantitatively. Qualitative analysis was conducted to analyze the types of elements contained in the deposits and quantitative analysis was carried out to determine the concentration of elements in the borehole results. Classification of Ni and Fe elements in laterite profiles in the study area was divided into three, namely low, moderate and high (Table 1).

Table 1. Classification of Ni and Fe elements in laterite

Element	grade (%)		
	low	moderate	high
Ni	< 1.2	1.2 - 1.4	> 1.4
Fe	<45	45 - 50	> 50

The results of geochemical analysis of the limonite layer have a high content of Fe, which is> 35% and low Ni elements <1.2%. Based on the description of the core obtained a minimum thickness of 0.5 meters at the drill point 113 and the maximum thickness is 6.85 meters at the drill point 25. The thickness difference in the limonite layer is caused by various aspects, such as bedrock, rock structure, topography, climate, subsurface water and surface water [3] [4].

In the saprolite layer, the minimum thickness is 1.09 meters at borehole 29 and the maximum thickness is 11.5 meters at the borehole 25. The difference in saprolite thickness is due to the lack of fractures in the bedrock so that the weathering intensity is low. In addition it is also influenced by the sloping topography so that the thickness of the saprolite is thinned because the chance of surface water to seep into the soil is not optimal. Bedrock is a rock that is still fresh and weathering processes are still very small. Observation of peridotite bedrocks in the core, shows blackish green color, holocrystalline crystallinity, phaneritic granularity, equigranular relations, subhedral-anhedral, olivine, and pyroxene forming minerals, and serpentine.

The thickness of laterite in the study area was classified into three, namely thin (<7.2 meters) medium (7.2 - 10 meters) and thick (> 10 meters). On the map of the relationship between slope and laterite thickness, it can be seen that the thick distribution of sludge is generally at a slope of 0 - 5%, although there is also a slope of 5-10% in the southwest. The spread of the thickness of the precipitate is moderate at a slope of 0-10%. Whereas the thickness of thin deposits is spread in a slope of 10-15%. The development of intensive laterite deposits on gentle slopes in the slope is 0-10%, so the weathering process is relatively the same in the area. Similarly, the absorption of rainwater to the bedrock will be intensive which causes chemical weathering greater than physical weathering.

3.5 Influence of Slope with Lateritic Thickness and Ni and Fe grade

The influence of topography on thickness in the study area can be analyzed from overlays of laterite thickness data and slope maps. The thickness of laterite at slopes of 0 - 5% and 5-10% is generally higher, because the area has a flat topography and also not steep, so that surface water moves slowly and will have enough time to penetrate deeper through rock fractures. This penetration causes the mobile elements to be carried along with the flow of water and eventually accumulates to form thick laterites. In contrast to the slopes of 10-15% and > 15% which are generally thinner, which is caused by a steep topography causing surface water does not have enough penetration time to the surface. So that the weathering process is less than optimal. The slope of the slope has a large influence on the level of weathering and accumulation. However, in some locations the slope is not found thick laterite deposits, it can also be caused by the influence of the structure is not optimal to form fractures in rocks [2]. Protolith also plays a role in the formation of laterite nickel deposits, namely harzburgite [6] peridotite, whereas in the study area it is dominated by lherzolite peridotite.

Nickel distribution data associated with slope, the distribution of nickel with grade values> 1.4% and 1.2% - 1.4%, generally located in topography with slopes of 0 - 5% and 5-10%. In this topography, groundwater does not move quickly so that nickel deposits that are mobile are not displaced. Nickel distribution of slope <1.2%, generally located in topography which has slopes of 10-15% and> 15%. In these slopes, surface water does not have enough time to penetrate downwards so that the formation of laterite and nickel is less effective and if penetration occurs to the bedrock, it can cause the nickel element to be carried away by the subsurface water flow which is relatively faster.

Distribution of iron ore levels high to moderate are in all slope values in the study area but generally are at slope percentages of 0-5%, 5-10%, and > 15%. While the distribution of iron ore with low levels is at every percentage slope. So that in the study area shows that the influence of topography on iron ore distribution is not significant. It is also related to high Fe content compared to Ni in thick to thin laterite deposits. Serpentinite rock that is widely distributed in peridotite units is a rock originating from deposits containing high levels of Fe. Petrographic analysis shows that magnetite minerals are scattered and some form veins in serpentine rocks and serpentinized peridotite rocks. Magnetite is a mineral that is responsible for high Fe levels compared to Ni.

In the process of chemical weathering, groundwater rich in CO_2 comes from the air and decomposition of plants decomposes unstable minerals such as olivine and pyroxene in ultra-alkaline rocks, which will produce Mg, Fe, soluble Ni and Si tend to form colloids form very fine silica particles. In solution, Fe is oxidized and settles as ferric hydroxide. Iron deposits are concentrated in the limonite zone. In this zone is dominated by goethite, hematite, magnetite and clay minerals.

Factors that greatly affect the formation and distribution of iron ore are chemical weathering, which break down and dissolve primary minerals. Another factor that is also very supportive is that groundwater will leach minerals to the limit between limonite and saprolite; other factors can be in the form of pH, and topography. Laterite type iron ore is generally found in the peak hills which are relatively sloping or have a slope of less than 10%, making it one of the main factors of iron ore distribution.

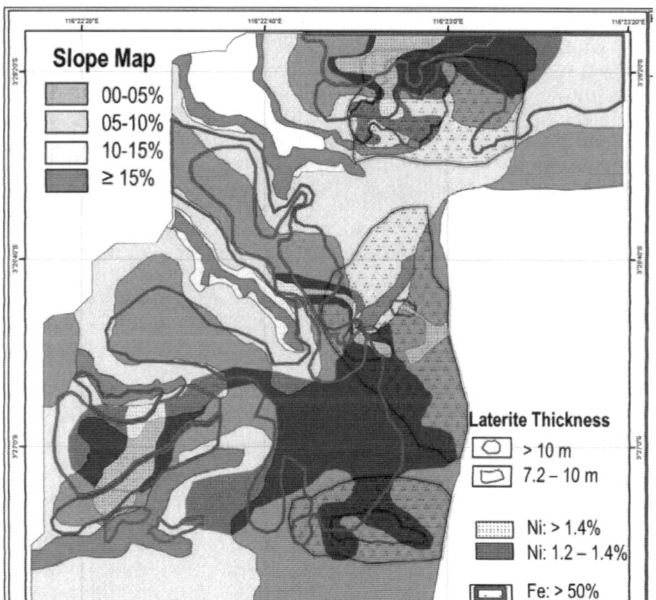

Figure 9. The distribution of thickness and grade of Ni-Fe on slope map

4. Conclusion

The results of the study were in the Mangguruh area of Pulau Sebuku sub-district, Kota Baru district, South Kalimantan Province, as follows:

1. The topographic slope of the study area is divided into four (4), namely topography with a large slope of 0-5%, topography with a slope of 5-10%, topography with a slope of 10-15%, and topography with a large slope > 15%.
2. Moderate to high laterite thickness in topography with slopes of 0-5% and 5-10%, while topography with a slope of 10-15% and> 15% has a thickness of thin laterite; this shows that topography is one of the dominant factors affecting the thickness of laterite.
3. Topography is the dominant factor affecting the distribution of nickel in the study area.
4. Distribution of iron ore is dominant in slopes of 0-5%, 5-10%, and> 15%, while low iron ore is at each percentage slope, then topography is an insignificant factor in the distribution of iron ore.

References

[1] Nurhakim, M. U Dwiatmoko, N H Romla and Adip M 2011 Identification of potential laterite ore deposits in the central part of Sebuku Island in South Kalimantan Province (In Bahasa: Identifikasi potensi endapan bijih laterit di bagian tengah Pulau Sebuku Provinsi Kalimantan Selatan Jurnal) Infoteknik Vol. 12 No. 2 48-53
[2] Thorne R. Roberts S. and Herrington R. 2012 The formation and evolution of the Bitincke nickel laterite deposit, Albania Miner Deposita, Vol. 47, Issue 8, 933–947
[3] Ilyas A, and Koike K 2012 Geostatistical modeling of ore grade distribution from geomorphic characterization in a laterite nickel deposit Natural Resource Research, Volume 21, Issue 2, pp 177–191
[4] Ilyas A Kashiwaya K and Koike K 2016 Ni grade distribution in laterite characterized from geostatistics, topography and the paleo-groundwater system in Sorowako, Indonesia Journal Geochemical Exploration, Vol. 165, 174-188.
[5] Rustandi E. Nila E S Sanyoto P and Margono U 1995 Geological Map of The Kotabaru Sheet, Kalimantan, Scale 1:250,000 Geological Research and Development Centre.
[6] de-Benavent C V Domènech C Tauler E Galí S Tassara S Proenza Joaquín 2017 Fe–Ni-bearing serpentines from the saprolite horizon of Caribbean Ni-laterite deposits: new insights from thermodynamic calculations Miner Deposita 52:979–992.

Acknowledgments

The author would like to thank PT. Sebuku Iron Lateritic Ores, especially the Geodev division which has provided opportunities for data collection and permission to publish it. Thanks also to Directorate of Research and Community Service, Directorate General of Research and Development Strengthening Ministry of Research, Technology and Higher Education of the Republic of Indonesia for decentralization grants, the year 2018.

Calcareous Nannofossil Assemblages of Tonasa Formation Palakka Area, South Sulawesi: Implication of Paleoenvironmental application

M Farida[1], A Jaya[1], T Sato[2]

[1]Geological Engineering Department, Hasanuddin University, Indonesia
[2]Department of Earth Resource Science, Faculty of International Resource Sciences, Akita University, Japan

The corresponding author's e-mail: meutia.nugraha@gmail.com

Abstract. Tonasa limestone formation in this area consists of interbedded between marl and limestone, which were developed intensively in the marine environment from the Late Eocene to Oligocene. One important component of this rock is Calcareous nannofossil as the primary production of CaCO3. The aim of this study are to obtain data from the distribution of calcareous nannofossil, assemblages and early investigation about the implications for the paleoenvironment. Calcareous nannofossil is a good indicator for determining paleoenvironmental conditions such as modern coccolithophores that show seawater stratification. The material under study came from the outcrops of the Barru River, which were collected by measuring section method, then the samples were prepared using the smearslide technique. The result are the presence some marker species such as *Sphenolithus distentus, Distyococcites bisectus, Zygrhablithus bijugatus, Phontosphaera enormis, Cyclicarg, olithus abisectus, Sphenolithus predistentus, Discoaster saipanensis, Spenolithus pseudoradians*. The presence of Discoaster as typical warm water organism characterize the deep photic zone, therefore indicates stratified water colomn as well as deepened nutricline.

1. Introduction

Sulawesi is one of the most interesting places for investigated. its K-shape which was the result of the collision of the cratonic Asia with the micro continent causing this region have complex geological condition. An example of this unique area is the western arm of Sulawesi has a different geological framework to the east (figure 1). The Tertiary sediment is widespread on the western Sulawesi, and Tonasa limestone is one of the most thickness of the sedimentary rock, which composed by coral limestone, bioclastic limestone and marl interbedded, and redeposit limestone [1],[2]. Previous study suggest that the Tonasa limestone deposited on relatively stable shallow-water, known as the Tonasa carbonate platform [2]. many authors have conducted research on the Tonasa Formation, such as lithology in general, stratigraphic sequence, and biostratigraphy of foraminifera, however no studies of calcareous nannofossil. The detail foraminifera study correlated with the paleotemperature also conducted in Palakka area [3].

In this paper we results investigation of calcareous nannofossil assemblages from the limestone and marl interbeds of Barru River section. And the purpose of this research are to analyze correlation between the assemblages of calcareous nannofossil and their implication to paleoenvironmental

IOP Conf. Series: Materials Science and Engineering **619** (2019) 012016 doi:10.1088/1757-899X/619/1/012016

identification. This is a preliminary study of calcareous nannofossil in this area, therefore the investigation has not been in quantitative research.

Many authors have made it clear that calcareous nannofossils have broader applications, as well as good tools for paleoenvironmental analysis, age determination, including paleoecologic and paleoclimatic conditions [4, 5, 6, 7, 8]. The quantitative research of calcareous nannofossils have also conducted, for instance its relationship with productivity [7, 9,10,11]. *Discoaster* species abundantly in the Paleogene to Neogene intervals and had been interpreted as a typical warm water species [12], and the relationship between coccolith size and Discoaster abundance are useful to reconstruc the changes of the sea surface stability [13, 14, 15]

2. Geological Setting

Sulawesi Island has been complex tectonic in the world, which is situated at juntion three major plate which an almost complete Tertiary sequence [16]. The Tertiary carbonate was developed extensively in Southeast Asia, one of which is the Tonasa Formation with a thickness of up to thousands of meters. Interbeds of the limestone and marl are well exposed in Barru River and surrounding areas. Tonasa Limestone Formation extends from the north of Barru area to the Jeneponto in the south. The Tonasa Formation is a deposit of carbonate rock from the late Eocene to the Middle Miocene [1], moreover this formation was deposited as a shallow - water and outer ramp / endinal precipitate of the carbonate platform, particularly in the Pangkajene and Jeneponto area [17].

Stratigraphically, Balangbaru Formation is the oldest sedimentary rock unconformably overlies the basement complex of the western, which composed by deep marine sediments, Overlying of Balangbaru Formation is angular unconformably of Mallawa Formation that characterized by deposits of marginal transitions from Paleocene age, and it is overlain by carbonate deposits, formed initially a transgressive sequence [1, 17, 18]. The study area is located at Barru River, Barru Regency South Sulawesi Province (figure 1). This area consists of interbedded carbonate rock deposits.

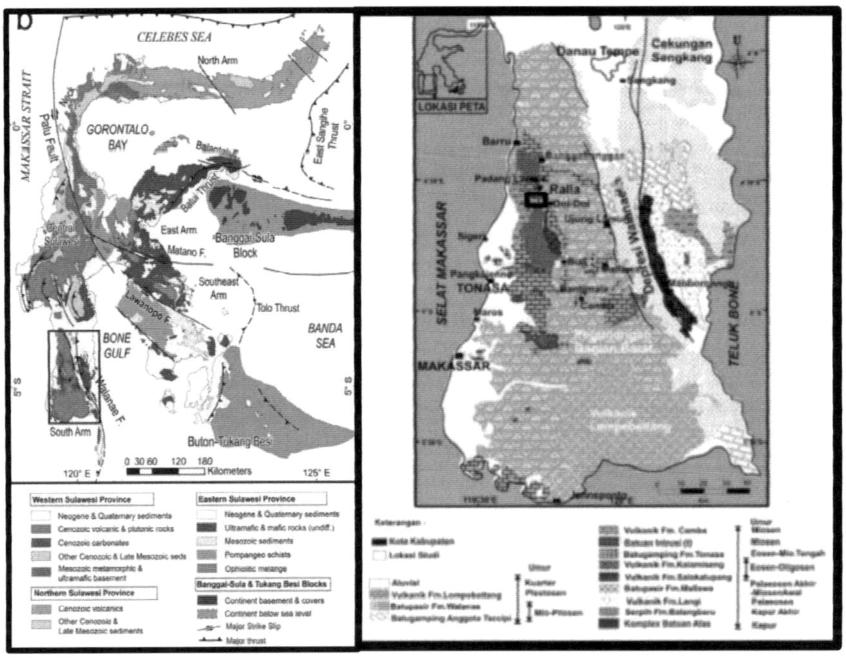

Figure 1. Lithotectonic setting of Sulawesi Island and Geological map of South Sulawesi (after [1,2,18,19] modified)

3. Samples and Methods

Samples used in this study we collected from the outcrop of Barru River with measuring section method. A total of 40 layers of marl and limestone interbeds from this section were taken to be analyzed nannofossils. All samples were processed by the smear slides method. As much as 100-300 grams of dried sediment are made into powder, then a small sample is sprinkled sprinkled on a 24 x 24 mm glass cover and dripped with water using a pipette. The samples were flattened on all cover glass surfaces using a toothpick and then dried on a hot plate at 70 ° C. Then observed species and taken nannofossils photographed under a light microscope with 1000x magnification. In order to describe the assemlages of nannofossil we identified all species across all slide view.

4. Results

Barru area is one of the locations where Tonasa limestone is well exposed, and the Barru River section shows marl and limestone interbeds and is a good for stratigraphic measuring section (figure 2). Marls dominated in this section with thickness of layers 1,5 – 40 cm, and thickness of limestone around 1 – 60 cm. Strike/dip of layers generally have orientation N 325°E/34°. Length of the measuring section is 353,96 meter and total thickness is around 23 m.

Figure 2. Photograph of outcrop between marl (X) and limestone (Y) interbeds in Barru River

4.1. Distribution of Calcareous Nannofossils

The distibution of Calcareous nannofossils from bottom to the top of the Barru River section shown that species are increasingly diverse, while discoaster are almost found in each layer. The photograps from the selected taxa are illustrated in figure 3 and figure 4. Generally, calcareous nannofossils in this section is fairly abundant and moderately preserved, however some samples are poorly preserved. There are 2 layers of 40 layers that are not found nannofossils, this may be due to the diagenesis processes in limestone. As many as 15 genera and 25 species are found,

Reworked specimens have been identified from this section, they are mainly Paleocene – early Eocene age. The presence of this species may be due to the transport of sedimentary material from older sedimentary rocks. This fact indicates that there was marine sedimentary rock older than the Tonasa Formation.

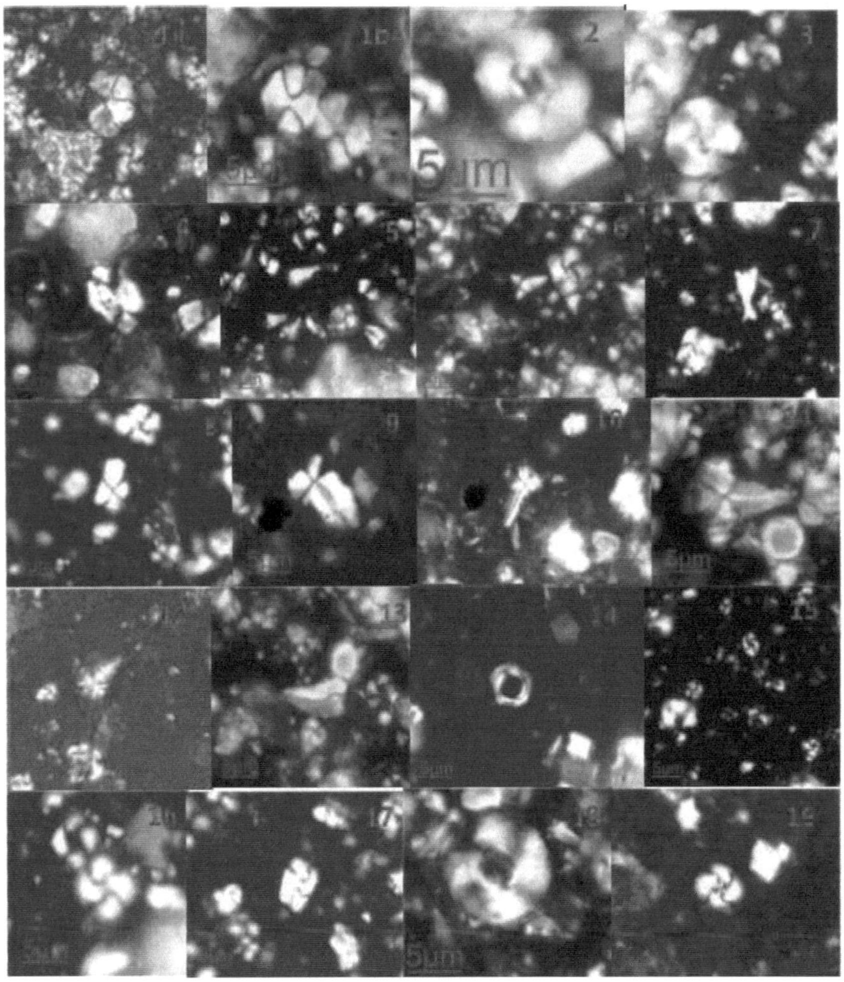

Figure 3. Photomicrograph of nannofossils all in CN (Cross Nicol), **1a,b**. *Braarudosphaera bigelowii* (Gran & Braarud) Deflandre, **2**. *Cyclicargolithus abisectus* (müller) Wise, **3**. *Dictyococcites bisectus* (Hay, Mohler & Wade, 1966) Bukry & Percival (1971), **4**. *Pontosphaera enormis* Locker (1967) Perch-Nielsen (1984a), **5**. *Sphenolithus predistentus* Bramlette & Wilcoxon (1967), **6**. 7, *Sphenolithus distentus* (Martini, 1965) Bramlette & Wilcoxon (1967), **8.9.10.11.12**, *Sphenolithus pseudoradians* Bramlette & Wilcoxon (1967), **13**. *Zygrhablithus bijugatus* (Deflandre in Deflandre & Fert, 1954) Deflandre (1959), **14**. *Calcidiscus protoannulus* (Gartner, 1971) Loeblicb & Tappan (1978), **15**. *Dictyococcites* sp., **16.17**. *Dictyococcites scrippsae* Bukry & Percival (1971), **18**. *Reticulofenestra umbilica* (Levin, 1965) Martini & Ritz.kowski {1968), **19**. *Cribrocentrum reticulatum* (Gartner & Smith, 1967) Perch-Nielsen (1971d).

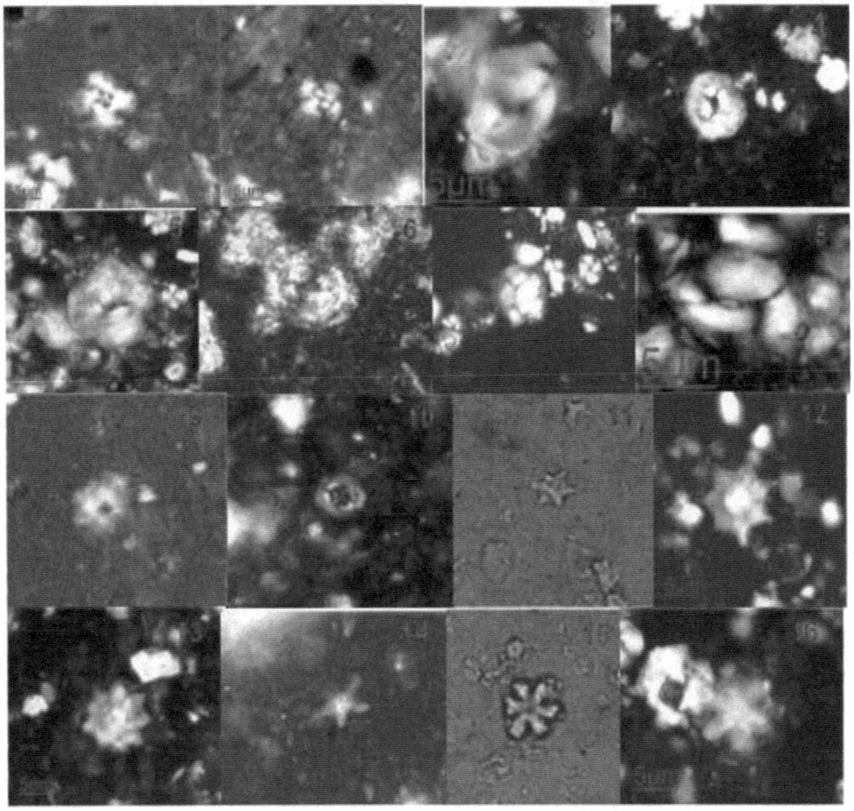

Figure 4. Photomicrograph of nannofossils : CN (Cross Nicol), PN (Parallel Nicol), **1,2**. *Cribrocentrum reticulatum* (Gartner & Smith, 1967) Perch-Nielsen (1971d), CN, **3**. *Ellipsolithus macellus* (Bramlene & Sullivan, 1961) Sullivan (1964), CN, **4,5,6**. *Coccolithus eopelagicus* (Bramlette & Riedel, 1954) Bramlette &:Sullivan (1961), CN, **7**. *Coccolithus pelagicus* (Wallicb, 1877) Schiller (1930), CN, **8**. *Helicosphaera* sp., CN, **9**. *Markalius inversus* (Deflandre in Deflandre & Fert, 1954) Bramlette & Martini (1964), CN, **10**. *Chiasmolithus altus* Bukry & Percival (1971), CN, **11**. *Discoaster saipanensis* Bramlette & Riedel (1954), PN, **12**. *Discoaster saipanensis* Bramlette & Riedel (1954), CN, **13**. *Discoaster barbadiensis* Tan (1927), CN, **14**. *Discoaster sublodoensis* Bramlette & Sullivan (1961), CN, **15**. *Discoaster defiandrei* Bramlette & Riedel (1954) PN, **16**. *Discoaster* sp.

4.2. Biostratigraphy

The age of the sample studied from a total of 25 calcareous nannofossil species was determined from the First Occurence (FO) and Last Occurence (LO) of marker species, and based on the result identified marker species, they are FO of *Sphenolithus pseudoradians* Bramlette & Wilcoxon, LO of *Discoaster saipanensis* Bramlette & Riedel, LO of *Discoaster barbadiensis* Tan, also FO and LO of *Sphenolithus distentus* (Martini) Bramlette & Wilcoxon, LO of *Sphenolithus predistentus* Bramlette & Wilcoxon, *Pontosphaera enormis* Locker, Perch-Nielsen, and LO of *Dictyococcites bisectus* (Hay, Mohler & Wade) Bukry & Percival, and the LO of *Zygrhablithus bijugatus* (Deflandre) Deflandre, therefore the age of rocks in the Barru River outcrop was determined by according to the first and last occurence of marker species based on the standard zonation of Martini [20] is NP 16 – NP 24 and standard zonation by Okada and Bukry [21] is CP 14a – CP 19a or age equivalent as the Middle Eocene to Oligocene.

4.3. Paleoenvironmental Analysis
The presence some species indicating specific environment could be identified. Paleoenvironment of Palakka area especially upstream of Barru River Section is reconstructed based on calcareous nannofossils data. From the results of the identification of nannofossils, Discoaster species are found in almost every rock layer, the presence of these species as typical warm water that lived in the low photic zone.

A living nannofossil species, such as *Florisphaera profunda*, presently found in lower photic zone, which is a proxy to reconstruct the stability of the Quaternary sea surface condition with nutricline and thermocline. This specimen is not found in the Pliocene to older age, therefore it is difficult to reconstruct the sea surface conditions during the Neogene. Instead it was Discoaster that lived on Paleogene to Neogene (Tersier).

References
[1] Sukamto R 1982 The geology of the Pangkajene and western part of Watampone , Sulawesi (Quadrangels series) scale 1:250.000, 1 sheet *Geological Research and Development Centre* Bandung
[2] Wilson M E J & Bosence D W J 1996 The Tertiary evolution of South Sulawesi: a record in redeposited carbonates of the Tonasa Limestone Formation *Geological Society Special Publication* no. 106, ed R Hall and D Blundell pp 365 – 389
[3] Farida M, Pratiwi, Husain R 2014 Paleotemperature of Middle Eocene based on foraminifera at Palakka area South Sulawesi *Int. J. Of Eng. & Sci. Appl.* vol. 1 pp 77 – 84
[4] McIntyre A 1967 Coccoliths as Paleoclimate Indicators of Pleistocene Glaciation *Science* vol. 158 1314 – 17
[5] Bukry D 1971 Discoaster evolutionary trends *Micropaleontology* vol. 17 pp 43 – 52
[6] Haq B U 1980 Biogeographic history of Miocene calcareous nannoplankton and paleoceanography of the Atlantic Ocean *Micropal.* vol. 26 pp 414 – 443
[7] Sato T and Chiyonobu S 2009 Cenozoic paleoceanography indicated by size change of calcareous nannofossil and *Discoaster* abundance *Fossils, The Palaentological Soc. of Japan* vol 86 pp 12 – 19
[8] Adeigbe O C, Adebowale M, Ola-Buraimo A O, Bankole S A and Ayinla H A 2014 Calcareous nannofossils and foraminifera studies: an integrated approach to the depositional environmental study and biostratigraphy of Dep-1 well offshore eastern Dahomey Basin, Southwestern Nigeria *British. J.of appl. Sci.& Technol.* vol 4(2) pp 319 – 337
[9] Stoll H M, Shimizu N, Archer D, and Ziveri P 2007 Coccolithophore productivity response to greenhouse event of the Paleocene-Eocene thermal maximum. *Earth & Planet. Sci. Letters* vol 258 pp 192 – 206
[10] Farida M, Imai R, and Sato T 2012 Miocene to Pliocene paleoceanography of the western equatorial Pacific Ocean based on calcareous nannofossils ODP hole 805B *Open J. Geol.* vol 2 pp 72 – 79
[11] Imai R, Farida M, Sato T, and Iryu Y 2015 Evidence for eutrophication in the northwestern Pacific and eastern Indian Oceans during the Miocene to Pleistocene based on the nannofossil accumulation rate, Discoaster abundance, and coccolith size distribution of Reticulofenestra *Mar. Micropal.* vol 116 pp 15 – 27
[12] Bukry D 1973 Low-Latitude Coccolith Biostratigraphic Zonation *Initial Reports of the Deep Sea Drilling Project* vol 15 pp 685 – 703
[13] Sato T and Chiyonobu S 2009 Cenozoic Paleoceanography Indicated by Size Change of Calcareous Nannofossil and *Discoaster* Abundance *Fossils, the Palaen. Society of Japan* vol 86 pp 12 – 19
[14] Farida M, Imai R and Sato T 2012 Miocene to Pliocene Paleoceanography of the Western Equatorial Pacific Ocean Based on calcareous nannofossils, ODP Hole 805B *Open Journal*

of Geology vol 2 pp 72 – 79

[15] Imai R, Farida M, Sato T & Iryu Y 2015 Evidence for eutrophication in the northwestern Pacific and eastern Indian oceans during the Miocene to Pleistocene based on the nannofossil accumulation rate, Discoaster abundance, and coccolith size distribution of Reticulofenestra vol 116 pp 15 – 27

[16] Wilson M E J & Bosence D W J 1996 The Tertiary evolution of South Sulawesi : a record in redeposited carbonates of the Tonasa Limestone Formation *Geol. Society Spec.Publ. no.* 106, ed R Hall and D Blundell pp 365 – 389

[17] Wilson M E J & Bosence D W J 1997 Platform–top and ramp deposits of the Tonasa carbonate platform, Sulawesi, Indonesia *Petrol. Geol. of SE Asia* no. 126, ed A J Fraser, S J Matthews and R W Murphy pp 247 – 279

[18] Van Leeuwen T M, Susanto E S, Maryanto S, Hadiwisastra S, Sudijono, Muhardjo and Prihardjo 2010 Tectonostratigraphic evolution of Cenozoic marginal basin and continental margin successions in the Bone mountains, Southwest Sulawesi, Indonesia *J. of Asian Earth Sci.* vol 38 pp 233 – 254

[19] Jaya A, Nishikawa O and Hayasaka Y 2017 LA-ICP-MS zircon U-Pb and muskovite K-Ar ages of basement rock from the south arm of Sulawesi, Indonesia *Lithos* vol 292 – 293 pp 96 – 110

[20] Martini E 1971 Standard Tertiary and Quaternary calcareous nannoplankton zoantion. *In* Farinacci, A., *eds*., *Proc. II Planktonic Conf., Roma*, vol 2 pp 739 – 785

[21] Okada H & Bukry D 1980 Supplementary modification and introduction of code numbers to the low-latitude coccolith biostratigraphic zonation (Bukry, 1973; 1975). Mar. Micropaleontol., vol 5 (3) pp 321 – 5

Acknowledgments

We would like to thank the Ministry of Research, Technology and Higher Education of the Republic of Indonesia, and also thank the Research and Community Services of Hasanuddin University for giving us the opportunity to fund our research. Hopefully the results of our research contribute to the development of science, especially studies in the field of geology. And we don't forget to thank the local government for giving permission to study in the Barru area of South Sulawesi Province.

Beneficiation of Lateritic Iron Ore from Malili Area, South Sulawesi, Indonesia Using Magnetic Separator

Sufriadin*, S Widodo and M Trianto

Department of Mining Engineering, Hasanuddin University, Engineering Campus, Jl. Poros Malino, KM.6 Bontomarannu Gowa 92171, Sulawesi Selatan, Indonesia

*Corresponding author's, e-mail: sufri.as@unhas.ac.id

Abstract. The increasing demand of sponge iron as raw materials in steel industry nowdays has led to increase the effort in obtaining such materials from various sources. This study aims to characterize and beneficiate of a lateritic iron ore sample from Malili area of South Sulawesi. Mineralogy of the ore was investigated by means of X-ray diffractometry (XRD) and microscopy techniques; whereas chemical composition of the sample was determined by using X-ray fluorescence (XRF) spectroscopy. Result of mineralogical analysis indicated that ore sample consists of goethite, magnetite and chromite. The ore shows fine grained size with accicular texture. Chemical analysis reveals that sample is composed of Fe_2O_3 32.8%, Al_2O_3 35.5%, SiO_2 21.0%, Cr_2O_3 7.2%, MnO 1.5%, and NiO 1.2%. Sieving of 20 kg iron ore sample was carried out to separate into five size fractions: (+212)μm, (-212+180)μm, (-180+150)μm, (-150+90)μm and (-90)μm with mass distribution of 12.0kg; 2.8kg; 0.9kg; 1.8kg and 1.6kg respectively. Each feed fraction was further treated by using dry low-intensity drum magnetic separator to produce concentrate. Result of each fraction having the mass percentage of 15.5%; 31.7%; 30.7%; 32.0% and 40.0% with the respective iron content (Fe_2O_3) of 45.0%; 47.9%; 43.9%; 46.7% and 47.2%. The Fe_2O_3 grade of concentrate in all fractions have narrow range values, but the mass percentage of concentrate increase significantly with finer grain size fraction. The highest Fe_2O_3 recovery was found to be 89.82% with -90 μm size fraction.

Keywords: iron ore, beneficiation, magnetic separator, goethite, concentrate.

1. Introduction

Lateritic ore is usually formed by chemical weathering containing oxides and hydroxides of iron and aluminum which found in the tropical regions [1]. The estimated resources of lateritic iron ore in Indonesia in measured resource category may reach up to 1.307 billion tons, while its proved reserve is about 297.25 million tons [2]. This kind of ore is dispersed in Lampung, West Java, South Kalimantan, South Sulawesi, Southeast Sulawesi, North Maluku, and Papua. In addition, there are also resources of iron sand, primary iron, and sedimentary iron. One of the lateritic iron deposit in South Sulawesi is located in Malili Area of East Luwu Regency. This deposit is predicted to have about 20 million tons of resource which has been mined by PT Panca Digital Solution in 2010 with the grade of 0.62-0.94 % Ni and 36.2-41.36 % Fe [3].

The iron ore which was formerly economic to be mined has a grade of 60 % Fe. Nonetheless, it has not been longer viable since the 2000s. Nowadays by using the new and simple technology, an iron ore with the grade of 30 % Fe has sufficient for iron ore to be economic mined. In China and India, iron ore

IOP Conf. Series: Materials Science and Engineering **619** (2019) 012017 doi:10.1088/1757-899X/619/1/012017

is processed using simple technology, such as magnetic separator, gravity concentration and blast furnace for iron producing which has already been used by manufacturing industry [4].

Processing of minerals is strongly influenced by its properties, both physical characteristics such as mineralogy, specific gravity, electrostatic, magnetism, and chemical nature of the ore [5]. Consequently, selection of appropriate methods depend on the properties of the. For example, ore with high content in goethite (FeOOH) should be heat-treatment to transform into magnetite before the application of magnetic separation [6-7]. This research aim to examine the mineralogical and chemical composition of lateritic iron ore sample from Malili area and beneficiation of such ore using magnetic separator.

2. Sample and Methods

This research consists of two main stages, there are field and laboratory activity. However, it begun with a literature review on the Malili area, followed by lateritic iron ore sampling in Malili, East Luwu Regency, South Sulawesi Province, exactly in Lampia Village with coordinate locality lies at 2°44'25.6'' latitude south and 121°05'16.9'' longitude east. Sampling was conducted by scraping method. Sample was taken from lateritic soil profile which has indication of iron-rich material. Sample preparation was begun with air-drying floor or indoor drying method at room temperature for two weeks until the surface water content decreases.

Initial sample of lateritic iron ore was characterized by mineralogical analysis using X-Ray Diffraction (Shimadzu X-7000). Sample preparation was started with grinding using agate mortar until particles passed 200 mesh sieve. Powder sample was pressed to find flat surface on the holder to be tested with X-ray fluorescence (XRF) spectrometer for chemical determination.

In supporting the mineralogical data, microscopic analysis was carried out using a polarized microscope (Nikon Eclipse LV100). For this purpose, sample was prepared by mixing with powder resin and then it was heated and pressed with a mounting press at a certain temperature and pressure. Polishing was done until the sample surface was found to be flat and then it was observed under a microscope. The chemical analysis was conducted with X-Ray Fluorescence (Shimadzu) whereby the sample was prepared first into pellets. It is a kind of sample specimen in the form of a pressed powder by a PVC ring until it solidified.

Beneficiation process for a 20 kg sample of lateritic iron ore was ground using ball mill then it was sieved into five grain size fractions: (+212)μm; (-212+180)μm; (-180+150)μm; (-150+90)μm; and (-90)μm. Each fraction was screened and weighed to determine the grain size distribution and mass of the feed respectively. The feed of each size fraction was beneficiated using magnetic separator. Drum-type magnetic separator is the most widely applied as compared to other types, such as magnetic pulley, plate magnet, grate magnet, and suspended magnet [8]. In this study, we use drum magnetic separator with low intensity. The product of concentrate (magnetic rich material) and tailing (magnetic poor material) resulting from beneficiation were further weighed and chemically analyzed in order to find out recovery.

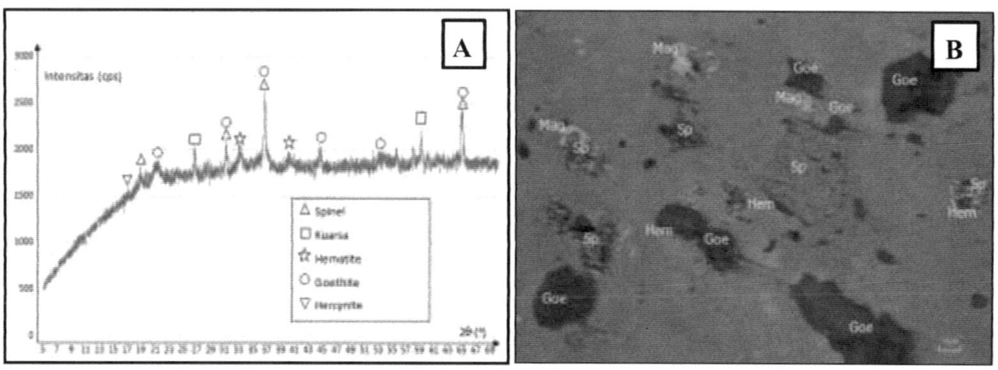

Figure 1. XRD pattern of a lateritic iron ore sample from Malili (A); Microscopic appearance (Goe: goethite, Hem: hematite, Chr: chromite, Mag: magnetite, Sp: spinel), (B).

IOP Conf. Series: Materials Science and Engineering **619** (2019) 012017 doi:10.1088/1757-899X/619/1/012017

3. Results and Discussions

The result of mineralogical analysis using the XRD method exhibit that initial sample of lateritic iron ore is mainly composed by goethite (FeO.OH). It is indicated with the presence of reflection intensity with d-spacing of 4.18Å and and 2.69Å. Spinel group mineral such as spinel ($Al_{0,39}Cr_{1,58}Mg_{0,5}O_4$) itself, magnetite ($Fe_3O_4$) and hercynite ($Al_{2,1}Fe_{0,1}O_4$) are also present in minor quantity, whereas hematite (Fe_2O_3) and quartz (SiO_2) are two oxide minerals which occur in the sample in small amount. Figure 1-A shows the diffractogram of a lateritic iron ore sample.

X-ray diffraction is a reliable and fast method for identifying minerals. The result of mineralogical analysis also showed that mineralogy of the lateritic iron ore has generally poor crystallinity, which is represented by a widening peak shape. The degree of crystallinity of a mineral is equivalent to the 2-theta degree [9]. This is supported by microscopic observation such as shown by goethite (FeO.OH) with fine crystalline size and acicular texture. Spinel group minerals ($Al_{0,39}Cr_{1,58}Mg_{0,5}O_4$), magnetite ($Fe_3O_4$) and hematite ($Fe_2O_3$) were detected to be associated with goethite. Goethite is formed by weathering process in which iron oxide is converted to iron hydroxide [10].

Figure 1-B shows the microscopic appearance of lateritic iron ore. Goethite show unhedral with acicular texture; whereas spinel exhibits euhedral to subhedral with grain size ranging between 100 and 300 microns. Result of chemical analysis using X-ray fluorescence method is given in Table 1. It is shown that iron, aluminum and silicon are the three elements dominant with >90 wt% in total.

Table 1. Chemical composition of lateritic iron ore feed

Major Oxide	**Composition (wt%)**
Fe_2O_3	32.845
Al_2O_3	35.596
SiO_2	21.133
Cr_2O_3	7.202

The concentration Fe_2O_3 can be derived from minerals such as goethite, magnetite, hematite, hercynite, and other minerals that have iron element composition. The availability of Al_2O_3 could be derived from the spinel group minerals and some of them are likely associated with goethite.

Beneficiation process was started with classifying the initial sample (feed) based on grain size using a multilevel sieve. The sieving result of the dominance of coarse grain size is shown in Figure 2. There is 60% lateritic iron ore sample from Malili area was included in the size of +212 μm.

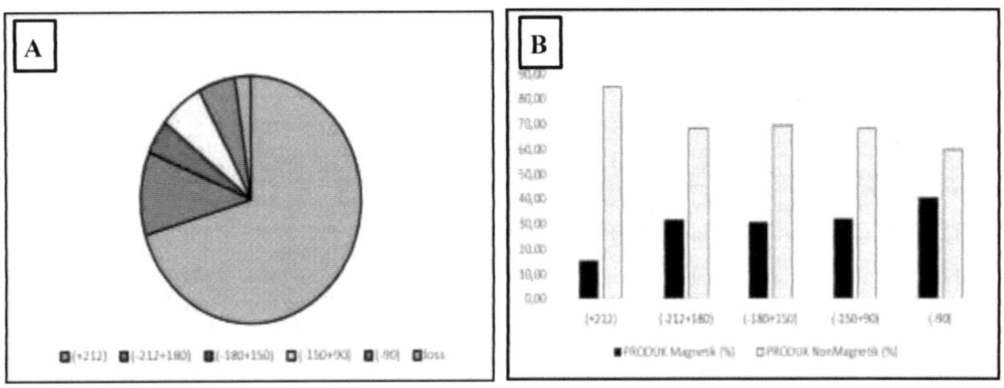

Figure 2. Mass percentage of iron ore feed (A) and comparison of products percentage between magnetic and non-magnetic materials (B).

The beneficiation process using magnetic separator produces a concentrate that increases with the smoothness of the grain size of the fraction. The smaller the size of the feed particles indicate the easier of magnet attracts to them. This also proves that the finer grain size of the feed, showing the higher degree of mineral liberation. The degree of liberation is the amount of weight of a valuable mineral that is completely free, commonly stated in percent, compared to the amount of all minerals contained in the ore, either free or still bind. The magnetic separation result is presented in Figure 2B.

The color of a concentrate is darker than a tailing product. Based on mineralogical analysis on the sample of concentrate and tailing, there are diffractogram differences indicated the variation of the minerals composition. Figure 3 shows the comparison of diffractograms which were obtained from mineralogical analysis using X-Ray Diffraction method.

Based on the diffractogram, it can be interpreted that the magnetic minerals are the minerals within the poor crystallinity, meanwhile the non-magnetic minerals has a good crystallinity. This was shown on the widened peak shape until no peak shaped, such as at the angle of 2θ = 65°, 58°, 36°, 30°, and others.

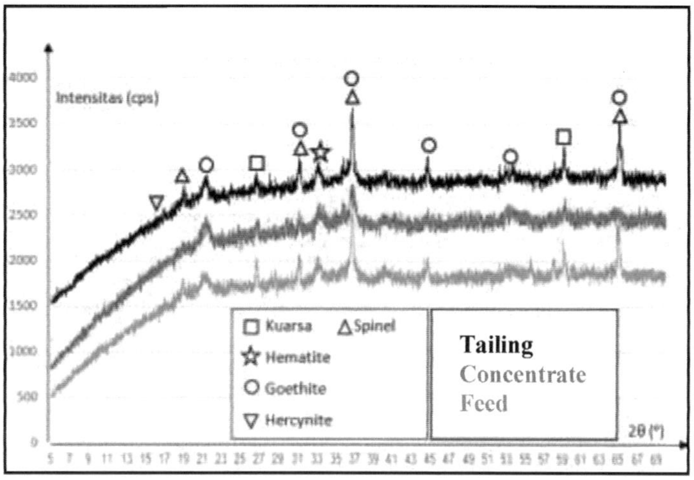

Figure 3. Comparison of diffractogram patterns of the sample before and after separation.

Based on chemical analysis using X-Ray Fluorescence method, lateritic iron ore experienced an increase in Fe_2O_3 grade from 32% in the feed to an average of 46% in the concentrate, after passed a magnetic separation process. The increase of Fe_2O_3 grade per fraction has an identical value with moderately small difference value, between 43% and 47%. The level of magnetism depends on the Fe^{3+} ion content.

Table 2. Chemical composition of product concentrates (wt%)

Size Fraction (μm)	Fe_2O_3	SiO_2	Al_2O_3	Cr_2O_3
Feed	32.8	21.1	35.5	7.2
(+212)	45.0	31.1	18.2	2.8
(-212+180)	47.9	17.8	24.2	7.3
(-180+150)	43.9	16.3	28.5	8.9
(-150+90)	46.6	16.4	26.5	8.2
(-90)	47.2	20.3	26.1	3.9

This suggests that goethite (FeO.OH) has a fairly good magnetic susceptibility, consistent with XRD analysis that the concentrate is mainly composed of goethite with poor crystallinity. Goethite is the most dominant mineral found in lateritic iron ore [11]. Whereas the grade of non-magnetic Al_2O_3, SiO_2, and Cr_2O_3 decreases. For more details, the chemical composition of the concentrate is shown in Table 2.

In tailing products, the percentage of Fe_2O_3 decreased, with the Fe_2O_3 grade from 25.4 to 39.0 wt% the lowest was 25.2%. Even as Al_2O_3 and SiO_2 increased due to its non-magnetic. Table 3 shows the transformation of grade in tailing.

The recovery is obtained from the ratio between the concentrate mass multiplied by its grade and the feed mass multiplied by its grade. It is caused of the processing of minerals never goes perfectly (perfect means all the valuable minerals can be obtained) which means that there is some valuable minerals going into the tailing that are eventually discarded. If the recovery is equal to 100% (perfectly) then the weight of the concentrate is probably equal to the weight of the incoming feed, consequently the concentrate must be equal to the feed grade, meaning that no process occurs. The recovery formulation can be written as follows:

$$R = [(C.c)/(F.f)]*100\%$$

Table 3. Chemical composition of product tailing (wt%)

Fraction (µm)	Fe_2O_3	SiO_2	Al_2O_3	Cr_2O_3
(+212)	39.0	22.2	31.1	3.3
(-212+180)	25.4	19.8	41.1	11.0
(-180+150)	25.2	16.6	43.1	12.7
(-150+90)	25.4	15.5	43.4	13.6
(-90)	39.0	21.3	30.1	7.1

Results of calculation of recovery of Fe_2O_3 along with graph showing the concentrate of grade and recovery against size fraction are presented in Table 4 and Fig. 5 respectively.

Table 4. Recovery of Fe_2O_3.

Fraction (µm)	Concentrate		Feed		***RECO-VERY (%)***
	Fe_2O_3 (%)	Mass (gr)	Fe_2O_3 (%)	Mass (gr)	
(+212)	45.0	1821.0	32.8	12081.3	**20.6**
(-212+180)	47.9	854.0	32.8	2785.2	**44.7**
(-180+150)	43.9	424.8	32.8	895.1	**63.5**
(-150+90)	46.6	803.3	32.8	1880.3	**60.7**
(-90)	47.2	986.3	32.8	1579.4	**89.8**

4. Conclusion

The lateritic iron ore from Malili area is composed of minerals with poor crystallinity mainly goethite (FeO.OH) which is characteristic of minerals that is formed by weathering. Spinel ($Al_{1.641}Cr_{0.244}Mg_{1.115}O_4$), hematite ($Fe_2O_3$), magnetite ($Fe_2O_3$), and hercynite ($Al_{2,1}Fe_{0,1}O_4$) were also identified.

Lateritic iron ore sample from Maili area consists of 33%Fe_2O_3, 21%SiO_2, 35%Al_2O_3, 7%Cr_2O_3, 1.5%MnO, and 1.2%NiO. The concentrate product by magnetic separator has an average grade of 46%Fe_2O_3, 16%SiO_2, 24%Al_2O_3, 6%Cr_2O_3, 1%MnO, and 1.2%NiO. The tailing product shows an average grade of 25%Fe_2O_3, 20%SiO_2, 38%Al_2O_3, 9,5%Cr_2O_3, 1.1%MnO, and 1%NiO.

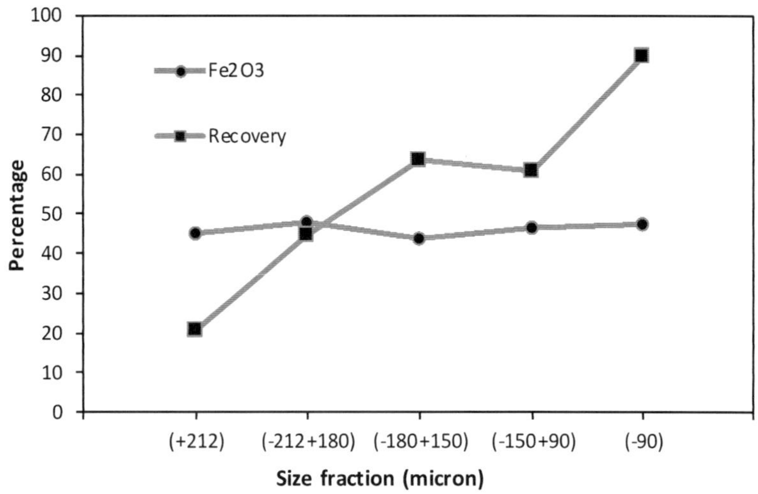

Figure 4. Graph showing distribution of size fractions with Fe_2O_3 recovery and grade of concentrate.

Result of iron ore beneficiation using magnetic separator indicates an increasing amount of concentrate with the finer fraction of the grain size of the feed. Consequently, the highest Fe_2O_3 recovery was about 89.83% and it was found at (-90) μm size fraction of the feed.

References

[1] Irabor, E I and Okolo P O 2010 Chemical and mineralogical characteristics of lateritic iron ore deposit at Iyuku, Etsako West Local Government area of Edo State, Nigeria. *Global Journal of Pure and Applied Science,* 16, pp 309 – 312

[2] The Balance of Energy, Coal and Geothermal Resources (in Bahasa) 2018 PSDG Kementerian ESDM RI (http://psdg.geologi.esdm.go.id/index.php?option, accessed on 28 October, 2018.

[3] Dinas Energi dan Sumber Daya Mineral Kabupaten Luwu Timur, 2011., http://luwutimurkab.go.id/dinasesdm/file/iup2011/pdf. accessed on August, 12, 2017.

[4] Sembiring S F 2009 Jalan Baru untuk Tambang: Mengalirkan Berkah bagi Anak Bangsa., Jakarta: *Elex Media Komputindo.* p319

[5] Wills B and Finch J A 2006 *Mineral Processing Technology: An Introduction to the Practical Aspects of Ore Treatment and Mineral Recovery*. New York: Pergamon Press, p550

[6] O'Connor O, Cheung W H, Valix M, 2006 Reduction roasting of limonite ores: effect of dihydroxylation, *Int. J. Mineral Process*, 80, pp 88- 99

Oh Jang K, Venkata R M, Hapugoda S, Nguyen A V, Buckard W J 2014 Chemical and mineral transformation of a low grade goethite ore by dehydroxylation, reduction roasting and
[8] magnetic separation, *Mineral Engineering*, 60, pp 14 – 22.

Svoboda J 2004 *Magnetic Techniques for The Treatment Materials.* Kluwer Academic
[9] Publisher. New York, p 642

Herdianita N R 1999 Pengukuran Kristalinitas Silika Berdasarkan Metode Diffraktometer Sinar-
[10] X., Proceedings. ITB, 31, pp 41-47

Ahmad W 2009 *Nickels laterite: fundamentals of chemistry, mineralogy, weathering process,*
[11] *formation and exploration*, Unpub. Training Manual VITSL Sorowako p 320

Wang B, Guo Q, Wei G, Zhang P, Qi T, 2012 Characterization and Atmospheric, Hydrochloric Acid Leaching of Limonite Laterite from Indonesia *Hydrometallurgy*, 129-130, pp 7-13

Acknowledgements
Thanks to the head of LP2M UNHAS (Institute for Research and Community Services of Hasanuddin University) who has provided research grant through the PTUPT scheme to SF. Thanks also to Mr. Ade, Laboratory Assistant of the Geological Engineering Department who has assisted in the polished sections preparation for microscopic analysis.

On some examples of BEM solution to elasticity problems of isotropic functionally graded materials

S. Hamzah[1], **M. I. Azis**[2,*], **E. Syamsuddin**[3]
[1]Department of Civil Engineering, Hasanuddin University, Makassar, Indonesia
[2]Department of Mathematics, Hasanuddin University, Makassar, Indonesia
[3]Department of Physics, Hasanuddin University, Makassar, Indonesia
E-mail: mohivanazis@yahoo.co.id (*Corresponding author)

Abstract. A boundary element method is derived for the solution of static elasticity problems of quadratically graded isotropic materials. Some particular problems are considered to illustrate the application of the method.

1. Introduction
Applications of the boundary element method (BEM) cover many kind of engineering problems such as deformation of elastic materials, pollutant transport and so on in either isotropic or anisotropic and either homogeneous or non-homogeneous materials. Examples of early work considering for homogeneous media includes papers by Haddade, et. al [1] and Azis, et. al [2, 3] in which BEM was used to find numerical solution to pollutant transport problems for homogeneous anisotropic materials.

For inhomogeneous media paper [4] considered the use of BEM to solve deformation problems for inhomogeneous isotropic materials, the work in [5] solved heat conduction problems, and paper [6] considered transient heat conduction problems, papers [7, 8] studied a BEM for a class of elliptic boundary value problems of functionally graded media and paper [9] worked on the diffusion convection reaction equation. Recent work on deformation problems for inhomogeneous materials was reported in [10] and in [11].

The current study employs a displacement based method to solve static deformation problems of quadratically graded isotropic materials. Some specific problems will be solved numerically. They are problems of a constrained slab, extension of a slab with a crack inclusion, compression of a slab with and without a hole inclusion.

2. Governing equations
Using a Cartesian frame $Ox_1x_2x_3$ we consider the governing equations

$$\sigma_{ij,j} = \mathbf{0} \tag{1}$$

where σ_{ij} for $i, j = 1, 2, 3$ denotes the stress tensor, the indexed commas indicate partial differentiation with respect to the spatial coordinates x_j and the repeated suffix summation convention (summing from 1 to 3) is employed. The stress-displacement relations are

$$\sigma_{ij} = \alpha\delta_{ij}u_{k,k} + \beta\left(u_{i,j} + u_{j,i}\right) \tag{2}$$

IOP Conf. Series: Materials Science and Engineering **619** (2019) 012018 doi:10.1088/1757-899X/619/1/012018

where u_k for $k = 1, 2, 3$ is the displacement and δ_{ij} the Kronecker delta. Also in (2) $\alpha(\mathbf{x})$ and $\beta(\mathbf{x})$ with $\mathbf{x} = (x_1, x_2, x_3)$ denote the Lamé parameters which are taken to be twice differentiable functions of the spatial variables x_1, x_2 and x_3. Substitution of (2) into (1) yields

$$\left[\alpha\delta_{ij}u_{k,k} + \beta\left(u_{i,j} + u_{j,i}\right)\right]_{,j} = \mathbf{0} \tag{3}$$

3. Statement of the boundary value problem

An inhomogeneous isotropic elastic material occupies the region Ω in R^3 with boundary $\partial\Omega$ which consists of a finite number of piecewise smooth closed surfaces. On $\partial\Omega_1$ the displacement u_i is specified and on $\partial\Omega_2$ the stress vector $P_i = \sigma_{ij}n_j$ is specified where $\partial\Omega = \partial\Omega_1 \cup \partial\Omega_2$ and $\mathbf{n} = (n_1, n_2, n_3)$ denotes the vector of normal to $\partial\Omega$ pointing outward. We seek a solution to (3) for any point inside the domain Ω which satisfies the specified boundary conditions on $\partial\Omega$.

4. Transformation to a constant coefficient equation

The coefficients $\alpha(\mathbf{x})$ and $\beta(\mathbf{x})$ are assumed to be

$$\alpha(\mathbf{x}) = \alpha^{(0)}b(\mathbf{x}) \quad \beta(\mathbf{x}) = \beta^{(0)}b(\mathbf{x}) \tag{4}$$

where $\alpha^{(0)}$ and $\beta^{(0)}$ are constants. Substitution of (4) into (3) yields

$$\left\{b\left[\alpha^{(0)}\delta_{ij}u_{k,k} + \beta^{(0)}\left(u_{i,j} + u_{j,i}\right)\right]\right\}_{,j} = \mathbf{0} \tag{5}$$

Let

$$\psi_i(\mathbf{x}) = b^{1/2}(\mathbf{x})\, u_i(\mathbf{x}) \tag{6}$$

Adopting the procedure developed in Manolis and Shaw [12], the transformation (6) will be used to transform (3) to a constant coefficients equation. Therefore (5) becomes

$$\left\{b\left[\alpha^{(0)}\delta_{ij}\left(b^{-1/2}\psi_k\right)_{,k} + \beta^{(0)}\left(\left(b^{-1/2}\psi_i\right)_{,j} + \left(b^{-1/2}\psi_j\right)_{,i}\right)\right]\right\}_{,j} = \mathbf{0}$$

Thus

$$\alpha^{(0)}\left[b\left(b^{-1/2}\psi_k\right)_{,k}\right]_{,i} + \beta^{(0)}\left[b\left(b^{-1/2}\psi_i\right)_{,j}\right]_{,j} + \beta^{(0)}\left[b\left(b^{-1/2}\psi_j\right)_{,i}\right]_{,j} = \mathbf{0} \tag{7}$$

Now

$$\begin{aligned}
&\left[b\left(b^{-1/2}\psi_k\right)_{,k}\right]_{,i} \\
&= \frac{1}{4}b^{-3/2}b_{,i}b_{,k}\psi_k - \frac{1}{2}b^{-1/2}b_{,ki}\psi_k - \frac{1}{2}b^{-1/2}b_{,k}\psi_{k,i} + \frac{1}{2}b^{-1/2}b_{,i}\psi_{k,k} + b^{1/2}\psi_{k,ki} \\
&= -b^{1/2}_{,ki}\psi_k + b^{1/2}\psi_{k,ki} - \frac{1}{2}b^{-1/2}b_{,k}\psi_{k,i} + \frac{1}{2}b^{-1/2}b_{,i}\psi_{k,k}
\end{aligned} \tag{8}$$

Similarly

$$\left[b\left(b^{-1/2}\psi_i\right)_{,j}\right]_{,j} = -b^{1/2}_{,jj}\psi_i + b^{1/2}\psi_{i,jj} \tag{9}$$

$$\left[b\left(b^{-1/2}\psi_j\right)_{,i}\right]_{,j} = -b^{1/2}_{,ij}\psi_j + b^{1/2}\psi_{j,ij} - \frac{1}{2}b^{-1/2}b_{,i}\psi_{j,j} + \frac{1}{2}b^{-1/2}b_{,j}\psi_{j,i} \tag{10}$$

Use of (8), (9) and (10) into (7) gives

$$b^{1/2}\left[\alpha^{(0)}\delta_{ij}\psi_{k,k}+\beta^{(0)}\left(\psi_{i,j}+\psi_{j,i}\right)\right]_{,j}-\left[\alpha^{(0)}\psi_k b^{1/2}_{,ki}+\beta^{(0)}\psi_i b^{1/2}_{,jj}+\beta^{(0)}\psi_j b^{1/2}_{,ij}\right]$$
$$-\left(\alpha^{(0)}-\beta^{(0)}\right)\left[\frac{1}{2}b^{-1/2}\right]\left[b_{,k}\psi_{k,i}-b_{,i}\psi_{k,k}\right]=\mathbf{0} \tag{11}$$

If $b(\mathbf{x})$ assumes the form

$$b(\mathbf{x})=\left(\gamma_0+\gamma_1x_1+\gamma_2x_2+\gamma_3x_3\right)^2 \tag{12}$$

where γ_t, $t=0,1,2,3$ are constants and also

$$\alpha^{(0)}=\beta^{(0)} \tag{13}$$

thus $\alpha(\mathbf{x})=\beta^{(0)}\left(\gamma_0+\gamma_1x_1+\gamma_2x_2+\gamma_3x_3\right)^2=\beta(\mathbf{x})$ then (11) reduces to

$$\left[\alpha^{(0)}\delta_{ij}\psi_{k,k}+\beta^{(0)}\left(\psi_{i,j}+\psi_{j,i}\right)\right]_{,j}=\mathbf{0}\quad(\text{with } \alpha^{(0)}=\beta^{(0)}) \tag{14}$$

From (2) the corresponding stresses are given by

$$\sigma_{ij}=-\psi_k\,\sigma^{[b]}_{ijk}+b^{1/2}\,\sigma^{[\psi]}_{ij}$$

where

$$\begin{aligned}\sigma^{[b]}_{ijk}&=\alpha^{(0)}\delta_{ij}\,b^{1/2}_{,k}+\beta^{(0)}\left(\delta_{ki}\,b^{1/2}_{,j}+\delta_{kj}\,b^{1/2}_{,i}\right)\\ \sigma^{[\psi]}_{ij}&=\alpha^{(0)}\delta_{ij}\,\psi_{k,k}+\beta^{(0)}\left(\psi_{i,j}+\psi_{j,i}\right)\end{aligned}$$

and the stress vector

$$P_i=-\psi_k P^{[b]}_{ik}+b^{1/2}P^{[\psi]}_i \tag{15}$$

where

$$\begin{aligned}P^{[b]}_{ik}&=\sigma^{[b]}_{ijk}n_j\\ P^{[\psi]}_i&=\sigma^{[\psi]}_{ij}n_j\end{aligned} \tag{16}$$

A boundary integral equation for the solution of (14) is

$$\lambda(\mathbf{x}_0)\psi_j(\mathbf{x}_0)=\int_{\partial\Omega}\left[\Gamma_{ij}(\mathbf{x},\mathbf{x}_0)\,\psi_i(\mathbf{x})-\Phi_{ij}(\mathbf{x},\mathbf{x}_0)\,P^{[\psi]}_i(\mathbf{x})\right]ds(\mathbf{x}) \tag{17}$$

where$\lambda=0$ if $\mathbf{x}_0\notin\Omega\cup\partial\Omega$, $\lambda=1$ if $\mathbf{x}_0\in\Omega$ and $\lambda=\frac{1}{2}$ if $\mathbf{x}_0\in\partial\Omega$ and $\partial\Omega$ has a continuously turning tangent at $\mathbf{x}_0$. The Φ_{ij} in (17) is any solution of the equation

$$\left[\alpha^{(0)}\delta_{ij}\Phi_{km,k}+\beta^{(0)}\left(\Phi_{im,j}+\Phi_{jm,i}\right)\right]_{,j}=-\delta_{im}\delta(\mathbf{x}-\mathbf{x}_0)$$

where δ_{im} is the Kronecker delta and the Γ_{ij} is given by

$$\Gamma_{im}=\left[\alpha^{(0)}\delta_{ij}\Phi_{km,k}+\beta^{(0)}\left(\Phi_{im,j}+\Phi_{jm,i}\right)\right]n_j$$

For the three dimensional case

$$\begin{aligned}
\Phi_{ij} &= \frac{1}{16\pi\beta^{(0)}(1-\nu)d}\left[(3-4\nu)\delta_{ij}+d_{,i}d_{,j}\right]\\
\Gamma_{ij} &= -\frac{1}{8\pi(1-\nu)d^2}\\
&\left[\frac{\partial d}{\partial n}\left\{(1-2\nu)\delta_{ij}+3d_{,i}d_{,j}\right\}+(1-2\nu)\left(n_i d_{,j}-n_j d_{,i}\right)\right]
\end{aligned}$$

and for two dimensional case

$$\begin{aligned}
\Phi_{ij} &= \frac{1}{8\pi\beta^{(0)}(1-\nu)}\left[(3-4\nu)\log\frac{1}{d}\delta_{ij}+d_{,i}d_{,j}\right]\\
\Gamma_{ij} &= -\frac{1}{4\pi(1-\nu)d}\\
&\left[\frac{\partial d}{\partial n}\left\{(1-2\nu)\delta_{ij}+2d_{,i}d_{,j}\right\}+(1-2\nu)\left(n_i d_{,j}-n_j d_{,i}\right)\right]
\end{aligned}$$

where $d=\|\mathbf{x}-\mathbf{x}_0\|$, $\nu=\alpha^{(0)}/(2(\beta^{(0)}+\alpha^{(0)}))$ and $\partial d/\partial n = d_{,k}n_k$.

Substitution of (6) and (15) into (17) yields

$$\lambda(\mathbf{x}_0)\,b^{1/2}(\mathbf{x}_0)\,u_j(\mathbf{x}_0)=\int_{\partial\Omega}\left\{u_i(\mathbf{x})\left[b^{1/2}(\mathbf{x})\,\Gamma_{ij}(\mathbf{x},\mathbf{x}_0)-P_{ki}^{[b]}(\mathbf{x})\,\Phi_{kj}(\mathbf{x},\mathbf{x}_0)\right]-\right.$$
$$\left.P_i(\mathbf{x})\left[b^{-1/2}(\mathbf{x})\,\Phi_{ij}(\mathbf{x},\mathbf{x}_0)\right]\right\}ds(\mathbf{x})$$

This equation provides a boundary integral equation for determining u_i and σ_{ij} at all points inside Ω.

5. A perturbation method

The coefficients $\alpha(\mathbf{x})$ and $\beta(\mathbf{x})$ are assumed to take the form

$$\alpha(\mathbf{x}) = \alpha^{(0)}b(\mathbf{x})+\epsilon\alpha^{(1)}(\mathbf{x}) \tag{18}$$
$$\beta(\mathbf{x}) = \beta^{(0)}b(\mathbf{x})+\epsilon\beta^{(1)}(\mathbf{x}) \tag{19}$$

with

$$\alpha^{(0)}=\beta^{(0)} \quad \text{and} \quad b^{1/2}_{,ij}=0$$

and where $\alpha^{(1)}$ and $\beta^{(1)}$ are twice differentiable functions. Thus from (3)

$$\left\{b\left[\alpha^{(0)}\delta_{ij}u_{k,k}+\beta^{(0)}(u_{i,j}+u_{j,i})\right]\right\}_{,j}=-\epsilon\left[\alpha^{(1)}\delta_{ij}u_{k,k}+\beta^{(1)}(u_{i,j}+u_{j,i})\right]_{,j}$$

Use of the (6) and following the analysis used to derive (11) from (5) gives

$$\left[\alpha^{(0)}\delta_{ij}\psi_{k,k}+\beta^{(0)}(\psi_{i,j}+\psi_{j,i})\right]_{,j}=-\epsilon b^{-1/2}\left[\alpha^{(1)}\delta_{ij}u_{k,k}+\beta^{(1)}(u_{i,j}+u_{j,i})\right]_{,j} \tag{20}$$

A solution to equation (20) is sought in the form

$$\psi_i(\mathbf{x})=\sum_{r=0}^{\infty}\epsilon^r\psi_i^{(r)}(\mathbf{x}) \tag{21}$$

From (6) and (21) the displacement u_k may be written in a series form

$$u_k(\mathbf{x}) = \sum_{r=0}^{\infty} \epsilon^r u_k^{(r)}(\mathbf{x}) \tag{22}$$

where $u_k^{(r)}$ corresponds to $\psi_k^{(r)}$ according to the relationship

$$\psi_k^{(r)} = b^{1/2}\, u_k^{(r)}$$

Substitution of (21) into (20) and equating the coefficients of powers of ϵ gives

$$\left[\alpha^{(0)}\delta_{ij}\psi_{k,k}^{(r)} + \beta^{(0)}\left(\psi_{i,j}^{(r)} + \psi_{j,i}^{(r)}\right)\right]_{,j} = h^{(r)} \qquad \text{for } r = 0, 1, \ldots, \tag{23}$$

where

$$\begin{aligned} h^{(0)}(\mathbf{x}) &= \mathbf{0} \\ h^{(r)}(\mathbf{x}) &= -b^{-1/2}\left[\alpha^{(1)}\delta_{ij}u_{k,k}^{(r-1)} + \beta^{(1)}(u_{i,j}^{(r-1)} + u_{j,i}^{(r-1)})\right]_{,j} \qquad \text{for } r = 1, 2, \ldots \end{aligned} \tag{24}$$

The integral equation for (23) is

$$\begin{aligned} \lambda(\mathbf{x}_0)\,\psi_j^{(r)}(\mathbf{x}_0) &= \int_{\partial\Omega}\left[\Gamma_{ij}(\mathbf{x},\mathbf{x}_0)\,\psi_i^{(r)}(\mathbf{x}) - \Phi_{ij}(\mathbf{x},\mathbf{x}_0)\,P_i^{[\psi^{(r)}]}(\mathbf{x})\right] ds(\mathbf{x}) \\ &\quad + \int_{\Omega} h_i^{(r)}(\mathbf{x})\,\Phi_{ij}(\mathbf{x},\mathbf{x}_0)\,dS(\mathbf{x}) \qquad \text{for } r = 0, 1, \ldots \end{aligned} \tag{25}$$

where

$$P_i^{[\psi^{(r)}]} = \left[\alpha^{(0)}\delta_{ij}\psi_{k,k}^{(r)} + \beta^{(0)}\left(\psi_{i,j}^{(r)} + \psi_{j,i}^{(r)}\right)\right] n_j$$

Also

$$P_i^{[\psi^{(r)}]} = b^{1/2}P_i^{(r)} + u_k^{(r)}P_{ik}^{[b]} \qquad \text{for } r = 0, 1, \ldots$$

where

$$P_i^{(r)}(\mathbf{x}) = \left[\alpha^{(0)}\delta_{ij}u_{k,k}^{(r)} + \beta^{(0)}\left(u_{i,j}^{(r)} + u_{j,i}^{(r)}\right)\right] n_j$$

and $P_{ik}^{[b]}$ is given by (16). Thus the integral equation (25) may be written in the form

$$\begin{aligned} \lambda(\mathbf{x}_0)\,b^{1/2}(\mathbf{x}_0)\,u_j^{(r)}(\mathbf{x}_0) &= \int_{\partial\Omega}\left\{u_i^{(r)}(\mathbf{x})\left[b^{1/2}(\mathbf{x})\,\Gamma_{ij}(\mathbf{x},\mathbf{x}_0) - P_{ki}^{[b]}(\mathbf{x})\,\Phi_{kj}(\mathbf{x},\mathbf{x}_0)\right]\right. \\ &\quad \left. -P_i^{(r)}(\mathbf{x})\left[b^{1/2}(\mathbf{x})\,\Phi_{ij}(\mathbf{x},\mathbf{x}_0)\right]\right\} ds(\mathbf{x}) \\ &\quad + \int_{\Omega} h_i^{(r)}(\mathbf{x})\,\Phi_{ij}(\mathbf{x},\mathbf{x}_0)\,dS(\mathbf{x}) \end{aligned} \tag{26}$$

Now, the corresponding value of P_i may be written as

$$P_i = bP_i^{(0)} + \sum_{r=1}^{\infty}\epsilon^r\left(bP_i^{(r)} + G_i^{(r)}\right) \tag{27}$$

where

$$G_i^{(r)}(\mathbf{x}) = \left[\alpha^{(1)}\delta_{ij}u_{k,k}^{(r-1)} + \beta^{(1)}(u_{i,j}^{(r-1)} + u_{j,i}^{(r-1)})\right] n_j$$

IOP Conf. Series: Materials Science and Engineering **619** (2019) 012018 doi:10.1088/1757-899X/619/1/012018

To satisfy the boundary conditions in Section 3 it is required that

$$\begin{aligned} u_i^{(0)} &= u_i && \text{on } \partial\Omega_1 \\ P_i^{(0)} &= b^{-1} P_i && \text{on } \partial\Omega_2 \end{aligned}$$

where u_i takes on its specified value on $\partial\Omega_1$ and P_i takes on its specified value on $\partial\Omega_2$. It then follows from (22) and (27) that for $r = 1, 2, \ldots$

$$\begin{aligned} u_i^{(r)} &= \mathbf{0} && \text{on } \partial\Omega_1 \\ P_i^{(r)} &= -b^{-1} G_i^{(r)} && \text{on } \partial\Omega_2 \end{aligned}$$

The integral equation (26) may now be used to find the numerical values of the unknowns on the boundary $\partial\Omega$ and the numerical values of $u_i^{(r)}$ and derivatives in the domain Ω for $r = 0, 1, \ldots$.

6. Numerical results

Some problems will be considered in this section. They are plane strain and plane stress problems. The integral equations obtained in Sections 4 and 5 will be used to solve them numerically. The integrals in the integral equations are calculated using Gaussian quadrature (see Abramowitz and Stegun [13]).

For the chosen elastic parameters of the forms (18) and (19), the right hand side of (24) is assumed to be so small as it is only necessary to retain two terms in the expression (22).

6.1. Example 1: Extension of a slab

Consider the boundary value problem given in Figure 1 for a material with elastic coefficients

$$\alpha(\mathbf{x}) = 1.4\widehat{\alpha}(1 + 0.1x_1')^2 \qquad (28)$$

$$\beta(\mathbf{x}) = 1.2\widehat{\alpha}(1 + 0.1x_1')^2 \qquad (29)$$

where $\widehat{\alpha}$ is a reference elastic modulus and $x_1' = x_1/l$.

The elastic coefficients (28) and (29) are of the forms (18) and (19) with $b(\mathbf{x}) = (1 + 0.1x_1')^2$, $\beta^{(0)} = \alpha^{(0)} = 1.2\widehat{\alpha}$, $\alpha^{(1)} = \widehat{\alpha}(1 + 0.1x_1')^2$, $\beta^{(1)} = 0$ and $\epsilon = 0.2$. The boundary conditions (see Figure 1) are

$$\begin{aligned} &P_1/\widehat{P} = 0 \quad u_2/\widehat{u} = 0 \quad \text{on AB} \\ &P_1/\widehat{P} = 1 \quad P_2/\widehat{P} = 0 \quad \text{on BC} \\ &P_1/\widehat{P} = 0 \quad u_2/\widehat{u} = 0 \quad \text{on CD} \\ &u_1/\widehat{u} = 0 \quad u_2/\widehat{u} = 0 \quad \text{on AD} \end{aligned}$$

where $\widehat{u}$ is a reference displacement and $\widehat{P} = \widehat{\alpha}\widehat{u}/l$. The analytical solution is $u_1/\widehat{u} = x_1'/[3.8(1 + 0.1x_1')]$, $u_2 = 0$ and $\sigma_{11}/\widehat{P} = 1$, $\sigma_{12}/\widehat{P} = 0$ and $\sigma_{22}/\widehat{P} = 0.3864$.

Table 1 – Table 4 show the BEM and analytical solutions for some points in the domain Ω and for the cases when the boundary $\partial\Omega$ is divided into 80, 160 and 320 elements. The BEM solutions converge to the known solution as the number of elements increases. The displacement displays fourth figure and the stress third figure accuracy when 320 boundary elements are used.

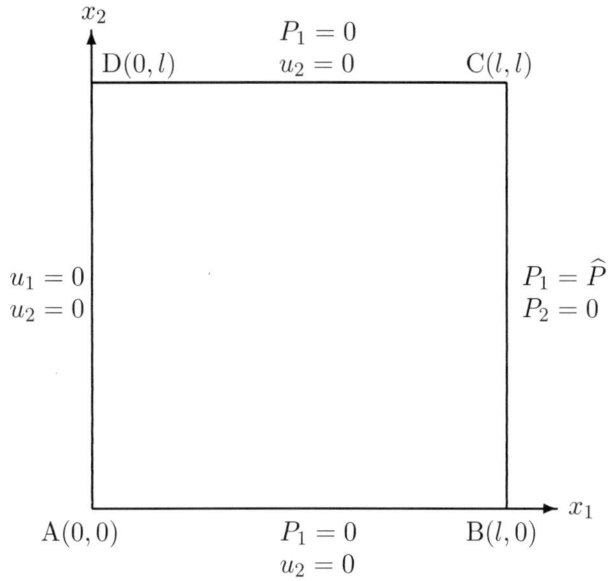

Figure 1. The geometry for Example 1

Table 1. Displacement solutions for Example 1

Position	BEM 80 elements		BEM 160 elements		BEM 320 elements	
(x'_1, x'_2)	$u_1/\widehat{u}$	$u_2/\widehat{u}$	$u_1/\widehat{u}$	$u_2/\widehat{u}$	$u_1/\widehat{u}$	$u_2/\widehat{u}$
(.1,.5)	.0258	.0000	.0259	-.0000	.0260	-.0000
(.3,.5)	.0763	.0000	.0765	-.0000	.0766	-.0000
(.5,.5)	.1248	.0000	.1251	-.0000	.1252	.0000
(.7,.5)	.1715	-.0000	.1719	.0000	.1720	.0000
(.9,.5)	.2165	-.0000	.2169	.0000	.2171	.0000

Table 2. Displacement solutions for Example 1

Position	Analytical	
(x'_1, x'_2)	$u_1/\widehat{u}$	$u_2/\widehat{u}$
(.1,.5)	.0261	.0000
(.3,.5)	.0766	.0000
(.5,.5)	.1253	.0000
(.7,.5)	.1722	.0000
(.9,.5)	.2173	.0000

Table 3. Stress solutions for Example 1

Position	BEM 80 elements			BEM 160 elements		
(x'_1, x'_2)	$\sigma_{11}/\widehat{P}$	$\sigma_{12}/\widehat{P}$	$\sigma_{22}/\widehat{P}$	$\sigma_{11}/\widehat{P}$	$\sigma_{12}/\widehat{P}$	$\sigma_{22}/\widehat{P}$
(.1,.5)	.9987	.0000	.3694	.9994	-.0000	.3689
(.3,.5)	.9983	.0000	.3710	.9992	-.0001	.3697
(.5,.5)	.9988	.0000	.3714	.9995	-.0001	.3699
(.7,.5)	.9993	-.0000	.3715	.9998	-.0002	.3700
(.9,.5)	.9983	-.0004	.3724	.9996	-.0020	.3704

Table 4. Stress solutions for Example 1

Position	BEM 320 elements			Analytical		
(x_1', x_2')	$\sigma_{11}/\widehat{P}$	$\sigma_{12}/\widehat{P}$	$\sigma_{22}/\widehat{P}$	$\sigma_{11}/\widehat{P}$	$\sigma_{12}/\widehat{P}$	$\sigma_{22}/\widehat{P}$
(.1,.5)	.9998	-.0000	.3687	1.0000	.0000	.3684
(.3,.5)	.9997	-.0000	.3691	1.0000	.0000	.3684
(.5,.5)	.9998	.0000	.3692	1.0000	.0000	.3684
(.7,.5)	.9999	.0001	.3692	1.0000	.0000	.3684
(.9,.5)	.9996	-.0002	.3695	1.0000	.0000	.3684

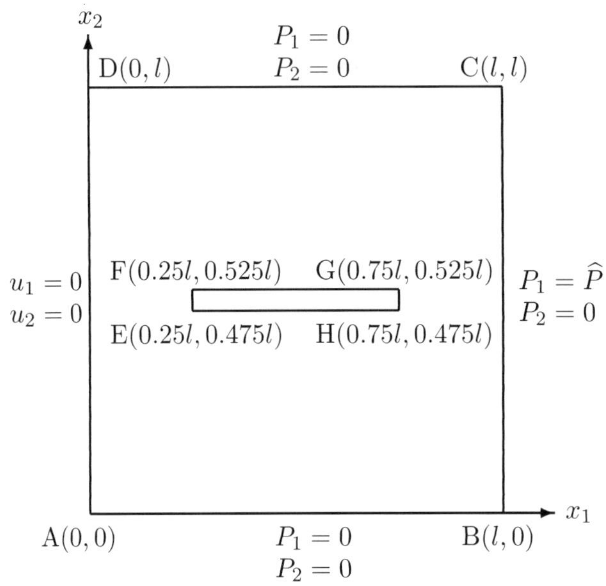

Figure 2. The geometry for Example 2

6.2. Example 2: Extension of a slab with a crack inclusion

Consider the boundary value problem for a material with a crack inclusion as shown in Figure 2 and with elastic coefficients given in (28) and (29). The boundary conditions (see Figure 2) are

$$
\begin{aligned}
&P_1/\widehat{P} = 0 \quad P_2/\widehat{P} = 0 \quad \text{on AB} \\
&P_1/\widehat{P} = 1 \quad P_2/\widehat{P} = 0 \quad \text{on BC} \\
&P_1/\widehat{P} = 0 \quad P_2/\widehat{P} = 0 \quad \text{on CD} \\
&u_1/\widehat{u} = 0 \quad u_2/\widehat{u} = 0 \quad \text{on AD} \\
&P_1/\widehat{P} = 0 \quad P_2/\widehat{P} = 0 \quad \text{on EFGH}
\end{aligned}
$$

The outer boundary is divided into 320 elements and the inner boundary (the crack) is divided into 84 elements.

Table 5 shows the BEM solution for the displacements at some points in the domain Ω. And Figures 3 and 4 show respectively the deformation along the outer boundary of the slab and along the inclusive crack boundary. The new coordinate system $OX_1'X_2'$ in Figures 3 and 4 is

Table 5. Displacement solutions of Example 2

Position	Homogeneous		Inhomogeneous	
(x'_1, x'_2)	$u_1/\widehat{u}$	$u_2/\widehat{u}$	$u_1/\widehat{u}$	$u_2/\widehat{u}$
(.9,.1)	.2764	.0507	.2771	.2022
(.9,.3)	.2780	.0279	.2229	.1880
(.9,.5)	.2853	.0061	.2477	.1753
(.9,.7)	.2803	-.0148	.2901	.1854
(.9,.9)	.2757	-.0374	.2540	.1741

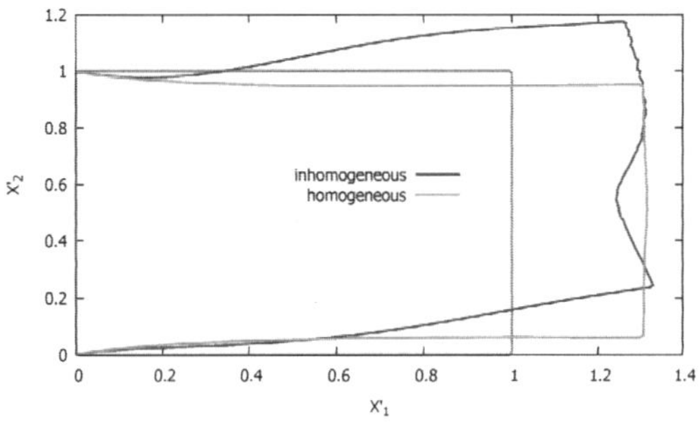

Figure 3. Deformation along the outer boundary of the slab of Example 2

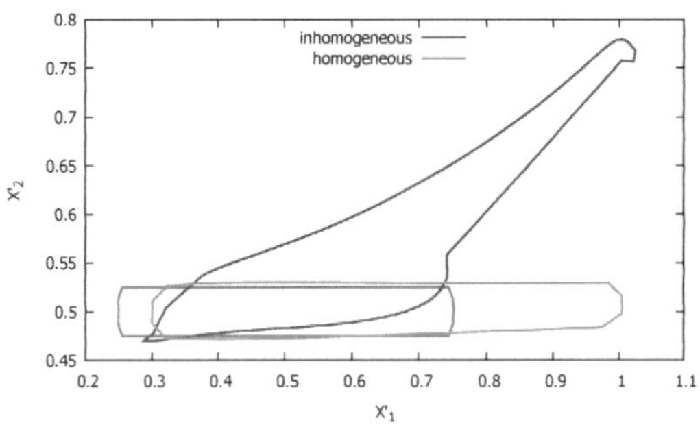

Figure 4. Deformation along the inclusive crack boundary $-.125 \leq x'_1 \leq .125$, $-.125 \leq x'_2 \leq .125$ inside the slab of Example 2

the system for deformed object, where the coordinate variables are defined by $X'_i = x'_i + u'_i$ for $i = 1, 2$.

In Table 5 and Figures 3 and 4 two cases, that is when the slab is assumed to be homogeneous (ie. $b(\mathbf{x}) = 1$) and inhomogeneous (ie. $b(\mathbf{x}) = (1+0.1x'_1)^2$), are considered. Table 5, Figure 3 and Figure 4 indicate the effect of the inhomogeneity function $b(\mathbf{x})$ especially on the displacement $u_2/\widehat{u}$ near the side BC.

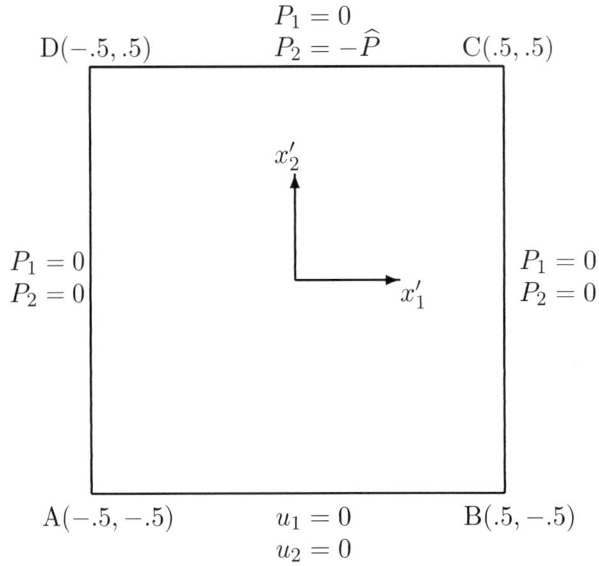

Figure 5. The geometry for Example 3

6.3. Example 3: Compression of a slab

Consider the boundary value problem as shown in Figure 5. The elastic coefficients of the material are given by

$$\alpha(\mathbf{x}) = 1.5\widehat{\alpha}(1 + rx'_1 + sx'_2)^2 \quad (30)$$
$$\beta(\mathbf{x}) = \widehat{\alpha}(1 + rx'_1 + sx'_2)^2 \quad (31)$$

and the boundary conditions are

$$\begin{array}{lll} u_1/\widehat{u} = 0 & u_2/\widehat{u} = 0 & \text{on AB} \\ P_1/\widehat{P} = 0 & P_2/\widehat{P} = 0 & \text{on BC} \\ P_1/\widehat{P} = 0 & P_2/\widehat{P} = -1 & \text{on CD} \\ P_1/\widehat{P} = 0 & P_2/\widehat{P} = 0 & \text{on AD} \end{array}$$

The elastic coefficients (30) and (31) take the form (18) and (19) with $b(\mathbf{x}) = (1 + rx'_1 + sx'_2)^2$, $\alpha^{(0)} = \beta^{(0)} = \widehat{\alpha}$, $\alpha^{(1)} = \widehat{\alpha}(1 + rx'_1 + sx'_2)^2$, $\beta^{(1)} = 0$ and $\epsilon = 0.5$. If $\widehat{\alpha} = 2.49 \times 10^3$ ksi then the material's elastic coefficients under consideration are the coefficients for a magnesium alloy.

Four cases concerning the elastic coefficients will be considered. The case of homogeneous materials is the case when $r = s = 0$. The other cases are for inhomogeneous cases which are when $r = 0, s = 0.1$; $r = 0.1, s = 0$; $r = 0.1, s = 0.1$.

Figure 6 shows the deformation along the outer boundary of the compressed slab and Figure 7 shows the deformation along the inner region $-0.25 \leq x'_1 \leq 0.25$, $-0.25 \leq x'_2 \leq 0.25$ inside the slab. Again, the figures indicate the effect of the inhomogeneity function $b(\mathbf{x})$.

6.4. Example 4: Compression of a slab with a hole inclusion

Consider the problem for a compression of an inhomogeneous slab of square shape with a hole inclusion. The coefficients are given by equations (30) and (31) and the boundary conditions

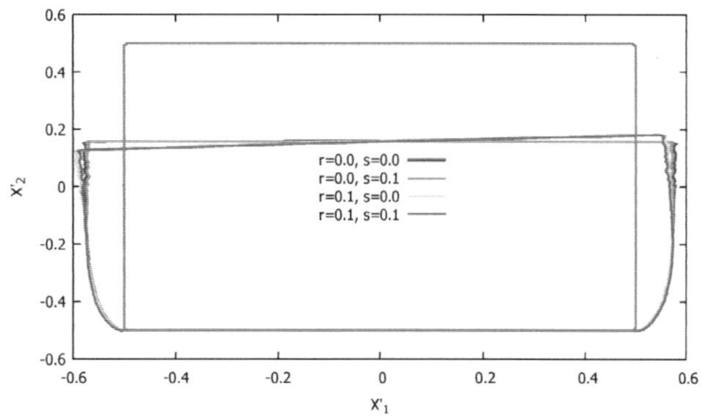

Figure 6. Deformation along the outer boundary of the compressed slab of Example 3

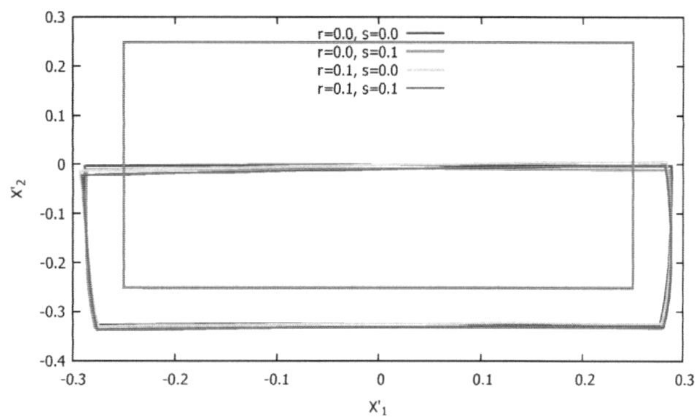

Figure 7. Deformation along the inner region $-.25 \leq x'_1 \leq .25$, $-.25 \leq x'_2 \leq .25$ inside the compressed slab of Example 3

are (see Figure 8)

$$\begin{aligned}
&u_1/\widehat{u} = 0 \quad u_2/\widehat{u} = 0 \quad \text{on AB} \\
&P_1/\widehat{P} = 0 \quad P_2/\widehat{P} = 0 \quad \text{on BC} \\
&P_1/\widehat{P} = 0 \quad P_2/\widehat{P} = -1 \quad \text{on CD} \\
&P_1/\widehat{P} = 0 \quad P_2/\widehat{P} = 0 \quad \text{on AD} \\
&P_1/\widehat{P} = 0 \quad P_2/\widehat{P} = 0 \quad \text{on EFGH}
\end{aligned}$$

Again, four cases concerning the material's elastic coefficients α and β will be considered. The values of the parameters a and b are taken to be same as in Section 6.3 for the four cases. Figure 9 shows the deformation along the outer boundary of the compressed slab and Figure 10 shows the deformation along the inclusive hole boundary occupying the region$-.125 \leq x'_1 \leq .125$, $-.125 \leq x'_2 \leq .125$. Once again, these two figures show the effect of the inhomogeneity function $b(\mathbf{x})$.

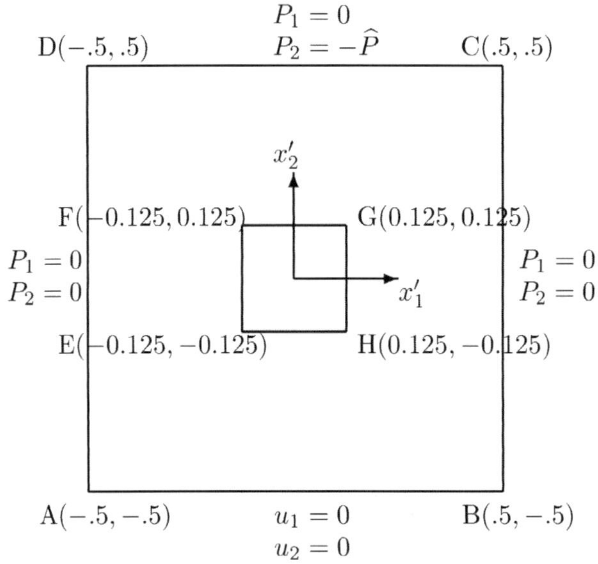

Figure 8. The geometry for Example 4

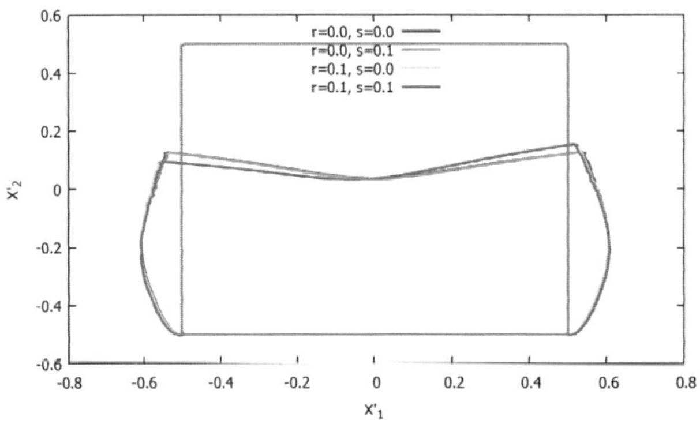

Figure 9. Deformation along the outer boundary of the compressed slab of Example 4

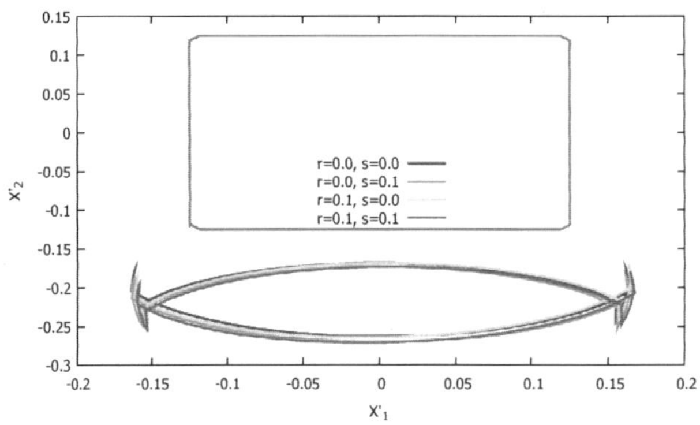

Figure 10. Deformation along the inclusive hole boundary $-.125 \leq x'_1 \leq .125$, $-.125 \leq x'_2 \leq .125$ inside the compressed slab of Example 4

7. Conclusion

The integral equations (17) and (25) were derived from this governing equation (3) and a BEM was then constructed for calculation of numerical solutions to the problems for isotropic functionally graded media specifically quadratically graded media. The results also show that the inhomogeneity of the loaded material will certainly yield an effect on its deformation. In general, homogeneous loaded materials will deform symmetrically, while inhomogeneous loaded materials will not.

Acknowledgements

The author would like to thank The Hasanuddin University and The Ministry of Research, Technology and Higher Education of the Republic of Indonesia for the provided support.

References

[1] Haddade A, Salam N, Khaeruddin and Azis M I 2017 A boundary element method for 2D diffusion-convection problems in anisotropic media *Far East Journal of Mathematical Sciences* **102**(8) 1593

[2] Azis M I, Kasbawati, Haddade A and Thamrin S A 2018 On some examples of pollutant transport problems solved numerically using the boundary element method *Journal of Physics: Conference Series* **979**(1)

[3] Azis M I, Asrul L, Khaeruddin and Paharuddin 2018 BEM solutions for unsteady transport problems in anisotropic media *JP Journal of Heat and Mass Transfer* **15**(4) 915

[4] Clements D L and Azis M I 2000 A Note on a Boundary Element Method for the Numerical Solution of Boundary Value Problems in Isotropic Inhomogeneous Elasticity *Journal of the Chinese Institute of Engineers* **23**(3) 261

[5] Azis M I and Clements D L 2008 Nonlinear transient heat conduction problems for a class of inhomogeneous anisotropic materials by BEM *Engineering Analysis with Boundary Elements* **32**(12) 1054

[6] Azis M I and Clements D L 2014 A Boundary Element Method for Transient Heat Conduction Problem of Nonhomogeneous Anisotropic Materials *Far East Journal of Mathematical Sciences* **89**(1) 51

[7] Salam N, Haddade A, Clements D L and Azis M I 2017 A boundary element method for a class of elliptic boundary value problems of functionally graded media *Engineering Analysis with Boundary Elements* **84**(3) 186

[8] Azis M I 2019 Numerical solutions to a class of scalar elliptic BVPs for anisotropic exponentially graded media *Journal of Physics: Conference Series* **1218** 012001

[9] Azis M I 2019 Standard-BEM solutions to two types of anisotropic-diffusion convection reaction equations with variable coefficients *Engineering Analysis with Boundary Elements* **105** 87

[10] Azis M I, Toaha S, Bahri M and Ilyas N 2018 A boundary element method with analytical integration for deformation of inhomogeneous elastic materials *Journal of Physics: Conference Series* **979**(1) 012072

[11] AAzis M I and Clements D L 2014 On some problems concerning deformations of functionally graded anisotropic elastic materials *Far East Journal of Mathematical Sciences* **87**(2) 173

[12] Manolis R P and Shaw R P 1996 Green's function for the vector wave equation in a mildly heterogeneous continuum *Wave Motion* **24** 59

[13] Abramowitz M and Stegun I A 1972 *Handbook of mathematical functions: with formulas, graphs and mathematical tables*, Dover Publications, Washington

Garbage Compaction Mechanism of Three-Wheels Electrical Motorcycle

Alfaroq[1] and Andi Cahya Adi Putera[1], Abdul Rahman[1], Abdul Kadir Muhammad*[1], Rafiuddin Syam[2]

[1] Center for Mechatronics and Control System, Mechatronics Enginering, State Polytechnic of Ujung Pandang

[2]Departerment of Mechanical Enginering, University of Hasanuddin

[1,2]Makassar, Indonesia

Email: paroimran17@gmai.com, cahya393@gmail.com, abdulrahman@poliupg.ac.id, kadir.muhammad@poliupg.ac.id, rafiuddinsyam@gmail.com

Abstract. This research aims to design a trash in a three-wheels electric motor with a compaction mechanism so that it can increase the volume of the trash than usual and the jack mechanism that reduces human power for waste management. This developed also includes a garbage cover with a cover mechanism that can block the smell of garbage when open or closed. The mechanism is controlled using a microcontroller and the actuator used a stepper motor. This tool has been tested using garbage and the equipment is running well.

1. Introduction

The increasing purchasing power of the various kinds of staples, research and technology support activities of a region's economic growth by increasing consumer dynamics that impacted negatively on the exhaust of waste products or so-called junk tend to show increased rates.

In the middle of current technological developments in human civilization, garbage management has reached the recycling utilization to change shape into electrical energy. As the case in many developed countries that manage the garbeage that can benefit return.

In 2016 by JaoeryJoenedhilAkib, Festo Andre Hardinsi&Akri, alumni of Polytechnic of Ujung Pandang in 2016 has made Prototype Motorcycle Electric Tricycle, and in 2017 by Rusliadi, Eddy KurniawanPasaribu and Harbin, alumni of Polytechnic of Ujung Pandang 2017 already develop Motorcycle Electric Tricycle Carrier Trash. Three Wheel Electric Motorcycles Waste Transporters has a specification that is,BLDC electric motors 350 [Watt] high torque, the battery is 48V. [1]

The final project is made aims to:

1. Develop draft trash on the electric motor tricycle.
2. Designing and building a trash compacting mechanism in the tub electric three-wheeled motorcycle with the addition of the volume of waste by 25%.
3. Designing and building systems on garbage jack electric motor tricycle.

2. Garbage Compaction

Garbage is unwanted residual material after the end of a process. Trash is defined by humans according to the degree of wear ability, in natural processes is not actually a trash concept, only products produced after and during the natural process takes place. However, because the human defined the concept of environment then waste can be divided according to the types.According to the definition of the World Health Organization (WHO) trash is something that is not being used, not used, not liked or something dumped derived from human activities and does not happen by itself [2][11].

Waste management is all the activities carried out in dealing with garbage since brought to the final disposal. Broadly speaking, the activities in waste management includes control of waste generation, waste collection, transfer and transport, treatment and disposal.

2.1. Components

*2.1.1. Microcontroller.*microcontroller family name which was originally created by the company Smart Projects. One of the figures is the creator Massimo Banzi. These boards are hardware devices that are open source system[3].

Microcontroller we used is a microcontroller board based ATmega328. This microcontroller has a lot of input / output pins between pin which can be used as PWM output, analog input, 16 MHz crystal oscillator, a USB connection, a power jack, ICSP head and the reset button. Microcontroller is able to support the microcontroller, can be connected to a computer using a USB cable. Microcontroller is an electronic circuit that is open source, and has hardware and software that is easy to use. Microcontroller can recognize the surrounding environment through various types of sensors and can control lights, motors, and various other types of actuators. Microcontroller has many types, including Uno, Mega 2560, Fio and others[4].

Figure 1. Board Microcontroller

2.1.2. Step Down. Step down is a component which acts as a regulator in the electrical circuit. In the present state of this module is used to regulate the voltage into the circuit and is supplied to the controller component that is Microcontroller.

Figure 2. Step Down Module lm2596

2.1.3. Stepper Motors. Stepper motors are one type of dc motor controlled by digital pulses[5]. stepper motor is a digital actuator whose input is in the form of programmed energization of the stator windings and whose output is in the form of discrete angular rotation [6].

Figure 3. Stepper Motor

2.1.4. Batteries. Batteries are energy storage medium that is capable of converting chemical energy into electrical energy through an electrochemical oxidation-reduction reaction or so-called redox reactions. Unlike heat engines (in motorcycles or engine coolant) which has a low efficiency limitation as the second law of thermodynamics consequence, the battery has a higher efficiency because the mechanism of energy conversion takes place electrochemically [7]. Lithium Ion Batteries able to provide greater power, can be made with small size and light weight[8].

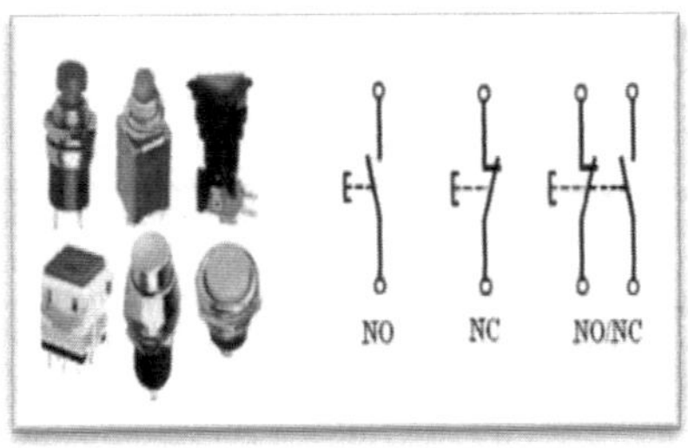

Figure 4. Symbol Push Button

2.1.5. Button. Button is to control the ON or OFF condition of an electric circuit, especially in the control. The working principle of working momentary pushbutton is meant when a button on tap for a moment then the contact of the button will return to its original position [9].

2.1.6. Lead Screw. Leadscrew is a mechanical component that serves as the successor to the rotational force of the motor into a translational force to push the burden of the plate in a compacting mechanism or mechanisms such as levers in the trash. Lead screw are used in various motion delivery systems where power is transmitted by converting rotary to linear motion[10].

IOP Conf. Series: Materials Science and Engineering **619** (2019) 012019 doi:10.1088/1757-899X/619/1/012019

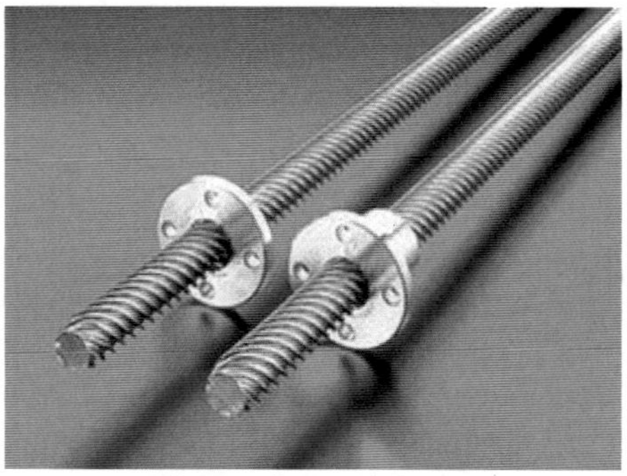

Figure 5. Lead Screw

3. Methodology

3.1. Mechanical design

The basic point that many changes or development is the mechanism of compaction itself contained in the basin and consists of two mechanisms to be split into two parts, the mechanism of the jack which exists between the tub with the framework of the motorcycle, and the cover mechanism which is at the top of the tub with an additional some parts to support additional forms on the cover mechanism.

3.1.1. Compaction mechanism. In this mechanism, the use of stepper motors and transmission systems leadscrew or screw shaft we use to support the movement of plate compactor that will encourage and solidify the waste that goes into the tub. The materials that we use into a compactor plate is acrylic and buffer the motor and the leadscrew mechanism is incorporated strip plate welding method.

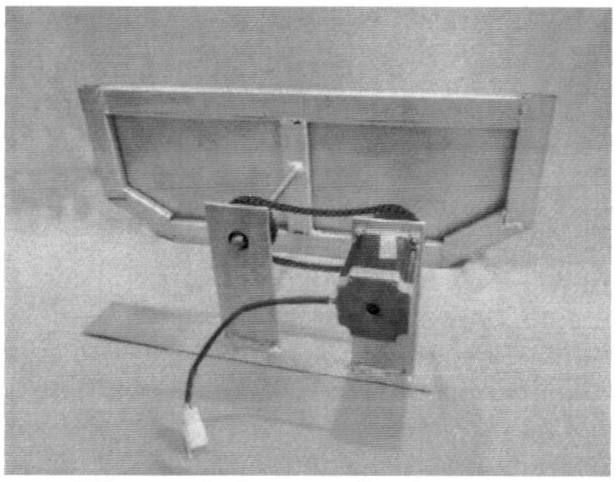

Figure 6. Compaction Mechanism

3.1.2. Jack Mechanism.
Supporting one of the trashes that we have developed is the existence of a mechanism whereby the jack can maximize processing waste that has meet the bath and then be discarded.

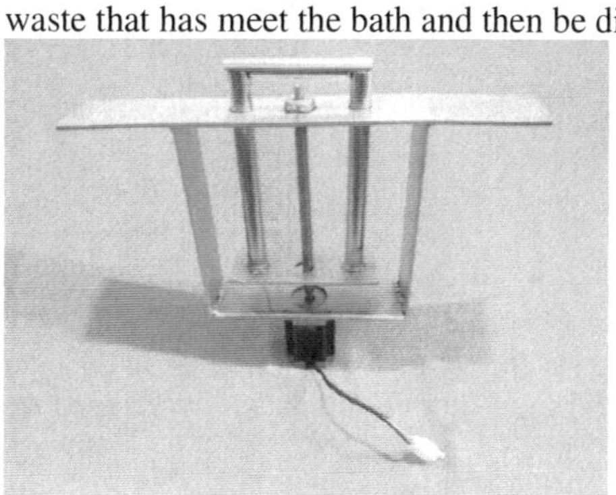

Figure 7. Jack Mechanism

As with the compactor mechanism, in this mechanism also uses a stepper motor actuators and leadscrew transmission system in order to support the process of appointing the trash to the disposal process.

3.1.3. Closing Mechanism.
At the closing mechanism in the series this thesis intends to eliminate odors that might come out of the tub when the tub is opened to insert the trash. As we know that the problems that often arise in waste transportation that smell out of the landfill due to release of the smell that came through the door of the shelter that does not have a mechanism to prevent the release of doors from the tub.

Therefore, we make the closing mechanism bins that are likely able to block the smell out of the bath when done transporting waste. The method we use is by using two half-cylinder composite tub as initial shelter before it is put into a tub. Then the next half-cylindrical containers that are smaller will be closed first so that the position is still in a state tub covered by a half-cylinder container larger. And in the end after the first container is closed, then the container is then sealed as well as its original state so that trash will never be in the open state and the smell would not be exposed out.

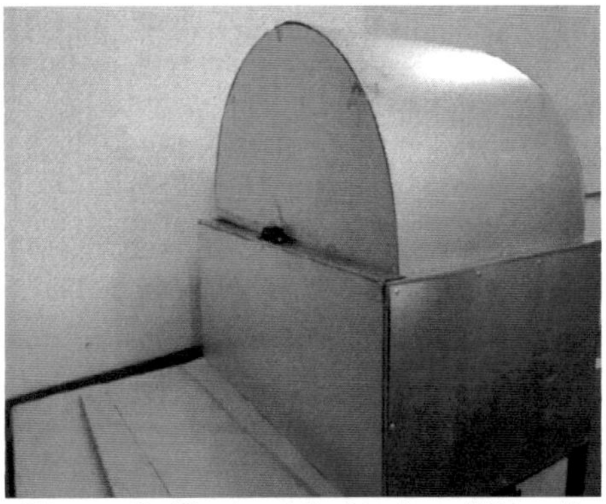

Figure 8. Closing Mechanism (Closed)

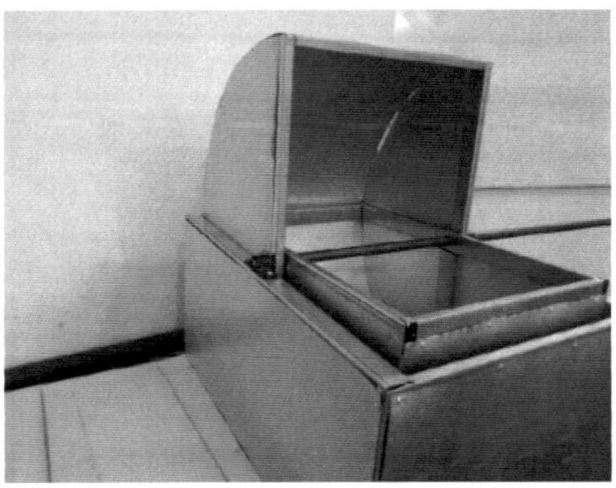

Figure 9. Closing Mechanism (Open)

3.2. Electronic Scheme
In this final project using the panel box / box control that serves as a control center and other electrical circuits. With the assistance from the voltage source 12 V battery and connecting the small wires, circuit consists of an Arduino Mega, motor drivers, bride board, and step down are packed in a box of acrylic.

IOP Conf. Series: Materials Science and Engineering **619** (2019) 012019 doi:10.1088/1757-899X/619/1/012019

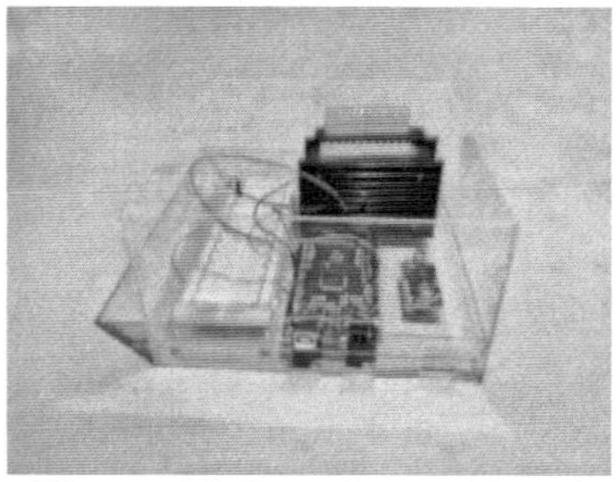

Figure 10. Box Control Panel

3.3. Control Scheme

3.3.1. Block Diagram. Based on the depiction on the block diagram below is known through the groove feedback to control system to run the mechanism in an automatic way[11][12]. So just by pressing the push button mechanism operates during all cycles.

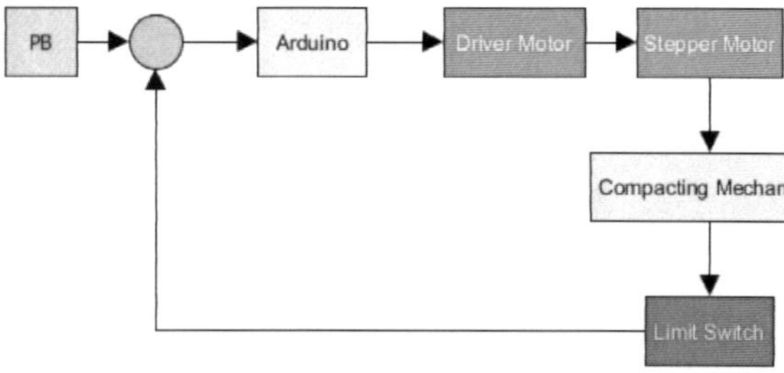

Figure 11. Block Diagram Control

3.3.2. Algorithm. The control scheme on this tool uses Arduino software that is inherited from the Arduino hardware as a whole with the provisions of the device used is the Arduino Mega. Used in its use logic function "if", "do", and "while" for the selection decision to execute the program. The completeness of the other program uses a simple formula language C.

Figure 12. Program on software

3.4. Experiments Setup

In collecting data on the mechanism of this final project is done by inserting a container of waste in starting a closure mechanism, and garbage will be entered into concurrently with closing the tub retraced to the early part of the cover. Furthermore trash into the tub and the plate will push the waste

IOP Conf. Series: Materials Science and Engineering **619** (2019) 012019 doi:10.1088/1757-899X/619/1/012019

towards the back of the tub. And so it goes waste from many and indirectly the compactor mechanism will solidify the waste that goes into the tub.

4. Result and Discussion

4.1. Method of Manufacture

Starting from the improvements that will be implemented tub compacting mechanism therein with another plate is coated on the inside and a little hollow as the framework for the second floor. Compactor mechanism is the next stage, which is by stringing together several plate strips made by the method of welding as a buffer from the motor and leadscrew by the two systems as needed in a bath which is divided into two floors. In this same mechanism we use a chain and gear system to transfer power from the motor to the leadscrew. Liaison between the motor shaft to the gear used clutch made using the method of turning. Continued jack mechanism that the mechanical system consisting of a strip plate frame, stepper motors as actuators driving mechanism,

4.2. Efficiency place on the vessel

In the data collection has been done, it is known that basically the trash compaction mechanism has a significant role in the efficiency of space, whether it is at the bottom of the tub and the top of the trash. Can be seen in the table below of the results of data collection that we do show the numbers prove that in the tub where maximum efficiency when using a compaction mechanism.

Table 1. Comparison of Volume With and Without junkie

No.	Body parts	Without compactors (cm3)	Using compactors (cm3)
1.	Floor 1 (Bottom)	51 840	64 800
2.	Level 2 (Top)	34 560	41 480

Based on the above calculation of volume expansion bathtub, obtained information that the additional volume of garbage if using a compacting mechanism that is 25% lower and 20% at the top, and if the calculated overall efficiency of trash that some 22.5%.

4.3. Voltage Graphic

Based on the results of measurements of the amount of usage of each trial voltage operation of the compacting mechanism, it is known that the use of voltage is not too large due to the need for motor actuator which is also not too great to see the difference between the initial voltage and a final voltage generated from the data that has been taken.

Table 2. Consumption Tension Experiment and movement of compactors

Movement	Initial Voltage (V_o)		Final Voltage (V_t)
1	24.06		24.01
2	24.01		23.94
3	23.94		23.88
4	23.88		23.82
5	23.81		23.75

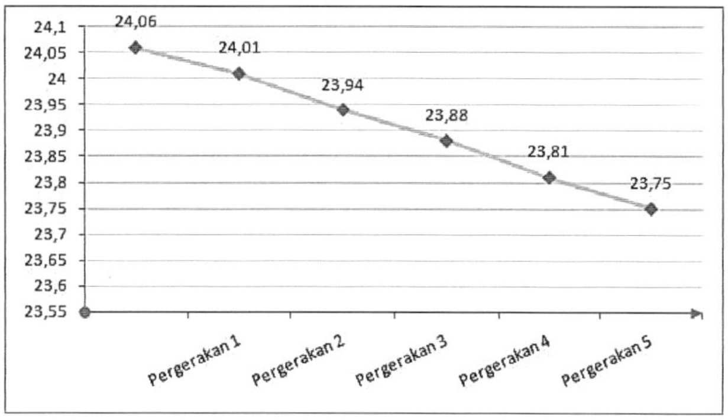

Figure 13. Voltage Graphic

5. Conclusion

Based on the results and discussion on this research can be concluded as follows:

1) Garbage bins designed development on three-wheeled electric motorcycle successfully made using economical materials and in accordance with the initial planning research manufacture.
2) Making the compaction mechanism has been created and prove additional volume of waste can be increased by 25%, as it is attached to the results.
3) Making the jack mechanism successfully fabricated using stepper motors as actuators and leadscrew.

References

[1] Rusliadi, Eddy Kurniawan Pasaribu, and Harbin. 2017. Development of Electric Motorcycles Tricycle Carrier Trash. Makassar. Department of Mechanical Engineering Polytechnic of Ujung Pandang.

[2] Basahona, Ato. 2016. Definition of Waste According to experts, (Online) (http://www.atobasahona.com/2016/10/ pengertian-sampah-menurut-para-ahli.html), Accessed February 6, 2018.

[3] Kadir, Abdul. 2015. From Zero to A Pro Arduino. Yogyakarta; CV. Andi Offset.

[4] Arduino Home Page. 2012. Arduino Mega, (Online) (https://www.arduino.cc/en/Main/arduinoBoardMega), Accessed February 7, 2018.

[5] Faisal Ahmad. Stepper Motor 2016, (Online) (https://gudangfaisal.blogspot.co.id/2016/06/motorstepper.html?m=1), Retrieved January 18, 2018.

[6] Athani, V.V. 2005. *Stepper Motor : Fundamentals, Aplications And Design*. New Delhi : New Age International (P) Ltd.

[7] Wiryawan. Lithium Ion Battery 2015, (Online) (http: //wiryawangpblog.blogspot.com / 2015/04 / battery-lithium.html), Retrieved on February 17, 2018.

[8] Enterprise, Jubilee. 2010. *Teknik Menghemat Baterai*. Jakarta : PT. Gramedia.

[9] Azizan, Akhdan. 2014 Push Button (Original in Bahasa: Tombol Tekan), (Online) (https://akhdanaziza.com/tombol-tekan-push-button/), Retrieved on February 6, 2018.

[10] Vahid-Araghi, Orang. 2011. *Friction-Induced Vibration in Lead Screw Drives.* Springer : New York.

[11] R Syam, K Watanabe, K Izumi, 2007,An Adaptive Actor-Critic Algorithm With Multi-Step Simulated Experiences For Controlling Nonholonomic Mobile Robots, Soft Computing 11 (1), 81-89, Springer Verlag Germany.

[12] R Syam, K Watanabe, K Izumi , 2005, A study on constructing a neuro-interface using the concept of a virtual master–slave system Artificial Life and Robotics,Vol. 9, 1, pp 51-57

Control Design for Preventing Dengue Infections using Input-Output Linearization Method

Kasbawati*, Firman, Waode Ainun Apriyani Amrin

Department of Mathematics, Hasanuddin University
Jl. Perintis Kemerdekaan Km. 10 Tamalanrea Makassar, Indonesia

*Corresponding author: kasbawati@gmail.com

Abstract. This study deals with a theoretical study of control design to prevent transmission of dengue fever viruses. The input-output linearization method is a feedback control method that can be applied to design the control functions so that the transmission of dengue viruses can be disrupted. Fogging is considered as an eradication effort and vaccination is considered as a prevention effort. Both treatments are considered as the input controls of the method while infected human variable is considered as the output control. By treating eradication effort and prevention effort as the control inputs we design controls that can stabilize the system and lead the infected population going to zero. Some numerical schemes are presented to confirm the theoretical results and the analytical results.

1. Introduction

Infectious diseases such as dengue fever are mainly found in tropical and subtropical regions such as Indonesia. From 1968 to 2009, the World Health Organization (WHO) records that Indonesia is a country with the highest cases of dengue fever in Southeast Asia where the first case was found in Surabaya at 1968 [1]. Since the disease can cause organ damage even death, early treatment should be arranged to reduce the mortality rate caused by this disease.

Mathematical modelling is one of the powerful tools to study dengue fever problem especially to analyse different strategies that might be useful in controlling the disease. This theoretical approach can help us to understand the transmission dynamics of the viruses that can involve human and mosquito population. Various mathematical modelling and mathematical theory have been formulated and developed to study the transmission of dengue viruses in human and mosquito populations. For instance Nuraini et al., 2009 [2] and Chávez et al., 2017 [3] that examined mathematical model of dengue to identify main factors that influence the spread of the disease; Diekmann et al. [4] and Driessche & Watmough [5] that developed mathematical tools called the next generation matrix and basic reproduction number to quantify the multiplication factor of the primary infectious cases to be the secondary infectious cases. Some researchers also combined mathematical model with optimal control theory to study effects of controls such as vaccination, space spraying, and treatment cycle in reducing infectious both in human and mosquito populations [6-10]. In this research we also study the transmission of dengue viruses by combining mathematical model of the disease with the input-output linearization method to observe the dynamics of the system before and after control. This study will give another point of view regarding control study of the transmission of dengue viruses such that strategies can be identified to overcome the outbreak of disease.

This paper is organized as follow. In Section 1, we present an introduction about dengue fever and some mathematical models about this disease. In Section 2 we present derivation of linear control using input-output linearization method and in the last section we present some numerical simulations and discussion about the simulation results.

2. Control Design Using Input-Output Linearization Method

Before deriving the input control, firstly let us consider the formulation of the mathematical model of dengue disease. Formulating the mathematical model of the spread of dengue infections involves the interactions of two population, i.e. human population and mosquitos' population. The two populations have an important role in the spread of dengue viruses. In the previous study, Ningsih (2017) [10] has formulated the mathematical model for dengue by considering an integrated vector control for reducing the transmission of dengue. It assumed that the integrated vector control for reducing the transmission viruses is a combination of vaccination for susceptible human and fogging for mosquito population. By using the similar model, we have the following dimensionless system [10],

$$\dot{\mathbf{x}}(t)=\begin{bmatrix} \mu_h-\left(B\beta_{mh}\dfrac{N_m}{N_h}x_6(1-u_1)+\mu_h\right)x_1+\theta x_3 \\ B\beta_{mh}\dfrac{N_m}{N_h}x_1x_6(1-u_1)-(\eta_h+\mu_h)x_2 \\ \eta_h x_2-(\theta+\mu_h)x_3 \\ \varphi\left(1-\dfrac{N_m}{kN_h}x_4\right)(x_5+x_6)-(\eta_A+\mu_A)x_4 \\ \eta_A x_4-(B\beta_{hm}x_2+\mu_m+u_2)x_5 \\ B\beta_{hm}x_2x_5-(\mu_m+u_2)x_6 \end{bmatrix}, \tag{1}$$

Table 1. Variables and parameters of the model (1).

Parameter	Description	Value	Ref.
$1/\mu_h$	average lifespan of humans (days)	$71*365$	[6]
N_h	total of human population	48000	Assumed
N_m	total of mosquito population	500000	Assumed
B	average number of bites on humans by mosquitoes, per day	0.8	[6]
β_{mh}	transmission probability from infected human (per bite)	0.35	[6]
β_{hm}	transmission probability from infected mosquito (per bite)	0.375	[6]
θ	decreasing rate of human immunity per day	0.05	Assumed
$1/\eta_h$	mean of viremia period (in days)	3	[6]
η_A	maturation rate from larvae to adult (per day)	0.08	[6]
φ	number of eggs at each deposit per capita (per day)	10	Assumed
k	number of dismissed larvae per human	6	Assumed
$1/\mu_A$	natural mortality of larvae (in day)	4	[6]
$1/\mu_m$	average lifespan of adult mosquitoes (in days)	10	[6]

where $\boldsymbol{x_1(t)}$ represents the proportion of susceptible humans at time t, $\boldsymbol{x_2(t)}$ is the proportion of infected humans with dengue virus at time t, $\boldsymbol{x_3(t)}$ is the proportion of humans recovered from dengue hemorrhagic fever at time t, $\boldsymbol{x_4(t)}$ is the proportion of mosquitoes in the water phase including eggs, larvae, and pupae, $x_5(t)$ is the proportion of susceptible mosquitoes that do not carry dengue virus, and $x_6(t)$ is the proportion of dengue virus-infected mosquitoes that can spread the virus to human. Definitions of all variables and parameters of the model (1) are given in Table 1.

Next, we will derive the input control for vaccination and fogging by defining the infected population as the output control, $y = x_2$.

2.1. Control design for effect of vaccination strategy (u_1)

For designing the vaccination control, we assume that $u_2 = 0$. By using that assumption we can rewrite system (1) becomes

$$\dot{\mathbf{x}} = \mathbf{f}(\mathbf{x}) + \mathbf{g}(\mathbf{x})u_1,$$

with

$$\mathbf{f}(\mathbf{x}) = \begin{bmatrix} \mu_h - B\beta_{mh}\dfrac{N_m}{N_h}x_1x_6 - \mu_h x_1 + \theta x_3 \\ B\beta_{mh}\dfrac{N_m}{N_h}x_1x_6 - (\eta_h + \mu_h)x_2 \\ \eta_h x_2 - (\theta + \mu_h)x_3 \\ \varphi\left(1 - \dfrac{N_m}{kN_h}x_4\right)(x_5 + x_6) - (\eta_A + \mu_A)x_4 \\ \eta_A x_4 - (B\beta_{hm}x_2 + \mu_m)x_5 \\ B\beta_{hm}x_2x_5 - \mu_m x_6 \end{bmatrix}, \quad \mathbf{g}(\mathbf{x}) = \begin{bmatrix} B\beta_{mh}\dfrac{N_m}{N_h}x_1x_6 \\ -B\beta_{mh}\dfrac{N_m}{N_h}x_1x_6 \\ 0 \\ 0 \\ 0 \\ 0 \end{bmatrix}.$$

For $i = 1$, we have the Lie derivative (see [11-12] for detail about Lie derivative) as follows,

$$L_g L_f^{1-1}h(\mathbf{x}) = L_g L_f^{1-1}h(\mathbf{x}) = L_g h(\mathbf{x}),$$

with

$$L_g h(\mathbf{x}) = \frac{\partial h}{\partial \mathbf{x}} \cdot \mathbf{g}(\mathbf{x}) = \frac{\partial x_2}{\partial \mathbf{x}} . \mathbf{g}(\mathbf{x}) = \begin{bmatrix} 0 & 1 & 0 & 0 & 0 & 0 \end{bmatrix} \cdot \begin{bmatrix} B\beta_{mh}\dfrac{N_m}{N_h}x_1x_6 \\ -B\beta_{mh}\dfrac{N_m}{N_h}x_1x_6 \\ 0 \\ 0 \\ 0 \\ 0 \end{bmatrix}$$

$$= -B\beta_{mh}\frac{N_m}{N_h}x_1x_6 .$$

Since $L_g h(\mathbf{x}) \neq 0$ then we have the relative degree for system (1) with output control $y = x_2$ is equal to 1. Therefore we have the transformation system as follow

$$T(x) = \begin{bmatrix} z_1 \\ \phi_1 \\ \phi_2 \\ \phi_3 \\ \phi_4 \\ \phi_5 \end{bmatrix}, \tag{2}$$

with $z_1 = x_2$ and $\phi_1,\ldots,\phi_5$ that fulfils

$$\frac{\partial \phi_i}{\partial \mathbf{x}} \cdot \mathbf{g}(\mathbf{x}) = 0, \quad i = 1,\ldots,5. \tag{3}$$

Suppose we chose functions $\phi_1 = x_3, \phi_2 = x_4, \phi_3 = x_5, \phi_4 = x_6$, and $\phi_5 = x_1 + x_2$. It is easy to show that the chosen functions fulfil equation (3) such that we get the transformation system,

$$T(x) = \begin{bmatrix} z_1 \\ \phi_1 \\ \phi_2 \\ \phi_3 \\ \phi_4 \\ \phi_5 \end{bmatrix} = \begin{bmatrix} x_2 \\ x_2 \\ x_4 \\ x_5 \\ x_6 \\ x_1 + x_2 \end{bmatrix} \tag{4}$$

From equation (4) we get

$$\begin{aligned} \dot{z}_1 &= v_1, \\ \dot{\phi}_1 &= \dot{x}_3, \\ \dot{\phi}_2 &= \dot{x}_4, \\ \dot{\phi}_3 &= \dot{x}_5, \\ \dot{\phi}_4 &= \dot{x}_6, \\ \dot{\phi}_5 &= \dot{x}_1 + \dot{x}_2, \end{aligned} \tag{5}$$

where v_1 is the new input, $v_1 = -c_0 z_1$ with positive parameter c_0. If the system (1) is substituted to equation (5) then we get

$$\begin{aligned} \dot{z}_1 &= v_1 \\ \dot{\phi}_1 &= \eta_h x_2 - (\theta + \mu_h) x_3, \\ \dot{\phi}_2 &= \varphi\left(1 - \frac{N_m}{kN_h} x_4\right)(x_5 + x_6) - (\eta_A + \mu_A) x_4, \\ \dot{\phi}_3 &= \eta_A x_4 - (B\beta_{hm} x_2 + \mu_m) x_5, \\ \dot{\phi}_4 &= B\beta_{hm} x_2 x_5 - \mu_m x_6, \\ \dot{\phi}_5 &= \mu_h - \mu_h x_1 + \theta x_3 - (\eta_h + \mu_h) x_2. \end{aligned} \tag{6}$$

If we rewrite system (6) using the transformation variables in (2) then we have the normal system of (1) with output control $y = x_2$ as follow

$$\begin{aligned} \dot{z}_1 &= v_1, \\ \dot{\phi}_1 &= \eta_h z_1 - (\theta + \mu_h)\phi_1, \\ \dot{\phi}_2 &= \varphi\left(1 - \frac{N_m}{kN_h}\phi_2\right)(\phi_3 + \phi_4) - (\eta_A + \mu_A)\phi_2, \\ \dot{\phi}_3 &= \eta_A \phi_2 - (B\beta_{hm} z_1 + \mu_m)\phi_3, \\ \dot{\phi}_4 &= B\beta_{hm} z_1 \phi_3 - \mu_m \phi_4, \\ \dot{\phi}_5 &= \mu_h - \mu_h(\phi_5 - z_1) + \theta\phi_1 - (\eta_h + \mu_h) z_1, \\ y &= h(x) = z_1. \end{aligned} \tag{7}$$

If $z_1 = 0$ then we get the zero dynamics as follow

$$\begin{aligned}
\dot{\phi}_1 &= -(\theta + \mu_h)\phi_1, \\
\dot{\phi}_2 &= \varphi\left(1 - \frac{N_m}{kN_h}\phi_2\right)(\phi_3 + \phi_4) - (\eta_A + \mu_A)\phi_2, \\
\dot{\phi}_3 &= \eta_A\phi_2 - \mu_m\phi_3, \\
\dot{\phi}_4 &= -\mu_m\phi_4, \\
\dot{\phi}_5 &= \mu_h - \mu_h\phi_5 + \theta\phi_1.
\end{aligned} \tag{8}$$

By using linearization method, it is easy to show that the zero dynamic is a stable system. Furthermore, we can formulate the control function for u_1 using the following term [11],

$$u_1 = \frac{1}{L_g h(\mathbf{x})} \cdot \left[-L_f h(\mathbf{x}) + v_1\right].$$

Consider

$$L_g h(\mathbf{x}) = \frac{\partial h}{\partial \mathbf{x}} \cdot \mathbf{g}(\mathbf{x}) = \frac{\partial x_2}{\partial \mathbf{x}} \cdot \mathbf{g}(\mathbf{x}) = \begin{bmatrix} 0 & 1 & 0 & 0 & 0 & 0 \end{bmatrix} \cdot \begin{bmatrix} B\beta_{mh}\dfrac{N_m}{N_h}x_1x_6 \\ -B\beta_{mh}\dfrac{N_m}{N_h}x_1x_6 \\ 0 \\ 0 \\ 0 \\ 0 \end{bmatrix} = -B\beta_{mh}\frac{N_m}{N_h}x_1x_6,$$

$$L_f h(\mathbf{x}) = \frac{\partial h}{\partial \mathbf{x}} \cdot \mathbf{f}(\mathbf{x}) = \frac{\partial x_2}{\partial \mathbf{x}} \cdot \mathbf{f}(\mathbf{x}) = \begin{bmatrix} 0 & 1 & 0 & 0 & 0 & 0 \end{bmatrix} \cdot \begin{bmatrix} \mu_h - B\beta_{mh}\dfrac{N_m}{N_h}x_1x_6 - \mu_h x_1 + \theta x_3 \\ B\beta_{mh}\dfrac{N_m}{N_h}x_1x_6 - (\eta_h + \mu_h)x_2 \\ \eta_h x_2 - (\theta + \mu_h)x_3 \\ \varphi\left(1 - \dfrac{N_m}{kN_h}x_4\right)(x_5 + x_6) - (\eta_A + \mu_A)x_4 \\ \eta_A x_4 - (B\beta_{hm}x_2 + \mu_m)x_5 \\ B\beta_{hm}x_2x_5 - \mu_m x_6 \end{bmatrix}$$

$$= B\beta_{mh}\frac{N_m}{N_h}x_1x_6 - (\eta_h + \mu_h)x_2.$$

Therefore we get the control function for u_1,

$$u_1 = \frac{B\beta_{mh}\dfrac{N_m}{N_h}x_1x_6 - (\eta_h + \mu_h)x_2 + c_0 z_1}{B\beta_{mh}\dfrac{N_m}{N_h}x_1x_6}. \tag{9}$$

2.2. *Control design for effect of fogging strategy* (u_2)

By assuming $u_1 = 0$ and using the similar way as before, we get the relative degree as follow. For i=1, we get

IOP Conf. Series: Materials Science and Engineering **619** (2019) 012020 doi:10.1088/1757-899X/619/1/012020

$$L_g L_f^{1-1} h(\mathbf{x}) = L_g h(\mathbf{x}),$$

with

$$L_g h(\mathbf{x}) = \frac{\partial h}{\partial \mathbf{x}} \cdot \mathbf{g}(\mathbf{x}) = \frac{\partial x_2}{\partial \mathbf{x}} . \mathbf{g}(\mathbf{x}) = \begin{bmatrix} 0 & 1 & 0 & 0 & 0 & 0 \end{bmatrix} \cdot \begin{bmatrix} 0 \\ 0 \\ 0 \\ 0 \\ -x_5 \\ -x_6 \end{bmatrix} = 0.$$

Since we have $L_g h(\mathbf{x}) = 0$, it means that we have to continue our quantification. For $i = 2$

$$L_g L_f^{2-1} h(\mathbf{x}) = L_g L_f h(\mathbf{x}),$$

with

$$L_f h(\mathbf{x}) = \frac{\partial h}{\partial \mathbf{x}} \cdot \mathbf{f}(\mathbf{x}) = \frac{\partial x_2}{\partial \mathbf{x}} \cdot \mathbf{f}(\mathbf{x}) = \begin{bmatrix} 0 & 1 & 0 & 0 & 0 & 0 \end{bmatrix} \cdot \begin{bmatrix} \mu_h - B\beta_{mh} \frac{N_m}{N_h} x_1 x_6 - \mu_h x_1 + \theta x_3 \\ B\beta_{mh} \frac{N_m}{N_h} x_1 x_6 - (\eta_h + \mu_h) x_2 \\ \eta_h x_2 - (\theta + \mu_h) x_3 \\ \varphi \left(1 - \frac{N_m}{kN_h} x_4 \right) (x_5 + x_6) - (\eta_A + \mu_A) x_4 \\ \eta_A x_4 - (B\beta_{hm} x_2 + \mu_m) x_5 \\ B\beta_{hm} x_2 x_5 - \mu_m x_6 \end{bmatrix}$$

$$= B\beta_{mh} \frac{N_m}{N_h} x_1 x_6 - (\eta_h + \mu_h) x_2 .$$

So that we get

$$L_g L_f h(\mathbf{x}) = \frac{\partial L_f h(\mathbf{x})}{\partial \mathbf{x}} \cdot \mathbf{g}(\mathbf{x}) = \frac{\partial \left(B\beta_{mh} \frac{N_m}{N_h} x_1 x_6 - (\eta_h + \mu_h) x_2 \right)}{\partial \mathbf{x}} \cdot \mathbf{g}(\mathbf{x})$$

$$= \begin{bmatrix} B\beta_{mh} \frac{N_m}{N_h} x_6 & -(\eta_h + \mu_h) & 0 & 0 & 0 & B\beta_{mh} \frac{N_m}{N_h} x_1 \end{bmatrix} \cdot \begin{bmatrix} 0 \\ 0 \\ 0 \\ 0 \\ -x_5 \\ -x_6 \end{bmatrix} = -B\beta_{mh} \frac{N_m}{N_h} x_1 x_6 .$$

Since $L_g L_f h(\mathbf{x}) \neq 0$ then we get relative degree with output control $y = x_2$ is equal to 2. Because of the relative degree of the system is equal to 2 then we have the transformation system

$$T(x) = \begin{bmatrix} z_1 \\ z_2 \\ \phi_1 \\ \phi_2 \\ \phi_3 \\ \phi_4 \end{bmatrix},$$

where $\phi_1, \dots, \phi_4$ satisfies equation (3). Suppose we chose $\phi_1 = x_1, \phi_2 = x_3, \phi_3 = x_4, \phi_4 = x_1 + x_2$. It is easy to show that the chosen functions fulfil equation (3) such that we get the transformation system,

$$T(x) = \begin{bmatrix} z_1 \\ z_2 \\ \phi_1 \\ \phi_2 \\ \phi_3 \\ \phi_4 \end{bmatrix} = \begin{bmatrix} x_2 \\ B\beta_{mh}\dfrac{N_m}{N_h}x_1x_6 - (\eta_h + \mu_h)x_2 \\ x_1 \\ x_3 \\ x_4 \\ x_1 + x_2 \end{bmatrix}. \tag{10}$$

Using equation (10) we have

$$\begin{aligned} \dot{z}_1 &= z_2, \\ \dot{z}_2 &= v_2, \\ \dot{\phi}_1 &= \dot{x}_1, \\ \dot{\phi}_2 &= \dot{x}_3, \\ \dot{\phi}_3 &= \dot{x}_4, \\ \dot{\phi}_4 &= \dot{x}_1 + \dot{x}_2. \end{aligned} \tag{11}$$

where $v_2 = -c_0z_1 - c_1z_2$ is the new input control with positive parameters c_0 and c_1. If we substitute equation (1) into equation (11) then we get

$$\begin{aligned} \dot{z}_1 &= z_2, \\ \dot{z}_2 &= v_2, \\ \dot{\phi}_1 &= \mu_h - \left(B\beta_{mh}\frac{N_m}{N_h}x_6 + \mu_h\right)x_1 + \theta x_3, \\ \dot{\phi}_2 &= \eta_h x_2 - (\theta + \mu_h)x_3, \\ \dot{\phi}_3 &= \varphi\left(1 - \frac{N_m}{kN_h}x_4\right)(x_5 + x_6) - (\eta_A + \mu_A)x_4, \\ \dot{\phi}_4 &= \mu_h - \mu_h x_1 + \theta x_3 - (\eta_h + \mu_h)x_2. \end{aligned} \tag{12}$$

If we rewrite system (12) using the transformation variables in (10) then we have the normal system of (1) with output control $y = x_2$ as follow

$$
\begin{aligned}
\dot{z}_1 &= z_2, \\
\dot{z}_2 &= v_2, \\
\dot{\phi}_1 &= \mu_h - \mu_h \phi_1 + \theta \phi_2, \\
\dot{\phi}_2 &= \eta_h z_1 - (\theta + \mu_h)\phi_2, \\
\dot{\phi}_3 &= \varphi \left(1 - \frac{N_m}{kN_h} \phi_3 \right)(a) - (\eta_A + \mu_A)\phi_3, \\
\dot{\phi}_4 &= \mu_h - \mu_h \phi_1 + \theta \phi_2 - \eta_h z_1 - \mu_h z_1, \\
y &= h(x) = z_1.
\end{aligned} \tag{13}
$$

If $z_1 = z_2 = 0$ then we get the zero dynamics as follow

$$
\begin{aligned}
\dot{\phi}_1 &= \mu_h - \mu_h \phi_1 + \theta \phi_2, \\
\dot{\phi}_2 &= -(\theta + \mu_h)\phi_2, \\
\dot{\phi}_3 &= \varphi \left(1 - \frac{N_m}{kN_h} \phi_3 \right)(a) - (\eta_A + \mu_A)\phi_3, \\
\dot{\phi}_4 &= \mu_h - \mu_h \phi_1 + \theta \phi_2.
\end{aligned} \tag{14}
$$

By using linearization method, it is easy to show that the zero dynamic (14) is a stable system. Therefore we can formulate the control function for u_2 using the following function [11],

$$
u_2 = \frac{1}{L_g L_f h(\mathbf{x})} \cdot \left[-L_f^2 h(\mathbf{x}) + v_2 \right] \tag{15}
$$

with

$$
L_g L_f h(\mathbf{x}) = \frac{\partial L_f h(\mathbf{x})}{\partial \mathbf{x}} \cdot \mathbf{g}(\mathbf{x}).
$$

Consider that

$$
L_f h(\mathbf{x}) = \frac{\partial h}{\partial \mathbf{x}} \cdot \mathbf{f}(\mathbf{x}) = \frac{\partial x_2}{\partial \mathbf{x}} \cdot \mathbf{f}(\mathbf{x}) = B\beta_{mh} \frac{N_m}{N_h} x_1 x_6 - (\eta_h + \mu_h) x_2.
$$

Therefore we have

$$
L_g L_f h(\mathbf{x}) = \frac{\partial L_f h(\mathbf{x})}{\partial \mathbf{x}} \cdot \mathbf{g}(\mathbf{x}) = \frac{\partial \left(B\beta_{mh} \frac{N_m}{N_h} x_1 x_6 - (\eta_h + \mu_h) x_2 \right)}{\partial \mathbf{x}} \cdot \mathbf{g}(\mathbf{x})
$$

$$
= \left[B\beta_{mh} \frac{N_m}{N_h} x_6 \quad -(\eta_h + \mu_h) \quad 0 \quad 0 \quad 0 \quad B\beta_{mh} \frac{N_m}{N_h} x_1 \right] \cdot \begin{bmatrix} 0 \\ 0 \\ 0 \\ 0 \\ -x_5 \\ -x_6 \end{bmatrix}
$$

$$
= -B\beta_{mh} \frac{N_m}{N_h} x_1 x_6
$$

and

$$L_f^2 h(\mathbf{x}) = \frac{\partial L_f h(\mathbf{x})}{\partial \mathbf{x}} \cdot \mathbf{f}(\mathbf{x}) = \frac{\partial \left(B\beta_{mh} \frac{N_m}{N_h} x_1 x_6 - (\eta_h + \mu_h) x_2 \right)}{\partial \mathbf{x}} \cdot \mathbf{f}(\mathbf{x})$$

$$= \begin{bmatrix} B\beta_{mh} \frac{N_m}{N_h} x_6 & -(\eta_h + \mu_h) & 0 & 0 & 0 & B\beta_{mh} \frac{N_m}{N_h} x_1 \end{bmatrix} \cdot \begin{bmatrix} \mu_h - B\beta_{mh} \frac{N_m}{N_h} x_1 x_6 - \mu_h x_1 + \theta x_3 \\ B\beta_{mh} \frac{N_m}{N_h} x_1 x_6 - (\eta_h + \mu_h) x_2 \\ \eta_h x_2 - (\theta + \mu_h) x_3 \\ \varphi \left(1 - \frac{N_m}{kN_h} x_4 \right) (x_5 + x_6) - (\eta_A + \mu_A) x_4 \\ \eta_A x_4 - (B\beta_{hm} x_2 + \mu_m) x_5 \\ B\beta_{hm} x_2 x_5 - \mu_m x_6 \end{bmatrix}$$

$$= B\beta_{mh} \frac{N_m}{N_h} x_6 \left(\mu_h - B\beta_{mh} \frac{N_m}{N_h} x_1 x_6 - \mu_h x_1 + \theta x_3 \right) - (\eta_h + \mu_h) \left(B\beta_{mh} \frac{N_m}{N_h} x_1 x_6 - (\eta_h + \mu_h) x_2 \right)$$

$$+ B\beta_{mh} \frac{N_m}{N_h} x_1 (B\beta_{hm} x_2 x_5 - \mu_m x_6).$$

Suppose $L_f^2 h(\mathbf{x}) = f_2$ then we have

$$u_2 = \frac{f_2 + c_0 z_1 + c_1 z_2}{B\beta_{mh} \frac{N_m}{N_h} x_1 x_6}. \tag{16}$$

3. Numerical Results and Discussion

In this section we present numerical simulation of system (1) by considering different cases to evaluate effect of the controls applied to the system. For simulation purposes we use parameter values listed in Table 1. We assume that the initial conditions of the system in dimensionless term are $x_1(0) = 0.9450$; $x_2(0) = 0.0050$; $x_3(0) = 0.005$; $x_4(0) = 0.0001$; $x_5(0) = 0.0001$; $x_6(0) = 0.001$ with $N_h = 48000$ and $N_m = 500000$. Furthermore we chose $c_1 = c_2 = 1$ and the final time observation is $t_f = 100$ days. Solution of system (1) is approximated using Runge-Kutta fourth order numerical scheme (see [13] for detail about the method). We consider three cases: $u_1 \neq 0, u_2 = 0$; $u_1 = 0, u_2 \neq 0$; and $u_1 \neq 0, u_2 \neq 0$.

Case 1: $u_1 \neq 0, u_2 = 0$

In this case, we assume that control is only applied to the human population to quantify effect of vaccination without controlling mosquito population. Since u_1 is defined as effect of vaccination to the susceptible population (in dimensionless unit), we restrict its value in the bounded region $0 \leq u_1(t) \leq 1$. By setting $u_2 = 0$ we get the simuation results for human population depicted in Figure 1 and for human mosquito population depicted in Figure 2. In Figure 1-(a) we can observe that effect of vaccination is relatively small in reducing number of infected population. For instance without control, the maximal infected population about 25% while after control the maximal number of infected population decreases only about 1% (see Figure 1-(b)). So does for the recovered population, it only decreases around 1%. Conversely, effects of vaccination can be observed in the mosquito populations. Figure 2-(a) shows that number of susceptible mosquito decreases from 32% becomes 25%. While in Figure 2-(b) we can observe that number of infected mosquito also decreases about 2% at 40 days.

Case 2: $u_1 = 0, u_2 \neq 0$

In this case we assume that control only applies for mosquito population. We assume that there is no vaccination for susceptible human. Therefore we have $u_1 = 0$, $0 \leq u_2(t) \leq 1$. Numerical results for this case are given in Figure 3 and 4. Figure 3 presents the proportion of susceptible human, the proportion of infected human, and the proportion of recovered human. We can observe that control applied for mosquito population significantly affects the reduction of infected human. The maximum number of infected human significantly decreases. As time is increased the number of infected human goes to zero. Conversely, as time is increased the number of susceptible human is also increase implying that almost all susceptible humans are free from infection when mosquito population is controlled. In Figure 4 we can observe that the same interpretation is applied to the mosquito populations. When mosquito population is controlled using fogging treatment, effects of this treatment significantly decrease not only infected mosquito but also susceptible mosquito and its larvae.

Case 3: $u_1 \neq 0, u_2 \neq 0$

In case three, both populations are controlled to quantify effect of simultaneous control. Here we assume that $u_1 \neq 0, u_2 \neq 0$. Simulation results for this case are depicted in Figure 5 and 6. We can observe that the simulation results in this case are not different from the previous results in case 2. Vaccination did not give significant effect to reduce infection in both human and mosquito populations. It means that the best effective way to reduce the number of infection caused by dengue viruses is by controlling the mosquito population only. This is due to control of the human population through vaccination does not have a significant effect in reducing the number of infected humans. This is also become the conclusion of this study that can be recommended to the policy maker such that outbreak of dengue fever can be avoided.

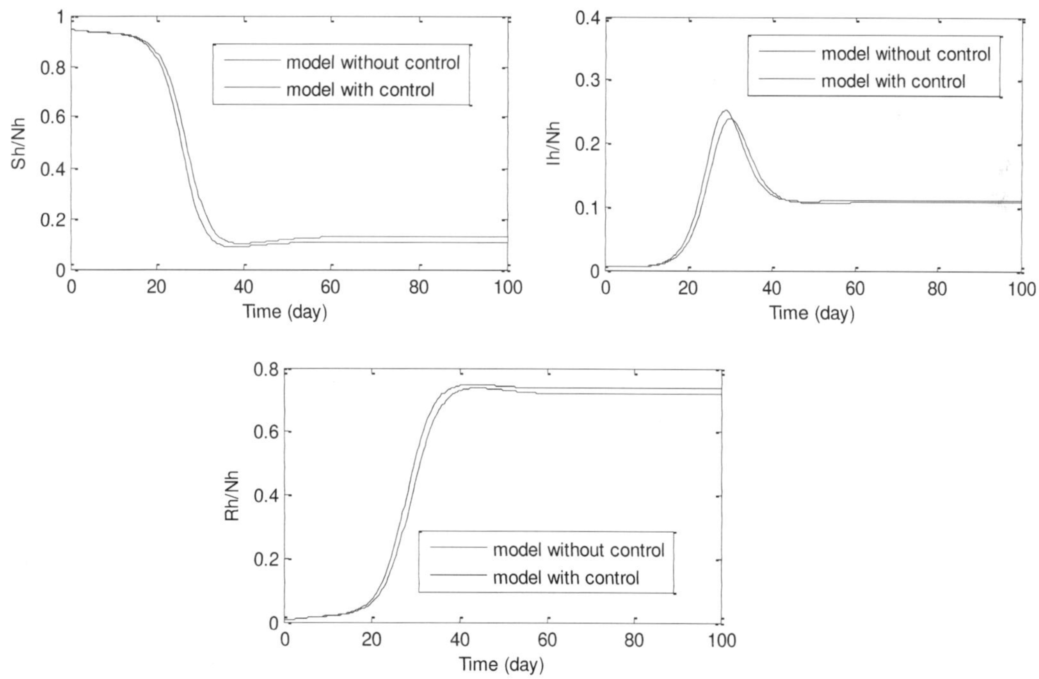

Figure 1. Numerical solutions for susceptible human (top-left), infected human (top-right) and recovered human (bottom) with and without vaccination control (case 1).

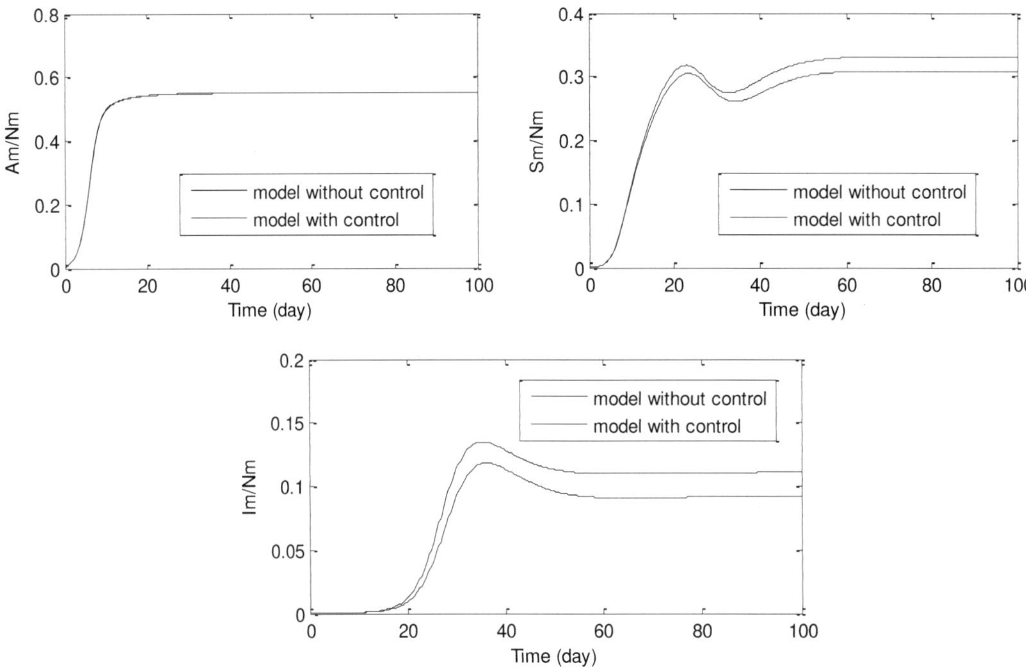

Figure 2. Numerical solutions for larvae (top-left), susceptible mosquito (top-right), and infected mosquito (bottom) with and without vaccination control (case 1).

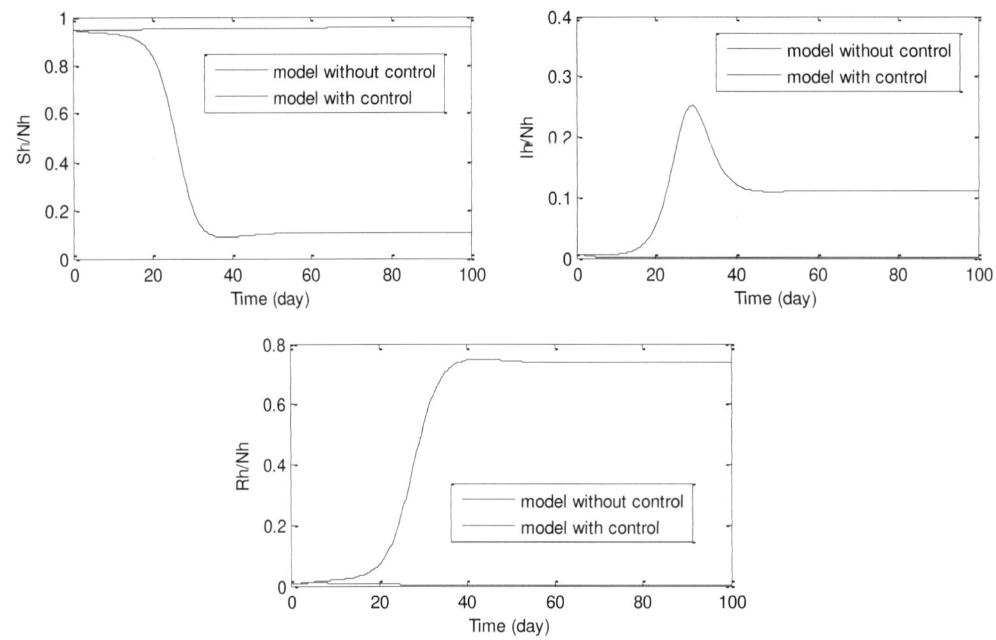

Figure 3. Numerical solutions for susceptible human (top-left), infected human (top-right) and recovered human (bottom) with and without fogging control (case 2).

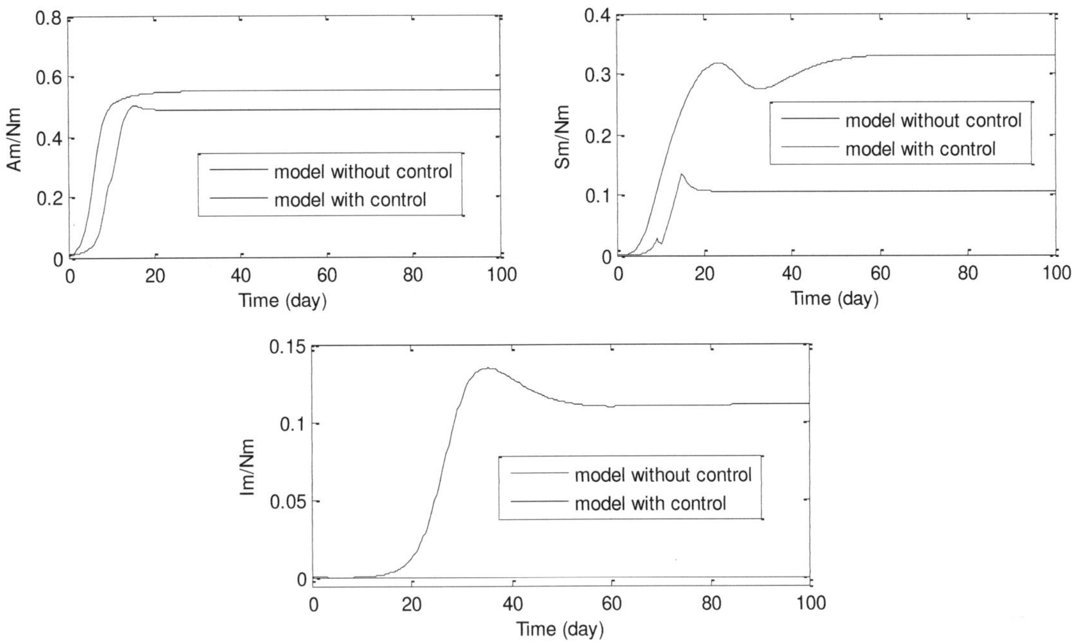

Figure 4. Numerical solutions for larvae (top-left), susceptible mosquito (top-right), and infected mosquito (bottom) with and without fogging control (case 2).

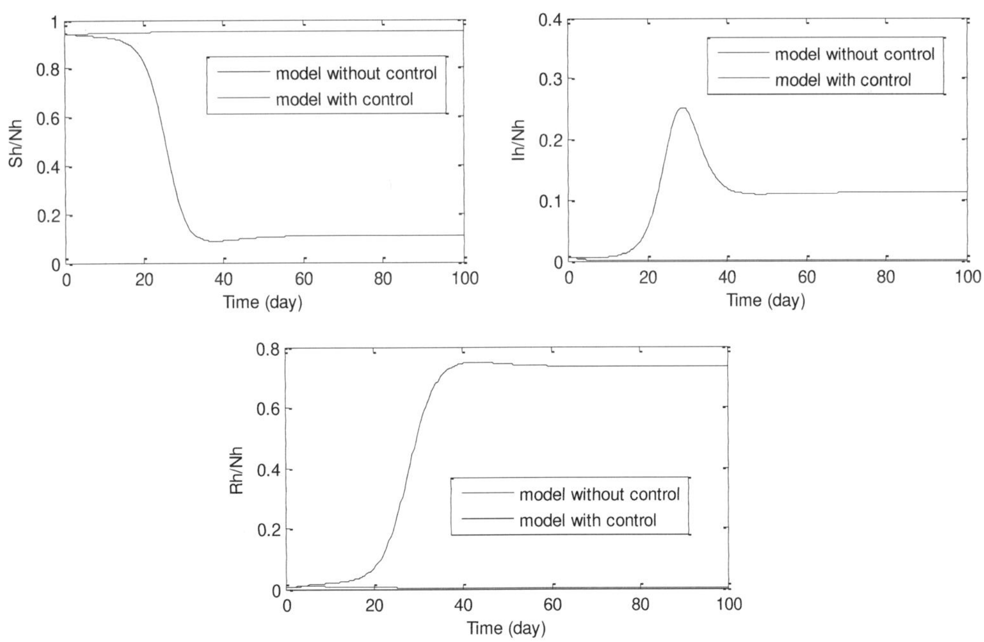

Figure 5. Numerical solutions for susceptible human (top-left), infected human (top-right) and recovered human (bottom) with and without vaccination and fogging controls (case 3).

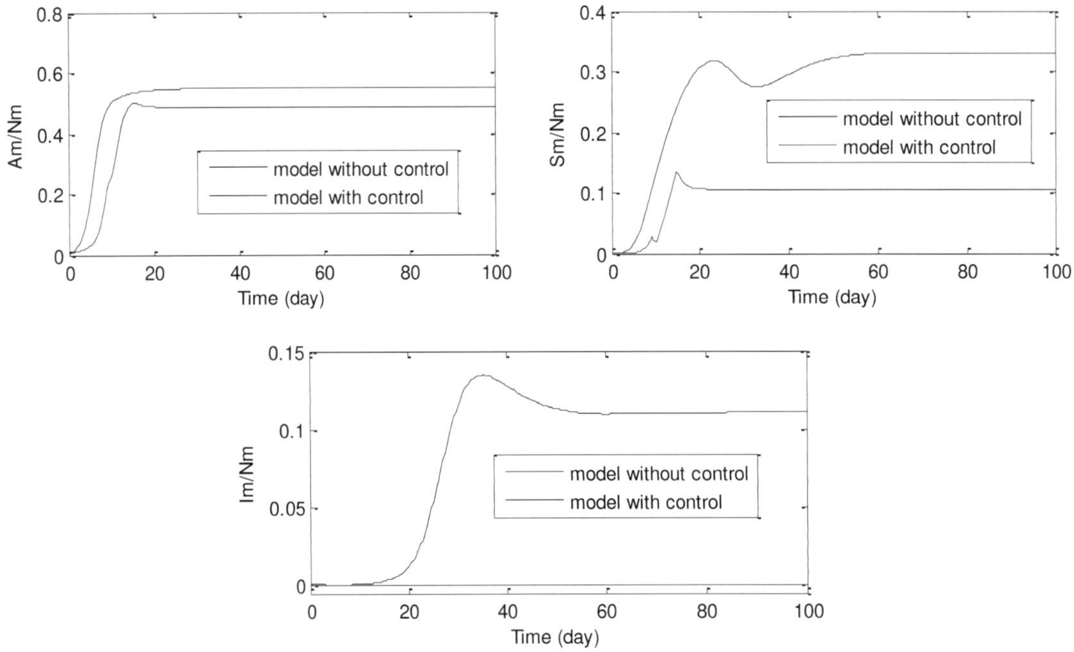

Figure 6. Numerical solutions for larvae (top-left), susceptible mosquito (top-right), and infected mosquito (bottom) with and without vaccination and fogging controls (case 3).

Conclusions

In this paper we derived a mathematical model describing the transmission of dengue infections that involved two population, i.e. human and mosquito populations. For controlling the spread of dengue viruses, an integrated vector control was taken into consideration in the model by combining effect of vaccination for susceptible human and fogging for mosquito population. The two controls were designed by using a feedback control as the input-output linearization method. Fogging was considered as an input control with an eradication effort and vaccination was also considered as an input control with prevention effort. On the other hand, the infected human variable was considered as the output control that should be minimized. Analytically, we found the closed form of the controls in term of functions in time that depended on the model parameters. Numerically, we presented simulations for different cases and we found that vaccination did not give significant effect to reduce the infection in human population. Based on the simulation results we concluded that the best effective way to reduce the number of infection caused by dengue viruses was by controlling the mosquito population only. By controlling the mosquito population, outbreak of dengue fever can be avoided.

References

[1] WHO 2017 *Emergencies preparedness, response: Dengue fever* (available from: http://www.who.int/csr/don/archive/disease/dengue_fever/en/ accesed on 10th August, 2018).
[2] Nuraini N, Tasman H, Soewono E and Sidarto K A 2009 *Math. Comput. Model* **49** p. 1148–1155.
[3] Chávez J P, Götz T, Siegmund S and Wijaya K P 2017 *Math. Biosci.* **289** p. 29–39.
[4] Diekmann O, Heesterbeek J A P and Metz J A P 1990 *J. Math. Biol* **28** p. 503-522.
[5] van den Driessche P and Watmough J 2002 *Math. Biosci.* **180** p. 29-48.
[6] Rodrigues H S, Monteiro M T T and Torres D F M 2010 *Math. Comput. Model* **52** p. 1667–1673.

[7] Aldila D, Götz T and Soewono E 2013 *Math. Biosci.* **242** p. 9–16.
[8] Rodrigues H S, Monteiro M T T and Torres D F M 2014 *Math. Biosci.* **247** p. 1–12.
[9] Christofferson R C and Mores C N 2015 *Vaccine* **33** p. 7069–7074.
[10] Ningsih S 2017 *Optimal Vaccination and Fogging on Controlling the Spread of Dengue Fever (in Bahasa: Kontrol Optimal Penyebaran Penyakit Demam Berdarah Dengue melalui Vaksinasi dan Fogging)* (Thesis: FMIPA UNHAS).
[11] Isidori A 1998 *Nonlinear Control Systems* (New York: Springer-Verlag).
[12] Khalil H 2002 *Nonlinear Control System* Third Edition (New Jersey: Prentice Hall).
[13] Lenhart S and Workman J T 2007 *Optimal Control Applied to Biological Models* First Edition (Chapmanand Hall/CRC).

Acknowledgment
The authors would like to thank LP2M Unhas that supported this research through Penelitian Benua Maritim Spesifik (BMIS) 2018.

IOP Conf. Series: Materials Science and Engineering **619** (2019) 012021 doi:10.1088/1757-899X/619/1/012021

Car Driver Drowsiness Recognition Android-Based System

Indrabayu[1], Siti Khumaerah Mufti[2], Intan Sari Areni[3]

[1,2]Department of Informatic Engineering, Faculty of Engineering, Hasanuddin University, Indonesia

[3]Department of Electrical Engineering Faculty of Engineering, Hasanuddin University, Indonesia

Email: indrabayu@unhas.ac.id, khumufti@gmail.com, intan@unhas.ac.id

Abstract. — The growth of the number of vehicles in Indonesia especially in terrestrial mode is also offset by the increasing number of accidents. One of the main causes of traffic accidents is due to drivers who driving in drowsiness. This condition leads to loss of concentration and even awareness of the driver. The drowsiness recognition system built into android smartphone platforms which owned by almost all walks of life, can help the community and government in providing direct warning especially to motorists while driving in a drowsy state. The method used is the Haar-Cascade Classifier to detect the driver's face and utilize the detectable face size ratio to create ROI in the mouth area. Then the using of the FindContour algorithm to find contours in dark areas when the mouth is open because it evaporates. The size of the contour will be compared to the Mouth ROI size for the drowsy driver's decision. System creation using Android Studio with C ++ and Java programming languages, as well as OpenCV libraries for Haar-Cascade and FindContour. Based on the test to 10 respondents, the accuracy of the system recognizes sleepiness on the driver reached 88.7% during the day.

1. Introduction

Current transportation means has become a major need for human beings especially in Indonesia. According to data from the Central Bureau of Statistics from 2003 to 2011, there has been a surge in the number of motor vehicles, both passenger cars, buses and trucks reaching 12-15% per year, and is forecast to continue increasing [1]. However, the number of accidents occurring in land transportation in Indonesia is quite high. According to data from the Police Force of the Republic of Indonesia, tiredness and drowsiness are the main causes of traffic accidents in 2016 which until July reached 41% [2]. Specifically, accidents caused by drivers are not ready in response to traffic and sleepy situations are closely related to the driver's vision. This condition not only endangers drivers but also other road users. Therefore, it is advisable to rest for a while when getting sleepy. However, there are still many drivers who ignore the safety protocol, one of them because of the lack of direct warning [3]. Several studies have provided solutions by creating a detection system that can recognize drowsiness by image processing and machine learning using a camera mounted on the steering wheel, as well as a car dashboard. Therefore, it takes a more practical and affordable system, which can recognize drowsiness and can alert the driver before it actually loses concentration. One of them is, a system built in the smartphone android platform.

Smartphones are almost owned by all walks of life. Utilizing a smartphone camera to detect and recognize sleepiness in the driver can be an alternative to help provide immediate alert when the driver is sleepy. Therefore, in this study, made an android-based system that can recognize sleepiness in car drivers. Using smartphone's camera to record the driver's face in real-time. The system will then process image data using Haar-Cascade Classifier to detect faces. Then the FindContour algorithm takes advantage of contour sizes when the mouth opens because it evaporates to recognize sleepiness in the driver.

2. Parameters to detect drowsiness on the driver

Drowsiness is one of the symptoms caused by fatigue that reduces strength, reaction speed, coordination ability, balance and also the ability of a person in making decisions. Therefore, when drowsy, it is not recommended to perform activities that require high concentrations with dangerous consequences such as driving [4]. Please follow these instructions as carefully as possible so all articles within a conference have the same style to the title page. This paragraph follows a section title so it should not be indented. According to Narender Kumar, the driver's sleepiness condition can be seen from the psychological condition, behaviour, and visual appearance of the driver. Furthermore, to make the driver drowsiness system can be designed with techniques to analyse visual appearance such as eyes and mouth [6]. One of them can be done based on the condition of the mouth when yawning.

The yawning mouth can be used as a parameter to detect sleepiness in the driver. One of the studies conducted by Kumar and Barwar in 2014 that decides someone is yawning or not by checking the size of the mouth. If the height of the mouth is greater than a certain threshold, it means that someone is yawning, a mistake then appears when the system fails to distinguish the yawning condition by laughing, talking and singing [7]. According to the National Sleep Foundation basically humans yawning under the following conditions [8]:

- Drowsiness
- Seeing others yawn
- Bored
- Disease
- The brains working hardly
- Body is having less movement

However, under the normal conditions humans will yawn for about six seconds [9].

3. System development

The aim of this work is to recognize sleepiness or drowsiness in driver's yawn to give the driver a direct alert.

3.1. Capturing driver image

First step is to record driver's face using smartphone front camera. The shooting angle is set between -30 to 30 degrees, with 30 frame per second, mp4 format for video, and RGB format for image frame. This will be the input for the system development.

3.2. Face detection

Face on the recording data is detected using Haar Cascade Classifier. A complex algorithm to detect faces on image using a square feature that gives a specific indication on a picture or image, called Haar-Like Feature. The image that is processed is the image in the grayscale form. So, the RGB format will be converted to grayscale first. The grayscale frame will be detected using the haar-like feature that traces the image using a box-shaped sliding window. After the sliding window ends, the box size will be minimized and sliding back through the image. And so on until box size cannot be minimized anymore. So it will be a lot of repetitive detection results.

IOP Conf. Series: Materials Science and Engineering **619** (2019) 012021 doi:10.1088/1757-899X/619/1/012021

Figure 1. Illustration of the grayscale frame with red-box sliding windows with haar-like feature inside.

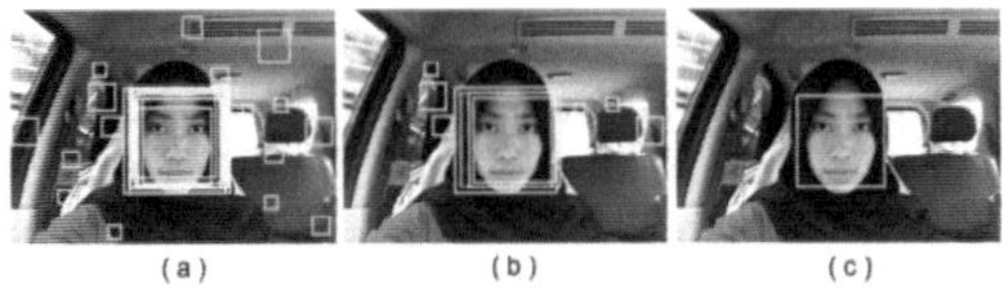

Figure 2. Illustration comparison of detection results with modified minNeighbours: (a) value 0; (b) value 3; (c) value 8.

System creation uses modified value of minNeighbours to 8, to minimize the possibility of the system detecting facial-like features on non-face objects. By modifying minNeighbours to 8, then only features (or surroundings) with at least eight times detection will be classified as a strong classifier as shown on Figure 2.

3.3. Creating MouthROI

The next stage is to determine the ROI (Range of Interest) on the mouth of the face. By determining the ROI, it can be localized to certain parts that are desired to be processed, so it can be more targeted and accurate. ROI is set at the mouth position, the center-bottom of the face. ROI size is also set as not too big and not too fit on the size of the mouth to anticipate the size of the mouth when it opens. MouthROI determined MouthROI using these following stages:

3.3.1. Creating FaceROI. A previously detected face, converted to ROI under the name FaceROI which will be the place of the process to determine MouthROI by utilizing coordinate value x and y as well as the height (h) and width (w) from FaceROI.

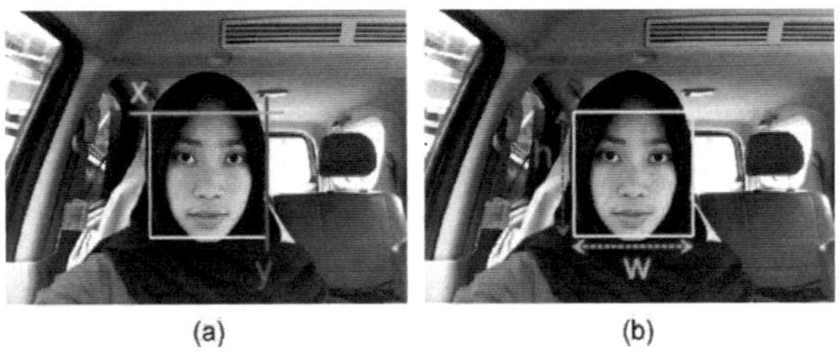

Figure 3. (a) Illustration of FaceROI coordinate; (b) Illustration of FaceROI size

3.3.2. Specifying the coordinate and determining the size of MouthROI. Coordinates will determine the position of MouthROI in FaceROI. The desired coordinates are shown in illustration figure 3. Coordinate of MouthROI area are marked with red dots and named(x', y'). To find the value of (x', y') using following equation:

$$x' = \left(r_x + \frac{1}{3}\, w_r\right) \tag{1}$$

$$y' = \left(r_y + 1.3\, h_r\right) \tag{2}$$

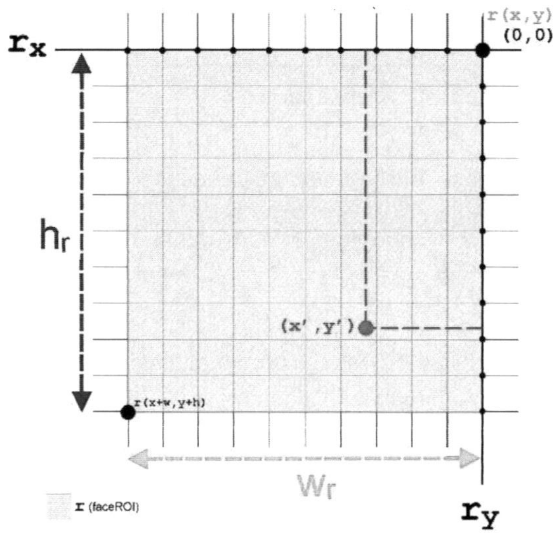

Figure 4. Illustration of desired coordinates of MouthROI within the FaceROI.

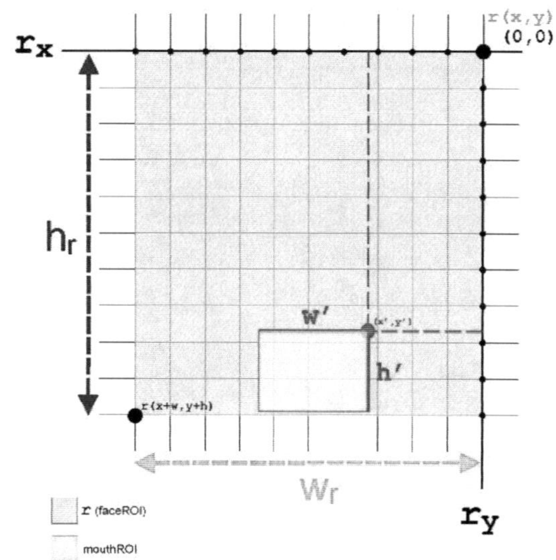

Figure 5. Illustration of desired size of MouthROI

Next, the desired size of MouthROI is a rectangle consisting of a height of $(h')$and width (w'). The desired size is displayed using a yellow rectangle in figure 5. To find the size of MouthROI using the following equation:

$$h' = \frac{1}{4.5} h_r \tag{3}$$

$$w' = \frac{1}{3.18} w_r \tag{4}$$

And then the MouthROI we had made, will displayed using red bounding box.

3.4 Finding contour

The next stage is the contour finding for the dark areas contained in the area of MouthROI. The contour will be used for decision making process. The flow of the contour search process is shown in Figure 8 below:

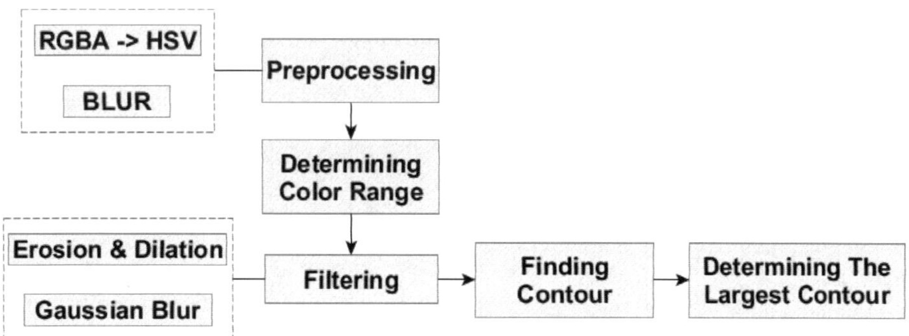

Figure 6. Contours finding process flow.

The process flow described as follows:

(1) Pre-processing. Stage to prepare the image for further process. The following steps in pre-processing are shown in table 1.

IOP Conf. Series: Materials Science and Engineering **619** (2019) 012021 doi:10.1088/1757-899X/619/1/012021

Table 1. Pre-processing stages

	Proses	Illustration	Description
	Image input from MouthROI	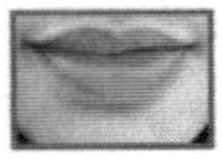	Image in RGB format
1	Converting RGB image to HSV	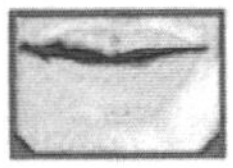	Image is converted into HSV format, because HSV is better at distinguishing colours (better than RGB).
2	Blur (size kernel 4x4)	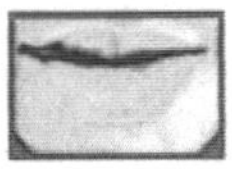	Blur used to smooth out the border and reduce the noise on the image

(2) *Determining colour range.* Colour range is five-collected colours that are considered to represent a variety of colours in dark areas when the mouth is open. The colour value of HSV in OpenCV is [0-180, 0,255, 0,255]. To convert the initial value of HSV to be compatible with OpenCV, use the following equations:

$$H_{OpenCV} = Hue \times 0{,}5 \quad (5)$$

$$S_{OpenCV} = Saturation \times 2{,}55 \quad (6)$$

$$V_{OpenCV} = Value \times 2{,}55 \quad (7)$$

Table 2. Colour conversion table

Colour (in RGB)	Hex Code	RGB	HSV	HSV-OpenCV
A	#000000	[0, 0, 0]	[0, 0, 0]	[0, 0, 0]
B	#374A2B	[55, 74, 43]	[96, 42, 29]	[48, 207, 74]
C	#290001	[41, 0, 1]	[360, 100, 16]	[180, 255, 40]
D	#754447	[117, 68, 71]	[356, 42, 46]	[178, 107, 117]
E	#BF7575	[191, 117, 117]	[0, 39, 75]	[0, 99, 191]

Furthermore the color will be defined using inRange function of OpenCV. One colour consists of one edge of the upper threshold and one edge of lower threshold.

IOP Conf. Series: Materials Science and Engineering **619** (2019) 012021 doi:10.1088/1757-899X/619/1/012021

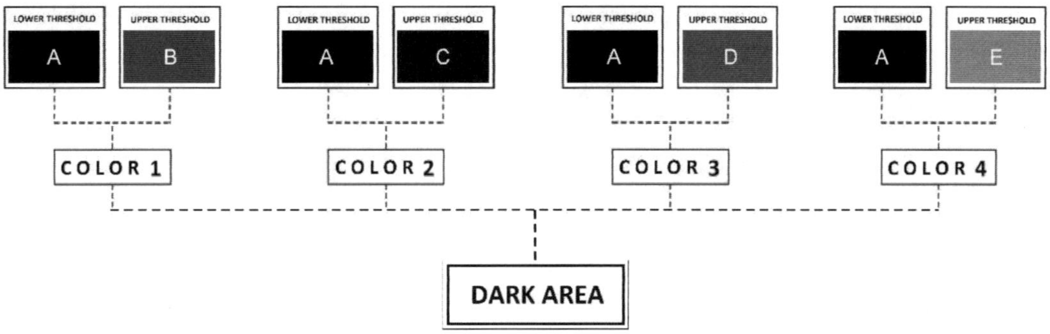

Figure 7. Illustration of colour definition for colour range of dark area.

(3) Filtering. This stage aims to improve the image quality. There are two types of process shown in Table 3.

Table 3. Filtering process illustration

Proses	Illustration	Description
Input for filter process is the image that contains colours that has been defined		Filtered image in binary form
1. Dilation (5 iterations) Erosion (10 iterations)		Also knows as closing operation that useful to smooth the image and remove small holes (noise)
2. *Gaussian Blur* (kernel 9x9)		Gaussian blur operation is used to smooth the image. Compared to other blur techniques, Gaussian blur provides the most tolerant output for colour differences.

(4) Contour finding. Contour finding based on pre-defined colours that has been filtered before, so our objects have less noise and smoother edges. Using the FindContour function from OpenCV, using CV_RETR_External mode which only looks for the outer contours of the object and CV_CHAIN_APPROX_SIMPLE method, the process will be using minimal computing and limit energy sources.

(5) Determining the largest contour. Inconsistencies of light and conditions around *MouthROI* can cause a color discoloration and cause a shadow on a particular part of the image. The shadow can add dark areas to the mouth, so that miscalculation can affect decision making. For that of all the contour areas that have been detected, will be selected one of the largest.

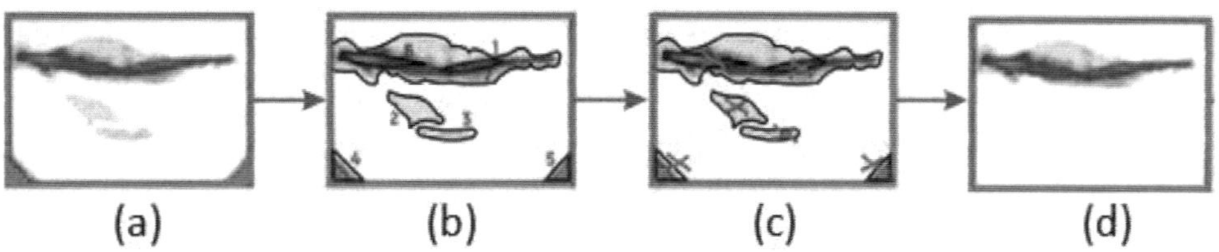

Figure 8. Illustration of determining the largest contours process. (a) the contours detected; (b) there are 7 contour detected; (c) the entire contours are compared to each other and the small ones are eliminated; (d) one of the largest contours saved as the output

The output contour will be given a red bounding box. Both the contour and the bounding box will be displayed on the system view.

Figure 9. Screenshot of system view. The green line is the contour and the red-rectangle around the contour is bounding box (bb) contour**.**

4. Decision making

The decision-making process is the process of determining the condition of the driver based on predefined parameters. Basically decision making is divided into 2 namely: open mouth decision; and drowsy driver. In decision-making using the comparison between MouthROI and BB contours is illustrated in figure 10.

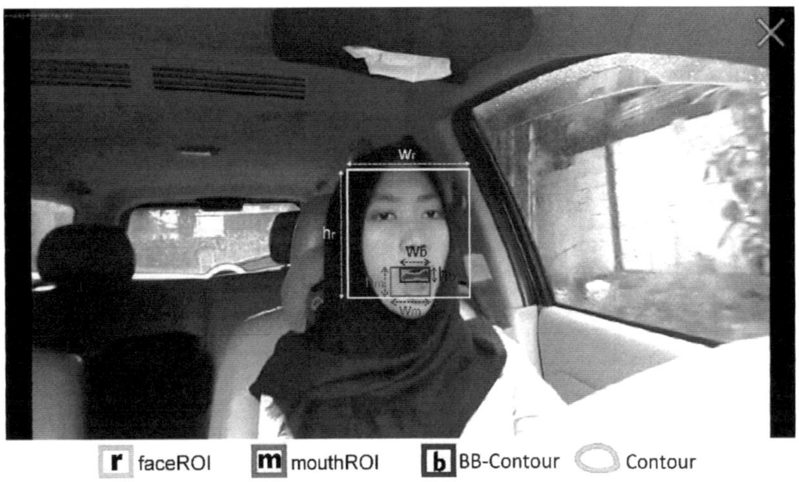

Figure 10. Illustration of comparison instruments for decision making. (The initial shown on the picture, will be used for the rest of the paper).

4.1. Pre-experiment

Before setting the parameter for decision making, there are several pre-experiments to perform the basis form for determining the way to placing smartphone camera and decision-making parameter. Through the system that can processing 10 frames per second, there are three pre-experiment described as follows:

(1) Limit Angle of Shooting Tolerance: The first test is to determine the tolerance of the proper shooting angle, the camera position from the driver's face. Tests were performed on two video data taken at different angles. To get the average value (Nt), the equation is:

$$Nt = \frac{Na}{Nb} \times 100 \qquad (8)$$

where Na is the total number of frames with the face detected, and Nb is the total number of frames in the video. So the results of the test are shown in Table 4 as follows:

Table 4. Result of tolerance angle of shooting for the best picture search testing

Vid	Angle	Duration (Seconds)	Number of Frames with detectable face (a)	Average
1	<30°	15	131	87%
2	> 30°	15	88	58%

Based on Table 4, with a shooting angle of more than 30 degrees, the system tends to fail to detect the driver's face. So the tolerance of shooting angle used is -30 to 30 degrees.

(2) *Threshold Mortality Size Limit.* The test is performed to obtain a size comparison to differentiate the yawning condition with other open mouth conditions. Research conducted by Kumar and Barwar in 2014 decided the parameters of a person to yawn can be seen from the size of the mouth height is greater than a certain threshold. However, errors occur when the system fails to distinguish the yawning condition with the condition when the respondent laughs, speaks and sings [7]. Therefore, for pre-experimental testing used the comparison of some conditions:

- **A** is the condition when the driver speaks and the contour is detected
- **B** is the condition when the driver evaporates and the contour is detected
- **C** is the condition when the driver laughs and the contour is detected
- **D** is the condition when the driver sings and contours Detected

Tests were performed on three video data with the primary goal of obtaining the threshold with the minimum possible size, but with a high tolerance level to distinguish the condition of yawning with others. Therefore that compared not only high-size, but also volume. The estimated threshold used is shown in figure 11, the orange shading indicates the range of sizes that are the parameters of whether a threshold is grouped as **X**, **Y**, or **Z**.

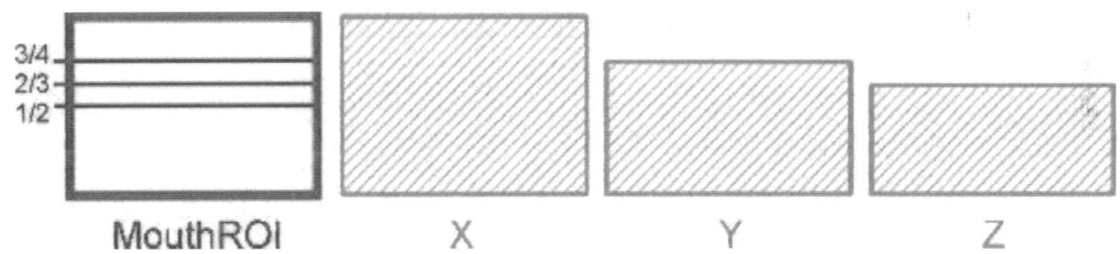

Figure 11. Illustration of size **m (red box)** and **b (blue box)**

- **X threshold** occurs if and only if the frame detected size **b is** greater than or equal to 3/4 size **m**
- **Y threshold** occurs if and only if the frame detected size **b** greater than or equal to 2/3 the size of **m,** but smaller than 3/4 size **b**
- **The Z threshold** occurs if and only if the frame detected size **b is** greater than or equal to 1/2 size **m,** but less than 2 / 3 sizes of **b**

the test results of three video data testing shown in table 5 below:

Table 5. Best size of threshold testing result

Vid	Condition	Seconds	Frame Detected					
			Volume Threshold			High Threshold		
			X	Y	Z	X	Y	Z
1	A	3	0	0	2	0	0	2
	C	5	0	1	4	0	0	2
	B	8	1	3	32	12	23	31
2	B	8	0	3	6	22	22	12
	C	10	0	0	3	0	0	13
3	C	5	0	0	2	1	0	5
	D	6	1	0	3	0	0	1

Based on Table 5 it can be concluded that the evaporative condition (**b**) is most detected at the **Z** threshold.

(3) *Best Frame Limit Threshold*: The best frame number threshold search is done to distinguish yawning from sleepiness with other conditions. After testing of 3 video data then got the result as follows:

Table 6. Best frame threshold searching test result.

Vid	Condition	Sec	Total Frame		Average (%)
			Volume	High	
1	B	7	29	58	43
2	B	7	9	53	30.5
3	B	7	21	27	30
AVERAGE (%)					34.5

So the best value as a parameter for decision making is the detection of open mouth 35 times on the frame for 7 seconds (35 of 70 frames).

4.2. *Decision taking*

Using three pre-experiment results, decision taking was created and described as follows:

(1) *Open mouth detection.* Open mouth detection will detect condition that approach the condition when mouth is opened because of yawning. The size of **m** and **b** will be compared. If the volume **b** or height **b** is greater than half of the volume or height of **m,** it will be defined as a mouth open. Every single mouth open is defined as one counter and starting the counter.

(2) *Drowsiness recognition.* Basically humans will yawning for about 6 seconds [9], that's why the timer set for 7 second that will be used to count the duration of the mouth open during yawning. System could process 10 frames per second, it means the 7-seconds timer is equal to 70 frames. In this section, system will recognize sleepiness or drowsiness in driver's yawn – as National Sleep Foundation has claimed that sleepy is not the only reason of human's yawning, it could means bored, disease, imitating, hot brain, and body less movement [8] –. When counter (mouth open detected) in 35 frames or more as long as the 7-seconds timer starts hitting 5 until 7, the alert will ring.

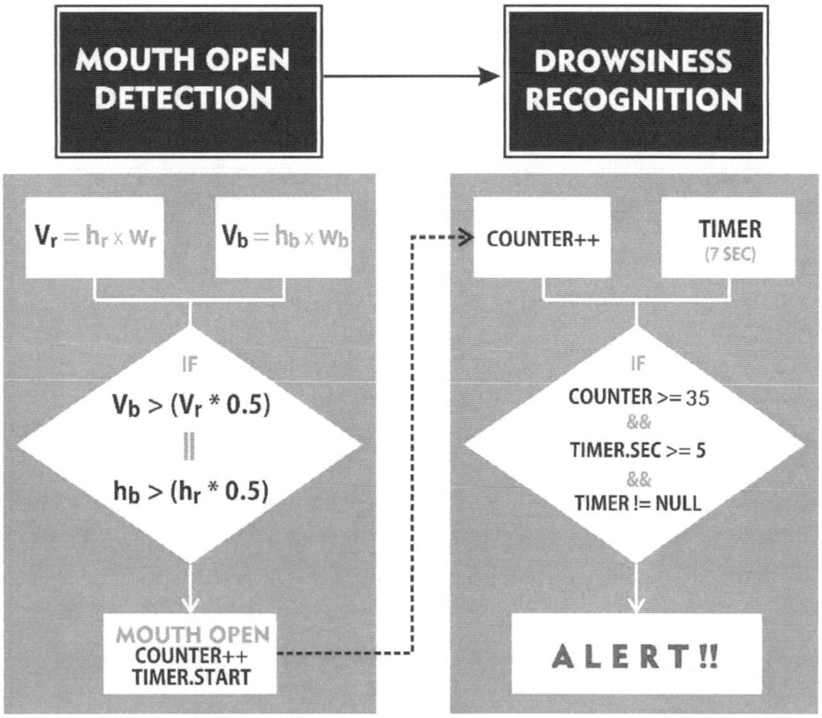

Figure 12. The flow of decision-taking system

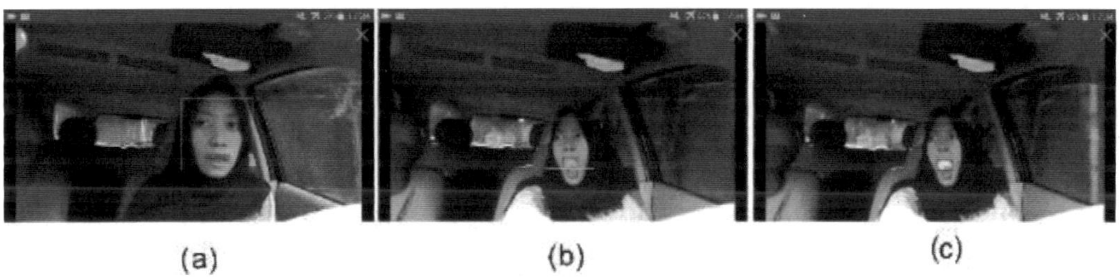

Figure 13. Screenshot of system detection result: (a) normal condition; (b) mouth open detected; (c) drowsiness recognised.

5. Experimental results

The performance of drowsiness recognition of this system has been measured using 10 respondents in real-time in the middle of the day. The smartphone was installed on car dashboard under 30 degrees from driver's face. To determine the accuracy of the system, data measured using confusion matrix variables.

Table 7. Confusion matrix

	Result of Detection (Drowsiness)	Result of Detection (Not Drowsy)
Actual Condition (Drowsy)	True Positive (TP)	False Negative (FN)
Actual Condition (Not Drowsy)	False Positive (FP)	True Negative (TN)

Table 7 is described as follows:

- TP = The actual condition is drowsy and there is a warning by the system.
- FN = The actual condition is sleepy and there is no warning by the system.
- FP = The actual condition is not sleepy and there is a warning by the system.
- TN = The actual condition is not sleepy and there is no warning by the system.

Therefore, to determine the level of accuracy (A) of drowsy detection, the equation follow is used:

$$\boldsymbol{A} = \frac{\boldsymbol{TP+TN}}{\boldsymbol{TP+TN+FP+FN}}\ \boldsymbol{x\ 100} \tag{9}$$

Screenshots of system measured using confusion matrix through 10 respondents are presented in pictures below:

Figure 14. True positive (TP); respondents is yawning and alerts appear

Figure 15. True negative (TN); condition when driver is not yawning and no alert shown. Mouth open conditions like laughing, singing, and speaking are grouping in this true negative situation.

Figure 16. False positive (FP) system error and fails to alerts driver while yawning.

Figure 17. False negative (FN) condition when respondents yawn but no alert shown. Error detection could cause by Haar-Cascade fails to detect face because there are things on driver's face, or respondents yawn under 5 seconds.

6. Conclusions

Utilizing *Range of Interest* (ROI) after Haar-Cascade algorithm had detected face, could elevate the system for further feature-face detection, as well as the mouth detection. The dark feature that made when mouth is open could be taking into account using FindContour algorithm to track the movement of the dark areas in mouth. If the contour size was bigger than half of ROI size, it will be considered as an *open* mouth. To recognize the sleepiness in it, the number of frame is measured that if in 5-7 seconds, system detected at least 35 frames contained *open* mouth, it will be decided as yawning and notification will appear. The result reaches 88.7% accuracy.

References

[1] (2011) Central Bureau of Statistics. [On line]. Available: https://www.bps.go.id/
[2] (2016) Victim of Accidents [Online]. Available: www.menyusurijalan.com/
[3] Kusyono, "Several Ergonomic Factors Associated with Low Back Pain Complaints at Angkot Drivers Department Gunungsari-Celancang (PP) Cirebon," thesis, UNDIP, Bandung, 2012.
[4] Anriani, Rina, "Recognition of Human Face Features for Drowsy Driver Detection, "Hasanuddin University, Makassar, 2017.
[5] (2013) Kompas Cyber Media [Online]. Available: https://lifestyle.kompas.com/
[6] Kaplan, S., Guvensan, MA, Yavuz, AG, Karalurt, Y., "Driver Behavior Analysis for Safe Driving: A Survey," IEEE Trans. Intell.Transp. Syst. 16, 3017-3032. doi: 10.1109 / TITS.2015.2462084, 2015.
[7] Kumar, Narender, Dr. NC Barwar, "Detection of Eye Blinking and Yawning for Monitoring Driver's Drowsiness in Real Time," International Journal of Application or Innovation in Engineering and Management (IJAIEM), Vol 3, Issue 11, Nov. 2014.
[8] (2016) National Sleep Foudandation [Online]. Available: www.sleep.org/
[9] Underwood, Emily. (2016) Science Magazine: The Bigger Your Brain, The Longer You Yawn. [On line]. Available: www.sciencemag.org/
[10] Turiyanto MD., Djoko P., Dikairono R., "Implementation of Facial Recognition Technique Based on LBP Feature on Food Introduction Robot," thesis, Surabaya Institute of Technology, Surabaya.

Acknowledgments

I would like to express our deepest appreciation to all those who provided us the possibility and assistance. Also, for all of our respondents who had spent their precious time to help us conducted this research.

Numerical Examination of Metamaterial Concentrator Design for High Efficient Thermal Energy Harvester

Zaenab Muslimin[1], Rapha Nichita Kaikatui[2], Elyas Palantei[3], Intan Sari Areni[4], Syafruddin Syarif[5], Ansar Suyuti[6], Indrabayu[7], Merandy Pala'biran[8]

[1,2,3,4,5,6] Department of Electrical Engineering, Hasanuddin University
[7] Department of Informatic Engineering, Hasanuddin University
[8] Department of Mechanical Engineering, Bandung Institute of Technology

naiirapha26@gmail.com

Abstract. Nowadays, research on energy harvesting is highly interest to harvest more waste thermal energy which produced by any kind of thermal engine in our daily life. Thermal metamaterials, designed by transformation thermodynamics are artificial structure that can actively control heat flux at a continuum scale. However, right now the problem of using metamaterial on thermal harvesting is to know what is the best shape design, configuration, and best material to obtain highly efficient on thermal energy harvesting because of the number of research on this field is limited. Flower-shaped metamaterial thermal concentrator has developed and it showed that the design can performed well on directing heat flux to specific location. However, the Flower-shaped metamaterial design that has been conducted before not optimized yet. In the current paper, the numerical simulation and analysis for several metamaterial and shape construction variations were performed using Ansys Workbench™ to characterize the best configuration to optimizing flower-shaped thermal concentrator. A number of interesting numerical computation presented on this research manuscript including total heat flux, temperature distribution, and thermal concentrator efficiency on the surface of the constructed concentrator. The aim of this paper is to provide more reference on metamaterial design for thermal harvesting.

Keywords: Energy Harvesting, Thermal Concentrator, and Metamaterial

1. Introduction

Energy is one of the most important need for human daily activity and also to drive and powering mechanical and electrical devices on industrial activity. However, according to US Department of Energy it has estimated that as much as 20 to 50% of the industrial activity total energy consumption discharged as waste heat from various device. While some waste heat losses from industrial processes are inevitable, facilities can reduce these losses by improving equipment efficiency or installing waste heat recovery technologies. To harvesting thermal energy from waste heat, it is needed a medium that can be used as a collector of heat from a flow of heat flux. A collector theoretically can be more efficient to harvesting thermal energy whenever the heat flux from arbitrary direction can be concentrated to some core or a specific location. Harvesting thermal energy through heat flux manipulation and focusing has been proposed by researchers as a practical extension of work on thermal metamaterials, and has practical applications for autonomously powered sensors or self-actuated devices (Dede et al, 2016).

One of the research on thermal concentrator has been proposed. The type of thermal concentrator proposed by Liu et al (2017) is configured as called Flower-shaped thermal energy harvester made by metamaterials. The design has been examined numerically and experimentally and provide excellent concentration efficiency with remarkable energy density increase and the temperature gradient intensifies in the target core, while temperature profile keeps undisturbed. However, Liu et al (2017) just examined one type of flower-shaped thermal concentrator configuration design, with one type of metamaterial configuration.

Therefore, on this research numerical study is conducted with several design flower-shaped like thermal concentrator design, with each design simulated with 4 type of metamaterial configuration. The objective of this research is to find the optimum result of flower-shaped-like thermal concentrator configuration that can provide more efficiency on flower-shaped metamaterial thermal concentrator.

2. Method

In this research, the model of thermal concentrator is design with took basis on the thermal concentrator that was developed by Liu et al (2017), with 3 type of improved design configuration and simulated with 4 type of material configuration. The numerical simulation in this research use transient thermal analysis features from Ansys Workbench™ software. For the measurement of the concentration effect, the temperature of plane at hot section and cool section is set as T_1=150°C and T_2=70°C. In this research, the simulation is set with neglecting the effect of convection heat transfer phenomenon occur. The result of numerical simulation on this research presented on graph of Temperature vs location, with plotted location on each design. The graph that contain all type of metamaterial configuration temperature vs location graphic then analyzed to find the optimum configuration of flower-shaped like thermal concentrator design by measuring the thermal efficiency of the concentrator and then making comparison with the thermal concentrator that has been designed by Liu et al (2017).

3. The Concept of Thermal Energy Harvesting Concentrator

The basic concept of the design in this research is using flower-shaped thermal concentrator that has been developed by Liu et al (2017). The concept of directing heat flux has been explained well by Bandaru et al (2015) with the equation of heat flux angle below.

$$\emptyset = \mathrm{Tan}^{-1}\left[\frac{(-1+c)\sin(\theta)\cos(\theta)}{cos^2(\theta)+csin^2(\theta)}\right], \text{with } c = \frac{\left[1+\left(\frac{k_1}{k_2}\right)\right]^2}{4\cdot\frac{k_1}{k_2}}$$

The equation then become the basis of flower-shaped configuration by using some configuration of metamaterial that has been arranged to get heat flux angle of direction that can direct to the center of the design to made a thermal concentrator. This design has been performed well by Liu et al (2017) on their research for thermal concentrator.

With the basis of design that performed by Liu et al (2017), then on this research the aim is to improve and optimize flower-shaped metamaterial thermal concentrator design. The idea to improve the design is first, using triangular fin-shaped design (three-dimension perspective) on the petal section of flower-shaped metamaterial thermal concentrator because the lower rate of triangular fin-shaped on dissipated heat to its environment compare than rectangular fin-shaped that Liu et al (2017) used. The reason of this idea because on thermal concentrator application, the aim of petals section is just to transfer and concentrating heat to the specific location, or in this research is the core section of flower-shaped metamaterial, not to dissipate it to the environment like application of fin-shaped on heat sink and the other cooler devices. The idea of fin-shaped design on heat transfer application has been widely broad for many application, especially for triangular fin-shaped because the lower of total volume needed of triangular fin-shaped design on heat transfer application, but have lower heat transfer application compared with rectangular fin-shaped. Several authors has been investigated the efficiency of triangular fin-shaped and compare it to the rectangular fin-shaped (Mirapalli and Kishore (2015), Pirompugd and Wongwishes (2013), etc), it is can be concluded that triangular fin-shaped design have lower Biot number on transferring heat compare with rectangular fin-shaped design because triangular fin-shaped have less surface area. Otherwise, in this research the concept of the material that used in this research is using combination of three materials with different thermal conductivity to directing heat flux to specific location, this concept then become thermal metamaterial. The concept of metamaterial in thermal directing application has been explained and applied by several author (Yang et al (2014), Chen et al (2015), Han et al (2014), Peralta & Fachinotti (2017)). Based on that, in this research the material that used is the material with that have higher thermal conductivity as a heat flux director, and lowest thermal conductivity as a substrate between director to keep the heat flux flow in the director material, and material with mid thermal conductivity like stainless steel as a background and core section of flower-shaped metamaterial to be an intake section and the collector of the heat flux. In the concern of daily application and feasible of manufacturing or fabrication, in this research the configuration of metamaterial just varied in the substrate section with four type lower conductivity material which easy to get, and keep the background, core, and petals section in one type of material which also easy to get. The list of all material thermal properties in this research can be seen in the Table 1.

Table 1. List of all material thermal properties in this research

Materials	Density ($kg\ m^{-3}$)	Specific Heat ($J\ kg^{-1}\ C^{-1}$)	Thermal Conductivity ($W\ m^{-1}\ C^{-1}$)
Copper	8933	385	400
Stainless Stell	7750	480	15.1
PDMS (Polydimethylsiloxane)	1460	970	0.15
Polyimide	1400	1150	0.25
Thermal Epoxy	1900	1150	0.249

The base design that has been designed by Liu et al (2017) and all the design with several material configuration of metamaterial thermal concentrator that has been designed in this research is shown in the Figure 1. The thickness of all the metamaterials design that designed in this research is 0.5 mm, similar with the thickness of metamaterial that designed by Liu et al (2017).

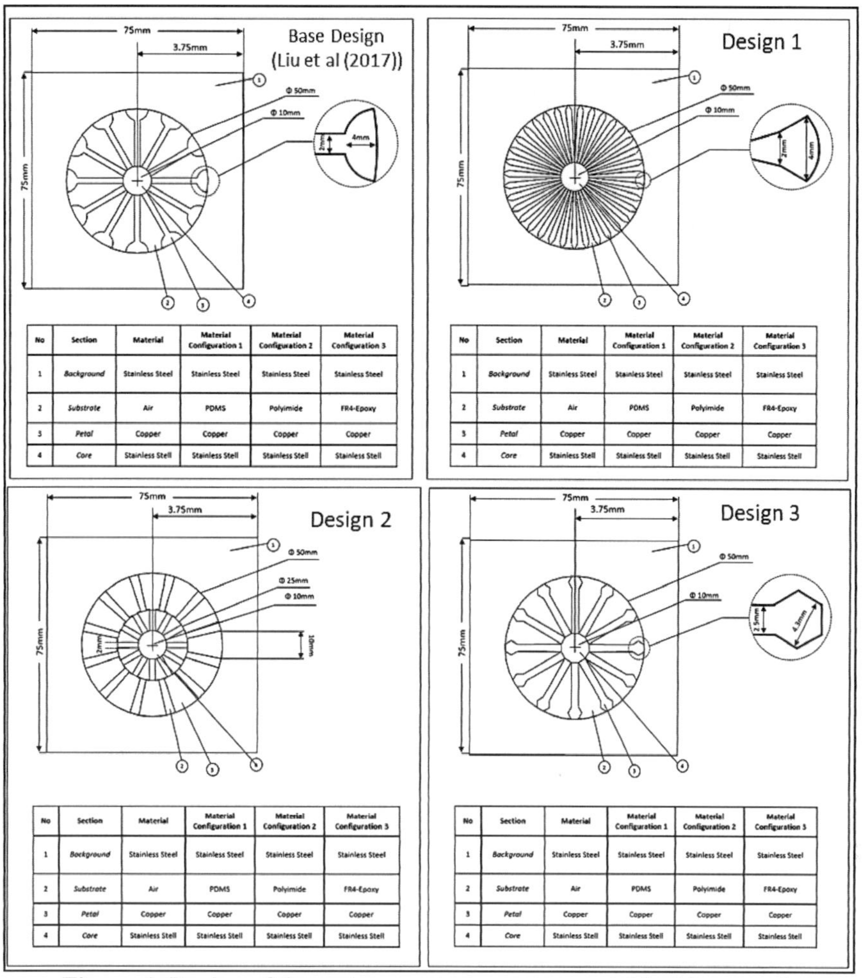

Figure 1. Design of thermal concentrator metamaterial in this research

4. Result

4.1 . Effect of Material configuration

Based on the result of numerical simulation, the effect of the material configuration that used in this research is less to the thermal distribution and concentration in the metamaterial. This result can be seen in the Figure 2 that shows the temperature distribution of metamaterial concentrator with using Design 1 as an example of the result.

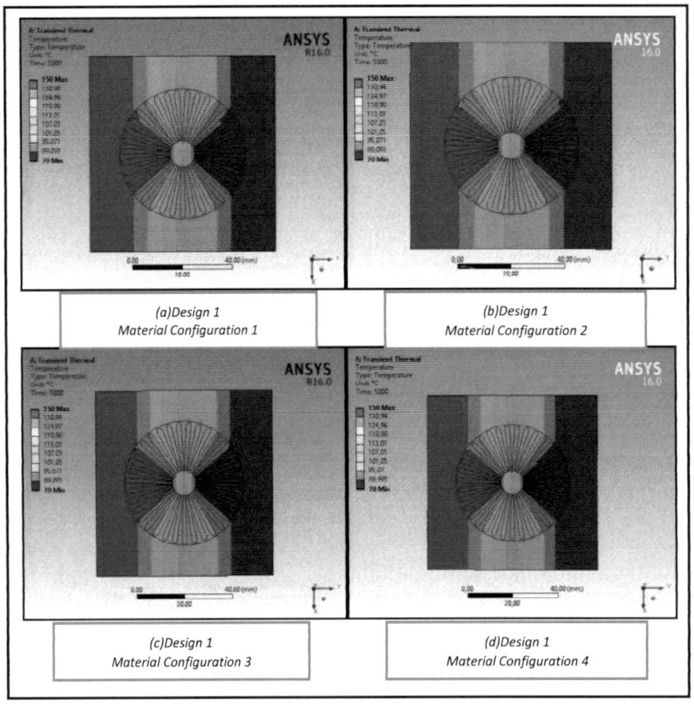

Figure 2. The Temperature distribution of metamaterial concentrator for every material configuration with using design 1 as an example of the result

The same result is occurs for other design that used in this research. More detail of temperature distribution comparison for all material type that used in this research across the Design 1 can be seen in the Figure 3.

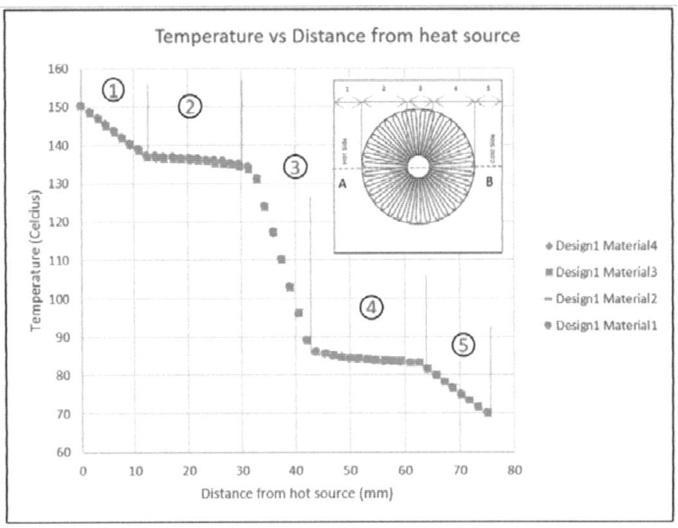

Figure 3. Temperature profile along Design 1 with different material configuration

The reason of this phenomenon is because the differential of material configuration in this research is just on variation in the substrate section design. Furthermore, the difference of thermal conductivity for each material that fill the substrate section is small. But from the analysis of concentrator efficiency that showed in Table 2, the lower of substrate section thermal conductivity, higher concentrator efficiency. As we can see the design that using air as the substrate (k = 0.026 W $m^{-1}C^{-1}$) is more efficient than PDMS (k = 0.15 W $m^{-1}C^{-1}$) and so on.

4.2. *Effect of Improved Design*

The idea of improved design in this paper is based on the concept of heat transfer phenomenon in the triangle fin-shaped for Design 1 and Design 2. The concept of heat transfer with triangle fin-shaped has been explained in the Design Concept section. The Design 1 is designed with small triangle shape-like on the outer circle of concentrator and then with more narrow triangle fin shaped-like petals, and as the result, the Design 1 have small petals with more petals than the other design, with total 35 petals. The Design 2 is designed with more broad triangle fin shaped-like, and the shape is build from the outer circle with length 25 mm, and then continued with rectangle fin-shaped like petals to the core outer diameter. The Design 3 is The Design 3 is one of the design that designed with arbitrary design on the metamaterial concentrator petal section. This design was purposed to be design that to be reference for the design that designed with arbitrary shape compared with concepted design. The result of transient thermal and total heat flux concentrated numerical simulation for each design with using example of material configuration 2 (Using PDMS as substrate, with k = 0.15 W $m^{-1}C^{-1}$) can be seen in the Figure 4 and Figure 5.

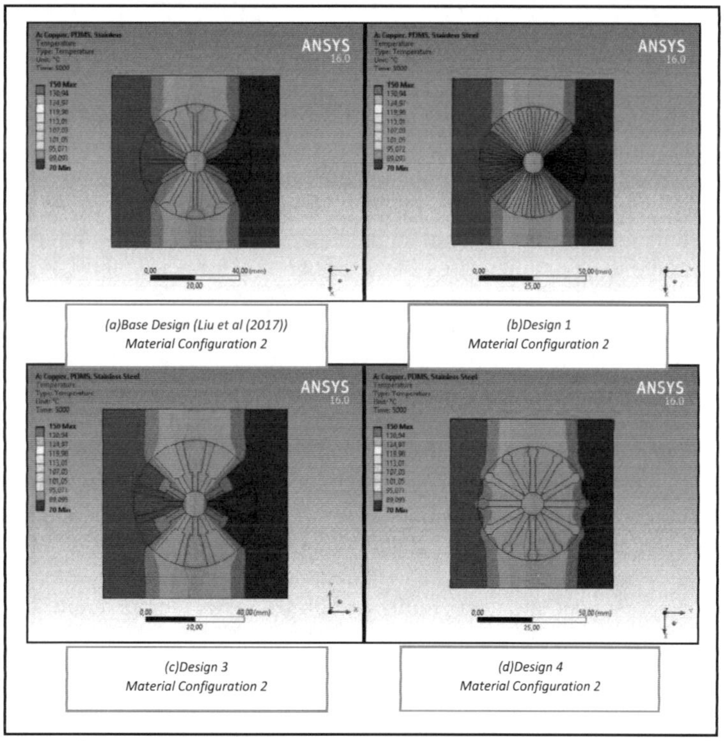

Figure 4. Temperature distribution on each design 2D plane, with using Material Configuration 2 as comparison

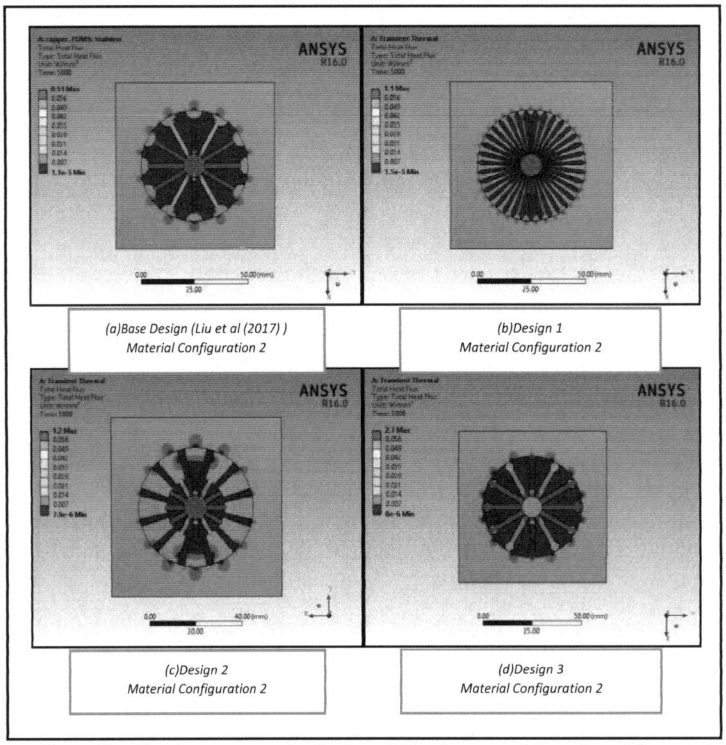

Figure 5. Total heat flux distribution on each design 2D plane, with using Material Configuration 2 as comparison

From Figure 4 and Figure 5, we can see the result of simulation showing that the Design 1 on figure (b) have more sophisticated result on the thermal concentration function compared with other design. The design can harvest heat flux from wider area of heat source and then concentrated it well on the core section of concentrator, indicated by highest temperature indicator and highest total heat flux collected (red scale) that can be narrow to the core with more area on the core surface with highest total heat flux the design can get in core is 1.1 W/ mm^2. The design from Liu et al (2017) that showing by base design on figure (a) of Figure 4 and Figure 5, showed that higher temperature and higher total heat flux collected (red scale) cannot concentrated well on core section of metamaterial design, with higher temperature range just directed by one petals, and have a lot of heat loss to the director material surround, so the maximum of heat flux the design can get just 0.51 W/ mm^2. The idea of making triangular fin-shaped like more broad with less amount of petals section on the Design 2 also have less efficient thermal concentration behavior, it is can be seen on the figure (c) of Figure 4 and Figure 5, but even better than Base Design by Liu et al (2017). The concentration on higher temperature in initial phase can be directing by broad triangular fin-shaped-like, but then the temperature just drop to lower temperature range scale at the point where the rectangular fin-shaped applied, and so the total heat flux (red scale) cannot distributed well on the core section like Design 1, even it is have slightly more higher value of total heat flux compared with Design 1, but the difference just 0.1 W/ mm^2. The Design 3 is the worst result among all, it is can be seen on figure (d) of Figure 4 and Figure 5, it has been predicted because the design was designed with arbitrary shape design. The temperature profile along all the metamaterial on this research is showed on the Figure 6 below. The temperature profile is taken from the point A to B as it shown on the Figure 6 legend.

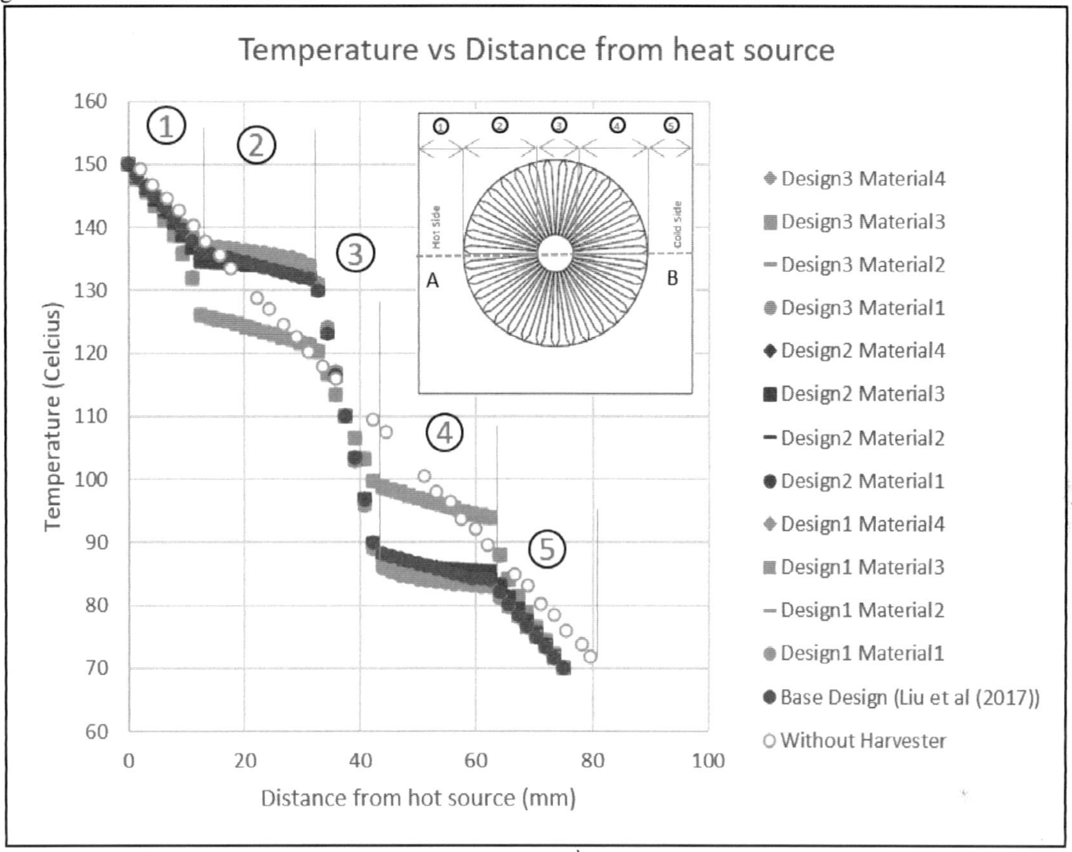

Figure 6. Temperature profile along all the metamaterial design on this research

The zone of temperature profile is divided into five section, that is the number 1 is background section zone, from the edge of hot side until boundary of concentrator outer circle, number 2 is petals section which is from the boundary of concentrator outer circle into boundary of concentrator inner circle, number 3 is along diameter of core, number 4 is from the right side of inner core boundary into the right side of outer circle boundary, and the number 5 is from outer circle boundary to edge of cold side material. From the chart on Figure 6 we can see that the metamaterial with Base Design, Design 1, and Design 2 can concentrated heat well on the edge of core section so the temperature remain high, compared with temperature profile of material with no thermal concentrator or harvester. The more specific value of thermal concentrator comparison in this research is can be seen on the Table 2, which is showed the efficiency of all metamaterial thermal concentrator that studied in this research.

Table 2. Concentrator efficiency of all the metamaterial design on this research

Design Compartment		Thermal Efficiency
Base Design (Liu et al (2017)	Material	88.14%
Design 1	Material Configuration 1	89.86%
	Material Configuration 2	89.68%
	Material Configuration 3	89.62%
	Material Configuration 4	89.57%
Design 2	Material Configuration 1	88.66%
	Material Configuration 2	88.49%
	Material Configuration 3	88.35%
	Material Configuration 4	88.29%
Design 3	Material Configuration 1	81.20%
	Material Configuration 2	81.16%
	Material Configuration 3	81.13%
	Material Configuration 4	81.11%

From the table 2 we can see that the best design among all is the Design 1, with the most sophisticated result of concentrator efficiency is by Design 1 with Metamaterial Configuration 1, with efficiency is 89.86 %. But for practical application, may the Design 1 with Metamaterial Configuration 2 is more applicable which is using PDMS as substrate with efficiency 89.68 %, different with Metamaterial Configuration 1 which is using air as substrate. It is because air is an unstable matter, which the properties can be varied, depend by environment condition.

5. Conclusion

From the result on this research, it is can be concluded that metamaterial with flower-shaped like design can harvest thermal to the specific location, and the flower-shaped metamaterial that developed first by Liu et al (2017) can be optimized with using improved design like Design 1 with triangular fin-shaped like in petals section with less Biot number compared than rectangular fin-shaped of director material. Also, from the result of comparison between Design 1 and Design 2 we also can get that the more petals also can get higher thermal concentrator efficiency.

The incremental of thermal concentrator efficiency which achieved with Design 1 up to 1.72 % and the incremental of total heat flux can be harvested is 116% (two times as larger as), compared with the Base Design which designed by Liu et al (2017). Otherwise, in this research is also studying about several material configuration with just varying substrate section with four type material, the result is like in the table 2, the incremental of concentrator efficiency is very low, that is because the difference of thermal conductivity is also less. But from the table 2 we can conclude that lower the substrate thermal conductivity, the more efficiency of thermal concentrator we can get.

The limitation on this research is the design that conducted in this research have not been fabricated yet, so further study needed to fabricated the design, so it is can be see the design performance as physically.

In the future research it is can suggest that more study to exploring new design configuration or material configuration with recent material technology that can have much higher or even lower thermal conductivity to applied for thermal concentrator is very needed to gain more thermal concentrator efficiency than this research can gain. Otherwise, the design was studied in this research just take design in millimeter scale of size, it is maybe can be conducted research that perform analysis of metamaterial thermal concentrator which has smaller size until nano-sized metamaterials. Furthermore, further study and development of the current research activity is the best reference to contribute the manufacturing of metamaterial to support highly thermal energy harvesting efficiency such as applied on thermoelectric generator, and has practical applications for autonomously powered sensors or self-actuated devices compact power supply of low power consumption appliances and so on.

6. References

[1] Bandaru P R, K P Vemuri, F M Canbazulo, R S Kapadia 2015. Layered Thermal Metamaterials For Directing And Harvesting Of Conductive Heat. *Applied Physics* Letters. **5** 053403

[2] Anonymous 2008. B Sc Incorporated *Waste Heat Recovery*: Technologies and Opportunities in U.S Industry. U.S Department of Energy: Industrial Technology Program

[3] Celik, H K, Y S Tan, R Seviour, A E W Rennie, 2009. Effect of Thermal and Mechanical Deformation of Metamaterial FDM Components. *US – TURKEY Workshop On Rapid Technologies*. September 24 – 24

[4] Chen Fei and Dang Yuan Lei 2015. Experimental Realization of Extreme Heat Flux Concentration with Easy-to-Make Thermal Metamaterials. *Scientific Reports* **5.** 11552

[5] Dede, Ercan M. et al 2016. Collection of Low-Grade Waste Heat for Enhanced Energy Harvesting. *AIP Advances* **6.** 055113

[6] Han, Tiancheng, Xue Bai, John T. L. Thong , Baowen Li , and Cheng-Wei Qiu 2014 Full Control and Manipulation of Heat Signatures: Cloaking, camouflage and Thermal Metamaterials. *Adv. Mater* **26**, 1731–1734

[7] J B Pendry, D Schurig, D R 2006, Controlling Electromagnetic Fields. *Science* **312**, 1780

[8] Liu, Wenmei, Chuwen Lan, Muwei Ji, Jitan Yao, Jin Wang, Bo Li, and Ji Zhou 2017 A Flower-Shaped Thermal Energy Harvester Made by Metamaterials. *Global Challenges* **1**, 1700017

[9] Mirapalli, Sandhya Kishore P 2015. Heat Transfer Analysis on a Triangular Fin *International Journal of Engineering Trends and Technology (IJETT)* – Volume 19 Number **5**, ISSN: 2231-5381

[10] Narayana, Supradeep Savo S, and Sato Y, 2013 Transient Heat Flux Shielding Using Thermal Metamaterials. *Applied Physics Letters* **102**, 201904

[11] Peralta Ignacio and Víctor D. Fachinotti. 2017 Optimization-Based Design Of Heat Flux Manipulation Devices With Emphasis On Fabricability. *Scientific Reports* **7**, 6261

[12] Pirompougd, Woraschest & Somchai Wongwishes 2013. Partially wet fin efficiency for the longitudinal fins of rectangular, triangular, concave parabolic, and convex parabolic profiles. *Journal of the Franklin Institute Vol 350 Issue 6. Elsevier*

[13] Shen, Xiangying Chaoran Jiang, Ying Li, and Jiping Huang 2016. Thermal Metamaterial For Convergent Transfer Of Conductive Heat With High Efficiency. *Applied Physics Letters* **109** 201906

[14] Shi, Y et al. 2014 Energy. *Convers Manage.* **80**, 110

[15] Tianzhi Yang Krishna P Vemuri, and Prabhakar R Bandaru 2014. Experimental Evidence For The Bending Of Heat Flux In A Thermal Metamaterial *Applied Physics Letters* **105**, 083908

[16] Vemuri K, P Canbazoglu, F M, and Bandaru P R, 2014. Guiding Conductive Heat Flux Through Thermal Metamaterials *Applied Physics Letters*. **105** 193904 B.Sc Incorporated.2008. Waste Heat Recovery: Technologies and Opportunities in U.S Industry. U.S Department of Energy: Industrial Technology Program

Acknowledgments
This work was supported by Telecommunication Radio and Microwave Laboratory, Departement of Electrical engineering, Hasanuddin University.

IOP Conf. Series: Materials Science and Engineering **619** (2019) 012023 doi:10.1088/1757-899X/619/1/012023

Evaluation of Finite Control Set - Model Predictive Control for Dynamic Resources Management in A Shared-Resources System

D Aryani[1], L Wang[2], N A Hidayatullah[3]

[1] Department of Electrical Engineering, State Polytechnic of Ujung Pandang, Indonesia
[2] School of Engineering, RMIT University, Australia
[3] Department of Electrical Engineering, State Polytechnic of Madiun, Indonesia

E-mail: dharma.aryani@poliupg.ac.id

Abstract.

This paper presents an evaluation of Finite Control Set-Model Predictive Control (FCS-MPC) performance for dynamic resources provisioning. Realization of a shared-resources environment is experimented in software system network by virtualization technique. Control parameter regulates the resources provisioning among the users to guarantee the achievement of performance metrics in the network. The objective preferences are maintained in relative term by manipulating a finite set of operating points as the control states. Moreover, an optimization procedure is implemented based on receding horizon prediction of MPC in a feedback control scheme. Further investigation is conducted to examine the performance of optimal FCS and constrained FCS for dynamic system management. The experimental results demonstrate that constrained FCS system provides a better response stability than the optimal FCS. Correspondingly, the range and quantity of input states are significantly influential to the flow of control action. A wider range and longer control state sequence yields to a smoother control signal stream.

1. INTRODUCTION

Optimization scheme in a multi-users networked system is aimed to preserve satisfactory performance objectives by managing the infrastructure provisioning. In a shared resources environment such software system, the principal properties of performance quality are response time and service differentiation. It has been practically confirmed that the settings of output preferences and workload uncertainty are main challenges that could cause performance shortcomings in software system. Therefore, a self-adapting ability is very crucial to be integrated into control system procedure in order to enable dynamics adjustment in unpredictable circumstances [1, 2].

One of predictive control approaches that has been used extensively to control dynamic system with certain states of operating points is Finite Control Set-Model Predictive Control (FCS-MPC). Studies by [3] and [4] can be referred as successful implementation of FCS-MPC to handle the issue of dynamic resource management. The signature of FCS control method is its algorithm simplicity and straightforward application. Earlier findings by [5] and [6] have shown that FCS-MPC is a well-proven predictive techniques for power converters and drives control. In addition, based on [6], FCS-MPC is a preferable control for power electronics. The FCS-MPC technique was initially exposed in [6] and [7]. In these studies, FCS-MPC was analyzed with PWM and the hysteresis method. Similar methods have

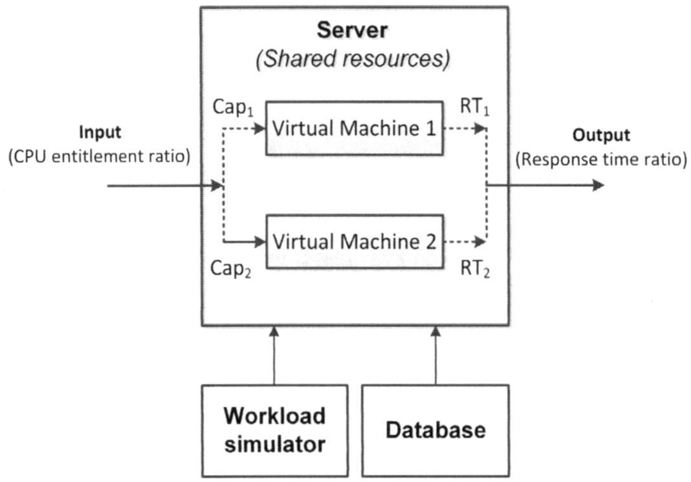

Figure 1: Structure of Virtualized software system

been investigated for various applications, such as power converters ([8, 3]), and the Permanent Magnet Synchronous Motor (PMSM) ([9, 10, 11]).

In FCS-MPC, a control action is selected from optimal switching states based on error value between the predicted and real output value. FCS-MPC calculates the output prediction, and computes cost function based on the finite control states. The optimal solution is the switching state which minimizes prediction error in objective function. This method features simple implementation and leads to a robust performance. In regards to relative management scheme of software system, control parameter is CPU capacity of the server which needs to be managed dynamically in a certain degree. The relative relationship is interpreted by pairing consecutive clients to get the ratio value.

The main contribution of this paper is the implementation of Finite Control Set-Model Predictive Control (FCS-MPC) to optimize dynamic management of a shared-resources system. Two schemes are evaluated, the optimal formulation and constrained FCS. Experiments are carried out in a two-class virtual machines environment for an equal performance differentiation setting. The results exhibit control system robustness with a better system's stability when the constrained FCS algorithm is implemented into the feedback control loop.

The remaining topics of this paper are presented into five sections. Section 2 covers the description of virtualized software system and its model structure. Section 3 presents the formulation of FCS-MPC formulation which comprises of optimal FCS-MPC and constrained FCS-MPC. The experimental results are evaluated in Section 4, with a brief conclusion in Section 5.

2. System Dynamic and Characterization

2.1. Virtualized software System

A test bench for experimentation of a shared-resources environment is attained in a software system network based on *RUBiS* application. *RUBiS* is a site benchmark of e-commerce for multi-tiers applications to emulate the system dynamics of *ebay.com*. The physical structure comprises of a server, database, and workload simulator as depicted in Figure 1. Two virtual machines (VM_1 and VM_2) are installed in the server and act as the clients that will utilize all resources in the server. For the purpose of virtualization, $Xen2.6$ hypervisor is deployed by considering its credit-based scheduler which can be used to manipulate the CPU allocation proportionally to each VM. To execute an automated control in a feedback control method, an actuator is included to actuate the system with the estimated control signal. In addition, a sensor is integrated to measure the time that server needs in order to respond the incoming requests of each client.

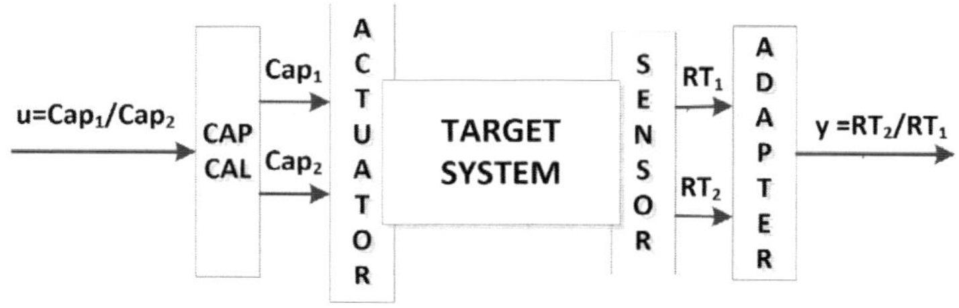

Figure 2: Relative Management Scheme for Two Users

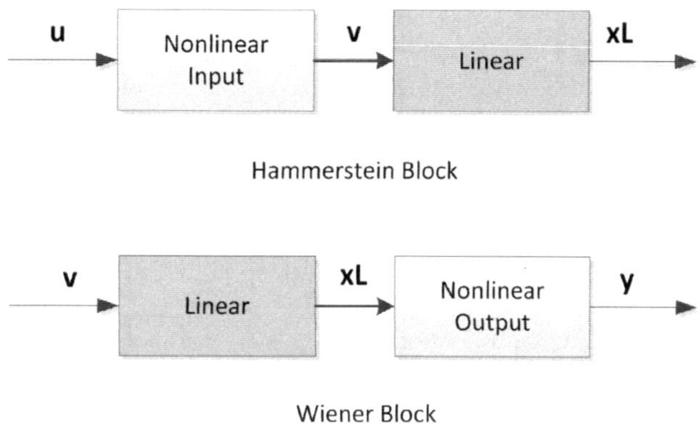

Figure 3: Hammerstein-Wiener Structure

Maintaining the performance objective in relative mode requires a dynamic resources management. In a condition where the level of CPU allocation for a client is higher than the other, it means that more resources have been provisioned to serve the workload (request) from this client. The CPU capacity entitlement ratio for VM_1 over VM_2 ($u(k) = \frac{Cap_1(k)}{Cap_2(k)}$) is input variable, while the full CPU capacity is numerically quantified as 100%. Furthermore, output parameter is response time ratio which is expressed as $y(k) = \frac{RT_1(k)}{RT_2(k)}$, where $RT_1(k)$ and $RT_2(k)$ are the measured time of the server response to the client requests. Figure 2 illustrates the framework of a relative management objectives for two clients.

2.2. Dynamic Model Estimation

Model estimation using a system identification approach refers to the study in [12] and [13]. Characterization of virtualized software system obtained by a system identification for nonlinear dynamics in a form of block-oriented model. Figure 3 shows a Hammerstein-Wiener structure to formulate the dynamic model estimation.

Nonlinear models in input and output element are identified in their inverse function in order to use the models as compensator for the nonlinear characteristic. Then, the rest of system dynamic is taken as linear function. System identification is proceeded by using a set of input and output data which is generated from the experimental testbed. A multi-level sinusoidal input with minimum 100 requests/s workloads for each client class is given into the system input, and the output response for 400 data samples are measured (see Figure 4).

For Hammerstein block, an inverse static input nonlinearity model is a polynomial function from least squares optimization. It is a function of $u(k) = f^{-1}(v(k))$ which is estimated by setting $v(k)$ values in

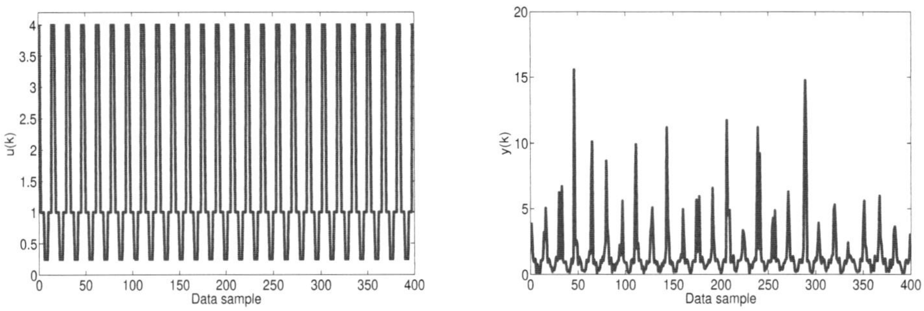

Figure 4: Input and Output Data for System Identification

range $v_{min} = -15$, $v_{max} = 15$ and $\delta v = 0.5$. The model is formulated as in eq. 1.

$$u(k) = 4.17e^{-7}v(k)^5 + 9.34e^{-6}v(k)^4 + 1.018e^{-4}v(k)^3 + 0.0028v(k)^2 + 0.08v(k) + 1.0045 \quad (1)$$

Subsequently, a linear model ($xL(k) = f(v(k))$) is formulated in an ARX (Auto-Regressive eXogenous) model,

$$xL(k+1) = 0.4201xL(k) + 0.0430v(k) \quad (2)$$

In Wiener block, another inverse nonlinear model is approximated as a function of $xL(k) = f^{-1}(y(k))$ which is estimated in B-spline curve function $B(y(k))$ with parameter N_i as basis function, the curve order ρ , and Cp_i as control points of the curve. The mathematical equation is written as follows,

$$xL(k) = \sum_{i=0}^{s} N_{i,\rho}((y(k))Cp_i \quad (3)$$

3. Finite Control Set - Model Predictive Control

Control signal in FCS-MPC utilizes a finite number of input states. In each sampling time, the input state that minimizes cost function is selected and implemented as the control action. Cost function (J) is evaluated for each input state. Then, the optimal control signal is the one with the least error in cost function optimization.

3.1. Optimal Finite Control Set (OFCS)

The basic algorithm of FCS is the a receding horizon approach with one-step ahead prediction and on-line optimization. The value of $y(k)$ for prediction uses the updated value from current measurements of output response. This updating procedure is closed-loop feedback mechanism in FCS method. The discrete model of an SISO system is defined as,

$$\begin{bmatrix} y_1(k+1) \\ y_2(k+1) \end{bmatrix} = A \begin{bmatrix} y_1(k) \\ y_2(k \end{bmatrix} + B \begin{bmatrix} u_1(k) \\ u_2(k) \end{bmatrix} \quad (4)$$

and the cost function to be solved as optimization problem is the squared error between reference and the output values:

$$J = \begin{bmatrix} r_1(k) - y_1(k+1) & r_2(k) - y_2(k+1) \end{bmatrix} \begin{bmatrix} r_1(k) - y_1(k+1) \\ r_2(k) - y_2(k+1) \end{bmatrix} \quad (5)$$

Furthermore, by defining $f(k)$ as,

$$\begin{bmatrix} f_1(k) \\ f_2(k) \end{bmatrix} = \begin{bmatrix} r_1(k) \\ r_2(k) \end{bmatrix} - A \begin{bmatrix} y_1(k) \\ y_2(k \end{bmatrix} \quad (6)$$

IOP Conf. Series: Materials Science and Engineering **619** (2019) 012023 doi:10.1088/1757-899X/619/1/012023

the objective function can be rewritten in the following notation:

$$J = \left(\begin{bmatrix} f_1(k) & f_2(k) \end{bmatrix} - B \begin{bmatrix} u_1(k) & u_2(k) \end{bmatrix} \right) x \left(\begin{bmatrix} f_1(k) \\ f_2(k) \end{bmatrix} - B \begin{bmatrix} u_1(k) \\ u_2(k) \end{bmatrix} \right) \tag{7}$$

Transforming the J function into a quadratic form yields to:

$$J = \begin{bmatrix} f_1(k) & f_2(k) \end{bmatrix} \begin{bmatrix} f_1(k) \\ f_2(k \end{bmatrix} - 2 \begin{bmatrix} u_1(k) & u_2(k) \end{bmatrix} B^T \begin{bmatrix} f_1(k) \\ f_2(k \end{bmatrix} + \begin{bmatrix} u_1(k) & u_2(k) \end{bmatrix} B^T B \begin{bmatrix} u_1(k) \\ u_2(k) \end{bmatrix} \tag{8}$$

Consequently, the system optimization can be solved as a least squares problem. The first derivative of J function from (8) in respect to control signal u is

$$\frac{\delta J}{\delta u_{12}(k)} = -2B^T \begin{bmatrix} f_1(k) \\ f_2(k) \end{bmatrix} + 2B^T B \begin{bmatrix} u_1(k) \\ u_2(k) \end{bmatrix} \tag{9}$$

where the minimum value of J is acquired when (9) is equal to 0. Control signal with the minimum value of J is written as follows,

$$\begin{bmatrix} u_1(k) \\ u_2(k) \end{bmatrix} = (B^T B)^{-1} B^T \begin{bmatrix} f_1(k) \\ f_2(k) \end{bmatrix} \tag{10}$$

Next, the selected control action is obtained by substituting (6) into (10). It leads to the function below:

$$\begin{bmatrix} u_1(k)^{opt} \\ u_2(k)^{opt} \end{bmatrix} = B^{-1} \left(\begin{bmatrix} r_1(k) \\ r_2(k) \end{bmatrix} - A \begin{bmatrix} y_1(k) \\ y_2(k) \end{bmatrix} \right) \tag{11}$$

This $u(k)$ value is the optimal solution of one-step ahead prediction algorithm in FCS.

The derived algorithms in all sections throughout this paper represent multi input-multi output system. In the structure of relative performance management of a virtualized software system in Figure 2, the input and output signal are formulated in the ratio value from the two virtual machines. Therefore, the identified dynamic model is a single input-single output function (SISO), as follows

$$y(k+1) = ay(k) + bu(k) \tag{12}$$

Furthermore, all corresponding equations for the controller design should be treated as a SISO system. For instance, the optimal control signal from (11) is simplified to,

$$u(k)^{opt} = b^{-1}(r(k) - ay(k)) \tag{13}$$

and the cost function for SISO system is:

$$J = (r(k) - y(k+1))^2 \tag{14}$$

Control law for Optimal FCS-MPC (OFCS) is sorted in the following steps:

(i) At sampling time k, find the reference value $r(k)$ and measure output $y(k)$.

(ii) Calculate the control solution for all the optimal control signal candidates v_n, $n = 1, 2, 3, \ldots, N$ using (13), where N is the total number of control states.

(iii) Compute the objective function J based on the squares of error between the predicted value $y(k)$ and the reference value $r(k)$ using (14).

(iv) Find J_n, which has the smallest value and get its corresponding index n. The control signal with this index (v_n) is the optimal control solution to be implemented in the feedback control system at sampling time k.

(v) Repeat from step 1 for the next sampling time $(k+1)$.

3.2. Candidates of Control States

In this section, the effect of control states length is examined using Optimal FCS. The feedback control system is evaluated with two different length of input states. The available resources are assumed to be 100% for the total capacity and the minimum allocation for each virtual machine is 20%. By setting $\delta u = 1$ for each Cap_1 and Cap_2, the input signal ($u = \frac{Cap_1}{Cap_2}$) can be arranged in a sequence of 61 input operating points [13].

$$u = \left\{ \frac{20}{80}, \frac{21}{79}, \ldots, \frac{50}{50}, \ldots, \frac{79}{21}, \frac{80}{20} \right\} \tag{15}$$

For Hammerstein-Wiener feedback control structure, input signal is the intermediate variable v. The values of v are assigned as the set of possible control states in FCS control scheme. To transform the input u so that the operating points are equally spaced, the intermediate variable $v(k)$ values are set arbitrarily by $v_{min} = -15$, $v_{max} = 15$ where $\delta v = 0.5$. Thus, the number of v operating points is 61 with the sequence of $v = \{-15, -14.5, \ldots, 0, \ldots, 14.5, 15\}$. The sequence of v based on the selection of δv determines the length of input states. The smaller the δv value, the longer the operating input set.

The OFCS algorithm is implemented to control the system with reference value $r = 1$ and nominal workloads of 100 requests/second for both clients. The control scheme is in Hammerstein-Wiener structure which is designed based on the linear model in eq. 2. Then, using eq. 12, parameter a and b are determined, $a = 0.4201$, and $b = 0.0430$.

The values of $\delta v = 0.5$ for 61 points and $\delta v = 0.1$ for 301 points are chosen. It can be seen that the performance of the FCS control system with 301 control states is significantly better than the one with 61 states. The output responses are presented in Figures 5 and 6. The output response from the controlled

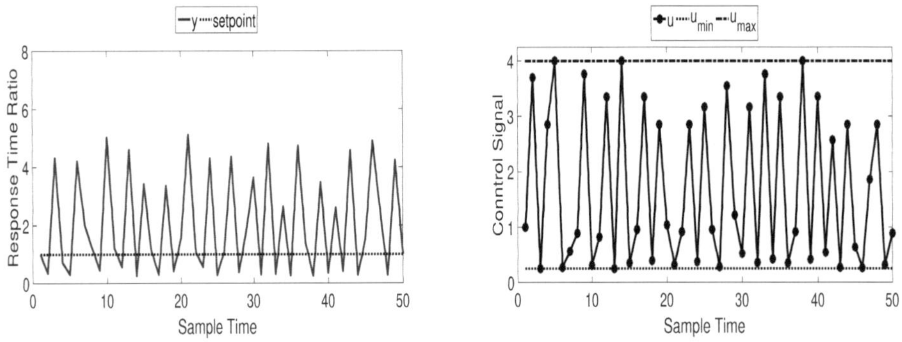

Figure 5: Output response and input signal u of OFCS-MPC for 61 control states of w

system with a longer sequence of control states has a less steady state error and the change of control signal trajectory is more stable than the one with fewer control states. It can be analyzed that with the longer control states, the rate of change in control signal will be smoother because the system is provided with a wide range of control state candidates.

3.3. Constrained Finite Control Set (CFCS)

This section presents the procedure for finding the actual control signal $u(k)$ among the available candidates. The optimal control signal is not automatically equal to one of the control signal candidates. Therefore, constraint is deployed in the prediction of the OFCS control system to revise the searching procedure to calculate the actual control signal. The objective function for the constrained FCS-MPC

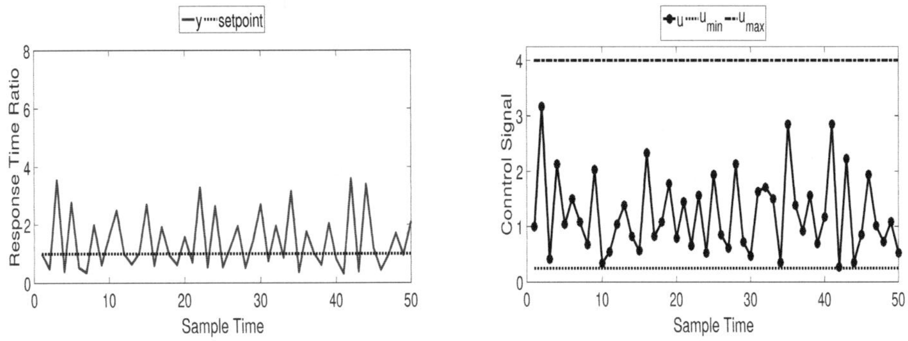

Figure 6: Output response and input signal u of OFCS-MPC for 301 control states of w

(CFCS) is converted to:

$$J = \left(\begin{bmatrix} u_1(k) \\ u_2(k) \end{bmatrix} - \begin{bmatrix} u_1(k)^{opt} \\ u_2(k)^{opt} \end{bmatrix} \right)^T (B^T B) \left(\begin{bmatrix} u_1(k) \\ u_2(k) \end{bmatrix} - \begin{bmatrix} u_1(k)^{opt} \\ u_2(k)^{opt} \end{bmatrix} \right) \tag{16}$$

This function is derived based on the value of the optimal control signal as in (11). Then, the objective function J can be presented in the formulation below:

$$J = B^2(u_1(k) - u_1(k)^{opt})^2 + B^2(u_2(k) - u_2(k)^{opt})^2 \tag{17}$$

It is sufficient to evaluate this function by using only:

$$J = (u_1(k) - u_1(k)^{opt})^2 + (u_2(k) - u_2(k)^{opt})^2 \tag{18}$$

Furthermore, to apply a constrained FCS-MPC method in relative performance management, the algorithm of CFCS should be adjusted in compatible formulations for a single input-single output function (see (12) and (13 for the output function and optimal control signal respectively). Likewise, the cost function from (18) becomes

$$J = (u(k) - u(k)^{opt})^2 \tag{19}$$

The control procedures of the constrained optimal finite control set are summarized as follows:

(i) At sampling time k, measure output $y(k)$ and get the reference value $r(k)$.

(ii) Calculate the optimal control signal $u(k)^{opt}$ using (13).

(iii) Compute the objective function J_n for all the control candidates ($n = 1, 2, \cdots, N$) using (19), where N is the total number of control states.

(iv) Find J_n, which has the smallest value, and get its corresponding index n. The control signal with this index (v_n) is the optimal control solution to be implemented in the feedback control system at sampling time k.

(v) Repeat from step 1 for the next sampling time $(k+1)$.

4. Evaluation of Optimal FCS and Constrained FCS

In this section, the performance of Constrained FCS-MPC and Optimal FCS-MPC are evaluated in an experiment where the reference value is $r = 1$. Feedback controller with an addition of pre-input and post-output nonlinear compensator is presented in Figure 7.

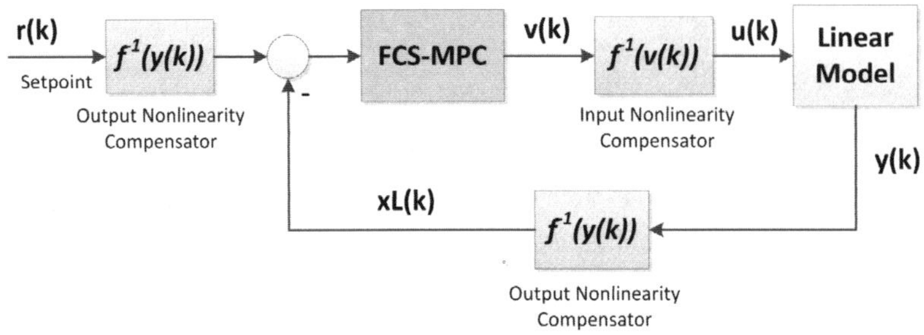

Figure 7: Feedback control loop with nonlinear compensations

301 control states are used for this experiment. The disturbance for the control system is the workload changes of $client_1$ and $client_2$. During the first 40 samples, workloads for all clients are 100 requests/second, and then they are increased simultaneously to 200 requests/second. Experimental results for OFCS and CFCS are given in Figures 8 and 9. In addition, the statistics for output response are presented in Table 1.

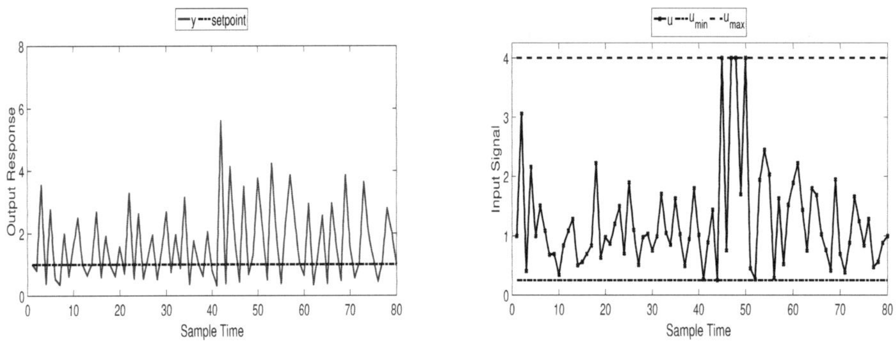

Figure 8: Output response and control signal of OFCS

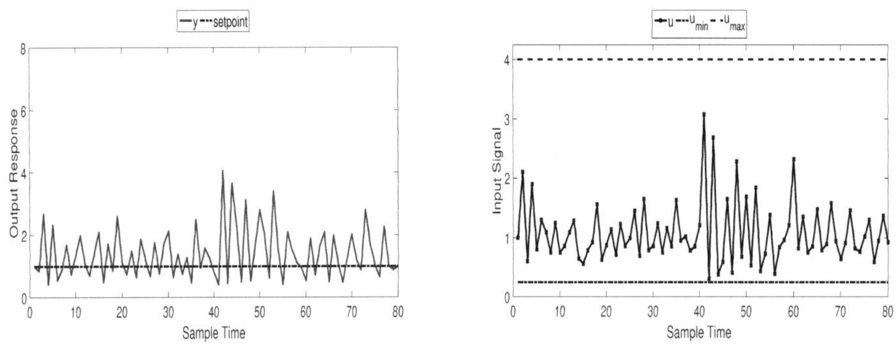

Figure 9: Output response and control signal of CFCS

Table 1: Output response statistics of OFCS and CFCS

Experiment	MSE	Minimum value	Maximum value
OFCS	1.8095	0.32	5.59
CFCS	0.8217	0.41	4.05

It can be inferred from the results that the improved procedure of control state searching by adding constraint in the CFCS algorithm makes the movement of control signals less aggressive than in OFCS. Likewise, the output response of CFCS has a smaller steady state error compare to the OFCS response. The MSE value of OFCS is 1.8095, which is two times bigger than the CFCS. This error caused by the oscillations of output response in OFCS with a higher magnitude of overshoots compare to the CFCS response.

5. CONCLUSIONS

FCS-MPC has been studied and implemented for dynamic resource provisioning in one type of shared-resources environment. Based on the experimental results, it can be concluded that a wider range of states selection provides a smoother dynamic of control signal. Moreover, constrained FCS-MPC demonstrates a better performance where its output response has smaller steady state error than the optimal FCS-MPC.

References

[1] Patikirikorala T, Wang L, Colman A and Han J 2012 *Control Engineering Practice* **20** 49–61
[2] Padala P, Shin K G, Zhu X, Uysal M, Wang Z, Singhal S, Merchant A and Salem K 2007 Adaptive control of virtualized resources in utility computing environments *ACM SIGOPS Operating Systems Review* vol 41 (ACM) pp 289–302
[3] Vargas R, Ammann U and Rodriguez J 2009 *IEEE Transactions on Power Electronics* **24** 894–902
[4] Quevedo D E, Aguilera R P, Perez M A, Cortés P and Lizana R 2012 *IEEE Transactions on Power Electronics* **27** 3128–3136
[5] Cortés P, Kazmierkowski M P, Kennel R M, Quevedo D E and Rodríguez J 2008 *IEEE Transactions on industrial electronics* **55** 4312–4324
[6] Kouro S, Cortés P, Vargas R, Ammann U and Rodríguez J 2009 *IEEE Transactions on Industrial Electronics* **56** 1826–1838
[7] Rodriguez J, Pontt J, Silva C A, Correa P, Lezana P, Cortés P and Ammann U 2007 *IEEE Transactions on Industrial Electronics* **54** 495–503
[8] Vargas R, Cortés P, Ammann U, Rodríguez J and Pontt J 2007 *IEEE Transactions on Industrial Electronics* **54** 2697–2705
[9] Morel F, Lin-Shi X, Rétif J M, Allard B and Buttay C 2009 *IEEE Transactions on Industrial Electronics* **56** 2715–2728
[10] Chai S and Wang L 2012 Finite control set model predictive control of 2lvsi-pmsm using interpolated switching states *IECON 2012-38th Annual Conference on IEEE Industrial Electronics Society* (IEEE) pp 1799–1804
[11] Preindl M and Bolognani S 2013 *IEEE Transactions on Power Electronics* **28** 1007–1015
[12] Aryani D, Wang L and Patikirikorala T 2017 *IFAC-PapersOnLine* **50** 7795–7800
[13] Aryani D, Wang L and Patikirikorala T 2016 *International Journal of Systems Science* **48** 1146–1161

Analysis of Natural Light Distribution in the Building

Nurul Jamala[1], Ramli Rahim[2], Rosady Mulyadi[3]

[1,2,3] Department of Architecture, Faculty of Engineering, Hasanuddin University, Indonesia

E-mail: nuruljamala@yahoo.co.id

Abstract. Global warming affect the condition of the sky and also affect the temperature on the earth surface. Changes in sky conditions affect the distribution of light into the building. Sunlight is a natural light source in the morning, afternoon and evening. Utilization of natural lighting into buildings, can reduce energy use as an artificial lighting source. Phinisi Tower Building has a unique form of hyperbolic paraboloid, so it can be a guide in designing buildings. This building needs to be analyzed, how the distribution of light into the building, so it can become a guide in buildings design. The purpose of this study is how the influence of position, floor height and window openings on the building envelope to the distribution of natural light. Quantitative research method is to analyze the level of illumination inside and outside the building in the form of tables and graphs. The result of research is illuminance level of daylight depends on the sky condition and it have affect the light distribution into the building. The increased light distribution at the height of the floor. As well as of window opening on the building envelope (top lighting and side lighting) have effect on the level of illuminance in the building. The distribution of natural light will increase during the day, if the sky condition does not change in one day.

Keywords: light distribution, level illuminance, natural light

1. Introduction

The concept of energy-efficient architecture is to optimize the lighting system by considering the integration between natural lighting (daylight) and artificial (lights). The use of energy in buildings, can be reduced if it can maximize the distribution of natural light, but by considering the negative effects include: thermal, glare and brightness.

Phinisi Tower of Makassar State University is the Center Academic Building. This building implements the Hyperbolic Paraboloid Facade which is a futuristic application of scientific and technological sophistication. This building is a 3-storey podium and 12-story tower. This form of building is a screen metaphor of phinisi boat like picture 1 below.

The building lighting design is designed according to the illumination standard recommended by SNI 03-6575-2001 [1]. Level of Illuminance at work space of 350 lux, meeting room of 300 lux and hallway as connecting space of 150 lux. Recommendations standar of illuminance level for guideline to be reference for lighting design in building namely for the standard of inspiration need to be a reference in designing buildings namely Illuminating Engineering Society [2] and Illuminating Engineering Society of Nort America [3]. Light distribution in buildings, can be influenced by sky conditions, position, wide and shape window [4].

Based on this recommendation, the designer can determine amount of armatures required on space, according to the activities in the space. Design of building although it does not meet these standards, can be running well, the activities of space. Based on this, it is necessary to study the distribution of natural light on selected objects [5].

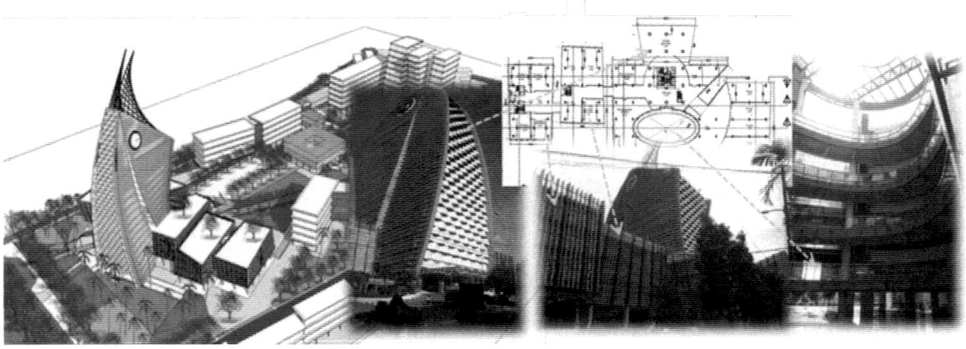

Figure 1. Phinisi Tower of Makassar State University

2. Literature Study

The Phinisi tower building uses a diagonal and horizontal facade on the building's envelope, so there is no glare and brightness in space. This building is one example of building in designing high building. In terms of aesthetics, building facade is a positive value but needs to be considered the visual comfort of space users so that work productivity can increased [6]. Sky condition have classified into three groups sky condition i.e. overcast, clear and intermediate sky, trying to define the luminance distribution of the intermediate sky [7][8]. According to other researcher concluded that sky conditions based on the number and type of clouds can be grouped into: (a) Clear sky is the sky without clouds; (b) Overcast sky is a sky completely covered in white or gray white clouds or partial or complete partial clouds; (c) intermediate sky is a partially clouded sky with a variety of darker clouds and numbers: and (d) Uniform sky is the sky with the same luminance in all positions not dependent on the geographic latitude and the height of the sun [9].

Local Government Rules No. 38 (2012) explained that the building envelope is a building element that surrounds the building. They are transparent or not transparent walls or the roof. Skylight from the roof operates as lighting lamps that emit light directly with the direction of light down the hole while the light from the side using vertical openings to take advantage of daylight. The recommendation standard of illuminance levels in the office room are base on recommendation value by CIE (Commision Internationalde I'Elaire) and IES (Illuminating Engineers Society). These recommendations are national and international standards of illuminance levels for design of lighting design [10].

Indonesia is a tropical humid region with high relative humidity characteristics (60% -80%), high radiation (80% per year), and unstable velocity (urbanized between 0-> 30 m / seconds) [11]. The distribution of light lumination in open plan office by using general lighting and suitability of light lumination distribution [12]. The researchers concluded that examined quantitative guidance of illuminance, luminance and ratio values in office buildings [13]. Utilization of natural lighting is closely related to the geographical position of a building because the relative movement of the sun on each coordinate on earth is different. For that needed a solar diagram that can help the observation and estimate the amount of sunlight that enters into a building [14]. Within a year there is a change in altitude angles in the sun and the sun's rotation changes the orientation of the building's orientation. In the movement of the sun there are four angle changes that is between March, June, September and December.

3. Methodology

Measurements are done for 3 days, 19, 20 and 21 May 2018. The measurements are done in the morning (08.00-09.00), noon (12.00-13.00) and afternoon (16.00-17.00). Measurements begin by measuring the level of illuminance to determine the condition of the sky at the time of measurement.

The measurement results are made in the form of tables and graphs to analyze the distribution of natural light in the hall area of the connecting several room. Selected lobby area is Zone A and B located on the 1st, 2nd and 3rd floors. The orientation of building in zone A1 is east-west, while zone B is *east-west*. The point notation uses code A-B and 1-3, for example A1, A3, B2, C3 and so on, as in Figure 2 below.

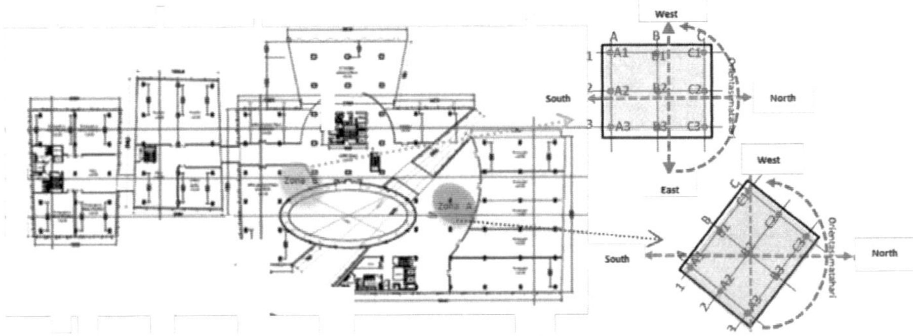

Figure 2. The position of the hall and the placement of the measuring point of the research object (Zones A and B)

4. Results and Discussions

4.1 Analysis of the sky conditions at the time of measurement

Makassar city is located in Latitude -5,1° and Longitudinal 119,5° (+8,0). Field measurements were carried out for 3 days in the morning, afternoon and evening. The condition of the sky at the time of measurement of the difference of the sky conditions of planning. The first measurement on 19 May 2018 on outside the building has a clear sky condition of 51,000 lux (morning), 94,200 lux (noon) and 32,100 lux (afternoon). The second and third measurements were conducted on 20 and 21 May 2018 on intermediate sky conditions, as shown in Figure 3 below.

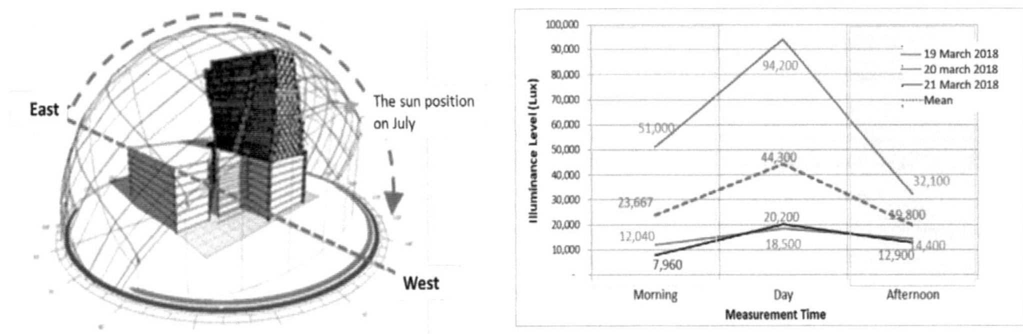

Figure 3. Sky Condition in the Morning, Noon and Afternoon

Figure 3 shows the position of the sun in May and the measurement on graph for 3 days. This gafik shows that the highest illumination rate is 94,200 lux while the lowest is 7960 lux. The average measurement results outside the building is 23,667 lux (morning), 44,300 lux (afternoon and 19,800 lux (afternoon) .This mean value shows that the highest level of illumination outside the building is during the day The results show that the increase of natural light distribution into the building, will increase if the sky conditions increase.

4.2 Analysis of first measurement results in the morning (Zone A floors 1,2 and 3)

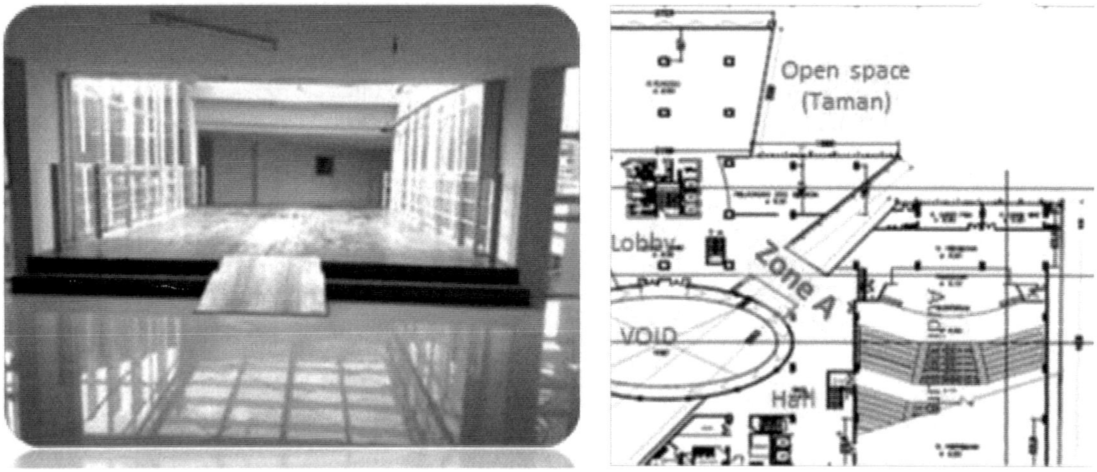

Figure 4. The position Zone A in the Podium Makassar State University

In Phinisi building there are several hall area as interconnecting work space. One of the selected objects is Zone A on the 1st, 2nd and 3rd floors. This area is directly related to the building envelope, but have the different models or positions. Zone A on the 3rd floor, directly related to the sunlight from the top and side of the building, while Zone B is only on the side envelope of the building. The average value of the first measurement results shows the illumination level as in the following table 1.

Table 1. The results of measurement of iluminance level in the morning on floors 1, 2 and 3

	Measurement Point	1	2	3
1st Floor	A	164	328	690
	B	62	167	137
	C	311	315	280
2nd Floor	A	267	665	956
	B	142	194	192
	C	186	525	417
3rd Floor	A	1465	6200	6350
	B	3810	7150	6020
	C	4390	7250	5050

Table shows the illuminance level on the 3rd floor is the highest compared to the 1st and 2nd floors, because on Zone A on the 3rd floor is surrounded by the opening of the building envelope in the side and top lighting. The highest of illuminance level in this area is 7250 lux (C2) and the lowest is 3810 lux (B1). The illuminance level on the 2nd floor is the highest of 690 lux (A3) and the lowest is 62 lux (B1), while on the 3rd floor is the highest 956 lux (A3) and the lowest 142 lux (B1).

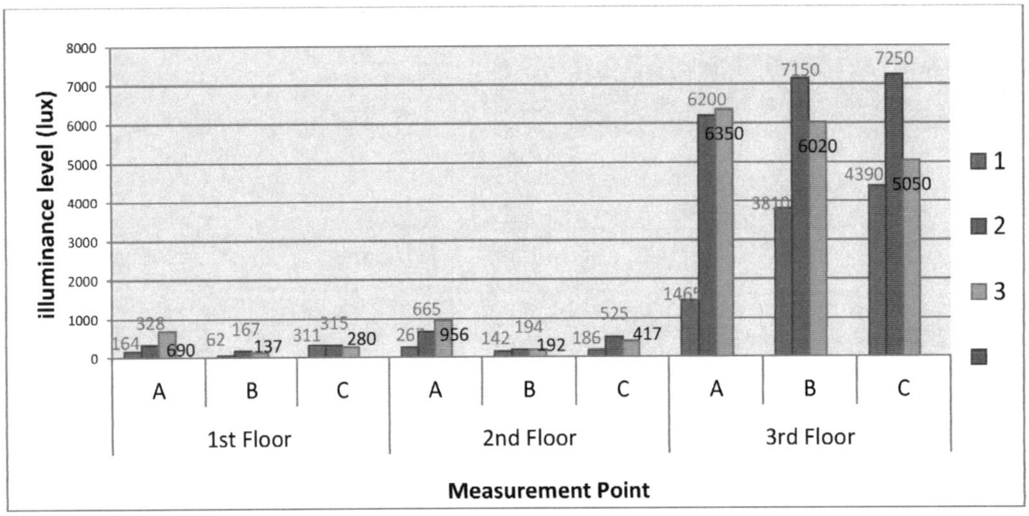

Figure 5. illuminance Level of measurement point A1-C3 on floors 1, 2 and 3.

Figure 5 shows a graph of the first measurements in the morning on floors 1,2 and 3. This graph shows that the distribution of light is higher, when the openings on the building envelope are wider. The results of the analysis show that the open area effect on the distribution of natural light into the space.

Furthermore, in order to simplify the analysis of illumination values on Zone A of floor 1, 2 and 3, it is calculated the mean value at the point of A, B and C as shown in the following Fig. 6 and 7.

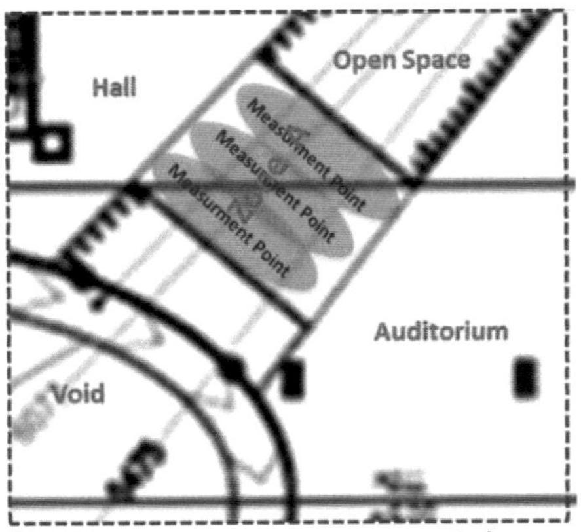

Figure 6. Block of Measurement Point at A1-A3, B1-B3 and C1-C3

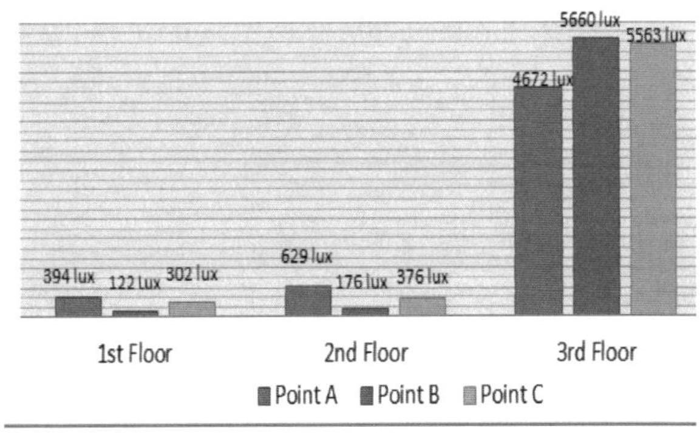

Figure 7. Average value of illumnance level on floors 1, 2 and 3 in the morning

The mean values in Fig. 7 show the distribution of light at the points A, B and C on the 3rd floor is very high at 4672 lux (point A), 5563 lux (point B) and 5660 lux (point C). This description shows that the middle area of space has the highest illumination level is at point B. In Zone A floor 3 has very high illumination level, because the light distribution in this Zone has openings of building envelope that is side lighting (right and left) and Top Lighting. On floors 1 and 2 show that the distribution of natural light is lower than that of the 3rd floor Zone, since it only has a building envelope opening on the side lighting (right and left). The results of the analysis show that the area not on the building envelope affects the level of illumination in the building.

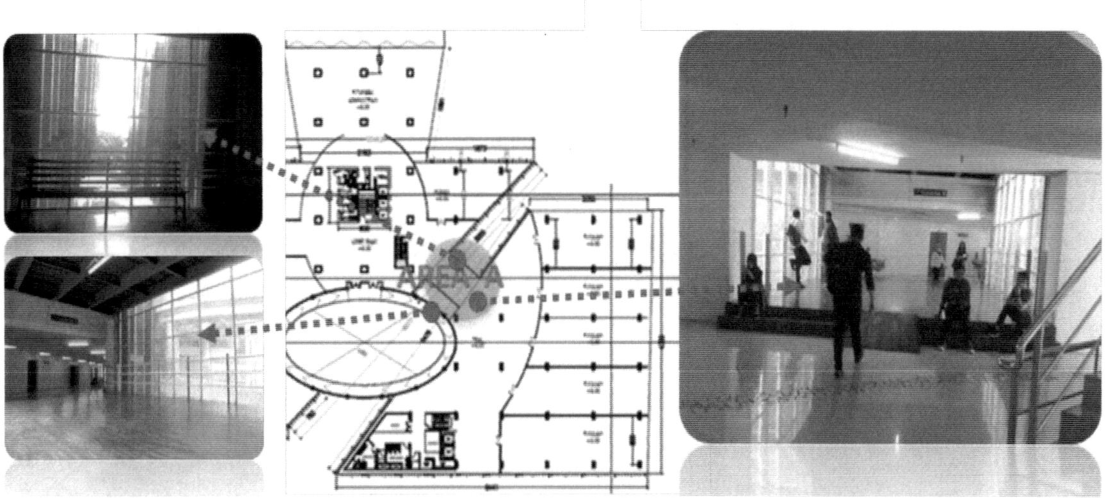

Figure 8. The position side Lighting (right and left) on the building envelope

4.3. Average value analysis for 3 days measurement in the morning

Measurements were made in the morning for 3 days with different sky conditions ie 51,000 (first) lx, 12,040 lx (second) and 7960 (third) with a mean value of 23,667 lux. Measurements made on the 1st, 2nd and 3rd floors are to know how the influence of floor height on natural light distribution. Mean value for 3 times measurement indicate that level of illumination at zone A floor 1 equal to 114-251 lux, floor 2 equal to 198-569 lux and floor 3 equal to 298 5-k5256 lux. Graph of measured average value in the morning can be seen in the following picture.

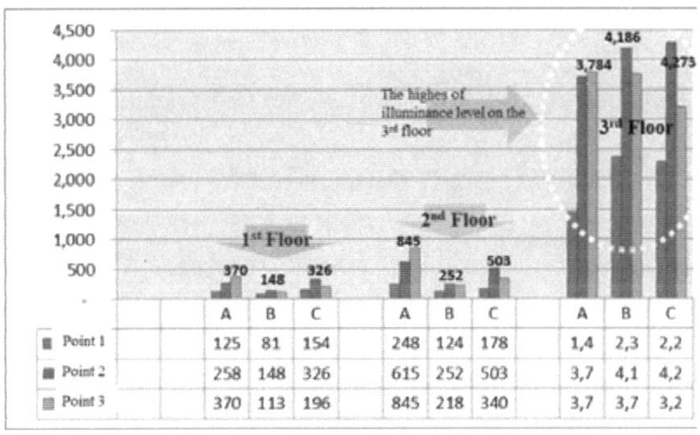

Figure 9. Average value for three days measurement in the morning

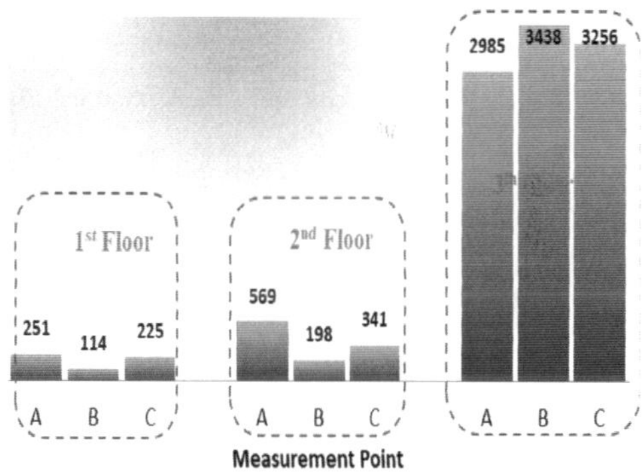

Figure 10. Average value of point area for three days measurement in the morning

The picture above shows the graph of light distribution that goes into buildings on floors 1,2 and 3 with different percentage of different light distribution. The level of illumination in the Zone A area of the 4th floor is 2985 (average point A), 3438 lux (average point B) and 3256 lux (mean of point C). this graph shows that the higher the position of the space then the level of illuminance is increasing. For example: comparison between the Zone A of the 1st and 2nd floors, where this area has a building envelope opening with the same area and shape. The level of illuminance on the 2nd floor is higher than the 1st floor with the average value of 281 lux (1st floor) and 533,3 lux (2nd floor). This shows that the distribution of natural light affects the height of the building.

Figure 11. The position of the Zone A area on the 1st, 2nd and 3rd floor

4.4. Average value analysis for 3 times measurements during the day

The average value of level illuminance on floors 1,2 and 3, maximum of 18,706 lux (measuring point B2, floor 3) and minimum 154 lux (B2 measurement point, 1st floor). This graph can be seen in Figure 15 below.

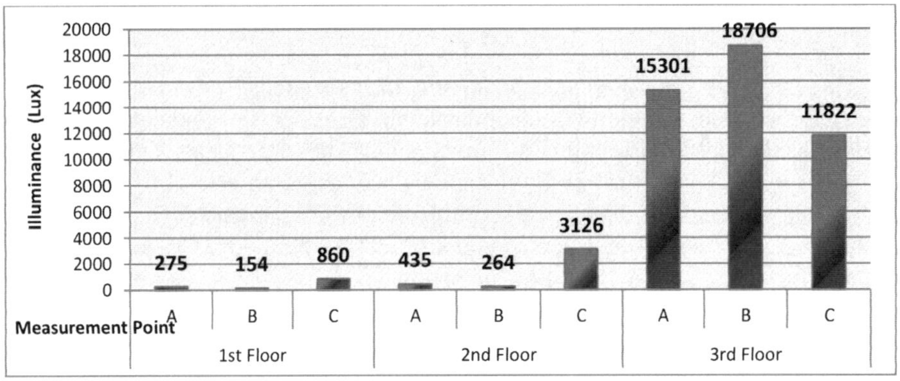

Figure 12. Average value of illuminance level on during the day

The sky conditions during the day were highest at 94,200 lux (first day), 18,500 lux (second day) and 20,200 lux (third day) so that the average value was 44,300 lux. The highest of illuminance level in the area Zone A is 860 lux (1st floor), 3126 lux (2nd floor) and 18.706 lux floor 3. The level of illuminance on the 3rd floor is the highest, because there are openings in the building envelope on top and side of the building. This indicates that the distribution of natural light into buildings influences the height of the floor and the width of the building envelope.

4.5 Average value analysis for 3 times measurement in the afternoon

Measurements in the afternoon conducted for 3 days, so that the measurement results are mean to determine the effect of the illuminance level on certain variables.

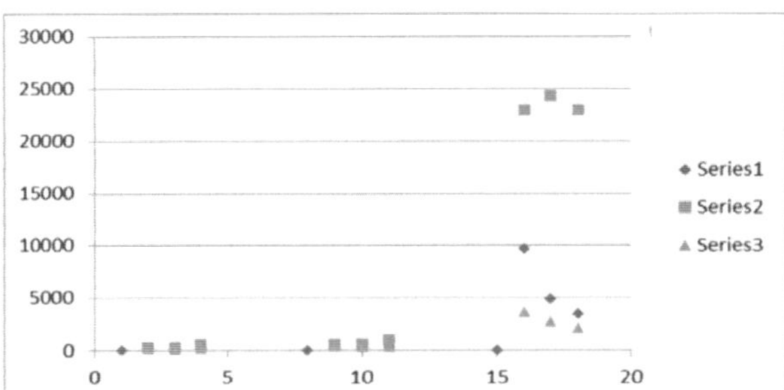

Figure 13. Measurement result of illuminance level in the afternoon

Figure 14 shows the average value of illuminance in the afternoon. The average value of the illuminance in the afternoon has the highest value on the 3rd floor, because the distribution of light on the top of building envelope is not blocked, so that the level illuminance is very high. The level of illuminance in the Zone A of 3rd floor is at least 12,220 lux (group A) and at least 9522 lux (group C). the graph shows a very high degree of level illuminance on the 3rd floor, because the building cover is bigger than the 1st and 2nd floors. The average value on the 1st and 2nd floors has the highest value in area group C (front of the building envelope) which is 347 lux (group C, floor 1) and 531 lux (group C, 2nd floor). The level of illuminance in this area is higher than others, because the building faces to the west and the sun sets in the afternoon

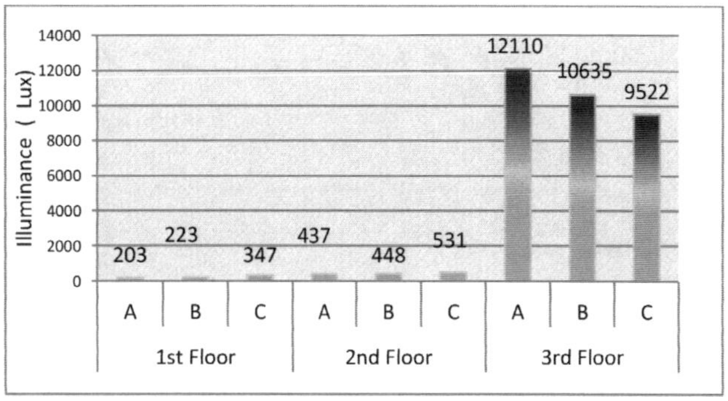

Figure 14. Average value in the afternoon for three days

4.6. Illumination value analysis on Zone A floor 1,2 and 3 in the morning, day and afternoon

The average value of the levels illuminance in Zone A on the floors 1, 2 and 3 shows the ratio of illuminance levels in the lobby area in the morning, afternoon and evening. The measurement results show that the highest sky conditions that is during the day, with the average value of illuminace rate of 44,300 lux. The distribution of light in the morning, afternoon and evening can be seen in Figure 17 below.

	Point	Morning	Day	Afternoon
1st Floor	Point A	251	275	203
	Point B	114	154	223
	Point C	225	860	347
2nd Floor	Point A	569	435	437
	Point B	198	264	448
	Point C	341	3126	531
3rd Floor	Point A	2985	15301	9980
	Point B	3438	18706	10635
	Point C	2890	11822	9522

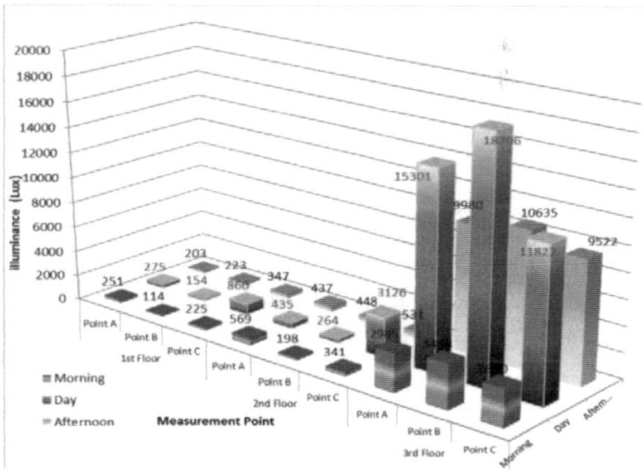

Figure 15. Measurement result on Zone A for three time

The average value of illuminance level in zone A area 1, 2 and 3 shows that the illuminance level during the day is higher than in the morning and afternoon. The analysis of the condition of the sky, also shows that the illuminance level at noon is higher than the morning and evening. Based on this, it can be seen that the distribution of light into the building is related to the sky condition. The level of illuminance in the zone A zone fluctuates, so it only analyzes on the 4th floor, because it has the highest of level illuminance compared on the 1st and 2nd floors.

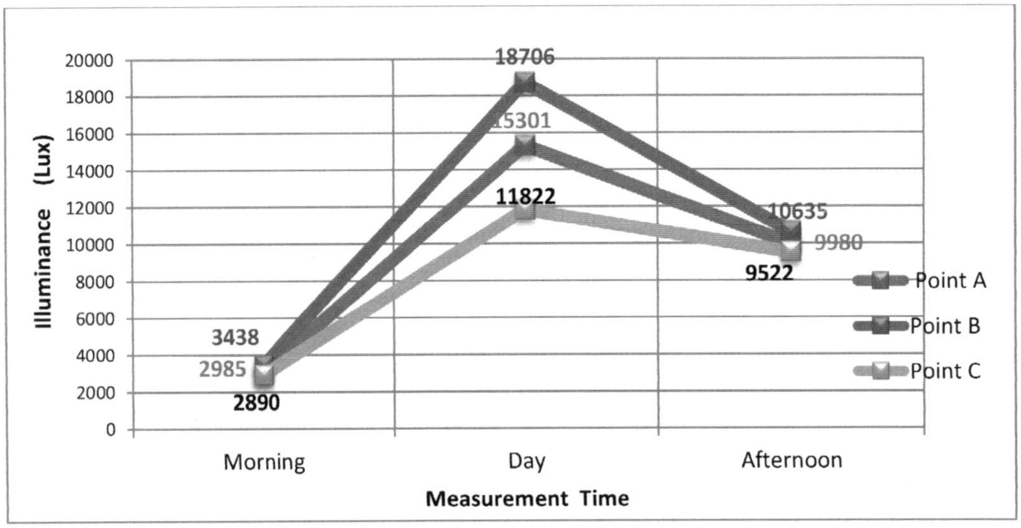

Figure 16. Average value of measurement result on Zone A area

Figure 16 shows the average value of illumination level on the 4th floor. The analysis of the light distribution on the 4th floor shows that the highest illuminance level occurs at noon with a value of 18706 lux (point A), 15302 lux (point A) and 1822 lux (point C). the lowest illumination level is in the morning at 3438 lux (point B), 2985 lux (point A) and 2890 lux (point C).

4.7 Illuminance level analysis in Zone B

Zone A and B area conditions are closely related to having openings on both the left and right sides of the building envelope. The orientation of the window is different ie Zone A on east-west whereas Zone B on Southwest-Northeast (figure 17)

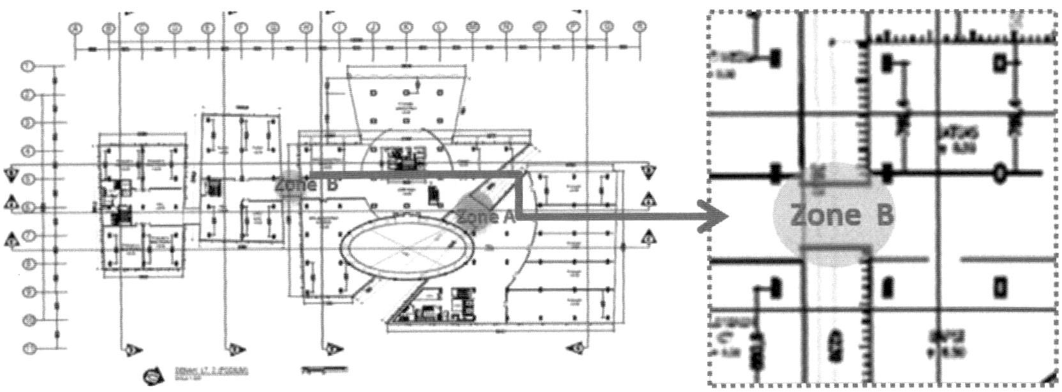

Figure 17. The position of the Hall in Zone B

Measurements conducted for 3 days in the morning, *during the day* and afternoon. the average value of the measurement results can be seen in Figure 19 below. Level of illuminance In Zone B has the highest value on the 3rd floor compared to the level of illumination on the 1st and 2nd floors. The measurement results show that the highest illumination level is *during the day* time is 323 lux (floor),

559 lux (2nd floor) and 4240 lux (3rd floor). in the Zone B has the lowest of the illuminance level at noon that is 300 lux (1st floor), 371 lux (2nd floor) and 1306 lux (3rd floor).

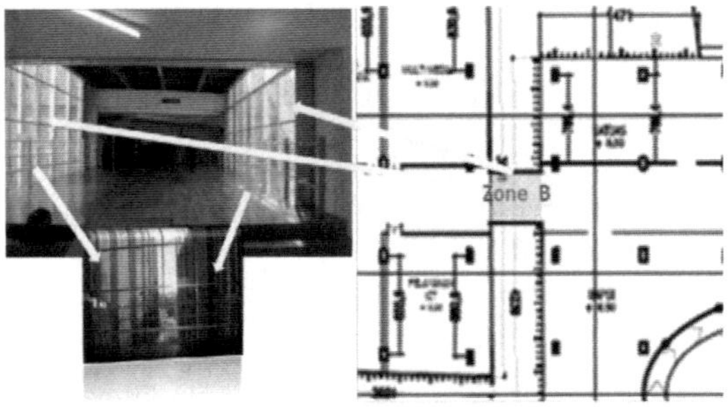

Figure 18. Window opening on the building envelope (Side Lighting)

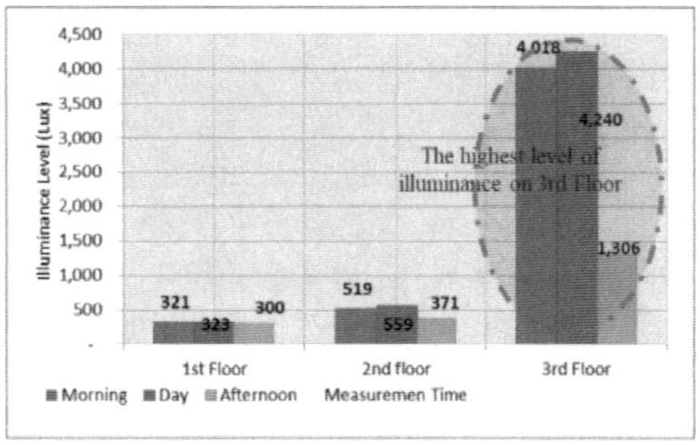

Figure 19. Average level of illuminance on 1st,2nd and 3rd Floor

4.8 Results of comparison analysis of natural light distribution in Area A and B

The results of the analysis show that there is no difference in the rules of light distribution in Zone A and B, although the orientation of the building envelope is not the same but resembles to have a similar light distribution (figure 20).

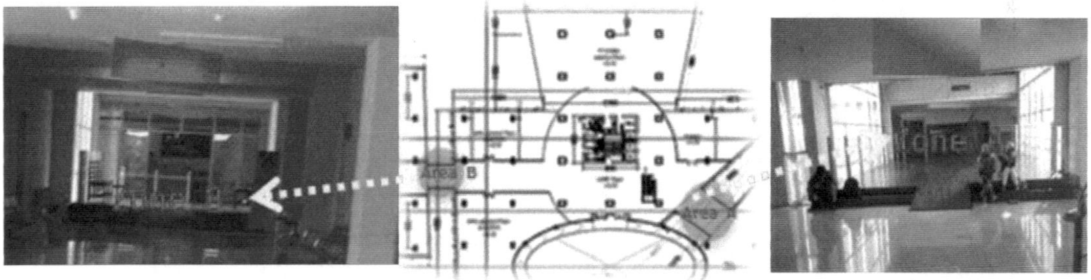

Figure 20. Window opening on the building envelope in Zone B area

Overall it shows that the Zone A has a higher of illuminance level than the Zone B, since the Zone A positions have partially openings of the building envelope surrounded by open space and serve as a void. In the Zone B area has openings on the building envelope from the left and right, but is blocked by the building resulting in a reduction in the distribution of natural light. This can be seen in Figure 21 below.

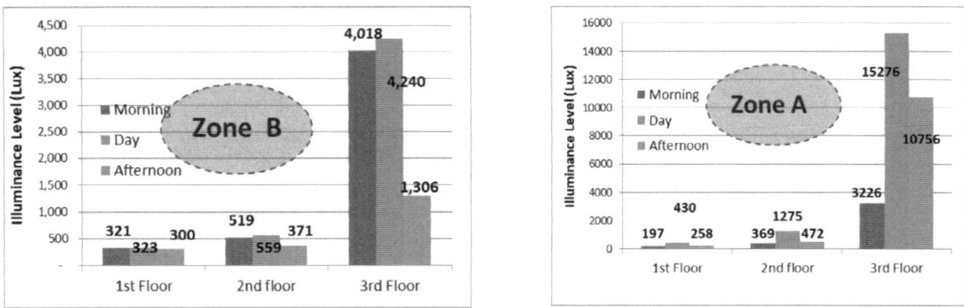

Figure21. . Illuminance Levels in Zone A and B

Figure 21 Shows that Overall Illumination Level In Zone A is higher than Zone B. The level of iluminance on Zone A of floor 4 is between 3226-15276 lux while in Zone B is between 1306-4240 lux. Overall, the illumination level in Zone A on floors 1, 2 and 3 are higher than the Zone B on Floors 1,2 and 3. The illumination level in Zone A is between 282-4240 lux whereas in Zone B is between 197-15267 lux. The level of illuminance in the Zone A on floors 1, 2 and 3 are the same that is the highest during the day, as well as in the B on floors 1, 2 and 3.

5. Conclusion

The level of illuminance outside the building depends on the condition of the sky (clear sky, intermediate sky and overcast sky) The sky conditions affect the distribution of light into the building and the distribution of light that increasing in the height of the floor of the building. As well as of wide opening on the building envelope (top lighting and side lighting) have effect on the level of illuminance in the building. Example in this research, that the level of illuminance in Phinisi building hall area is higher on the 3rd floor compared to the 1st and 2nd floor.

The opening position on the building envelope also affects the distribution of natural light. If there is no building or other near the building envelope, then the distribution of light can be maximally in to the building. For example: the ratio of illuminance levels in the Zone A is higher than the illumination level in the Zone B

The results of the analysis concluded that the distribution of natural light will increase during the day, if the condition of the sky does not change in one day. Zone A and B area conditions are close to the same i.e. have openings on the left and right side of the building envelope and orientation of window opening is different i.e. area Zone A east-west direction while Zone B direction of Southwest-Northeast, but the procedure of distribution of light is same.

6. References

[1] Indonesian Nasional Standard 03-6575-2001 *The Design of The System of Artficial Light in The Building* (Original in Indonesia language)
[2] Illuminating Engineering Society (IES) 1977 *code for Interioir Lighting* (London : IES Lighting Handbook)
[3] Illuminating Engineering Society of Nort America (IESNA) 2004 *American National Standars Practice for Office Lighting (*New York*:ANSI/IESNA-RP*)
[4] Indonesian Nasional Standard 03-2396-2001 *The Design of The System of Natural Light in The Building* (Original in Indonesia language)
[5] Jamala N 2012 *Visual Comfort on Drawing Studio Room Using the Echotect Program* Journal Gelagar Vol 26-1 (Original in Indonesia language)
[6] Jamala, N 2017 *The Effect building Façade on Natural lighting (Case Study: Building Of Phinisi Tower UNM)* Proceeding by AIP Publishing 978-0-7354-1499-0 (American Institute of Physics) Volume 1831- 02006
[7] Nakamura H et al 1985 *Luminance Distribution of Intermediate Sky* Journal of light and Vis. Environment 9 (1) p n6-13
[8] Rahim R 2009 *Theory and Application of sky Luminance Distribution in* Indonesia National Library (Original in Indonesia language)
[9] Sugianto 1998 *Building in Indonesia with a Tropical Climate Humidity from Aspect Building Physics* Jakarta (Original in Indonesia language)
[10] United Nations Environment Programmed (UNEP) 2006 *Energy Efficiency guide for industry in Asia* (India: *www.energyefficiensiasia.org*) p. 1-43.
[11] Satwiko, P., (2009). Building Physics. Yogyakarta, Andy Publisher (Original in Indonesia language)
[12] Tommy 2002 *Preferred Luminance Distribution in Working Area*
[13] Newsham G R et al 2004 *Lighting Design for Open Plan Offices* (Canada: Construction Technology

[14] Gunawan Ryani 2011 *Simulation of the Opening of Direct Sunlight Lighting Design* Katholik Parahyangan University

IOP Conf. Series: Materials Science and Engineering **619** (2019) 012025 doi:10.1088/1757-899X/619/1/012025

A class of variable coefficient elliptic equations solved using BEM

N. La Nafie[1], N. Ilyas[2], M. I. Azis[2,*] A. K. Amir[2]

[1]Department of Chemistry, Hasanuddin University, Makassar, Indonesia
[2]Department of Mathematics, Hasanuddin University, Makassar, Indonesia

E-mail: mohivanazis@yahoo.co.id (*Corresponding author)

Abstract. In this paper a BEM is used to solve a class of variable coefficient elliptic equations numerically. Some examples are considered to show the convergence, consistency, and accuracy of the numerical solutions.

1. Introduction

BEM solutions to a number of problems for homogeneous media governed by several types of equations have been obtained (see for examples [1, 2, 3, 4]. However, BEM has not been used extensively to solve problems of inhomogeneous media. Progress towards this end has been shown, works in [5, 6, 7, 8, 9, 10, 11, 12] are the examples. Specifically, the governing equation considered by Salam et. al in [11] takes the form

$$\frac{\partial}{\partial x_i}\left[\kappa_{ij}(x_1,x_2)\frac{\partial\mu(x_1,x_2)}{\partial x_j}\right]=0 \qquad (1)$$

where the coefficients κ_{ij} depend on x_1 and x_2 and the repeated summation convention (summing from 1 to 2) is employed.

This paper is intended to extend the work by Salam et. al [11] for problems with governing equation (1) to for 2D boundary value problems governed by another type of (dimensionless) elliptic equation of the form

$$\frac{\partial}{\partial x_i}\left[\kappa_{ij}(x_1,x_2)\frac{\partial\mu(x_1,x_2)}{\partial x_j}\right]+\beta(x_1,x_2)\,\mu(x_1,x_2)=0 \qquad (2)$$

Equation (2) may be used to model a variety of problems such as steady infiltration problems (when $\beta < 0$, see for examples [13, 14]), acoustic problems (when $\beta > 0$, see for examples [15, 16]), and antiplane strain in elastostatics and plane thermostatic problems (when $\beta = 0$).

Equation (2) will be transformed to a constant coefficient equation from which a boundary integral equation will derived. It is necessary to place some constraint on the class of coefficients κ_{ij} and β for which the solution obtained is valid. The analysis of this paper is purely formal; the main aim being to construct effective BEM for class of equations which falls within the type (2).

2. The boundary value problem

Referred to a Cartesian frame Ox_1x_2 a solution to (2) is sought which is valid in a region Ω in R^2 with boundary $\partial\Omega$ which consists of a finite number of piecewise smooth closed curves. On $\partial\Omega_1$ the dependent variable $\mu(\mathbf{x})$ ($\mathbf{x} = (x_1, x_2)$) is specified and on $\partial\Omega_2$

$$P(\mathbf{x}) = \kappa_{ij}\left(\partial\mu/\partial x_j\right) n_i \tag{3}$$

is specified where $\partial\Omega = \partial\Omega_1 \cup \partial\Omega_2$ and $\mathbf{n} = (n_1, n_2)$ denotes the outward pointing normal to $\partial\Omega$.

For all points in Ω the matrix of coefficients $[\kappa_{ij}]$ is a real symmetric positive definite matrix so that throughout Ω equation (2) is a second order elliptic partial differential equation. Therefore equation (2) may be written explicitly as

$$\frac{\partial}{\partial x_1}\left(\kappa_{11}\frac{\partial\mu}{\partial x_1}\right) + 2\frac{\partial}{\partial x_1}\left(\kappa_{12}\frac{\partial\mu}{\partial x_2}\right) + \frac{\partial}{\partial x_2}\left(\kappa_{22}\frac{\partial\mu}{\partial x_2}\right) + \beta\mu = 0$$

Further, the coefficients κ_{ij} and β are required to be twice differentiable functions of the two independent variables x_1 and x_2.

The method of solution will be to obtain boundary integral equations from which numerical values of the dependent variables μ and P may be obtained for all points in Ω. The analysis is specially relevant to an anisotropic medium but it equally applies to isotropic media. For isotropy, the coefficients in (2) take the form $\kappa_{11} = \kappa_{22}$ and $\kappa_{12} = 0$ and use of these equations in the following analysis immediately yields the corresponding results for an isotropic medium.

3. The boundary integral equation

The boundary integral equation is derived by transforming the variable coefficient equation (2) to a constant coefficient equation. The coefficients κ_{ij} and β are required to take the form

$$\kappa_{ij}(\mathbf{x}) = \overline{\kappa}_{ij}g(\mathbf{x}) \tag{4}$$

$$\beta(\mathbf{x}) = \overline{\beta}g(\mathbf{x}) \tag{5}$$

where the $\overline{\kappa}_{ij}$ and $\overline{\beta}$ are constants and g is a differentiable function of $\mathbf{x}$. Use of (4) and (5) and in (2) yields

$$\overline{\kappa}_{ij}\frac{\partial}{\partial x_i}\left(g\frac{\partial\mu}{\partial x_j}\right) + \overline{\beta}g\mu = 0 \tag{6}$$

Let

$$\mu(\mathbf{x}) = g^{-1/2}(\mathbf{x})\,\psi(\mathbf{x}) \tag{7}$$

so that (6) may be written in the form

$$\overline{\kappa}_{ij}\frac{\partial}{\partial x_i}\left[g\frac{\partial\left(g^{-1/2}\psi\right)}{\partial x_j}\right] + \overline{\beta}g^{1/2}\psi = 0$$

That is

$$\overline{\kappa}_{ij}\left[\left(\frac{1}{4}g^{-3/2}\frac{\partial g}{\partial x_i}\frac{\partial g}{\partial x_j} - \frac{1}{2}g^{-1/2}\frac{\partial^2 g}{\partial x_i\partial x_j}\right)\psi + g^{1/2}\frac{\partial^2\psi}{\partial x_i\partial x_j}\right] + \overline{\beta}g^{1/2}\psi = 0 \tag{8}$$

Use of the identity

$$\frac{\partial^2 g^{1/2}}{\partial x_i\partial x_j} = -\frac{1}{4}g^{-3/2}\frac{\partial g}{\partial x_i}\frac{\partial g}{\partial x_j} + \frac{1}{2}g^{-1/2}\frac{\partial^2 g}{\partial x_i\partial x_j}$$

permits (8) to be written in the form

$$g^{1/2}\overline{\kappa}_{ij}\frac{\partial^2\psi}{\partial x_i\partial x_j}-\psi\overline{\kappa}_{ij}\frac{\partial^2 g^{1/2}}{\partial x_i\partial x_j}+\overline{\beta}g^{1/2}\psi=0$$

It follows that if g is such that

$$\overline{\kappa}_{ij}\frac{\partial^2 g^{1/2}}{\partial x_i\partial x_j}=0 \tag{9}$$

then the transformation (7) carries the variable coefficients equation (6) to the constant coefficients equation

$$\overline{\kappa}_{ij}\frac{\partial^2\psi}{\partial x_i\partial x_j}+\overline{\beta}\psi=0 \tag{10}$$

And if g is such that

$$\overline{\kappa}_{ij}\frac{\partial^2 g^{1/2}}{\partial x_i\partial x_j}-\overline{\beta}g^{1/2}=0 \tag{11}$$

then

$$\overline{\kappa}_{ij}\frac{\partial^2\psi}{\partial x_i\partial x_j}=0 \tag{12}$$

Also, substitution of (4) and (7) into (3) gives

$$P=-P_g\psi+P_\psi g^{1/2} \tag{13}$$

where

$$P_g(\mathbf{x})=\overline{\kappa}_{ij}\frac{\partial g^{1/2}}{\partial x_j}n_i \qquad P_\psi(\mathbf{x})=\overline{\kappa}_{ij}\frac{\partial\psi}{\partial x_j}n_i$$

A boundary integral equation for the solution of (10) and (12) is given in the form

$$\eta(\mathbf{x}_0)\psi(\mathbf{x}_0)=\int_{\partial\Omega}\left[\Gamma(\mathbf{x},\mathbf{x}_0)\psi(\mathbf{x})-\Phi(\mathbf{x},\mathbf{x}_0)P_\psi(\mathbf{x})\right]ds(\mathbf{x}) \tag{14}$$

where $\mathbf{x}_0=(a,b)$, $\eta=0$ if $(a,b)\notin\Omega\cup\partial\Omega$, $\eta=1$ if $(a,b)\in\Omega$, $\eta=\frac{1}{2}$ if $(a,b)\in\partial\Omega$ and $\partial\Omega$ has a continuously turning tangent at (a,b).

The so called fundamental solution Φ in (14) is any solution of the equation

$$\overline{\kappa}_{ij}\frac{\partial^2\Phi}{\partial x_i\partial x_j}+\overline{\beta}\Phi=\delta(\mathbf{x}-\mathbf{x}_0)$$

and the Γ is given by

$$\Gamma(\mathbf{x},\mathbf{x}_0)=\overline{\kappa}_{ij}\frac{\partial\Phi(\mathbf{x},\mathbf{x}_0)}{\partial x_j}n_i$$

where δ is the Dirac delta function. Following Azis in [17], for two-dimensional problems Φ and Γ are given by

$$\begin{aligned}\Phi(\mathbf{x},\mathbf{x}_0) &= \begin{cases}\frac{K}{2\pi}\ln R & \text{if } \overline{\beta}=0\\ \frac{\imath K}{4}H_0^{(2)}(\omega R) & \text{if } \overline{\beta}>0\\ \frac{-K}{2\pi}K_0(\omega R) & \text{if } \overline{\beta}<0\end{cases}\\ \Gamma(\mathbf{x},\mathbf{x}_0) &= \begin{cases}\frac{K}{2\pi}\frac{1}{R}\overline{\kappa}_{ij}\frac{\partial R}{\partial x_j}n_i & \text{if } \overline{\beta}=0\\ \frac{-\imath K\omega}{4}H_1^{(2)}(\omega R)\overline{\kappa}_{ij}\frac{\partial R}{\partial x_j}n_i & \text{if } \overline{\beta}>0\\ \frac{K\omega}{2\pi}K_1(\omega R)\overline{\kappa}_{ij}\frac{\partial R}{\partial x_j}n_i & \text{if } \overline{\beta}<0\end{cases}\end{aligned} \tag{15}$$

IOP Conf. Series: Materials Science and Engineering **619** (2019) 012025 doi:10.1088/1757-899X/619/1/012025

where

$$
\begin{aligned}
K &= \ddot{\tau}/\zeta \\
\omega &= \sqrt{|\overline{\beta}|/\zeta} \\
\zeta &= \left[\overline{\kappa}_{11} + 2\overline{\kappa}_{12}\dot{\tau} + \overline{\kappa}_{22}\left(\dot{\tau}^2 + \ddot{\tau}^2\right)\right]/2 \\
R &= \sqrt{(\dot{x}_1 - \dot{a})^2 + (\dot{x}_2 - \dot{b})^2} \\
\dot{x}_1 &= x_1 + \dot{\tau}x_2 \\
\dot{a} &= a + \dot{\tau}b \\
\dot{x}_2 &= \ddot{\tau}x_2 \\
\dot{b} &= \ddot{\tau}b
\end{aligned}
$$

where $\dot{\tau}$ and $\ddot{\tau}$ are respectively the real and the positive imaginary parts of the complex root τ of the quadratic

$$\overline{\kappa}_{11} + 2\overline{\kappa}_{12}\tau + \overline{\kappa}_{22}\tau^2 = 0$$

and $H_0^{(2)}$, $H_1^{(2)}$ denote the Hankel function of second kind and order zero and order one respectively. K_0, K_1 denote the modified Bessel function of order zero and order one respectively, $\imath$ represents the square root of minus one. The derivatives $\partial R/\partial x_j$ needed for the calculation of the Γ in (15) are given by

$$
\begin{aligned}
\frac{\partial R}{\partial x_1} &= \frac{1}{R}(\dot{x}_1 - \dot{a}) \\
\frac{\partial R}{\partial x_2} &= \dot{\tau}\left[\frac{1}{R}(\dot{x}_1 - \dot{a})\right] + \ddot{\tau}\left[\frac{1}{R}\left(\dot{x}_2 - \dot{b}\right)\right]
\end{aligned}
$$

Use of (7) and (13) in (14) yields

$$
\begin{aligned}
\eta(\mathbf{x}_0)\, g^{1/2}(\mathbf{x}_0)\, \mu(\mathbf{x}_0) &= \int_{\partial\Omega} \left\{ \left[g^{1/2}(\mathbf{x})\,\Gamma(\mathbf{x},\mathbf{x}_0) - P_g(\mathbf{x})\,\Phi(\mathbf{x},\mathbf{x}_0)\right]\mu(\mathbf{x}) \right. \\
&\quad \left. - \left[g^{-1/2}(\mathbf{x})\,\Phi(\mathbf{x},\mathbf{x}_0)\right] P(\mathbf{x}) \right\} ds(\mathbf{x}) \qquad (16)
\end{aligned}
$$

This equation provides a boundary integral equation for determining μ and P at all points of Ω.

4. Numerical examples

Some particular boundary value problems will be solved numerically by employing the integral equation (16). The main aim is to show the validity of the analysis for deriving the boundary integral equation (16) and the appropriateness of the BEM in solving the problems through the derived boundary integral equation (16). Standard boundary element method is employed to obtain numerical results. The integrals in equation (16) are evaluated numerically using the Bode's quadrature (see Abramowitz and Stegun in [18]).

4.1. Examples with analytical solutions

In order to see the convergence and accuracy of the BEM we will consider some examples of problems with analytical solutions. Three possible multiparameter forms of function $g(\mathbf{x})$ are quadratical (satisfying (9)), exponential and trigonometrical (satisfying (11) with $\overline{\beta} > 0$ and $\overline{\beta} < 0$ respectively). For all problems considered in this section we take

$g(\mathbf{x}) = [A(\alpha_0 + \alpha_1 x_1 + \alpha_2 x_2)]^2$ for quadratically graded materials
$g(\mathbf{x}) = [A \exp(\alpha_i x_i)]^2$ for exponentially graded materials
$g(\mathbf{x}) = \{A[\cos(\alpha_i x_i) + \sin(\alpha_i x_i)]\}^2$ for trigonometrically graded materials

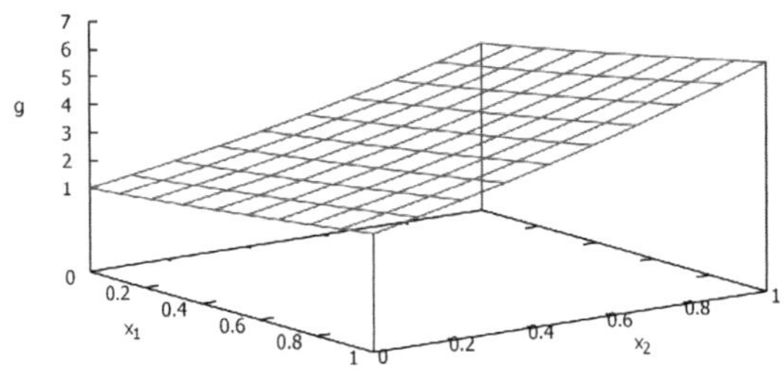

Figure 1. A quadratic inhomogeneity function $g(\mathbf{x}) = (1 + 0.5x_1 + x_2)^2$

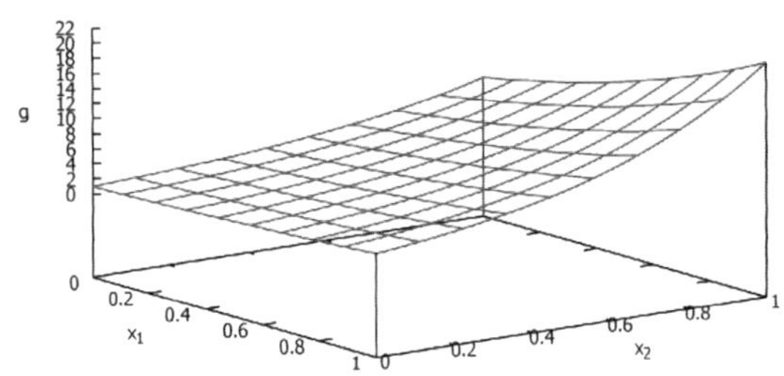

Figure 2. An exponential inhomogeneity function $g(\mathbf{x}) = [\exp(0.5x_1 + x_2)]^2$

with $A = 1, \alpha_0 = 1, \alpha_1 = 0.5, \alpha_2 = 1$. Plots of $g(\mathbf{x})$ are shown in Figures 1–3. The geometry of the region Ω and the boundary conditions are as depicted in Figure 4. The values of the constant coefficients $\overline{\kappa}_{ij}$ for the governing equation (2) are

$$\overline{\kappa}_{11} = 1, \overline{\kappa}_{12} = 0.5, \overline{\kappa}_{22} = 1$$

Also, the parameters for the analytical solutions are taken to be

$$B = 1, \gamma_0 = 1, \gamma_1 = 0.5, \gamma_2 = 0.5$$

4.1.1. Quadratically graded media For quadratically graded media the function $g(\mathbf{x})$ satisfies (9). Therefore equation (10) has to be the corresponding constant variable coefficient equation in which $\overline{\beta} > 0$, $\overline{\beta} < 0$ or $\overline{\beta} = 0$.

Problem 4.1.1.1: Case $\overline{\beta} > 0$ in equation (10)
We take analytical solutions

$$\begin{aligned} \psi(\mathbf{x}) &= B[\cos(\gamma_i x_i) + \sin(\gamma_i x_i)] \quad \text{thus } \overline{\beta} = \overline{\kappa}_{ij}\gamma_i\gamma_j = 0.75 \\ \mu(\mathbf{x}) &= B[\cos(\gamma_i x_i) + \sin(\gamma_i x_i)] / [A(\alpha_0 + \alpha_1 x_1 + \alpha_2 x_2)] \end{aligned}$$

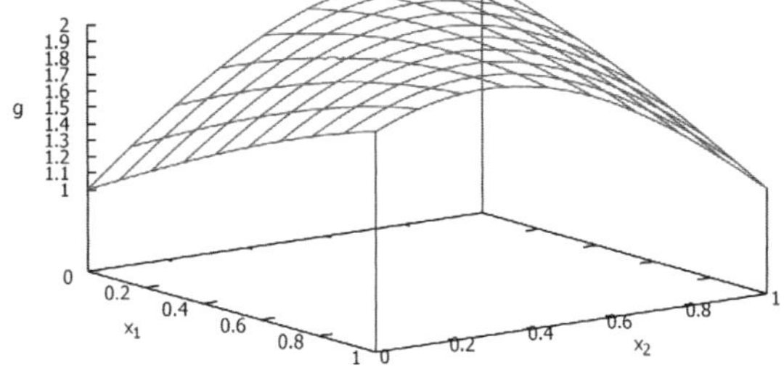

Figure 3. A trigonometrical inhomogeneity function $g(\mathbf{x}) = [\cos(0.5x_1 + x_2) + \sin(0.5x_1 + x_2)]^2$

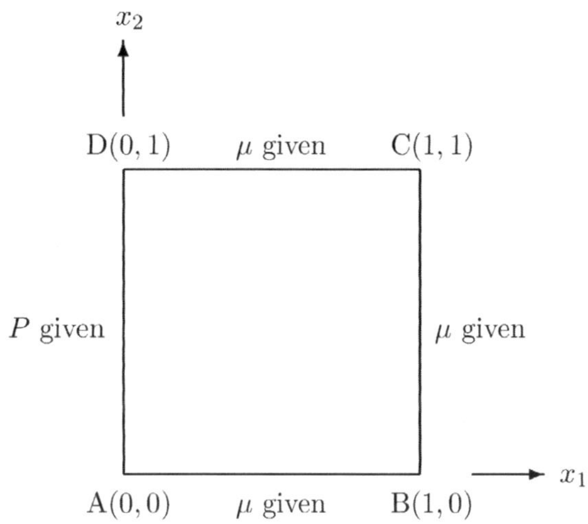

Figure 4. The geometry of all problems in Section 4.1

Table 1 shows the results of the analytical and BEM solutions with 80, 160 and 320 segments of equal length. The BEM solution converges to the analytical solution as the number of segments increases.

Problem 4.1.1.2: Case $\overline{\beta} < 0$ in equation (10)
Analytical solutions are

$$\begin{aligned} \psi(\mathbf{x}) &= B\exp(\gamma_i x_i) \quad \text{thus } \overline{\beta} = -\overline{\kappa}_{ij}\gamma_i\gamma_j = -0.75 \\ \mu(\mathbf{x}) &= B\exp(\gamma_i x_i) / [A(\alpha_0 + \alpha_1 x_1 + \alpha_2 x_2)] \end{aligned}$$

The results are shown in Table 2. Again, the BEM solution converges to the analytical solution as the number of segments increases.

Problem 4.1.1.3: Case $\overline{\beta} = 0$ in equation (10)

Table 1. BEM and analytical solutions for Problem 4.1.1.1

(x_1, x_2)	μ	$\partial\mu/\partial x_1$	$\partial\mu/\partial x_2$	μ	$\partial\mu/\partial x_1$	$\partial\mu/\partial x_2$
	BEM 80 segments			BEM 160 segments		
(.1,.5)	.8070	-.0480	-.3075	.8070	-.0477	-.3076
(.3,.5)	.7942	-.0800	-.3197	.7942	-.0798	-.3200
(.5,.5)	.7753	-.1079	-.3289	.7754	-.1078	-.3291
(.7,.5)	.7512	-.1324	-.3355	.7513	-.1325	-.3356
(.9,.5)	.7226	-.1536	-.3399	.7226	-.1541	-.3397
(.5,.1)	.9261	-.0982	-.4409	.9263	-.0986	-.4414
(.5,.3)	.8452	-.1011	-.3732	.8453	-.1011	-.3736
(.5,.7)	.7128	-.1162	-.2984	.7128	-.1160	-.2985
(.5,.9)	.6554	-.1246	-.2765	.6554	-.1245	-.2766
	BEM 320 segments			Analytical		
(.1,.5)	.8070	-.0476	-.3077	.8070	-.0475	-.3078
(.3,.5)	.7942	-.0797	-.3201	.7942	-.0796	-.3202
(.5,.5)	.7754	-.1078	-.3292	.7754	-.1078	-.3293
(.7,.5)	.7513	-.1325	-.3356	.7513	-.1326	-.3357
(.9,.5)	.7226	-.1542	-.3397	.7226	-.1544	-.3396
(.5,.1)	.9264	-.0986	-.4417	.9266	-.0988	-.4420
(.5,.3)	.8454	-.1012	-.3738	.8455	-.1012	-.3740
(.5,.7)	.7128	-.1160	-.2986	.7128	-.1159	-.2987
(.5,.9)	.6554	-.1244	-.2767	.6554	-.1244	-.2768

Table 2. BEM and analytical solutions for Problem 4.1.1.2

(x_1, x_2)	μ	$\partial\mu/\partial x_1$	$\partial\mu/\partial x_2$	μ	$\partial\mu/\partial x_1$	$\partial\mu/\partial x_2$
	BEM 80 segments			BEM 160 segments		
(.1,.5)	.8716	.1537	-.1252	.8712	.1541	-.1258
(.3,.5)	.9047	.1777	-.0945	.9044	.1779	-.0952
(.5,.5)	.9427	.2020	-.0659	.9424	.2020	-.0666
(.7,.5)	.9856	.2270	-.0388	.9852	.2266	-.0394
(.9,.5)	1.0336	.2532	-.0127	1.0331	.2523	-.0130
(.5,.1)	.9994	.1298	-.2364	.9997	.1297	-.2386
(.5,.3)	.9626	.1710	-.1373	.9626	.1709	-.1385
(.5,.7)	.9352	.2277	-.0112	.9348	.2277	-.0116
(.5,.9)	.9375	.2507	.0331	.9371	.2506	.0329
	BEM 320 segments			Analytical		
(.1,.5)	.8710	.1543	-.1261	.8709	.1545	-.1264
(.3,.5)	.9043	.1780	-.0955	.9041	.1781	-.0959
(.5,.5)	.9423	.2019	-.0670	.9421	.2019	-.0673
(.7,.5)	.9851	.2265	-.0396	.9849	.2263	-.0399
(.9,.5)	1.0329	.2519	-.0131	1.0327	.2516	-.0132
(.5,.1)	.9998	.1297	-.2396	.9999	.1296	-.2407
(.5,.3)	.9625	.1708	-.1391	.9625	.1708	-.1397
(.5,.7)	.9346	.2276	-.0118	.9344	.2276	-.0120
(.5,.9)	.9368	.2505	.0328	.9366	.2505	.0327

Table 3. BEM and analytical solutions for Problem 4.1.1.3

(x_1, x_2)	μ	$\partial\mu/\partial x_1$	$\partial\mu/\partial x_2$	μ	$\partial\mu/\partial x_1$	$\partial\mu/\partial x_2$
	BEM 80 segments			BEM 160 segments		
(.1,.5)	.8390	.0516	-.2177	.8388	.0518	-.2181
(.3,.5)	.8487	.0456	-.2103	.8486	.0458	-.2107
(.5,.5)	.8573	.0409	-.2032	.8572	.0409	-.2036
(.7,.5)	.8651	.0370	-.1966	.8650	.0368	-.1969
(.9,.5)	.8722	.0340	-.1906	.8720	.0334	-.1906
(.5,.1)	.9625	.0142	-.3403	.9627	.0139	-.3416
(.5,.3)	.9032	.0315	-.2586	.9032	.0314	-.2594
(.5,.7)	.8208	.0459	-.1638	.8207	.0460	-.1641
(.5,.9)	.7911	.0486	-.1349	.7909	.0486	-.1350
	BEM 320 segments			Analytical		
(.1,.5)	.8388	.0519	-.2183	.8387	.0520	-.2185
(.3,.5)	.8485	.0458	-.2110	.8485	.0459	-.2112
(.5,.5)	.8572	.0408	-.2039	.8571	.0408	-.2041
(.7,.5)	.8649	.0366	-.1971	.8649	.0365	-.1972
(.9,.5)	.8719	.0331	-.1906	.8718	.0329	-.1907
(.5,.1)	.9628	.0138	-.3423	.9630	.0137	-.3429
(.5,.3)	.9032	.0313	-.2598	.9032	.0312	-.2601
(.5,.7)	.8206	.0460	-.1642	.8205	.0460	-.1644
(.5,.9)	.7908	.0486	-.1351	.7907	.0487	-.1352

Now we choose analytical solutions

$$\begin{aligned}\psi(\mathbf{x}) &= B(\gamma_0 + \gamma_1 x_1 + \gamma_2 x_2) \quad \text{with}\, \overline{\beta} = 0\\ \mu(\mathbf{x}) &= B(\gamma_0 + \gamma_1 x_1 + \gamma_2 x_2) \,/\, [A(\alpha_0 + \alpha_1 x_1 + \alpha_2 x_2)]\end{aligned}$$

The results are shown in Table 3. Once again, the BEM solution converges to the analytical solution as the number of segments increases.

4.1.2. Exponentially graded media

Problem 4.1.2.1

A BVP of anisotropic FGM will be considered as a test problem. The function of inhomogeneity $g(\mathbf{x})$ satisfying equation (11) is taken to be (see Figure 2)

$$g(\mathbf{x}) = [A\exp(\alpha_i x_i)]^2$$

Therefore ψ must satisfy (12). We choose analytical solution

$$\mu(\mathbf{x}) = B(\gamma_0 + \gamma_1 x_1 + \gamma_2 x_2) \,/\, [A\exp(\alpha_i x_i)]$$

Table 4 shows the results of the analytical and BEM solutions with 80, 160 and 320 segments of equal length. The BEM solution converges to the analytical solution as the number of segments increases.

Table 4. BEM and analytical solutions for Problem 4.1.2.1

(x_1, x_2)	μ	$\partial\mu/\partial x_1$	$\partial\mu/\partial x_2$	μ	$\partial\mu/\partial x_1$	$\partial\mu/\partial x_2$
	BEM 80 segments			BEM 160 segments		
(.1,.5)	.7502	-.0867	-.4610	.7501	-.0866	-.4613
(.3,.5)	.7310	-.1046	-.4691	.7309	-.1045	-.4695
(.5,.5)	.7087	-.1180	-.4717	.7086	-.1181	-.4720
(.7,.5)	.6840	-.1279	-.4697	.6840	-.1280	-.4700
(.9,.5)	.6578	-.1347	-.4638	.6576	-.1349	-.4641
(.5,.1)	.9157	-.1057	-.5613	.9159	-.1055	-.5624
(.5,.3)	.8077	-.1151	-.5179	.8077	-.1153	-.5186
(.5,.7)	.6190	-.1161	-.4252	.6189	-.1160	-.4253
(.5,.9)	.5385	-.1114	-.3797	.5384	-.1108	-.3801
	BEM 320 segments			Analytical		
(.1,.5)	.7501	-.0866	-.4614	.7500	-.0865	-.4616
(.3,.5)	.7309	-.1044	-.4697	.7309	-.1044	-.4698
(.5,.5)	.7086	-.1181	-.4722	.7085	-.1181	-.4724
(.7,.5)	.6839	-.1281	-.4700	.6839	-.1282	-.4702
(.9,.5)	.6575	-.1352	-.4641	.6575	-.1354	-.4641
(.5,.1)	.9160	-.1056	-.5631	.9161	-.1057	-.5638
(.5,.3)	.8077	-.1153	-.5189	.8077	-.1154	-.5193
(.5,.7)	.6188	-.1160	-.4253	.6188	-.1160	-.4254
(.5,.9)	.5383	-.1108	-.3800	.5383	-.1108	-.3800

4.1.3. Trigonometrically graded media

Problem 4.1.3.1

The function of inhomogeneity $g(\mathbf{x})$ satisfying equation (11) is taken to be (see Figure 3)

$$g(\mathbf{x}) = \{A\left[\cos(\alpha_i x_i) + \sin(\alpha_i x_i)\right]\}^2$$

The analytical solution is

$$\mu(\mathbf{x}) = B(\gamma_0 + \gamma_1 x_1 + \gamma_2 x_2) / \{A\left[\cos(\alpha_i x_i) + \sin(\alpha_i x_i)\right]\}$$

Table 5 shows the results of the analytical and BEM solutions with 80, 160 and 320 segments of equal length. The BEM solution converges to the analytical solution as the number of segments increases.

4.2. Examples without analytical solutions

In this section we will consider some examples of problems without simple analytical solutions. We setup some problems for a homogeneous isotropic material by taking $g(\mathbf{x}) = 1, \overline{\kappa}_{11} = \overline{\kappa}_{22} = 1, \overline{\kappa}_{12} = 0$ and with symmetrical boundary conditions. This function $g(\mathbf{x})$ satisfies equation (9) thus we will take $\psi(\mathbf{x})$ that satisfies (10) to be put into the integral equation (14). The aim is to see the consistency of the BEM of whether it produces symmetrical solutions.

Problem 4.2.1: Case $\overline{\beta} > 0$ in equation (10)

For this problem we take $\overline{\beta} = 1$ and the boundary conditions are as shown in Figure 5. Table 6 shows the results of the BEM solution using 80, 160, 320 and 640 segments of equal length. As

Table 5. BEM and analytical solutions for Problem 4.1.3.1

(x_1, x_2)	μ	$\partial\mu/\partial x_1$	$\partial\mu/\partial x_2$	μ	$\partial\mu/\partial x_1$	$\partial\mu/\partial x_2$
	BEM 80 segments			BEM 160 segments		
(.1,.5)	.9457	.2493	.1383	.9455	.2498	.1375
(.3,.5)	.9994	.2883	.2221	.9992	.2885	.2214
(.5,.5)	1.0616	.3351	.3174	1.0614	.3350	.3168
(.7,.5)	1.1341	.3917	.4286	1.1339	.3913	.4281
(.9,.5)	1.2191	.4613	.5614	1.2188	.4604	.5611
(.5,.1)	1.0134	.1545	-.0789	1.0136	.1543	-.0803
(.5,.3)	1.0180	.2418	.1212	1.0180	.2416	.1203
(.5,.7)	1.1474	.4536	.5500	1.1471	.4536	.5494
(.5,.9)	1.2874	.6240	.8705	1.2870	.6240	.8700
	BEM 320 segments			Analytical		
(.1,.5)	.9454	.2500	.1372	.9453	.2502	.1369
(.3,.5)	.9992	.2887	.2210	.9991	.2888	.2207
(.5,.5)	1.0614	.3350	.3165	1.0613	.3350	.3162
(.7,.5)	1.1338	.3911	.4279	1.1337	.3910	.4276
(.9,.5)	1.2187	.4600	.5609	1.2186	.4596	.5608
(.5,.1)	1.0137	.1542	-.0810	1.0138	.1541	-.0817
(.5,.3)	1.0180	.2416	.1198	1.0180	.2415	.1194
(.5,.7)	1.1470	.4536	.5492	1.1469	.4536	.5489
(.5,.9)	1.2868	.6240	.8698	1.2867	.6240	.8695

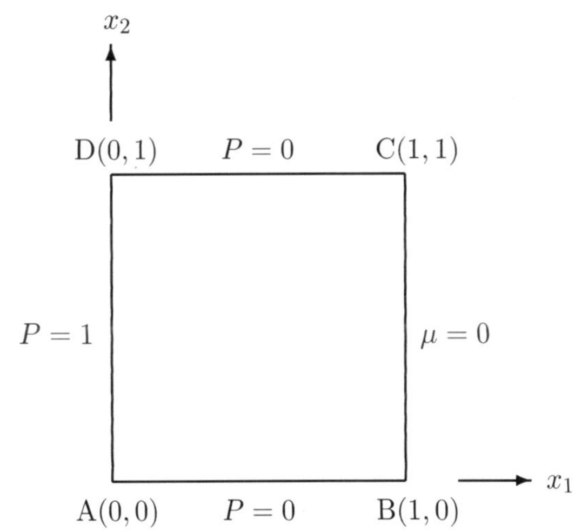

Figure 5. The geometry of Problem 4.2.1 and Problem 4.2.2

expected, the results converge as the number of segments increases and also they are symmetrical about the axes $x_2 = 0.5$.

Problem 4.2.2: Case $\overline{\beta} < 0$ in equation (10)

We take $\overline{\beta} = -1$ and boundary conditions are as shown in Figure 5. Table 7 shows the results of the BEM solution using 80, 160, 320 and 640 segments of equal length. The results converge

Table 6. BEM solution for Problem 4.2.1

(x_1, x_2)	μ	$\partial\mu/\partial x_1$	$\partial\mu/\partial x_2$	μ	$\partial\mu/\partial x_1$	$\partial\mu/\partial x_2$
	BEM 80 segments			BEM 160 segments		
(.1,.5)	1.4447	-1.1500	-.0000	1.4475	-1.1502	-.0000
(.3,.5)	1.1875	-1.4138	-.0000	1.1902	-1.4148	-.0000
(.5,.5)	.8830	-1.6213	-.0000	.8854	-1.6231	-.0000
(.7,.5)	.5433	-1.7639	-.0000	.5453	-1.7664	-.0000
(.9,.5)	.1821	-1.8358	-.0000	.1835	-1.8392	-.0000
(.5,.1)	.8829	-1.6223	.0000	.8853	-1.6235	-.0000
(.5,.3)	.8829	-1.6216	.0002	.8854	-1.6232	.0001
(.5,.7)	.8829	-1.6216	-.0002	.8854	-1.6232	-.0001
(.5,.9)	.8829	-1.6223	-.0000	.8853	-1.6235	.0000
	BEM 320 segments			BEM 640 segments		
(.1,.5)	1.4487	-1.1503	.0000	1.4493	-1.1504	.0000
(.3,.5)	1.1913	-1.4152	.0000	1.1919	-1.4154	.0000
(.5,.5)	.8864	-1.6237	.0000	.8869	-1.6240	.0000
(.7,.5)	.5462	-1.7674	.0000	.5466	-1.7678	.0000
(.9,.5)	.1842	-1.8406	.0000	.1845	-1.8411	.0000
(.5,.1)	.8864	-1.6239	-.0000	.8869	-1.6241	-.0000
(.5,.3)	.8864	-1.6238	.0000	.8869	-1.6240	-.0000
(.5,.7)	.8864	-1.6238	-.0000	.8869	-1.6240	.0000
(.5,.9)	.8864	-1.6239	.0000	.8869	-1.6241	.0000

Table 7. BEM solution for Problem 4.2.2

(x_1, x_2)	μ	$\partial\mu/\partial x_1$	$\partial\mu/\partial x_2$	μ	$\partial\mu/\partial x_1$	$\partial\mu/\partial x_2$
	BEM 80 segments			BEM 160 segments		
(.1,.5)	.6628	-.9270	-.0000	.6642	-.9279	-.0000
(.3,.5)	.4896	-.8114	-.0000	.4907	-.8125	-.0000
(.5,.5)	.3361	-.7289	-.0000	.3370	-.7300	-.0000
(.7,.5)	.1961	-.6756	-.0000	.1968	-.6767	-.0000
(.9,.5)	.0640	-.6494	-.0000	.0645	-.6505	-.0000
(.5,.1)	.3360	-.7292	.0000	.3369	-.7301	.0000
(.5,.3)	.3360	-.7289	.0001	.3370	-.7300	.0000
(.5,.7)	.3360	-.7289	-.0001	.3370	-.7300	-.0000
(.5,.9)	.3360	-.7292	-.0000	.3369	-.7301	-.0000
	BEM 320 segments			BEM 640 segments		
(.1,.5)	.6647	-.9283	.0000	.6650	-.9285	.0000
(.3,.5)	.4912	-.8130	.0000	.4914	-.8132	.0000
(.5,.5)	.3374	-.7304	.0000	.3375	-.7306	.0000
(.7,.5)	.1971	-.6771	.0000	.1972	-.6773	.0000
(.9,.5)	.0647	-.6510	.0000	.0648	-.6512	.0000
(.5,.1)	.3373	-.7304	-.0000	.3375	-.7306	-.0000
(.5,.3)	.3373	-.7304	.0000	.3375	-.7306	.0000
(.5,.7)	.3373	-.7304	-.0000	.3375	-.7306	-.0000
(.5,.9)	.3373	-.7304	.0000	.3375	-.7306	.0000

as the number of segments increases and also they are symmetrical about the axes $x_2 = 0.5$.

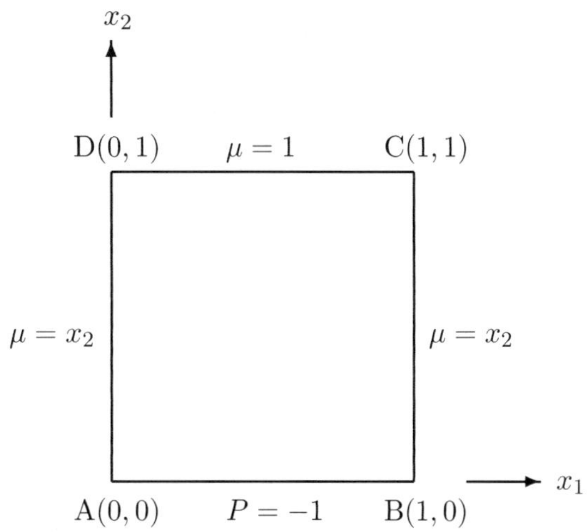

Figure 6. The geometry of Problem 4.2.3

Problem 4.2.3: Case $\overline{\beta} = 0$ in equation (10)
We consider a problem with $\overline{\beta} = 0$ and the boundary conditions are as shown in Figure 6. Table 8 shows the results of the BEM solution using 80, 160, 320 and 640 segments of equal length. The results converge as the number of segments increases and also they are symmetrical about the axes $x_1 = 0.5$.

Table 8. BEM solution for Problem 4.2.3

(x_1, x_2)	μ	$\partial\mu/\partial x_1$	$\partial\mu/\partial x_2$	μ	$\partial\mu/\partial x_1$	$\partial\mu/\partial x_2$
	BEM 80 segments			BEM 160 segments		
(.1,.5)	.5001	.0014	1.0001	.5001	.0006	1.0001
(.3,.5)	.5004	.0007	1.0003	.5002	.0004	1.0002
(.5,.5)	.5004	.0000	1.0004	.5002	.0000	1.0002
(.7,.5)	.5004	-.0007	1.0003	.5002	-.0004	1.0002
(.9,.5)	.5001	-.0014	1.0001	.5001	-.0006	1.0001
(.5,.1)	.1005	-.0000	.9993	.1002	.0000	.9996
(.5,.3)	.3004	-.0000	.9998	.3002	.0000	.9999
(.5,.7)	.7006	.0000	1.0012	.7003	.0000	1.0006
(.5,.9)	.9009	-.0000	1.0019	.9005	-.0000	1.0010
	BEM 320 segments			BEM 640 segments		
(.1,.5)	.5000	.0003	1.0000	.5000	.0002	1.0000
(.3,.5)	.5001	.0002	1.0001	.5000	.0001	1.0001
(.5,.5)	.5001	-.0000	1.0001	.5001	-.0000	1.0001
(.7,.5)	.5001	-.0002	1.0001	.5000	-.0001	1.0001
(.9,.5)	.5000	-.0003	1.0000	.5000	-.0002	1.0000
(.5,.1)	.1001	-.0000	.9998	.1001	-.0000	.9999
(.5,.3)	.3001	-.0000	.9999	.3000	-.0000	1.0000
(.5,.7)	.7001	-.0000	1.0003	.7001	-.0000	1.0002
(.5,.9)	.9002	-.0000	1.0005	.9001	-.0000	1.0002

5. Conclusion

The scalar elliptic governing equation (2) is used for modelling physical problems such as steady infiltration problems (when $\beta < 0$), acoustic problems (when $\beta > 0$), and antiplane strain in elastostatics and plane thermostatic problems (when $\beta = 0$). The boundary integral equation (16) was derived from this governing equation (2) and straight from (16) a BEM was then constructed for calculation of numerical solutions to the problems for anisotropic functionally graded media specifically including quadratically, exponentially and trigonometrically graded media. The results show the convergence, consistency, and accuracy of the BEM solutions. Together with its ease in implementation, it may be concluded that BEM is a good numerical method for solving such kind of problems.

Acknowledgements

The author would like to thank The Hasanuddin University and The Ministry of Research, Technology and Higher Education of the Republic of Indonesia for the provided support.

References

[1] Azis M I, Kasbawati, Haddade A and Thamrin S A 2018 On some examples of pollutant transport problems solved numerically using the boundary element method *Journal of Physics: Conference Series* **979**(1)

[2] Azis M I, Asrul L, Khaeruddin and Paharuddin 2018 BEM solutions for unsteady transport problems in anisotropic media *JP Journal of Heat and Mass Transfer* **15**(4) 915

[3] Haddade A, Salam N, Khaeruddin and Azis M I 2017 A boundary element method for 2D diffusion-convection problems in anisotropic media *Far East Journal of Mathematical Sciences* **102**(8) 1593

[4] Azis M I 2019 Numerical solutions for the Helmholtz boundary value problems of anisotropic media *Journal of Computational Physics* **381** 42

[5] Cheng A H-D 1984 Darcy's Flow with Variable Permeability: A Boundary Integral Solution *Water Resources Research* **20** 980

[6] Clements D L and Azis M I 2000 A Note on a Boundary Element Method for the Numerical Solution of Boundary Value Problems in Isotropic Inhomogeneous Elasticity *Journal of the Chinese Institute of Engineers* **23**(3) 261

[7] Azis M I and Clements D L 2014 On some problems concerning deformations of functionally graded anisotropic elastic materials *Far East Journal of Mathematical Sciences* **87**(2) 173

[8] Azis M I, Toaha S, Bahri M and Ilyas N 2018 A boundary element method with analytical integration for deformation of inhomogeneous elastic materials *Journal of Physics: Conference Series* **979**(1) 012072

[9] Azis M I and Clements D L 2014 A Boundary Element Method for Transient Heat Conduction Problem of Nonhomogeneous Anisotropic Materials *Far East Journal of Mathematical Sciences*, **89**(1) 51

[10] Azis M I and Clements D L 2008 Nonlinear transient heat conduction problems for a class of inhomogeneous anisotropic materials by BEM *Engineering Analysis with Boundary Elements* **32**(12) 1054

[11] Salam N, Haddade A, Clements D L and Azis M I 2017 A boundary element method for a class of elliptic boundary value problems of functionally graded media *Engineering Analysis with Boundary Elements* **84**(3) 186

[12] Azis M I 2019 Standard-BEM solutions to two types of anisotropic-diffusion convection reaction equations with variable coefficients *Engineering Analysis with Boundary Elements* **105** 87

[13] Clements D L and Lobo M 2010 A BEM for time dependent infiltration from an irrigation channel *Engineering Analysis with Boundary Elements* **34** 1100

[14] Solekhudin I and Ang K-C 2012 A DRBEM with a predictor-corrector scheme for steady infiltration from periodic channels with root-water uptake *Engineering Analysis with Boundary Elements* **36** 1199

[15] Barucq H, Bendali A, Fares M, Mattesi V and Tordeux S 2017 A symmetric Trefftz-DG formulation based on a local boundary element method for the solution of the Helmholtz equation *Journal of Computational Physics* **330** 1069

[16] Loeffler C F, Mansur W J, Barcelos H D M and Bulcão A 2015 Solving Helmholtz problems with the boundary element method using direct radial basis function interpolation *Engineering Analysis with Boundary Elements* **61** 218

[17] Azis M I 2017 Fundamental solutions to two types of 2D boundary value problems of anisotropic materials *Far East Journal of Mathematical Sciences* **101**(11) 2405

[18] Abramowitz M and Stegun I A 1972 *Handbook of mathematical functions: with formulas, graphs and mathematical tables*, Dover Publications, Washington

IOP Conf. Series: Materials Science and Engineering **619** (2019) 012026 doi:10.1088/1757-899X/619/1/012026

Audial comfort optimalization of The Lanraki Biringkanaya Makassar Church room

Muhammad Taufik Ishak[1], Asniawaty[2], Dwi Retno Sulistyowati[3]

[1,2,3] Department of Architecture, Faculty of Engineering, Hasanuddin University, Jalan Poros Malino Km. 6, Gowa, 92119, Sulawesi Selatan, Indonesia

E-mail: d.retno84@gmail.com

Abstract. Church is one of the unique worship buildings which is full of audial activities such as speech, songs, and music. The purpose of this research is to evaluate and give redesign suggestions so that the audial comfort in the church room can be achieved. Quantitative research method by analyzing data with the statistically-descriptive. Reverberation time was calculated by the Sabine formula, redesign using Autodesk Ecotect Analysis 2011 program, and the questionnaire processed by IBM SPSS Statistics 25 program. The results conclude that the sound pressure level distribution has been spread evenly, while the background noise and the reverberation time in the church room still exceed the standard. But overall, the user has felt comfort with the condition of the audial church room. So, we made change of materials and furniture that may affect the value of reverberation time and background noise in a room. We hope this research can be a reference to design the next church or similar buildings.

Keywords: audial of the church, background noise, Ecotect, reverberation time, sound pressure level.

1. Introduction

Unlike the auditorium, in the church building, the voice of worship leader needs a touch to be more authoritative and interesting [1]. Because the flows of the sound will determine the quality and quantity of the sound moreover, it is very important to match the wishes of the sound recipient.

The performance of the good audial quality in a room is influenced by the objective factors of the interior design of its enclosure fields (floors, barriers, and ceilings) and dimensions influenced by user capacity. In addition, the using and placement of coating materials is absorptive or reflective that enclose them. So, to achieve the good acoustics, we must pay attention about the material selection and application in a space because both of them playing an important role in creating good audial in a space.

The Lanraki Biringkanaya Church is located at Lanraki Street, Makassar, Indonesia. Nearby Kibaid Church of Batubara Daya and Nurul Iman Mosque and there is Tongkonan above the pulpit in the church room. This make it unique to be research because of the potential noise from the crowd and its material using and complaints by the users that they cannot hear the speech properly so that the further research is needed to evaluate the audial condition in the church room and its redesign suggestions.

Figure 1. Interior Lanraki Biringkanaya Church, Makassar.

2. Literature Study

Most of influential parameters in acoustic room design is reverberation time [2-5]. Reverberation Time (RT) is time range to take a sound energy to decay up to one millionth of its original energy, it is 60 dB. A room that have a long RT will cause decreasing speech intelligibility, because the direct sound is still strongly influenced by the sound reflection. The room with a too short's RT will impress the "dead room" [6]. According to Egan [7], the optimal RT value for the speech condition in the church is 0,5 – 1,4 seconds and in the music condition is 1,4 – 2,6 seconds. According to Cowan [8], the optimal RT value for contemporary church is 1,4 – 1,6 seconds.

Table 1. Sound absorption coefficient material [9]

Num	Material	Sound absorption coefficient				
		125	250	500	1000	2000
1	Brick plaster	0,01	0,02	0,02	0,03	0,04
2	Ceramic tiles floor	0,01	0,01	0,01	0,02	0,03
3	Gypsum board	0,29	0,10	0,05	0,04	0,07
4	Concrete + paint	0,09	0,07	0,01	0,01	0,01
5	Wood ¼" thickness	0,42	0,21	0,10	0,08	0,06
6	Glasses	0,35	0,25	0,18	0,12	0,07
7	Thick glass	0,18	0,06	0,04	0,03	0,02
8	Solid wood	0,14	0,12	0,08	0,08	0,07
9	Plastic	0,40	0,34	0,25	0,19	0,15
10	Speaker	0,60	0,54	0,45	0,39	0,35
11	Steel	0,40	0,34	0,25	0,19	0,15

This is manual calculation of the reverberation time in the church room with the Sabine formula [10]:

$$RT = \frac{0{,}161.V}{A} \qquad A = \sum S.\alpha \tag{1}$$

Where,
RT = reverberation time (sec)
V = volume (m^3)
A = total absorbing (m^2 sabins)
S = surface material area (m^2)
α = material absorption coefficient

Most of large rooms have problems with Background Noise Level (BNL) exceeding the required noise criteria that affects the acoustic performance of the space [11]. Parker said, "noise is sound which is unwanted" [12]. Noise acceptable so that activities and functions in a room is not disturbed, especially in the church room is 35-40 dB [13]. In the worship building requires good acoustics to get a solemn atmosphere [14]. To reach noise acceptable requirement, the research field measurement, analytical calculation, and subjective analysis are necessary. The noise subjectivity depends on things like environment and conditions, social-culture, and hobbies [15]. The intensity of presence affects the subjectivity of the listener as it relates to adaptation to the conditions [16]. More and more subjects are in certain conditions, the subject will adapt to these conditions. Factors that can affect human hearing's subjectivity are age [17], gender [18], occupation type, listener position, etc.

The placement of the loudspeaker also greatly affects the sound distribution in the space that is not disturb the user's hearing. To know that sound distribution is good in a room, we can measure sound pressure level. Sound Pressure Level (SPL) is the logarithmic ratio of the energy of a sound source with the reference energy source, measured in dB (deciBel). The requirement that the listener can hear the information even in a different position is the difference between the farthest and closest sound level no more than 6 dB. If in a relatively small space where sound sources with normal sound levels have been able to reach the farthest listener, it is almost certain that the sound distribution of the room is evenly distributed [6].

3. Methodology

This study used measurement methods, calculations, and questionnaires. Measurements used Sound Level Meter (SLM) to determine the sound pressure level in a point recorded with a sound effect (so that the measured sound source is stable), and to determine the background noise level in quiet conditions (when there are no people and vehicles around) and crowded conditions (when there are lots of people and vehicles passing by and rain). This is intended to make maximum measurable noise.

The study was conducted in the main room of the Lanraki Biringkanaya Church (figure 1). The area of the church is 29 m x 16.10 m and a capacity of 400 people. The measuring point is divided into 16 points on the ground floor and 9 points on the upper floor (balcony) based on the sitting position of the church in the church room. Specifically for background noise, 17 additional measuring points are located at a location near the opening with a distance of 1 meter from the opening (door, window, stained glass).

Reverberation time is calculated manually with the Sabine formula. Closed questionnaires are distributed according to the measurement point with 5 respondents as sample per measurement point. The contents of the questionnaire include the age, gender of the respondent, and the type of work as these factors can influence the subjectivity of listeners.

The data analysis uses quantitative-descriptive method so that the results of the questionnaire data are used to explain the relationship with the measurement data. After that, the design improvement suggestions are simulated using the Autodesk Ecotect Analysis 2011 program.

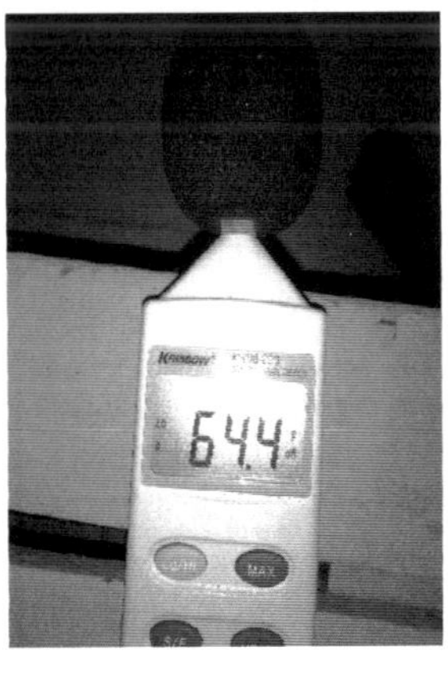

Figure 2. Sound Level Meter

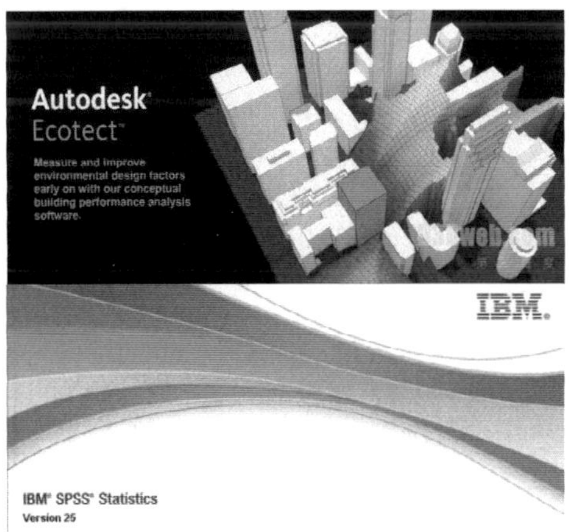

Figure 3. Autodesk Ecotect Analysis 2011 and IBM SPSS Statistics 25 program.

4. Results and Discussions

4.1 Sound Pressure Level (SPL)

The shape of the church is symmetrical so that the measured SPL value at the measuring point 3 and 14 is same because the distance to the loudspeaker, the distance to the wall, and the distance to the podium (altar) are same. Similarly at the measuring point 4 and 13, the measuring points 6 and 12, the measuring points 7 and 10, etc. Here is the picture of the symmetrical of that measuring points.

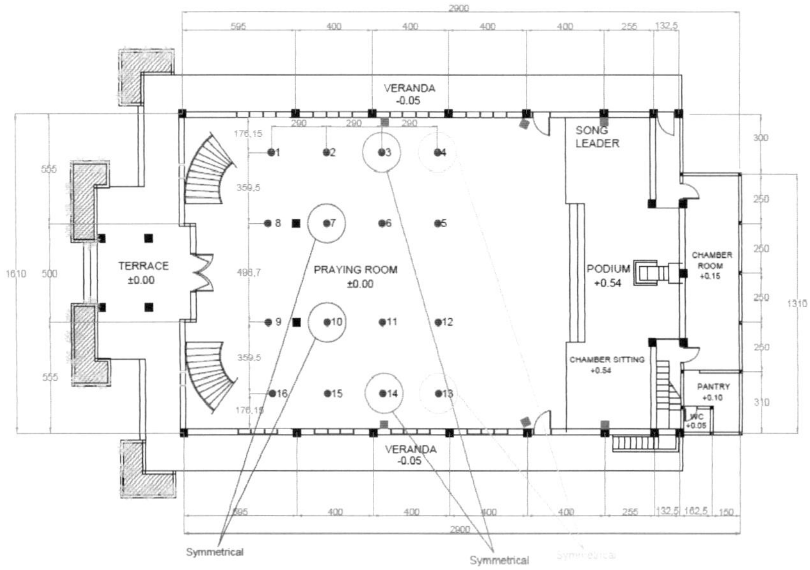

Figure 4. Symmetrical of the measuring points.

SLM that available for this research only two tools so that the measurement have been doing by putting one SLM closest to the sound source (loudspeaker) and the other was placed in the measuring point. Although on each measuring point can not be measured in the same time, but it is considered more efficient because the measuring result can be matched by adjusting it according to the measured SPL on the loudspeaker. After the sound source has been matched, we can compare it. For example on speaker 1 measured 50 dB, at the same time in measuring point 1 measured 40 dB. While on speaker 2 measured 45 dB, at the same time in measuring point 2 measured 30 dB. So, to make it same with the first data, adding the difference between the SPL of the speakers by adding 5 dB so that on the measuring point 2 the SPL is 35 dB if it is assumed that the sound source is 50 dB.

After measuring on each loudspeaker to check which is loudspeaker dominated the sound that the listener can hear at each measuring point, evidently that only 1 loudspeaker can not be used as a benchmark in measurement because the measured SPL reached only 83,4 dB while the measuring result on 25 measuring point shows the SPL that exceed 100 dB. It means that the loudspeaker did not dominated the sound that the listener can hear on the closest measuring point but the captured sound came from another loudspeaker that it close to the measuring point. So, measurement with this guns sound effect has been doing only on loudspeaker that it close to the measuring point (dominated) except one loudspeaker does not dominated. Here is the SPL measurements with the guns sound effect with the result of its equation to the sound source.

Point	Speaker	D to MP (m)	SPL of speaker (dB)	+ / -	SPL on measuring point (dB)	
					Before	After +/-
1	2	7,08	120,2	0,0	106,8	106,8
2	2	4,89	118,9	1,3	108,4	109,7
3	2	3,83	118,8	1,4	108,6	110,0
4	3	4,74	121,2	-1,0	108,3	107,3
5	3	6,77	121,6	-1,4	106,5	105,1
6	2	6,23	118,6	1,6	106,8	108,4
7	2	6,93	118,8	1,4	105,7	107,1
8	2	8,75	118,2	2,0	104,1	106,1
9	6	8,75	116,5	3,7	102,5	106,2
10	6	6,93	116,4	3,8	103,2	107,0
11	6	6,23	115,9	4,3	104,1	108,4
12	7	6,77	121,8	-1,6	106,7	105,1
13	7	4,74	121,1	-0,9	108,2	107,3
14	6	3,83	117,1	3,1	107,0	110,1
15	6	4,89	115,8	4,4	105,3	109,7
16	6	7,08	116,4	3,8	103,2	107,0
17	5	11,45	110,3	9,9	95,7	105,6
18	5	11,57	112,4	7,8	95,3	103,1
19	5	12,30	112,4	7,8	94,8	102,6
20	5	9,19	111,7	8,5	97,2	105,7
21	5	8,20	112,5	7,7	98,4	106,1
22	5	7,99	113,1	7,1	99,5	106,6
23	5	4,56	110,8	9,4	100,6	110,0
24	5	4,92	111,3	8,9	100,1	109,0
25	5	6,46	112,0	8,2	99,8	108,0

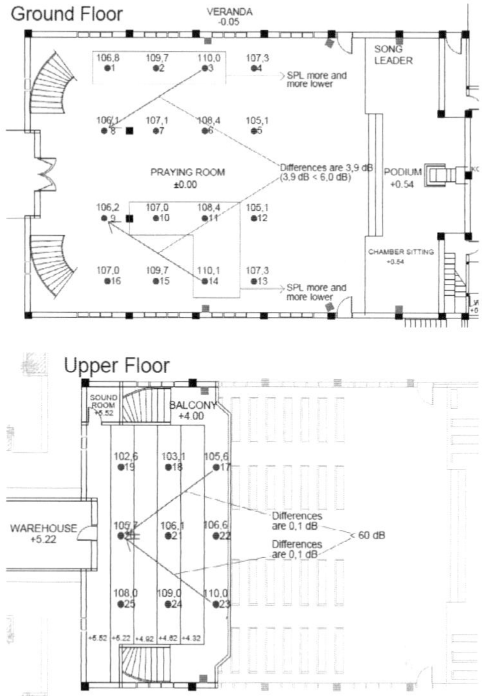

Figure 5. SPL comparison on farthest and nearest measuring point by the loudspeaker with guns sound effect.

From the table can be easy to compare between the measuring points because it is considered that the sound source was same, 120,2 dB. For checking, it could be seen by SPL on symmetrical measuring point. There was still a little difference due to the time difference as one of the reason so that maybe the SLM shifted by vibration. Although the measurement has been doing on the loudspeakers that dominated, it did not rule out the other loudspeakers also come out the sound so it can affect the value of measurement on the measuring point.

To determine whether the SPL distribution in the space is evenly distributed or not, the farthest and the nearest distance from the sound source were determined. Figure 5 (right side) shows the farthest and closest point by the loudspeaker and the measurement results. The comparison shows that the SPL distribution on the ground floor and balcony of the church was evenly distributed because the difference SPL between the farthest and closest point from the sound source was less than 6 dB.

Figure 5 also shows that on point 3 which is in the distance 3,83 meter measuring 110 dB, on the point 2 in 4,89 meter measuring 109,7 dB, and the point 1 in 7,08 meter measuring 106,8 dB. Similarly, on the measuring point 14, 11, 10 that SPL more lower. More far away from the sound source (loudspeaker), the SPL more lower. It means that the distance affects the SPL, in line with Mills (1976) opinion who told about designing of the acoustic system by setting the listener's distance to the sound source.

Listener response to the audial condition in the Lanraki Biringkanaya Church room can be a comparative data on audial condition previously discussed mathematically so that the questionnaire was used. Here is the comparison of the respond by respondents about the sound condition's comfort of this church room and the audio repair needed.

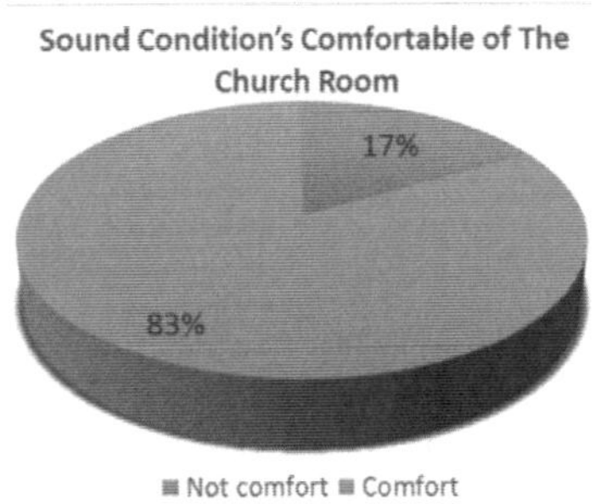

Figure 6. Comparison graphic of the respondents about the sound condition's comfortable of the church room.

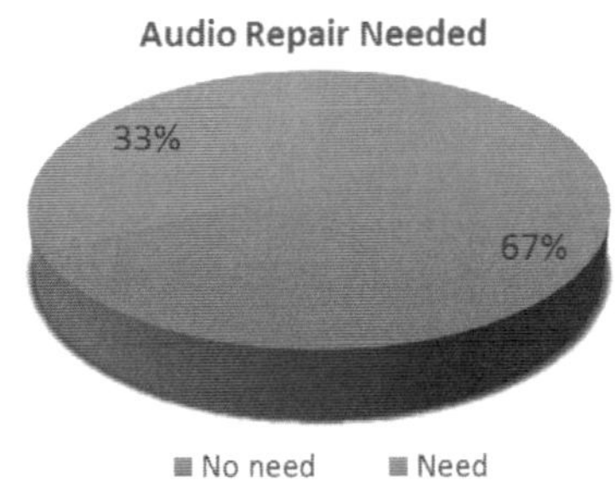

Figure 7. Comparison graphic of the respondents about an audio repair needed.

Figure 6 shows that respondents who already comfortable about sound condition of this church were 83% or 104 of 125 respondents. While who feel not comfortable just a little, it were 17% or 21 of 125 respondents. It means that users felt comfort about the audial condition of the church room. Figure 7 shows 67% or 84 of 125 respondents said they were not need audio repair and only 33% or 41 of 125 respondents who said needs audio repair in the church room. It means that the users was comfortable with the audio condition of the church room. Here are the cases that arised regarding to the distribution of SPL in the church.

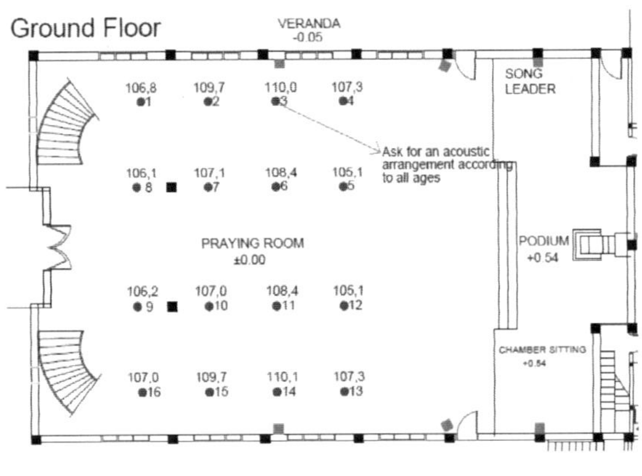

Figure 8. On the ground floor, there was respondent who requested to do acoustic arrangement that be adapted to all ages.

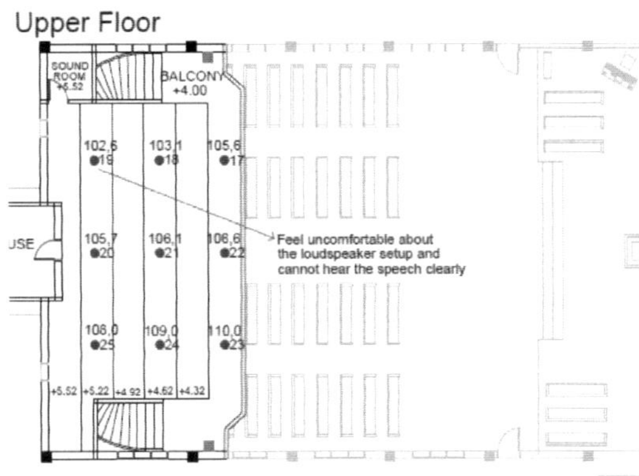

Figure 9. On the upper floor, there were respondents who can not hear the speech clearly.

Respondent on point 3 was 80 years old, male, and retired. According to the measurement result that on the measuring point 3 the highest SPL (reached 110 dB) and in real condition (before the SPL in all sound sources was equated) the SPL is 108,6 dB. While 105 dB only, can damage the ears within 1 hour. Elderly people was lost their hearing ability in old age or we call presbycusis and the incidence of hearing loss for men is three times higher than for women, it makes sense for respondents to complain that they can not hear well.

Figure 9 shows that on the point 17, 18, and 19 SPL measurement, resulting lower values than measuring point of 20 to 25. Evidently that only respondents in the 19 felt uncomfortable about the loudspeaker arrangement and could not hear the sound of the speech clearly. The measuring point 19 in the rear of the upstairs (balcony) and the measured SPL on the point 19 is the lowest where the real condition (before the SPL in all sound sources was equated) is 94,8 dB. But the lowest SPL on this point was not reason to make the clearless of the speech because 60 dB only (equivalent to general conversation) could be heard by normal humans. According to the responses of the respondents, the intrusion of the speech heard on this point was influenced by outdoor noise and echo speech sound.

4.2 Background Noise Level (BNL)

At the time of worship, sometimes the condition around the Lanraki Biringkanaya Church are very crowded by children who not following worship and playing, vehicles that passing, and buy-selling transactions. This church does not use sound proofing material so that allow for leaks of sound through the pores of space-form elements such as wall and ceiling and there was expansion slits on windows and doors.

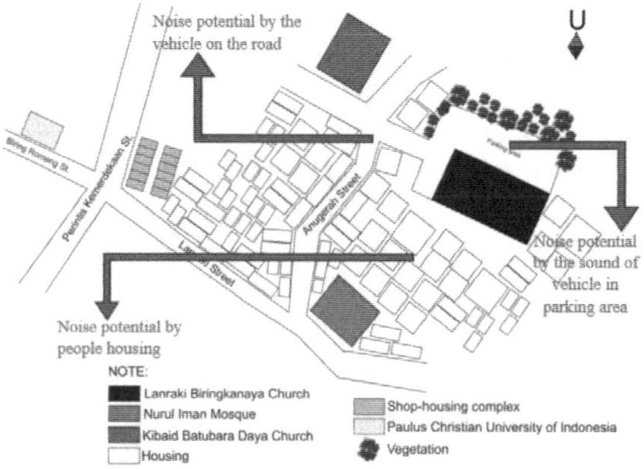

Figure 10. Layout of the Lanraki Biringkanaya Church with the analysis of its condition.

Figure 11. Slits on the doors, windows, and installation of stained glass became the potential for sound leaks.

Here are the comparison graphic between the BNL average in crowded condition and quiet condition near the opener of church room and on the seats area. According to Koenigsberger (1973), the noise acceptable for the church room is 35 dB - 40 dB. The value of 69,8 dB exceed 40 dB and the difference between the crowded and the quiet conditions is 54,2 dB so there was a temporary suspicion that the church has a sound leaks.

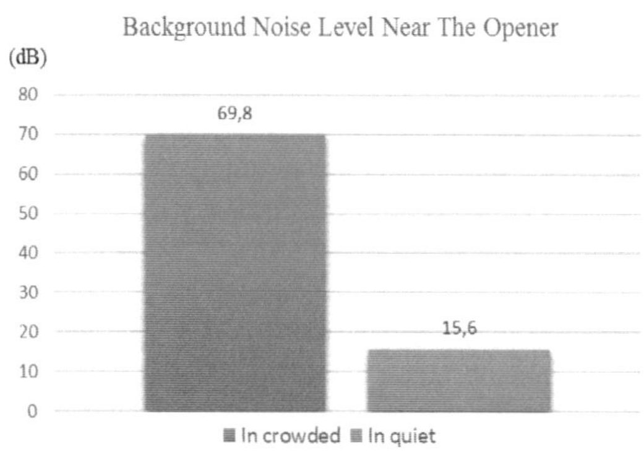

Figure 12. Comparison graphic between the BNL average in crowded condition and quiet condition near the opener of church room.

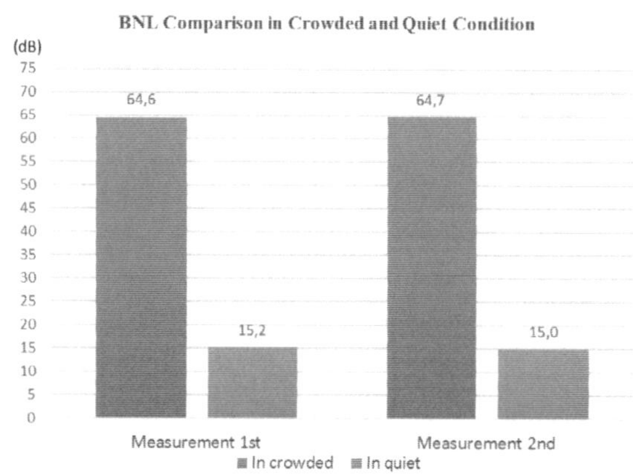

Figure 13. Comparison graphic of BNL in crowded and quiet condition on the seats area.

Although the measurement was also doing when quiet condition, but it did not rule out the possibility there was sound like mechanical electrical where the church using AC so that the BNL was not 0 dB. It shows that noise was not only from the outside of the room, but there was also noise in the room. Figure 13 shows that the average BNL in crowded condition was 64,6 dB at the first measurement and 64,7 dB at the second measurement. According to Koenigsberger (1973), the noise acceptable for the church room is 35 dB - 40 dB. This indicates that in crowded condition, the noise was exceed of the comfort standard.

The average of BNL on the seats area in quiet condition is 15,2 dB at the first measurement and 15,0 dB at the second measurement. The sound in the room does not exceed the comfort standard, but the outside sound has been the potential to cause noise for the users. This is seen at the first measurement, the BNL 64,6 dB and 15,2 dB was very far difference, it was 49,4 dB. At the second measurement, the BNL 64,7 dB and 15,0 dB was very far difference, it was 49,7 dB. The difference that too far shows that the noise is caused by the sound leaks so that the sound from outside can enter into the church room.

Potential noise from outside could come from the sound of vehicles on the road, parking area, rain, and settlements. Here are the results of the simulation using the Autodesk Ecotect Analysis 2011 application to explain it.

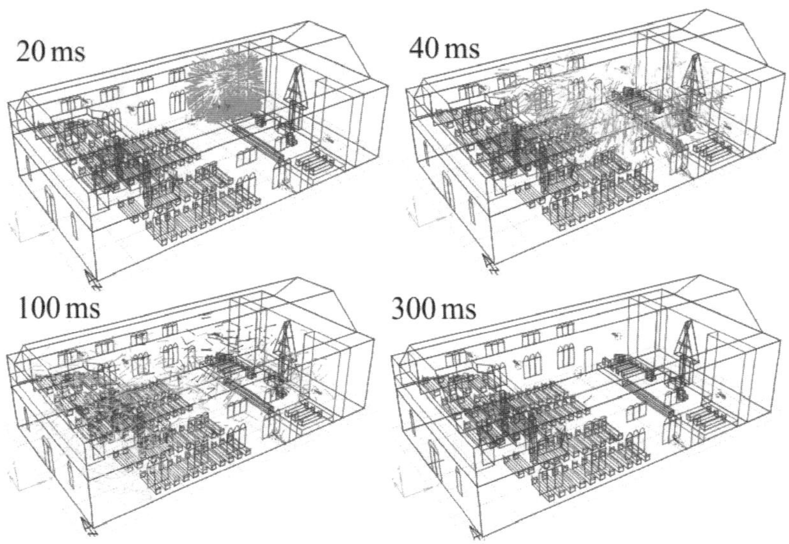

Figure 14. The rays of sound simulation that reflected to the surface with the *Ecotect* program

Figure 14 shows the sound rays sourced from the loudspeaker, proceed until it reflected to the wall, windows, doors, and stained glass at 40 ms. The reflection result continued until the sound bounced to the surface again on different point. This happened repeatedly until the sound reflection were lost on 300 ms.

It should be noted that a small gap in the wall and ceiling can be formed due to lack of perfect mixture of materials and methods of installation so that the density formed is also less. In the installation of windows, doors and stained glass, cracks are indeed made for the expansion chamber, whereas in the following simulation images there were no gaps in the wall and ceiling or expansion gap. So the results of this simulation shows that if there are no gaps in the wall, windows, doors, and installation of stained glass, a sound leak will not occur. So to insulate sound leaks to reduce background noise, it could be doing by covering the cracks on wall, ceiling, windows, doors, installation of stained glass, or other gaps in the church. Here is the description about questionnaire results compared to BNL.

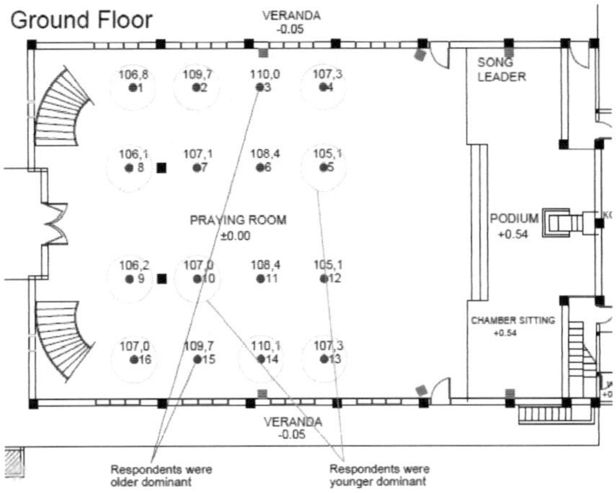

Figure 15. People who was disturbed by the noise from outside of the room based on the questionnaire result (ground floor).

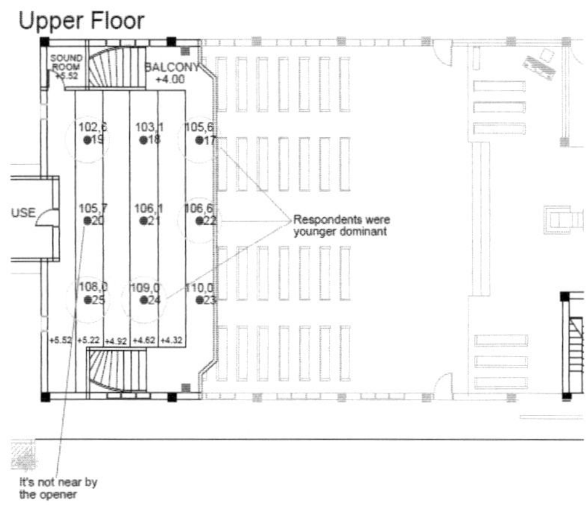

Figure 16. People who was disturbed by the noise from outside of the room based on the questionnaire result (upper floor).

The inside was not close to the wall is also disturbed, not only on the edge that heard noise from the outside of the room. This is according to the BNL measurement result that each measuring point of the BNL exceed of the standard. On measuring point 5, 10, 17, 22, and 24 were disturbed by outside sound because the respondents were young dominant so their hearing is still very sensitive. On the measuring point 3 and 15 did not felt disturbing by the sound of the outside room because the respondents were elder dominant so that hearing became less. On the measuring point 20 did not felt disturbing by the sound of the outside room although the respondents is dominant still young because around the measuring point 20 there is no window that can be a potential noise and behind the measuring point was still there room so that barried noise from outside.

Table 2. Manual calculation of reverberation time in Lanraki Biringkanaya Church room

Num	Object	Material	Total area (m^2)	Coef.	A
1	Wall	Brick plaster	658,95	0,02	13,18
2	Floor	Ceramic 60x60 cm wood fiber motif	510,40	0,01	5,10
		Ceramic 60x60 cm red blood color	23,82	0,01	0,24
3	Ceiling	Gypsum board	2173,16	0,05	108,66
4	Column	Concrete + paint	10,08	0,01	0,10
5	Main door	Wood + politur	6,85	0,08	0,55
6	Side door	Wood + politur	5,14	0,08	0,41
7	Glass door	Clear glass	7,56	0,04	0,30
8	Window	Wood frame	28,80	0,10	2,88
		Cloudy glass	31,20	0,18	5,62
9	Stained glass	Wood frame	25,92	0,10	2,59
		Carving glass	13,20	0,04	0,53
10	Pulpit	Solid wood	14,15	0,08	1,13
11	MC pulpit	Solid wood	1,68	0,08	0,13
12	Congregation seat	Solid wood	384,12	0,08	30,73
13	Assembly seat	Solid wood	17,46	0,08	1,40
14	Singer seat	Solid wood	17,46	0,08	1,40
15	Musician seat	Plastic + steel	0,80	0,25	0,20
16	Speaker	Speaker	25,60	0,45	11,52
17	Musical instrument	Plastic +steel	2,56	0,25	0,64
		Total A			187,31
		Volume			2417,32
		Reverberation Time (RT)			2,08

However, the noise disturbance did not interfere the users in the church room, as evidenced by the dominant response of felt comfort to the sound condition in the church room and in the comments column only 4% of respondents complain the issue of outside noise. It means they could tolerated such discomfort due to the subjectivity of respondents.

4.3 Reverberation Time (RT)

Here are the tables about manual calculation and computerization result of the reverberation time. Input materials of this computerization based on the materials of the church. Computerization make us be able and easy to see the simulation model and visualization of the church room.

Table 3. Material input based on Lanraki Biringkanaya Church room material

Num	Object	Material	Koefisien absorpsi bunyi						
			63 Hz	125Hz	250 Hz	500 Hz	1 kHz	2 kHz	3 kHz
1	Wall	Brick Plaster	0,10	0,07	0,03	0,12	0,02	0,02	0,03
2	Floor	Concrete slab tiles	0,01	0,01	0,01	0,01	0,02	0,03	0,07
3	Ceiling	Gypsum plaster board	0,33	0,28	0,10	0,05	0,03	0,07	0,09
4	Column	Concrete block plaster	0,12	0,09	0,07	0,01	0,01	0,01	0,02
5	Main door	Glass standard	0,51	0,35	0,25	0,18	0,12	0,07	0,04
6	Side door	Solid core pine timber	0,17	0,14	0,12	0,08	0,08	0,07	0,06
7	Window	Single glazed-timber frame	0,17	0,14	0,09	0,02	0,00	0,00	0,00
8	Staired glass	Single glazed-timber frame	0,17	0,14	0,09	0,02	0,00	0,00	0,00
9	Pulpit	Solid timber	0,17	0,10	0,07	0,07	0,08	0,05	0,04
10	MC pulpit	Solid timber	0,17	0,10	0,07	0,07	0,08	0,05	0,04
11	Seats	Solid timber	0,17	0,10	0,07	0,07	0,08	0,05	0,04
12	Musician seat	Plastic	0,43	0,40	0,34	0,25	0,19	0,15	0,14
13	Speaker	Speaker	0,56	0,60	0,54	0,45	0,39	0,35	0,34
14	Musical instrument	Plastic	0,43	0,40	0,34	0,25	0,19	0,15	0,14

Table 4. The simulation result of RT using Ecotect

Volume: 2417.32 m^3
Seating: 400
Occupancy: 90%

Frequency	Total absorption	RT(60) Sabine
63 Hz	169,426	1,87
125 Hz	154,352	2,01
250 Hz	102,192	1,92
500 Hz	86,455	2,04
1 kHz	53,913	2,43
2 kHz	59,003	2,06
4 kHz	99,451	1,76
8 kHz	98,447	1,46
16 kHz	121,693	1,38

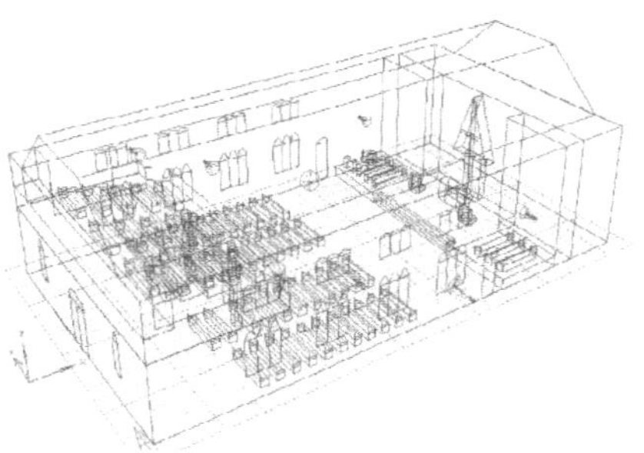

Figure 17. Model of simulation.

Figure 18. Visualization indoor of the church room.

The results of the RT value with manual calculation and computerization has not too different the calculation of manual obtained 2,08 seconds and computerization obtained 2,04 seconds. Thus, these two RT values were matched with the condition in the real building.

Optimal RT value for church in speech condition is 0,5 – 1,4 seconds and in music condition is 1,4 – 2,6 seconds. This shows that the condition of music in the church has met the standard, while the speech condition still needs to be improved. The questionnaire that have been given to church users contains the echo of speech problem. This can be compared to the RT high value in the church room. Here is the description.

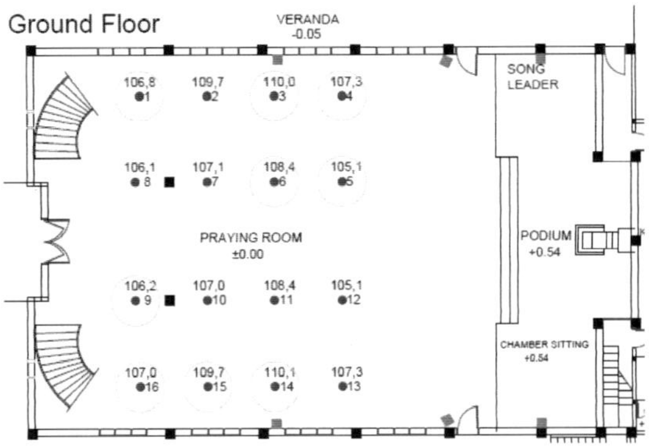

Figure 19. Position of the respondents who felt disturbed by the speech's echo (ground floor).

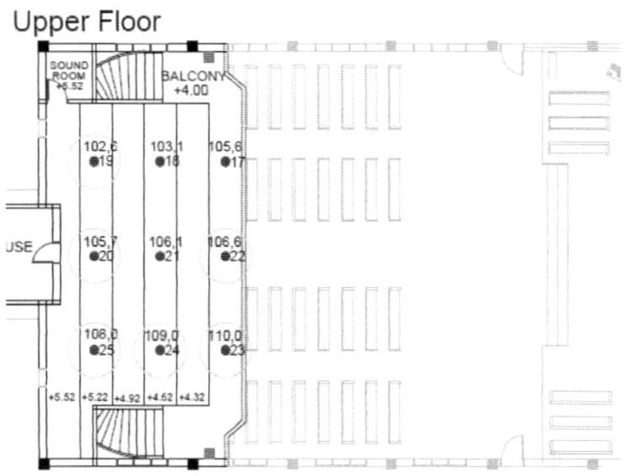

Figure 20. Position of the respondents who felt disturbed by the speech's echo (upper floor).

The green circle is the respondents who felt disturbed by the speech's echo. The figure shows the pattern formed by the respondent responses of the echo of speech disturbance was almost circular, so that almost every edge of the church room was echo. This maybe due to the reflection of sounds that can be caused by the shape of the space and the surface material of the building elements of the church room. This is also in line with the calculation of the RT value in that room that exceeds the standard of the speech condition so that when the preacher speaks for the next, the sound received by the listener has not finished decaying so that the words of the preacher sound mixed and made the speech was not clear.

Although the respondents said that they were felt comfort about the sound condition in the church room, but the respondents gave complains when looking at certain factors. This happened because the subjectivity of the users as the listeners so they can tolerate the uncomfort they felt.

To achieve audial comfort in the Lanraki Biringkanaya Church room, we could reduce the Reverberation Time (RT) values of that room. Reducing RT value could be doing by modified or added material that has small absorption coefficient with material that has large absorption coefficient. Acoustic simulation using Ecotect program has been doing to get the optimal redesign and according to RT value standard. Various alternatives are suggested so the best alternative can be chosen as an optimize solution and could be applied to the church, which is adaptable and flexible acoustic design. Variables that would be modified or added are materials, dimension of the space-form elements, and furnitures. This is simulation results from several alternative redesign for the church.

The optimal RT value for church in speech is 0,5 – 1,4 sec and in music conditions is 1,4 – 2,6 sec. The intersection of these values is 1,4 sec as to determine the comfortable RT in the church room and achieve the maximum RT value in both functions (speech and music). From the table it can be seen that to achieve a value of 1,4 seconds (no more and no less) is by combine the variables to be modified or added. We could use carpet, plywood, and rockwool as materials that could be reduce RT value. We could change the thickness of the wall and add more seats as dimension and furniture could be reduce RT value too. This shows that the material, the dimension of the space-form elements, and the using of furniture in a room can affect the value of the reverberation time of that room.

Table 5. Recapitulation of redesign alternatives to achieve optimal RT value

Variable	Redesign alternative	Absorption total	RT (sec)
Material	1. Adding carpet 15 mm on the ground floor.	167,645	1,43
	2. Adding carpet 15 mm on the ground floor and changing the solid wood doors with the sliding glass doors.	168,696	1,43
	3. Adding plywood 3/8 inch to the interior wall of the room.		
	4. Adding rockwool 5 mm to the ceiling.	171,020	1,43
		170,884	1,42
Dimension of space-form elements	1. Adding the thickness of the wall 1,5 times become 27,5 cm.	175,994	1,39
Furniture	1. Adding the number of seats, 60 seats.	86,455	1,39
	2. Adding upholstered on the seats.	168,432	1,43
Combination	1. Adding carpet 15 mm on the ground floor and plywood 6 mm on the part of podium wall and the back wall of congregation.	173,545	1,40
	2. Adding rockwood 5 mm on ceilings and plywood 5 mm on the part of podium wall.	174,664	1,40
	3. Adding carpet 15 mm on the ground floor and 6 seats.		
		167,745	1,40

5. Conclusion and Suggestions

Regarding to the acoustic condition of the Lanraki Biringkanaya Church room, the distribution of sound pressure levels in the church room was evenly distributed, while background noise level and reverberation time in the church room has not achieved the comfort standards yet. Regarding the user's response to the audial comfort in the church, the users felt comfortable overall. But when looking at the bakcground noise factor and the reverberation time factor, users still complain about the noise from outside of the church room and speech's echo.

Redesign to create a good design concept in the Lanraki Biringkanaya Church room to achieve audial comfort can be doing by insulating background noise, by covering the cracks on the wall, ceiling, windows, doors, installation of stained glass, or other gaps in the church. Reverberation time insulation can be carried out with a combination of material, the dimension of the space-form elements, and the using of furniture in a church room.

Distance can affect the sound pressure level. In this church, more far away from the sound source, the SPL more lower. The difference of listener position in a room can affect the subjectivity of human hearing.

When we want to build a church or similar buildings, we should be considered the various aspects of audial comfort and potential disturbances that may occur. Using reflective and absorber materials of the sound must be balanced in a room that required optimum audial. The absorption coefficient of the material can be adjusted to the market conditions in order to insert into the Ecotect application so that the simulation results are closer to the truth of redesign. Further research can include variables and parameters that have not been included in this study such as air density that affected by differences of air humidity, air density that affected by differences of height, differences of frequency, early decay time, and clarity.

6. References

[1] Doelle, Leslie L. (1972). *Environmental Acoustics*. New York: McGraw-Hill.
[2] Errede, S. (2013). *Acoustical Physics of Music*. University of Illinois: Department of Physics.
[3] Kahle, E., 2013. *Room Acoustical Quality of Concert Halls: Perceptual Factors and Acoustic Criteria - Return From Experience*. Toronto, Canada.

[4] Kuttruff, H. (2009). *Room Acoustics*. London: Spon Press.
[5] Svensson, C. & Nilsson, E. (2008). *Optimum Room Acoustic Comfort TM (RACTM) can be achieved by using a selection of appropriate acoustic descriptors*. Euronois.
[6] Indrani, H. C. (2007). *Analisis Kinerja Akustik Ruang Auditorium Universitas Kristen Petra Surabaya*. Dimensi Interior, (pp. 109-121).
[7] Egan, M. D. (1988). *Concepts in Architectural Acoustics*. New York: McGraw-Hill.
[8] Cowan, J. (2007). *Building Acoustics. In: Springer Handbook of Acoustics*, ed. T. D. Rossing. New York: Springer, (pp. 387-425).
[9] Satwiko, P., 2009. *Building Physics*. Yogyakarta: Andi Offset.
[10] Sabine, W. C. (1993). *Design for Good Acoustics. Collected Papers on Acoustics*. Los Altos, U.S: Trade Cloth ISBN 0-9321 Peninsula Publishing.
[11] Legoh, F. (1993). *Acoustic Design and Scale Model Testing a Multi Purpose Auditorium*. U.K: The University of Salford.
[12] Parker, S. P. (1994). *McGraw-Hill Dictionary of Scientific and Technical Terms*. New York: McGraw-Hill.
[13] Koenigsberger, et al (1973). *Manual of Tropical Housing and Building*. Bombay: Orient Longman.
[14] Steemers, K. &. Steane, M. Ann (2006). *Environmental Diversity in Architecture*. London: Spon Press.
[15] Mediastika, E. C. (2008). *Akustika Bangunan*. Jakarta: Penerbit Erlangga.
[16] Poetiray, et al (2015). *Pengukuran Kebisingan Bangunan Gereja Terbuka Studi Kasus; Gereja Puh Sarang, Kediri*. ECO-Architecture Conference. Wonosobo.
[17] Ballenger, John Jacob (1977). *Disease of The Nose, Throat, and Ear*. Philadelphia: Lea & Febiger.
[18] Nelson, D. I., et al. (2005). *The Global Burden of Occupational Noise-Induced Hearing Loss*. American Journal of Industrial.

IOP Conf. Series: Materials Science and Engineering **619** (2019) 012027 doi:10.1088/1757-899X/619/1/012027

Experiments on Motion and Oscillation Controls of a Gantry Crane System Using Parallel Proportional Controllers

A K Muhammad, R Tandioga, S Ka'ka, L Lewi, A Rahman and A Taufik

Center for Mechatronics and Control Systems, Mechanical Engineering Department, State Polytechnic of Ujung Pandang, Makassar, Indonesia

kadir.muhammad@poliupg.ac.id

Abstract. The purposes of this research are to formulate the mathematical model of the system, to propose an effective control scheme and to perform experiments on a laboratory scale. The system used in this paper consists of a trolley, a pendulum as load and a dc motor to drive the trolley. Mathematical model of trolley position and load angle were formulated considering voltage of the motor and damping constant of the air calculated specifically using energy balance. Two proportional (*P*) controllers were designed in parallel in such a way to drive the trolley and to reduce oscillation of the load. The system and the proposed control scheme were confirmed through experiments. The experimental results revealed that the motion and oscillation of the system can be controlled effectively.

1. Introduction

The use of gantry crane systems for transporting payload is very common in engineering constructions, industrial applications and trading. For instance, a gantry crane is used for lifting containers from a ship onto a truck. The crane should move the load as fast as possible without having any excessive payload motion at the final position. However, moving the load using the crane is not an easy task. Most of the common gantry cranes result in a swing motion when payload is suddenly stopped after a fast motion [1]. So, the gantry crane needs a skillful operator to manually control it and to immediately stop the swing at the right position. Furthermore to unload, the operator has to wait until the load stops swinging [2]. It is often time consuming process which eventually affects the productivity in transporting payload. These problems can be solved using a controlled gantry crane system. Its controller system should deal with not only its motion but also oscillation of the payload.

The purposes of this research are to formulate the mathematical model of the system, to propose an effective control scheme and to perform experiments on a laboratory scale. The system used in this paper consists of a trolley, a load and a dc motor to drive the trolley. Mathematical model of trolley position and load angle were formulated considering voltage of the motor and damping constant of the air calculated specifically using energy balance. Two proportional (*P*) controllers were designed in parallel in such a way to drive the trolley and to reduce oscillation of the load.

2. Modeling

2.1. Mathematical Model

Figure 1 shows the dc motor used in this study. Electrical part of the motor consists of resistor, *R* and conductor, L. Voltage of *V* is the source of power. Inverse voltage of the motor and its constant are denoted by *e* and *K* respectively. Torque constant of the motor is denoted by K_t. Mechanical part of the motor has mass moment of inertia, *J* , viscous damping constant, *b* and motor torque *T*. Radius of gear, angular position and velocity of the motor are denoted by *r*, θ and $\dot{\theta}$. Transfer function of voltage of the motor to position of the trolley *X* and is expressed by [3]

$$\frac{X(s)}{V(s)} = \frac{r.K_t}{s\left((Ls+R)(Js+b)+KK_t\right)} \tag{1}$$

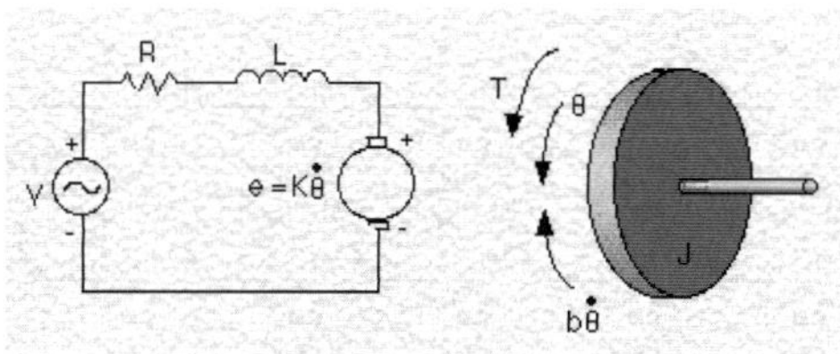

Figure 1. DC Motor

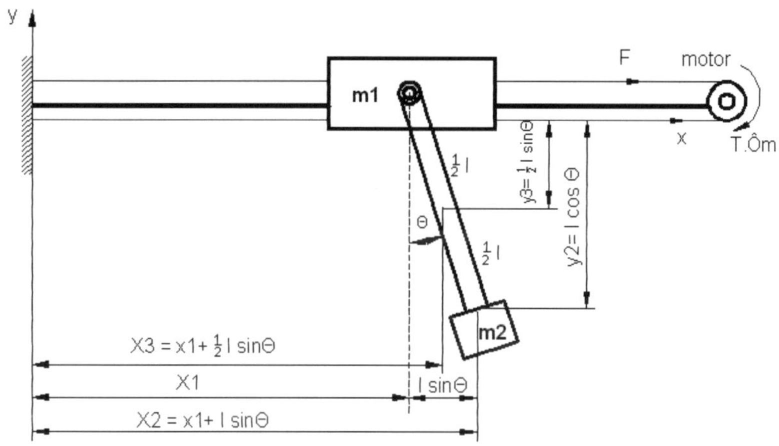

Figure 2. Design of the gantry crane system

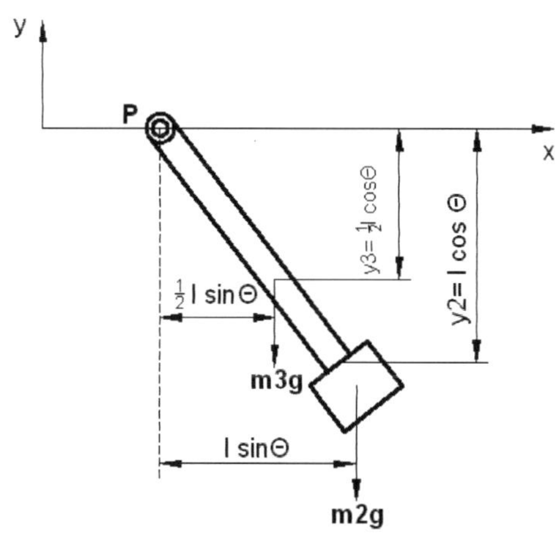

Figure 3. Free body diagram of the load

Figure 2 and figure 3 show the gantry crane system and free body diagram of the load. Based the figures, transfer function of position of the trolley *to* angle of the load α and is expressed by [3]

$$\frac{\theta(s)}{X_1(s)} = \frac{2.22.s^2}{s^2 + 1.25D_f.s + 21.8} \tag{2}$$

where D_f is damping constant of the air calculated specifically using energy balance.

2.2. Experimental Model

Figure 4 shows the experimental model of the gantry crane system. The gantry crane consists of the frame, the trolley, the load, the dc motor and the base. The motor is fixed in one side of the frame. The trolley is connected to the motor through a belt transmission system. An encoder is attached in the other side of the motor. IMU (inertia measurement unit) sensor is attached to the load. The frame is mounted to the base.

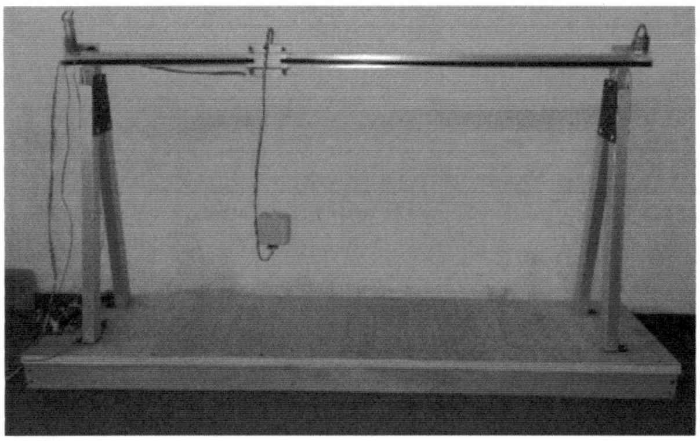

Figure 4. Experimental model of the gantry crane system

3. Control Scheme

A control scheme to drive the trolley and to reduce oscillation of the load was designed for the dc motor. Two *P*-controllers were designed in parallel for the system. The controlled torque of the motor are defined as follows

$$\tau = k_{p1}(x_d - x) + k_{p2}(\theta_d - \theta) \tag{3}$$

where k_{p1} and k_{p2} are proportional gains for position and angle of the load, respectively. Desired and actual positions of the trolley are denoted by x_d and x, respectively.

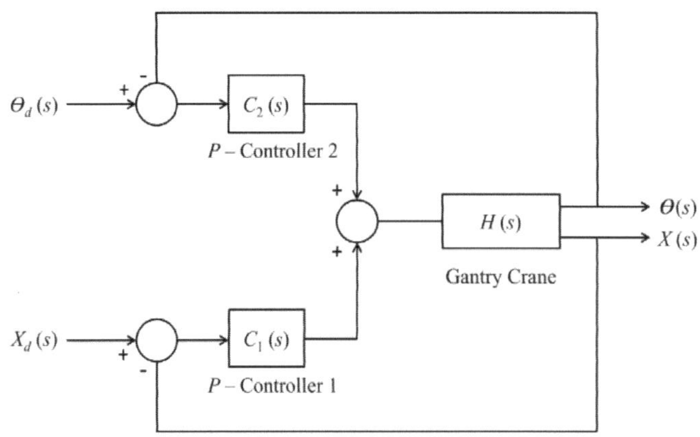

Figure 5. Block diagram of the proposed control scheme for the gantry crane system

Desired and actual angles of the load are denoted by θ_d and θ, respectively. θ_d is set to zero. The block diagram of the controlled system is shown in figure 5.

4. Experiment

4.1. Experimental Set-up

In order to investigate the validity of the proposed control scheme, an experimental set-up was designed. The set-up is shown in figure 6.

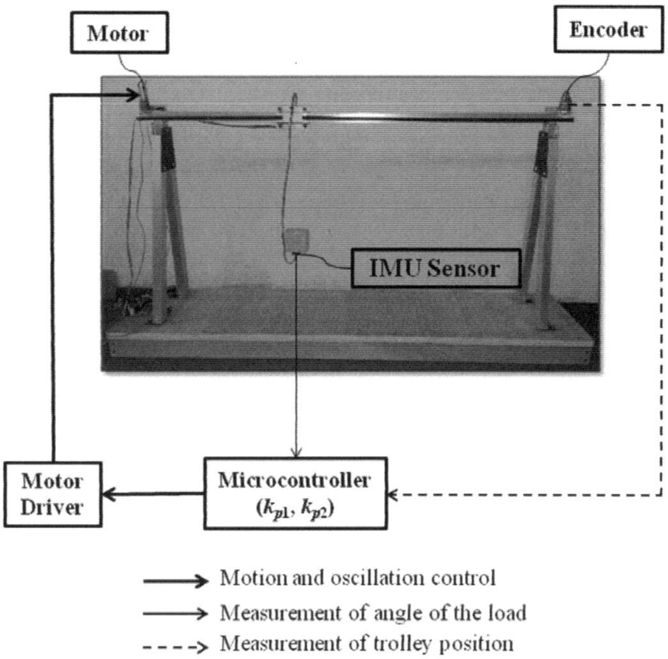

Figure 6. Schematics of measurement and control systems

The gantry crane system consists of the frame, the trolley, the load, the dc motor, the encoder, the IMU sensor and the base. The encoder and the IMU sensor provide information of trolley position and angle of the load. The both information were processed by a microcontroller using two *P*-controllers. The actuating signal sent to a motor driver to drive the motor and control oscillation of the load in the motor voltage range.

4.2. Experimental Method

Position of the trolley was set to 20 [cm]. Actuation signal were obtained from the first *P*-controller based on error of the trolley position and by the second one based on angle of the load. The proportional control scheme was implemented using arduino nano.

4.3. Experimental Results

Experimental time history responses of the trolley position and angle of the load on the controlled system were measured under the control scheme as shown in figure 7 and figure 8. Examining several gains of the *P*-controllers leaded to $k_{p1} = 3$ and $k_{p2} = 4$ as the better ones for the system. Figure 7 shows the uncontrolled and controlled time history responses of trolley position. The trolley position of 20 [cm] was achieved for 33 [s] and 20 [s] for uncontrolled and controlled system. Figure 8 shows the uncontrolled and controlled time history responses of angle of the load. Maximum angle of the load was

11 [degrees] for the uncontrolled system and could be reduced after 50 [s]. Maximum angle of the load was 4 [degrees] for the controlled one and could be reduced after 25 [s].

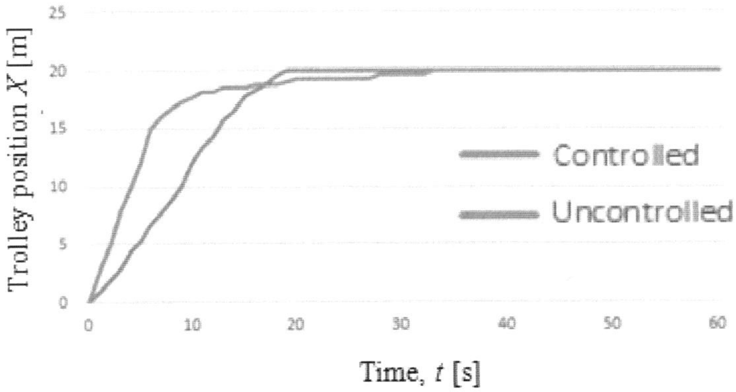

Figure 7. Time history responses of the trolley position

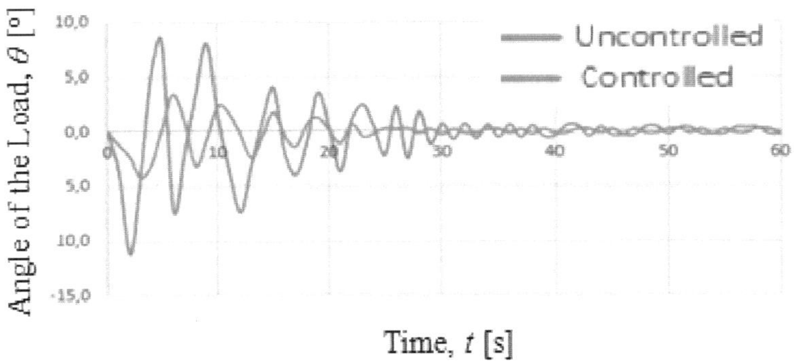

Figure 8. Time history responses of the load angle

5. Conclusion

The Mathematical model of the gantry crane system had been formulated considering voltage of the motor and damping constant of the air calculated specifically using energy balance. Two proportional (*P*) controllers were designed in parallel in such a way to drive the trolley and to reduce oscillation of the load. The proposed control scheme was examined through the experiments. The experimental results have been revealed that the position of the trolley and oscillation of the load can be controlled effectively.

References

[1] Omar H M 2003 *Control of Gantry and Tower Crane* MS Thesis Virginia Tech
[2] Wahyudi et al 2007 Control Startegy for Automatic Gantry Crane Systems; a Practical and Intelligent Aproach *Int. J. Advanced Robotics System* **4 No. 4** 447
[3] Muhammad A K 2011 Modeling and a Control Strategy for Gantry Crane Systems *J. Teknologi* **13 No. 4** 201

Optimization of stirring parameters on the rheocast microstructure and mechanical properties of aluminum alloy ADC12

Syaharuddin Rasyid[1,2], **Ilyas Renreng, Effendy Arif**[1], **Hairul Arsyad**[1], **and Muhammad Syahid**[1]

[1] Department of Mechanical Engineering, Faculty of Engineering, Hasanuddin University, Makassar, 90245, Indonesia
[2]Department of Mechanical Engineering, Politeknik Negeri Ujung Pandang, Makassar, 90245, Indonesia.
E-mail: syaharuddinrasyid@poliupg.ac.id

Abstract. This paper presents an experimental design approach for the optimization process parameters for the aluminum solid alloy casting of ADC12 alloys. To achieve this goal, the stirring time parameter and stirring speed are selected and two levels of this parameter are considered. Design of expert (DOE) of tests was used for experimental design and analysis of results. The aluminum ADC12 slurry is stirred by a mechanical stirrer (straight plate stirrer models) at 610°C with a variation of speed 250, 300, 350 rpm for 20, 40, 60 seconds. Furthermore, the aluminum slurry of ADC12 is poured on a metal mold with temperature 600°C. The microstructure characteristics were examined by direct observation using optical microscopy, secondary α-Al phase dendrite arm spacing and shape factor were identified. The mechanical properties were investigated by the hardness test and tensile test. The microstructure and mechanical properties of aluminum alloys ADC12 made with semi-solid rheocasting casting technology using several parameters stirring have been studied, and the results obtained can be synergized as follows. The highest mechanical properties (Hardness 122,5 HB, Tensile strength 238.2 N / mm2) occur at 300 rpm stirring speed and 20 seconds stirring time. Secondary Dendrite Arm Spacing (SDAS) lowest occurred at 25 rpm stirring speed and 20 seconds stirring time of 13.93 μm. The highest shape factor occurs at 300 rpm stirrer spin and 20 seconds stirring time of 0.47. The optimal stirring parameters recommended for producing a hardness response of 122,071 HRC, tensile strength 236,271 N / mm2, Shape Factor (SF) 0.38994, and Secondary Dendrite Arm Spacing (SDAS) are 317 rpm (stirring speed) and 20 seconds (stirring time).
Keywords: Optimization, DOE, ADC12.

1. Introduction

The semi-solid metal forming technology has evolved since the early 1970s, and leader in the metal processing technology in the 21st century [1-3]. Conventional casting, dendrites micro-structure formed naturally may have disabilities, such as depreciation and gas shocks that weaken the mechanical properties [4]. In the process of semi-solid, micro-globular structures formed by shear rate applied to the molten metal which is in the range of compaction [5]. The solid fraction with a high percent (± 60%), solids still has good mobility and can be formed by conventional forming processes [6].

IOP Conf. Series: Materials Science and Engineering **619** (2019) 012028 doi:10.1088/1757-899X/619/1/012028

Some researchers have explained the dendrites transformation into spheroidal morphology with a stirring mechanism. The shear rate causes the dendrite to bend and break so that it decreases surface energy [7, 8]. Dendrite melting results at the root locally due to the stirring process [9-10]. Changes in the dendritic transformation into spheroidal morphology due to solute convection stirring at [11]. In recent years, many methods have been introduced for the production of the semisolid slurry as scientifically sound and viable industry with the preferred microstructure is called a thixotropic microstructure as a raw material. One of them is mechanical stirring. A number of researchers have investigated the formation of semi-solid metal with mechanical stirring. Rashid et al. [12] have improved the mechanical properties of aluminum alloy ADC12 with mechanical stirring molten alloy. Fan et al. [13] by using a mechanical stirrer, can produce thin shielding parts that cannot be produced by the conventional die-casting process. Barabazon et al. [14] have investigated the effects of mechanical stirring on microstructure and mechanical properties of Aluminum Alloys. Niroumand et al. [15-17] to evaluate the microstructure rheocast by creating new mechanisms. They are Al-7.1% Si alloys that are engineered under different stirring speeds and solid fractions. they show that the newly proposed mechanism for the formation and evolution of micro rheocast can explain and justify experimental results.

Some researchers have used DOE to optimize various types of manufacturing processes [18-22], In this study, efforts have been made to develop models to predict optimal stirring parameters in the ADC12 aluminum alloy semi-solid casting process.

In this work by using a mechanical stirrer slurry maker constructed by the authors, several experiments have been conducted on the effects of stirring parameters such as stirring time and stirring speed. For all these experiments, investigated on the microstructure and mechanical properties of aluminum alloys ADC12 in the semi-solid foundry.

2. Experimental Procedures

In this study, the secondary ADC12 alloy was used. The chemical composition of this alloy is shown in Table 1. The liquidus temperature of this alloy is 582°C. The chemical composition of the aluminum alloy ADC12 is seen in Table 1.

Table 1. The composition of ADC12 aluminum alloys

ADC12	Weight %											
Aluminum	Si	Cu	Fe	Mn	Mg	Zn	Ti	Cr	Ni	Pb	Sn	Al
Alloys	9.55	2.01	0.91	0.16	0.22	1.31	0.03	0.02	0.14	0.11	0.02	85.49

2.1. Preparation of SSR slurry.

Metal molds are prepared and heated to a temperature of ±250°C. Aluminum alloy material ADC12 is prepared (±280 gram). The alloy material of aluminum ADC12 is melted to 650°C temperature using gas furnace. Metal mold temperature measurements using infrared temperature gauges and aluminum fluid measurements using a thermocouple gauge. In a temperature of 605 ^{O}C, the aluminum alloy liquid or slurry of ADC12 is stirred with a mechanical stirrer straight plate stirrer models of stirring speed of 250, 300, and 350 rpm. Time stirring 20, 40, and 60 seconds. The liquid or aluminum alloy of ADC12 aluminum is poured into the metal mold. The specimens of casting result are made of the tensile specimen, hardness, and microstructure. In this study, the effect of time of stirring and stirring speed on size, the main feature of ADC12 rheocast, and recognizing stirring requirements for significant changes in dendritic microstructure was examined.

2.2. Mechanical testing and microstructure analysis.

The mechanical properties of the foundry are investigated experimentally, including the nature of hardness and tensile properties. Hardness is evaluated by a Brinnel hardness tester, where steel ball indenter is used at 613 N load for 5 s. Tensile properties are checked at room temperature using a universal screw driven type screw machine with a capacity of 100 kN. The test specimens were designed based on ASTM B557. Characteristics of microstructures are examined by optical

IOP Conf. Series: Materials Science and Engineering **619** (2019) 012028 doi:10.1088/1757-899X/619/1/012028

microscopy (MO). The secondary dendritic arm spacing of the α-al (SDAS) phase and the size of the eutectic base phase Si were measured using image analysis.

Image analysis techniques were used to investigate the microstructure of rheocast samples following standard metallographic procedures and. average particle size and average dendritic arm spacing (DAS) dendrite were measured using a linear intercept method. The equation used to calculate the particle form factor (SF) of the particle is [23]:

$$SF = \frac{4.\pi.A}{P^2} \quad (1)$$

Where P and A are the perimeter and area of individual particles respectively. A perfectly spherical particle would have a shape factor value of unity, and an infinitely long needle-like particle would have a shape factor equal to zero.

2.3. Experimental design and statistical analysis.

To explore the effect of the operational factors on the response in the region of investigation, a DOE at two levels was performed. Stirring speed (rpm, A) and stirring time (min, B) were selected as independent factors. The range of values and coded levels of the factors are given in Table 2.

A polynomial equation (Eq. 2) was used to predict the response as a function of independent factors and their interactions. An interaction is the failure of the one factor to produce the same effect on the response at different levels of another factor [24]. In this work, there were four independent factors; therefore, the response for the quadratic polynomials becomes:

Table 2. Independent Factors and their Levels for DOE of semi solid casting process.

Independent Factors	Unit	Level		
		-1	0	1
Stirring Speed (A)	(rpm)	250	300	350
Stirring Time (B)	(second)	20	40	60

$$Y = \beta_0 + \sum \beta_i x_i + \sum \beta_i x_i^2 + \sum\sum \beta_{ii} x_i x_j \quad (2)$$

where b_0, b_i, b_{ii}, b_{ij} are the constant, linear, square and interaction regression coefficient terms, respectively, and xi and xj are the independent factors (A and B).

Design-Expert 6 software was used for multiple regression analysis, analysis of variance (ANOVA), and analysis of ridge maximum of data in the response surface regression (RSREG) procedure. The goodness of the model was evaluated by the coefficient of determination R^2 and its statistical significance was checked by the F-test.

3. Result and Discussion

This study demonstrates the effect of stirring speed and stirring time for the optimization of the semi solid casting route. The design is used to obtain 9 design points within the whole range of two factors for experiments. The designs and the response are given in Table 3. Following the experiments, the response surface is approximated by DOE.

Table 3. Design layout and experimental results.

Std	Stirring Speed (rpm)	Stirring Time (second)	Coded A	Coded B	Hardness (HBN)	Tensile Strength (N/mm^2)	SDAS (μm)	SF
1	250	20	-1	-1	120.6	227.4	14.99	0.47
2	300	20	0	-1	122.5	235.1	13.93	0.43
3	350	20	1	-1	121.8	231.8	16.99	0.38
4	250	40	-1	0	120.7	225.7	21.63	0.38
5	300	40	0	0	122.2	231.8	16.77	0.30

6	350	40	1	0	121.9	229.9	17.62	0.36
7	250	60	-1	1	120.3	220.5	24.43	0.36
8	300	60	0	1	121.7	226.7	17.89	0.26
9	350	60	1	1	121.4	225.0	20.67	0.32

3.1. Hardness

Results for hardness at stirring speeds and stirring time show that it fits the quadratic model. The ANOVA for the hardness data is given in Table 4. Having its Prob>F of much less than 0.01, the quadratic model is valid. As for the coefficients, the stirring speeds and stirring time was considered a significant factor.

Table 4. ANOVA with CI = 95% for model and factors of the hardness.

Source	Sum of Squares	DF	Mean Square	F Value	Prob > F	
Model	4.58	5	0.92	33.54	0.0078	significant
A	2.04	1	2.04	74.75	0.0033	
B	0.37	1	0.37	13.73	0.0342	
A^2	2.07	1	2.07	75.68	0.0032	
B^2	0.094	1	0.094	3.44	0.1608	
AB	2.500E-003	1	2.500E-003	0.092	0.7820	
Residual	0.082	3	0.027			
Cor Total	4.66	8				

The obtained empirical equation of the hardness in the form of an actual factor is as stated in equation (1),

$$Hardness = 122.28 + 0.58A - 0.25B - 1.02A^2 - 0.22B^2 - 0.025AB \quad (1)$$

Where A is stirring speed (rpm) and B is stirring time (s).

For convenience, the equation can be displayed as response surface contour as well as three-dimensional surfaces, as shown in Figure 1.

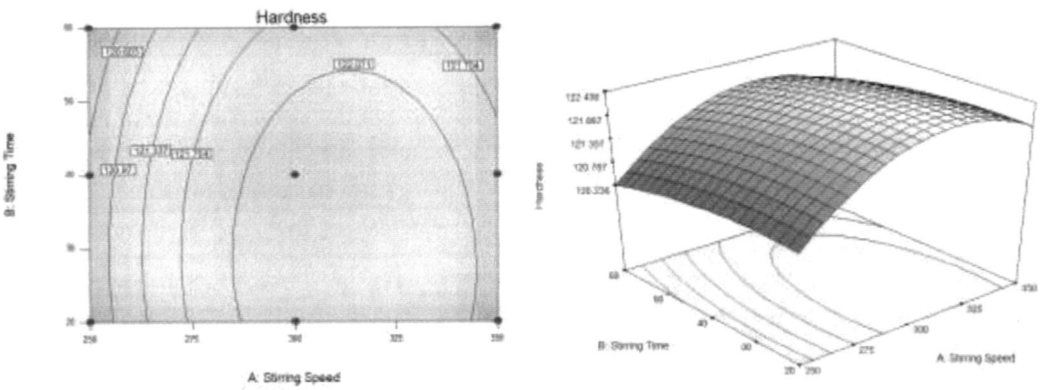

Figure 1. Response surface graph of (a) contours and (b) 3D Surface for hardness.

(Figure 1 shows the hardness of casting results with variations of stirrer rotation (250, 300, and 350 rpm) and stirring time 20, 40, and 60 seconds. The graph shows the hardness of the ADC12 material of semisolid casting. The longer the stirring time the hardness value decreases, but on the increase of stirrer rotation, the hardness value increases until 300 rpm rotation and the hardness value decreases at 350 rpm rotation.)

3.2. Tensile Strength

Results for tensile strength at various stirring speeds and stirring time show that it fits the quadratic model. The ANOVA for the tensile strength data is given in Table 5. Having its Prob>F of much less than 0.01, the quadratic model is valid. As for the coefficients, both of the stirring speeds and stirring time was considered a significant factor. Tensile strength was insensitive to the change in stirring speeds and stirring time.

Table 5. ANOVA with CI = 95% for model and factors of the tensile strength.

Source	Sum of Squares	DF	Mean Square	F Value	Prob > F	
Model	154.03	5	30.81	87.58	0.0019	significant
A	28.60	1	28.60	81.31	0.0029	
B	81.40	1	81.40	231.41	0.0006	
A^2	40.20	1	40.20	114.28	0.0017	
B^2	3.83	1	3.83	10.88	0.0458	
AB	2.500E-003	1	2.500E-003	7.107E-003	0.9381	
Residual	1.06	3	0.35			
Cor Total	155.09	8				

The obtained empirical equation of tensile strength in the form of an actual factor is as stated in equation (2),

$$Tensile\ Strength = 232.12 + 2.18A - 3.68B - 4.48A^2 - 1.38B^2 + 0.025AB \quad (2)$$

here A is stirring speed (rpm) and B is stirring time (s).

For convenience, the equation can be displayed as response surface contour as well as three-dimensional surfaces, as shown in Figure 2.

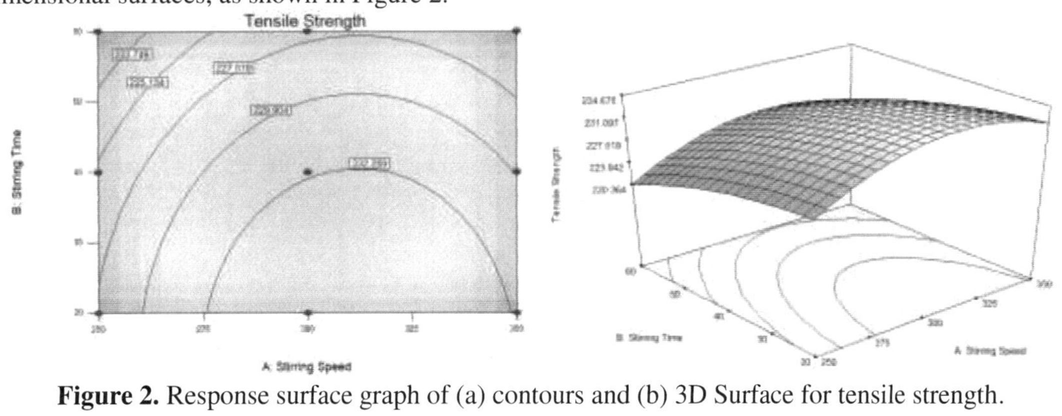

Figure 2. Response surface graph of (a) contours and (b) 3D Surface for tensile strength.

Figure 2 shows the tensile stress values of casting results with variations of stirrer rotation (250, 300, and 350 rpm) and stirring time (20, 40, and 60 sec.) The graph shows the tensile stress values of the semisolid-coated ADC12 material. then the value of tensile stress decreases, but at the increase of stirrer rotation, tensile stress value increases up to 300 rpm rotation and tensile stress value decreases at 350 rpm rotation.

3.3. Secondary Dendrite Arm Spacing (SDAS)

Results for secondary dendrite arm spacing at various stirring speeds and stirring time show that it fits the quadratic model. The ANOVA for the secondary dendrite arm spacing data is given in Table 6. Having its Prob>F of much more than 0.01, the quadratic model is valid. As for the coefficients, both of the stirring speeds and stirring time was considered a significant factor. Secondary arm spacing was insensitive to the change in stirring speeds and stirring time.

Table 6. ANOVA with CI = 95% for model and factors of secondary dendrite arm spacing.

Source	Sum of Squares	DF	Mean Square	F Value	Prob > F	
Model	83.39	5	16.68	8.88	0.0510	Not significant
A	5.55	1	5.55	2.96	0.1841	
B	48.62	1	48.62	25.90	0.0147	
A^2	20.37	1	20.37	10.85	0.0459	
B^2	0.55	1	0.55	0.29	0.06266	
AB	8.29	1	8.29	4.42	0.01263	
Residual	5.63	3	1.88			
Cor Total	89.02	8				

The obtained empirical equation of secondary dendrite arm spacing in the form of an actual factor is as stated in equation (3),

$$SDAS = 16.55 - 0.96A + 2.85B + 3.19A^2 - 0.52B^2 - 1.44AB \tag{3}$$

here A is stirring speed (rpm) and B is stirring time (s).

For convenience, the equation can be displayed as response surface contour as well as three-dimensional surfaces, as shown in Figure 3.

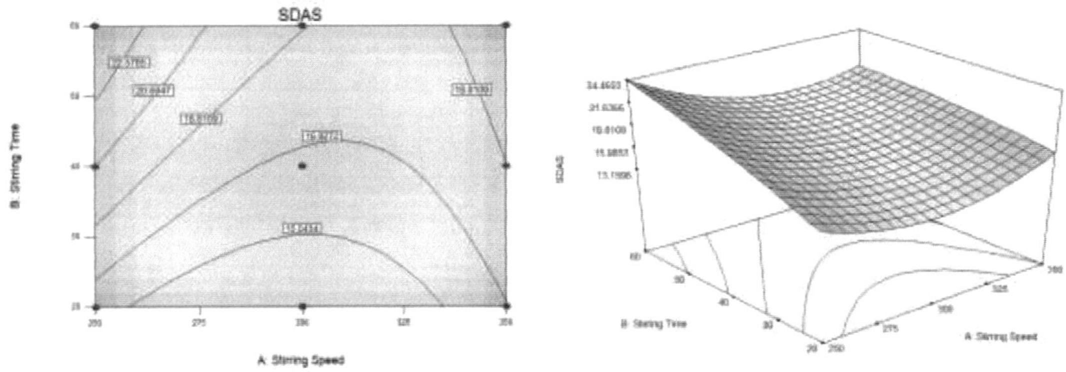

Figure 3. Response surface graph of (a) contours and (b) 3D Surface for secondary dendrite arm spacing.

Figure 3 shows the grade of secondary dendrite arm spacing with mixed rotation variations (250, 300, and 350 rpm) and stirring time (20, 40, and 60 sec.) The graph shows the secondary dendrite arm spacing of ADC12 semisolid foundry. The longer the agitation time the secondary dendrite arm spacing value is higher, but in the stirrer rotation increase, the secondary dendrite arm spacing value decreases to 300 rpm and the secondary dendrite arm spacing value increases at 350 rpm rotation.)

3.4. Shape Factor (SF)

Results for shape factor at various stirring speeds and stirring time show that it fits the quadratic model. The ANOVA for the shape factor data is given in Table 7. Having its Prob>F of much less than 0.01, the linear model is valid. As for the coefficients, both of the stirring speeds and stirring time was considered a significant factor. Shape factor was insensitive to the change in stirring speeds and stirring time.

Table 7. ANOVA with CI = 95% for model and factors of shape factor.

Source	Sum of Squares	DF	Mean Square	F Value	Prob > F	
Model	0.023	2	0.012	6.95	0.0274	significant
A	3.750E-003	1	3.750E-003	2.26	0.1831	
B	0.019	1	0.019	11.63	0.0143	
Residual	9.939E-003	6	1.656E-003			
Cor Total	0.033	8				

The obtained empirical equation of secondary arm spacing in the form of an actual factor is as stated in equation (4),

$$SF = 0.36 - 0.025A - 0.057B \tag{4}$$

here A is stirring speed (rpm) and B is stirring time (s).

For convenience, the equation can be displayed as response surface contour as well as three-dimensional surfaces, as shown in Figure 4.

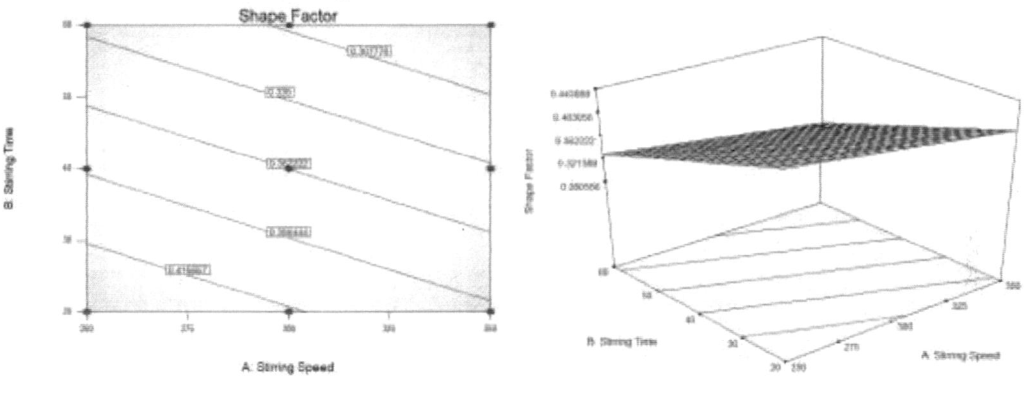

Figure 4. Response surface graph of (a) contours and (b) 3D surface for shape factor.

Figure 4 shows the shape factor values of casting results with variations of stirrer rotation (250, 300, and 350 rpm) and stirring time (20, 40, and 60 seconds). The graph shows the shape factor value of the ADC12 material of semisolid foundry. The longer the stirring time the shape factor value decreases, but in the stirrer rotation increase, shape factor value increases until 300 rpm rotation and shape factor value decreases at 350 rpm rotation.)

3.5. Optimization

Now that the empirical model for all casting responses as a function of stirring speed and stirring time has been obtained, the selection of the optimal casting parameter setting can be performed. One can adjust the expected range of each casting response and the range of stirrer speed and stirring time in line with expectations for all foundry responses can be determined. For example, that in order to obtain optimal mechanical properties, minimum SDAS, and optimum shape factor, the stirring parameters should be carried out at 300-325 rpm rotation range and 20 seconds stirring time. To achieve this criterion, the stirrer speed range and the timing of the sweep must be within the yellow plot of the overlay (Figure 5) of all casting responses.

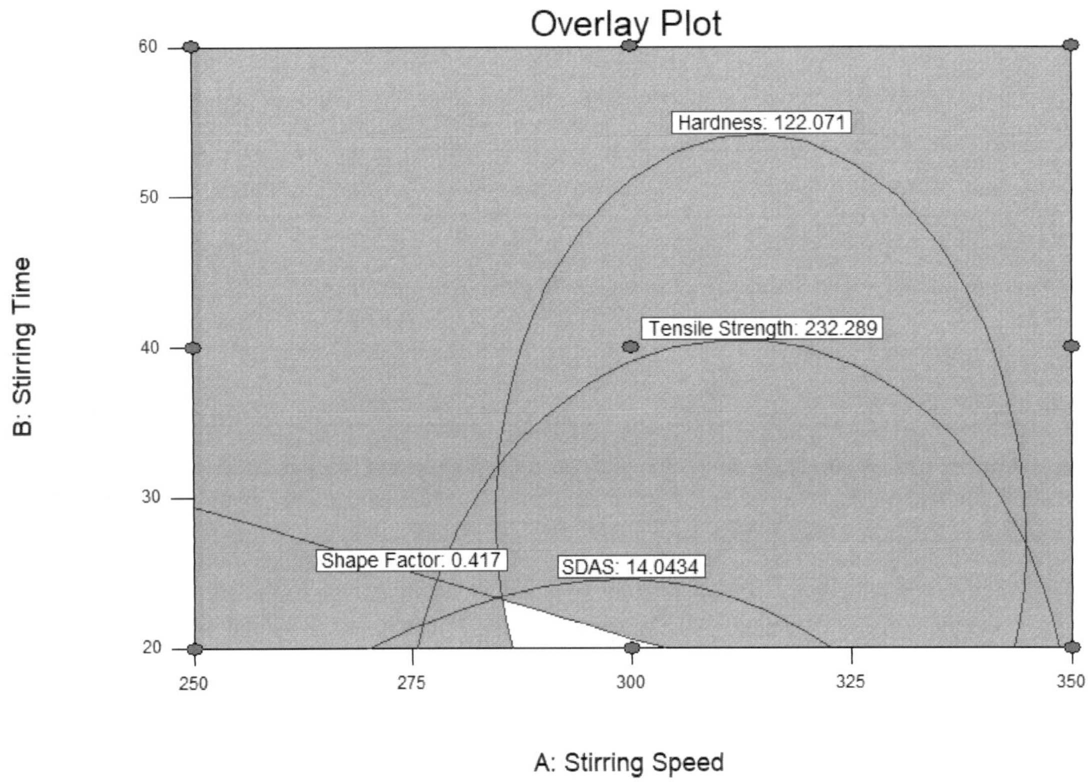

Figure 5. Overlay plot of the input factors for the predetermined response criteria of a minimum of 122.071 HRC hardness, 232.289 N/mm^2 tensile strength, 0.417 shape factor, and 14.0434 μm SDAS.

4. Conclusion

The microstructure and mechanical properties of aluminum alloys ADC12 made with semi-solid rheocasting casting technology using several parameters stirring have been studied, and the results obtained can be synergized as follows. The highest mechanical properties (Hardness 122,5 HB, Tensile strength 238.2 N/mm^2) occur at 300 rpm stirring speed and 20 seconds stirring time. Secondary Dendrite Arm Spacing (SDAS) lowest occurred at 25 rpm stirring speed and 20 seconds stirring time of 13.93 μm. The highest shape factor occurs at 300 rpm stirrer spin and 20 seconds stirring time of 0.47. The optimal stirring parameters recommended for producing a hardness response of 122,071 HRC, tensile strength 236,271 N/mm^2, Shape Factor (SF) 0.38994, and Secondary Dendrite Arm Spacing (SDAS) is 317 rpm (stirring rate) and 20 seconds (stirring time).

References

[1] Flemings J 1991. Behavior of Metal Alloys in The Semisolid State. Met Mat Trans. A 22A:957
[2] Kiuchi S A, Augiyama 1995. Behavior of A356 Alloy in Semi-Solid State Produced by Mechanical Stirring. J. Inter. 24(1), p. 32.
[3] Sukumaran K, Pai B B, Chakraborty M 2004. The effect of isothermal mechanical stirring on an Al–Si alloy in the semisolid condition. J. Mat. Sci. and Eng., A369, p. 275.
[4] Seijiro M, Yasunori H, Ken-ichiro M, Hiroyasu M. 2002. Application of resistance heating technique to mushy state forming of aluminium alloy. J. of Mat. Proc. Tech. 125-126, p. 477 482.

[5] Lee B S, Joo D H, Kim H M 2005. Extrusion behavior of Al-Cu alloys in the semisolid state. Mat. Sci. and Eng. A 402, p. 170–176.
[6] Zhou B F and Yan H. 2011. Study on Semi-Solid Magnesium Alloy Produced by Mechanical Stirring. Adv. Mat. Res. Vol. 146-147, p. 1723-1728.
[7] Vogel Y and Zhang K. 2003. The formation of rosette phase structural evolution during the reheating and semisolid casting of AlSi7Mg Alloy. J. of Mat. Proc. Tech. 137, p. 195-200.
[8] Vogel A, Doherty R.D, Cantor B. 1979. Stir-cast microstructure and a low crack growth. Proc. of Int. Con. on Sol. Met. Soc. London, 518.
[9] Hellawel T, Stjohn D H, Steinberg T 2000. The shear behaviour of partially solidified Al–Si–Cu alloys. Mat. Sci. and Eng. A286, p.18-29.
[10] Hellawell A, 1996. Grain evolution in conventional and rheocasting. Conf. Shef. UK,40.
[11] Mullis A M 1999. Growth induced dendritic bending and rosette formation during solidification in a shearing flow. J. Acta Mater, 47:1783.
[12] Rasyid S, Arif E, Arsyad H and Syahid M 2018. effect of mechanical stirrer and pouring temperature on semi solid rheocasting of ADC12 al alloy: mechanical properties and microstructure J. of Eng. and App. Sci. 13, p. 2032 2037.
[13] Fan Z 1981. Semi-solid Metal Processing. Proc.of Ninth North Am. Man. Res. Con. p. 140-144.
[14] Barabazon D, Browne D J, Carr A J 2002. Mechanical stir casting of aluminium alloys from the mushy state: process, microstructure and mechanical properties. Mat. Sci. and Eng. A.236, p. 370–381.
[15] Niroumand B, Xia K 2000. 3D study of the structure of primary crystals in a rheocast Al-Cu alloy. Mat. Sci. and Eng. A283, p. 70–75.
[16] Falak P, Niroumand B 2005. Rheocasting of an Al-Si alloy. Scr.Mat. 53, p. 53–57.
[17] Mirzadeh H, Niroumand B 2009. Fluidity of Al-Si semisolid slurries during rheocasting by a novel process. J. of Mat. Proc. Tech., 209, p. 4977–4982.
[18] Khosravi H, Eslami FR, Askari PM 2014. Modeling and optimization of cooling slope process parameters for semi-solid casting of A356 Al alloy, T. Nonferr. Metal. Soc. 24 p. 961-968.
[19] Khosravi H, Eslami F R, Askari P M 2017. Optimization of SiC particle distribution during compocasting of A356-SiCp composites using D-optimal experiment design. J. of Par. Sci. and Tec. 3 145-154.
[20] Sultan A Z, Sharif S, and Kurniawan D 2015. Statistical Analysis and Mathematical Modelling to Determine Minimum Surface Roughness Value and Cutting Force when Drilling AISI 316L. Int. Con. on Adv. Man. Tech. 20-22.
[21] Zheng CQ, Samuel E, Laplum F 2011 Optimization of the process parameters and tooling improvement for the rheocasting of high quality aluminum components using the SEED process. The Min, Met and Mat Soc. P. 273-277.
[22] Rasyid S, Arif E, Arsyad H, and Syahid M. 2018. Effects of stirring parameters on the rheocast microstructure and mechanical properties of aluminum alloy ADC12 Mat. Web of Conf. 197, 12004
[23] Dahle A K and Suery. M 2004. Deformation Behavior of Aluminum Alloys during Solidification Proc. of Int. Con. on Sem. Proc. of All. and Com.
[24] Montgomery D C. 1997 Design and Analisis of Experiments. Wiley, New York.
[25] Reisi, M., Niroumand, B. 2009. Effects of Stirring Parameters on The Rheocast Structure of Al–7.1wt.%Si Alloy. J. of Alloys and Compo.470 (2009) 413–419.

Acknowledgments
This article can be completed in cooperation between the lead author and lecturer of the Department of Mechanical Engineering, Hasanuddin University. Thanks to the Department of Mechanical Engineering, Ujung Pandang State Polytechnic, and Geology Department of Hasanuddin University for the provision of research facilities.

Performance analysis of rack type solar dryers with mass variations of dried material and types of fins

Jamal Jamal[a], Sri Suwasti[b], and Sukma Abadi[c]
Mechanical Engineering Department, Politeknik Negeri Ujung Pandang, Makassar, Indonesia
E-mail:[a]jamal_mesin@poliupg.ac.id, esspoltek@gmail.com,
[c]sukabadi2000@yahoo.com,

Abstract. The purpose of this study was to determine the performance of solar dryers based on effect of variations on mass of dried material and effect of types of fins. The air flow in the solar dryer is a natural convection. Solar dryers used in this study were four solar dryers of rack type which has three stacking shelves with different type of fin respectively. Testing method that used was direct simultaneous testing. The material that was dried was corn. Testing of the first day was conducted with an initial material mass of 500 grams/rack while the second day used 900 grams/rack. Testing were conducted in the city of Makassar, South Sulawesi, Indonesia, from 10:00 am until 16:00 pm. Testing with the initial mass of material of 900 grams/rack had greater drying rate and efficiency but the decline in water content was lower than the testing with initial mass of material of 500 grams/rack. Based on the decrease in water content, drying rate and efficiency, the use of a large wave fin had the best performance and almost the same with the zigzag fin. The use of a small wave fin had the lowest performance while the plate fin slightly better than the small one.

Keywords: fin, the mass of material, solar dryer, rack type, natural convection

1. Introduction

Plantation products require post-harvest handling. The post-harvest handling is needed to extend life of plantation products to prevent rapid deterioration [1]. One of methods of the post-harvest handling is drying process. Without the drying process farmers can get loses up to 40% [2].

There are several energy resources used in the drying process. One of the energy resources is widely used is solar energy. Solar energy is a renewable alternative energy [3] so that the utilization of solar energy grows fastly mainly in areas that have high solar energy. Indonesia is a tropical country that has a very large potency of solar energy [4].

Utilization of solar energy for the drying process requires solar dryer equipment. One of models of the solar dryers is a rack type [5-8]. Advantage of the solar dryer of rack type is capable for drying materials in large quantities. There are many efforts to improve the performance of solar dryer rack type, including the use of gravels [5], fins [1] and plates [7] as heat storage. Performance improvement can also be done by providing a forced convection flow [6] and the use of wind ventilator [8]. All agricultural products can be dried in a solar dryer rack type, in many studies, various agricultural products that have been dried are chili [5], eggplant [6], maize [7, 1], bananas [9] etc.

One of factors that affect the performance of the solar dryer rack type is convection air flow in the drying chamber. There are two types of convection air flow namely natural convection [10, 11] and forced convection [12, 13]. Forced convection flow can increase the drying rate [14] but it requires electrical power whereas natural convection flow does not require electrical power but the performance is not optimum. This study seeks to optimize the flow of natural convection to increase the rate of the dryer.

The use of fins is one way to improve the performance of solar dryers. Fin serves to increase heat absorption area [15] without increasing dimensions of the solar dryer. Research that conducted was an attempt to optimize the performance of the solar dryer rack type with the influence of the mass of material that was dried and the use of various types of fin.

2. Methodology

The material that be dried in this study was corn. Tests on the first day were done with the initial mass of material be dried was 500 grams/rack or 1.5 kg for each solar dryer with the initial moisture content of 78.41%. Tests on the second day were done with the initial mass of material be dried was 900 grams/rack or 2.7 kg for each solar dryer with the initial moisture content of 79.10%.

Fins used in this study were made by zinc material with a thickness of 1 mm. The tests used four types of fin namely small wave, big wave and zigzag plate. The types of fin can be seen in Figure 1. All fins dimensions were 600 mm x 1090 mm except the height of the fins. The similarity of the four types of the fin was the same dimensions when stretched to form of sheets straight plate.

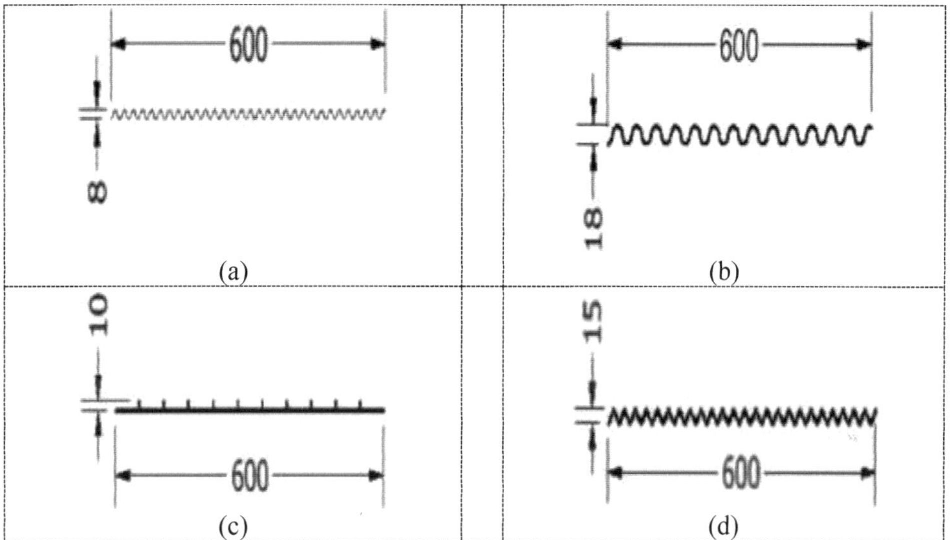

Figure 1. Various types of fins (a) small wave (b) big wave (c) plate and (d) zigzag.

Testing scheme can be seen in Figure 2. The intensity of solar radiation entering the room through the glass cover of solar dryers so that temperature of the air in the drying chamber increased. The increase in temperature caused the air had a lower density so that the air becomes lighter [15] and moved towards the drying chamber. The movement of air due to the temperature difference is called the natural convection [15]. The air that moved up and replaced by outside air that entered through the channel of inlet. The outside air was dry air. The hot air in the drying chamber caused the drying process occurred in the material so that the water vapor came out through the material caused the air becomes humid and its density is greater. The humid air would drain out through the channel of water vapor. The air that was on the top of the drying chamber would come out through the air outlet so that the natural air flow could be occurred which increased the heat transfer rate [15].

There are four innovations in this study which were effort to increase the drying rate namely the use of fins to increase the heat transfer area so that accelerated the rate of heat transfer. The use of the inlet channel to keep the air in the drying chamber remained the dry air. The use of the channel of water vapor to dispose of moist air so the air remained the dry air. The use of the channel of outlet kept the natural convection flow persisted.

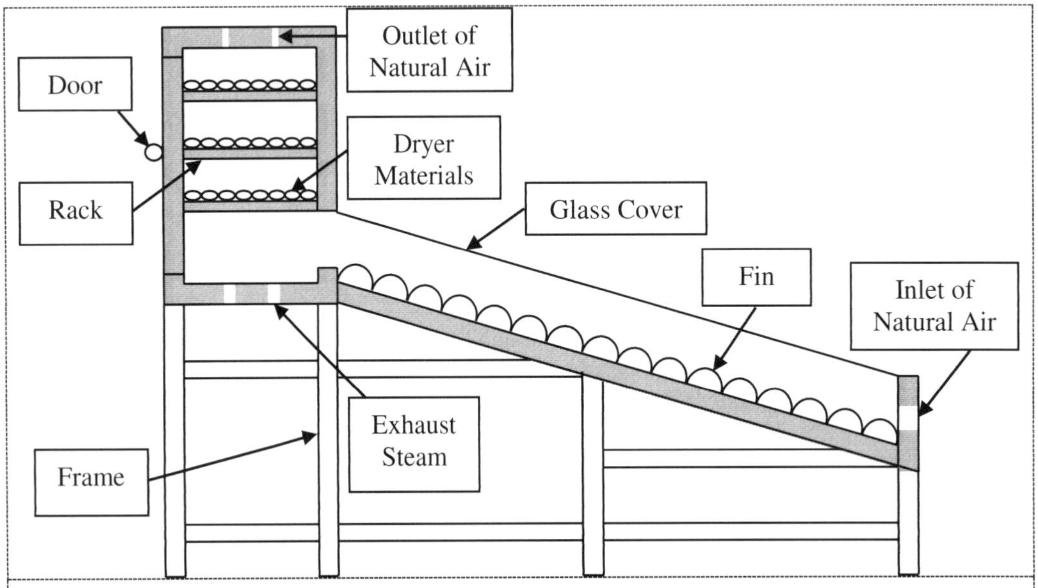

Figure 2. Experimental Scheme.

Figure 3. Testing of the solar dryer of rack type with four types of fins.

Tests of the solar dryer rack types were done simultaneously for the four types of fins that can be seen in Figure 3. The tests were done in order to obtain accurate results. As for test with various of the initial mass of material performed on different days with sunny weather conditions. The tests conducted in Makassar, South Sulawesi, Indonesia. The tests were conducted at 10:00 am until 16:00 pm. Data were collected at the beginning of the test and every 1 hour.

3. Results and discussion

Intensity of the solar radiation and ambient temperature at the time of the testing can be seen in Figure 4. The intensity of the solar radiation on the first day (testing with a mass of material 500 grams/rack) was slightly larger than the intensity in the second day (testing with a mass of material 900 grams/rack), as well as the ambient temperature on the first day was slightly larger than the second day. Maximum intensity of the solar radiation and ambient temperature occurred at 12:00 until 13:00 pm.

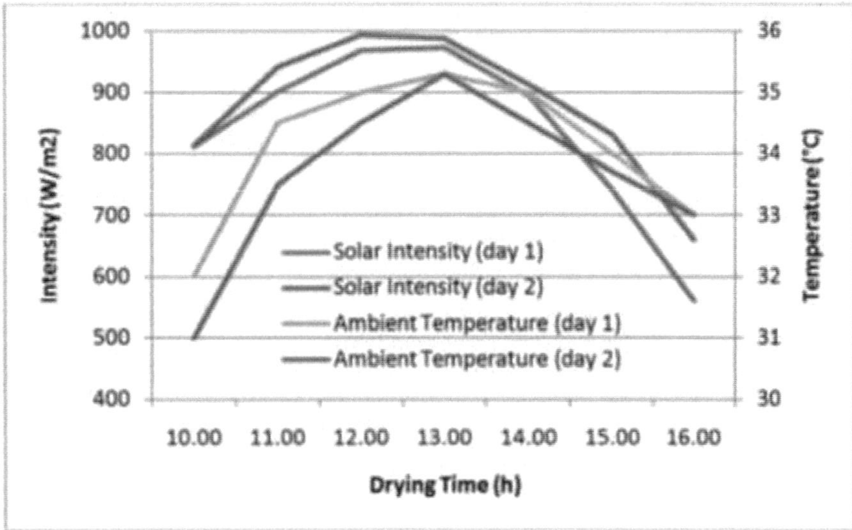

Figure 4. Solar intensity and ambient temperature when testing.

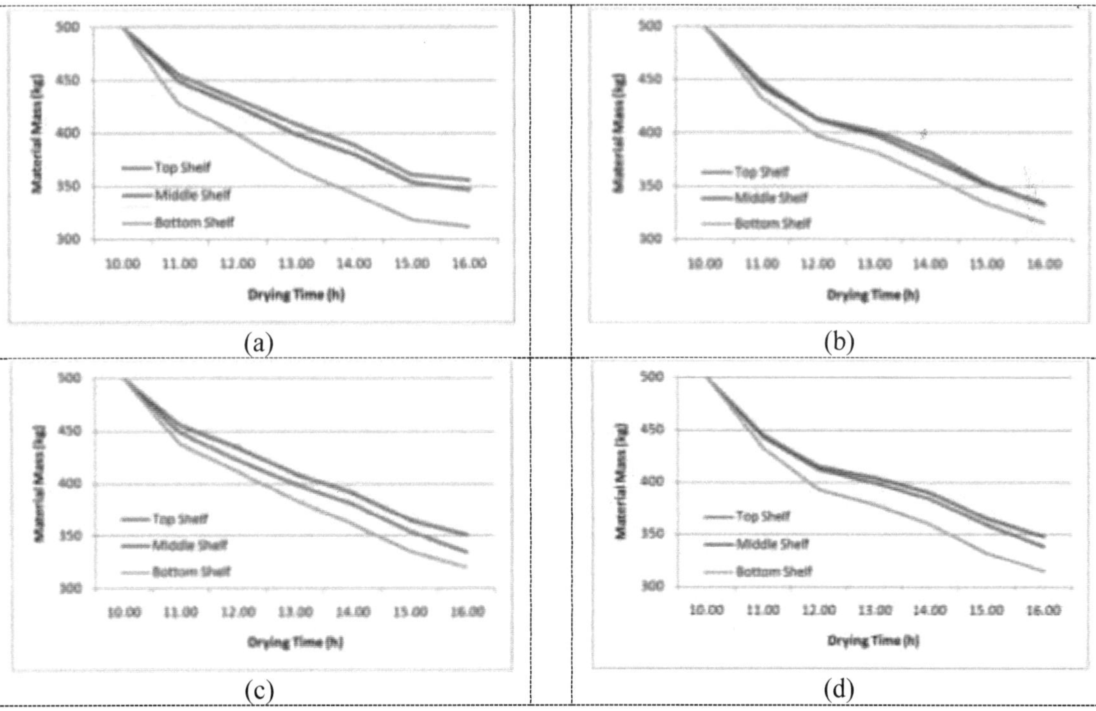

Figure 5. Decreasing of the mass of material on each shelf with initial mass of 500 g/rack using various types of fins (a) Small Wave (b) Big Wave (c) Plate (d) Zigzag.

IOP Conf. Series: Materials Science and Engineering **619** (2019) 012029 doi:10.1088/1757-899X/619/1/012029

Figure 5 shows that the decreasing of the mass of material in the drying process with the initial mass of 500 grams/rack. The biggest material mass reduction occurred on the bottom shelf because the bottom shelf of the rack firstly received the hot air. The material mass reduction on the top shelf was greater than on the middle shelf when the fins of plate and zigzag were used. Furthermore, in the use of the small wave fins, decreasing of the mass of material on the middle shelf was greater than the top shelf. The use of the large wave fins provided the same reduction in the mass of material between the top shelf and the middle shelf. It can be also seen in Figure 5 that the use of the large wave fins, plate and zigzag fins provided more uniform drying on all the shelves compared to the use of a small wave fin.

Figure 6 shows that the decreasing of the mass of material in the drying process with the initial mass of 900 grams/rack. The biggest material mass reduction occurred on the bottom shelf because the bottom shelf of the rack firstly received the hot air. Furthermore, decline of the masses of material on the top and middle shelf were almost the same except in testing with the big wave fins which was the decreasing of the mass of material on the middle shelf is greater than the top shelf. It can be also seen in Figure 6 that uneven drying was happened. Drying rate on the bottom shelf was much greater than on the middle shelf and the top shelf.

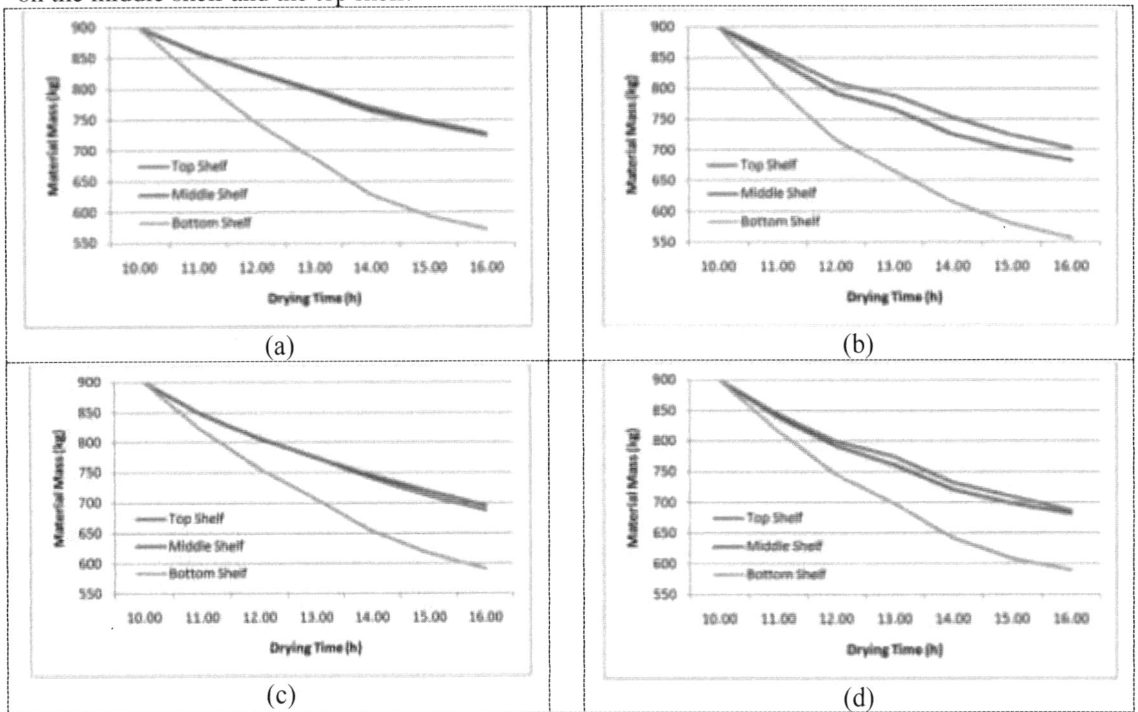

Figure 6. Decreasing of the mass of material on each shelf with initial mass of 900 g/rack using various types of fins (a) Small Wave (b) Big Wave (c) Plate (d) Zigzag.

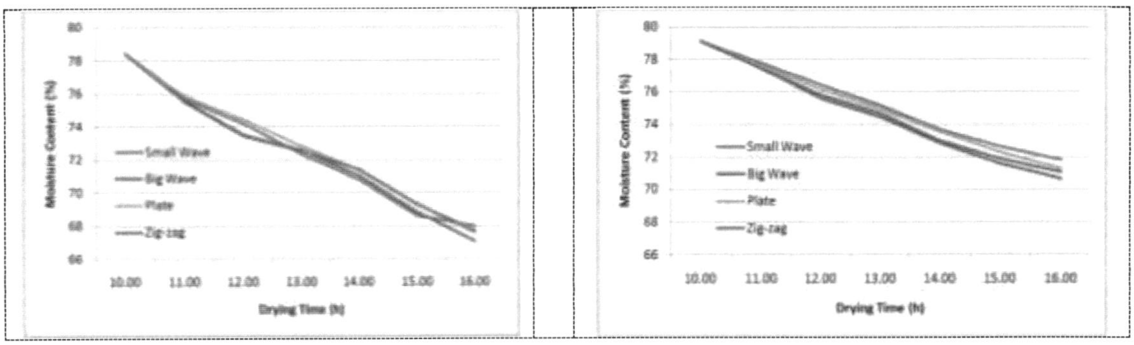

(a) (b)

Figure 7. The influence of the type of fin to decrease water content with initial masses (a) 500 g/rack, (b) 900 g/rack.

Figure 7 shows that the testing of the initial mass of 500 grams/rack (Figure 7.a) had a decreasing of water content of materials that was greater than the test with initial mass of 900 grams/rack (Figure 7.b). It can be also seen in Figure 7 that the big decreasing of water content happened in the use of the large wave, zigzag, plate and small waves fins.

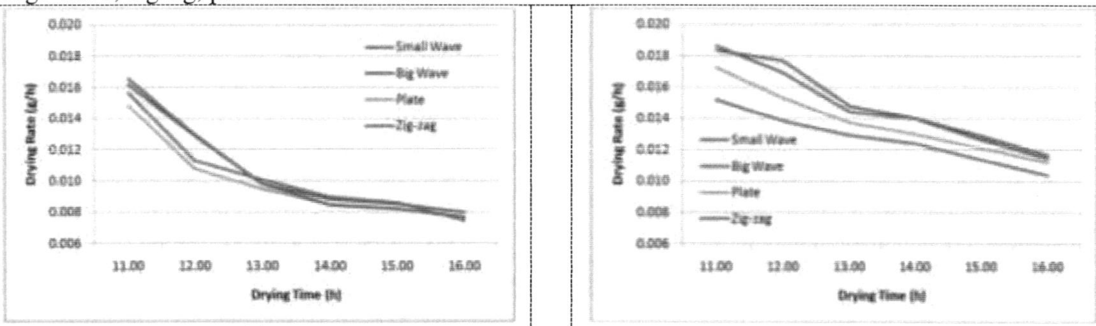

(a) (b)

Figure 8. The influence of the types of fin on the drying rate with initial masses (a) 500 g/rack, (b) 900 g/rack

The drying rate of material in the testing with the initial mass of 900 grams/rack shown in Figure 8.b was greater than the drying rate of material in the testing with initial mass of 500 grams/rack as shown in Figure 8.a. It can be also seen in Figure 8 that the biggest drying rate of material generated by the use of large wave fin even though its value is almost equal to the use of zigzag fin. Furthermore, the use of the small wave fins had the lowest drying rate of material.

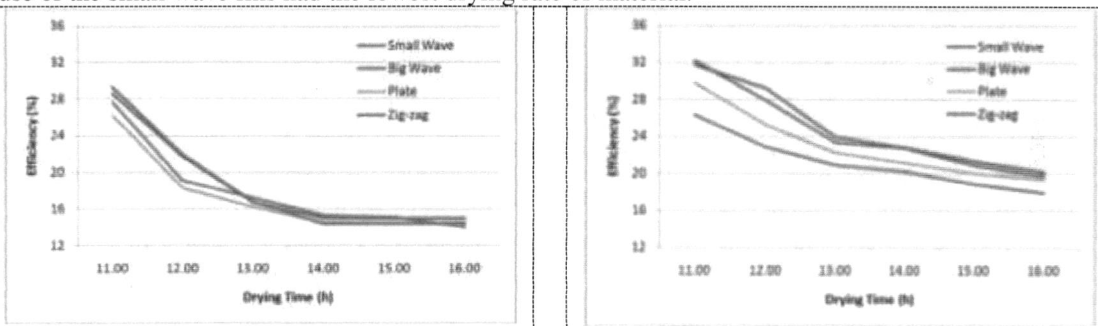

(a) (b)

Figure 9. The influence of the types of fin on the efficiency with initial masses (a) 500 g/rack, (b) 900 g/rack

Testing with the initial mass of 900 grams / rack shown in Figure 9.b obtained greater efficiency of testing compared to the testing with the initial mass of 500 grams / rack shown in Figure 9.a. The use of a large wave fin had the greatest efficiency even though its value was almost equal to the use of zigzag fin as shown in Figure 9 while the use plate fins produced greater efficiency compared to the use of small wave fins.

4. Conclusions

Based on the reduction in moisture content, result of the testing with the initial mass of 500 grams/rack was better than testing with initial mass of 900 grams/rack. Based on the drying rate and efficiency of solar dryers, result of the testing with the initial mass of 900 grams/rack was better than testing with initial mass of 500 grams/rack.

The use of a large wave fins provides moisture reduction, the rate of drying and the greatest efficiency although its performance was similar to the use of the zigzag fin. The use of small wave fins has the lowest performance while the plate fins was slightly better than the use of small wave fins.

5. References

[1] Suwasti, S., Abadi, S., Shiddiq Yunus, A. M., & Jamal. (2015). Impacts of Fin Variation on the Performance of Shelf Type Solar Dryer. *Journal of Advanced Research in Fluid Mechanics and Thermal Sciences*, *13*, 17-21.

[2] El-Sebaii, A. A., & Shalaby, S. M. (2012). Solar drying of agricultural products: A review. *Renewable and Sustainable Energy Reviews*, *16*(1), 37-43.

[3] Devabhaktuni, V., Alam, M., Depuru, S. S. S. R., Green II, R. C., Nims, D., & Near, C. (2013). Solar energy: Trends and enabling technologies. *Renewable and Sustainable Energy Reviews*, *19*, 555-564.

[4] Handayani, N. A., & Ariyanti, D. (2012). Potency of solar energy applications in Indonesia. *International Journal of Renewable Energy Development*, *1*(2), **33-38**.

[5] Kamble, A. K., Pardeshi, I. L., Singh, P. L., & Ade, G. S. (2013). Drying of chilli using solar cabinet dryer coupled with gravel bed heat storage system. *J Food Res Technol*, *1*(2), 87-94.

[6] Azimi, A., Tavakoli, T., Beheshti, H. K., & Rahimi, A. (2012). Experimental study on eggplant drying by an indirect solar dryer and open sun drying. *Iranica Journal of Energy and Environment*, *3*(4), **347-353.**

[7] Folaranmi, J. (2008). Design, construction and testing of simple solar maize dryer. *Leonardo Electronic Journal of Practices and Technologies*, *1*(13), **122-130**.

[8] Bolaji, B. O., Olayanju, T. M., & Falade, T. O. (2011). Performance evaluation of a solar wind-ventilated cabinet dryer.

[9] Hassanain, A. A. (2009). Simple solar drying system for banana fruit. *World journal of agricultural sciences*, *5*(4), **446-455.**

[10] Pangavhane, D. R., Sawhney, R. L., & Sarsavadia, P. N. (2002). Design, development and performance testing of a new natural convection solar dryer. *Energy*, *27*(6), **579-590**.

[11] Abdullahi, Y., Momoh, M., Garba, M. M., & Musa, M. (2013). Design and construction of an adjustable and collapsible natural convection solar food dryer. *Int J Comput Eng Res*, *3*(6), **1-8**.

[12] Mohanraj, M., & Chandrasekar, P. (2009). Performance of a forced convection solar drier integrated with gravel as heat storage material for chili drying. *Journal of Engineering Science and Technology*, *4*(3), **305-314**.

[13] Tchaya, G. B., Kamta, M., & Kapseu, C. (2014). Improvement of an indirect solar dryer with forced convection by variation of airflow mode. *International Journal of Emerging Technology and Advanced Engineering*, *4*(1), **518-522**.

[14] Babagana, G., Silas, K., & Mustafa, B. G. (2012). Design and construction of forced/natural convection solar vegetable dryer with heat storage. *ARPN Journal of Engineering and Applied Sciences*, *7*(10), **1213-1217**.

[15] Bergman, T. L., Incropera, F. P., Lavine, A. S., & Dewitt, D. P. (2011). *Introduction to heat transfer*. John Wiley & Sons.

Acknowledgments
Authors would like to thank to Research, Technology and Higher Education Ministry of Indonesia for funding support through competitive research grant scheme 2014 - 2015.

IOP Conf. Series: Materials Science and Engineering **619** (2019) 012030 doi:10.1088/1757-899X/619/1/012030

MJO Anomalies relationship with Volcanic Eruption in Indonesian Archipelago

Achmad Yasir Baeda[1], Dharmawan Abdullah[1], Chairul Pao'tonan[2]

[1]Marine Disaster and Weather Anomalies Research Laboratory, Ocean Engineering Department, Faculty of Engineering, Universitas Hasanuddin, Indonesia
[2]Ocean and Coastal Geotechnical Laboratory, Ocean Engineering Department, Faculty of Engineering, Universitas Hasanuddin, Indonesia

Abstract. Madden-Julian Oscillation (MJO) is a large-scale atmospheric phenomenon that crosses the equator with a propagation area ranging from 15° North to 15° South, and moves from west to east, precisely from the Indian Ocean to the Pacific. MJO propagation is usually characterized by the rising of Sea Surface Temperature (SST) in Indian Ocean respectively. MJO anomaly is closely related to the variability of weather in the area it passes, including the Indonesian Archipelago. The impact of MJO anomaly in Indonesian Archipelago itself is suspected to cause not only excessive rainfall, but also a very long consecutive No Rain Days (NRD), with in the end can also trigger massive crop failures. In addition to SST, there are also external events that can affect the duration of the MJO, i.e. volcanic eruptions. Therefore, the study aims to determine the effect of volcanic eruption events on MJO anomalies, through the SST variable as an MJO anomaly parameter and also the occurrence of volcanic eruptions in the same time span. Data from both variables were taken over a period of 35 years, from 1982 to 2016. The SST data is a daily average data, obtained from ESRL NOAA, which is then filtered by the Bandpass Filter method to reinforce the oscillation of data. The MJO anomaly that seen from the SST fluctuations is then matched with time data and the duration of volcanic eruption events over the same time span. It found out that there are three MJO anomalies occurred during the time span of 35 years, and it is hypothesized that they were clearly in correspondence between the duration of volcanic eruption events with the incidence of MJO anomalies in Indonesian Archipellago.

1. Introduction

Equatorial regions, especially Indonesia, are influenced by very complex atmospheric and oceanographic phenomena. This phenomenon has variations in space and time varies, one of which is the intra-seasonal cycle. Viewed from its geographic position, Indonesia is surrounded by two large continents (Asia and Australia) and two large oceans (the Pacific and the Indies), and it's a centre of mass water movement at various depths. The territory of Indonesia has a complex topography that adds to the sea-atmosphere variability in the Indonesian Ocean [1]. One of the atmospheric phenomena that occur is Madden-Julian Oscillation (MJO). MJO was first discovered by Madden-Julian [2]. MJO is one of the dominant oscillations in equatorial region [3]. Another important aspect of the MJO is its time scale, related to the average period of each occurrence of 45 days. Seto [4] states that the MJO in the active phase has a correlation of high rainfall intensity to the area it passes. MJO controls a considerable fraction of the total precipitation over Indonesia and that the rainfall variability over the surrounding ocean is more clearly controlled by the MJO compared to over the large land masses [5].

The measurement of the Outgoing Longwave Radiation (OLR) variant in the convection region will read a larger signal than the red noise so it can show the MJO signal [6]. OLR actually is the size or value of Earth radiation that has a long wavelength detected from outer space. This detection is usually done with satellite equipment, and measured value illustrates the extent of the inhibition of the Earth's radiation; which is a negative value indicating the magnitude of the obstacle. The smaller the value of OLR on a negative scale indicates the greater the obstacles that can be visualized as the higher clouds inhibit those that are usually convective clouds. In general, the OLR pattern illustrates the pattern of potential convective areas [7].

Volcanic eruptions can inject into the stratosphere tens of *teragrams* of chemically and microphysically active gases and solid aerosol particles, which affect the Earth's radiative balance and climate, and disturb the stratospheric chemical equilibrium. The volcanic cloud forms in several weeks by SO2 conversion to *sulfate* aerosol and its subsequent microphysical transformations [8] [9]. The resulting cloud of sulfate aerosol particles, with an *e-folding decay time* of approximately 1 year [10], has important impacts on both shortwave and longwave radiation. The resulting disturbance to the Earth's radiation balance affects surface temperatures through direct radiative effects as well as through indirect effects on the atmospheric circulation [11].

Figure 1 indicates the major radiative processes resulting from the stratospheric aerosol cloud from a major volcanic eruption. The most obvious and well-known effect is on solar radiation. Since the sulfate aerosol particles are about the same size as visible light, with a typical effective radius of 0.5 □m, but have a singlescatter albedo of 1, they strongly interact with solar radiation by scattering. Some of the light is backscattered, reflecting sunlight back to space, increasing the net planetary albedo and reducing the amount of solar energy that reaches the Earth's surface. This backscattering is the dominant radiative effect at the surface and results in a net cooling there. Much of the solar radiation is forward scattered, resulting in enhanced downward diffuse radiation that somewhat compensates for a large reduction in the direct solar beam [11].

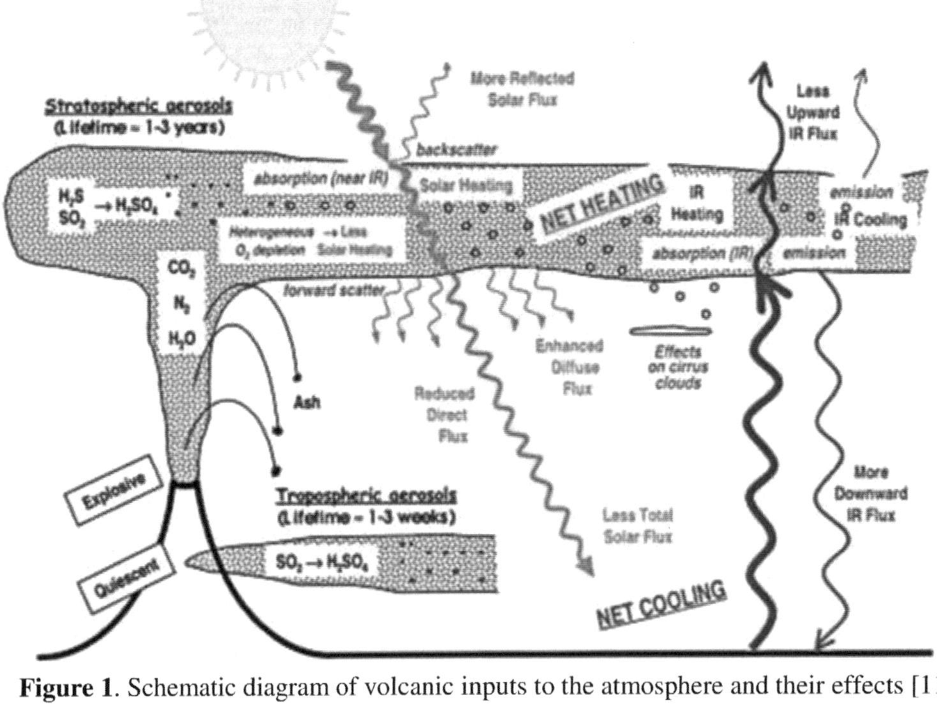

Figure 1. Schematic diagram of volcanic inputs to the atmosphere and their effects [11].

Table 1. Effects of Large Explosive Volcanic Eruptions on Weather and Climate [11]

Effect	Mechanism	Begins	Duration
Reduction of diurnal cycle	Blockage of shortwave & emission of longwave radiation	Immediately	1-4 days
Reduce tropical precipitation	Blockage of shortwave radiation, reduced evaporation	1-3 months	3-6 months
Summer cooling of NH continents	Blockage of shortwave radiation	1-3 months	1-2 years
Stratospheric warming	Stratospheric absorption of shortwave & longwave radiation	1-3 mon ths	1-2 years
Winter warming of NH continents	Stratospheric absorption of shortwave & longwave radiation, dynamics	$\frac{1}{2}$ year	One or two winters
Global cooling	Blockage of shortwave radiation	Immediately	1-3 years
Global cooling from multiple eruptions	Blockage of shortwave radiation	Immediately	10-100 years
Ozone depletion, enhanced UV	Dilution, heterogeneous chemistry on aerosols	1 day	1-2 years

The study aims to determine the relationship between MJO anomalies that occurred in Indonesian Archipelago with volcanic eruption events, through the SST variable as an MJO anomaly parameter and also the occurrence of volcanic eruptions in the same time span.

2. Research Method

2.1. Data

The data used in this study were downloaded from the Earth System Research Laboratory of the National Oceanic and Atmospheric Administration (ESRL NOAA)[12] for Sea Surface Temperature (SST) data with 0.25° x 0.25° spatial resolution & daily temporal resolution taken from 1982 to 2016 (35 years), and from the National Geophysical Data Center / World Data Service (NGDC/WDS)[13] for volcano eruption data. The SST data is then filtered with Bandpass Filters using Ferret.

2.2. Bandpass Filtering

The MJO phenomenon has a dominant period ranging from 40 to 60 days or classified as an intra-seasonal period, so to reinforce the oscillations of data during that period it is necessary to filtered them with an intra-seasonal frequency range [1]. Data filtering is expected to limit the impact of seasonal, annual, or inter-annual oscillation phenomena. Filtering data to be applied in this research is bandpass filter with period of 20-100 days. The cutoff value of the frequency used as the input of the bandpass filter function is 0.01 and 0.05. The low frequency value (100 days) is represented by 0.01 and the high frequency value is represented by 0.05 (20 days). The bandpass filter discards the signal oscillations with periods below 20 days and above 100 days. The input data of the X_t variable is filtered by the Lanczos equation and produces the time series data Y_t. The time series equation is used as follows [14]:

$$Y_t \sum_{k=-\infty}^{k=\infty} \overline{w_k}\, X_{t-k} = \overline{w_k} \tag{1}$$

$$\overline{W_k} = \left(\frac{sin2\pi fc_1k}{\pi k} - \frac{sin2\pi fc_2k}{\pi k}\right)\sigma, k = -n, \ldots, 0, \ldots, n \quad (2)$$

where:
$\overline{W_k}$ = signal weight at 95% confidence interval;
fc_1 = *cut off* of 1st frequency;
fc_2 = *cut off* of 2nd frequency which give "0" respon of *Nyquist* frequency;
σ = sigma factor.

3. Result and Discussion

The daily data of sea surface temperatur (SST) is filtered with Bandpass Filter method, to reinforce the oscillation of data during the MJO period. The filtered data then be taken average value for each year. The annual average SST data shown in figure 2.

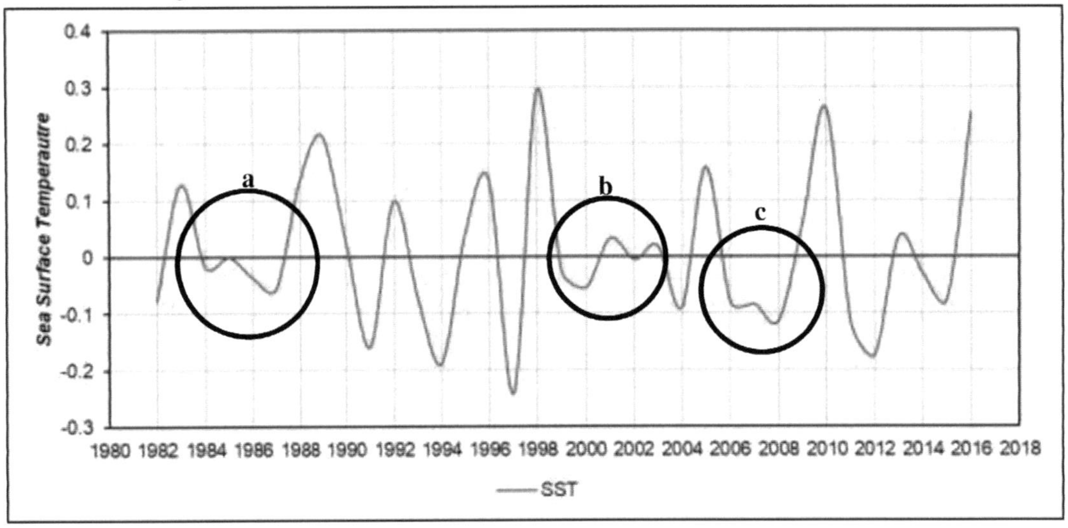

Figure 2. Annual average SST in Indonesian Waters (1982 – 2016)

In figure 2, there are 3 MJO anomalies occurred for 35 years (1982 – 2016), seen from the abnormal pattern of SST (circled areas).

Large volcanic eruptions inject sulfate gas into the stratosphere and produce aerosols clouds in the stratosphere, then cause the earth surface cooling but produce warming in the stratosphere[11]. This is occurs because the solar flux will be blocked by the aerosol cloud. Some flux will be reflected, so the flux that arrives on the surface becomes less. These occurences can be seen in figure 1. From these statements, it is possible that SST are related with volcanic eruptions, if we match the anomaly with the eruption data that occurs, there is a volcanic eruption that occurs before and at the intervals of the anomaly occurs.

In 1982 there was an eruption of El Chicón volcano in Chiapas, Mexico, with a scale of 5 VEI (Volcanic Explosivity Index), and with 800 DVI (Dust Veil Index). The eruption was often associated with the powerful El Niño incident of 1982 - 1983. Based on satellite measurements, it is estimated that the El Chicón eruption of 1982 spewed 7 metric tons of SO2 (Sulfur Dioxide) in the air [11]. The eruption is also suspected to have a relationship with the occurrence of SST anomalies that occurred around 1983 to 1984 (Fig 2 in the a circle).

Furthermore, during time span of 1999 to 2001, while Indonesian waters experiencing SST anomalies, there were also series volcanic eruptions occurred at the same region. At least 24 eruptions recorded during that period, with the largest eruption were at Karangetang mountain in Siau, North Sulawesi in 1999, with a VEI scale of 3. Meanwhile, there was also a major eruption that occurred before the anomaly occurred, i.e. Soufriere Hills, in the Caribbean Islands, with a VEI scale of 4. For the anomaly occurred in 2007 - 2008, there was an eruption in form of long ash emission (from mid 2007 to early 2008, with maximum 4VEI) of Rabaul volcano in Papua New Guinea, before the anomaly occurred. Some incident data the above mentioned volcanic eruption is suspected to be associated with the occurrence of SST anomalies.

Table 2. Volcanic eruptions from 1982 to 2016 which are related to the SST anomaly [13]

No	Year	Month	Day	Name	Location	Country	Elevation	Type
1.	1982	3	29	Chichon, El	Mexico	Mexico	1150	Tuff cone
2.	1982	4	4	Chichon, El	Mexico	Mexico	1150	Tuff cone
3.	1982	5	17	Galunggung	Java	Indonesia	2168	Stratovolcano
4.	1982	5	27	Chichon, El	Mexico	Mexico	1150	Tuff cone
5.	1983	7	23	Colo [Una Una]	Sulawesi-Indonesia	Indonesia	507	Stratovolcano
6.	1983	10	3	Miyake-jima	Izu Is-Japan	Japan	815	Stratovolcano
7.	1984	9	9	Mayon	Luzon-Philippines	Philippines	2462	Stratovolcano
8.	1984	10	16	Etna	Italy	Italy	3350	Stratovolcano
9.	1985	5	10	Semeru	Java	Indonesia	3676	Stratovolcano
10.	1985	11	13	Ruiz	Colombia	Colombia	5321	Stratovolcano
11.	1986	8	21	Oku Volc Field	Africa-W	Cameroon	3011	Maar
12.	1986	11	15	Oshima	Izu Is-Japan	Japan	758	Stratovolcano
13.	1987	1	25	Pacaya	Guatemala	Guatemala	2552	Complex volcano
14.	1999	1	20	Soufriere Hills	W Indies	Montserrat	915	Stratovolcano
15.	1999	8	15	Shiveluch	Kamchatka	Russia	3283	Stratovolcano
16.	1999	10	5	Guagua Pichincha	Ecuador	Ecuador	4784	Stratovolcano
17.	1999	10	16	Tungurahua	Ecuador	Ecuador	5023	Stratovolcano
18.	2000	7	27	Semeru	Java	Indonesia	3676	Stratovolcano
19.	2000	8	23	Arenal	Costa Rica	Costa Rica	1657	Stratovolcano
20.	2001	5	14	Etna	Italy	Italy	3350	Stratovolcano
21.	2002	1	3	Karangetang [Api Siau]	Sangihe Is-Indonesia	Indonesia	1784	Stratovolcano
22.	2002	8	28	Etna	Italy	Italy	3350	Stratovolcano
23.	2002	11	3	Reventador	Ecuador	Ecuador	3562	Stratovolcano
24.	2002			Semeru	Java	Indonesia	3676	Stratovolcano
25.	2005	1	27	Manam	New Guinea-NE of	Papua New	1807	Stratovolcano
26.	2005	10	1	Santa Ana	El Salvador	El Salvador	2365	Stratovolcano
27.	2006	7	18	Tungurahua	Ecuador	Ecuador	5023	Stratovolcano
28.	2006	8	17	Tungurahua	Ecuador	Ecuador	5023	Stratovolcano
29.	2007	3		Manam	New Guinea-NE of	Papua New	1807	Stratovolcano
30.	2008	8	8	Kasatochi	Aleutian Is	United State	314	Stratovolcano
31.	2008	8	22	Tungurahua	Ecuador	Ecuador	5023	Stratovolcano

4. Conclusion

This research concluded clearly that there is a relationship between SST anomalies and volcanic eruptions that occurred in Indonesian Archipelago; in form of ash emissions spewed in the air which can effect the solar radiation intensity. Unfortunately, the amount of ash emission produced by that volcanic eruptions, still not calculated properly in every event that occurred. Which then lead to a question, "Is it possible to predict the active period of MJO if volcanic eruptions occur on a large scale?"

5. References

[1] Wu, C. &. (2009). Topographic influence on the MJO in the maritime continent. *J. Climate, 22*, 5433-5448.

[2] Madden, R.A., & P. Julian. (1971). Detection of a 40-50 dat oscillation in the zonal wind in the tropical Pacific. *J. Atmos Sci, 28*, 702-708.

[3] Madden R.A., & P. Julian. (1994). Observations of the 40-50 day tropical oscillation. *Month Weather Rev, 122*, 814-837.

[4] Seto, T. (2004). Effect of Madden-Julian Oscillation on Variability of Tropical Rainfall (in Bahasa Indonesia: Pengaruh Osilasi Madden-Julian terhadap Variabilitas Curah Hujan Tropis), *Jurnal Sains & Teknologi Modifikasi Cuaca I(V)*, 55-58.

[5] Hidayat, R., & S. Kizu. (2010). Influence of the Madden–Julian Oscillation on Indonesian rainfall variability in austral summer. *Int. J. Climatol, 30*, 1816-1825.

[6] Geerts, B., & M. Wheeler. (1998). *The Madden-Julian Oscillation*. Retrieved October 24, 2018, from http://www.das.uwyo.edu/~geerts/cwx/notes/chap12/mjo.html

[7] Aldrian, E. (2000). Monthly average rainfall pattern in Indonesia, review of the results of the contour data with ECHAM T-42 resolution (in Bahasa Indonesia: Pola hujan rata-rata bulanan wilayah Indonesia, tinjauan hasil kontur data penakar dengan resolusi ECHAM T-42), *Jurnal Sains & Teknologi Modifikasi Cuaca. 1(2)*, 113-123.

[8] Pinto, J.P., R.P. Turco, & O.B. Toon. (1989). Self-limiting physical and chemical effects in volcanic eruption clouds. *J. Geophys. Res., 94*, 11,165-11,174.

[9] Zhao, J., R.P. Turco, O.B. Toon. (1995). A model simulation of Pinatubo volcanic aerosols in the stratosphere. *J. Geophys. Res., 100*, 7315-7328.

[10] Barnes, J. E., & D. J. Hoffman. (1997). Lidar Measurement of stratospheric aerosol over Mauna Loa Observatory. *Geophys. Res. Lett., 24*, 1923-1926.

[11] Robock, A. (2000). Volcanic Eruptions and Climate. *Reviews of Geophysics, 38, 2*, 191-219.

[12] *NOAA High Resolution SST data provided by the NOAA/OAR/ESRL PSD, Boulder, Colorado, USA.* (n.d.). Retrieved from Earth System Research Laboratory, Physical Sciences Data: https://www.esrl.noaa.gov/psd/

[13] *National Geophysical Data Center / World Data Service (NGDC/WDS): Significant Volcanic Eruptions Database.* (n.d.). Retrieved from National Geophysical Data Center, NOAA: https://www.ngdc.noaa.gov/

[14] Duchon, C. E. (1979). Lanczos Filtering in One and Two Dimensions. *Journal of Applied Meteorology, 18*, 1016-1022.

IOP Conf. Series: Materials Science and Engineering **619** (2019) 012031 doi:10.1088/1757-899X/619/1/012031

Experimental study on the influence of draft relative due to wave transmission energy coefficient on hanging sheet pile breakwater

Chairul Paotonan[1], Sabaruddin Rahman[1], Hasdinar Umar[1], Taufiqur Rachman[1], Achmad Yasir Baeda[1], Agustinus M[2] and Alif Hidayat[2]

[1]Lecturer of Ocean Engineering Department, Faculty of Engineering, Universitas Hasanuddin, Indonesia
[2]Student of Ocean Engineering Department, Faculty of Engineering, Universitas Hasanuddin, Indonesia

Email: paotonan_ch@yahoo.com, sabarahman5@gmail.com, hasdinar.umar@gmail.com, ocean_d321@gmail.com, baedayasir@gmail.com,agus.mellolo@gmail.com alifhidayatk@gmail.com

Abstract. Rubble mount breakwateris recently become the most widely used breakwater, consist both of natural and artificial stone. It required a large dimension of structure so that the investment costs become more and more expensive. One of the solutions that can be implemented is Hanging Sheet Pile breakwater. Unfortunately, the performance of this type of breakwater has not been known yet. Performance of this breakwater represented by wave energy transmission coefficient (KE_t). Therefore, this study aims to examine the effect of relative draft (s/d) and relative wave height (H/s) to KE_t. This study conducted in Ocean Technology Laboratory of Universitas Hasanuddin. The parameters studied are structure dimensions and wave parameters. Wave parameters consist of wave height (H) and wave period (T), while the structure parameter is the draft structure (D). The result of this study shows that the greater value of both incident wave steepness (Hi/gT^2) and structure relative draft (s/d), then the smaller of KE_t value will be achieved. Also, the greater relative H/s, the smaller KE_t will be achieved. For all cases of study, the transmission coefficient ranges from 0.122 – 0.786.

1. Introduction

The type of breakwater that is widely applied so far to protect the beach and harbour pool is the rubble mound breakwater, both with natural and artificial stone. These breakwaters have disadvantages due to its large dimensions and development costs. One potential type of breakwaters to be developed in order to reduce building dimensions and large development costs is the Hanging Sheet Pile. This type of breakwater is similar to a vertical pole breakwater. The first research on pole breakwater was conducted by Wiegel and reported that the transmission coefficient was only influenced by the distance between the cylindrical pole and the cylinder diameter [1]. Hayashi and Kano conducted a study of cylindrical pole breaks that focused on moment distribution on cylinder poles and wave transmission through breakwaters [2]. Hayasi and Kano then conducted and developed the theory about the effect of water spray through cylindrical pole breakwater. Furthermore, they also conducted experiments to verify the solutions developed earlier, and they reported that there is a difference between the developed analytical solutions with experimental results. Hayashi and Kano reasoned that this happened because the dissipation of energy in front of cylindrical poles, had been neglected. Furthermore, Hayashi et al revised the results of their research applying the assumption of waves in the shallow water, and then comparing it with the experimental results. Hayashi et al. Reported that there was a lag between theory and experimental results and also reported that the greater the distance between the cylinders the greater the waves are transmitted. Truit and Herbich conducted a model test to determine the transmission of waves on vertical cylindrical poles and by varying the distance of the cylinder pile, their diameter and also with using irregular waves [3].

Truit and Herbich compared their research results with theories developed by Hayashi et al and reported that experiments have value closely with analytical solutions. Truit and Herbich therefore concluded that the formula developed by Hayashi et al yielded a corresponding result for irregular waves. Furthermore, Truit and Herbich investigated the effect of wave height and water depth on wave transmission and reported that wave parameters are important parameters affecting wave transmission, but the wave breaking geometry also plays an important role in wave transmission phenomena.

Herbich and Douglas continue the previous research that has been done by Truit and Herbich by conducting a study of vertical cylindrical pole breakwaters made two lines in the direction of wave propagation [4]. Results obtained from a two-row vertical cylindrical split are compared with one-line vertical cylindrical pole breakers. Herbich and Douglas reported that two-row vertical cylinders can reduce 15% of the transmission wave for the value of b/D = 0.2, where b is the distance between the cylindrical gaps while D is the diameter of the cylinder. For b/D = 0.1, a wave breaker with two vertical cylinder lines can reduce the transmission wave height by 10%. Herbich and Douglas also examined the effects of wave periods, wave heights and water depth and reported that the transmission waves increased with increasing of water depth due to wave heights (d/H) and transmission waves decreasing with increasing wave steepness (ratio between wave height and wavelength, H/L). Kukano and Liu examined the spread of waves through a single vertical cylindrical pole through theoretical development to solve the vertical cylinder spreading effect on the waves by modelling the flow around the vertical cylinder [5]. Kukano and Liu consider the energy dissipation between the cylindrical poles consisting of two shapes of pole, i.e. rectangular and round. Kukano and Liu reported that the results of his research can be trusted for certain cases, but further research is needed by varying the shape of vertical piles and wave characteristics. Mani and Jayakumar examined the transmission of waves on cylindrical poles hanging in one line [6]. Mani and Jayakumar vary the cylinder loads submerged in water (s), the spacing between the cylinder pole (b) and the wave parameters (height and period) and report that for b/d = 0.22 and y/h = 0.46, can reduce the coefficients transmission up to 0.5. For H_i/gT^2 greater than 0.008, breakwaters can reduce up to 50% and for H_i/gT^2 between 0.005 to 0.008 breakwaters can reduce up to 40% of the incident wave.

Rao and Satyananayana conducted a model study that focused on wave transmission through a two-row cylindrical pole breaker [7]. Rao and Satyananayana studied the effect of water depth, the steepness of the incident wave, the distance between the poles, and the distance between the pole-breaking pole lines to the wave transmission. Rao and Satyananayana also compared the solid type and hollow type of pole.

In conclusion the results of Rao and Satyananayana's research is the effect of water depth on the transmission of wave can be neglected both for breakwaters and perforated cylinders, the greater the wave steepness the smaller the wave transmission will be. The greater the inter-pole distance, the greater the transmission wave will be. The use of the breaker in a two-line pole, causes a decrease in the transmission wave. There is no difference between the two-line breakwaters for the hollow and solid poles; and the difference in continuing waves between the hollow and solid pole breakwater is negligible. The Koraim & Salem study undertakes to see the performance of a horizontal semi-circular split pipe supported by a vertical cylinder mantle in reflecting, transmitting and transmitting waves [8]. Koraim & Salem reported that the breakwaters would be effective if the semi-circular pipe is placed horizontally, the diameter of the semi-circle pole increases, the semicircular pole break splitting is 45°, the wavelength is greater than half the water depth, and the wavelength is greater than twice of water depth. Koraim [9] and Koraim, et al [10] examine the performance of hanging breakwaters made of L-shaped and C-shape bars that are laid flat with the support of vertical poles, through theoretical and experimental approaches for regular waves. The parameters studied were wavelength, L-shape and C-shape bars, the distance between L-shape and C-shape bars, diameter and spacing between vertical supports. Koraim reported that the comparison between experimental results and theoretical predictions for the reflection, transmission and dissipation coefficients is excellent.

Paotonan conducts study on a pilecap type breakwater with an analytic approach to find the magnitude of wave height passing through the wave breaker represented by the value of the transmission coefficient [11]. This study also find that the construction cost of pilecap type breakwater is much cheaper than conventional breakwaters. Investment cost savings can be up to 72% more than conventional type breakwaters. Paotonan and Suyatno conducted research on stresses and deformations in sheet pile breakers by varying the gap width between sheet pile and constant wave parameters using ANSYS [12]. They also reported that the greater the width of the gap between the sheet pile, the less force, tension and deformation occur. The greater the stress, then the greater deformation will occurs, in nonlinear relationship. Previous research has not specifically examined the wave transmission on the breaking sheet pile. Therefore, this study focused to examine the effect of wave velocity (H_i/gT^2) and relative depth (s) of sheet pile breakwaters to wave energy transmission coefficient (KE_T).

2. Theoretical Basis

The waves that travel from the deep sea to the coast and are blocked by buildings, some of their energy is reflected, some transmitted and others are dissipated. The magnitude of the reflected wave energy, dissipated and transmitted depends on the characteristics of the coming wave (period, height, and water depth), the type of building (smooth or rough surface, water pass or waterproof) and building geometry (slope, elevation and width of building height). The illustration of fluid interaction with sheet pile breakwaters can be seen in Figure 1.

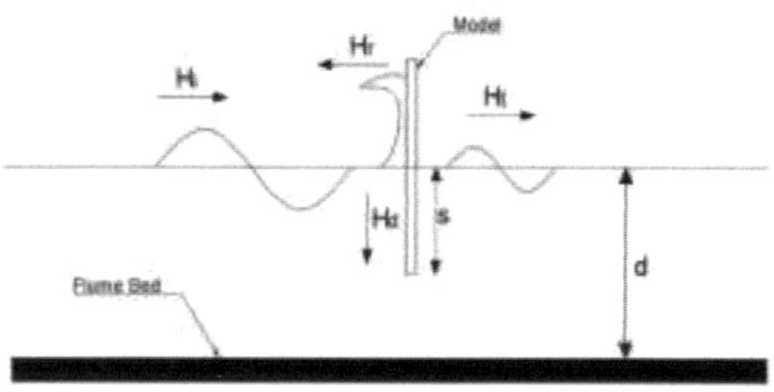

Figure 1. Illustration of fluid interaction with sheet pile breakwaters

Figure 1 shows that H_i, H_r, H_d and H_t are the incident, reflection, dissipation & transmission wave height respectively. In the context of wave energy, the illustration in Fig. 1 can be formulated as follows;

$$E_i = E_r + E_d + E_t \tag{1}$$

If both sides of Eq. 1 divided by E_i, then;

$$\frac{E_r}{E_i} + \frac{E_d}{E_i} + \frac{E_t}{E_i} = 1 \tag{2}$$

Based on linear wave theory, wave energy is formulated with;

$$E = \frac{\rho g H^2}{8} \tag{3}$$

If eq. 3 is substituted to eq. 1, the equation will be;

$$\left(\frac{H_r}{H_i}\right)^2 + \left(\frac{H_d}{H_i}\right)^2 + \left(\frac{H_t}{H_i}\right)^2 = 1 \tag{4}$$

or

$$(K_r)^2 + (K_d)^2 + (K_t)^2 = 1 \tag{5}$$

IOP Conf. Series: Materials Science and Engineering **619** (2019) 012031 doi:10.1088/1757-899X/619/1/012031

By measuring the wave height in front of the model (H_i) and the wave height behind the hanging sheet pile (H_t) model, the wave energy transmission coefficient $KE_t = (K_t)^2$ can be calculated. The energy transmission coefficient of this wave will be influenced by two non-conventional parameters of wave steepness and relative draft; and can mathematically written as follows:

$$KE_t = f\left(\frac{H_i}{gT^2}; \frac{s}{d}\right) \tag{6}$$

3. Experimental Setup

This research was conducted in the Ocean Technology Laboratory, Faculty of Engineering, Universitas Hasanuddin; through physical experiment using flume tank equipped with sophisticated wave generator. The wave generator used is a piston type wave maker consisting of actuators and electric motors. The wave flume is 24 meters long, with 1 meter width and 1.22 meters of height of probes. The model is placed inside a wave flume and reinforced with a wood and iron base. The sketch and laying of models within the wave flume can be seen in Fig. 2, 3 and 4.

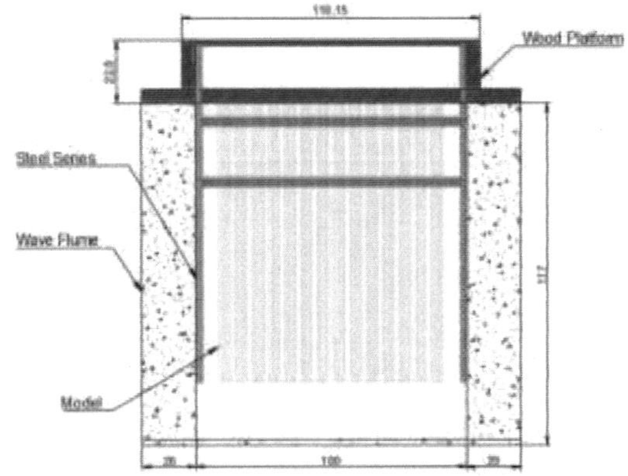

Figure 2. Sketch of the breakwater

Figure 3. Model position inside the flume

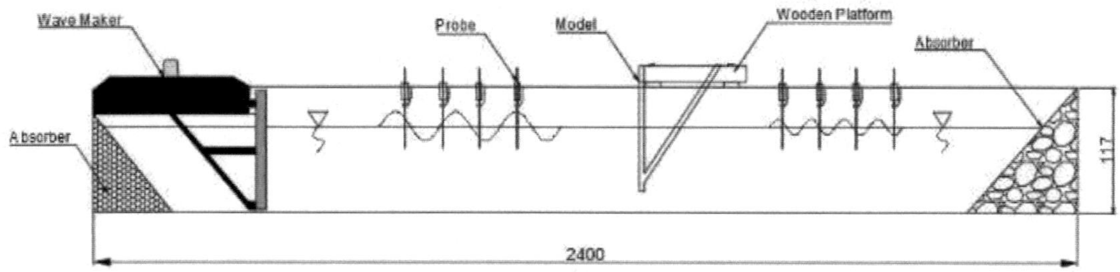

Figure 4. Sketch of Hanging Sheet Pile Breakwater position inside *wave flume*

Depth of water used is d = 0.6 m. The wave period (T) was varied 5 times, i.e. 0.8 seconds, 1.2 seconds, 1.6 seconds, 2 seconds and 2.4 seconds. While the wave height (H) was varied as much as 5 times per period change (T) above, i.e. 0.032 m, 0.064 m, 0.096 m, 0.128 m and 0.160 m. Variable draft of sheet pile set s_1 = d, s_2 = 0.75d and s_3 = 0.50d or relative draft (s/d) consisting of 0.50; 0.75 and 1.00. The distance between the sheet pile at the width of the flume was set b = 0.005 m (0.5 cm).

4. Result and Discussion

By varying wave height, H, the wave period, T, at a water depth of 0.60 m, and varying the structure draft (s) that immersed in water and also specifying the gap width or spacing of the sheet pile, b = 0.005 m, and measuring of the wave height in front and behind the model, then H_i as incident wave height, and wave transmission height, H_t, can be calculated. Totally there are 8 probes used in this study, i.e. 4 in front of the model and 4 behind the model. When wave comes in, H_i is calculated using data on the probe placed in front of the model. While the transmission wave height, H_t is calculated using the data on the probe placed behind the model. The formula for calculating H_i and H_t is as follows.

$$H_i = \left[\frac{H_{max} + H_{min}}{2}\right]_{Front} \tag{7}$$

$$H_t = \left[\frac{H_{max} + H_{min}}{2}\right]_{Back} \tag{8}$$

With knowing H_i and H_t, KE_t values then can be calculated. Based on the incident wave height data, transmission wave height, wave period and full structure and water depth the steepness of wave, H_i/gT^2 and relative load (s/d) can be calculated. The value of T, H_i, H_t, KE_t, H_i/gT^2 and s/d can be seen in Table 1. Using the two values of non-dimensional parameters, KE_t values are presented in graphical form. The effect of H_i/gT^2 to KE_t value can be seen in Fig.5 to 7.

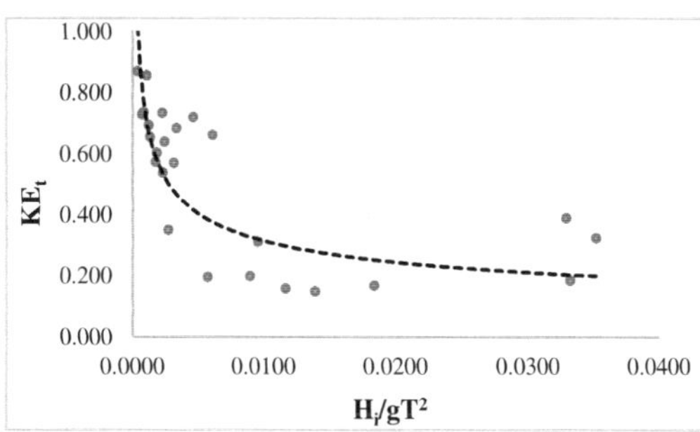

Figure 5. Effect of H_i/gT^2 on KE_t for s/d = 0.50 and width slit (b) = 0.005 m

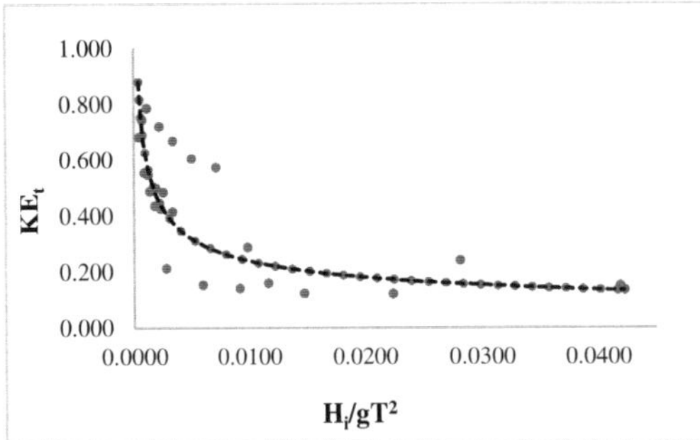

Figure 6. Effect of H_i/gT^2 on KE_t for s/d = 0.75 and width slit (b) = 0.005 m

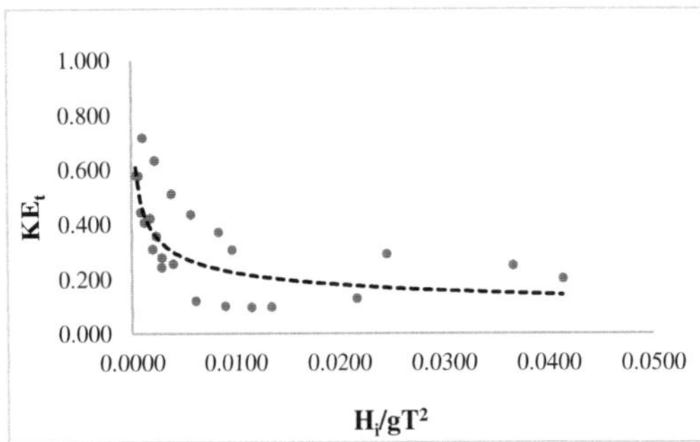

Figure 7. Effect of H_i/gT^2 on KE_t for s/d = 1.00 and width slit (b) 0.005 m

Table 1. Value of H_i, H_t, H_i/gT^2 and KE_t for various value of s/d

T(S)	Hi	Hi	Hi	Ht	Ht	Ht	Hi/gT²	Hi/gT²	Hi/gT²	KEt	KEt	KEt
	s/d=1	s/d=0.75	s/d=0.5	s/d=1	s/d=0.75	s/d=0.5	s/d=1	s/d=0.75	s/d=0.5	s/d=1	s/d=0.75	s/d=0.5
0.80	0.06	0.06	0.06	0.03	0.03	0.03	0.01	0.01	0.01	0.31	0.29	0.31
0.80	0.12	0.14	0.14	0.05	0.05	0.05	0.02	0.02	0.02	0.17	0.12	0.13
0.80	0.21	0.18	0.15	0.09	0.09	0.08	0.03	0.03	0.02	0.19	0.24	0.29
0.80	0.22	0.27	0.23	0.13	0.10	0.12	0.04	0.04	0.04	0.32	0.14	0.25
0.80	0.21	0.26	0.26	0.13	0.10	0.12	0.03	0.04	0.04	0.39	0.15	0.20
1.20	0.04	0.04	0.04	0.02	0.02	0.02	0.00	0.00	0.00	0.35	0.21	0.24
1.20	0.08	0.08	0.09	0.04	0.03	0.03	0.01	0.01	0.01	0.20	0.15	0.12
1.20	0.13	0.13	0.13	0.06	0.05	0.04	0.01	0.01	0.01	0.20	0.14	0.10
1.20	0.17	0.16	0.16	0.07	0.07	0.05	0.01	0.01	0.01	0.16	0.16	0.10
1.20	0.20	0.21	0.19	0.08	0.07	0.06	0.01	0.01	0.01	0.15	0.12	0.10
1.60	0.03	0.03	0.03	0.03	0.02	0.02	0.00	0.00	0.00	0.86	0.79	0.72
1.60	0.06	0.05	0.05	0.05	0.05	0.04	0.00	0.00	0.00	0.73	0.72	0.63
1.60	0.08	0.08	0.09	0.07	0.07	0.07	0.00	0.00	0.00	0.68	0.67	0.51
1.60	0.12	0.12	0.14	0.10	0.10	0.09	0.00	0.00	0.01	0.72	0.60	0.44
1.60	0.15	0.18	0.21	0.12	0.13	0.13	0.01	0.01	0.01	0.66	0.57	0.37
2.00	0.03	0.03	0.03	0.02	0.02	0.02	0.00	0.00	0.00	0.73	0.74	0.58

T(S)	Hi	Hi	Hi	Ht	Ht	Ht	Hi/gT²	Hi/gT²	Hi/gT²	KEt	KEt	KEt
	s/d=1	s/d=0.75	s/d=0.5	s/d=1	s/d=0.75	s/d=0.5	s/d=1	s/d=0.75	s/d=0.5	s/d=1	s/d=0.75	s/d=0.5
2.00	0.05	0.05	0.05	0.04	0.03	0.03	0.00	0.00	0.00	0.69	0.55	0.41
2.00	0.07	0.07	0.08	0.06	0.05	0.04	0.00	0.00	0.00	0.60	0.50	0.31
2.00	0.10	0.10	0.11	0.08	0.07	0.06	0.00	0.00	0.00	0.64	0.48	0.28
2.00	0.12	0.13	0.15	0.09	0.08	0.08	0.00	0.00	0.00	0.57	0.41	0.26
2.40	0.02	0.02	0.02	0.02	0.02	0.02	0.00	0.00	0.00	0.87	0.68	0.58
2.40	0.05	0.05	0.05	0.04	0.04	0.03	0.00	0.00	0.00	0.74	0.55	0.44
2.40	0.07	0.07	0.07	0.06	0.05	0.05	0.00	0.00	0.00	0.66	0.49	0.41
2.40	0.10	0.10	0.10	0.08	0.07	0.06	0.00	0.00	0.00	0.57	0.44	0.42
2.40	0.13	0.13	0.13	0.09	0.08	0.08	0.00	0.00	0.00	0.54	0.43	0.36

Fig. 5 to 7 shows that an increasing wave steepness, (H_i/gT^2), will cause a decreasing value of the transmission wave energy coefficient, (KE_t). This indicates that the steeper the waves come in front of the sheet pile breakwaters model, will cause a shrinking ratio of wave height behind and the front of the model. This phenomenon occurs because a wave with a large steepness occurs significant turbulence in front of the model and causing a decrease in wave height passing the model relative to the incident wave height. But note that the distribution of data for all is relatively large. For all the values of s/d are shown in Figures 5 to 7, for the H_i/gT^2 less than 0.01, the decreasing of KE_t due to the increasing of H_i/gT^2, is significant. Meanwhile on the value of H_i/gT^2 greater than 0.01, the increasing of H_i/gT^2 is relatively have a small effect on the changing of KE_t.

The second objective of this study was to determine the draft relative effect (s/d) on the transmission energy coefficient. Therefore, by combining the graphs in Fig. 5 to 7 visualization of the effect s/d graphically can be done. The effect of s/d against KE_t for various values H_i/gT^2 can be seen in Figure 8.

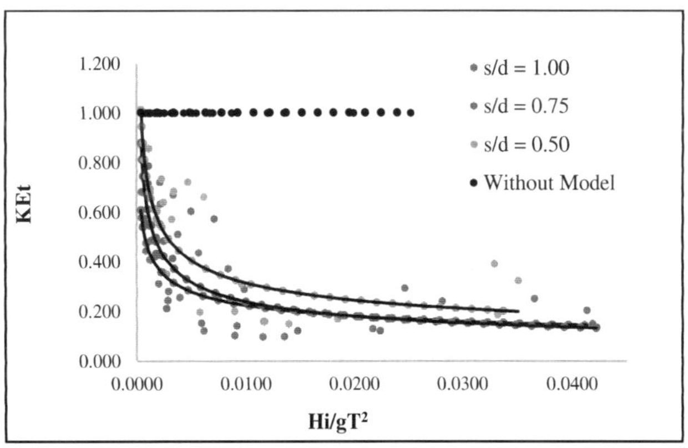

Figure 8. Effect of s/d due to KE_t for several value of H_i/gT^2

Figure 8 shows that when experiments in no-model scheme, the energy transmission coefficient is close to 1. This means, there is no wave energy damping occurs. This can be seen on the flat KE_t graphical line as H_i/gT^2 function at the same ordinate value of 1. Figure 8 also shows that the KE_t graph for the hanging sheet pile model with s/d 0.50 is above the KE_t graph for the model with s/d 0.75 and 1.00. This means that for the constant value of H_i/gT^2, the greater value of s/d, the smaller of the KE_t will be, and it means the more effective the model for reducing the incoming wave energy. Nevertheless, the difference value of KE_t for model with s/d equal to 0.50 and 0.75, is relatively small. Even for the value of H_i/gT^2 equal or greater than 0.01, the KE_t graphical line for s/d 0.50 and 0.75 is almost coincide. This means that when the value of H_i/gT^2 is greater or equal to 0.01, it will not be much benefit, although it enlarges the value s/d from the point of view of wave energy damping by the sheet pile breakwaters. The dissemination phenomena of KE_t as a considerable Hi/gT^2 function, may be caused by several things:

a. In certain wave simulations (high and large wave periods), the model becomes unstable and tends to move following the wave motion. So that in certain periods, the KE_t value increases as the resonant effect of the transmission wave and the reflection of the wave absorbers at the flume end
b. The wave absorber that is located at the end of the flume is less effective in reducing the reverse wave, which result an increasing wave height behind the model.

5. Conclusion

Based on the above discussions, the following concluded several things, namely:

1. The damping of wave energy through the hanging sheet pile breakwater is influenced by the draft structure (s), width of the structure gap (b), the incident wave height (H) and the wave period (T)
2. The greater value of wave steepness (H_i/gT^2) and the draft relative (s/d), then the smaller transmission energy coefficient (KE_t) will be.

6. References

[1] Wiegel, (1960), Transmission of waves past a rigid vertical thin barrier. Coastal Engineering
[2] Hayashi, T., Kano, T. (1966). Hydraulic research on the closely spaced pile breakwater. Coastal Engineering Proceedings, 1(10).
[3] Truitt, C., Herbich, J. (1987). Transmission of random waves through pile breakwaters. Coastal Engineering Proceedings, 1(20).
[4] Herbich, J., Douglas, B. (1989). Wave transmission through a double-row pile breakwater. Coastal Engineering Proceedings, 1(21).
[5] Kakuno, S., & Liu, P. L. (1993). Scattering of water waves by vertical cylinders. Journal of Waterway, Port, Coastal and Ocean Engineering, (3), 302.
[6] Mani, J. S., & Jayakumar, S. (1995). Wave transmission by suspended pipe breakwater. Journal of Waterway, Port, Coastal & Ocean Engineering,121(6), 335.
[7] Rao, S., Rao, N. B. S., & Sathyanarayana, V. S. (1999). Laboratory investigation on wave transmission through two rows of perforated hollow piles. Ocean Engineering, 26(7), 675-699.
[8] Koraim, A. S. & Salem, T. N. (2012). The hydrodynamic characteristics of single suspended row of half pipes under reguler waves. Coastal Engineering, 50(0), 1-9.
[9] Koraim, A. S. (2014). Hhydraulic characteristics of pile-supported L-shape bars used as screen breakwater. Coastal Engineering, 83(0), 36-51.
[10] Koraim, A. S., Iskander, M. M., & Elsayed, W. R. (2014). Hydrodynamic performance of double rows of piles suspending horizontal c shaped bars. Coastal Engineering, 84(0), 81-96.
[11] Paotonan, C. (2015). Analytical approach waves through on pilecap breakwater (Original in Bahasa : Pendekatan analitis gelombang melalui pemecah gelombang tipe pilecap). Prosiding Seminar Nasional Teknik Sipil Universitas Muhammadiya Surakarta.
[12] Paotonan, C., dan Suyatno, M.C.S. (2017). Stress and deformation of sheet pile breawater (Original in Bahasa : Tegangan dan deformasi pemecah gelombang sheet pile). Prosiding Seminar Nasional Sains dan Teknologi Fakultas Teknik Unhas ke-3.

Identification of Ground–fault Current in Various Transformer Connection and Generator Grounding using Symmetrical Components Method

A.R.Sultan[1], M.W.Mustafa[2], M.Saini[3]), A.Gaffar[4]

[1,3,4]Politeknik Negeri Ujung Pandang, [2]Universiti Teknologi Malaysia
Email: rizal.sultan@poliupg.ac.id

Abstract. Single line to ground-fault (SLG) fault is the record frequent fault probable to occur in an electric power system. The effect of ground-fault is determined by using grounding method and transformer connections. In this study, the simulation showed the performance of a simple power system within the SLG fault in the system at various transformer connection and generator grounding method. New approach the application of ground fault current calculation using symmetrical components method on unit generator transformers. A symmetrical components method applied with simulation was conducted in Matlab software. The results were then analyzed, presenting a comparison with the fault impact at various transformer connection and generator grounding method. The models of transformer connection for each side (primary and secondary) used were wye (Y), wye-grounded (Yn) and delta (Δ). It was clearly shown that the ground-fault currents at the generator bus had significantly an effect on the type of the transformer connections used during the ground-fault at the secondary side of the transformer.
Keywords; symmetrical components method, ground-fault, transformer connection

1. Intoduction

In general, a step up transformer at electric power station can be categorized as either unit generator-transformer configuration, unit generator-transformer configuration with generator breaker, cross-compound generator and generator involving a unit transformer [1, 2]. Ground-fault on the transmission line or busbar can affect the system configuration of the generator. Knowledge on ground-fault at transformer winding connection is essential in selecting the most suitable transformer for service requirement.

Several papers have deliberated the impact of variation transformer connections within a power system but no one has discussed, in particular, about the effect of ground fault current. The transformer configurations with the propagation of voltage sags [3] can influence the performance of voltage sags inside the industry facility, depending on the function of transformer connection used in the service transformer. It is known that the *Yn-Yn* transformer connected across the generator and the *Δ-Yn* transformer connected across the load will reduce the unbalance voltage[4]. Reference [5] describes the effect of the voltage transformers on the operating conditions of a ground-fault protection system for unit-connected generators. The different transformer connections was highly dependent on the unbalance voltage in the power system [6,7]. The influences of transformer connection on operation are discussed [8] focus on operation involved insulation stresses, autotransformer under various condition operations but not included for grounding.

There are many factors that can affect the level of ground-fault currents in unit generator transformer. The ground-fault current magnitude, particularly in the generator and transformers, are determined by analyzing the generator and transformer winding impedance [1, 9, 10]. This paper presents the effect of generator grounding method and three-phase transformer connections. The generator grounding method included solid grounding, resistance grounding and reactance grounding. Transformer connections were denoted as wye (*Y*), wye-grounded (*Yn*) and delta (*Δ*) in each primary and secondary side for a simple power system in unit generator-transformer configuration.

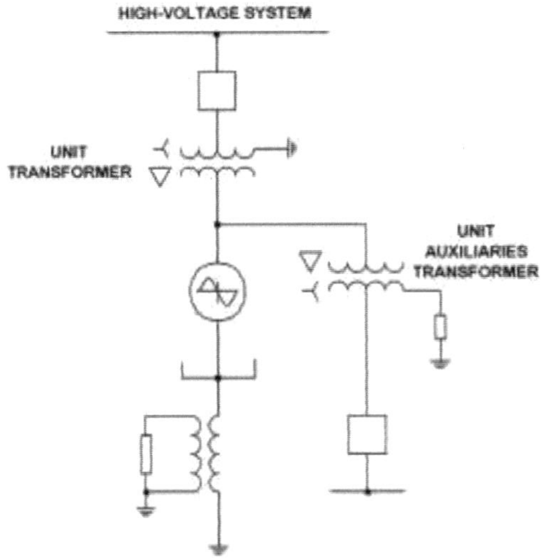

Figure 1. Unit generator-transformer configuration [1]

2. Principle

2.1 Transformer Connections

The primary and secondary winding of the transformer can be connected either in combinations of *Y*, *Yn* or *Δ* connections, which can result in nine possible connection combinations [9, 11] *Y-Y, Y-Yn, Y-Δ, Yn-Yn, Yn-Y, Yn-Δ, Δ-Y, Δ-Yn* and *Δ-Δ*.

The zero-phase sequence (ZPS) impedance of a transformer depends on the configuration of the winding. The ZPS impedance of a *Δ* winding is infinite, whereas the ZPS impedance of a Y-connected winding is a composite series with the ZPS impedance of a transformer and an impedance of any neutral grounding devices that might be present. Thus, an ungrounded Y-winding would present infinite zero-sequence impedance because of the absence of a neutral grounding connection that would appears as an open circuit in series with the ZPS impedance of the transformer winding itself [12, 13].

The sequence equivalent circuits of three-phase transformer depend on the connections of the primary and secondary winding. The different combinations of Y and Δ windings determine the configurations of the ZPS circuits. Figure 2 presents some of the connection diagram of a transformer connection and their ZPS equivalent network [9, 12]

2.2 Generator Grounding Method

Grounding systems can be categorized into four general groups : ungrounded, effectively grounded, low impedance, and high-impedance grounded. Ungrounded systems have no physical connection between the generator neutral and ground. Effectively grounded systems have a direct connection from the generator neutral to system ground. The phase-to-ground fault current on an effectively grounded system is very high; in many cases higher than that of a three-phase fault. A low-impedance grounded system has minimal impedance connected between the generator neutral and ground. This impedance limits the ground fault current to a value typically between 100 A and the magnitude of a three-phase fault. In a high-impedance grounded system, the ground-to-neutral impedance is large enough to limit the ground fault current to a value between 2 and 15 A [10]. For generator grounding, the grounding methods dominants are resistance grounding (39.4%), followed by solidly grounding (25.7%) then neutral reactor grounded (17.3 %) [13].

2.3 Symmetrical Components Method

Fortescue proposed the theory of symmetrical components in a famous paper in 1918. It was used to analyze the unbalanced short-circuit and open-circuit fault in practical power systems. This theory applies to a system of N unbalanced phasors, including the case of three unbalanced phasors representing a three-phase power system. The theory enables the transformation of three unbalanced phasors into a three set of balanced phasor, called the positive phase sequence (PPS), negative phase sequence (NPS), and zero-phase sequence (ZPS) [13].

The magnitude of a ground-fault current, especially at the generator and transformer are determined by generator and transformer winding impedance [1, 9, 10]. The ground-fault current magnitude is calculated by using the following equation [13, 14]:

$$I_{SLG} = \frac{3.V_f}{Z_0 + Z_1 + Z_2 + 3.Z_f} \quad \ldots (1)$$

Where V_f is the pre-fault voltage source. Z_1, Z_2, and Z_0 are the PPS, NPS, and ZPS impedances for network of different sequence respectively. Z_f represents the impedance of the ground return path including the fault arc, the grounding network, and the intended neutral impedance, when present. Generally, two settings of pre-fault voltages are assumed for fault currents calculation. Firstly, all the pre-fault voltages are assumed equal to 1 pu, i.e., the loads are neglected [9]. Secondly, the values of the pre-fault voltages are determined using load flow algorithms [14, 15]. The use of 1 pu makes the fault calculation simpler [16].

If one assumes Z_f=0 and Z_1=Z_2= Z_0, the magnitude of a ground-fault current will be :

$$I_{SLG} = \frac{V_f}{Z_1} \quad \ldots (2)$$

In resistance grounded systems (all reactance's set to zero), the ground resistor restricts the ground-fault current to a particular value regardless of the fault type or fault location. The ground-fault protective sensitivity can be relatively independent of continuous load current values and, therefore, can have a lower pick up setting than phase protective devices. The ground-fault currents are not transferred through system power transformers that are connected in Δ-Y or Δ-Δ, as the ground-fault. The Protection for each system voltage stage is independent of the protection at other voltage stages.

3. Research Method

This section presents the performance of various transformer connections and generator grounding method at the generator bus during SLG fault at each bus in the form of the single line diagram for analysis, as illustrated in Figure 2. Assume that the generators are operating at no-load and at their rated voltage at the occurrence of fault.

There are many factors affecting the level of ground-fault currents in unit generator transformer. Here, the influence of transformer configuration and generator grounding method are studied. Extensive simulation (a symmetrical components method) tests were carried out using Matlab Software. The adjusted parameters were as follows:

- The kind of generator grounding method was a change. The kinds of grounding method were solidly (0 ohm), reactance (all resistances set to zero) and resistance (all reactances set to zero) grounding.
- The type of two-winding transformer connection was changed. The types of transformer connection were *Y, Yn* and *Δ* as primary and secondary transformer connections with initial connection as *Yn-Yn* for transformer-1 (T_1) and *Yn-Yn* for T_2.
- Both the type of transformer connections and generator grounding method were changed.

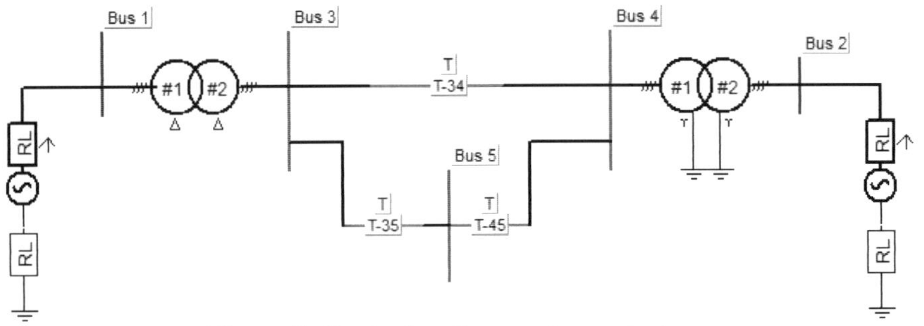

Figure 2. Single-line diagram for analysis

4. Analysis of simulation result

There are many factors that can affect the ground-fault current. Here, the effect of a transformer connection and generator grounding method are discussed. Simulations had been carried out using a symmetrical component method to study the effect of a generator grounding method and transformer connections onto the unit generator transformer. Simulation was conducted under three conditions: firstly was under the impact of a generator grounding method on ground-fault currents; secondly was under the effects of a transformer configuration distinct on ground-fault currents, and lastly was under the condition of combined effects of a generator grounding method and transformer configuration variations.

4.1 Effect of generator grounding method on ground-fault currents

Simulations were performed for solidly, resistance and reactance grounding method for *Yn-Yn* transformer connections using bolted fault impedance. This was done at two locations, namely SLG fault location nearby, and also far from the generating station.

4.1.1 Ground-faults closer to the generating station. For this condition, SLG fault was at Bus-1. The effect of grounding the generator through a solidly grounding, reactance and resistance grounding can be seen in Figure 3. This figure shows that, in solidly generator grounding (0 ohm), the fault current supplied to Bus-1 by Generator-1 was 5.0 per unit. In general, a large increase in the value of a resistance and reactance grounding would cause a decrease of the current. For the same impedance value, the value of a ground-fault current in resistance grounding was smaller than the reactance grounding. In contrast, the figure shows that the value under 26 ohms of the resistance grounding, asset value of ground-fault current was greater than the value of reactance grounding.

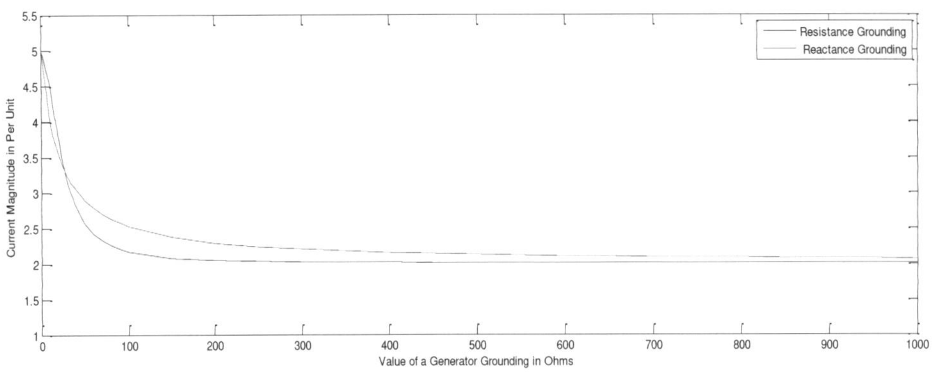

Figure 3. Magnitude of a generator current for SLG fault at Bus-1 for reactance and resistance grounding of the generator

4.1.2 Ground-faults away from the generating stations. In this condition, the SLG fault simulation was done at Bus-3 and Bus-5 as shown in Figure 4. The same conditions for SLG faults were closer to the generator. Using the identical value of the resistance / reactance, the resistance grounding had a lower magnitude of fault current. From Figure 4, it is apparent that the further the location of interference from Generator-1, the smaller current measure would be supplied by Generator-1 than when SLG fault occurred at Bus-3. This condition was influenced by reactance of the transmission line. The effects of disturbance would be smaller if the distance between generating station and fault location was further.

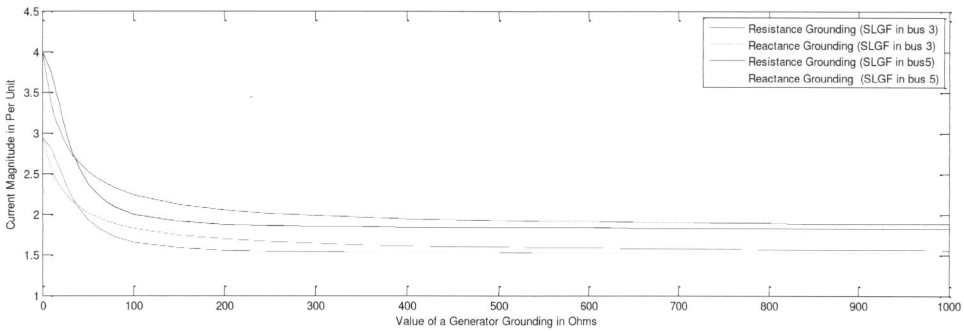

Figure 4. Magnitude of a generator current for SLG fault at Bus-3 and Bus-5 for reactance and resistance grounding of the generator

4.2 Effect of transformer connections on ground-fault current

Types of transformer connection used for this simulation were only *Yn, Y* and *Δ* at both primary and secondary of the transformer. According to the type of a ZPS transformer diagram, a ZPS circuit is similar for *Yn-Y* and *Y-Δ; Y-Yn* and *Δ*-Y, therefore, the transformer connection simulation was only performed on *Yn-Yn, Yn-Y, Y-Yn, Yn-Δ, Δ-Yn, Y-Y* and *Δ-Δ* transformer connections. Simulations were performed for solid generator grounding with bolted fault impedance. This was done at two locations, namely SLG fault location nearby, and far from the generating station.

4.2.1 Ground-faults closer to the generating station. For this condition, SLG fault at Bus-1. Figure 5 shows the comparison the effect of transformer connections on ground-fault current supplied by Generator-1 during SLG fault at various buses. Therefore, fault occurring at Bus-1, where the current supplied from the generator to the fault location at Bus-1 was equal to 5.0 per unit for all transformer connections, except for the *Yn-Δ* Connection transformer, in which the fault current supplied by the generator-1 was smaller with 4.5157 per unit. This was due to the ZPS reactance to Bus-1, where the value would become larger with the additional reactance of a transformer-1 at 0.05 per unit.

4.2.2 Ground-faults away from the generating stations. In this condition, the SLG fault simulation was done at Bus-2, Bus-3, Bus-4 and Bus-5. In the event of SLG fault at Bus-2, the magnitude of fault current supplied from the generator to the fault location for *Yn-Yn* and *Yn - Δ* transformer connections were 4.0 per unit and 3.4776 per unit respectively. Meanwhile, the magnitude of fault current for connection of other transformers was 1.8212 *pu*. Similarly, for disturbance at Bus-3, Bus-4 and Bus-5, transformer connections of *Yn-Yn* and *Δ-Yn* had the fault current greater than the other models of transformer connections. This was due to the ZPS diagrams for the transformer connections *Yn-Yn* and *Δ-Yn* which had the relationship between the primary and the secondary side of the transformer, as well as the additional reactance to ground at the bus of *Δ-Yn* transformer.

4.3 Combined effect of generator grounding method and transformer connections

For this simulation, the models being used for generator grounding were solidly grounding, resistance grounding and reactance grounding. Magnitude of reactance and resistance impedance for grounding was varied between 0 to 1000 ohms. Simulations were done for the transformer connection on *Yn-Yn, Yn-Y, Y-Yn, Yn-Δ, Δ-Yn, Y-Y* and *Δ-Δ* transformer connections. Simulations were also performed for generator grounding with solidly bolted fault impedance. This was done at two locations, namely SLG fault location nearby, and far from the generating station.

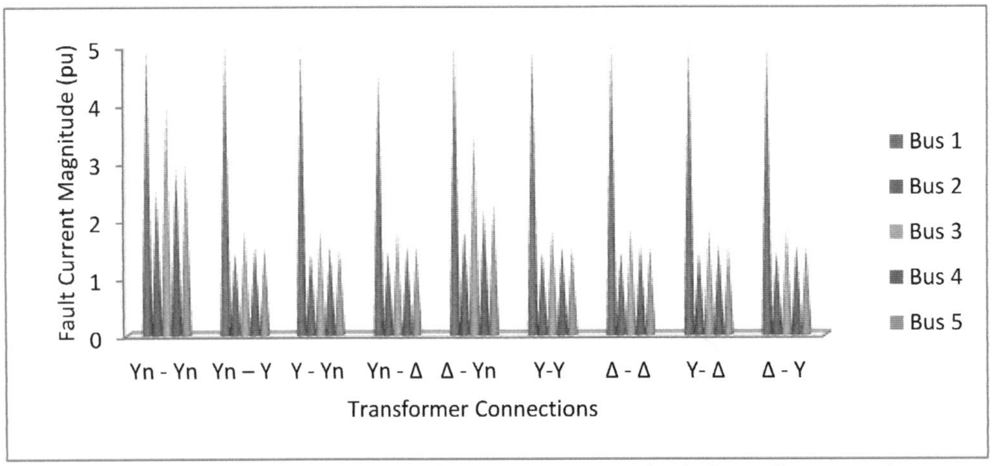

Figure 5. Comparison of fault current magnitude from Generator-1 for various locations of SLG fault

4.3.1 Ground-faults are closer to the generating station. For this condition, the SLG fault was at Bus-1. Figure 6 and Figure 7 present the current magnitude supplied from the generator to fault locations (Bus-1) in various grounding methods and transformer connections, based on the method of symmetrical components and data such as the appendix. It was clear that for solidly grounding, reactance grounding and resistance grounding, the *Yn-Δ* transformer connection produced the smallest ground-fault current compared to the others. The *Yn-Yn* transformer connection produced most of the fault currents. The *Yn-Y, Y-Yn* , *Δ-Yn, Y-Y* and *Δ-Δ* transformer connection had an identical ZPS circuit to produce the same ground-fault current.It can be seen that the highest ground-fault current for various transformer connection would be produced when the generator used was solid grounding. For the same value of impedance, the fault current of resistance grounding was greater than the reactance grounding.

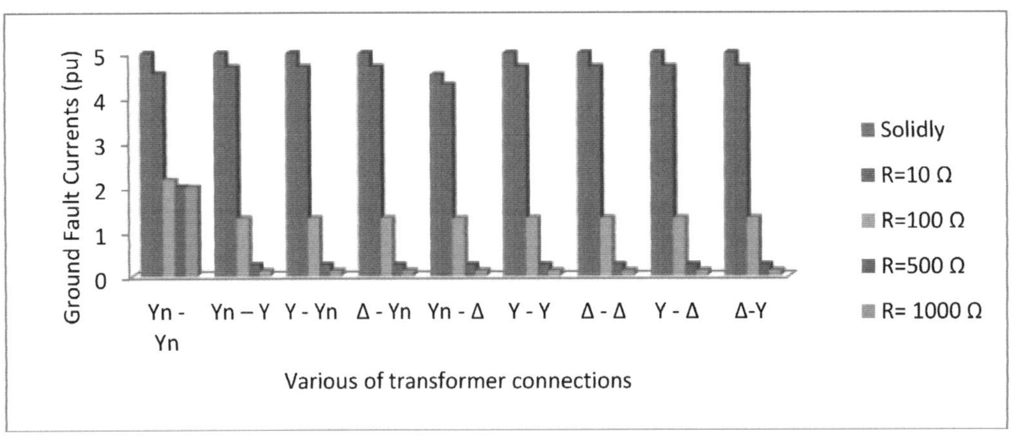

Figure 6. Magnitude of fault current in Generator-1 for distinct value of resistance grounding of a generator during SLG fault in Bus-1

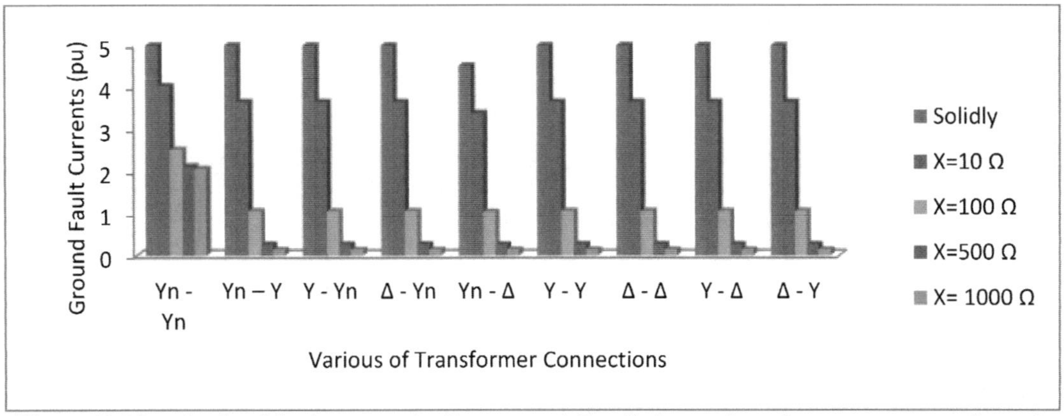

Figure 7. Magnitude of fault current in Generator-1 for distinct value of reactance grounding of a generator during SLG fault in Bus-1

4.3.2 Ground-faults away from the generating stations. In this condition, the SLG fault simulation was done at Bus-3 and Bus-5 separately. Table 1 and Table 2 show the comparison of magnitude of SLG fault current supplied by the generator to the fault location by applying symmetrical components method. From these tables, the *Yn-Y, Y-Yn, Δ-Yn, Y-Y* and *Δ-Δ* transformer connections had similar fault current for both of the faults through the generator grounding. This was because the isolation between Bus-1 and Bus-3 in the ZPS network led to 1.8212 *pu* at SLG fault at Bus-3 and 1.5162 *pu* at SLG fault at Bus-5. This value was not affected by the changes of the type and value of generator grounding. Similarly, for *Yn-Δ* transformer connection, the magnitude of fault current was equal to 3.4776 *pu* at SLG fault at Bus-3 and 2.2995 *pu* at SLG fault at Bus-5. As for the *Yn-Yn* transformer connection, a change of SLG fault current was affected by the value of generator grounding impedance. The higher value of the grounding impedance would produce the smaller of the ground-fault current.

It was also observed that if the fault location was far from the Generator-1, the ground-fault current would be smaller. At fault location of Bus-5, the ground-fault total was the combination of both from Generator-1 and Generator-2. When the source was multiplied, the fault current analysis became more complex.

Table 1 .Fault current of SLG at Bus-3 with various transformer connections and generator grounding

Grounding method	SLG fault current at various transformer connections (per unit)								
	Yn-Yn	*Yn – Y*	*Y - Yn*	*Δ-Yn*	*Yn - Δ*	*Y - Y*	*Δ-Δ*	*Δ-Y*	*Y-Δ*
Solidly	4.000	1.8212	1.8212	1.8212	3.4776	1.8212	1.8212	1.8212	1.8212
R=10 Ω	3.7533	1.8212	1.8212	1.8212	3.4776	1.8212	1.8212	1.8212	1.8212
R=100 Ω	1.9981	1.8212	1.8212	1.8212	3.4776	1.8212	1.8212	1.8212	1.8212
R=500 Ω	1.8290	1.8212	1.8212	1.8212	3.4776	1.8212	1.8212	1.8212	1.8212
R= 1000 Ω	1.8232	1.8212	1.8212	1.8212	3.4776	1.8212	1.8212	1.8212	1.8212
X=10 Ω	3.3540	1.8212	1.8212	1.8212	3.4776	1.8212	1.8212	1.8212	1.8212
X=100 Ω	2.2390	1.8212	1.8212	1.8212	3.4776	1.8212	1.8212	1.8212	1.8212
X=500 Ω	1.9199	1.8212	1.8212	1.8212	3.4776	1.8212	1.8212	1.8212	1.8212
X= 1000 Ω	1.8717	1.8212	1.8212	1.8212	3.4776	1.8212	1.8212	1.8212	1.8212

Table 2. Fault current when SLG at Bus-5 with various transformer connections and generator grounding

Grounding method	SLG fault current at various transformer connections (per unit)								
	Yn-Yn	*Yn – Y*	*Y - Yn*	*Δ-Yn*	*Yn - Δ*	*Y - Y*	*Δ-Δ*	*Δ-Y*	*Y-Δ*

Solidly	2.9412	1.5162	1.5162	1.5162	2.2995	1.5162	1.5162	1.5162	1.5162
R=10 Ω	2.8067	1.5162	1.5162	1.5162	2.2995	1.5162	1.5162	1.5162	1.5162
R=100 Ω	1.6512	1.5162	1.5162	1.5162	2.2995	1.5162	1.5162	1.5162	1.5162
R=500 Ω	1.5222	1.5162	1.5162	1.5162	2.2995	1.5162	1.5162	1.5162	1.5162
R= 1000 Ω	1.5177	1.5162	1.5162	1.5162	2.2995	1.5162	1.5162	1.5162	1.5162
X=10 Ω	2.5546	1.5162	1.5162	1.5162	2.2995	1.5162	1.5162	1.5162	1.5162
X=100 Ω	1.8180	1.5162	1.5162	1.5162	2.2995	1.5162	1.5162	1.5162	1.5162
X=500 Ω	1.5889	1.5162	1.5162	1.5162	2.2995	1.5162	1.5162	1.5162	1.5162
X= 1000 Ω	1.5535	1.5162	1.5162	1.5162	2.2995	1.5162	1.5162	1.5162	1.5162

5. Conclusion

This paper presents simulations using symmetrical components method to investigate the effect of transformer connections and generator grounding method on SLG fault at the generator-transformer unit. In the event of ground-fault at the primary side of a transformer, the transformer connection did not affect the magnitude of the SLG fault current. The ground-fault current was only affected by the generator grounding method. It was obviously shown that the solid grounding method produced the larger current than the other generator grounding methods. In the event of ground-fault at the secondary side of a transformer, it was clearly shown that the SLG fault at secondary side or line transmission depended on the transformer connection and generator grounding method used. All transformer connections, except for *Yn-Yn* with both neutral grounded, were produced in isolation between the primary and secondary in the ZPS circuit.

References:

[1] IEEE Std C37.102™-2006, IEEE Guide for AC Generator Protection.

[2] J.C.Das, "*Power System Relaying*", Wiley Encyclopedia of Electrical and Electronic Engineering, 1999.

[3] M.T.Aung and J.V.Milanovic, "The Influence of Transformer Winding Connections on the Propagation of Voltage Sags", *IEEE Transaction on Power Delivery*, Vol.21.No.1, pp.262-269, 2006.

[4] Zielichowski M and Fulczyk M., "Influence of Voltage Transformer on Operating Conditions of Ground-Fault Protection System for Unit-Connected Generator", *Electrical Power and Energy Systems*, Vol. 20, pp. 313-319,1998.

[5] Fulczyk.M and Bertsch.J, "Ground-Fault Currents in Unit-Connected Generators With Different Elements Grounding Neutral", *IEEE Transactions on Energy Conversion*, Vol. 17, No.I, pp.61-66, March 2002.

[6] Ying-Yi,H and Fu-Ming,W., "Investigation of Impacts of Different Three-phase Transformer Connections and Load Model on Unbalance in Power System by Optimization", *IEEE Transaction on Power Systems* ,Vol.12, No.2, pp.689-697, May 1997

[7] A.R.Sultan, M.W.Mustafa, M.Saini, 'Ground Fault Currents in Unit Generator-Transformer at Various NGR and Transformer Configurations' *2012 IEEE Symposium on Industrial Electronic and Applications (ISIEA 2012), September 23-26, 2012, Bandung- Indonesia*

[8] Blume, L. F., "Influence of Transformer Connection on Operation", *Journal of the American Institute of Electrical Engineers*, Vol XXXIII, Issue 1, pp.735-752, 1914

[9] Saadat H., "*Power System Analysis*", McGraw-Hill, 1999

[10] D.Reimert, "Protective Relaying for Power Generating Station", Chp.5 ,CRC Press. 2006

[11] C.Bayliss and B.Hardy, "Transmission and Distribution Electrical Engineering'', Newness. 2006.

[12] IEEE Std 142 -2007, IEEE Recommended Practice for Grounding of Industrial and Commercial Power Systems.

[13] Short T.A, "*Electric Power Distribution Handbook*", CRC Press LLC, USA, 2001.

[14] Tleis Nasser," Power Systems Modelling and Fault Analysis", Newness, Hungary, 2008

IOP Conf. Series: Materials Science and Engineering **619** (2019) 012032 doi:10.1088/1757-899X/619/1/012032

[15] Paap.G.G, "Symmetrical Components in The Time Domain and Their Application to Power Network Calculations", *IEEE Transactions Power Systems*, Vol. 15, No.2 pp. 522-528, May 2000.

[16] Abdel-Akher.M. and Nor, K. M, "Fault analysis of multiphase distribution systems using symmetrical components", *IEEE Transactions Power Delivery*, Vol. 25,No.4, pp. 2931-2939, Oct. 2010.

[17] Garcia, F.H, Ferraz.R.G, Bretas A.S and Leborgne R.C, "Single Line-to-Ground Faults on Distribution Systems: Effect of Pre-Fault Voltages on Fault Currents", *Power and Energy Society General Meeting, 2011 IEEE*

IOP Conf. Series: Materials Science and Engineering **619** (2019) 012033 doi:10.1088/1757-899X/619/1/012033

Application of Hybrid Solar and Wind Energy Generation for Paddle Wheel Aerator

Musrady Mulyadi[1], **A. M. Shiddiq Yunus**[2]
[1, 2] Mechanical Eng. Dept, Politeknik Negeri Ujung Pandang, Makassar 90245, Indonesia

The corresponding author's e-mail: shiddiq@poliupg.ac.id

Abstract. At the coastal areas of South Sulawesi, Indonesia, there are many of villagers work as fisherman, shrimps and fish farmers. Most of them use aerator paddle wheel that are operated by high wattage motors that are ranged from 750 Watt to 1000 Watt. Some others use gasoline engine for one aerator every day. Consequently, the use of high wattage motors will cost the farmers and the use of gasoline will contribute to the air pollution. Therefore, it is necessarily important to invent a renewable base energy source for aerator which in turn economically effective that in the same time could apply the energy diversification. Fortunately, solar and wind potency are quite promising in the area. In this paper, an investigation, examination and application of hybrid solar-wind turbine generator are conducted and applied to aerator paddle wheel for shrimp's pond. The required energy is 420 Wh/day. The measured average wind speed at Pancana Village is ranged from 3.25 m/s to 5.28 m/s. The averaged solar radiation potency in the village is 975 W/m^2. The hybrid system is a combination of two 150 Wp solar panels and 500 W wind turbine. The experimental result show that the solar panels could generate energy by about 370.54 Wh in average, while wind energy could generate 165.60 Wh energy every day. Therefore, it can be concluded that the required energy of the aerator at average paddle wheel of 80 rpm can be fulfilled although solar energy is dominant as energy supplier.

1. Introduction

The use of conventional energy could evoke the catastrophic impact to environment due to exhausted pollutant contents. Moreover, the deposits of conventional energy such as oil, gas and coal have been decreased from time to time [1]. This condition has force many countries including Indonesia to optimize the energy saving and the efforts to develop the new energy sources. Solar and wind energy are two popular renewable energy that have been implemented widely in the world [2]-[4]. Most of them are used for large scale to generate electric power that connected to the grid [5]-[7]. As a country that located in equator, Indonesia has a huge potency of solar and wind energy. This energy sources could be applied to many of farmers' equipment's Indonesia is a country of fishes and shrimp's producer. One of fishery equipment is aerator that is used to provide sufficient oxygen to the habitants of ponds. The aerator paddle wheel required high power that is range from 750 W to 1000 W. One of area in Sout Sulawesi, Indonesia that has potency of shrimp's production is Pancana. Pancana is a coastal village that is located at Barru District which land area of ponds and fishery is about 17.5 hectare. Majority of its community are work as shrimp farmers. The key factor that influences the aeration system at shrimp pond in order to improve the productivity of shrimp production is the control of flow direction and speed of air at the diffuser system. The rotation mechanism and energy source of rotation of paddle wheel are also factors that affecting the aerator performance system. Up today, shrimp farmers in Pancana are still using energy from the electrical grid and gasoline engine which in turn contribute to high cost and air pollution. Meanwhile, in Pancana village, the measured average wind speed is ranged from 3.25 m/s to 5.28 m/s and average solar radiation is about 975 W/m^2. Therefore, in this paper, an investigation, examination and application of hybrid solar-wind turbine generator are conducted and applied to aerator paddle wheel for shrimp's pond.

2. Experimental

2.1. Selection of Motor for Paddle Wheel

The selection of motor based on the paddle wheel requirements. Brushless DC Motor (BLDC) has specification with nominal voltage of $48V_{dc}$, rated current of 30A and maximum power of 350 W/1200W. Torque capability is up to 70 Nm with BLDC motor speed can be adjustable using twist throttle of maximum 30 Mph. In this paper, BLDC could operate the paddle wheel for 9 hours with constant speed of 80 rpm with consumed averaged power is 110 W. The paddle wheel speed is regulated at 80 rpm has been adjusted with the requirement of shrimp pond aeration need.

2.2. Battery Selection
The battery selection is based on the requirement of BLDC motor, if the paddle wheel operates in the night time, the energy will be supplied from the battery only because no solar radiance and limited wind speed. Based on energy calculation, the batteries capacity can be determined as 70W x 6h/12V = 35Ah to allow long duration life of batteries, 60% of maximum used is determined. Therefore, it can be obtained that the required battery capacity is 35Ah/60% = 58.33Ah. To match with the energy requirement of BLDC motor, 4 @12V, 50Ah battery that is connected in series are required to obtain 48V. The used average power for aerator is 70W that operate for 6 hours or equal to 420 Wh.

2.3. Solar Panel Selection
Solar energy is an alternative energy source that clean, abundantly available naturally and renewable. This solar energy is claimed to be a fast solution, measured, sustainable that could mitigate the climate change. According to the research that was conducted by Green Tech Media, solar panel installation has been increased about 34% in 2015 compared to prior year [8]. The solar panel configuration is a fixed slope where the fluctuation of solar radiance is considered against these parameters: Number of module, and angle of solar panels slope. Number of module is considered the voltage of Maximum Power Point Tracking (MPPT) of inverter and the voltage of module open circuit [9],[10]. In this paper, to fulfill the energy requirement on batteries charging, 2 x 150Wp (300 Wp). This selection is based on the nominal energy requirement of BLDC Motor 110W, easiness of it installation on the aerator, solar panel mechanical stability against the buoys. Two solar panels will give simplicity in configuring the panels whether in series or parallel. The specification of the used solar panel can be seen in Table 1.

Table 1. Solar Panel Specification

Output power	2 x 150 Wp
Cell type	Monocrystaline
Voltage at P_{max}	18.3V
Current at P_{max}	8.21A

2.4. Wind Turbine Selection
Generator selection is an important part for every design and construction of a wind turbine system. It is important to review the generator capability in generating voltage at low rpm. Magnet Permanent Alternator (PMA) is much efficient compared to DC generator. In this paper, wind turbine generator is selected based on daily average of wind speed at Pancana Village. Small Scale Wind Turbine type is suitable for low wind speed. Wind turbine i500 Ista Breeze is selected based on its suitability of the speed range at the location and generate power based on *IEC 61400-2* standard [11]. The detail specification of the wind turbine is shown in Table 2.

Table 2. Wind Turbine Specification

Output power	500 W
Wind speed start up	3 m/s
Rotor type	Permanent Magnet
Type	Brushless, Direct Drive, HAWT
Storage energy	Battery 4 x 12 VDC/ 2 x 24 VDC

2.5. Charge Controller
Charge controller is used to adjust the voltage and current from solar panel and wind turbine based on battery capacity. The operational algorithm of the controller allows the disconnection of current from

power supply to stop charging process. The disconnection is occurred when the battery voltage reaches 16V±1%V. On the other hand, when voltage battery drop up to 10.8V±1%, load current will be disconnected, therefore battery supply will be stopped simultaneously. In this state condition, hybrid charge controller is also work to control the charging and discharging of the batteries to avoid over charging process. The specification of the charge controller can be seen in Table 3. Additionally, this charge controller could detect the battery capacity. When battery is fully charged, automatically current charge process of both solar and wind turbine generators will be stopped [12].

Table 3. Charge Controller Specification

Type	Maximum Power Point Tracking (MPPT) eSmart 12V/24V/36V/48V-series
Efficiency	≥99.5%

2.6. Application of Hybrid Solar-Wind Turbine Based Paddle Wheel Aerator

Hybrid solar-wind turbine based paddle wheel aerator consisted of electrical and mechanical systems. Mechanical system is designed to consider the structural frame of solar panel stand based on size and weight of the panels. Stand for BLDC motor is designed to be robust, light, and capable to hold the vibration. Anti corrosion slide bearing is used to guarantee the rotation performance of the paddle wheel. To allow the mechanical power transmission from motor to drive shaft, an appropriate chain transmission is employed. To support the reliability of mechanical system, the mechanical structure is constructed using elbow-iron, long hollow shaft 200 mm, ϕ 1" *stainless steels*, slide bearing of ϕ 1" and paddle wheel with diameter of 620 mm. Number of blades is 8 and pontoon boat with size of 1760 mm, x 330 mm x 113 mm. The chain transmission is equipped with *Chain Guard Double Gear.* The construction frame is placed on the pontoon boat with aforementioned size of 1760 x 330 x 113 mm x 2 pieces with 1000 mm long of PVC pipe of 5". The heavy capacity of pontoon boat is 118.86 kg. It is important for the mechanical design to take into account some of parameters in order to achieve the stability consideration such as position of mass center, buoyancy point and metacentre point [13]. Based on calculation, the buoyancy point is 1792 N and gravity force of pontoon boat is 618.03 kg. The buoyancy point is larger than the gravity force of pontoon boat, therefore the pontoon boat could withstand with equipments weight of 61 kg where the stability is evenly stable. The design construction of the hybrid solar-wind turbine paddle wheel aerator can be seen in Fig. 1.

Figure 1. Hybrid Solar-Wind Turbine Paddle Wheel Aerator

3. Results and Discussion

3.1. Solar panel testing result

The testing of solar panel in area of shrimp ponds at Pancana Village is aimed to obtain the relationship between the output power and solar radiation per time unit of the solar panels. The result of the testing is shown in Fig. 2.

IOP Conf. Series: Materials Science and Engineering **619** (2019) 012033 doi:10.1088/1757-899X/619/1/012033

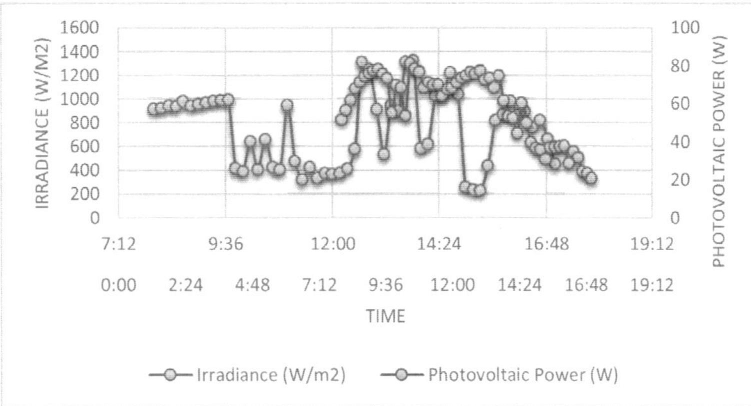

Figure 2. Solar (PV) Power Measurement

Fig. 2 shows the the relationship between the output power and solar radiation per time unit of the solar panels for series-batteries charging. The measurement is conducted from 08:00 am to 17:00 pm. Solar power characteristic is increased simultaneously with the increase of solar radiation that is ranged from 320 W/m^2 to 1123 W/m^2 with the range of solar power increase from 24.4 W to 77.3 W.

3.2. Wind Turbine testing result

The wind turbine is placed at the land that closes to the shrimp pond. The height of the wind turbine is 10 m. The test is conducted to investigate the relationship between wind speed variation on rotor speed and its generated power. The result is demonstrated in Fig. 3.

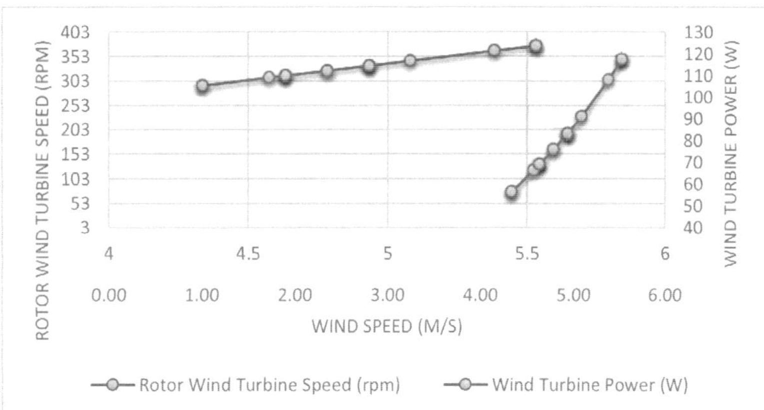

Figure 3. Wind Turbine Power Measurement

From Fig. 3, it can be seen that the wind speed is directly proportional to the generated power. The maximum rpm of the turbine is 374.11 rpm in maximum wind speed of 5.53 m/s. The maximum power generated is 117.11 W. Meanwhile, the minimum rpm (293.22 rpm) is reached at wind speed of 4.34 m/s. The generated power at this condition is only 56.46 W.

3.3. Hybrid Solar-Wind Turbine testing result

In this hybrid system, the average power generated of solar panels and wind turbine is 58.86 W and 82.80 W respectively. The generated power of the solar panels is continued based on the solar radiation from 08:00 am to 17:00 pm. However, wind turbine can only supply power momentary due to the intermittent characteristic of wind speed at the location. The total output power can be described from Fig. 4.

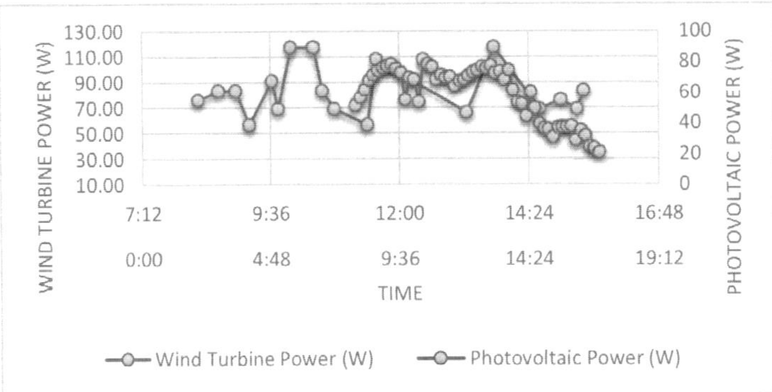

Figure 4. Output Power from the Hybrid System for Paddle Wheel Aerator

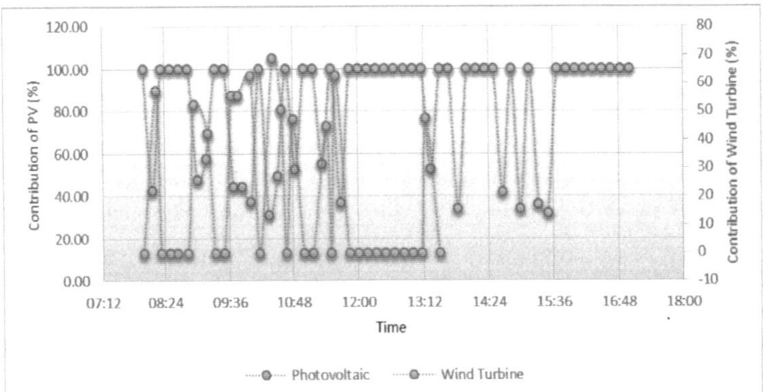

Figure 5. Energy Contribution by Wind Turbine - solar system for the Total Energy Production by the Hybrid System.

Fig. 5. exhibits the summary of energy contribution (energy share) of both solar and wind turbine. The total averaged generated power of the hybrid system is 526.90 Wh. From Fig. 5, it can be seen that solar panels contribute about 82.40% while the rest is by wind turbine. The diurnal contribution of solar panles to the hybrid system is 32.14% and it could supply maximum 100% in certain time. The maximum supply of solar panels occur during the wind turbine does not operate due to the very low wind speed. However, the contribution of wind turbine is varying from minimum 42.4% and maximum 68.67% depends on the wind speed condition at area.

4. Conclusion

The hybrid system has been installed and investigated for paddle wheel aerator application at shrimp's pond at Pancana Village. The use of average energy for the paddle wheel aerator application is 526.90 Wh. The amount of energy required to operate the BLDC motor for paddle wheel aerator is 420 Wh. Therefore, the required energy for the paddle wheel aerator using hybrid solar-wind turbine generator as the main energy sources can be fulfilled.

References

[1] Shiddiq Yunus, A.M., Abu-Siada, A., and Masoum, M.A.S., 2012, " Improving Dynamic Performance of Wind Energy Conversion System using Fuzzy-Based Hysteresis Current Controlled Superconducting Magnetic Energy Storage", IET Power Electronics, Vol 5, No.8, pp. 1305-1314.

[2] Yunus, A.M.S, Abu-Siada, A., and Masoum, M.A.S., 2011, "Effect of SMES on Dynamic Behaviors of Type-D Wind Turbine Generator-Grid Connected during Short Circuit", IEEE Power and Energy Society General Meeting, 6039276.

[3] Yunus, A.M.S, Abu-Siada, A., and Masoum, M.A.S., 2011, "Effect of SMES Unit on the Performance of Type-4 Wind Turbine Generator during Voltage Sag", IET Conference

Publication, 2011 (579 CP), pp. 94.
[4] Yunus, A.M.S and Saini, M., 2017, “Overview of SMES Units Application on Smart Grid Systems”, Proceeding- 2016 International Seminar on Intelligent Technology and Its Application, ISITIA 2016. 7828705, pp. 465-470.
[5] A.M.S. Yunus, A.Abu-Siada, M.A.S. Masoum, 2011, “Improvement of LVRT Capability of Variable Speed Wind Turbine Generators using SMES Unit”, Innovative Smart Grid Technologies Asia (ISGT), IEEE PES, pp. 1-7.
[6] M.Y. Khamaira, A.M.S. Yunus, A.Abu-Siada, 2013, “Improvement of DFIG-Based WECS Performance using SMES Unit”, Power Engineering Conference (AUPEC), Australasian Conference, pp. 1-5.
[7] A. M. S. Yunus and A. Wahdah, 2016, ”Modification and Testing of Wind Turbine with Double Savonius”, INTEK: Jurnal Penelitian 3(1), pp. 66-71.
[8] Banjarnahor D, Hanifan M, Budi E., 2017.Design of Hybrid Solar and Wind Energy Harvester for Fishing Boat.IOP Conf.Series:Earth and Enviromental Science 75(2017) 012007
[9] Barrass B and Derrett D, 2006, “Ship Stability” (Elsevier)
[10] Bhuiyan, M.M.H.; Asgar, M.A., 2003, “Sizing of a Stand-Alone Photovoltaic Power System at Dhaka”, Renew. Energy, 28, 929–938 .
[11] International Standard IEC 61400-2, 2006, “Part 2: Design Requirements for Small Wind Turbines”, International Electrotechnical Commision, Switzerland.
[12] Frank K and Gosmawi D, 2015, “Handbook of Energy Efficiency and Renewable Energy” (CRC Press).
[13] Munsell M, 2016, “ GTM research: global solar PV installations grew 34% in 2015 Greentech Media.

Acknowledgment
Authors would like to thanks Research, Technology, and Higher Education Ministry of Indonesia for funding the research through National Strategies Research Scheme 2018.

IOP Conf. Series: Materials Science and Engineering **619** (2019) 012034 doi:10.1088/1757-899X/619/1/012034

Fuel Production and Materials Synthesis by In-liquid Plasma

Shinfuku Nomura [1]
[1]Graduate school of Science and Engineering, Ehime University, Japan

Abstract. The in-liquid plasma process has been utilized to produce fuel gases from various liquids such as hydrocarbon and alcohol liquids, electrolytes, waste oil, biomass solutions, including glucose and cellulose, as well as to synthesize Zn, ZnO, WO_3, $Mg(OH)_2$ nanoparticles and diamonds. Using this technology, zero-emission hydrogen generation could become possible because it enables liquids to be directly broken down by plasma, making possible the decomposition of waste substances while producing useful materials. The application of the in-liquid plasma process for hydrogen production, decomposition of biomass, and synthesis of materials has already produced significant results. These highlight data will provide new information on "in-liquid plasma application.

1. Introduction

These Various methods have been proposed to date for discharge phenomena in liquids, such as DC / AC, pulse, high frequency, microwave, equilibrium plasma, non-equilibrium plasma, and so on. These discharge phenomena are called pulse or arc discharge in liquid, in-liquid plasma, solution plasma etc., depending on the discharge form and discharge method [1-4]. We call plasma in liquid "in-liquid plasma", using microwave (primarily 2.45 GHz MW) or high frequency radio generation (13.72 MHz or 27.12 MHz RF) [5].The generated plasma is called MW in-liquid-plasma and RF or High Frequency (HF) in-liquid plasma. In-liquid plasma is used broadly to refer to plasma in liquid. The reason we use plasma in these frequency bands is that they generate a gas phase plasma process for which most of the know-how regarding its application has been accumulated so far. Using this method, plasma generate bubbles within which the pressure is up to 200 times that of atmospheric while the surrounding liquid remains at room temperature and normal pressure [6-8]. In addition a supercritical state can be generated allowing its application range to be diversified. By using this reaction field, waste can be rendered harmless, decomposed, and materials can be synthesized.

Here, we will discuss a variety of applications of the in-liquid plasma method such as for hydrogen production from waste liquids or non-edible biomass, nanoparticle production for use in batteries and printed electronics.

2. Hydrogen production by the in liquid plasma process

2.1 Decomposition of hydrocarbon-based liquids

When plasma is generated in hydrocarbons, a chemical reaction is created that reaches 3500 K in atmospheric pressure which produces hydrocarbons hydrogen and solid carbon by thermal decomposition [9]. This technology can decompose various substances and recover hydrogen as a fuel gas. In the case of MW in-liquid plasma, it is relatively easy to generate in-liquid plasma simply by installing electrodes to increase the electric field strength in a conventional MW oven [9,10]. Table 1 shows the gas generation rate and the volume percentages of component gasses obtained by decomposition of hydrocarbon-based liquids using the in-liquid plasma process.

For hydrocarbons such as *n*-dodecane ($C_{12}H_{26}$), ideally it would be decomposed into hydrogen and solid carbon (13 moles hydrogen gas and 6 moles solid carbon), but in actuality, in addition to the target hydrogen gas, other hydrocarbon components such as methane gas and acetylene are generated. Since hydrocarbon-based liquids naturally contain a large amount of hydrogen, a method for safe storage and use of hydrogen is necessary. To that end, development of a compound called organic hydride is under way [11]. Oxides, such as CO and CO_2, are also generated when hydrocarbons are decomposed. The contents of the decomposition products depends on the components of the liquid, but for hydrocarbons or alcohols, 50% to 90% of the decomposition products is hydrogen gas, with the remainder being hydrocarbon components and solid carbon. While in-liquid plasma can be used as a recycling technology to make hydrogen from waste oil, energy is required for hydrogen production. From the viewpoint of extracting hydrogen from waste and using it as energy, EPR_{H2} the input energy to recoverable energy ratio [12,13], is found to be 0.16 for engine oil and 0.52 for methanol decomposition as is calculated by the following Eq.(1). Since energy is required to generate plasma, it naturally becomes 1 or less.

$$\mathrm{EPR}_{\mathrm{H2}} = \frac{n_{\mathrm{H2}} \times \Delta H_{H2}}{P} \quad (1)$$

Here, n_{H2} is the gas production rate of H_2 (mol / s), ΔH_{H2} is the heat of combustion of hydrogen (J / mol), P is input power of MW (W)

When water is pyrolyzed by plasma, hydrogen and oxygen are generated (part of the generated oxygen is dissolved in water, so that the ratio of the generated gas does not exactly become 2:1), the gas production rate υ is slower than other hydrocarbons and alcohol liquids. Comparisons of the gas production rate are shown in Table 1 where it can be seen that the in-liquid plasma method has a hydrogen gas production rate that is 30% - 100 % that of the electrolysis of alkaline water. [9,11-15]. Since the product gas from decomposition is not purely 100% hydrogen, a separation and recovery system is required for application to fuel cells. At present application of this method for large scale generation of hydrogen is not practical unless the gas production rate during decomposition is increased by several orders or the hydrogen purity is greatly improved. In order to apply this method to an actual machinery, it is necessary to develop a process such that total EPR becomes 1 or greater by advancing methods to recover usable solid carbon components or other valuable resources produced in the decomposition process.

2.2. Steam reforming experiment

The steam reforming method of natural gas is a technology established for commercially producing large amounts of hydrogen and is described by an endothermic reaction of Eq. (2) and a shift reaction of Eq. (3).

$$CH_4 + H_2O \rightarrow CO + 3H_2, \quad \Delta H_{298\,K} = +206\ \mathrm{kJ/mol} \quad (2)$$

$$CO + H_2O \rightarrow CO_2 + H_2, \quad \Delta H_{298\,K} = -41\ \mathrm{kJ/mol} \quad (3)$$

Table 1. Composition gasses and components for hydrogen production by various in-liquid plasma methods

Liquid	Method	Input power (W)	υ (ml/s)	Gas Yeild (%)							EPR_{H2}
				H_2	O_2	CO	CH_4	CO_2	C_2H_2	C_2H_4	
Engine oil	MW	700	10.9	82.0	0.2	0.6	1.8	0.1	13.4	1.9	0.16
n-dodecane[9]	MW	750	35.7	74.0	0.0	0.0	2.0	0.0	20.0	2.0	0.45
Methanol	RF	200	12.6	64.8	0.0	32.2	1.8	0.3	0.9	0.1	0.52
Cellulose (20 wt%)[15]	RF	150	0.3	45.6	13.1	17.4	1.2	18.1	0.3	4.3	0.01
Cellulose (40 wt%)[15]	RF	150	3.1	36.8	0	31.9	1.7	25.2	0.7	3.8	0.10
Pure water	RF	200	0.39	82.4	17.6	0.0	0.0	0.0	0.0	0.0	0.02
n-dodecane with Steam[13]	MW	250	12.5	73.8	0	12.9	2.5	0	6.8	3.9	0.47
Water-oil interface[14]	RF	250	5.3	62	0	23	3	2	6	4	0.32

From 1 mol of methane gas and 2 moles of water, 4 moles of hydrogen and 1 mol of carbon dioxide are produced. Therefore, the hydrogen production energy per 1 mol is significantly lower than that by electrolysis of water. By introducing water to oil decomposition when using the in-liquid plasma and inducing a steam reforming reaction, the amount of hydrogen generated is increased. The water may be introduced into the plasma reaction field directly from a cavity inside the electrode as shown in Fig. 1 (a) or by generating plasma near the interface between oil and water as shown in Fig. 1 (b). As shown in Fig. (a), water is supplied from the inside of the hollow electrode and becomes steam due the plasma which results in the generation of 12.9% of CO.

In addition, as shown in Fig. 2, this reaction increases the hydrogen gas production rate up to 1.4 times as compared to that without steam supply [13]. The comparison of with and without water steam in *n*-dodecane shown in Table 1 indicates that there was no difference in hydrogen purity, which means that the amount of hydrogen produced increased by steam introduction. However, for the method of Fig. 1, no shift reaction of CO and H_2O described in Eq. (3), which is found in the methane steam reforming method, was confirmed. Although the amount of steam supplied to the reaction field is small, MW absorbed by water as the amount of water passing through the electrode increases, the generation of plasma itself was difficult for 2.45 GHz MW plasma. On the other hand, in Fig. 2 (b), since RF of 27.12 MHz is used, there is no influence of water absorption, and when the tip of the electrode is installed on the lower water side, the influence of adhesion of carbon components to the electrode can be prevented. In this case, CO_2 is generated with hydrogen generation of about 60 to 71%, and the shift reaction can be confirmed. However, the amount of CO_2 generated is as small as 2%, conversely, it means that water is thermally decomposed, and the gas production rate during decomposition decreases.

Fig. 3 shows a chemical equilibrium diagram when the ratio of *n*-dodecane and water is (a) 10: 1 and (b) 1: 100. The gas temperature of the plasma at atmospheric pressure is about 3500 K. When water is not sufficiently supplied, only CO is generated, however, in (b), where the ratio of water is increased,

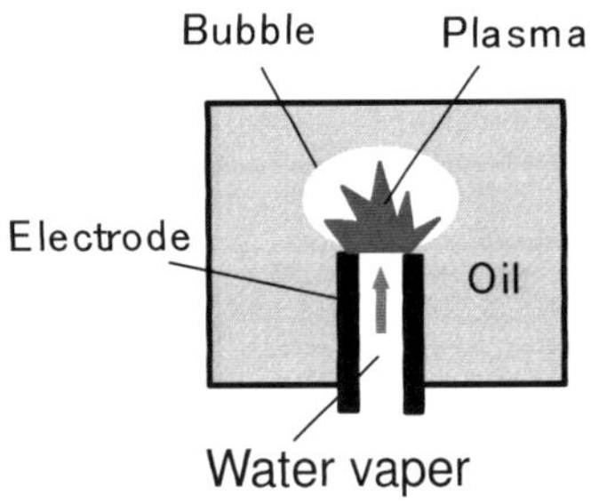

(a) Water vapor supply type plasma [13]

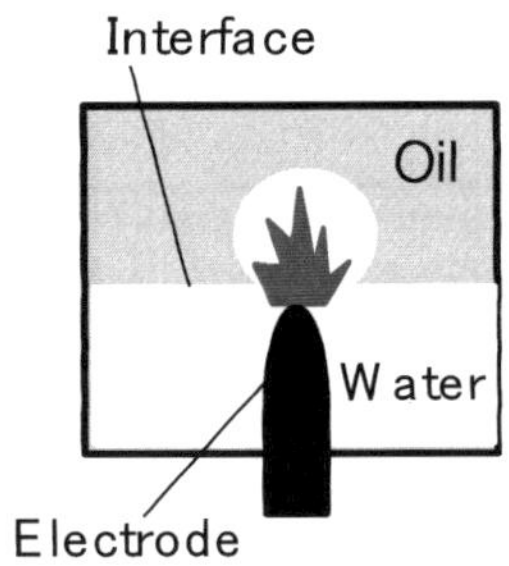

(b) Near oil water interface type [14]

Figure 1. Schematic diagram of plasma steam reforming in liquid

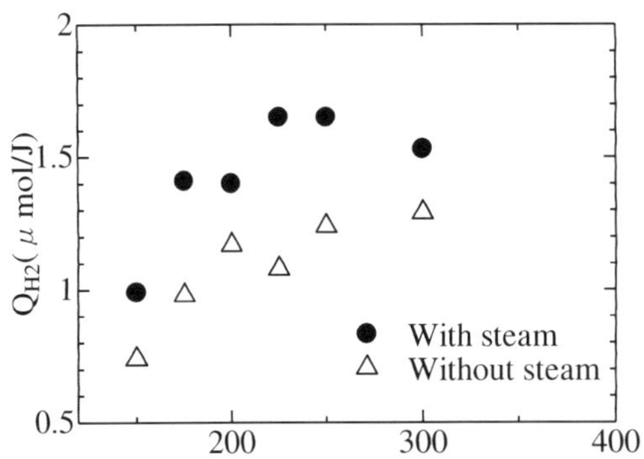

Figure 2. Hydrogen production efficiency by in-liquid plasma with steam reforming [13]

the generation of CO2 becomes conspicuous. Although the generation of CO and CO_2 depends on the amount of water in the reaction field, water is thermally decomposed within the plasma. When the amount of water is increased, energy is consumed for water decomposition. Therefore, in order to take advantage of the steam reforming reaction, it is necessary to adjust the amount of water supplied to the reaction field to optimize the steam reforming reaction.

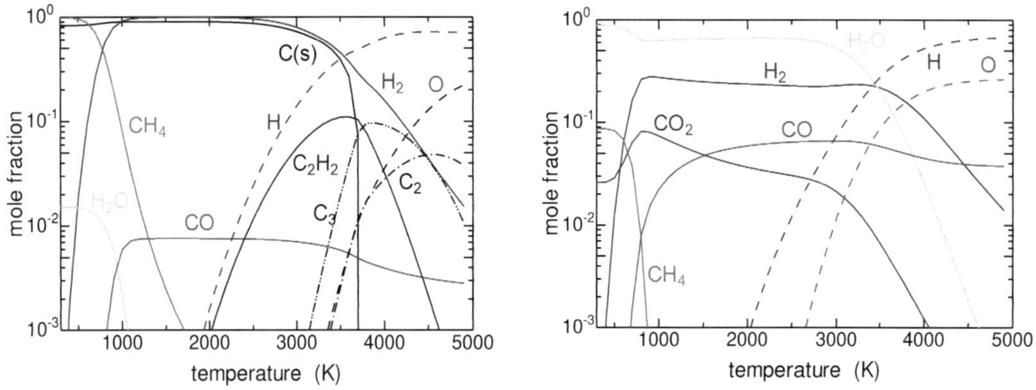

(a) $C_{12}H_{26}$:H_2O=10:1 (b) $C_{12}H_{26}$:H_2O=1:100

Figure 3. Thermal equilibrium composition of the mixture of *n*-dodecane and water [14]

2.3 Decomposition of cellulose

The results for the decomposition experiments carried out on cellulose powder are also shown in Table 1 and Fig. 4 for comparison. The amount of gas generated changes greatly depending on cellulose concentration. Since cellulose is insoluble, at a concentration of 27 wt% and below, the cellulose powder is dispersed in water. As the concentration of the cellulose increases, the relative amount of water decreases, so the diameter of the powder grains becomes smaller. Since cellulose is a nonvolatile substance, it does not enter into bubbles created by the heat of plasma evaporating the surrounding water. Therefore, decomposition of cellulose is an indirect decomposition of OH radicals existing around the bubble. At 30 wt%, the cellulose grains containing water (hydrated cellulose particles) are pushed out of the plasma region due to the pressure of bubble created by plasma generation and air gaps form in the plasma region without returning to their original positions due to the contact resistance between the particles with large diameter grains. Hence, the plasma cannot reach the cellulose grains containing water.

At 40 wt% and above, the cellulose enters a state of high viscosity, and the discharge of the plasma comes into direct contact with the cellulose powder, and the generation gas significantly increases. The fluidity disappears at 50 wt%, when the contact with the plasma is continued for an extended time resulting in the carbonization of the contact surface of cellulose and plasma. It may be possible to directly gasify cellulose itself using thermal plasma such as direct current arc discharge, however, the decomposition mechanism varies depending on the weight ratio of the substances dispersed in water for decomposition [12].

Figure. 5 shows a summary of the hydrogen production rate in the form of EPR_{H2} of Eq. (1) under various conditions. It is lowest in water, higher for cellulose and engine oil, and over 50% for methanol. Results for cellulose dispersed in an acid solvent such as diluted sulfuric acid or sodium hydroxide are shown in Fig. 5. In these solutions, the hydrogen production is larger than when cellulose powders are dispersed in water at the same weight ratio (20 wt%) [12]. The ions in the electrolyte collide with the electrode, resulting in emission of secondary electrons and the plasma region increases. The effect of sodium hydroxide is particularly remarkable.

Even with the same saccharide, monosaccharide glucose dissolves in the liquid, so it can be decomposed by plasma. When used in combination with ultrasound, decomposition is promoted and ultrasonic irradiation has a beneficial effect on plasma decomposition of saccharides [17].

2.4 Effect of catalyst introduction and type of catalyst

Because in-liquid plasma is generated within bubbles in the liquid, decomposition rates improve when bubbles are held in place by control plates installed above the electrodes as shown in Figs. 6 and 7. The effect of the shape and material of plates are summarized in Fig. 6. The use of copper or nickel for the electrode, and solid copper, solid nickel, porous ceramics, porous bronze and porous nickel for the control plate, have been investigated as to the effect of catalytic reaction. The gas generation rate for each combination of plate and electrodes have been compared. When the plate is placed above the electrode, the hydrogen production rate increases by 2.5 to 3.6 times [16]. The most effective device configuration in this experiment is a nickel electrode combined with a nickel porous plate. Separately, plasma decomposition of water was carried out under these configurations, however almost no differences were found for the combinations of these materials. The effect of nickel appears for methanol decomposition but not for water decomposition. However, the simple installation of a bubble control plate made of catalytic metal in in-liquid plasma reaction field improves the decomposition rate for compounds other than water. It is effective to construct the material of the electrode and plate from catalytic metals.

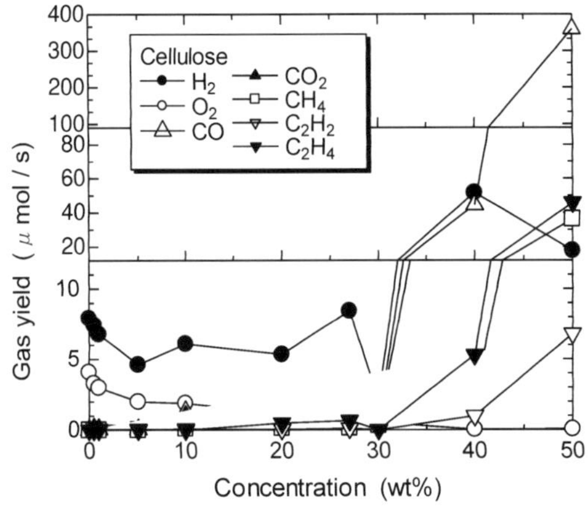

Figure 4. Gas production rate obtained by plasma decomposition of cellulose [15]

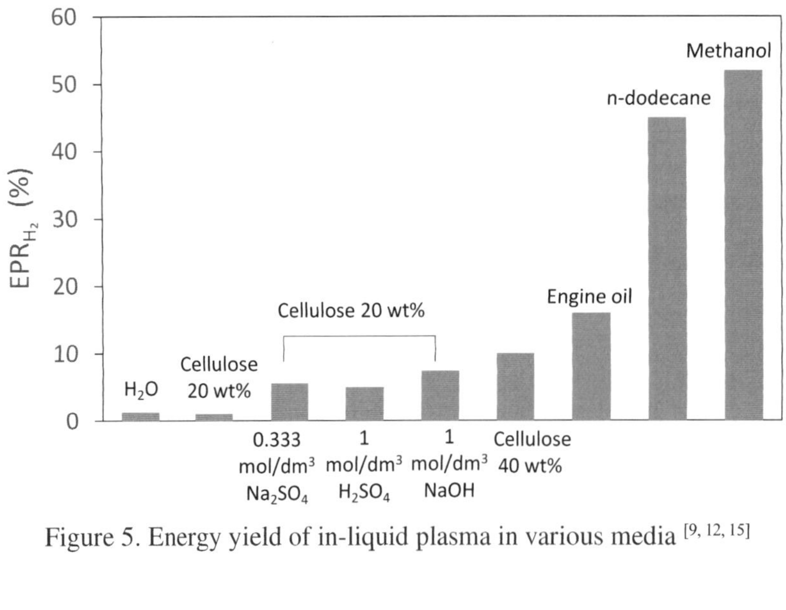

Figure 5. Energy yield of in-liquid plasma in various media [9, 12, 15]

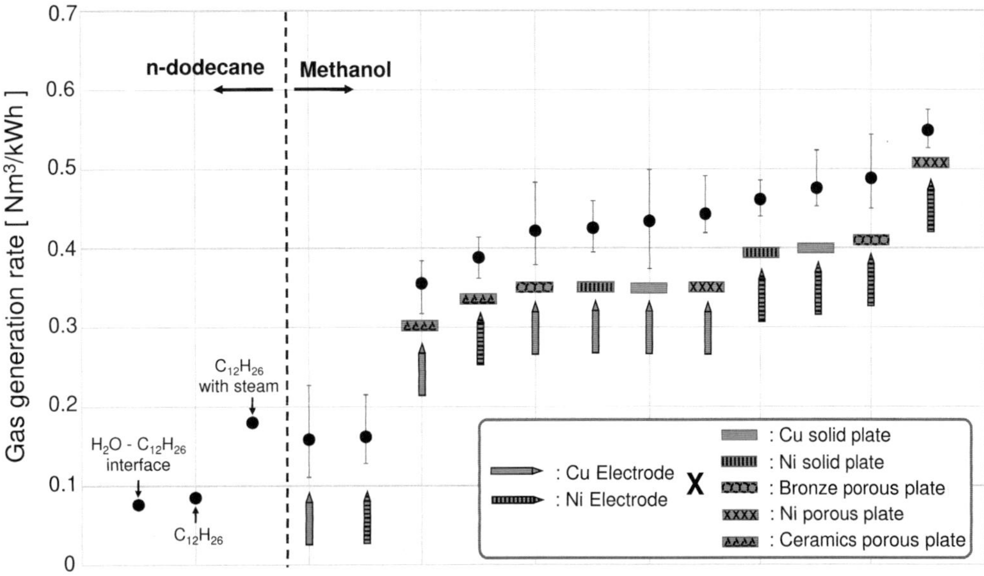

Figure 6. Comparison of gas production rate for each combination of electrode and plate [9, 12, 16]

3. Synthesis of metallic nanoparticles by in-liquid plasma

The synthesis mechanism of nanoparticles is obtained through processes of evaporation of metal by plasma which aggregates near the gas-liquid interface and precipitates out. At this time, the nanoparticles exist in a dispersed state in the liquid. The synthesis method of nanoparticles by in-liquid plasma is basically realized by inserting the raw metal to be converted into the plasma. If a metal with a low melting point is used for the plasma electrode, the electrode itself can be evaporated and it can be recovered, but then, since the condition of the plasma changes as the electrode is damaged, it is not suitable for synthesis of nanoparticles. Therefore, as shown in Fig. 7, there is a method for introducing metal as a thin wire into the plasma [18].

Nanoparticles of interest can be synthesized depending on the type of raw material and liquid utilized. The synthesis of nanomaterials by in-liquid plasma has been reviewed in detail in reference (3). For example, ZnO nanoparticles are synthesized by injecting zinc into plasma in pure water, whereas Zn particles are synthesized in plasma in ethanol [19]. It is generally known that ZnO nanoparticles become flower-type crystals [19], while Zn assumes a hexagonal crystal structure [20], however, after detailed analysis, Zn turned out to be cubic in liquid plasma (See Fig. 8).

It is believed that there is a severe quenching effect or intense temperature gradient in the reaction field of plasma submerged in liquid that makes it easier for a cubic structure to form, but this mechanism has yet to be clarified. Metal reduction by plasma can also be applied to recycle metal oxides after use such as Zn and Mg used as negative active materials in batteries [21]. However, the method for reducing a large amount of metal oxide depends on how best to bring these metals into contact with the plasma reaction field. Separately, there is already a method of reducing zinc oxide by using RF dielectric heating [22].

Table 2 summarizes the production of nanoparticles using in-liquid plasma including laser ablation. Compared with the laser ablation method, the in-liquid plasma method is much more effective, and it is convenient that nanoparticles can be recovered in a state of being dispersed in the liquid. A method for precisely controlling the particle diameter by controlling various plasma parameters has not yet been established. Synthesis of low-melting-point metal nanoparticles such as Sn or Cu by plasma in liquids has been reported [26].

4. Conclusion

At the gas-liquid interface exposed to the in-liquid plasma, active species are generated in which electrons are excited into various orbits. This environment can be used for a variety of purposes. For example, phenol can be synthesized in a single process by reacting benzene at the plasma interface in water [27]. This reaction is a simple one in which OH is bonded to the benzene ring. However, in the standard chemical process, this requires a plurality of processes using an acid catalyst and such. Whereas when in-liquid plasma is used, the radical reaction occurs as a simple chemical reaction. One of the advantages of decomposition and synthesis by plasma in liquid is that the expensive catalysts and microorganisms necessary for the chemical reaction processes for organic synthesis and decomposition of substances difficult to decompose are no longer required. This would eliminate the reliance on the use rare metals for radical reactions. We believe that it is important to be able to convert to a process that does not require rare metals even in fuel production and material synthesis processes using plasma in the future.

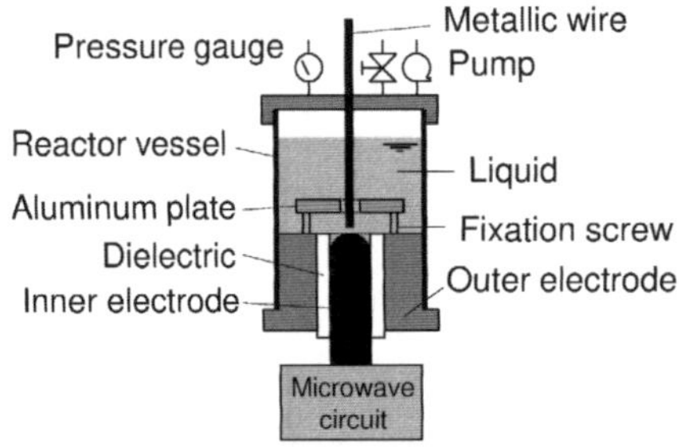

Fig. 7 Synthesis of nanoparticles by in-liquid plasma

Fig. 8 TEM image of zinc nanoparticles: (a) flower-type nanoparticles [19], (b) cubic zinc nanoparticles

Table 2

Particles	Target	Liquid	Productivity (g/hour)	Efficiency (μg/J)	Electrical source
$Mg(OH)_2$ [18]	Mg	Water	66	80.2	MW: 160W
Zn/ZnO [18]	Zn	Water	14	20.4	MW: 135W
Ag [18]	Ag	Water	0.8	0.74	MW: 220W
Zn [20]	Zn	Ethanol, Menthol	3.3	2.6	MW:250W
ZnO [24]	-	$Zn(CH_3COO)_2$	4.2	1	MW:800W
Zn/ZnO [25]	Zn	Tetrahydrofuran	0.016	5	YAG laser 125 mJ/pulse Repetition rate: 5kHz
WO_3 [23]	W	Water	15.5	13.65	MW : 220W

Synthesis of nanoparticles by plasma in liquid

References

[1] Bruggeman P J, *etal*, 2016 Plasma Sources Sci. Technol. **25** 053002
[2] Yang Y, Cho Y I , Freadman A, 2012 Plasma Discharge in Liquid, CRC Press, ISBN 9781138074958
[3] Horikoshi S and Serpone N, 2017 RSC Adv. **7** 47196
[4] Tsuchida A, Shimamura T, Sawada S, *etal*, 2018 Radiation. Phys. Chem. **147** 53
[5] Nomura S, Toyota H, Mukasa S, Takahashi Y, Maehara T, Kawashima A, and Yamashita H, 2008 Appl. Phys. Express, **1** 046002
[6] Nomura S, Mukasa S, Toyota H, Miyake H, Yamashita H, Maehara T, Kawashima A and Abe F, 2011 Plasma Sources Sci.Technol. **20** 034012
[7] Kawashima A, Nomura S, Toyota H, Takemori T, MukasaA and Maehara T, Nanotechnology, 2007 **18** 495603
[8] Rahim I, Nomura S, Mukasa S, Toyota H, 2015 Appl. Therm. Eng., **90** 120
[9] Nomura S, Toyota H, Mukasa S, Yamashita H, Maehara T, and Kawashima A, 2009 J. Appl. Phys., **106** 073306
[10] H. Toyota, S. Nomura, S. Mukasa, International Journal of Materials Science and Applications, **2** (2013) 83
[11] Alhumaidan F, Cresswell D, and Garforth A, 2011 Hydrogen Storage in Liquid Organic Hydride, Energy Fuels, **25** 4217
[12] Tange K, Nomura S, Mukasa S, Toyota H, Syahrial F, 2016 J. Jpn. Inst. Energy, **95** 1105
[13] Mochtar A A, Nomura S, Mukasa S, Toyota H, Kawamukai K, 2017 J. Jpn. Inst. Energy, **96** 86
[14] Yamane R ,Nomura S, Kawamukai K, Mukasa S, Toyota H, 2017 54th National Heat Transfer Symposium of Japan, CD-ROM H144 (in Japanese).
[15] Rahim I, Nomura S, Mukasa S, *etal*, 2015 Journal of Power and Energy Engineering, **3** 28
[16] Shiraishi R, Nomura S, Mukasa S, Nakano R, Kamatoko R, 2018 Int. J. Hydrog. Energy, **43** 4305
[17] Syahrial F, Nomura S, Mukasa S, Toyota H, Okamoto K, 2015 Int. J. Hydrog. Energy, **40** 11399
[18] Hattori Y, Nomura S, Mukasa S, Toyota H, Inoue T, Usui T, 2013 Journal of Alloys and Compounds, **578** 148
[19] Saito G, Nakasugi Y, Yamashita T, Akiyama T, 2014 Applied Surface Science, **290** 419
[20] Hattori Y, Mukasa S, Toyota H, Inoue T, Nomura S, 2010 Materials Letters, **65** 188
[21] Amaliyah N, Mukasa S, Nomura S, Toyota H, Kitamae T, 2015 Mater. Res. Express, **2** 025004
[22] Mukasa S, Udaka Y, Matsuzawa K, Doi N, Toyota H, Nomura S, 2017 J. Jpn. Inst. Energy, **96** 357
[23] Hattori Y, Mukasa S, Toyota H, Inoue T, Nomura S, 2011 Mater. Chem. and Phys., **131** 425
[24] Yonezawa T, Hyono A, Sato S, Ariyada O, 2010 Chem. Lett., **39**(7) 783
[25] Wagener P, Schwenke A, Chichkov B N, Barcikowski S, 2010 J. Phys. Chem. C, **114** 7618
[26] Saito G, Wan Omar Sidiq Bin Wan Mohd Azman, Nakasugia Y, Akiyama T, 2014 Advanced Powder Technology, **25** 1038
[27] Agung M , Nomura S, Mukasa S, Toyota H, Otsuka K, 2017 Plasma Sci.& Technol. **19**, 055503

IOP Conf. Series: Materials Science and Engineering **619** (2019) 012035 doi:10.1088/1757-899X/619/1/012035

Production of carbon dioxide, methane and nitrous oxide gas emissions in biodegradation of waste oil sludge

Fahruddin

Biology Department, Faculty of Mathematics and Natural Sciences, Hasanuddin University, Makassar, Indonesia

Email: fahruddin_science@unhas.ac.id

Abstract. Biodegradation Bioremediation of waste oil sludge through biodegradation using microorganisms still has an impact on the environment, because produces of gas emissions which causes to the global warming. The purpose of this study was to determine the production of carbon dioxide, methane, and nitrous oxide gas in biodegradation of waste oil sludge by using types of bioactivator from soil as the microbe inoculums sources was compost, mangrove and forest soil with adjusting the ratio of C:N:P of 360:60:1 to optimize the biodegradation condition. The results show that the addition of bioactivator on the treatment of waste oil sludge can increase the biodegradation process. The results show that the compost bioactivator treatment is 72 percent decrease in oil sludge, the forest soil bioactivator treatment of 58 percent and then the mangrove soil bioactivator of 46 percent. The observation of gas production obtained the highest concentration of carbon dioxide was produced on compost bioactivator treatment of 0.91 ppm on the 25 day, production of methane on compost bioactivator treatment was 0.66 ppm on the 25 day and production of nitrous oxide on the mangrove soil bioactivator treatment of 0.19 ppm on the 30 day; The decrease of hydrocarbon waste oil sludge and increased gas production are also associated with an increase in the population number of hydrocarbon degrading bacteria at all bioactivator treatments.

key words: gas emission, biodegradation, oil sludge

1. Introduction

One of the major problems faced by oil refineries is the safe disposal of oil sludge that is generated during processing of crude oil. Improper disposal of oil sludge leads to environmental pollution, particularly soil contamination and brings serious threat to the ground water. Many constituents of oil sludge are carcinogenic and potentially immunotoxicants [1]. Among many techniques used to decontaminate the affected sites, in situ bioremediation is the most widely used [2,3].

Bioremediation offers an alternative way that is not only effective in cost but also acceptable to the environment. Landfarming with using soil type is one of the advanced bioremediation methods that effective for contaminated soil of oil sludge. In this process, oil sludge from refinery waste is mixed with soil and subjected to enhance in situ bioremediation; inorganic fertilizer is applied to the site to provide fixed nitrogen, and phosphate and pulverized limestone ($CaCO_3$) are added to raise the pH of the soil-oily sludge mixture about 7.8[4,5].

However, in the biodegradation process of oil sludge, it will produce several gases formed by the decomposition both soil total organic from plant material and the components of hydrocarbon contained in oil sludge. Gas emission that is produced in the biodegradation process to CO_2, CH_4 and N_2O and many trace gases such as H_2, H_2S, OCS and NO [6].

CH_4 and CO_2 are the dominant gases as the product of organic matters decomposition, while emission of N_2O depends on the form of N that is found in soil, whereas the other trace gases are only intermediates and are released into the atmosphere [7,8]. Gases emission of CO_2, CH_4 and N_2O are included the main component that reduce the loss of outgoing infrared radiation [9].

Its Gases are relevant for atmospheric chemistry and climate that will lead to the higher temperatures at the earth surface leading to global warming [10,11].

Therefore, soil microbial activities are also the important source of the greenhouse gases [9, 12]. The third gases are interesting to be investigated in biodegradation process because they provide significant effect to the greenhouse effect.

2. Materials and Methods

2.1. Sampling.

There are three types of soil used as an inoculum source; soil forest; soil mangrove and compost. The soil was air dried, crushed, and sieved (mesh size, 2mm), and stored at 20^0C in plastic bag. Previously, the soil with pH 6.8 to 7.0 and with $Ca(OH)_2$ was used. oil sludge sample was collected from an open storage pit at the refinery at Unit Pertamina Indonesia, and it was stored in sealed plastic containers at a room temperature.

2.2. Characterization of soil and oil sludge

Physical and chemical properties of each sample of soil type was collected. Soil samples were analyzed for organic carbon, nitrogen, available phosphorus and moisture level [13]. The analysis of organic carbon was also conducted on oil sludge. Determination of chemical properties was to optimize the biodegradation condition.

2.3. Biodegradation experiments

Biodegradation treatments soil were mixed with oil sludge ratio 1:1 based on its C organic content. Nitrogen and phosphorus content of the mixture were adjusted to reach a ratio of C:N:P as 360:60:1 by adding NH_4NH_3 and K_2HPO_4 mineral nutrient respectively. Water content of the mixture was maintained at 90% of water holding capacity. This treatments were carried out in microcosm landfarming system [2].

Microcosm was created by placing the biodegradation treatments in flakes and was temporarily sealed with rubber stopper that was equipped with a rubber septum and air samples. Each microcosm consisted of 75 g biodegradation treatment. These treatments were incubated at room temperature for thirty days and there were three replications for each treatment. The biodegradation treatments were as follows: Concentration of oil sludge, gas emission, and hydrocarbon degrading microbes were determined every five days.

2.4. Determination of oil sludge concentration

Determination of oil sludge concentration of each treatment biodegradation were extracted by the following methods: 10 g of biodegradation treatments was mixed with 7 g of anhydrous Na_2SO_4 and extracted with diethyl ether in soxhlet extractor for 10 h. It was extracted and dried with an additional 0.1 g anhydrous Na_2SO_4 and transferred, with rinsing, to a tared vial. Evaporation was conducted at room temperature and the amount total of oil sludge hydrocarbon recovered was determined gravimetrically.

2.5. Gas emission analysis

Gas samples were collected using gas syringe. The samples were immediately injected into a gas chromotograph; CO_2, CH_4 and N_2O gas were detected with flame ionization detector. CH_4 and CO_2 were analyzed on an SRI GC (SRI Instruments, Torrance, Calif.), which was equipped with a flame ionization detector (FID) and a Hayesep D column (length, 2 m; 80/100 mesh). Helium was used as the carrier gas (flow rate, 20 ml. min^{-1}) and also synthetic air (250 ml min^{-1}) and H_2 (20 ml · min^{-1}) were used as burning gas. The oven temperature was 80^0C. Calibration was performed at each sampling occurrence by triplicate injection of 1,000 ppm of CH_4 in N_2 (Messer Griesheim). Standards were prepared by adding defined amounts to serum bottles of known volumes [7,10]. N_2O was analyzed with a gas chromatograph that was equipped with an electron capture detector and a Porapak Q-80/100 column (Supelco, Bellefonte, Pa). N_2O concentrations were expressed as the amount of N_2O in the gas phase plus the amount of N_2O [10].

2.6. Enumeration of microbial population

Enumeration of oil-degrading microbes used Bushnell-Hass agar (1.0 g KH_2PO_4, 1.0 g K_2HPO_4, 1.0 g NH_4NO_3, 0.2 g $MgSO_4$ 7 H2O, 0.05 g $FeCl_3$, 0.02 g $CaCl_2$, 20 g agar per liter). After steam sterilization, 5% of tween-80 was added into the medium of BH agar. Then, crude oil was mixed into BH agar and stirred by a magnetic stirrer. The medium of BH-oil agar was poured when it was cooled 50^0C. Serial dilution of sample were prepared and spread onto BH-oil medium plates. After incubation at 30^0C for 7 days, the colony was measured [14].

3. Results and Discussion

Determination of chemical properties on the soil sample including oil sludge was conducted to optimize the condition of biodegradation treatments, the results were analyzed in Table 1. such as; the content of nitrogen and phosporus of the mixture was adjusted to reach a ratio C:N:P as 360:60:1 by adding NH_4NH_3 and KH_2PO_4 mineral nutrient respectively.

Table 1. Chemical properties of each soil types and oil sludge sample

Soil type	Organic carbon	Nitrogen	Available phosporus
Soil of mangrove	31.9345	0.14	0.26
Compost	26.6766	0.08	0.45
Forest soil	34.6429	0.03	0.13
Oil sludge	62.8	0.067	0.19

From the analysis on total oil sludge hydrocarbon residue, it showed that all treatments had started to decrease slightly on day 5 where the highest was 14% at mangrove soil treatment. Then on day 10, it started experiencing significant concentration decrease especially at compost treatment compare with mangrove soil and forest soil treatment. The biodegradation process continued until the day 30 and remained on the compost treatment that showed a higher biodegradability indicated by the remaining oil sludge hydrocarbon residue of 28%. Then it was followed by mangrove treatment 42% and forest soil 63%, while on the control treatment there was a gradual decrease in the total hydrocarbon residue up to the 30th reaching 6% as shown in Figure 1.

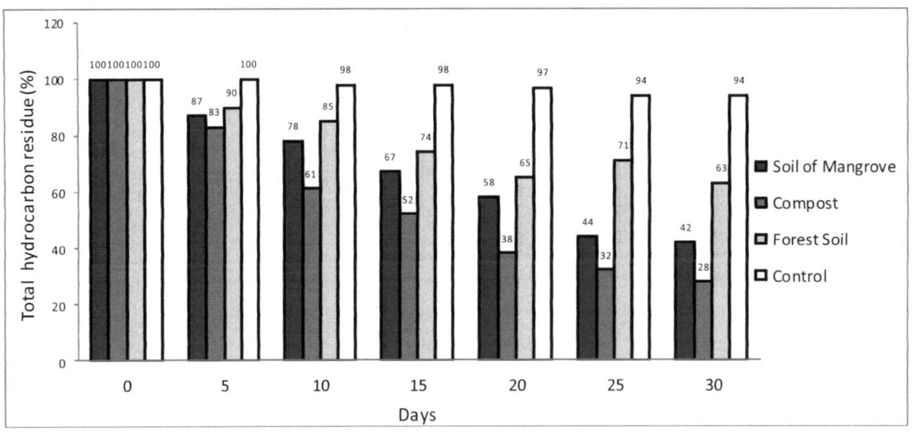

Figure 1. Total hydrocarbon residue in biodegradation of oil sludge with soil types are soil mangrove, compost and forest soil.

From the observation, it shows that on all treatments, biodegradation processes in oil sludge hydrocarbon occurred due to the presence of a number of microbes found in the soil type used as the inoculum source of degrading microbes [2]. The microbes present in this type of soil are easy to degrade the hydrocarbon oil sludge as petroleum hydrocarbon is similar to the molecular structure of the plant parts contained in the soil type. For example, hexane shares similarity with waxy substances in leaf, while aromatic hydrocarbon compound is similar to the lignin molecule structure in wood. This causes the microbes present in the soil type to be more enzymatically adaptive [15,16]. On the other hand, the loss of hydrocarbon residue in the control treatment occurred because abiotic loss process that is caused by environmental factors such as temperature, light or pressure that might cause hydrocarbon evaporation in oil sludge.

The addition of compost can give a great reduction of total hydrocarbon, followed by the addition of forest soil and mangrove soil. Production of CO_2 significantly increased at the 10th until 25th day after treatment. CH_4 gas was produced at the day and continuously increased up to the 30th day. While NO_2 gas that was treated especially with compost produced more gas than the other treatments that only produced gas at 20th day. Gas emission produced during biodegradation at forest soil of CO_2 was higher, which began at 10th day with 0.291 ppm and then decreased at 25th day. CH_4 was produced with high concentration at final incubation, but N_2O was lower than CH_4 as shown in Figure 2.

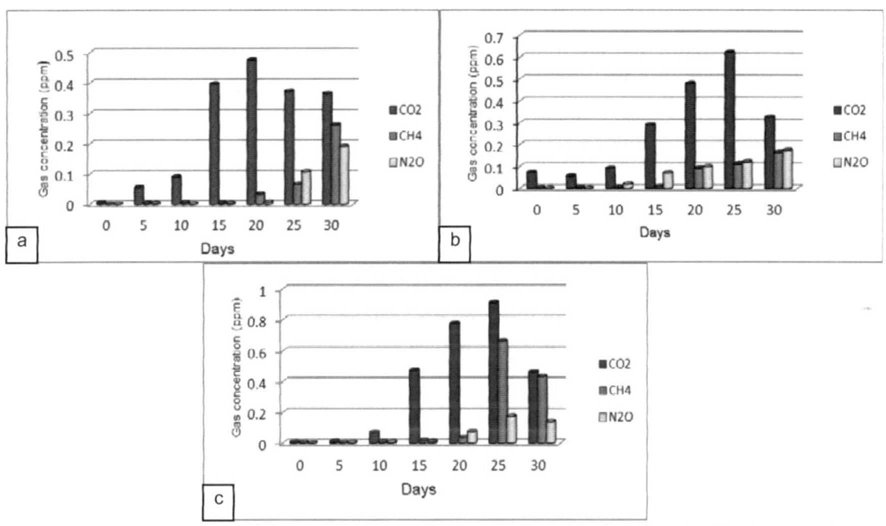

Figure 2. Production of CO_2, CH_4 and N_2O gas emissions in biodegradation treatment with (a) soil of mangrove. (b) forest soil; and (c) compost

The highest CO_2 gas was produced on the treatment of compost 0.91 ppm at the 25th day. The highest CH_4 gas on the compost treatment was 0.66 ppm at the 25th day and the highest N_2O was produced on the mangrove soil treatment was 0.19 at the 30th day. Observation of gas emission CO_2 of mangrove concentration that generated from 10th day to 25th day increased at 20th day. Even though the CO_2 was still the highest, by using compost treatment, CO_2 concentration increased at 15th day and N_2O increased at 30th day, so it was higher than mangrove soil and forest soil.

The process of hydrocarbon biodegradation of oil sludge is related to the growth of microbes and the resulting gases. The population of hydrocarbon degrading bacteria increased at 6th day after treatment. A significant increase in hydrocarbon degrading bacteria population was observed from the compost treatment up to the 30th day. While the treatment with mangrove soil and forest soil decreased at the 18th day as shown in Figure 3.

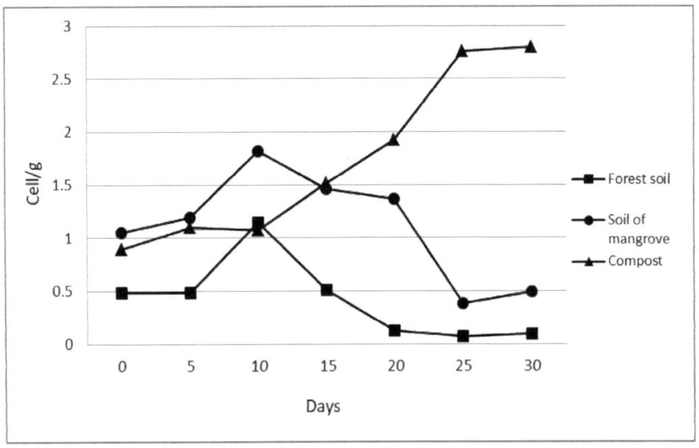

Figure 3. Number hydrocarbon degrading microbe of biodegradation treatment with forest soil, soil of mangrove and compost

The results of observation on all microbes that can degrade oil sludge hydrocarbon show that the increasing growth began at the 12 days in each treatment, except on forest soil that began to increase on the 6th day. It proves that microbe that can degrade oil sludge hydrocarbon need time to adapt with new subtract of oil sludge hydrocarbon. It also might occur due to the pressure to the microbe as the selection by the toxic component will decrease the population and diversity, so it gives the effect on diauxic growth of microbe [16,17].

Oil sludge is waste product of oil refinery activities that can badly affect the environment. Biodegradation is an effective method to solve this problem. In its application, soil is often being used as inoculum source. However, depending on vegetation growing on it, the content of organic material of each soil type has its own uniqueness. It affects the population and diversity of soil microbes in it [2,18]. The biodegradation process of oil sludge to produce CO_2, CH_4, N_2O and other gases emission is by decomposition both soil organic matter and hydrocarbon component contained in oil sludge [14].

Mineralization of organic component that comprises from inoculum source have been done by the microbe that can produce CO_2, CH_4, N_2O or any other gases. The gases produced are various depend on type of soil that is used as inoculum source. Different kind of soil contains unequal amount and type of microbe. In the other hand, they also include the compound of unequal organic composition. The simple organic molecules will crash sooner. Various inoculum source addition will affect the mineralization of oil sludge hydrocarbon [16]. Therefore, it will be seen from every type of various inoculum source addition and different gasses produced. The observation result showed that all treatments produced higher amount of CO_2. Here CH_4 and NO_2 gas were relatively in small amount, except in the second compost soil treatment in which these gases were shown in higher amount. It occurred because the compost soil contained a lot of microbe with many types. It also occurred on higher amount of nitrogen compound what was caused by higher level production. The higher level production of CH_4 gas was caused by anaerobic process that was made possible based on the compost soil texture itself.

The observation result on total residues of oil sludge hydrocarbon in each treatment of the added soil in oil sludge showed that the decreasing peak was at the 24th day of observation until the end stage of incubation especially on compost treatment. If it was closely related with gas production because it started on the 20th day. CO_2, CH_4 and NO_2 kept increasing during the observation. Even though it was not followed by the growth of microbe that could degrade oil sludge, the total amount of microbe still increased relatively. The possibility of gas production also came from the reduction of soil organic sources added, especially NO_2 gas source from denitrification of soil organic nitrogen [19,20].

Therefore, in this study, the highest emission gas is from the soil of compost inoculum and then, forest soil and mangrove soil are still relatively lower in gas production. However, in the other hand the ability of compost degradation is higher.

4. Conclusions

During the hydrocarbon oil sludge biodegradation process, several gases were produced is CO_2, CH_4 and N_2O. The increase in oil sludge biodegradation process was along with the increase in gas produced where CO_2 was the highest at compost treatment by 0.91 ppm, while CH_4 was 0.66 ppm on day 25, and N_2O was the highest on mangrove treatment by 0.19 ppm at day 30. The decrease of hydrocarbon oil sludge and increase the amount of gases generated are also accompanied by an increase in the number of hydrocarbon degrading microbes at all treatments.

References

[1] Mishra, S., J., Jyot, R.C., Kuhad, and B, Lal. 2001. Appl.Environ.Microbial 67 (4):1675 - 1681.
[2] Dibble, J.T. and R. Bartha, 1979. Appl.Environ.Microbial 37 : 729 – 737.
[3] Song, H. and R. Bartha, 1990. Appl.Environ.Microbial. 56(3) : 646 – 655.
[4] Eweis, J.B., S.J. Ergas, D.P.Y. Chang, and E.D. Schroeder, 1998. Bioremediation Principles. WCB, McGraw-Hill.Boston.
[5] Efsun, E. F. Olcay, T. Şağban and H. S. Başkaya, 2016. Turkish Journal of Agriculture and Forestry 40: 631-641.
[6] Shrestha M., W.R. Abraham, P.M. Shrestha, M. Noll and R. Conrad, 2008. Environ Microbiol. 10:400–412.
[7] Hardy K.R. and G.M. King, 2001. Appl. Environ.Microbial. 67 (8): 3671 – 3676.
[8] Lori, B, B.Peter, B.Philippe, D. Edward and P.Tans, 2017. Journal of Geophysical Research-Atmospheres 7 (122): 4070–4083.
[9] Trumbore S. 2006. Global Change Biol. 12:141–153.
[10] Matthies, C., A. Griebhammer and M.Schittroth, 1999. Appl. Environ.Microbiol. 65 (8): 3599 – 3604.
[11] Falkowski P.G, T. Fenchel and E.F. Delong, 2008. Science 320:1034–1039.
[12] Wang C., D.Y.F. Lai, J. Sardans, W.Wang, C. Zeng and J. Peñuelas, 2017. *PLoS ONE* 12(1): e0169254. https://doi.org/10.1371/journal.pone. 0169254.
[13] Nelson P.W. and C.E. Sommers, 1996. Total C, organic C and organic matter. In: PAGE, A.L. (Ed.). Methods of soil analysis. Part 2. Chemical methods. Madison: SSSA. pp. 539-579.
[14] Bagga J, M. Pandey and V. Pandey, 2015. International Journal of Innovative Research in Science, Engineering and Technology.4(10): 9993–10005.
[15] Christos G., M.C. Joanna, and J.S. Liz, 2014. J. Sci Food Agric. 94(12): 2362–2371
[16] Fahruddin and A.Abdullah, 2015. Asian Journal of Microbiology, Biotechnology and Environmental Sciences17(2): 321-324.
[17] Mamitha K.S., I. Grasian I. and Albert H.S., 2013. Bioinformation 9 (3):150–157.
[18] Langenheder S. and J.I. Prosser, 2008. Environ Microbiol. 10:2245–2256.
[19] Brüggemann, N., P.Rosenkranz, H. Papen, and K. Butterbach-Bahl, 2003. Geophys. Res. Abstr.,(5) 08463.
[20] Thomsen, T.R., Y. Kong and P.H. Nielsen, 2007. FEMS Microbiol Ecol. 60:370–382.

Acknowledgement

The writers would like to thank the Director research center of Hasanuddin University, Makassar, Indonesia on the implementation of this research.

Experimental Study for Measuring Density of slag melts by An Established Archimedean Double-bob Method

Arman[1] **and Hiromichi Takebe**[2]

[1] Department of Mechanical Engineering, Politeknik Negeri Ujung Pandang, Indonesia
[2] Department of Materials Science and Biotechnology, Graduate School of Science and Engineering, Ehime University, Japan

Abstract. The apparatus of Archimedean double-bob method for practicing accurate was established for density measurements of slag melts. The accuracy of density measurements at room temperature to establish measurements was determined in the following ways. Firstly, the densities of water were measured at 25, 30, 35, and 40 °C. Secondly, the densities of silicone fluids with the viscosities 0.01, 0.999, 4.850, and 29.44 Pa·s were measured at 25 °C. The measured densities have good agreement with the data of water standards with a precision of 0.07% and silicone fluid standards with a precision of 0.16%. The densities of Na_2O–SiO_2 slag melts with 60, 70 and 80 mol % SiO_2 were measured in the temperature range from 1350 to 1500 °C. The density decreases with increasing temperature and increases with increasing SiO_2 contents for all compositions. The reproducibility of this work is within the error estimation of density measurements. The density results of Na_2O–SiO_2 melts with 60, 70 and 80 mol % at 1400 °C show that the measurements data are within 0.23 % of the density values.

Keyword: Archimedean double-bob, density, slag melt, temperature, composition

1. Introduction

Physical property data e.g., density, viscosity, surface tension for coal slags are required in the design, operation, and the development of gasifier in the IGCC (Integrated Gasification Combined Cycle) [1]. The density of slag melts is one of fundamental and important properties for IGCC process as shown in Figure 1 [2,3]. However, limited studies on reliable data of density have been reported for coal slag melts [4]. Furthermore, the chemical compositions of coal slag melts from different origin countries can vary significantly and consequently the density of these coal slags can show appreciable variability [5].

Density of melts is necessary for the calculation of other properties, such as molar volume and coefficient of volume expansion in order to discuss the molecular structure of slag melts [9–11]. The coefficient of volume expansion also is one of important physical properties to control the melting process of slag melts [9].

The density measurements at high temperatures have been mainly carried out using maximum bubble pressure, pycnometer and Archimedean single- or double-bob methods [10]. The Archimedean methods give reproducible data sets with better precision [11]. Therefore, the Archimedean double-bob method is selected in this work. This convenient technique was established presently, and temperature dependence of the density was examined. The objective of this research describes an established Archimedean double-bob method for measuring the melt density of water, silicone fluid and slag melts in silicate system containing Na_2O–SiO_2 (mol%).

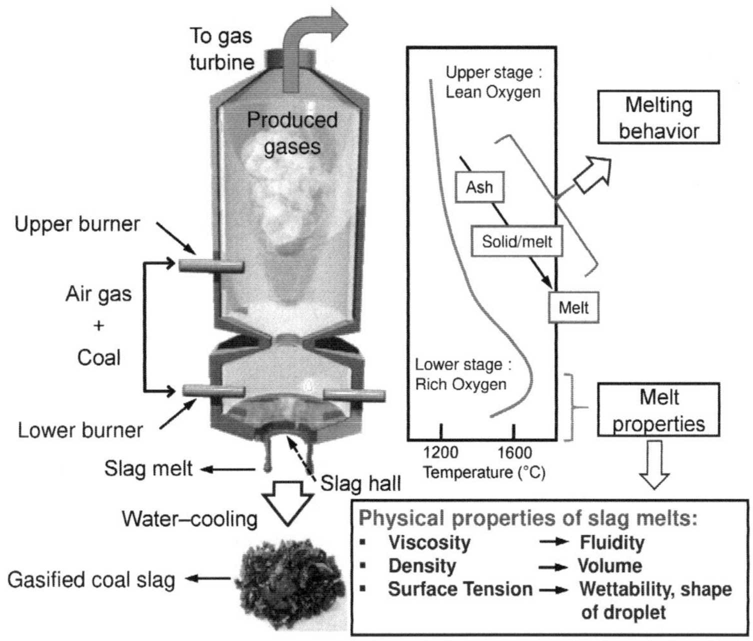

Figure. 1 The main physical properties of slag melts and schematic view of temperature distribution in the gasification furnace.

2. Principle of Measurements

Density of melts were measured using the Archimedean double-bob method. The Principle of the Archimedean double-bob method is to measured buoyancies of two platinum bobs with different volume immersed in melts as shown in Figure 2 The buoyancy *(W)* against the volume *(V)* of the bob immersed in a glass melt is determined by the following equation [12].

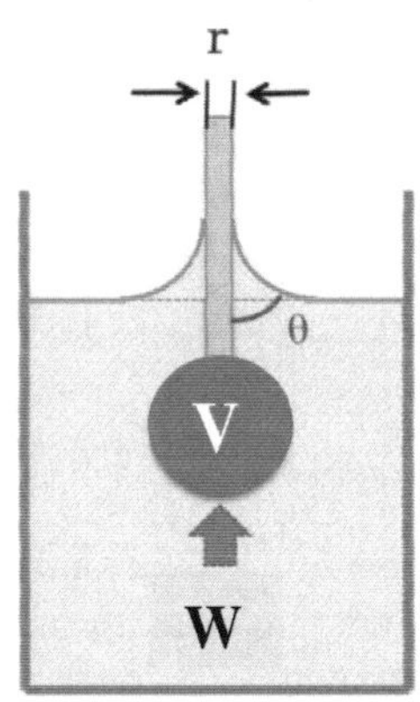

Figure. 2 Archimedean Method

$$W = V\rho - 2\pi r\gamma \cos\theta + A \qquad (1)$$

where.

W is the buoyancy of the bob immersed in the melt (10^{-3} kg)
V is the volume of the bob immersed in the melt (10^{-6} kg)
r is the radius of the suspending wire (10^{-3} kg/m^3)
ρ is the density of the melt (10^{-3} kg)
Υ is the surface tension of the melt (10^{-3} N/m)
θ is contact Angel between the suspending wire and the melt (°)
A is the constant that is related to the adhesion and convection of air (10^{-3} kg)

The buoyancy *(W_1)* against the volume *(V_1)* of the first (large) bob and the buoyancy *(W_2)* against the volume *(V_2)* of the second ball (small) bob immersed in a glass melt are determined by Figure 3 and the same equation.

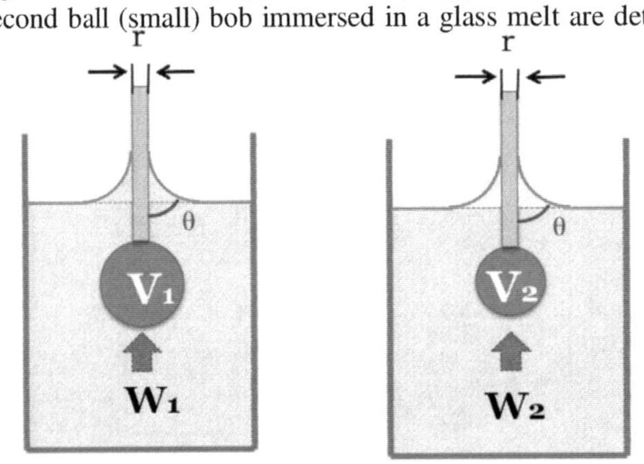

Figure. 3 Archimedean Double-bob Method

$$W_1 = V_1\rho - 2\pi r\gamma\cos\theta \quad (2)$$

$$W_2 = V_2\rho - 2\pi r\gamma\cos\theta + A \quad (3)$$

From Eq. 2 and 3, the density a glass melt can be expressed by the following equation.

$$\rho = \frac{(W_1 - W_2)}{(V_1 - V_2)} \quad (4)$$

where.

W is the buoyancy of the bob immersed in the sample (10^{-3} kg)
V is the volume of the bob immersed in the sample (10^{-6} kg)
r is the radius of the suspending wire (10^{-3} kg/m^3)
ρ is the density of the sample (10^{-3} kg)
θ is the acceleration of gravity (m/s^2)

Because the same suspension wire is used for each buoyancy measurement, the effect of surface tension, adhesion and convention on the suspension wire are identical, and can be eliminated. The volume of each bob at the measuring temperature was calculated from the following equation.

$$V'_1 = V_1 \times (1 + 3\alpha\Delta T) \quad (5)$$

$$V'_2 = V_2 \times (1 + 3\alpha\Delta T) \quad (6)$$

where.

V' is the volume of the bob at the measuring temperature (10^{-6} kg)
V is the volume of the bob at the room temperature (10^{-6} kg)
α is the thermal expansion coefficient of bob (1.0102 x 10^{-5} K^{-1}, 300-1273 K)
ΔT is the temperature difference between the measuring temperature and room temperature

The Volume of each bob at the room temperature was determined by measuring buoyancy of each bob in distilled water, of which the density is known, at the room temperature.

From Eq 1 and 2, the density a glass melt can be expressed by the following equation.

$$\rho = \frac{(W_1 - W_2)}{(V_1 - V_2)} = \frac{(W_1 - W_2)}{(1 + 3\alpha\Delta T) \times V_1 - V_2} \quad (7)$$

And then the final equation.

$$V_1 - V_2 = \frac{W_1 - W_2}{\rho_w} \quad (8)$$

where

V_1, V_2 are Volume of the large and small bobs immersed in the melt
W_1, W_2 are Buoyancy of the large and small bobs immersed in the melt
ρ_w is Density of purified water at room temperature

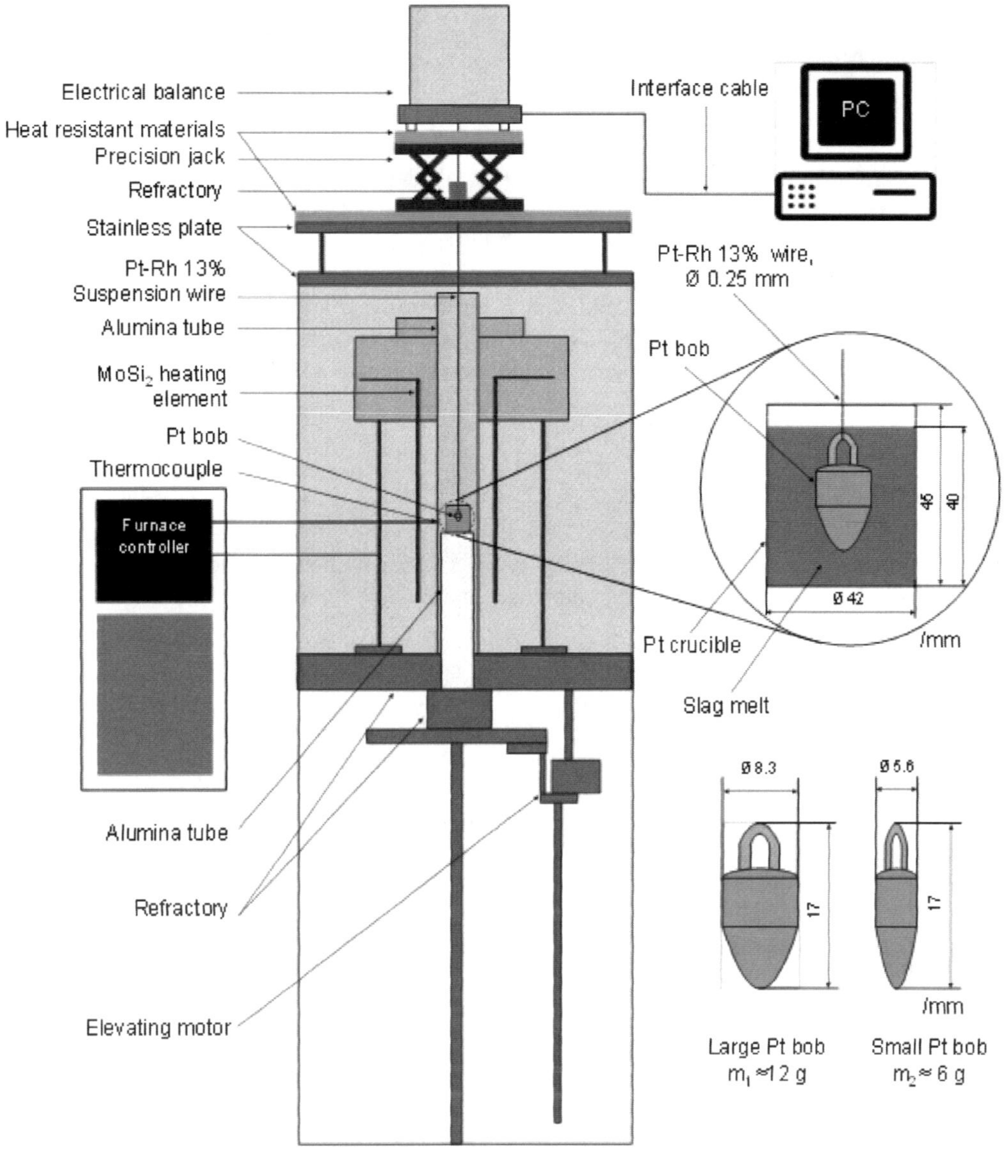

Figure. 4 Schematic apparatus for density measurements of slag melts: **(a)** whole diagram, **(b)** sample, and **(c)** large and small bobs.

3. Materials and Methods

The samples of density measurements were Na_2O-SiO_2 slag melts with 60, 70 and 80 mol % SiO_2. The samples were prepared from a mixture of raw materials of Na_2CO_3 and SiO_2 powders with 99.99 % purity. The mixture was melted at 1400 ºC for 2 hours in a Pt crucible placed in a $MoSi_2$ box furnace. The slag melt was poured onto iron mold for cooling as a slag sample. The sample was crushed and then re-melted in the air for density measurements in the temperature range from 1350 to 1550 °C.
The sample of density measurements of water and silicone fluid were measured at room temperature 25 to 40 °C. The samples of silicone fluid provided by the Silicone viscosity standard fluids of Brookfield. The chosen sample similar with the viscosity of slag melts of 0.01, 0.999, 4.850, and 29.44 Pa·s

The apparatus of Archimedean double-bob method used is similar to that[13], [8], [14] and is illustrated in Figure. 4. We have added some experimental techniques such us a computer interface to

IOP Conf. Series: Materials Science and Engineering **619** (2019) 012036 doi:10.1088/1757-899X/619/1/012036

see directly the graph buoyancy occurred in the monitor of PC and refractory under the electrical balances to avoid window blowing from the furnace.

The measuring temperatures were selected carefully that allows measure performed or until crystallization occurred in the sample based on fact sage data bank [15]. The buoyancies of the each bobs were measured by electrical balance with the addition a computer interface to connect of PC. The difference between two averages was entered into Eq. 9 to calculate melt density,

$$\rho = \frac{W_1 - W_2}{V_1 - V_2} \tag{9}$$

where W_1 and W_2 are the buoyancies and V_1 and V_2 are the submerged volumes of the large and small bobs, respectively. The thermal expansion of large and small bobs is corrected by **Eq. 10**, the data for thermal expansion of platinum (Waseda et.al., 1975) [6],

$$V_1 - V_2 = (1 + 3\alpha\Delta T).(V_1' - V_2') \tag{10}$$

where α is the thermal expansion of Pt, ΔT is temperature difference between measuring and room temperatures, and V_1' and V_2' are the volume of the large and small bobs at room temperature, respectively. Considering accuracy and reproducibility of the measurement were investigated density of water and silicone fluid as a standard sample measured in the room temperature.

4. Results and Discussion

4.1. Error Estimation

The error estimation of the above method was taken as the deviations in the determination of buoyancy, temperature, and electrical balance reading. The errors caused by temperature, viscosity and inhomogeneity of the sample in density measurements. The error range was depended on composition in slag melts [16]. The average value of error was 0.15% for all measured densities.

In order to confirm the reproducibility of density values, the standard sample was measured in the following ways. Firstly, the densities of water were measured 25, 30, 35, and 40 °C. Fig. 5 shows the densities of water were measured to be 0.9966, 0.9945, 0.9939, and 0.9912 at 25, 30, 35, and 40 °C, respectively. The measured densities have good agreement with data of previous study [17] with a precision of 0.07%.

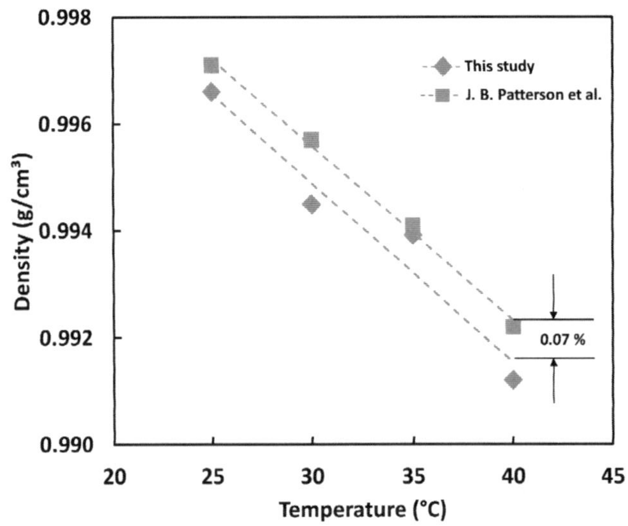

Figure 5 Variation of the density as a function of temperature in water. The dashed lines represent average linear fit of this study and previous study

Secondly, the density of silicone fluids with the viscosities 0.01, 0.999, 4.850, and 29.44 Pa·s were measured at 25 °C. Fig. 6 shows that the densities of silicone fluids were tested to be 0.941, 0.972, 0.973, and 0.974 g/cm^3, respectively. The measured densities have good agreement with the data of silicone fluid standards with a precision of 0.16%.

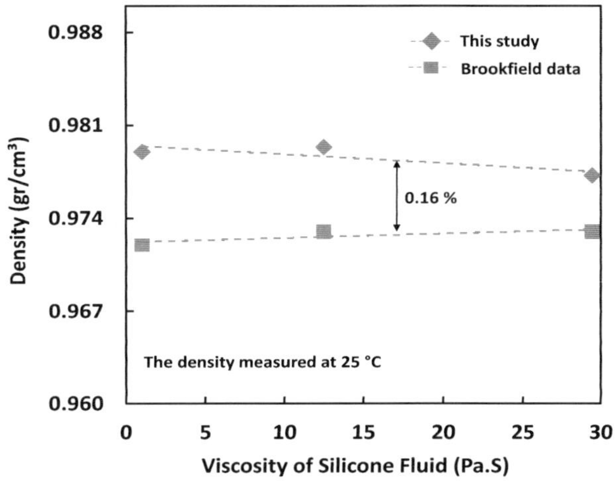

Figure 6 Variations of density with viscosity of silicone fluids. The dashed lines represent average linear fit of this study and previous study.

The results show that the precision density measurements of water and silicone fluid were 0.07 and 0.16 % of the density values, respectively. The accuracy and reproducibility of this work is within the error estimation of density measurements. Furthermore, the density of water and silicone fluid values are good agreement with previously reported values [18] within a deviation with the mean differences of 0.10%. The accuracy and reproducibility of density for water and silicone fluids support the reasonable results of density measurements in this work for measuring slag melts at high temperature.

4.2. Density of Na_2O-SiO_2

The difficulties of measuring density at high temperature are the consistency buoyancies of the bobs [12]. Data on the temperature dependence of the buoyancy of the Pt-wire and bobs immersed in the slag melts are shown in Fig. 7.

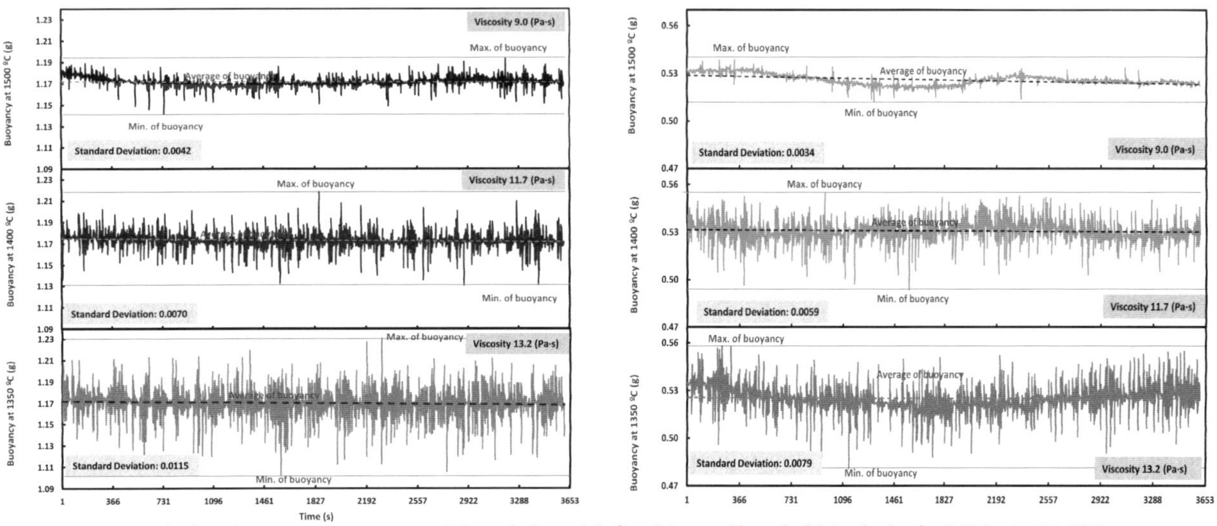

Figure. 7 Relation between buoyancy (g) and time (s) for (a) small and (b) Pt bobs in $20Na_2O$-$80SiO_2$ slag melt. The dashed lines represent average linear fit of the buoyancy.

The buoyancy data became stable within 30 min after the immersion in to the slag melts. The average of buoyancy during the last 30 min was used in density calculation.

Fig. 8 shows the temperature dependences of density for Na_2O-SiO_2 slag melts with 60, 70 and 80 mol % SiO_2. There are linear relationships between density and temperature for all Na_2O -SiO_2 slag melts between 1350 and 1500 °C. The densities of the NaO-SiO_2 slag melts decrease linearly with increasing temperature. The density varied from 2.213 to 2.238 g/cm^3. The density of Na_2O -SiO_2 slag melts decreases almost monotonically with increasing SiO_2 content.

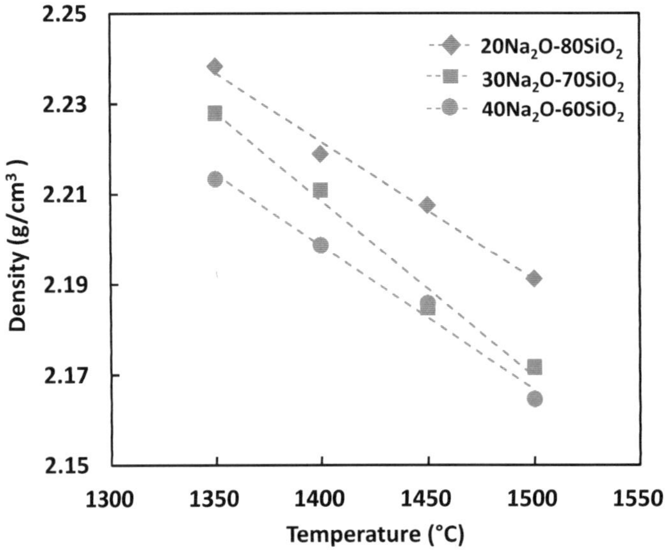

Figure. 8. Variations of density with temperature for NaO-SiO_2 slag melts with 60, 70 and 80 mol % SiO_2. The dashed lines are least square fit of this study

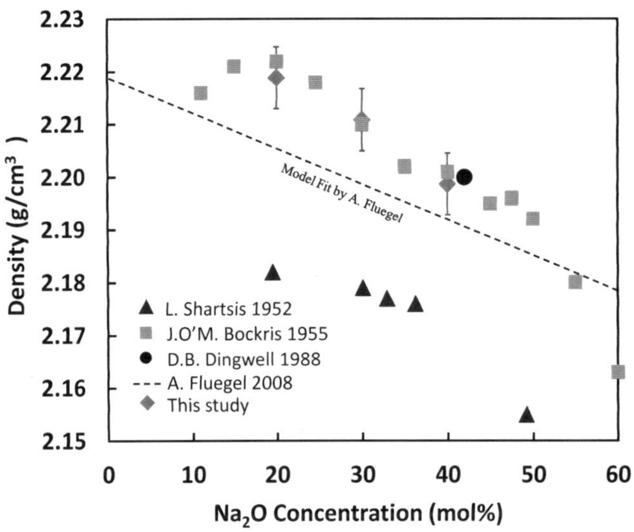

Figure. 9 Experimental data of previous and this study and model fit for the binary system Na_2O – SiO_2 at 1400 °C

Figure. 9 shows the experimental data with previous [13,19,20] and this study and the model fit [21] for the binary system Na_2O – SiO_2 at 1400°C. It can be seen that the experimental data by Shartsis et al. [19] and Bockris et al [13] are different. It is possible that Bockris in 1955 used the experimental data by Shartsis et al from 1951 for calibration. The measured densities data of this study have good agreement with the data of Bockris data with a precision of 0.23%.

From Fig. 9, it can be estimated that the error of repeated density measurements at 1400°C considering several investigators is approximately 0.007 g/cm^3. The error of repeated density measurements of one investigator using one experimental technique is approximately 0.001 to 0.01 g/cm^3 (4) for Na_2O – SiO_2 slag melts. The difference between the error of one investigator compared to the overall error of several investigators is caused by small systematic errors. Furthermore, the accuracy and reproducibility of density for Na_2O–SiO_2 (mol%) slag melts support the reasonable results of density measurements for gasified coal and synthesized slag melts in our previous study [4].

5. Conclusion

The apparatus of Archimedean double-bob method was established for density measurements of slag melts. The densities of water and silicone fluid were measured and have good agreement with the data of water standards with a precision of 0.07% and silicone fluid standards with a precision of 0.16%.
The temperature dependences of density for Na_2O-SiO_2 slag melts with 60, 70 and 80 mol % SiO2. There are linear relationships between density and temperature for all Na_2O-SiO_2 slag melts between 1350 and 1500 °C. The densities of the Na_2O-SiO_2 slag melts decrease linearly with increasing temperature. The density varied from 2.213 to 2.238 g/cm3. The density of NaO-SiO2 slag melts decreases almost monotonically with increasing SiO2 content.
The accuracy and reproducibility of density measurements for Na_2O-SiO_2 (mol%) slag melts support the reasonable results of density measurements for measuring gasified coal and synthesized slag melts

6. References

[1] Arman, Tsuruda A, Arma LH, Takebe H. Viscosity measurement and prediction of gasified and synthesized coal slag melts. Fuel 2017;200:521–8. doi:10.1016/j.fuel.2017.03.094.
[2] Aineto M, Acosta A, Rincon J, Romero M. Thermal expansion of slag and fly ash from coal gasification in IGCC power plant. Fuel 2006;85:2352–8. doi:10.1016/j.fuel.2006.05.015.
[3] Mondol JD, McIlveen-Wright D, Rezvani S, Huang Y, Hewitt N. Techno-economic evaluation of advanced IGCC lignite coal fuelled power plants with CO2 capture. Fuel 2009;88:2495–506. doi:10.1016/j.fuel.2009.04.019.
[4] Arman, Okada A, Takebe H. Density measurements of gasified coal and synthesized slag melts for next-generation IGCC. Fuel 2016;182:304–13. doi:10.1016/j.fuel.2016.05.117.
[5] Hsieh PY, Kwong KS, Bennett J. Correlation between the critical viscosity and ash fusion temperatures of coal gasifier ashes. Fuel Process Technol 2016;142:13–26. doi:10.1016/j.fuproc.2015.09.019.
[6] Mysen BO, Virgo D, Seifert FA. The structure of silicate melts: Implications for chemical and physical properties of natural magma. Rev Geophys Sp Phys 1982;20:353–83. doi:10.1029/RG020i003p00353.
[7] Mills KC. Estimation of physicochemical properties of coal slag and ashes. Miner Mater Ash Coal, Am Chem Soc Washinton, DC 1986;301:195–214. doi:10.1021/bk-1986-0301.ch015.
[8] Hwang C, Fujino S, Morinaga K. Density of Bi2O3–B2O3 binary melts. J Am Ceram Soc 2004;87:1677–82. doi:10.1111/j.1551-2916.2004.01677.x.
[9] Kuromitsu Y, Yoshida H, Takebe H, Morinaga K. Interaction between alumina and binary glasses. J Am Ceram Soc 1997;80 (6):1583–7. doi:10.1111/j.1151-2916.1997.tb03020.x.
[10] Sugawara T, Katsuki J, Shiono T, Yoshida S, Matsuoka J, Minami K, et al. High-temperature heat capacity and density of simulated high-level waste glass. J Nucl Mater 2014;454:298–307. doi:10.1016/j.jnucmat.2014.07.055.
[11] Linard Y, Nonnet H, Advocat T. Physicochemical model for predicting molten glass density. J

Non Cryst Solids 2008;354:4917–26. doi:10.1016/j.jnoncrysol.2008.07.013.

[12] Sasaki H, Tokizaki E, Terashima K, Kimura S. Density Variation of Molten Silicon Measured by an Improved Archimedian Method. Jpn J Appl Phys 1994;33:3803–7. doi:10.1143/JJAP.33.3803.

[13] Bockris JO, Tomlinson JW, White JL. The structure of the liquid silicates: partial molar volumes and expansivities. Trans Faraday Soc 1956;52:299. doi:10.1039/tf9565200299.

[14] Sukenaga S, Haruki S, Nomoto Y, Saito N, Nakashima K. Density and surface tension of CaO–SiO2–Al2O3–R2O (R=Li , Na , K) melts. ISIJ Int 2011;51 (8):1285–9. doi:10.2355/isijinternational.51.1285.

[15] www.factsage.com.

[16] Shiraishi Y, Ikeda K, Tamura A, Saito T. On the viscosity and density of the molten FeO–SiO2 system. Mater Trans JIM 1978;19 (5):264–74. doi:10.2320/matertrans1960.19.264.

[17] Patterson JB, Morris EC. Measurement of Absolute Water Density, 1 °C to 40 °C. Metrologia 2005;31:277–88. doi:10.1088/0026-1394/31/4/001.

[18] Laboratories BE. General purpose silicone fluids, 2010, p. 1–6.

[19] Shartsis L, Spinner S, Capps W. Density, expansivity, and viscosity of molten alkali silicates. J Am Ceram Soc 1952;35:155–60. doi:10.1111/j.1151-2916.1952.tb13090.x.

[20] Dingwell DB, Brearley M, Dickinson JE. Melt densities in the Na2O–FeO–Fe2O3–SiO2 system and the partial molar volume of tetrahedrally-coordinated ferric iron in silicate melts. Geochim Cosmochim Acta 1988;52:2467–75. doi:10.1016/0016-7037(88)90305-5.

[21] Fluegel A, Earl DA, Varshneya AK, Iii TPS. Density and thermal expansion calculation of silicate glass melts from 1000 ° C to 1400 ° C 2008;49:245–57.

Acknowledgments
This research is supported by New Energy and Industrial Technology Development Organization (NEDO) of Japan and Directorate General of Human Resource for Science, Technology and Higher Education (DIKTI) of Indonesia.

Analysis of ignition delay and burning rate of dammar resin

Jamal Jamal

Mechanical Engineering Department, Politeknik Negeri Ujung Pandang, Makassar, Indonesia

E-mail: jamal_mesin@poliupg.ac.id,

Abstract. This study aims to determine the ignition delay and the burning rate of dammar resin. The test is carried out by observing the process starting with the fall of the dammar resin above the furnace then starting to form until the combustion process ends. Observations were made using a Canon G15 digital camera that has the ability to record movies at speeds of 240 f/s. The analysis was carried out by first changing the recording results into an image using the program free video to jpg converter v.5.0.43 build 605. The results were obtained in the form of the mass relation of dammar resin material (g) to the ignition delay (s) in the equation$Y_I = -170{,}8.X_w^2 + 8{,}258.X_w + 1{,}631$. The mass relationship of dammar resin material (g) with the duration of combustion time (s) in the equation $Y_{the\ t} = 87{,}64.X_w + 1{,}373$. The mass relation of dammar resin material (g) to the burning rate (g/s) in the equations $Y_{vm} = -0{,}497.X_w^2 + 0{,}099.X_w + 0{,}004$. The relationship of the initial diameter of dammar resin (mm) on the flame distribution rate (mm²/s) in the equation $Y_{vD} = -0{,}051.X_D^2 + 0{,}306.the\ X_D + 2{,}368$.
Keywords: ignition delay, burning rate, dammar resin, Alternative Fuel

1. Introduction

Currently, the efforts continue to be made so that the world's energy needs can be met, including efforts to optimize the use of energy both existing energy such as water energy [1], solar energy [2-3] etc. Efforts to meet energy needs are also carried out by finding alternative and renewable energy [4 - 12]. The increasingly limited petroleum reserves and air pollution problems, so that it becomes an urgent need for the use of alternative fuels [4]. Therefore, joint efforts are needed to obtain alternative fuels that are equivalent and even superior and environmentally friendly [5].

Efforts to reduce exhaust emissions are also carried out by using alternative fuels from plant sources, including tobacco seed oil [6], tomato seed oil [7], vegetable oil [8] and soybean oil [9]. One of the plants that are also suitable as an alternative fuel is dammar resin (Dipterocarpaceae) because it contains hydrocarbons and other carbon compounds [10-11]. This content indicates that dammar resin has the potential to be developed as an alternative fuel. The feasibility of alternative fuels is determined by many factors, including the ignition delay and the burning rate [12]. As for research with alternative oil of jatropha oil [12] obtained results as shown in Figure 1.

This research was conducted to determine the flammability of dammar resin, i.e. the ignition delay and the burning rate of dammar resin.

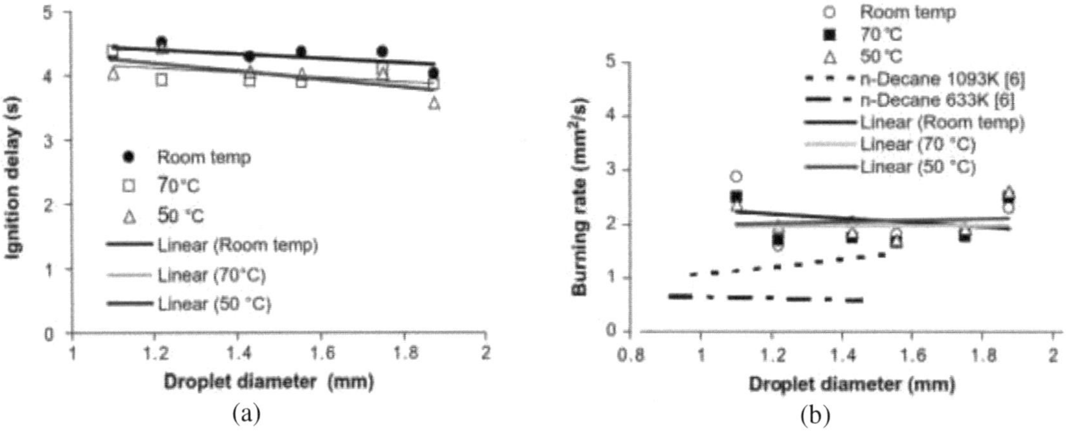

Figure 1. Effect of droplet diameter on (a) ignition delay and (b) burning rate.

2. Methodology

Testing of dammar resin combustion is done to determine the flame properties of dammar resin which is related to the function of dammar resin as an alternative renewable fuel.

The flame properties of the dammar resin observed were ignition time and ignition time. The equipment used is a furnace and a digital camera. The digital camera used is the Canon G15, this camera has the ability to record movies at speeds of 240 f/s. The experimental scheme can be seen in Figure 2. The samples observed were solid dammar resin. Before testing is done the selection of clean solid dammar resin.

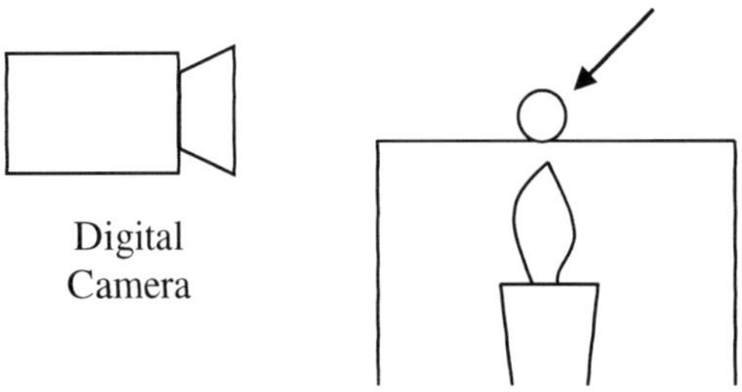

Figure 2. The experimental scheme of combusting low mass.

The test is done by turning on the furnace for 5 minutes, after that the digital camera with 240 f/s is turned on, then dropping the dammar resin over the furnace, shortly after the dammar resin turns on, turn off the furnace fire, observe the dammar resin ignition until it goes out, after dammar resin turns off the digital camera. The test done with a variation of the weight of the dammar resin is 0.02; 0.04; 0.06; 0.08 and 0.10 g and each weight variation was tested 5 times.

The time of dammar resin falls into the furnace until the combustion process occurs was called the ignition delay. The burning rate is measured by the mass of the material divided by the time of combustion. The burning time is measured from the start of the burning of the dammar resin on top of the furnace until it goes out. The results of digital camera readings are in the form of a video with MOV format; for processing video data then converted into an image using a program free video to jpg converter v.5.0.43 build 605.

3. Results and discussion

In dammar resin combustion with a material mass variation of 0.02 to 1.0 grams, when the dammar resin is dropped on the furnace, the dammar resin is not directly burned. The time of the dammar resin falls into the furnace until the combustion process occurs is called the ignition delay. The relation of the dammar resin mass (g) to the ignition delay (s) can be seen in figure 3. In the picture, the relationship that forms a curved line with the equation of the line is:

$$Y_I = -170{,}8.X_w^2 + 8{,}258.X_w + 1{,}631 \quad (1)$$

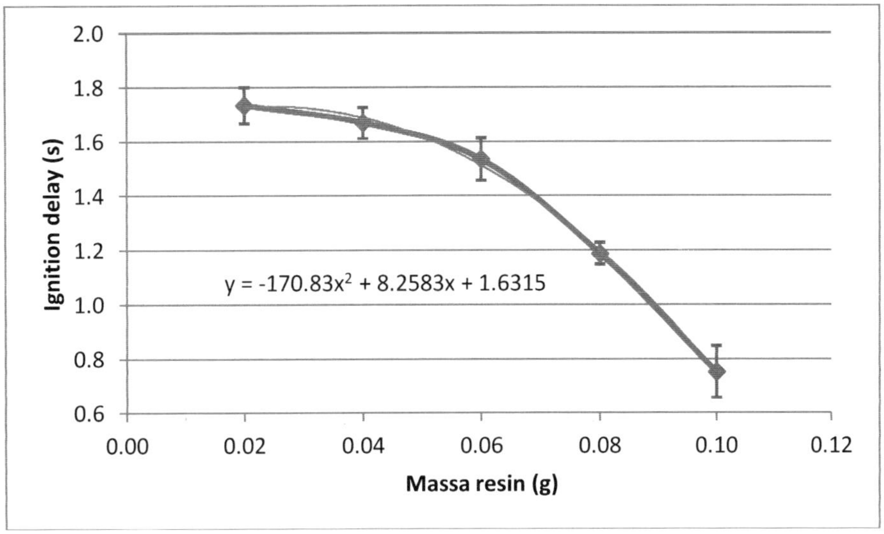

Figure 3. Effect of the dammar resin mass on the ignition delay (s) of dammar resin.

In Figure 3 shows that the larger the mass of dammar resin, the shorter the ignition delay, this is because the mass of the dammar resin, the large amount of gas formed from the heating results more and more so that a certain amount of gas is burned and forms a flame. The same tendency is also generated in the combustion of jatropha oil [12] as shown in Figure 1 (a) where the greater the diameter of the droplet (the greater the mass), the faster the ignition delay.

In dammar resin combustion, the ignition delay at 0.02 grams of material mass (equivalent to one drop of liquid fuel) is around 1.68 - 1.83 seconds (Figure 3), when compared with the results of jatropha oil burning [12] which has time wait for the flame to range from 3.6 to 4.4 seconds (Figure 1 (a)) then the dammar resin has a more flammable property than jatropha oil.

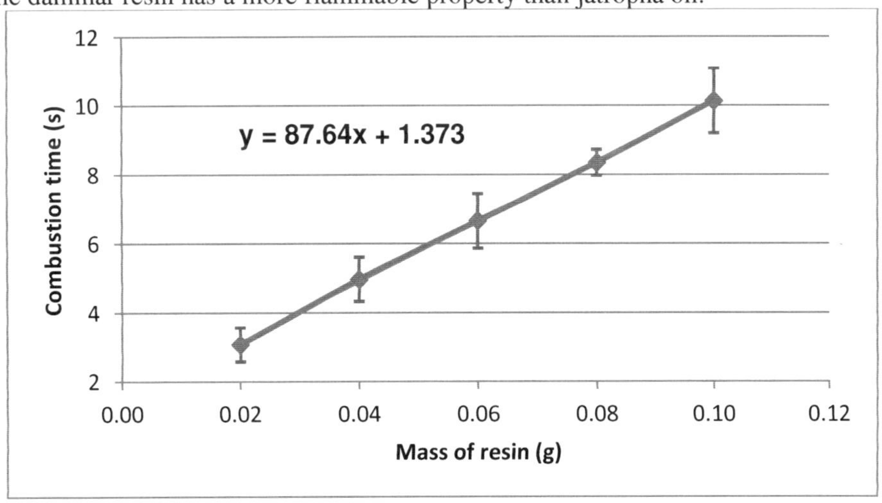

Figure 4. Effect of the dammar resin mass on combustion time.

In dammar resin combustion with a material mass of 0.02 to 1.0 grams, it can be obtained the results in the form of dammar resin mass relation (g) with the length of combustion time (s) can be seen in Figure 4. It can be seen that the relation of the dammar resin mass and combustion time is directly proportional and forms a linear line with the equation of the line as follows:

$$Y_t = 87{,}64. X_w + 1{,}373 \tag{2}$$

From the results in Figure 4, it is calculated the relationship of the mass of the material with the rate of mass burning which can be seen in Figure 5. The figure shows the relationship of the mass of the material with the burning rate in the form of parabolic. At the beginning of the test, there is an increase in the mass burning rate, the greater the mass of material increases the burning rate of the mass becomes smaller. This shows that one day the change in mass of material does not affect the rate of mass burning. In the figure the resulting graph forms the following curved line equation:

$$Y_{vm} = -0{,}497. X_w^2 + 0{,}099. X_w + 0{,}004 \tag{3}$$

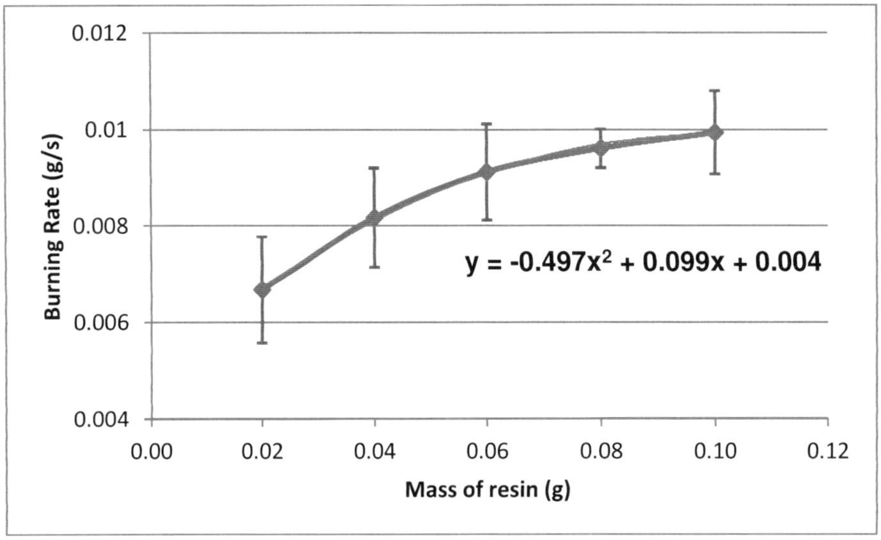

Figure 5. Effect of dammar resin mass on burning rate.

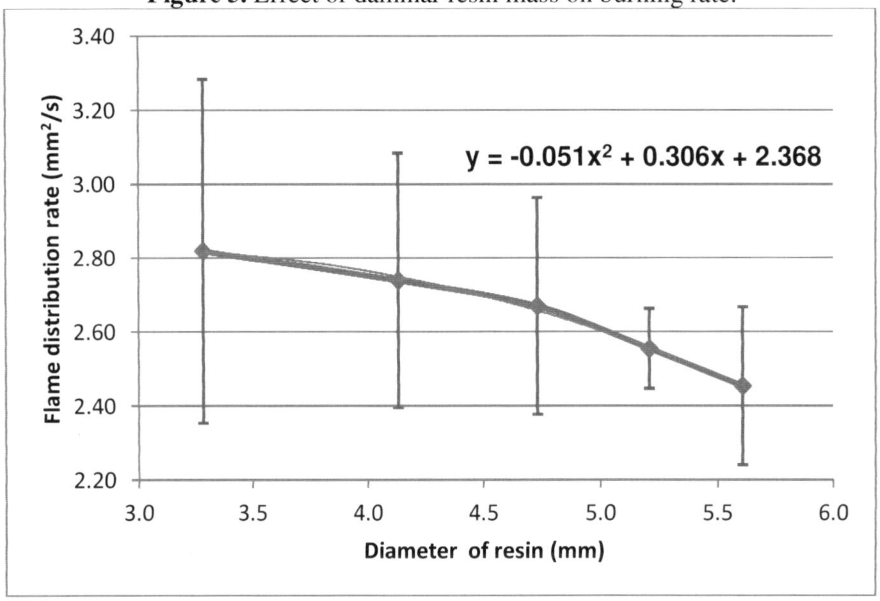

Figure 6. Effect of dammar resin mass on flame distribution rate.

The burning rate of dammar resin can also be described by the relationship of the initial diameter of the dammar resin to the flame distribution rate (mm²/s) as shown in Figure 6. It was obtained the dammar resin diameter value (mm) and the flame distribution rate (mm²/s) using the following equation:

$$X_D = 2\sqrt[3]{\frac{3}{4\pi}.\frac{X_w}{\rho}} \quad (4)$$

$$Y_D = \frac{\pi.X_D^2}{4.t} \quad (5)$$

In Figure 6, it can be seen that the relationship of the initial diameter and flame distribution rate is inversely proportional and parabolic in shape with the equation of the line is:

$$Y_{vD} = -0{,}051.X_D^2 + 0{,}306.X_D + 2{,}368 \quad (6)$$

In Figure 6, it shows that the larger the initial diameter of the dammar resin, the flame distribution rate (mm²/s) is slower. In combustion with a larger diameter, the surface area of the recipient of the flame energy is also greater. With the increased surface area of the recipient of the flame energy, the rate of spread of the flame is slower. The same tendency is also generated in the combustion of jatropha oil [12] as shown in Figure 1 (b) where the greater the diameter of the droplet, the flame distribution rate (mm²/s) will be smaller. In dammar resin combustion with a material mass of 0.02 grams (equivalent to 3.28 mm diameter) the flame distribution rate (mm²/s) is in the range of 2.62 - 3.44 mm²/s (figure 6), when compared to the oil burning distance [12] which has a flame distribution rate ranging from 1.6 - 2.9 mm²/s (figure 1 (b)) then the dammar resin has a flame distribution rate that is faster than jatropha oil.

4. Conclusions

Based on the results of the experiments was conducted, it can be concluded that:

1. The mass relationship of dammar resin material (g) to the ignition delay (s) is expressed by the equation $Y_I = -170{,}8.X_w^2 + 8{,}258.X_w + 1{,}631$.
2. The mass relationship of dammar resin material (g) with the duration of combustion time (s) is expressed by the equation $Y_t = 87{,}64.X_w + 1{,}373$.
3. The mass relationship of dammar resin material (g) to the burning rate (g/s) is expressed by the equation $Y_{vm} = -0{,}497.X_w^2 + 0{,}099.X_w + 0{,}004$.
4. The relationship of the initial diameter of dammar resin (mm) to the flame distribution rate (mm²/s) is expressed by the equation $Y_{vD} = -0{,}051.X_D^2 + 0{,}306.X_D + 2{,}368$.

5. References

[1] Jamal, J., & Lewi, L. (2018, June). Utilization of irrigation flow for the construction of micro-hydro power plant. In *AIP Conference Proceedings* (Vol. 1977, No. 1, p. 060018). AIP Publishing.

[2] Suwasti, S., Abadi, S., Shiddiq Yunus, A. M., & Jamal. (2015). Impacts of Fin Variation on the Performance of Shelf Type Solar Dryer. *Journal of Advanced Research in Fluid Mechanics and Thermal Sciences*, *13*, 17-21.

[3] Jamal, J., Tangkemanda, A., & Susanto, T. A. (2018, June). The effect of collector slope angle on the performance of solar water heater. In *AIP Conference Proceedings* (Vol. 1977, No. 1, p. 060019). AIP Publishing.

[4] Kazakov, A., Conley, J., & Dryer, F. L. (2003). Detailed modeling of an isolated, ethanol droplet combustion under microgravity conditions. *Combustion and Flame*, *134*(4), 301-314.

[5] Anand, K., Sharma, R. P., & Mehta, P. S. (2011). A comprehensive approach for estimating thermo-physical properties of biodiesel fuels. *Applied thermal engineering*, *31*(2-3), 235-242.

IOP Conf. Series: Materials Science and Engineering **619** (2019) 012037 doi:10.1088/1757-899X/619/1/012037

[6] Giannelos, P. N., Zannikos, F., Stournas, S., Lois, E., & Anastopoulos, G. (2002). Tobacco seed oil as an alternative diesel fuel: physical and chemical properties. *Industrial crops and products*, *16*(1), 1-9.

[7] Giannelos, P. N., Sxizas, S., Lois, E., Zannikos, F., & Anastopoulos, G. (2005). Physical, chemical and fuel related properties of tomato seed oil for evaluating its direct use in diesel engines. *Industrial crops and products*, *22*(3), 193-199.

[8] Demirbas, A. (2008). Relationships derived from physical properties of vegetable oil and biodiesel fuels. *Fuel*, *87*(8-9), 1743-1748.

[9] Moser, B. R., Williams, A., Haas, M. J., & McCormick, R. L. (2009). Exhaust emissions and fuel properties of partially hydrogenated soybean oil methyl esters blended with ultra low sulfur diesel fuel. *Fuel Processing Technology*, *90*(9), 1122-1128.

[10] Jamal, Wardana, I. N. G., Hamidi, N., & Widyanuriawan, D. (2014). Identification of Chemical Compound of Dammar Resin using Various Solvents with Gas Chromatographic--Mass Spectrometric Method. *Applied Mechanics & Materials*, (695).

[11] Jamal, Wardana, I. N. G., Hamidi, N., & Widyanuriawan, D. (2016). Thermal Analysis of Dammar Resin with Differential Scanning Calorimetry Method. In *Applied Mechanics and Materials* (Vol. 818, pp. 228-230). Trans Tech Publications.

[12] Wardana, I. N. G. (2010). Combustion characteristics of jatropha oil droplet at various oil temperatures. *Fuel*, *89*(3), 659-664.

Design and Development of Stirring and Cooking Machines for Palm sugar

Anwar Mazmur and Syarifuddin

Mechanical Engineering Department, Politeknik negeri Ujung Pandang, Jalan Perintis kemerdekaan Km. 10 Tamalanrea 90245 Makassar Indonesia

Email: mazmur.anwar@gmail.com, arifsyarif15@yahoo.com

Abstract. The production process of palm sugar is carried out by farmers or craftsmen with very simple equipment, namely cauldrons, stirrers and firewood stoves. Products was produced using this method are not good, short durability, require a long time, so it is inefficient and effective. The purpose of this study was to overcome worker fatigue, minimize work accidents and increase the production capacity of printed palm sugar and palm sugar. The research method that will be carried out is designing and making a stirrer and cooking machine for palm sugar palm, an electric motor drive system with several stages, namely the preparation stage in the form of a field survey of the conditions and production capacity of palm sugar, and preparation of materials and equipment required, the assembly phase of equipment components that have been made or purchased, and the Testing phase of the assembled equipment are tested. This testing phase will be repeated over and over until the data about the correct rotation is obtained to get the optimal amount of production. Specifications of the stirrer and cooking sugar ants are 155 cm high, 80 cm long, and 65 cm wide, stirring shaft diameter 1 inch and the power of the motor drive 1/2 hp, and a capacity of 29 kg per day.

1. Introduction

South Sulawesi, especially Enrekang and Pangkep districts, is one of the largest palm sugar producers in Eastern Indonesia. Palm sugar production in South Sulawesi from 2007 to 2011 experienced a significant increase, which is an average of 5% annually. The latest data collected shows the production of palm sugar in South Sulawesi for the year 20011 alone at 357,972 tons (BPS South Sulawesi, 2011). Generally, sugar production is marketed to urban areas, such as the city of Makssar and Ambon City.

The palm sugar industry or business as the location and object of research is located in Maiwa, Enrekang Regency, South Sulawesi Province. The industry is classified as a small industry and managed in a familial manner. Palm sugar has been known by the people of Indonesia as one of the sweeteners of food and beverages that can be substituted for sugar. Palm sugar is obtained from the tapping process of palm juice which is then reduced in water content until it becomes solid. This palm sugar product is in the form of printed sugar and ant sugar. Print sugar is obtained by cooking palm juice until it becomes thick like a cotton candy then printing it in a semicircular mold. As for the palm sugar of ant, the cooking process is longer that is until the palm sugar crystallizes, then dried until the water content is below 3%. This type of palm sugar has the advantage of palm sugar, which is longer durability, more hygienic and practical to use.

Based on the results of surveys and interviews with farmers and palm sugar craftsmen, they generally suggested that the quantity of production and the low quality of palm sugar caused by the

IOP Conf. Series: Materials Science and Engineering **619** (2019) 012038 doi:10.1088/1757-899X/619/1/012038

processing of palm sugar that had been done so far had not been optimal. This is because the palm sugar processing equipment and systems carried out by farmers or craftsmen of palm sugar have been very simple.

The process of producing palm sugar at the farmer or craftsman level is carried out with very simple equipment, namely using cauldrons, stirrers and firewood stoves. The products produced in this way are not good, besides the short durability also requires a long time, so it is not efficient and effective.

Another problem experienced by artisans is the lack of market certainty from the sugar product of the ant. The palm sugar they produce is mostly made on order and the amount is not much. So that most of the palm sugar crafters prefer to produce brown sugar as their main livelihood, because it is considered more promising and has market certainty in this case it was bought by the brown sugar collectors. To overcome the shortcomings and weaknesses described above, it is necessary to apply a mixing machine and a machine for cooking palm sugar ants that are driven by motorcycles. The machine intended is that which is easy to operate and can be made by a local welding workshop with the cost of making machines that can be reached by farmers or craftsmen of palm sugar.

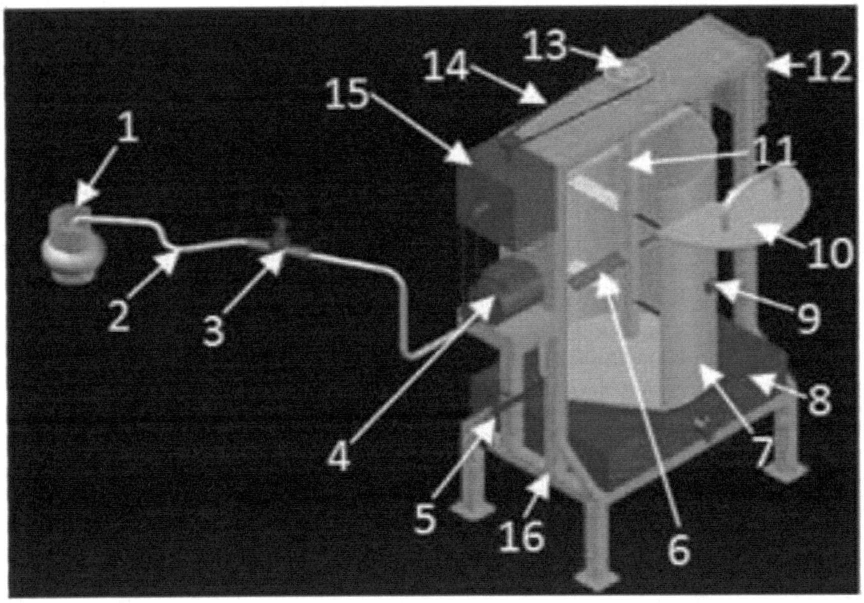

Figure 1. Design of stirring and cooking palm sugar machine

2. Research Method

2.1. Place and Time of Research

This research was carried out in Mechanical Workshop in Mein Engineering and Food Technology Laboratory in Chemical Engineering at Ujung Pandang State Polytechnic in May - October 2018.

Figure 2. Manufacturing process of stirring and cooking palm sugar machine

2.2. Sourcing data

Data sources used in the study are primary and secondary data. Primary data is obtained based on the results of direct interviews in the field and dissemination of information in the form of structured questions to respondents. Secondary data is obtained from library data as well as various data relating to palm sugar, data extraction in related departments such as agriculture, plantation services, and industry and trade services.

2.3. Collecting Data

Data collection method is done by interviewing with structured questions. Interviews are conducted in the form of deep interviews with the aim to explore hidden data that may be reluctant to be revealed by respondents

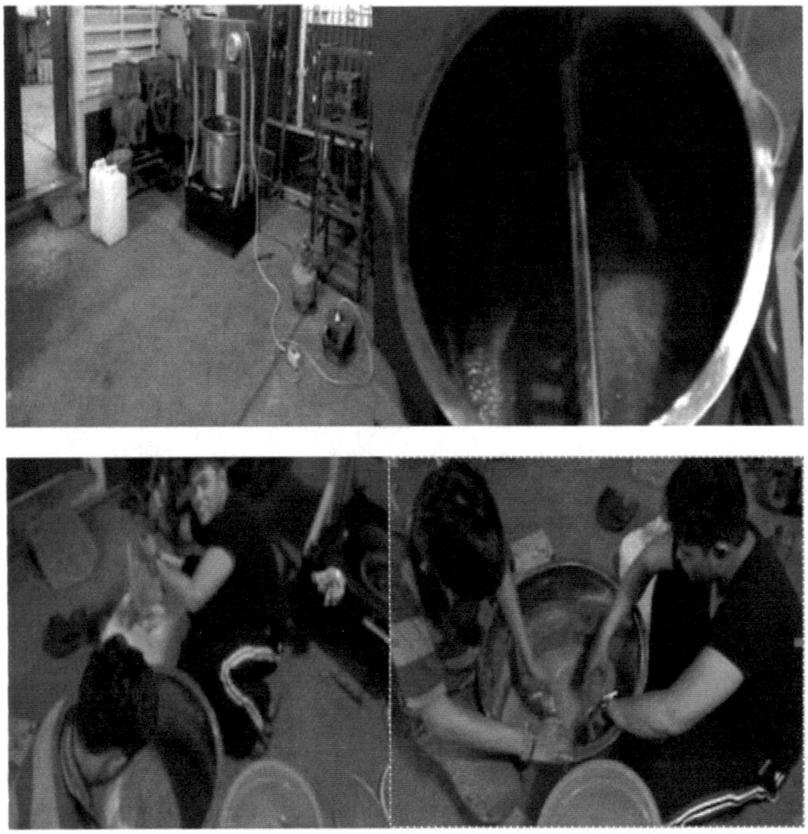

Figure 3. Testing Process of stirring and cooking palm sugar machine

2.4. Designing

In the introduction, it has been stated that the purpose of this paper is to plan and embody a stirring machine and a machine for cooking palm sugar, which is driven by a different motor than the previous stirring machine and palm sugar cooking machine. Hopefully the mixing machine and sugar cooking machine, which is driven by a motor which is made later, will have better and more efficient mixing and mixing results in the process. The design stages are as follows. Make a design (sketch drawing) of the components to be made. Designing is done by drawing on a computer using Autocad software as shown below:

2.5. Assembling

After all components are finished, the next stage is the assembly stage. The assembly process is the process of assembling or combining each component into a mutually supportive form so that a working mechanism is formed that matches what is desired. The assembly steps are as follows:

- The frame that has been made is placed on a flat floor.
- Attach the stove mount rail path to the top of the stove holder. In order for the rail to be strong, welding is needed with 90 A electric welding and 2.6 mm electrodes
- Insert the stove holder on the rail that has been installed in the frame.
- Install the bearing holder, motor holder, and reducer holder in the place that has been prepared in the drive chamber by welding 90 A electric welding and 2.6 mm electrodes.
- Install the pads on the stand that has been installed in the frame.
- Attach the shaft to bearing 1, simultaneously connecting the shaft to the gear between the bearings 1 and 2. Then press the shaft until it enters the bearing 2. 2. Tighten the bearing bolt.

- Connect the shaft with the stirring shaft by using a sheath and bolt 12.
- Paint parts of the stand frame and stove protector in black using spray-gun and orange supporting poles, home and motorbike protectors using spray-gun.
- Install the hard installation from the motor to the reducer through the pulley and belt on the drive chamber.
- Install the installation current from the motor to the socket.
- Install the cover of the drive chamber using a bolt.
- Install the stove on the stove holder on the stove holder on the frame.
- The tool is ready to be tested.

2.6. Testing Machine

To determine the performance of the palm sugar palm cooking and stirring machines, it was need to test the machine. The testing procedures of this machine are as follows:

- Put the juice in the tube while filtering to remove impurities. Turn on the stove on the cooking machine and stirrer of palm sugar.
- Cook the juice of the juice until it boils and produce yellow bubbles. Lift the yellow foam using a filter.
- Install the stirrer on the fixed shaft then tighten it with a bolt that has been prepared.
- After the juice of the juice is cooked marked by changing the color of the sugar to brown, add the pecan seeds that have been mashed, and then stir constantly when large bubbles appear turn off the stove.
- The juice concentration is left for ± 10 minutes.
- After silence, turn on the stirrer by pressing the ON button on the switch. Stir the sugar with a stirrer until it becomes sugar crystals or granules of raw material for palm sugar.
- After becoming a crystal, weigh the results using a scale.
- Clean the remnants of the raw material for palm sugar which is attached to the tube wall.

The results of the tests that have been carried out are as follows:

Table 1. Testing data for palm sugar making machines

No	Materials		Cooking Time		Time of Stirring	Results
	Palm sugar	Candlenut	Traditional	Machine		
1	50 litres	4 pieces	5 Jam	-	60 minutes	8,5 kg
2	50 litres	4 pieces	-	3 jam	15 minutes	8,5 kg

3. Results and Discussion

In traditional experiments the juice needed is 50 liters to produce 8.5 kg of raw material for palm sugar. This traditional method takes 5 hours to cook and 1 hour for the stirring process. So the total time taken is 6 hours. But sometimes it takes longer than that. This is caused by the uneven heat produced by firewood.

The stirring process also still uses human power so that the stirring process can cause work accidents in the form of fatigue and the palms of the workers sprained. In the traditional process can only be produced once a day so that it can only produce 8.5 kg / day.

In the experiment using a machine, the juice needed 50 liters produces 8.5 kg of raw material for ant palm sugar for 3 ¼ hours where, the cooking process uses a 3-hour machine, and the stirring process is 15 minutes.

At the time of cooking faster because the heat produced by the stove is more evenly distributed. In the second experiment the stirring process was easier, faster, and workers' work safety was guaranteed.

This is because they no longer use human power when mixing sugar to powder. Raw materials for palm sugar, which can be produced once is 14.5 kg because the capacity of the tube in this machine is 100 liters of juice. In production using machines can produce 2 times a day. So what can be produced in a day can reach 29 kg of raw material for palm sugar.

4. Conclusion

From the experimental data above, it can be concluded that:

- The mixing machine and cooking sugar of the ants are 155 cm high, 80 cm long, and 65 cm wide, the stirrer shaft diameter is 1 inc and the motor power is 1/2 hp.
- In the process of making raw materials for palm sugar using this machine, the results are more effective and efficient. Where in this process no longer use human power when stirring is replaced by mechanical power from an electric motor so that the safety of workers is guaranteed.
- With this tool can increase production capacity every day with a faster time. There was an increase in production compared to the traditional method from 8.5 kg to 29 kg per day due to the larger cooker tube capacity and production twice a day.

References

[1] Edwin. 2012.Perancangan Mesin Pengkristal Gula Jawa, Yogyakarta, Universitas Negri Yogyakarta.
[2] Ni Luh Desi In Diana Sari, I Nengah Sudika Negara, Alit Kumala Dewi. Ibm Kelompok Usaha Gula Semut Di Desa Besan Kecamatan Dawan – Klungkung Majalah Aplikasi Ipteks Ngayah : Volume 6, Nomor 1, Juli 2015
[3] Kartosapoetra. 1994. Teknologi Penanganan Pasca Panen Jakarta : Rineka Cipta.
[4] Khurmi, R.S. Gupta, J.K. 1982. Machine Design. Ed.3.New Delhi : Eursia Publishing House
[5] Santoso, Hieronymus Budi. 1995. Pembuatan Gula Kelapa. Penerbit Kanisius.Yogyakarta.
[6] Sularso. Kiyokatsu Suga, 1991. *Dasar Perencanaan dan Pemilihan Elemen Mesin.* Pradnya Paramita, Jakarta.

The Development Mechanism of Microcontroller-Based Water Level and Temperature Control

S Ka'ka[1,*], R Tandioga[1], Lewi[1], A K Muhammad[1], A Rahman[1] and A Taufik[1]

[1] Mechatronics Study Program, Mechanical Engineering Department, Ujung Pandang State Polytechnic, 90245, Makassar, South Sulawesi, Indonesia

[*] The Corresponding author E-mail address: kakasimon@ymail.com

Abstract. The development of Microcontroller-Based Water Level and Temperature Control is the development of Microcontroller-Based Water Level Control Design. This equipment is a tool that can be used to monitor and control the level and temperature of water. The aim of this research is to make a water level and temperature control device based on microcontroller, and to display the information of water level and temperature on the computer. The research methods used during the implementation process of this research are : literature study, hardware and software design, and data retrieval. The Microcontroller-based Water Level and Temperature Control uses two sensors, namely : ultrasonic sensor to detect water level, then the second sensor is a thermocouple sensor which is used to detect water temperature. This equipment uses a water pump to flow water from one tank to another, and the heater that used to heat water. The device has 2 buttons, where the first button is used to control the water level, and the second button is used to control the level and temperature of the water. Based on the results of the study it can be concluded that this water level and temperature control system works well as expected, where this equipment uses a multi position control system mode that has been programmed in the Arduino software and can display sensor readings in the form graphs using matlab software.
Keywords: Arduino Microcontroller, Ultrasonic Sensor, Thermocouple Sensor, Water Pump, Heater

1. Introduction

"Water Level Control (WLC)" is a tool that is used to identify water levels in a reservoir by utilizing a sensor as a detector of water level and has an actuator such as motor as a water pump. This system is widely used in the industrial world and in everyday life, because it is simple and easy maintenance. The control system is the process of controlling one or several quantities (variables, parameters) so that it is in a certain range. The combined work of various control devices in the production process is called the process control system, while all the equipment that forms the control system is called process control instrumentation [1]. According to [2], the control mode is divided into two types, namely the discontinuous control mode and the continuous control mode. The discontinuous control mode is divided into 2 types, namely two-position mode and multi-position mode. This two-position mode works only in two positions, that are on and off modes. On-off controllers are mostly used in simple control actions because their price are cheap and the control mode is the most basic. Multi-position mode is a two-position control mode that is equipped with several intermediate settings.

Microcontroller exists in a computer system that functions as a chip in the form of an IC (Integrated Circuit) that can receive input signals, process them and provide an output signal in accordance with the program loaded into it [1]. In other words, a microcontroller is a digital electronic device that has input and output as well as control with a program that can be written and deleted in a special way. The way the microcontroller works is actually reading and writing data.

The microcontroller is basically a computer on a single chip, in which there is a microprocessor, memory, Input/output (I/O) path and other complementary devices. The speed of data processing on the microcontroller is lower [3], when compared to a PC. In PC the speed of microprocessors used today has reached the order of GHz, while the speed operations of microcontroller generally range from 1 - 16 MHz Likewise, the RAM and ROM capacity on a PC can reach the order of Gbyte, compared to a microcontroller which only ranges in the order of byte/Kbyte.

One of the microcontrollers that is often used is Arduino which is an electronic circuit that is open source, and has hardware and software that is easy to use. Arduino can recognize the surrounding environment through various types of sensors and can control lights, motors, and other types of actuators. Arduino has many types, including Arduino Uno, Arduino Mega 2560, Arduino Fio, and others.

Figure 1. An AT Mega Arduino.

Arduino is a microcontroller board based on ATmega328. The Arduino has 14 input/output pins where 6 pins can be used as PWM outputs, 6 analog inputs, a 16 MHz crystal oscillator, USB connection, power jack, ICSP head, and reset button. Arduino is able to support the microcontroller and can be connected to a computer using a USB cable [4]. The standard statement form in the control system will be described in Figure 2.

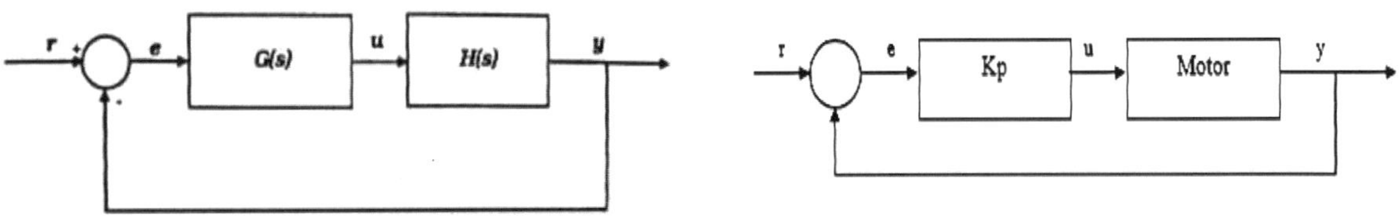

(a) (b)

Figure 2. Closed Loop Control and P-control

In the figure, r is the input, e is the error, u is the controller's output signal, G(s) is the controller, H(s) is the robot dynamic, and y is the output. Now the problem is how to design G(s).

Proportional control P applies if, $(s) = k$, which is k is a constant. If, $= G(s).e$, then the controller output equation is $u = kp.e$. Proportional control, P, for a motor can be illustrated as shown in Figure 2 (b). If K_p is a proportional constant, then K_p can be applied as gain without giving a dynamic effect on controller performance [5].

The use of P control has various limitations due to the non-dynamic control properties. However, in simple basic applications, this P control is quite capable of convergence even though the error in the steady state is relatively large. For example, in controlling the rotational speed of a dc motor, using proportional control, P, if the rotational speed is a function of time t (s), then the controller output u (t) is :

$$u(t) = Kp.(\Theta_{ref}(t) - \Theta_{act}(t)) \tag{1}$$

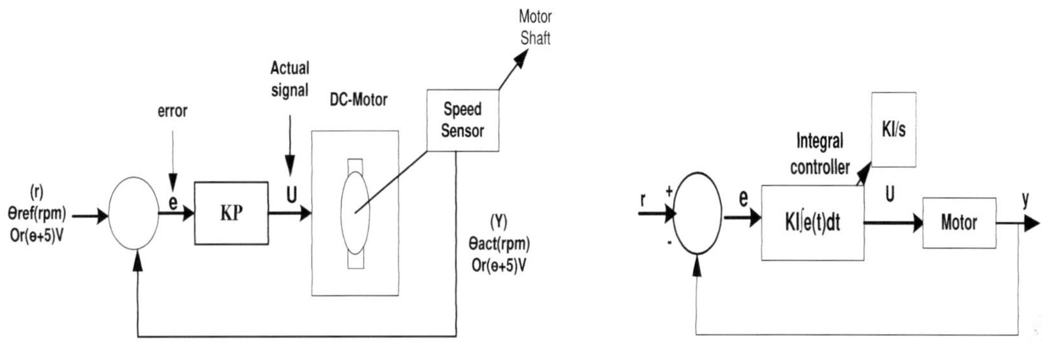

(a) (b)

Figure 3. The P and I control in a DC- motor.

The controller output value depends on the representation of errors, that is the reference speed minus the actual speed, which is less than the reference speed. If error is negative, it means that the actual speed is greater than the reference speed. If the actual speed is equal to the reference speed then the output signal will be zero because the error is zero. From the point of view of circuit, this output signal will cause the motor not to get an actuation signal again. As a result the motor shaft will stop rotating. Once the shaft rotation begins to decrease, the sensor will detect that the output speed is no longer equal to the reference speed. Actual speed will be smaller than the reference speed. The meaning of error is not the same as that zero, as a result the controller will start sending a positive actuating signal, u(t), again so that the motor will increase its speed again [6].

This is repeated so that the error at this P control cannot keep the error to always be zero, or in other words in a steady state, the error in control P cannot be zero. This is known as steady state error. The basic function of integral control, I, according to [7] is to reduce the stedy state error, while control I is rarely used alone in the application. It is usually combined with P control to improve the response to achieve a minimum error. The control I parameter can be illustrated in the motor control diagram as shown in Figure 3 (b).

Combination of I control and P control for a dc motor can be described as follows :

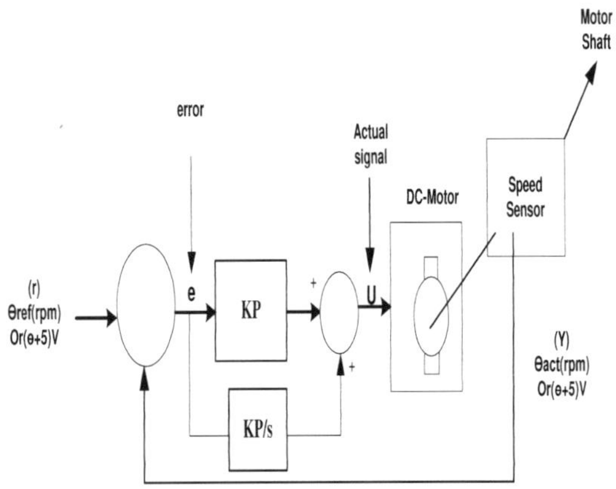

(a)

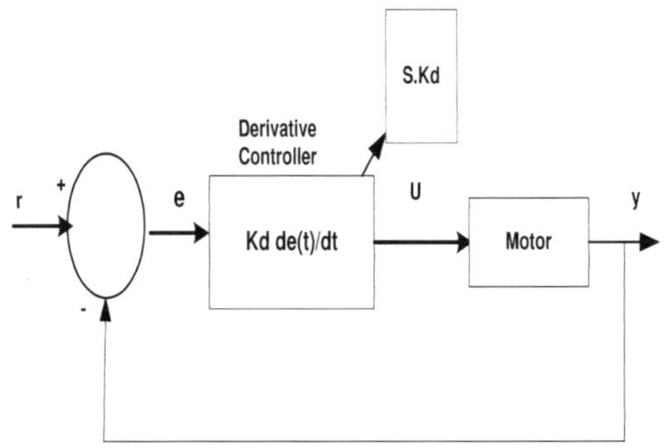

(b)

Figure 4 PI control and D control in DC Motor

The equation of the above controller output is :

$$u(t) = Kp.\left(\Theta_{ref}(t) - \Theta_{act}(t)\right) + Ki \int_0^t \left(\Theta_{ref}(t) - \Theta_{act}(t)\right) dt \qquad (2)$$

Parameter D works in the context of the rate/speed of the error so that it can reduce the effect of overshoot (over response) towards steady state. In other words, as if the control D is able to predict errors that will occur as an effect of the error speed that is calculated before.

The D control parameter can be illustrated in the motor control diagram as follows:

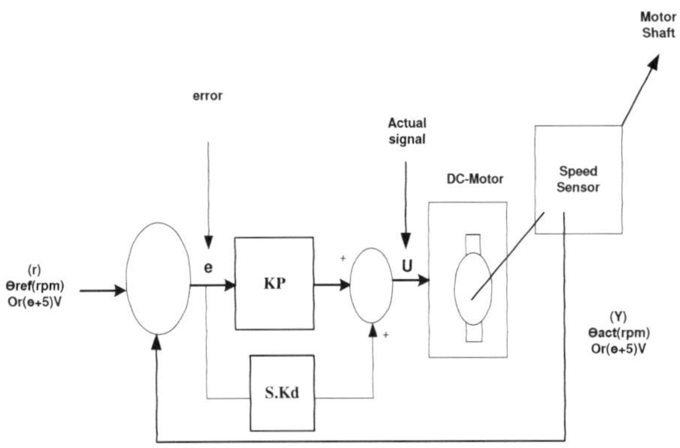

(a)

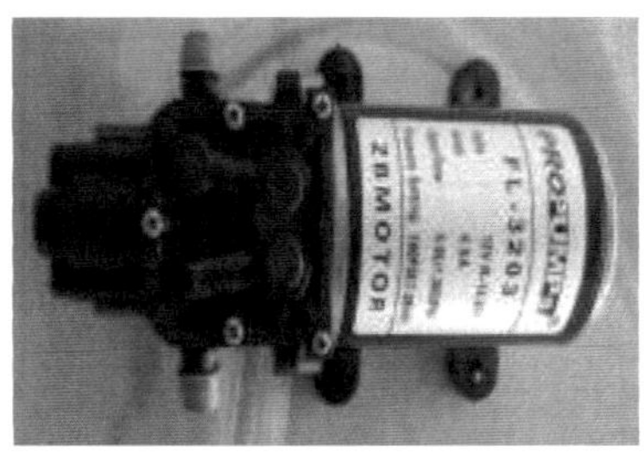

(b)

Figure 5. D and PD control in water pump dc motor.

Combination of D control and P control for a dc motor is described in Figure 5. So that the equation of controller output, u (t), is :

$$u(t) = Kp.\left(\Theta_{ref}(t) - \Theta_{act}(t)\right) + Kd.\frac{d(\Theta_{ref}(t) - \Theta_{act}(t))}{dt} \qquad (3)$$

Pump is one of the fluid machines included in the working machine class. The pump functions to flow the fluid from a lower place to a higher place or from a lower pressure place to a higher pressure place. In addition, the pump is also used to move the fluid from a higher place to a lower place, for example in a long and winding piping system.

Sensor is an equipment used to convert a physical quantity into an electrical quantity so that it can be analyzed with a certain electrical circuit [1]. Ultrasonic sensor is a sensor that operates based on the principle of sound wave reflection and is used to detect the presence of a particular object in front of it. Its working frequency is in the area above the sound wave from 40 kHz to 400 kHz. The ultrasonic sensor according to [8] is composed of two units, that are the transmitter and the receiver unit. The structure of the transmitter and receiver unit is very simple, where a piezoelectric crystal is connected to the mechanical anchor and is only connected to the vibrating diaphragm. Alternating voltage that has a working frequency in the range of 40 kHz - 400 kHz is given to the metal plate. The atomic structure of the piezoelectric crystal will contract (bind), expand or shrink based on the voltage polarity given and this is called the piezoelectric effect.

The contractions that occur are transmitted to the vibrating diaphragm so that ultrasonic waves emit into the air (surrounding area). The reflection of the ultrasonic waves will occur if there are certain objects and it will be received again by the receiving sensor unit. Then the receiving sensor unit will cause the vibrating diaphragm to vibrate and the piezoelectric effect produces an alternte voltage back with the same frequency. More details about the working principle of the ultrasonic sensor can be seen via the principle of the ultrasonic sensor in the following picture :

Figure 6. The working principle of an ultrasonic sensor

The amplitude of the electrical signal generated by the receiving sensor depends on the proximity of the detected object and the quality of the transmitter and receiver sensor. The process carried out on this sensor uses a reflection method to calculate the distance between the sensor and the target object.

The distance between the sensors is calculated by multiplying the half time used by the ultrasonic signal on its way from the sending circuit until it is received by the receiving circuit with the propagation speed of the ultrasonic signal on the air medium used [9]. The ultrasonic wave reflection can be used to measure the distance between the sensor and objects that can ideally be calculated by the following formula :

$$s = 0{,}5.v.t \tag{4}$$

Where, s is the distance between the object and the sensor (m) and v is the sound propagation speed on the medium which is 344 m/sec, t = traveling time (seconds). The reflection principle of this ultrasonic sensor can be seen in Figure 6. The Ultrasonic Receiver will receive ultrasonic signals emitted by ultrasonic transmitters with suitable frequency characteristics. The received signal will experience the frequency filtering process by using a band pass filter circuit, with the determined value of the pass-frequency.

Then the output signal will be amplified and passed to the comparator circuit with the reference voltage which is determined based on the amplifier output voltage when the distance between the mini vehicle sensor and the divider wall reaches a minimum distance to turn. It can an be assumed that the comparator output in this condition is high (logic 1) while a longer distance is low (logic 0). A temperature sensor of DS18b20 type is a digital temperature sensor which has an ADC (Analog to Digital Converter) in it with 12 bit resolution. This sensor has a fairly high level of accuracy that is ± 0.5 °C in the temperature range of -10 °C to + 85 °C and overall can measure temperatures from -55 °C to + 125 °C. This ADC resolution will be related to the temperature increase of each analog level, i.e. resolution of 9 bits has the temperature rise for each analog level of 0.5 °C, resolution of 10 bits has the temperature rise for each analog level of 0.25 °C, resolution 11 bits has the temperature rise for each analog level of 0.125 °C, and the resolution 12 bit has the temperature increase of each analog level of 0.0625 °C.

2. Research Methods

2.1. Flow Chart of Water Level and Temperature Control

Flow chart form of the mechanism of water level height and its temperture control using ultrasonic and temperature sensor is shown in Figure 7.

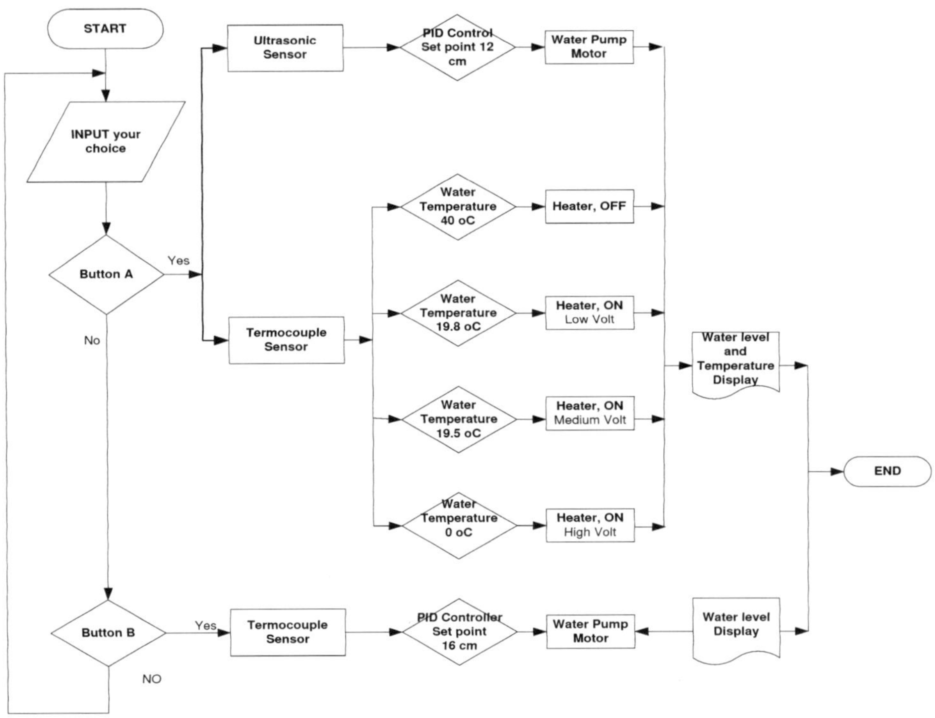

Figure 7. Flow chart of water level and temperature control.

2.2. Experiments

The Format of temperature registers on DS18b20 in Figure 8, is firstly calibrated to degrees Celsius, °C. In registers sizing of 2 bytes or 16 bits there is Sign-bits (S) which functions to indicate whether the temperature is in a positive state (getting hotter) or in a negative state (getting cooler).

IOP Conf. Series: Materials Science and Engineering **619** (2019) 012039 doi:10.1088/1757-899X/619/1/012039

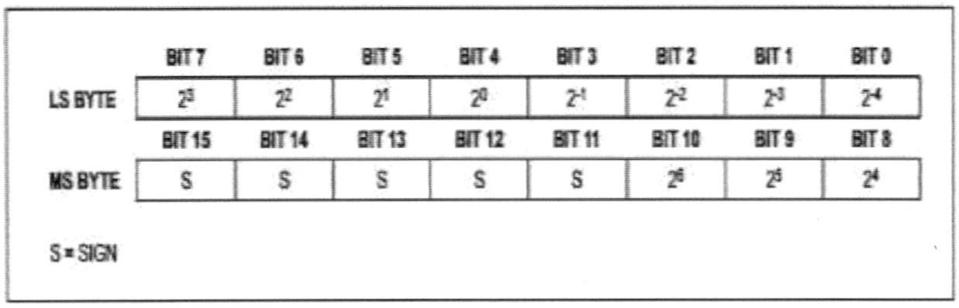

(a)

(b)

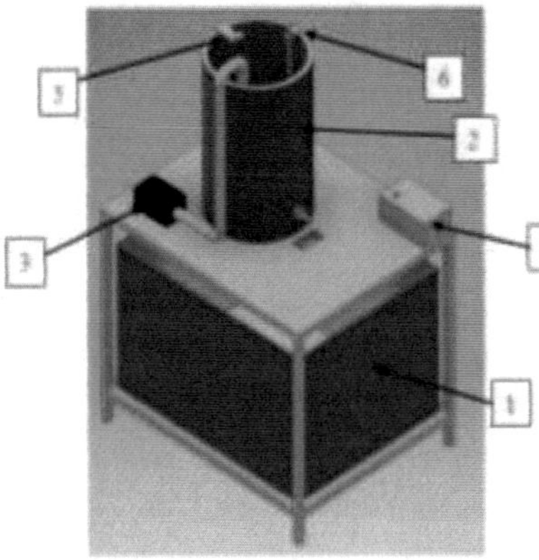

(c)

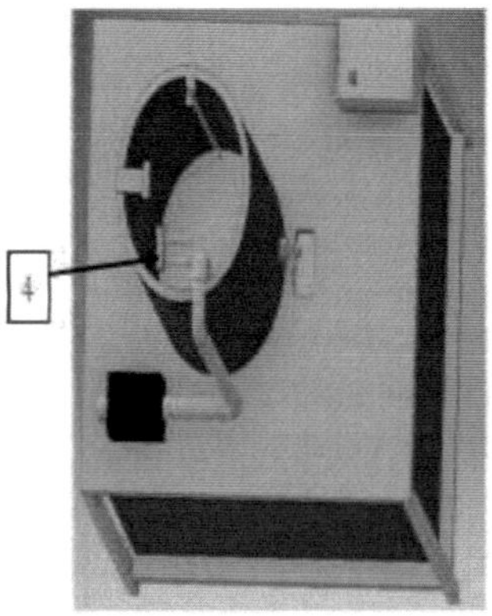

(d)

Caption :

1. Tank A
2. Tank B
3. Water Pump
4. *Heater*
5. Ultrasonic Sensor
6. Temperature Sensor
7. Motor *Driver* and *Heater*

Figure 8. Format of Register, temperature sensor, DS18b20, water level and temperature control device.

Table 1. The output data of digital sensor DS18b20

TEMPERATURE (°C)	DIGITAL OUTPUT (BINARY)	DIGITAL OUTPUT (HEX)
+125	0000 0111 1101 0000	07D0h
+85*	0000 0101 0101 0000	0550h
+25.0625	0000 0001 1001 0001	0191h
+10.125	0000 0000 1010 0010	00A2h
+0.5	0000 0000 0000 1000	0008h
0	0000 0000 0000 0000	0000h
-0.5	1111 1111 1111 1000	FFF8h
-10.125	1111 1111 0101 1110	FF5Eh
-25.0625	1111 1110 0110 1111	FE6Fh
-55	1111 1100 1001 0000	FC90h

Dimensions of stainless steel pipe : 0.6 cm diameter and 3.6 cm long. Jumper cable length : 91 cm with 0.4 cm diameter. Range of temperature measuring : -55 °C to + 125 °C. Accuracy : ± 0.5 °C in the temperature range of -10 °C to + 85 °C. Input voltage : 3.0 V - 5 V. Interface : Max 1-Wire Bus Protocol. Conversion time of 12-bit ADC analog to digital temperature data : 750 ms (maximum), and can use 127 DS18b20 at once in parallel.

2.3. *Collecting Data*
The digital data output from the DS18b20 sensor is shown in Table 1, and to communicate with the microcontroller, the DS18b20 sensor uses a 1-wire interface sensor. This means that it only uses 1 data path (and of course the ground) to send and receive data from the microcontroller.

2.4. *Data Analysis*
The simulation of water level control using MatLab/Simulink as shown in Figure 9, utilizes the setpoint at a height of 25 cm, while in the PID control set point control is made with $K_p = 95$, $K_I = 11$, and $K_D = 3.5$.

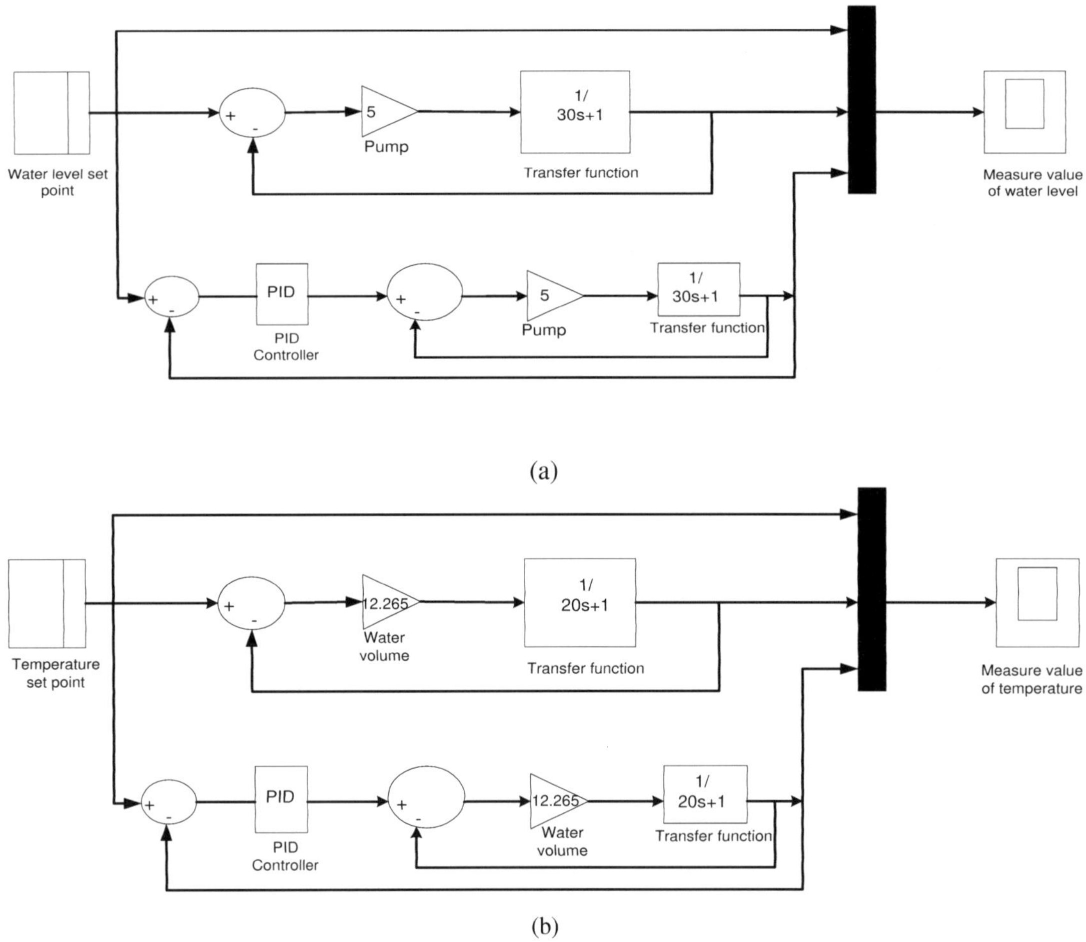

Figure 9. Block diagram of Mat Lab/ Simulink on water level and temperature control.

Shown in the graph that the green line is set point, blue line is a system with PID control, and red line is a system without PID control. The simulation results show that at the set point of 25 cm of water height, and the PID control is set at $K_p = 95$, $K_I = 11$, and $K_D = 3.5$, graphs is obtained as shown in Figure 10, i.e. with the PID control it is obtained the constant level of 25 cm, while controlling without using PID has a changing water level and is at a water level of less than 20 cm in an interval of 10 s.

3. Results And Discussion

3.1. Experiment of water level control

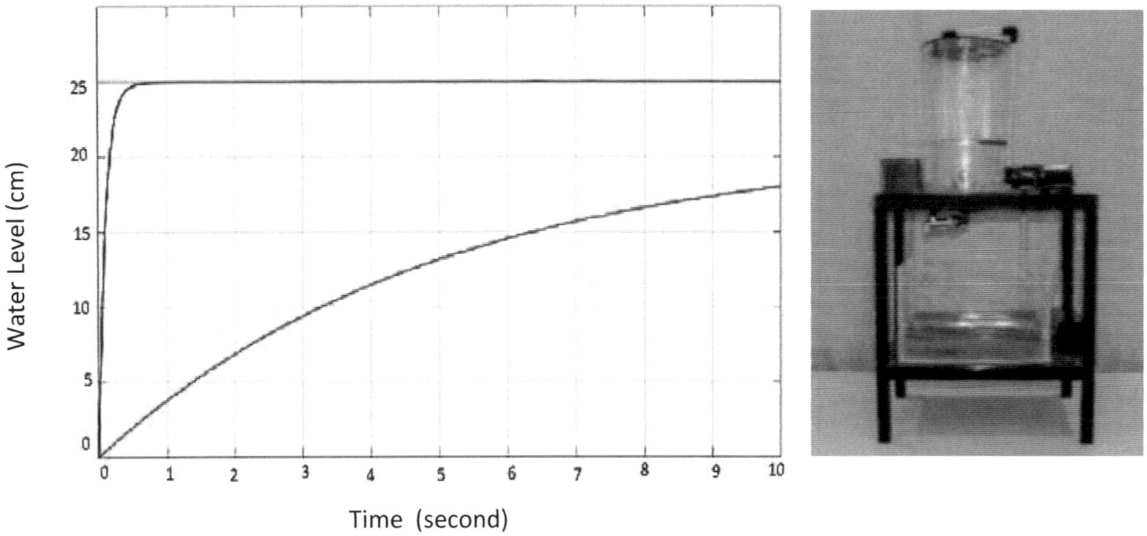

Figure 10. Graphs of Simulink program output for water level control.

3.2. Experiment of water temperature control

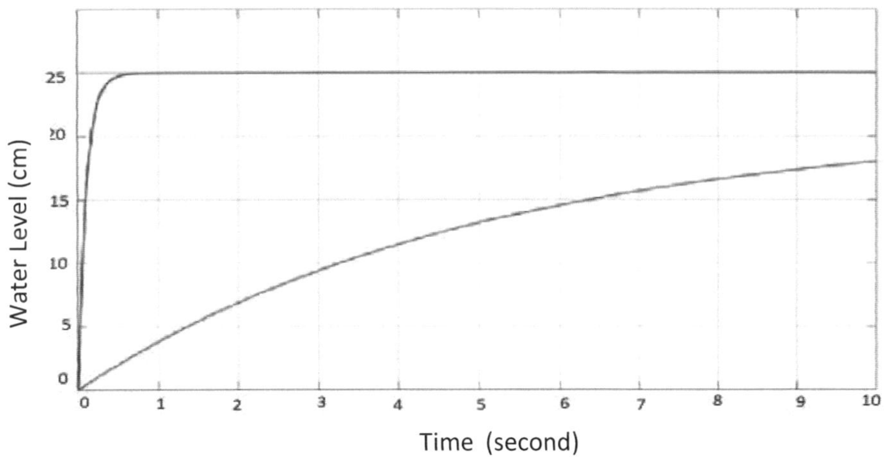

Figure 11. Graphs of Simulink program output for water temperature control.

In the graphs of Figure 11, if the set point is set at water temperature of 60 °C, and the PID control is set at K_P = 60, K_I = 16.50, and K_D = 2, then the maximum water level reaches 25 cm in t = 0.5 s, whereas controlling without PID results in the maximum value of the water level of 17.5 cm for 10 s.

3.3. Experiment of water heater control

Based on the heating experiments using heater in related to water levels and the duration of heating, then it is obtained the water temperature as shown in Table 2 and Figure 12.

Table 2. Temperature as a result of heating with heater.

Level (cm)	Time (s)					Temperature (degree celcius)						
	I	II	III	IV	V	t0	t1	t2	t3	t4	t5	Aver-age
12	60	120	180	240	300	25,69	25,82	26,27	27,29	28,53	30,02	0,86
10						26,02	26,49	27,25	28,18	29,26	30,45	0,88
8						26,17	26,86	27,69	28,78	29,96	31,25	1,01

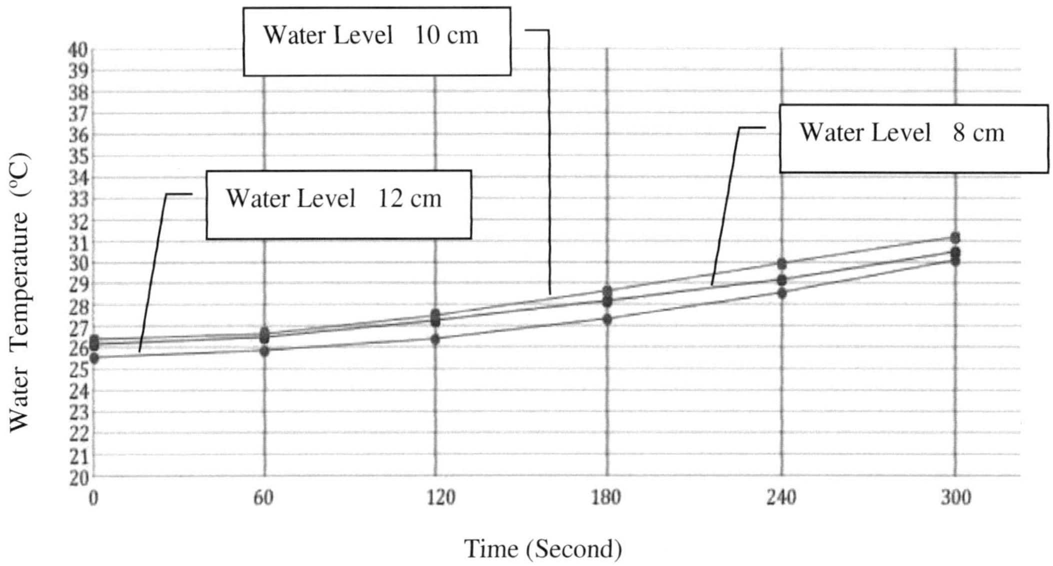

Figure 12. Graphs as a result of heating with Heater.

Based on the results of the heater test data, it is obtained that the heating power of the heater varies according to the voltage and water level in the tank. In Table 2, it can be seen that the time given in each experiment is the same i.e. 5 minutes, but the heat produced varies. When the water level is 8 cm and the voltage is 220 volt heater heats fastest, whereas at the water level 12 cm and the voltage given is 150 volts the heater heats quite long.

4. CONCLUSIONS

1. This water level and temperature control system has been proven as an alternative to controlling the level and temperature of water in the storage tank. This control system uses a microcontroller and ultrasonic sensor to detect water level, and the DS18B20 temperature sensor as a water temperature detector.
2. From the tests carried out, the control circuit of water pump dc motor functions well in controlling the level and temperature of water in the reservoir tank and can be monitored on a computer screen using MATLAB software.

REFERENCES

[1] N. E. Budiarso Z, "Microcontroller Based Damage Water Level Monitoring System, "Journal of Dinamika Informatika.," *Journal of Dinamika Informatika,* 2011.

[2] Simon Ka'ka[1,2*], Syukri Himran[2], Ilyas Renreng[2] and Onny Sutresman[2], "Pneumatic Actuator As Vertical Dynamic Load Simulator On The Suspension Mechanism Of A Quarter Vehicle Wheels," *ARPN Journal of Engineering and Applied Sciences,* **vol. 12,** 2017.

[3] I. d. M. A. Puluhulawa, "Design of Microcontroller Based Water Level Control Using Ultrasonic Sensors," 2016.

[4] H. Rianto, "Car Parking Safety uses a Microcontroller Based Ultrasonic sensor," 2010.

[5] L. Susanti, "Report on the Basic Design of Water Level ElectronicsControl" (Online)," 12 April 2017 2013.

[6] S. Deny, "Water Temperature Control Module Based on PID Using MATLAB, PID-BASED WATER TEMPERATURE CONTROL USING_MATLAB_SIMULINK (Online), (http://www.academia.edu/9629321/MODUL " 2013.

[7] M. N. K.G. Suastika, P. Yunus, "Ultrasonic Sensor As An Air Flow Meter In A Pipe," *Journal of Pendidikan Fisika Indonesia,* 2013.

[8] S. Ka'ka[1,2*], S. Himran, [2] Ilyas Renreng[2] and Onny Sutresman[2], "The Pneumatic Actuators As Vertical Dynamic Load Simulators On Medium Weighted Wheel Suspension Mechanism," *Journal of Physics,* **vol. 962**, p. 10, 2018.

[9] Q-ELECTRONICS, "How to use DS18B20 sensor with NodeMCU," *(http://www.qelectronics.id/tutorial-sensor-ds18b20-dengan-nodemcu/,* 2016.

ACKNOWLEDGEMENT

The authors would like to thank to the Study Program of Mechatronics Engineering, and also to the Department of Mechanical Engineering, State Polytechnic of Ujung Pandang for providing the facilities in Workshop and Laboratory.

Investigation of Creep Behaviour of Al-9Zn-4Mg and Al-9Zn-4Mg-5Cu Alloys

Muhammad Syahid[1], Bondan T. Sofyan[2]

[1] Department of Mechanical Engineering, Hasanuddin University,Indonesia
[2] Department of Metallurgy and Materials, University of Indonesia, Indonesia

Email: syahid.arsjad@gmail.com, bondan@eng.ui.ac.id

Abstract. Two aluminium alloys, Al-9Zn-4Mg and Al-9Zn-4Mg-5Cu was prepared for candidate material impeller turbine radial inflow system Organic Rankine Cycle (ORC) that works at 130 °C. The temperature is high for aluminium alloys so the creep behaviour of the alloys have to investigated. This study identified creep resistance of the alloys and also investigated the creep mechanism of the alloys. Creep test conduct at 50 MPa and variation temperature 150, 200 and 250 °C. The fracture surface and microstructure of subsurface was examinated by SEM. The creep test result showed both of alloys did not failure at 150 °C while Al-9Zn-4Mg-5Cu has a good creep strength compare with Al-9Zn-4Mg at temperature 200 and 250 °C. The fracture surface and microstructure of subsurface indicate the coble creep mechanism or difusional creep. The good creep strength of Al-9Zn-4Mg-5Cu due to $CuAl_2$ phase serves as a pin which inhibit the rate of diffusion at the grain boundaries.
Keywords: Creep, Al-Zn-Mg, Al-Zn-Mg-Cu, Impeller, Turbine ORC

1. Introduction

Al-Zn-Mg-Cu alloys have been widely used due to their low density, high strength to weight ratio high corrosion resistance. This kind of alloy is an ageing-hardening treated alloy and heat treatment processing has a great effect on its microstructures and properties. Over the years, new alloys have been developed with increasing total solute content and varying ratios between the three main alloying elements (Zn, Mg and Cu), in order to reach very high levels of strength [1,2]

Al-9Zn-4Mg and Al-9Zn-4Mg-5Cu was prepared for candidate material impeller turbine radial inflow system Organic Rankine Cycle (ORC) that works at 130 °C . Aluminium alloys have excellent mechanical properties of at temperatures between 25 and 100 °C but tensile strength sharply decreases with increasing at elevated temperatures [3]. However, creep resistance is one of the properties required for use of materials in elevated temperature applications. The rate of creep is a strong function of the applied stress and temperature and the resistance to this form of deformation is obviously of great importance when materials are used at elevated temperatures [4,5].

The mechanism of creep in aluminium and its alloys remains a controversial subject. For precipitation hardened aluminium alloys, the mechanism of creep is believed to consist of glide and diffusion assisted climb with the associated development of a dislocation substructure, the dislocation ± precipitate interactions are of primary importance. The creep rate will be determined by the rate at which dislocations interact with and ultimately overcome obstacles provided by the precipitates [5] . Therefore, creep resistance of precipitate hardenable aluminium alloys can be enhance by fine precipitates that provide obstacle to disclocation movement. In contrast, Rapid precipitate coarsening occurs together with loss creep resistance under high temperature applications [4]. The aim of this research is to investigate creep behaviour of Al-9Zn-4Mg and Al-9Zn-4Mg-5Cu that prepared for for candidate material impeller turbine radial inflow system Organic Rankine Cycle (ORC) that works at 130 °C .

2. Research Methodology

Al-9Zn-4Mg and Al-9Zn-4Mg-5Cu was used in this study. This material was manufactured by gravity die casting according to specimen of tension test. The material was heat treated with homogenization at 440± 5 °C for 24h and solution treatment at 460 ± 5 °C for 1 hour. Then, continue with artificial ageing at temperature 130 °C for 16 hours. Uniaxial tensile test was performed before uniaxial tension creep experiment was conducted. The uniaxial tension creep test was performed acording to standart of ASTME 139-06. Three different temperatures (150, 200, 250 °C) with constant stresses (50 Mpa). Surface and sub surface fracture of creep specimen was identified by Scanning Electron Microscope (SEM).

3. Result and Discussion

Specimen for creep test produced by gravity die casting, the microstructures can be seen at Figure 1a and 1b. The microstructures relatively same with impeller turbine was produced by investment casting (Figure 1.c) that used for impeller turbine radial inflow system Organic Rankine Cycle works at temperature 130 °C. Addition of 5 (wt.%) Cu increase second phase at grain boundary, commonly $MgZn_2$, $CuAl_2$ and $CuMgAl_2$ [2]. It can be seen that the second phase more thick in 5 wt.% Cu Alloys. The second phase of $MgZn_2$ and $CuAl_2$ have a good strength and can increase the properties of alloys. After heat treatment, the second phase play esential rule for precipitation hardening.

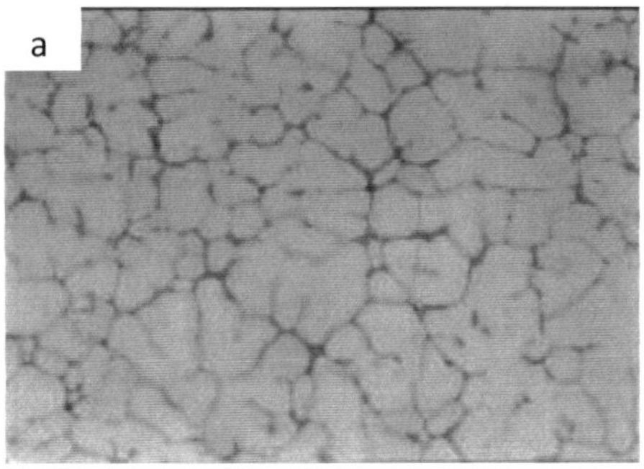

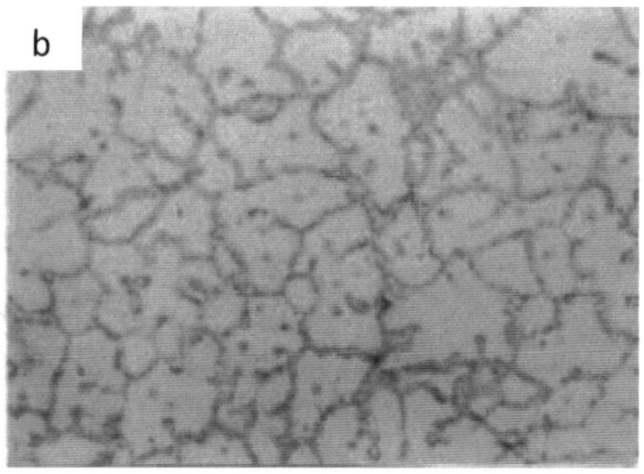

Figure 1. Microstructure of as cast specimen produced by die casting a) Al-9Zn-4Mg b) Al-9Zn-4Mg-5Cu

Figure 2a shows hardness value of alloy after heat treatment. Hardness value of Al-9Zn-4Mg-5Cu alloy achived 91 HRB higher than Al-9Zn-4Mg that only 67 HRB. Addition 5 wt. % Cu significantly increase the hardness of alloys due to $CuAl_2$ precipitate that formed through ageing proces. Figure 2b shows the ultimate strength and elongation of both alloys Al-9Zn-4Mg and Al-9Zn-4Mg-5Cu. It can be seen that ultimate strength of Al-9Zn-4Mg-5Cu at 179.59 MPa higher than Al-9Zn-4Mg that only 152.76 MPa. It also can be seen that the elongasion of Al-9Zn-4Mg-5Cu higher than Al-9Zn-4Mg. It means that $CuAl_2$ precipitate not only increase hardness value but also significantly increase strength, thoughness and ductility.

(a)

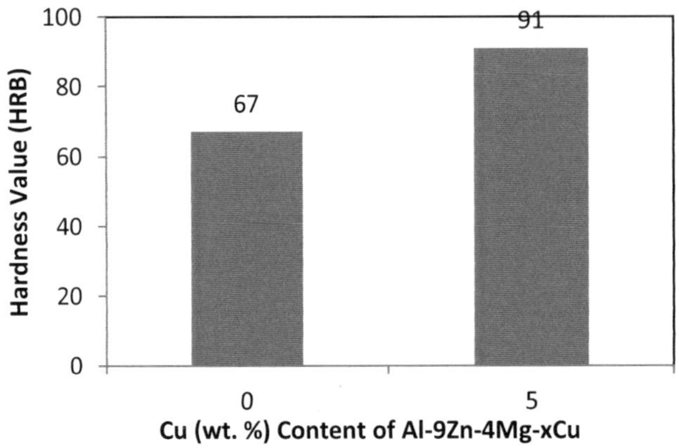

(b)

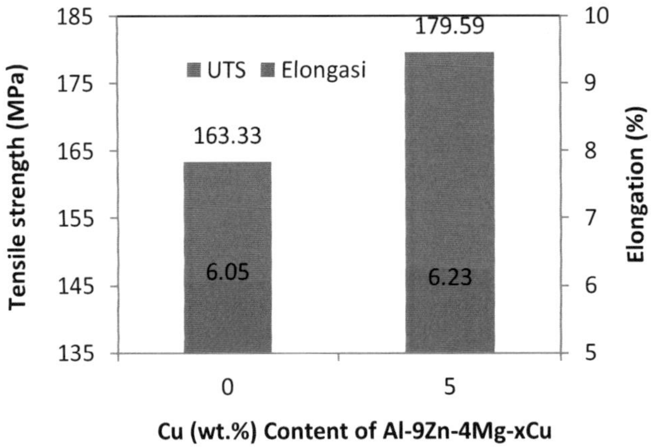

Figure 2. a.) Hardness value b.) Tensile strength and elongation of Al-9Zn-4Mg and Al-9Zn-4Mg-5Cu after heat treatment, homogenization at 440± 5 °C for 24h and solution treatment at 460 ± 5 °C for 1 hour. Then, continue with artificial ageing at temperature 130 °C for 16 hours

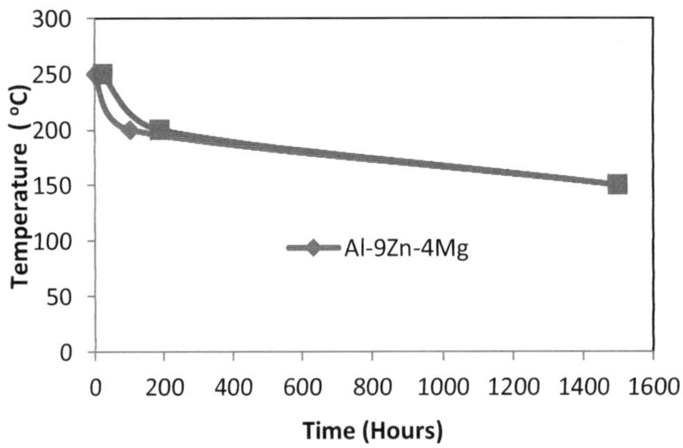

Note : Under temperature 150 °C, the alloy did not failure until 1500 hours

Figure 3. Tempcraturc vs time result of creep test Al-9Zn-4Mg dan Al-9Zn-4Mg-5Cu alloys with stressed 50 MPa

Figure 3 shows result of creep test of Al-9Zn-4Mg and Al-9Zn-4Mg-5Cu with stressed 50 MPa and variation temperature at 150, 200 and 250 °C. It can be seen that both of alloys failure when aplied temperature test at 250 °C and 200 °C. But both of alloys did not failure when temperature test at 150 °C until 1500 hours. It means that both of alloys can works under temperature 150 °C, save from creep failure. From Figure 3 also can be seen that addition 5 wt.% Cu increase creep resistance of alloys. Creep test at temperature 200 °C, Al-9Zn-4Mg failure at 104 hours and Al-9Zn-4Mg-5Cu failure at 190 hours. When temperature of creep test conducted at 250 °C, Al-9Zn-4Mg failure only at 1,5 hours and Al-9Zn-4Mg-5Cu failure at 25 hours. The increasing of temperature decrease creep resistance of alloy significantly. Increasing temperature decrease the ductility of alloys and became more brittle.

It can be seen at figure 4 the surface and sub surface facture of specimen after creep test. The surface fracture of Al-9Zn-4Mg-5Cu shows more dimple that indicate ductile fracture than Al-9Zn-4Mg. Sub surface fracture shows void at grain boundary. It is indicate the coble creep mechanism, where failure due to difusion of atom at grain boundary [4]. Void at grain boundary in Al-9Zn-4Mg more large than Al-9Zn-4Mg-5Cu, in Al-9Zn-4Mg-5Cu there are second phase at grain boundary that works as pin or obstacle and avoid the crack. The second phase was identified as $CuAl_2$. $CuAl_2$ not only play as precipitate that increase the strength and thoughness, but $CuAl_2$ which is not solute in the matrix and stay at grain boundary play as pin or provide effective obstacles to dislocation movement[5].

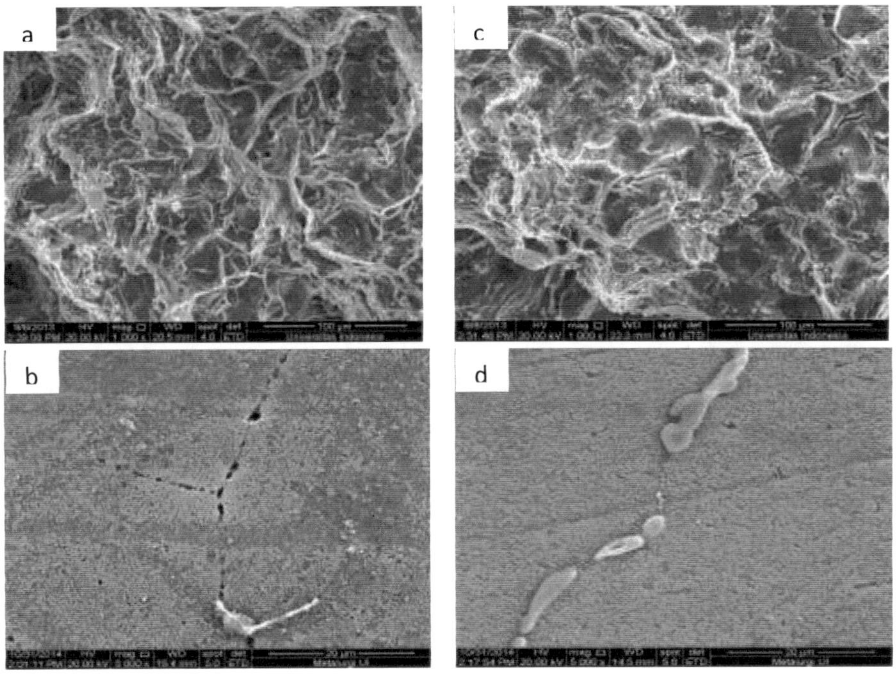

Figure 4. Surface fracture and sub surface fracture specimen after creep test, conducted at temperature 250 °C and stress 50 Mpa (a,b) Al-9Zn-4Mg and (c,d) Al-9Zn-4Mg-5Cu

4. Conclusion

The effects of Cu on the and creep resistance of the Al –ZnMg--Cu alloy were investigated. But both of alloys did not failure when temperature test at 150 °C until 1500 hours. It means that both of alloys can works under temperature 150 °C without creep failure. Creep resistance of Al-9Zn-4Mg-5Cu higher than Al-9Zn-4Mg due to $CuAl_2$ phase. $CuAl_2$ not only play as precipitate that increase the strength and thoughness, but $CuAl_2$ which is not solute in the matrix and stay at grain boundary play as pin or obstacle and works to avoid crack at the grain boundary in high temperature.

References

[1] ZANG Jin-xin, ZHANG Kun, DAI Sheng-long, Precipitation behavior and properties of a new high strength Al–Zn–Mg–Cu alloy, Trans. Nonferrous Met. Soc. China, **Vol 22**, 2012, pp. 2638–2644

[2] Chandan Mondal, A.K. MukhopadhyayOn the nature of $T(Al_2Mg_3Zn_3)$ and $S(Al_2CuMg)$ phases present in as-cast and annealed 7055 aluminum alloy, Materials Science and Engineering, **vol. A 391**, 2005, pp. 367–376

[3] ASM Metals handbook, Properties and Selection: Nonferrous Alloy and Special-Purpose Materials, **Vol. 2**, ASM International, USA, 1991.

[4] N. Mahathaninwong a , Y. Zhou , Creep rupture behavior of semi-solid cast 7075-T6 Al alloy, material science and engineering, **vol. A556**, 2012, pp 107-113

[5] J. S. Robinson, R. L. Cudd and J. T. Evans, Creep resistant aluminium alloys and their applications, Materials Science and Technology, **Vol.19**, pp. 143-155

Acknowledgments

The research was funded by the PUPT UI 2014. The authors would like to thanks also BATAN for facilitating creep test. MS is grateful for the provision of scholarship by BPP-DN.

Experimental study on machinability in sub-millimeter range of the Aluminum Alloy using Face Turning Method

Hairul Arsyad, Lukmanul Hakim, Nurul Annisa

Mechanical Engineering Department, University of Hasanuddin

The Corresponding Authors: arsyadhairul@yahoo.com

Abstract. One of the most factors that affect the quality of machining results is the machinability of materials. Machinability can affect the accuracy of dimension of the machining product, especially in producing small products. This study has purpose to obtain the circularity value and surface morphology of specimen as indicator of machinability of alloys Al 6020 by using face turning method in the range sub-millimeter. The effects of spindle speed and feed on circularity and surface morphology are the subject of the study. Three variations of spindle speed (440 rpm, 740 rpm and 1500 rpm) and six variations of feed (0.05 mm/rev, 0.1 mm/rev, 0.2 mm/rev, 0.3 mm/rev, 0.5 mm/rev and 1 mm/rev) were used. The circularity was calculated by image processing software and surface morphological was observed using SEM and optical microscope. Results show that the better circularity was found on smallest and largest feed for each spindle speed. The surface morphology shows some burr defect and also scatter.

Keywords. spindle speed, feed, circularity, surface morphology face turning

1. Introduction

The manufacture of engineering products with sub-millimeter sizes has earned many portions in recent years. The development of research in sub-millimeter area also led to small products improvement. The main goal of research for small product is stability and dimensional accuracy. The machining process also cannot be separated from the manufacture of products in this range. The word micro machining refers to the manufacture of the product by machining process which has dimension in range 1-999 micron [1-3]. However by reducing the size of the product in each manufacturing process is always accompanied by many problems. Some researchers have emphasized the change in material behavior during manufacturing of small product. Some of parameters have an effect on the fabrication of small metal products including size of the product, microstructure of raw materials, tools and also manufacturing conditions [4].

Several researchers have studied that in the range of sub-millimeter product, the raw materials shouldn't be considered as homogeneous materials. However, the raw materials should be stated as heterogeneous materials [4-5]. This is due to size effect phenomena, where the grain size of the material is close to the size of the product. Consequently the shape and final size of the product could be inaccurate.

In machining process, machinability has been known as a parameter that indicates the material ability to be manufactured. The machinability of material tends to decreases in the area of micromachining.

Produce component in very small size is highly dependent on microstructure conditions of raw materials such as grain size, grain boundary, crystallographic orientation, composition and also inclusion [6-7].

Face turning is one of turning operation that aims to make a smooth surface on cylindrical cross-section of machined products. Face turning is generally performed after major machining operations. The cutting process in the face turning is applied on two cylindrical ends of the workpiece. As the final cutting process, the facing process generally uses high rotation and small feeding depth and as a result is a smooth and fine cross-sectional surface. Several approaches to the machining process have been developed to examine the machinability for small products [2,3,9,10]. This research tried to use face-turning as a method to describe the machinability of Al 6020 in the range of submilimeter especially in circularity value and surface morphology

2. Experimental method

2.1 Material

The aluminum alloy was selected as testing material and the composition of the material shows in table1. The experimental set up was shown in figure1.

Table 1. The composition of raw material.

Element	Mg	Si	Fe	Pb	Zn	Cr	Ni	Cu	Al
%	1.91	0.564	0.248	-	-	-	-	0.173	Bal

2.2 *Face turning experiment set-up*

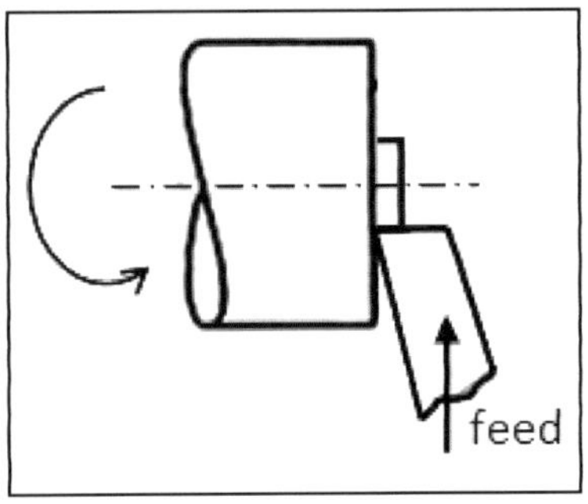

Figure 1. Experimental Setup of Face Turning method.

The experiments were carried out on conventional lathe (Emco Maier+Co). The tool used was HSS type.. The process parameters of face-turning was shows in table 2. Specimen produced by face turning then observed using optical microscopes and SEM.

Table 2. Process Parameters

Spindle speed (rpm)	Feed (mm/rev)					
440 740 1500	1	0.5	0.3	0.2	0.1	0.05

Circularity measurement is accomplish by photo of cross section of the workpieces and then processed using image-J as processing software. The measurement of circularity as machinability parameter is based on the outer circumference obtained from cross section of specimen after face turning process

3 Result and discussion

Figure 2 shows the example of specimen after face turning. In this study, the final diameter was setup to 0.5 mm for each feed and each rotational speed of spindle. After face turning process, the specimen then observed under optical microscope. The photo of circular cross section was taken for circularity measurement.

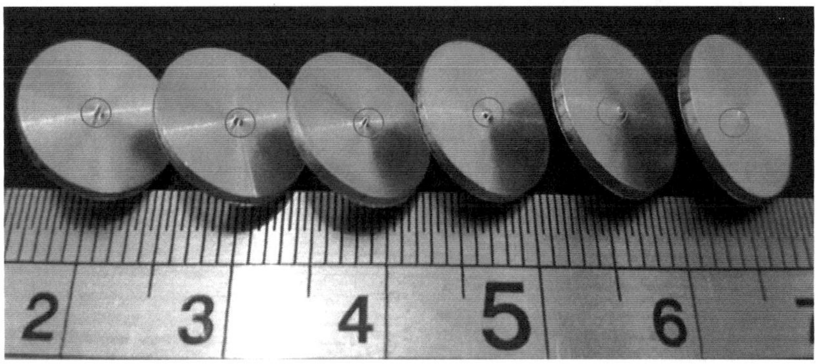

Figure 2 The Specimen produced by face turning process with different feed shows in red circle.

Figure 3 shows the circularity value with different feed in the experiment for each spindle speed. It is shows that at 400 rpm of spindle speed, the average circularity has a higher value at 1 mm/rev of feed but then decrease with decreases of feed (see figure 3a). However at 770 rpm of spindle speed (see figure 3b), the average value of circularity has the higher value at 0.1 mm/rev of feed and for 1500 rpm of spindle speed (see figure 3c) the higher value of circularity is at 0.05 mm/rev.

From figure 3 also shows that by increasing the spindle speed, the circularity tend to has higher inconsistency value, it can be seen from scattering data of circularity, especially at 1500 rpm. This indicates that the geometry of the final circular cross section is influenced by the feed and spindle speed. The value of the circularity can be obtained from the equation

$$Circularity = 4\pi \frac{Area}{Perimeter^2}$$

From the equation it is stated that the perfect circle occurs when the value of circularity equals to one.

The lower circularity indicated that a lot of chips are still attaching on circular cross section and reduce the quality of the cross section. This chip is also called uncut chip [8] which produces burr defects in the cutting process due to un-separated chips from workpiece. According to Wu that burr may be caused by orientation of the crystallographic orientation and it has a great influence on the micro cutting force [4]. The geometry of contact area between tool and workpiece will influence the geometry of chips and usually also produce segmented chips [5]. The geometry of contact area also affect the cutting force which is can create burr defect if the cutting force not successfully separate the chips.

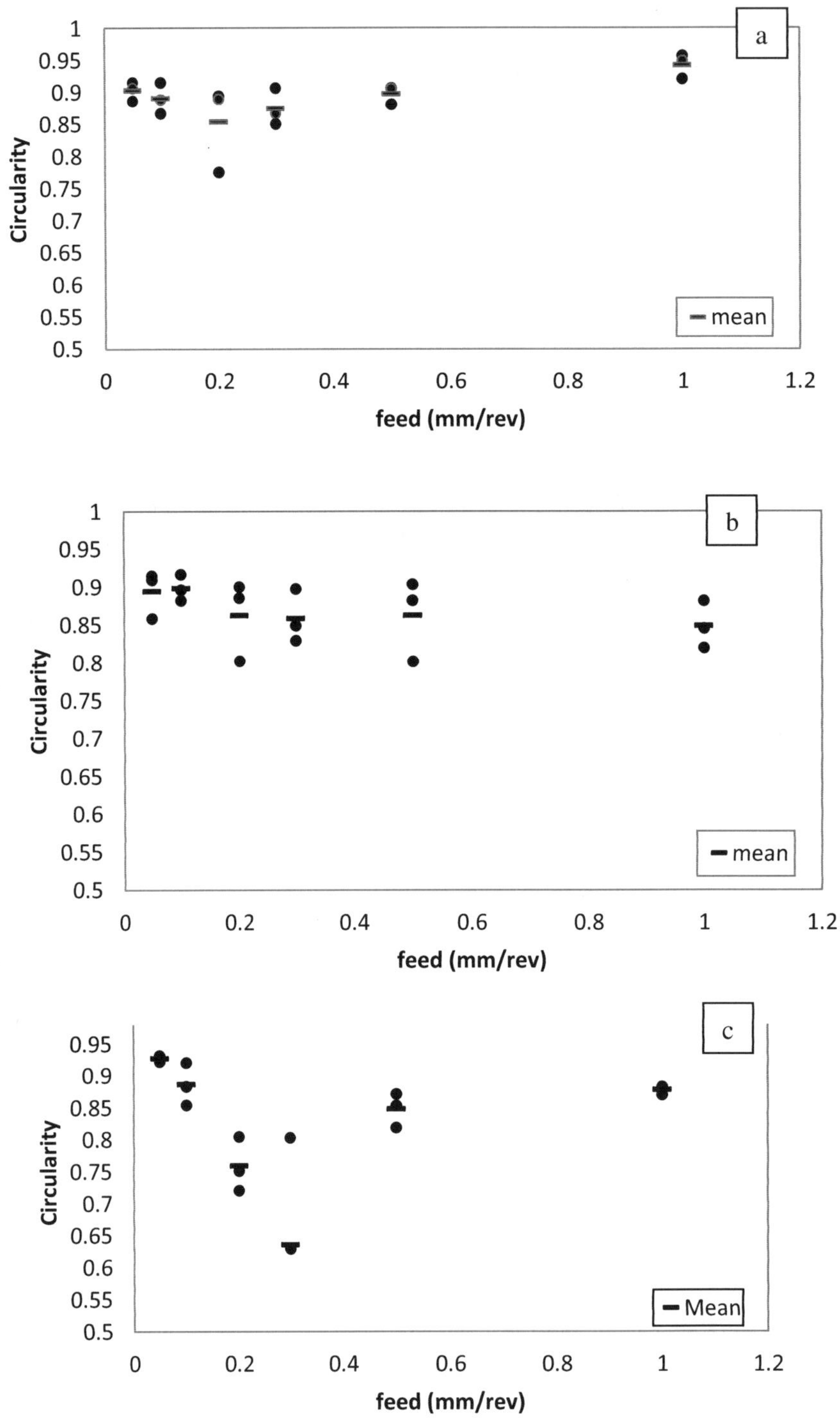

Figure 3. Influence of feed on the circularity value at different spindle speed: a. 440 rpm; b. 740 rpm; c. 1500 rpm

At 1500 rpm as seen in Figure 3c, show a low circular value in the range 0.1 to 0.5 mm of feed. In this range not only produces low circular values but also shows deviation. Some researchers have proven the effect of size that causes inconsistent data and produce low quality of cutting especially in

IOP Conf. Series: Materials Science and Engineering **619** (2019) 012041 doi:10.1088/1757-899X/619/1/012041

micro cutting. One of the causes of this size effect is the crystallographic orientation and size ratio between the microstructure and the cutting size [4]. For the small of feed the cutting process will produces a very thin chips which is the size of the chips could be close to the size of microstructure of the workpiece and the size effect can be affect the cutting process.

Figure 4 shows the average circularity value as a function of different spindle speed conditions. It is clear that the circularity value is higher at 440 rpm and 740 rpm compared to 1500 rpm. The effect of high rotation speed provides low circularity value at feeds 0.2 mm/rev, 0.3 mm/rev and 0.5 mm /rev, this is due to uncut chips. It seems that the circularity tends to has better value at lower spindle speed. It has been known that the cutting force tend to has lower value at high spindle speed. However small feed also will produce a small size of chip and reduce the defect of cutting.

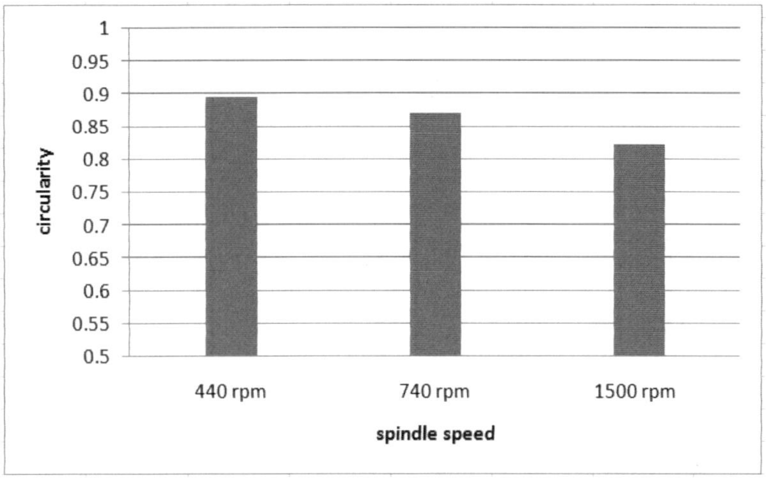

Figure 4. Average value of circularity in different spindle speed.

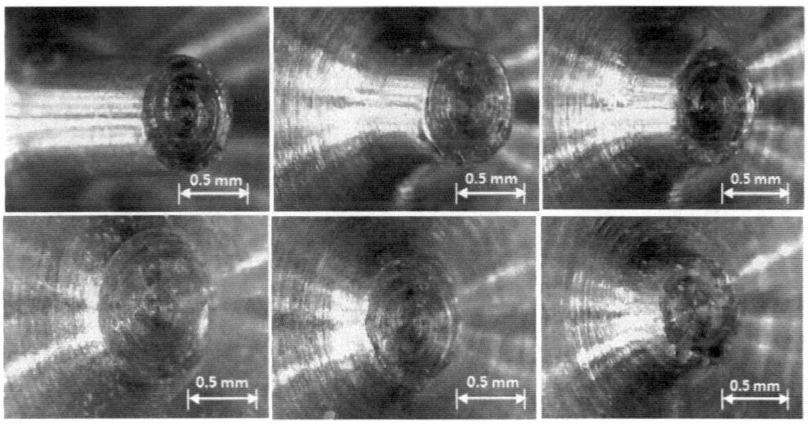

Figure 5. Result of face turning on surface morphology at different feed for 1500 rpm

This can be also attributed to the high rpm value involves high friction and heat. The presence of heat due to friction plays an important role in changing the behavior of shear deformation that occurs at the time of cutting. The heat of the workpiece due to high rotational speed causes the material to has lower deformation resistance. So that the material flow out more easily and create a permanent deformation and left on the surface of the specimen. The result is a defect in the form of burr formation. The presence of the burr defect causes the tolerance and geometry cannot be achieved [8].

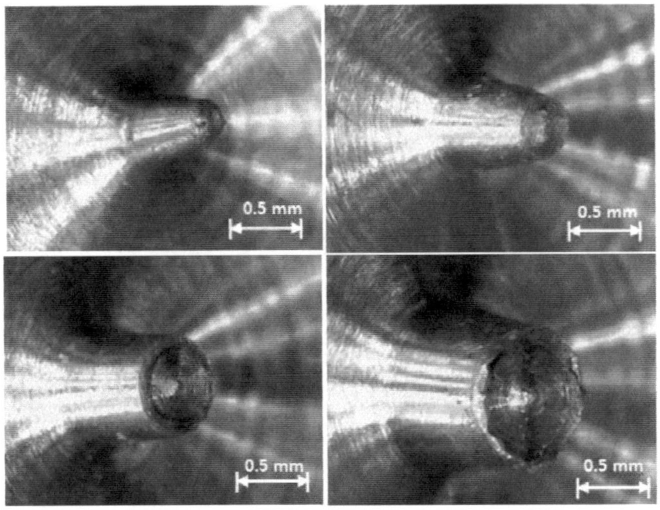

Figure 6. Result of face turning on surface morphology at different diameter of specimen for 1500 rpm with constant feed (0.5 mm/rev).

Figure 6 shows surface morphology with different diameter at 1500 rpm and 0.5 mm of feed. It shows that for small diameter, the diameter along specimen is tends to increases at the bottom of the specimen and show taper forms. This probably because of the tools geometry used in this study with high radius of nose. The face turning method used in this experiment shows the different result of circularity value and surface morphology. Its indicate that face turning method probably can be use to predict machinability in the range of micromachining.

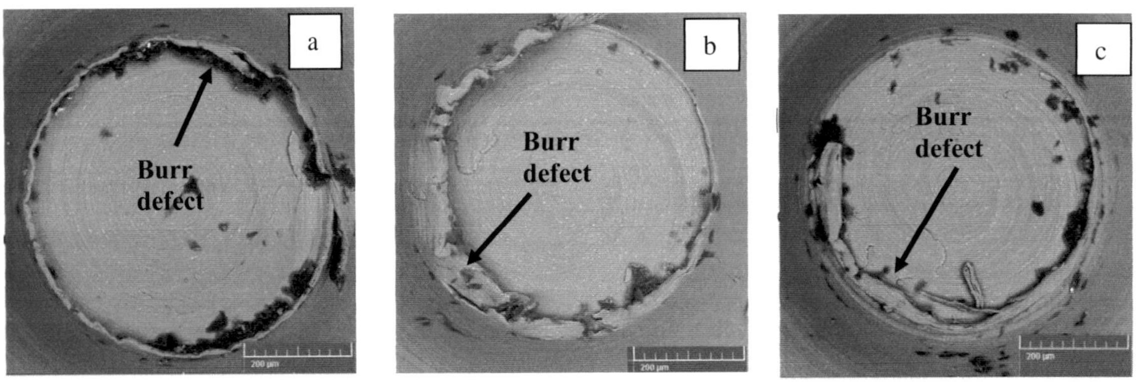

Figure 7. SEM of surface morphology at different spindle speed; a). 440 rpm b) 740 rpm, c) 1500 rpm

Figure 7 shows surface morphology at different spindle speed obtained from SEM, it shows that burr defect was take place for almost area at outer diameter. The burr defect is different in size and also in coverage area at outer diameter. However at 440 rpm, the width size of burr defect is lower than other spindle speed. As mention before that spindle speed affect the temperature of cutting due to friction between tool and workpiece. The mechanism of cutting is depend on shear deformation behavior that influence of chips formation. Uncompleted chips formation is a result of instability condition in cutting process and could be effect of increasing temperature due to high spindle speed

4. Conclusion

The following result can be conclude from this study: The high of circularity value was obtained from highest and smallest feed for each of spindle speed. However the highest circularity was obtained from 440 rpm of spindle speed,

1. At 1500 rpm of the spindle speed shows poor of circularity especially at 0.2 mm/rev, 0.3 mm/rev, and 0.5 mm/rev of feed. At this cutting parameter, shows large of burr defect at the surface of specimen.
2. The face turning method used in this experiment shows the different result of circularity value and surface morphology due to different condition of cutting process.
3. The face turning method probably can be used to study the machinability in the range of micromachining.

References

[1]. M. Prakash & M. Kanthababu & K. P. Rajurkar (2014), Investigations on the effects of tool wear on chip formation mechanism and chip morphology using acoustic emission signal in the microendmilling of aluminum alloy, Int J Adv Manuf Technol (2015) 77:1499–1511.

[2]. M. Azizur Rahman · M. Rahman · A. Senthil Kumar · H.S. Lim · A.B.M.A. Asad (2005), Development of micropin fabrication process using tool based micromachining, Int J Adv Manuf Technol (2006) 27: 939–944.

[3]. Jeong-Du Kim, Dong-Sik Kim, Development and Application of an Ultra-precision Lathe, International Journal of Advance Manufacturing Technology (1997) 13:164-171

[4]. Xian Wu, Liang Li, Ning He, Meng Zhao, Zhongbo Zhan (2015), Investigation on the influence of material microstructure on cutting force and burr formation in the micro cutting of copper, Int J Adv Manuf Technol (2015) 79:321–327.

[5]. JinshengWang, Yadong Gong, Gabriel Abba, Jean Francois Antoine, Jiashun Shi, Chip formation analysis in micromilling operation, Int J Adv Manuf Technol (2009) 45:430–447

[6]. E. Uhlmann, A. M. Zanatta, F. Mahr, J. O. Gomes (2013), Influence of inclusion contents on the micro-machinability of three plastic mold steels, Int J Adv Manuf Technol (2013) 68:2451–2460.

[7]. Dirk Biermann, Alexander Baschin, Influence of cutting edge geometry and cutting edge radius on the stability of micromilling processes (2009), Prod. Eng. Res. Devel. 3:375–380

[8]. Liu Zhanqiang, Shi Zhenyu, Wan Yi Definition and determination of the minimum uncut chip thickness of microcutting, International Journal Advance Manufacturing Technology (2013) 69:1219–1232.

[9]. G. Fromentin, J. Gasparoux, K. Agbeviade, J. H. Giovanola, Development of a precision machine to perform and study orthogonal micro-cutting, Production Engineering Research Development. (2016) 10:217–226.

[10]. Jeong-Du Kim, Dong-Sik Kim, Development and Application of an Ultra-precision Lathe, International Journal of Advance Manufacturing Technology (1997) 13:164-17.

Performance of Candlenut Shell (*Alleuretus moluccana*) Based Supercapacitor Electrode with Acid Electrolytes and Their Salts

M. Zakir[a], H. Kasim[a], I. Raya[a], Y. Lamba[a], Nuratisah[a], A. Belen Jorge[b]

[a]Department of Chemistry, Faculty of Mathematics and Natural Sciences, Hasanuddin University Jl. Perintis Kemerdekaan Km. 10 Makassar 90245, Indonesia
[b]Material Research Institute, School of Engineering and Material Sciences, Queen Mary University, Mile End Rd, E1 4NS London

[1]Corresponding author e-mail: muh.zakir@science.unhas.ac.id

Abstract. Candlenut shell was used as the precursor for activated carbon-based electrodes which prepared through two stages, namely carbonization and activation. The surface of the activated carbon was modified with HNO_3. Characterization of carbons and activated carbon was performed by SEM, BET, FT-IR test and capacitance analysis using Cyclic Voltammetry. The result of SEM analysis shows that carbon pore formation has occurred after activation with H_3PO_4. The surface area of carbon and activated carbon determined by the BET method is 125,828 m^2/g and 142,435 m^2/g, respectively. The results of FTIR and Boehm titration analysis showed an increase in oxygen-containing functional groups after modification with HNO_3. Based on the results of the analysis using Cyclic Voltammetry obtained the highest specific capacitance of unmodified activated carbon is 145.49 mF/g in H_2SO_4 0.5M electrolyte. In addition, the highest specific capacitance of activated carbon modified with HNO_3 is 155.92 mF/g obtained in HCl 0.1 M electrolyte. In the same condition, the contribution of salt electrolyte (Na_2SO_4 0.5 M and 0.1 M) in the specific capacitance of candlenut shell based supercapacitor electrodes is a little bit lower compared to the acid electrolytes.

1. Introduction

Technological advancements and an increasingly modern age demands enormous and growing energy consumption. Various technologies also require energy storage devices. Therefore, it takes a device that has a large energy and power and a shorter charging time to meet future technology needs. One of them is a supercapacitor [1,2].

Supercapacitors utilize electrode surfaces and electrolyte solutions to achieve greater capacitance. Supercapacitors are also more attractive because the main material used as an electrode is activated carbon. One of the raw materials for the manufacture of activated carbon is agricultural waste such as candlenut shell [3,4]. To maximize the function of activated carbon, a surface modification with HNO_3 to add active groups was reported to have an effect in increasing the specific capacitance of activated carbon as a super capacitor electrode material [5, 6].

Another way to improve the energy storage efficiency of a supercapacitor is the proper electrolyte selection. Commonly used electrolytes on supercapacitors include acid, base and salt electrolytes that have high conductivity, low resistance, easy wetting the electrode surface and lower cost [7]. Acid and alkaline electrolytes have compatibility with supercapacitors as they indicate the formation of rapid electric double coatings and salt electrolytes have high electrochemical potential stability that can increase the energy tension and density [7,8,9].

The use of acid and salt electrolytes with the same electrode material shows different values of different capacitances. Different concentrations also show different specific capacitances. This indicates that the type and concentration of the electrolyte affect the specific capacitance of the supercapacitor electrode [10,11].

Based on this, the research was conducted surface modification candlenut shell activated carbon with HNO_3 and determine the specific capacitance values using the electrolyte acids, bases and salts which, H_2SO_4, HCl, and Na_2SO_4.

2. Materials and Methods

2.1. Materials

Candlenut shell, aquades, H_3PO_4 85%, HNO_3 65%, H_2SO_4 0.1; 0.5; 1 M, HCl 0.1; 0,5; 1 M, Na_2SO_4 0,1; 0,5; 1 M, Na_2CO_3 0,05 N, $NaHCO_3$ 0,05 N, 0,05 N HCl, 0,05 N NaOH, copper wire, platinum wire, Whatman filter paper 42, paraffin, aluminum foil, universal pH paper, MO indicator, PP indicator, parafilm and tissue roll.

2.2. Methods

The samples of the candlenut shell were washed thoroughly and split into small pieces. Furthermore, dried under the sun.

Candlenut shell samples were clean and dry put in a porcelain dish and heated in a furnace at 400 ° C for 2 hours. This process will produce a carbon of candlenut shell. After carbonization, the resulting carbon is then cooled, smoothed, then sieved with a size of 100 mesh [12].

Candlenut shell carbon soaked with activator solution of H_3PO_4 85% by volume ratio H_3PO_4/carbon mass of 6:1 with a 24 hour soak time. Then the pecan shell carbon is filtered using a Buchner funnel and washed with distilled water until a neutral pH. Samples were obtained dried in an oven at 110 ° C for 1 hour. Activated carbon of hazelnut is cooled in desiccators [12].

Activated carbons are mixed with a chemical agent HNO_3 65%, with a mass ratio of 2:1 (volume: active carbon mass). Then, shaked at a constant rate (120 oscillations per minute) for 24 hours. Afterwards washed with aquades repeatedly until the pH is neutral. Then dried in the oven for 24 hours at 110 ° C [13].

A total of 0,25 g of activated carbon put into four 25-mL volumetric flask, each containing Na_2CO_3 0,05 N, 0.05 N $NaHCO_3$, NaOH 0.05 N and 0,05 N HCl and then the mixture allowed to stand for 24 hours. After that the solution is separated from the carbon by decantation. The separated solution, each taken 5 mL of Na_2CO_3, $NaHCO_3$ and NaOH solution, then added PP indicator, added excess HCl, then reversed using 0,05 N NaOH solution and for HCl solution was taken as much as 5 mL, added MM indicator and added excess NaOH, then titrated back using 0,05 N HCl [14].

The electrode body is made by connecting copper and platinum wires using steam solder. Once it is inserted into the pipette and glued using parafilm. The activated carbon shell of hazelnut is mixed with paraffin wax with carbon mass ratio / paraffin wax mass is 1:1 and stirred until homogeny. Subsequently, the carbon paste is fed into the electrode body by being pressed using a spatula to solidify and flatten [15,16].

This measurement uses three electrodes, Pt electrode, an electrode Ag/AgCl and carbon paste electrodes. Electrode test was performed with 50 mV/s scan rate using electrolyte solution of H_2SO_4, HCl, Na_2SO_4 with concentration 0,1 M, 0,5 M, and 1 M so as to obtain voltage and current voltammogram, then calculated value of specific capacitance of energy storage. The electrode body was prepared by connecting copper and platinum wires using a steam solder. The electrode body was then inserted into a tulip pipette and glued together using parafilm. CSAC sample was mixed with parafilm wax with a mass ratio of 1: 1 and stirred until homogeneous by using a spatula in a petri dish. The carbon paste was inserted into the electrode body by being pressed using a spatula for solidifying and flattening. Specific capacitance (Cs) of the carbon paste electrode was measured using Cyclic Voltammetry (CV) technique. This measurement employed three electrodes system i.e. Pt electrode, Ag/AgCl electrode and carbon paste electrode. Electrode testing was performed with scan rate of 50mV/s using electrolyte solution of H_2SO_4 0.1M to obtain voltage-current voltammogram. Based on that data, specific capacitance (Cs) was calculated.

3. Results and Discussions

3.1. Surface Analysis Activated Carbon of Candlenut Shell

The surface morphology and pore formation of activated carbon is shown from the results of the Scanning Electron Microscope (SEM) analysis with magnification of 3000, 7500 and 10000 times as shown in Figure 1.

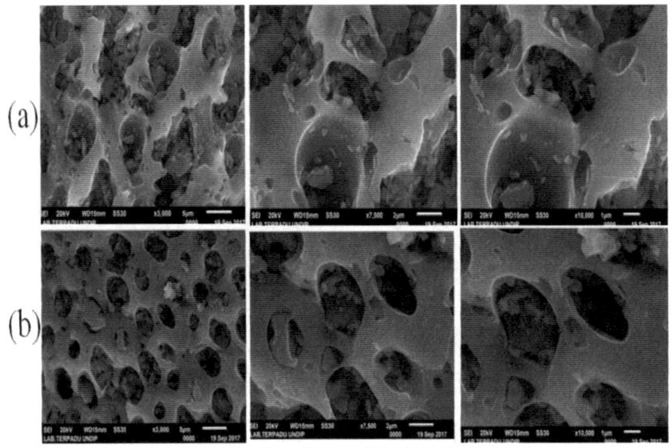

Figure 1. (a) SEM of carbon, (b) SEM of activated carbon

SEM analysis results show that the formation of carbon pores before irregular activation and many impurities that cover the carbon surface compared with after activation. The surface of the carbon pores after activation is cleaner and more open and the pore structure is formed more neatly and regularly.

The surface area, size and volume of activated carbon pores are confirmed from the measurements using the Brunauer-Emmett-Teller isotherm (BET) method shown in Table 1.

Table 1. Results of BET analysis of activated carbon

Sampel	S (m²/g)	V_{pori} (cc/g)	r_{pori} (A)
Carbon	125,828	0,116	16,072
Activated Carbon	142,435	0,121	17,134

Based on the results of SEM and BET analysis showed that the activation process with H_3PO_4 increased pore formation, surface area, volume and pore size of activated carbon candlenut shells. This is expected to facilitate the access of electrolyte ions to the supercapacitor.

Modification of the Activated Carbon Surface of Candlenut Shell

Modification of the surface of activated carbon with HNO3 is done to increase the group containing oxygen. The presence of this group on the carbon surface can increase the hydrophilicity of the carbon which supports the absorption of electrolyte ions into the pores of the carbon and can give the effect of pseudocapacitance [8].

The presence of oxygen-containing functional groups is determined by FTIR analysis as shown in Figure 2 and chemical characterization by the titration Boehm mehod shown in Table 2.

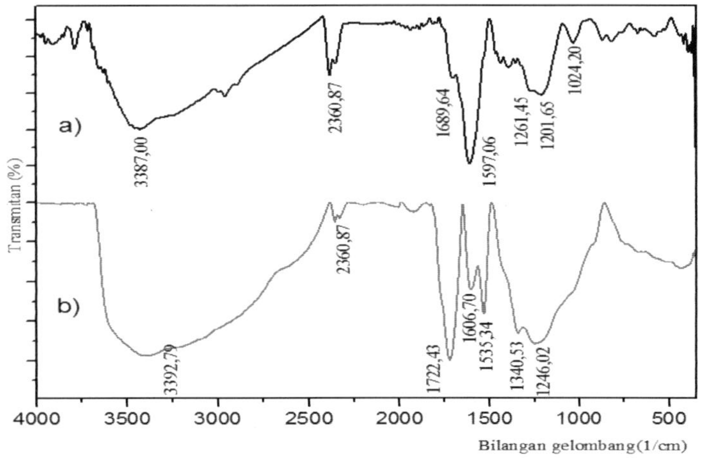

Figure 2. FTIR spectra of (a) activated carbon, (b) activated carbon modification

The FTIR results have a functional group absorption which is each functional group for -C-O, -C=O and -OH. This absorption shows the presence of carbonyl, hydroxyl, quinone and lactone groups. The results of the FTIR analysis of activated carbons before modification showed absorption bands in the 3387 cm-1 that indicated OH. The absorption band of 1597 cm-1 region indicates the presence of C=O group of quinone.

FTIR results after modification indicate an uptake in an area of 3392 cm-1 indicating the presence of OH from a carboxylic group reinforced by a C-O group in an area of 1246 cm1. The 1606 cm-1 absorption band indicates the presence of a C=O group of quinones. The absorption band of 1722 cm-1 indicates the presence of C=O of the lactone reinforced by the absorption area of 1340 cm-1 which is the C-O group.

Table 2. Active Carbon Functional Groups based on Boehm Titration

Sample	Acid groups (meq/g)				Base groups
	Carboxyl	Lakton	Phenol	Total	
Activared carbon	0,34	0,5	0	1,18	0,7218
Activared carbon modification	1,87	1,52	1,83	5,22	0,4392

Increased uptake from FTIR analysis and increased acid concentration from Boehm titration showed increased oxygen-containing groups after HNO_3 modification.

3.2. Determination of Specific Capacitance Using Electrolyte H_2SO_4

H_2SO_4 electrolyte is one of the commonly used acid electrolytes in the measurement of specific capacitance of supercapacitor. Figure 3 shows the voltammogram of the activated carbon candlenut shell electrode using H_2SO_4 electrolyte.

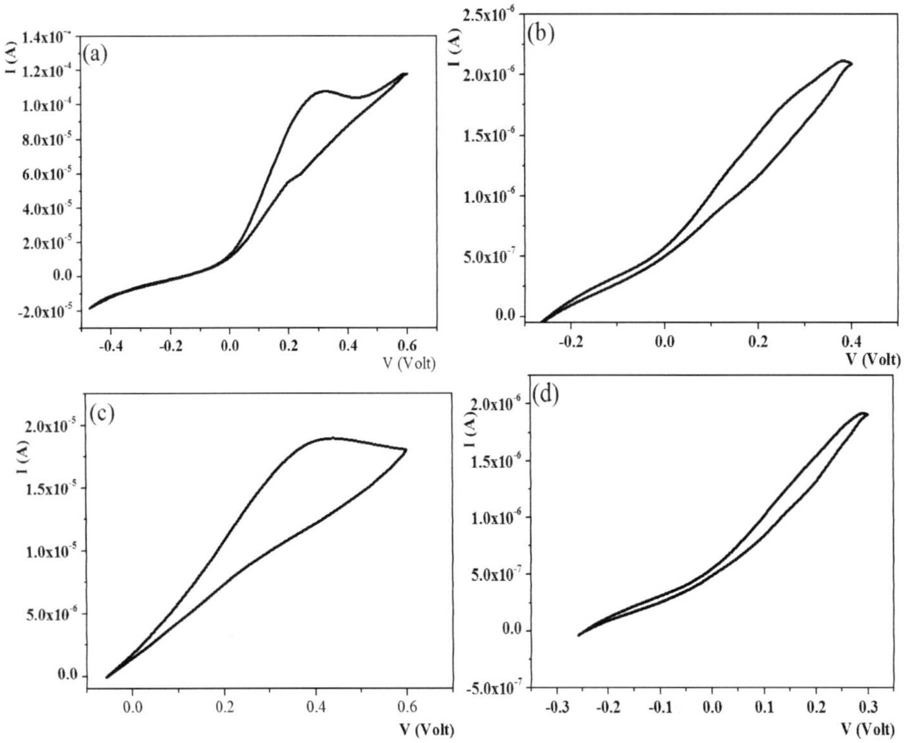

Figure 3. Voltammogram of the KATK electrode (a) before modification (b) after modification in H_2SO_4 electrolyte 0,1 M, (c) 0,5 M, (d) 1 M.

Table 3. Specific capacitance values of activated carbon using electrolyte H_2SO_4

Sample	Electrolyte concentration (M)	Cs (mF/g)
Before modification	0,5	145,49
After modification	0,1	2,26
	0,5	23,68
	1	2,37

The electrolyte concentration of H_2SO_4 0,5 M shows a high specific capacitance because the use of electrolytes equal to different concentrations is influenced by the velocity and amount of ions for the formation of electric double layers. If the concentration is low then the amount of ions is inadequate and if high concentrations can reduce ion mobility [7,11].

The decrease of specific capacitance after modification is predicted due to the decrease of the concentration of base group so that the use of acid electrolyte is not maximal in the formation of electric double layer.

3.3. Determination of Specific Capacitance Using HCl Electrolyte

HCl electrolyte is one of the acid electrolyte used in measuring the specific capacitance supercapacitor. Figure 4 shows a voltammogram of candlenut shell activated carbon paste electrode using a HCl electrolyte.

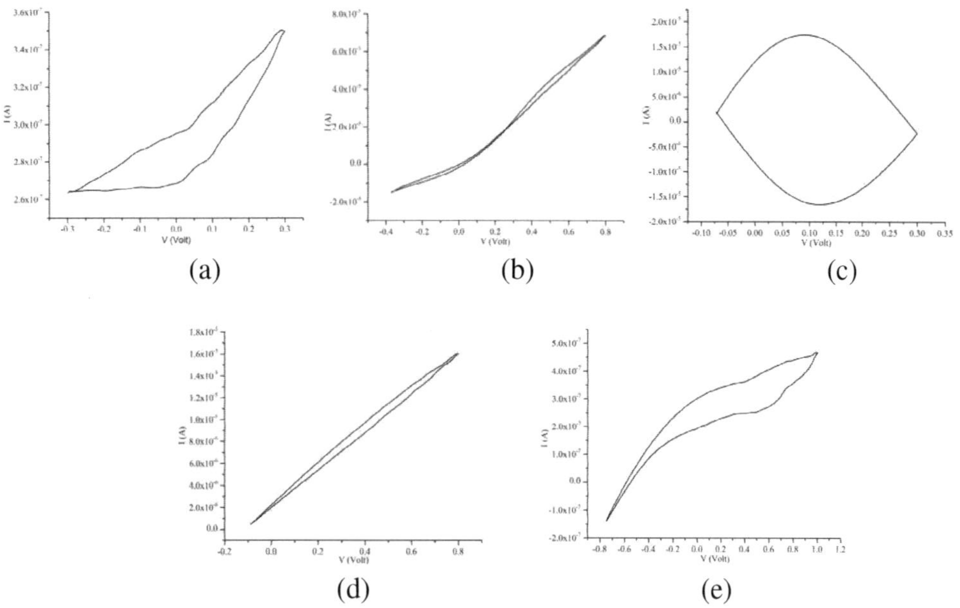

Figure 4. Voltammogram of the KATK electrode (a) before modification (b) after modification in HCl electrolyte 0.1 M, (c) 0.5 M, (d) 0.7 M and (e) 1 M.

The voltammogram of the carbon paste electrode using the HCl electrolyte exhibits a different cyclic pattern with the H_2SO_4 electrolyte. The curve with the HCl electrolyte exhibits a deviation or peak current caused by the high acid concentration of the carbon.

Table 4. Specific capacitance values of activated carbon using electrolyte NaOH

Sample	**Electrolyte concentration (M)**	**Cs (mF/g)**
Before modification	1	0.08
After modification	0.1	155.92
	0.5	35.88
	1	0.83

The concentration of HCl 0.1 M shows the highest specific capacitance probably because of the pseudocapacitance effect. Increased specific capacitances after modification indicate that surface modification with HNO_3 affects the specific capacitance of the supercapacitor electrode material with a particular electrolyte. This is because the modification of HNO_3 increases the acid group on carbon so that the electric double layer formation is faster by using HCl electrolyte. In addition, the peak current of the electrolyte voltammogram HCl indicates the effect of pseudocapacitance.

3.4. Specific Capacitance Determination Using Na_2SO_4 Electrolyte

The Na_2SO_4 electrolyte is one of the salt electrolytes which is widely studied as a supercapacitor electrolyte.

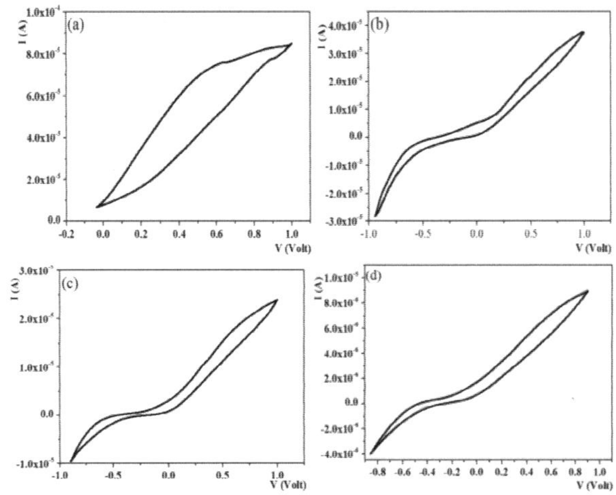

Figure 5. Voltammogram of the KATK electrode (a) before modification (b) after modification in Na_2SO_4 electrolyte 0,1 M, (c) 0,5 M, (d) 1 M.

Table 5. Specific capacitance values of activated carbon using electrolyte Na_2SO_4

Sample	Electrolyte concentration (M)	Cs (mF/g)
Before modification	0.1	78.20
After modification	0.1	67.21
	0.5	34.47
	1	12.92

The concentration of Na_2SO_4 0,1 M shows a high specific capacitance because the SO_4^{-2} anion in Na_2SO_4 is one of the largest molecules in the dissolved structure so that if the concentration is high then the electrolyte / electrolyte interface is saturated with ions which can disrupt the formation of the electric double layer.

The specific capacitance differences obtained from the use of electrolytes H_2SO_4, HCl and Na_2SO_4 indicate that not only the electrode material influences the specific capacitance of the supercapacitor. This is because the ion mobility and ion solubility of each electrolyte vary and the presence of an oxygen-containing group which can increase the specific capacitance for a particular electrolyte.

Based on the specific capacitance data, activated carbon of candlenut shell before and after modification has potential to be made of supercapacitor electrode material because its value is more than 1 mF/g [17].

4. Conclusions

Oxygen-containing group can affect the value of specific capacitance candlenut shell activated carbon based electrolyte. The number of concentration for acid and salt electrolytes influenced the ion mobility, existence of active group of electrode material and solubility of ion electrolyte. The highest specific capacitance of candlenut shell based supercapacitor electrode is obtained in HCl 0.1M electrolyte.

References

[1] Riyanto, A., 2014, Superkapasitor sebagai piranti penyimpanan energi listrik masa depan, Jurnal Ilmiah Pendidikan Fisika, 2(3), 1-8.

[2] Achmad, C., Surindra, M.D., and Prasetyo, B., 2014, Unjuk Kerja Sistem Turbin Angin Sumbu Horisontal Tipe TSD 500 Berdasarkan Nilai Tip Speed Ratio, Jurnal Teknik Energi, 10(2), 35-38.

[3] Chen, T., dan Dai, L., 2013, Carbon Nanomaterials For High Performance Supercapacitors, Materials Today, 16(7), 272-280.
[4] Muchammadsam, I.D., Taer, E. and Farma, R., 2015, Pembuatan dan Karakterisasi Karbon Aktif Monolit dari Kayu Karet dengan Variasi Konsentrasi KOH untuk Aplikasi Superkapasitor, JOM FMIPA, 2(1), 8-13.
[5] Tashima, D., Kurosawatsu, K., Sung, Y., Otsubo, M., and Honda, C., 2007, Surface Modification of Nanoporous Materials for Electric Double Layer Capacitors Application, Mater. Chem. Phys., 103, 158-161.
[6] Liu, Y., Hu, Z., Xu, K., Zheng, X., and Gao, Q., 2008, Surface Modification and Performance of Activated Carbon Electrode Material, Acta Phys. Chim. Sinica, 24(7), 1143-1148.
[7] Qun, P., Wenmao, T., Lan, D., and Guomi, M., 2012, Characteristics of Electric Double Layer in Different Aqueous Electrolyte Solutions for Supercapacitors, Journal of Natural Sciences, 17(3), 200-204.
[8] Demarconnay, L., Pinero, E.R., and Beguin, F., 2010, A Symmetric Carbon/Carbon Supercapacitor Operating At 1.6 V by using a Neutral Aqueous Solution, Electrochemistry Communications, 12, 1275-1278.
[9] Hsu, Y.K., Chen, Y.C., Lin, Y.G., Chen, L.C., and Chen, K.H., 2012, High Cell Voltage Supercapacitor of Carbon Nanotube/ Carbon Cloth Operating in Neutral Aqueous Solution, J. Mater. Chem., 22, 3383-3387.
[10] Bichat, M.P., Pinero, E. R., and Beguin, F., 2010, High Voltage Supercapacitor Built with Seaweed Carbons in Neutral Aqueous Electrolyte, CARBON, 48, 4351-4361.
[11] Tsay, K.C., Zhang, L., and Zhang, J., 2012, Effects of Electrode Layer Composition/Thickness and Electrolyte Concentration on Both Specific Capacitance and Energy Density of Supercapacitor, Electrochimica Acta, 60, 428–436.
[12] Surest, A.H., Kasih, J.A.F., and Wisanti, A., 2008, Pengaruh Suhu, Konsentrasi Zat Aktivator dan Waktu Aktivasi Terhadap Daya Serap Karbon Aktif dari Tempurung Kemiri, J. Tek. Kim., 15(2), 17-21.
[13] Ismanto, A.E., Wang, S., Soetaredjo, F.E., and Ismadji, S., 2010, Preparation of Capasitors Elektrode from Cassava Peel Waste, Bioresource Tech, 101, 3534-3540.
[14] Goertzen, S.L., Theriault, K.D., Oikcle, A.M., Tarasuk, A.C., and Andreas, H.A., 2010, Standardization of The Boehm titration. Part I. CO2 Expulsionand Endpoint Determination, Carbon, 48, 1252-1261.
[15] Vytras, K., Svancara, I. , and Metelka, R., 2009, Carbon Paste Electrodes in Electroanalytical Chemistry, J. Serb. Chem. Soc., 74(10), 1021-1033.
[16] Wachid, F.M. and Darminto, 2012, Analisa Fasa Karbon Tempurung Kelapa, Jurnal Teknik POMITS, 1 (1), 1-4.
[17] Nurdiansah, H. dan Susanti, D., 2013, Pengaruh Variasi Temperatur Karbonisasi dan Temperatur Aktivasi Fisika dari Elektroda Karbon Aktif Tempurung Kelapa dan Tempurung Kluwak terhadap Nilai Kapasitansi Electric Double Layer Capacitor (EDLC), JURNAL TEKNIK POMITS, 2(1), 2337-3539.

Acknowledgements

Grant No. 123/SP2H/PTNBH/DRPM/2018 provided by the Ministry of RISTEKDIKTI, Indonesia for this study is thankfully acknowledged.

Preliminary study on Acoustical Characteristics of Sago Midrib Composite

AsniawatyKusno[1], Sherly Asriani[2], KusnoKamil[3], Nurul Jamala[1]

[1] Department of Architecture, Faculty of Engineering, Hasanuddin University, Indonesia
[2] Study Program of Architecture, Engineering Faculty, Khairun University, Indonesia
[3] Mechanical Engineering Department, Engineering Faculty, Universitas Muslim Indonesia, Indonesia

Corresponding author's e-mail: kusno.kamil@umi.ac.id

Abstract. This research is a preliminary study on the use of sago midrib as an acoustic material. This study aims to determine the sago as acoustic material by measuring the absorption coefficient using (impedance tube method) of 100 Hz-1.6 KHz frequencies. The results of measurement were analyzed and compared with common acoustic material available on market as reference. From the measurement of the sago midrib composite, absorption coefficient has fluctuated from low to high frequencies as the absorption coefficient of reference material is found to be more stable.

Keywords: acoustical characteristics, absorption coefficients, impedance tube, sago midrib

1. Introduction

Sago trees/plants are found in many regions of Indonesia. Regions those are known as sago producers are Papua, Maluku (mainly Seram and Halmahera), Sulawesi, Kalimantan (especially West Kalimantan) and Sumatra (especially Riau). Large natural sago forests are found along the coastal lowlands and river mouths in Papua, Seram, Halmahera and Riau[1], [2]. In other areas the existing sago forests are mostly wild sago and become sago forests. In Java, sago is found limitedly in Banten and in several places along the north coast of Central Java[3], [4].

Sago trees are trees that produce primary and alternative food for people in Indonesia. According to a report[5] morphologically there are various types of sago plants, with high stems, wide trunk, wide leaves and long leaf bones. The leaf stalks from sago plants/trees are collected, stacked and crushed for many purposes.

Researches of sago trees had widely studied, ranged from the use of sago as food, sago fiber as composite material[6], [7], sago briquettes[8], sago piths[9] challenges and prospects for sago flour as a food source[10]–[12]. Although, foods taken from sago trunks are quite common, but not many reports of the midrib parts utilizations. Therefore this study will examine the opportunities for using sago midrib as building material alternative, specifically for its basic acoustical characteristics. It is expected in this initial study, new building materials with good acoustic characteristics will be introduced.

IOP Conf. Series: Materials Science and Engineering **619** (2019) 012043 doi:10.1088/1757-899X/619/1/012043

2. Research Method

This study has several stages, i.e. making materials, measuring acoustic characteristics, analysis, discussion and results as follows.

2.1. Material Making

Crushing. The cleaned sago midribs are cut into pieces about 100 cm long as seen Figure 1. The cut midrib then inserted into crusher so that the midrib becomes flat, opened, and reduced moisture contents, see Figure 2.

Drying. The post crushed material process is then dried. Drying process is carried out under direct sunlight until dry by water content below 10%. In this study, the drying process took about 1 month to reach dry with water content below 10%, as seen in Figure 3.

Cutting and compiling material. Dried material cut into 100 cm and 30 cm lengths. The material is then arranged layered longitudinally and transversely on steel mold, glued with Koyo Bond. Materials were stacked in several layers to get the desired thickness. Two types of material thickness were made, namely 2 cm and 3 cm, as seen in Figure 4.

Pressing. Material of flattened sago midrib after arranged transversely and longitudinally, then

Figure 1. Cleaned Sago midrib

Figure 2. Crushing process

Figure 3. Dried Sago midrib

Figure 4. Layered in longitudinal and transversal with glue

Figure 5. The material is pressed using a cold press

Figure 6. Sago midrib composite

covered with steel plate on top and clamped with bolts in several positions then pressed coldly with 2 Mpa pressure. Pressure is maintained for 24 hours until the material is formed into a sago midrib composite board as shown in Figure 5 and Figure 6.

Two types of sago midrib composite board thickness were 2 and 3 cm were measured. Composite board material is cut in a circular shape with a diameter of 10 cm as sample for measurement of absorption coefficient, see Figure 7. Measurement of plywood with thickness of 16 mm is also performed for comparison purposes. The absorption coefficient of the sago composite board is measured by two microphone method on impedance tube as shown in Figure 8. This method is a standard method commonly carried out by researchers or acousticians to get the absorption coefficient including the tube's sensor calibration [13]–[16]. This tube is intended to measure acoustic parameters by using a small sample placed on one end of the tube. A sound source emitted by computer speaker which is amplified then forwarded into the impedance tube which is then captured by the two microphones. Results are recorded and processed using the PULSE labshop software version 16.1. Data of recorded sound waves in impedance tube will produce absorption coefficient at frequency of 100 Hz to 1600 Hz. The record is then processed using KaleidaGraph program and displayed in curves.

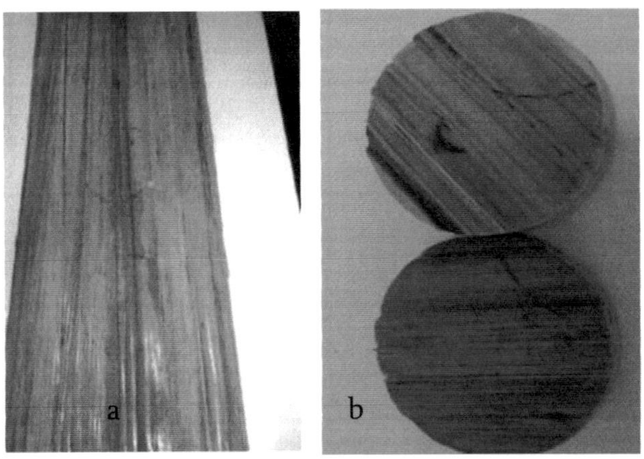

Figure 7. (a) Sago midrib composite ready to be cut;
(b) Cut Sago midrib composite

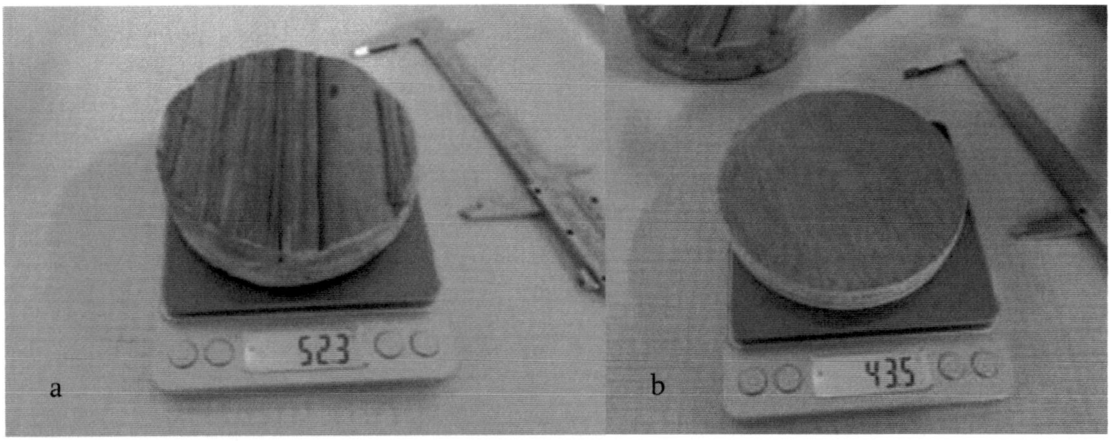

Figure 8. Weighting samples (a) Sago midrib and (b) Plywood for density measurements

3. Measurement Setup

Two types of sago midrib composite board thickness were 2 and 3 cm were measured. Composite board material is cut in a circular shape with a diameter of 10 cm as sample for measurement of absorption coefficient, see Figure 7. The circular cut Sago midrib material weighed and measured several times to determine the material density as seen Figure 8. Measurement of plywood with thickness of 16 mm is also performed for comparison purposes. The absorption coefficient of the sago composite board is measured by two microphone method on impedance tube as shown in Figure 9. This method is a standard method commonly carried out by researchers or acousticians to get the absorption coefficient including the tube's sensor calibration [13]–[16]. This tube is intended to measure acoustic parameters by using a small sample placed on one end of the tube. A sound source emitted by computer speaker which is amplified then forwarded into the impedance tube which is then captured by the two microphones. Results are recorded and processed using the PULSE labshop software version 16.1. Data of recorded sound waves in impedance tube will produce absorption coefficient at frequency of 100 Hz to 1600 Hz. The record is then processed using KaleidaGraph program and displayed in curves.

Data analysis method is by comparing the measurement results of the absorption coefficient of sago composite material (SM20 and SM30) with plywood (PW16) as reference material.The material is considered as easily found on market and also has the shape and dimensions that similar to SM20 and SM30materials. To see the effect of thickness on the absorption coefficients of material, then all dimension of samples are measured. Densities of specimen are also calculated to see whether it has effects on the specimen acoustical characteristics. Calculation of densities applied by the formula of

Figure 9. Absorption Coefficients measurement by using impedance tube

D = W / ($\pi \cdot r^2 \cdot t$), where D = density, W = weight, r = radius, and t = thickness of specimen. With diameter of all specimen is 10 cm, then the complete calculation are presented on Table 1.

Table 1. Dimension and density of specimens

Material	Thickness [m]	Diameter [m]	Weight [Kg]	Density [Kg/m3]
SM20	0.02	0.1	0.0524	83
SM30	0.03	0.1	0.0687	73
PW13	0.013	0.1	0.0435	107
PW16	0.016	0.1	0.0565	113

4. Results and Discussions

In Figure 10, it appears that α SM20 generally fluctuates below 0.4 with two peaks valued at 0.3 in the low frequency of 400 Hz and at high frequency of 1250 Hz, while the highest value 0.4 is reached at a frequency of 1550 Hz. The absorption coefficients of sago composite of 3 cm thickness [α SM30] are shown in Figure 11. This figure shows that α SM30 also has a fluctuating tendency as in αSM20. A SM30 generally fluctuates below 0.4 with a peak of 0.2 in the low frequency region of 400 Hz, at an intermediate frequency of 950 Hz and a height of 1550 Hz with a value of 0.4.Figure 12 shows the comparison of absorption coefficients of sago composite αSM20, αSM30 and plywood, αPW16. This comparison was made with the aim to be able to see the acoustic characteristics of sago midrib material compared to the acoustic characteristics of the widely available plywood material on the market. In this picture αSM20, αSM30 fluctuates with an average value (mean value) of 0.2 while αPW16 tends to be stable with a value of 0.05 starting at a low frequency area of 150 Hz to a frequency of 1,400 Hz.

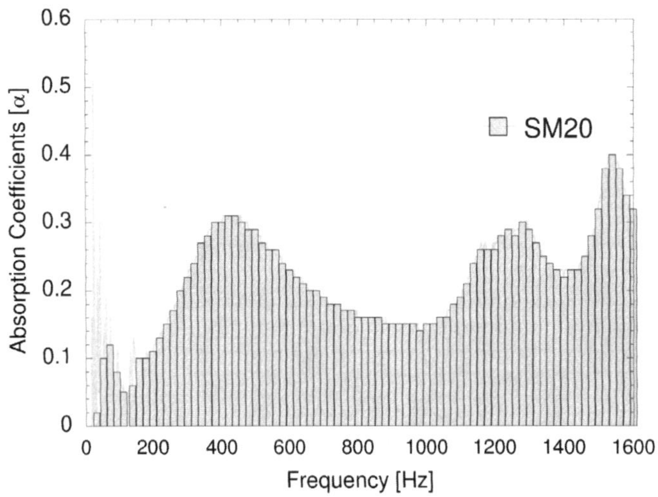

Figure 10. Absorption coefficients of Sago midrib 20 mm (SM20) with respect to frequency

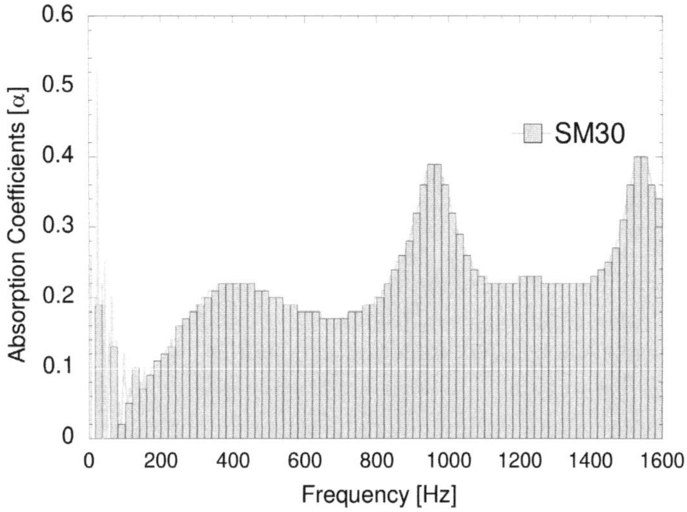

Figure 11. Absorption coefficients of Sago midrib 30 mm (SM30) with respect to frequency

Table 1 shows that additional 50% of specimen thickness from 2 cm to 3 cm has increased the weight of specimen by 25% from 0.0524 to 0.0687, making the density to become lower from 83 to 73 kg/m^3. Lower density give more porosity to specimen and perform better on absorption coefficient in the middle frequencies ranges of 800 – 1100 Hz, but relatively perform in the same value under 0.4 during high frequency range of 1400 to 1600 Hz.

Meanwhile, although different in thickness and densities, both specimen of plywood PW13 and PW16 were industrially produced then had almost similar performance of absorption coefficient throughout the range of experiment frequencies. From the explanation above, it is known that composite materials made from sago midribs have higher sound absorption compared to plywood. However, the ability to absorb sound on SM20 and SM30 material has fluctuated at several frequencies. This can occur due to the structural conditions of SM20 and SM30 materials which are

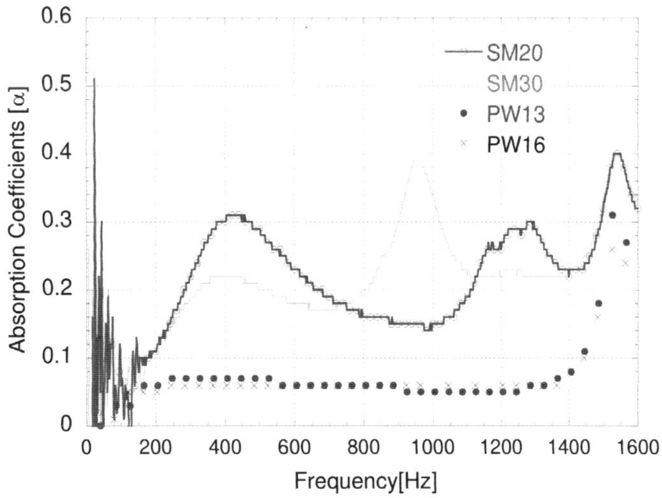

Figure 12. Comparison of absorption coefficients of Sago midribs and Plywoods with respect to

handmade and porous, so there is still a lot of uniformity in the structure and surface. Unlike PW16, although has lower absorption coefficients, thestructuresof material aremore solid with smoother surface so that measurement results are more stable.

5. Conclusions

Acoustic characteristics of sago midrib composite material in the form of absorption coefficients at frequencies of 100 Hz to 1.6 KHz show that although fluctuations still occur at certain frequencies, this composite material has absorptive properties or characters with a mean value of 0.2. This occurs in two types of material thickness measured, namely 2 cm and 3 cm. Based on the results of measurements and comparison of acoustic characteristics with plywood material it can be indicated that this sago midrib composite material has a good prospect as an acoustic material. Further research on the use of sago midrib as a building and acoustic material is still very necessary, especially in its durability.

References

[1] S. Abd-Aziz, "Sago starch and its utilisation," *J. Biosci. Bioeng.*, vol. 94, no. 6, pp. 526–529, 2002.
[2] P. A. Sopade and K. Kiaka, "Rheology and microstructure of sago starch from Papua New Guinea," *J. Food Eng.*, vol. 50, no. 1, pp. 47–57, 2001.
[3] E. Gusmayanti, T. Machida, and M. Yoshida, "A Simple Model to Simulate Sago Palm Growth," *World Conf. Agric. Inf. IT*, pp. 55–59, 2008.
[4] I. Mulyawanti, S. I. Kailaku, A. Nur, A. Syah, and E. Sjaifullah, "Pasting Properties Of Thermally Modified Sago Starch," pp. 469–474, 2014.
[5] A. A. Al-Hassan and M. H. Norziah, "Starch–gelatin edible films: Water vapor permeability and mechanical properties as affected by plasticizers," *Food Hydrocoll.*, vol. 26, no. 1, pp. 108–117, 2012.
[6] Mardin, I. N. G. Wardana, K. Kusno, and S. Wahyono, "Sea Water Effects on Surface Morphology and Interfacial Bonding of Sugar Palm Fiber to Sago Matrix," vol. 724, pp. 39–42, 2016.
[7] H. Mardin, I. N. G. Wardana, Pratikto, W. Suprapto, and K. Kamil, "Effect of Sugar Palm Fiber Surface on Interfacial Bonding with Natural Sago Matrix," *Adv. Mater. Sci. Eng.*, vol. 2016, 2016.
[8] R. S. Singhal, J. F. Kennedy, S. M. Gopalakrishnan, A. Kaczmarek, C. J. Knill, and P. F. Akmar, "Industrial production, processing, and utilization of sago palm-derived products," *Carbohydr. Polym.*, vol. 72, no. 1, pp. 1–20, 2008.
[9] A. A. Karim, A. P.-L. Tie, D. M. A. Manan, and I. S. M. Zaidul, "Starch from the Sago (*Metroxylon sagu*) Palm TreeProperties, Prospects, and Challenges as a New Industrial Source for Food and Other Uses," *Compr. Rev. Food Sci. Food Saf.*, vol. 7, no. 3, pp. 215–228, 2008.
[10] K. Israkarn, N. N. Nakornpanom, and P. Hongsprabhas, "Physicochemical properties of starches and proteins in alkali-treated mungbean and cassava starch granules," *Carbohydr. Polym.*, vol. 105, no. 0, pp. 34–40, 2014.
[11] A. R. Greenhill, W. A. Shipton, B. J. Blaney, B. Amoa, E. Kopel, D. Pelowa, M. Gena, and J. M. Warner, "Hazards and critical control points for traditional sago starch production in Papua New Guinea: Implications for food safety education," *Food Control*, vol. 21, no. 5, pp. 657–662, 2010.
[12] R. Shankar, M. S. Madihah, E. M. Shaza, N. A. K. O, A. A. Suraini, K. Kamarulzaman, S. R., M. M.S., S. E.M., N. A. K.O., S. A.A., K. K., R. Shankar, M. S. Madihah, E. M. Shaza, N. A. K. O, A. A. Suraini, and K. Kamarulzaman, "Application of different feeding strategies in

fed batch culture for pullulanase production using sago starch," *Carbohydr. Polym.*, vol. 102, no. 0, pp. 962–969, 2014.

[13] A. Kusno, M. Toyoda, K. Sakagami, T. Okuzono, T. Otsuru, K. Kamil, and V. Masgode, "Chicken feather: An alternative of acoustical materials," in *Proceedings of the 24th International Congress on Sound and Vibration, ICSV 2017*, 2017.

[14] N. B. C. Din, T. Otsuru, R. Tomiku, N. Okamoto, and K. Asniawaty, "Measurement method with a pressure-velocity sensor for measuring surface normal impedance of materials using ensemble averaging: Comparison with other methods and its geometrical configuration," *Acoust. Sci. Technol.*, vol. 33, no. 2, pp. 86–95, 2012.

[15] T. Otsuru, R. Tomiku, N. Okamoto, A. Kusno, and N. B. C. Din, "Ensemble averaged surface normal impedance measurement method in a reverberation room," *Acoust. Sci. Tech.*, vol. 32, no. 2, pp. 69–71, 2011.

[16] A. Kusno, T. Otsuru, R. Tomiku, N. Okamoto, and N. Bin Che Din, "The effect of humidity on the stability of pressure - Velocity sensors: A measurement method of absorption characteristics of building materials by using ensemble averaging," *AIJ J. Technol. Des.*, vol. 19, no. 41, 2013.

Acknowlegments

Appreciation is extended to Sherly Asriani on her work to prepare the samples from raw material to a ready-to-test specimen.

IOP Conf. Series: Materials Science and Engineering **619** (2019) 012044 doi:10.1088/1757-899X/619/1/012044

Analysis of Daylight Distribution on Building Height and Space Depth.

Nurul Jamala[1], Annajma Nurul Wika[2], Asniawaty Kusno[3]
[1,3]Department of Architecture, Engineering Faculty, Hasanuddin University, Indonesia
[2] Study Program of Architecture, Engineering Faculty, Hasanuddin University, Indonesia

Corresponding author's e-mail: nuruljamala@yahoo.co.id

Abstract. Lighting design is one of the priorities in designing buildings. Natural lighting can reduce energy consumption in buildings. Aesthetic levels are the basis of the planning concept, but need to pay attention to the concept of using natural lighting, in order to realize the energy-efficient building design. The purpose of the study was to determine how the level of illuminance affects the opening area, depth of space and height of the building. Quantitative research method is to analysis statistical data from measurements of illuminance levels in space. The results of the study concluded that the level of illuminance in the meeting room of the Center of Technology building in Hasanuddin University exceeded the illuminance standard. The extent of openings in the building envelope affects the illuminance level in space. Natural light distribution affects the height of the building's floor and the depth of space. Based on this conclusion, it is expected that in designing a building it is necessary to pay attention to the building facade, so that the distribution of natural light is not too high into the building and to avoid negative effects such as glare, high brightness and contrast.

1. Introduction

Global Warming is a natural phenomenon which causes an increase in temperature at the earth's surface. Global warming will be followed by climate change, namely increasing rainfall in some parts of the world which sometimes causes floods and erosion hazards. But otherwise found in other hemispheres will experience a prolonged dry season due to rising temperatures

Other changes that occur are increasing sunlight, which causes the level of illuminance in the sky to increase as well. The result of previous research concluded that illuminance level on palnning sky of 10,000 lux [1] and other research concluded that planning sky mean of 13.4 klux [2]. Change the illuminance level on the outside of the building improved in the last few years. This is caused by global warming, so that the light from the Sun is higher.The condition of the sky affects the distribution of light into the building and Light distribution affects the level of illuminance in space. This study will analysis the level of illuminance in the meeting room of Center of Technology Hasanuddin University building on clear sky conditions.

2. Literatur Study

Indonesia is a tropical region that has abundant availability of light, so it is necessary to think of Architects using this light as one of the factors in designing buildings, especially in many-storey buildings. integration between natural lighting (sunlight) and artificial (lighting) needs to be considered to optimize the lighting system in designing buildings. The use of natural lighting will minimize the use of energy as an artificial lighting source, but it is necessary to consider the negative effects including:

glare, brightness and bright. Indonesia is a humid tropical climate with high radiation characteristics (80% per year), high relative humidity (60% -80%) and unstable wind speed (velocity) (between 0-30 m / second) [3].

Orbits (circular paths of the earth) are not circular, but ellipses so that the distance between the earth and the sun varies with the rotation of the earth around the sun. This distance varies around 3,3 percent and this is what results in small changes in the sun's intensity of solar radiation received annually. The area of the building envelope which are equipped with wall openings of the very wide window, can be absorb natural light to the maximum so that the building has a relatively high level of illuminance, therefore occurs naturally light reflectance of 45,5%, but the negative effects occur namely glare and brightness so it is necessary to use a protective building façade as excessive sunlight [4]. SNI 03-2396-2001 explains that the daylight natural lighting factor is the ratio of the level of lighting at a point in the room to the level of flat lighting in an open field. This is a measure of the performance of the room's light hole [5]. SNI-6575-2001 that is about the procedure for designing artificial lighting systems in buildings [6]. This standard includes minimum requirements for artificial lighting systems in buildings. The preparation of this standard refers to the National Electrical Code (NEC), Illuminating Engineering Society (IES), International Electro Technical Commission (IEG) and Australian Standard Recommendation on illuminance level standard in meeting rooms is 300 lux, work space is 350 lux, archive warehouse is 150 lux and others [6].

The sun is the primary source of daylight. Howeve its light diffused and scattered by the earth's atmosphere, so that the whole sky becomes a secondary source. Daylight therefor is the sum of light which is produced directly by the sun (sunlight) and indirectlyby the scattering effect of the atmosphere (skylight). These two component of daylight are referred to separately although the term daylight is often use when referring to skylight being the main source of daylight in buildings. Both sunlight and skylight can be use to illuminate interiors directly through openings (windows, rooflights, etc) or indirectly by being reflected from external surfaces or devices (ground, building facades, shading control, etc) [7].

Utilizing sunlight as the main light source by creating access to various spaces in the building is one simple step but requires careful design consideration. Absolute through consideration is made in each design process so that the resulting building is not only environmentally friendly, but also comfortable to use and has a strong character and identity [8]. Sky condition have classified in to three groups condition i.e. overcast, clear and intermediate sky, trying to define the luminance distribution of the intermediate sky [9], [10]. Sunlight as a natural lighting source is expected to be able to enter the building to its full potential, but it should be noted that the negative effects. Brightness (brightness) is the subjective response of the eye to the light emitted / reflected by an object or the level of light of an object referring to human perception that observes the object. There is no special meaning of the brightness level as in lumination, so that the brightness level of an object cannot be measured (has no units), or is subjectively qualitative and generally high lumination has implications for high brightness as well. The distribution of light natural come in to the building is different for each building, so it necessary to analysis illuminance level in building.

3. Metodology

Quantitative descriptive research method was applied statistical data from the level of illuminance measurement at the measuring point inside the building. Data analysis using Excel program (bar and linear diagram) and Autodesk Ecotech program. The study was conducted at the meeting room of the building of the COT (Center of Technology) Faculty of Engineering Hasanuddin University. This building is the head office of the Faculty of Engineering, which consists of the Deans 'and deputy of Deans' workspace, meeting rooms, Quality Assurance rooms, administration rooms, and others. The meeting room at the COT building is located on floors 1,2 and 3.

This room is surrounded by a building glass opening in the form of massive glass. Orientation Building COT leads to the south while the envelope of meeting room buildings is oriented in two directions, namely Southeast and Southwest (figure 1). Measurements are made at the measuring point in the meeting room, which is designed using A-H notation and 1-8. The division of the measuring point zones is in the building veil left (Zone A) and right (Zone B), then the measuring point of the middle area of the space (Zone C) as shown in Figure 2 below.

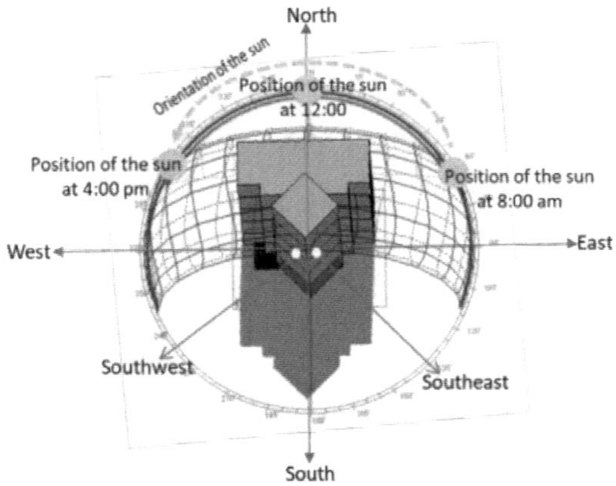

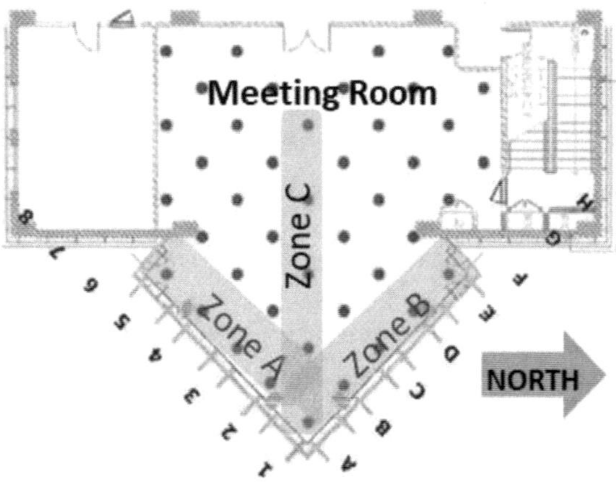

Figure 1. Position of the Sun **Figure 2.** Measurement point on zona A, B and C

Measurements are made in the morning (8:00 a.m. to 9:00 p.m.), noon (12:00 p.m. to 13:00 p.m.), and the afternoon (at 3 p.m. to 4 p.m.) Sky Intermediate and Sky Overcast. The rotation of the sun from east to west and the shadow of different buildings occur in the morning, afternoon and evening as shown in the following (figure 3).

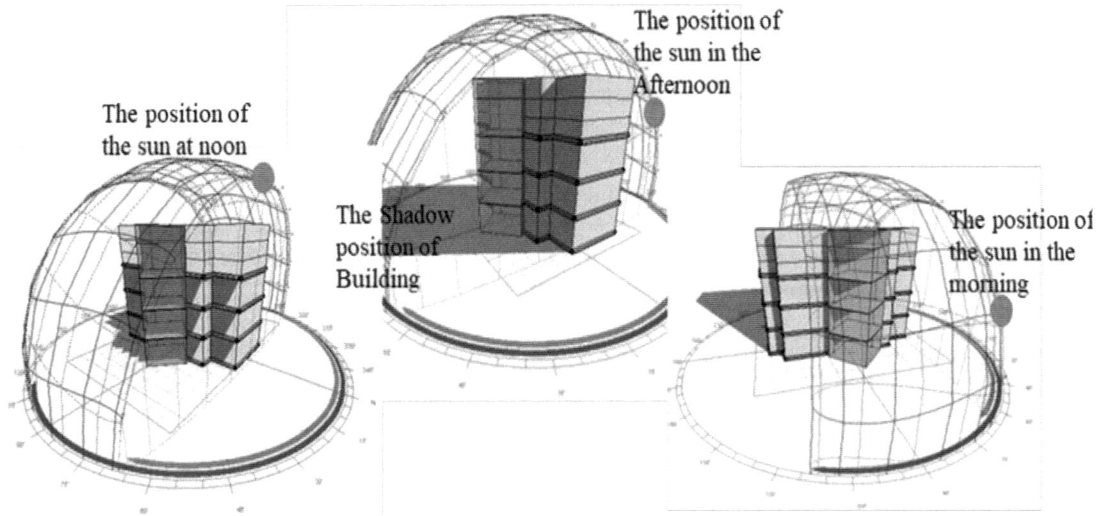

Figure 3. The shadow and daylight position of the building

4. Result and Discusion

4.1. Research object

The meeting room at the COT Building (Center of Technology) is located on floors 1,2 and 3. This room, directly related to the building envelope in the form of massive glass, as shown figure 4 below.

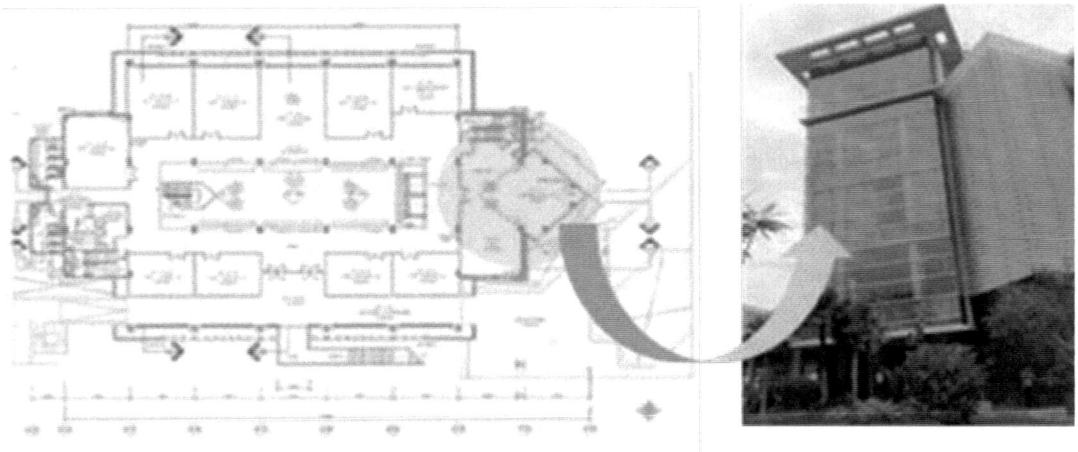

Figure 4. 1st Floor plan and building view

Meeting room on the 1st floor, functioned as a meeting room during guest visits from other universities. In addition, it also functioned as an internal meeting room for lecturers of the Faculty of Engineering, Department of Industry. The front view of this room, surrounded by a building envelope in the form of massive glass. The building envelope is angular so that the light distribution comes from the left and right sides of the building envelope as shown in Figure 5 below.

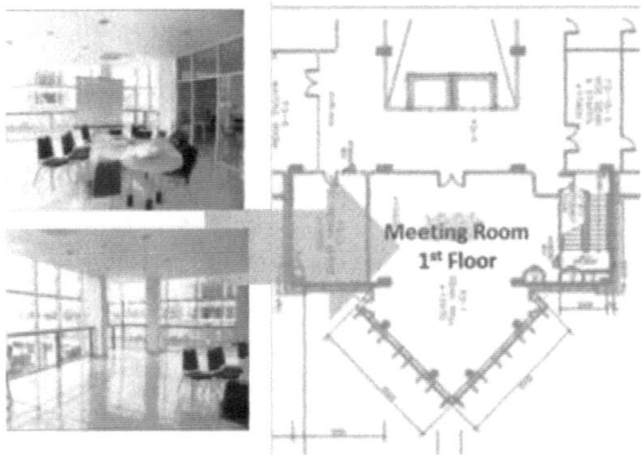

Figure 5. Meeting room on first floor

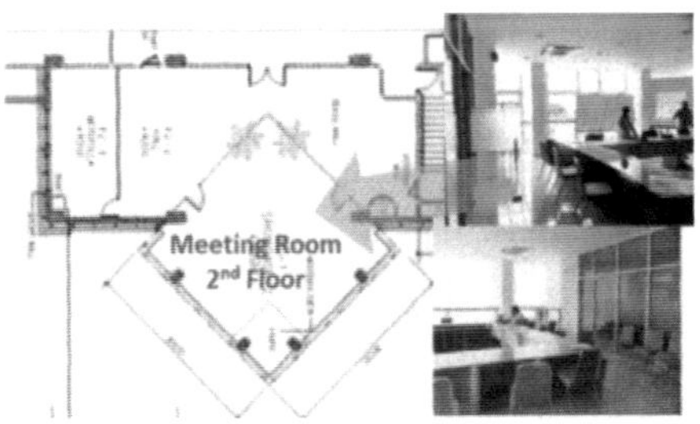

Figure 6. Meeting room on second floor

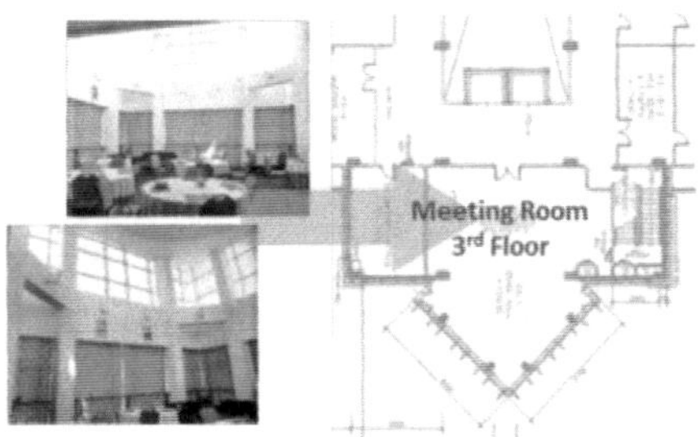

Figure 7. Meeting room on third floor

On the 2nd floor there is a meeting room for the lecturer of the Engineering department of the Machine department. In addition, this space is also used as a final exam room for Mechanical Engineering students. Meeting rooms on the 3rd floor are multi-functional, because this space is used as a gathering place for lecturers who have positions in the Faculty of Engineering of Hasanuddin University. In addition, this space also functions as a meeting room for visitors to the Faculty of Engineering of

IOP Conf. Series: Materials Science and Engineering **619** (2019) 012044 doi:10.1088/1757-899X/619/1/012044

Hasanuddin University. The 3rd floor is the highest floor in this building and the building envelope opening in this room is wider than the 1st and 2nd floors.

4.2 Analysis of measurement result on clear sky, intermediate and overcast

The condition of the sky on June 2017 is not constant, namely clear, intermediate and overcast sky. The illumination level measurements in meeting room conducted on 14 to 22 June 2019.
When the measurement level of illuminance , the sky conditions are 3250 lux to 24400 lux, so light distribution come into the room is very varied.

4.3. Analysis of Zone A Measurement Point (Left Area of Building Envelope)

Placement of the measuring point in Zone A is located on the left side area of the building envelope opening. This measurement is carried out to determine the percentage of sunlight distribution in the Zone A.

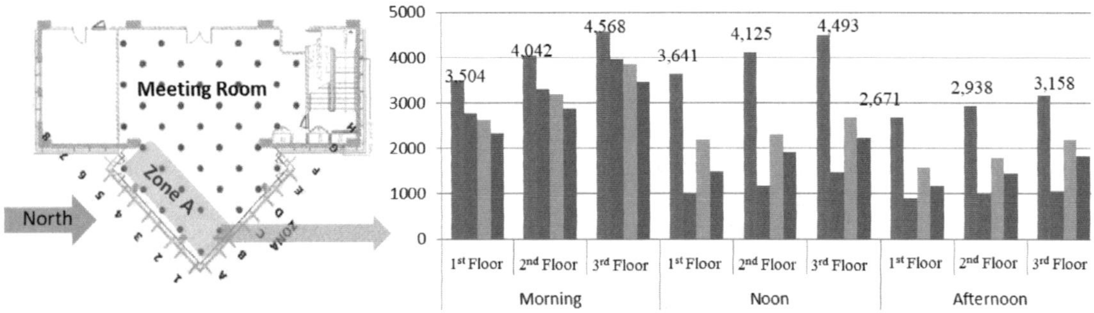

Figure 8. Illuminance level graph in zone A

The measurement results in the Zone A meeting room show that in general, the level of illuminance during the day is higher than in the morning and evening, as shown in Figure 7.

4.4. *Analysis of Zone B measuring points (left area of building envelope)*

The area in Zone B shows that the highest illuminance level occurs during the day. Zone A (AH point) is located in the building envelope area facing northwest, so that in the morning, afternoon and evening the distribution of light enters the room decreases from the measuring point A to D. This can be seen on the 1st, 2nd and 3rd floors in the morning, day and evening that has a level of illuminance going down from the front towards the back of the building envelope.

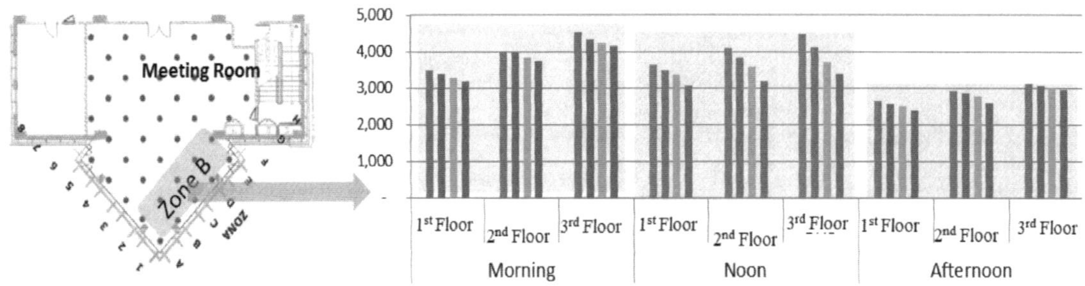

Figure 9. The level of illuminance on meeting room on 1st, 2st, and 3rd floor

4.5 Analysis of Zone C measuring points (middle area of space)

The measuring point in the middle area of space (zone C) is to find out how the influence of the depth of space on the level of illuminance. In area C zone has a light distribution from the left and right sides of the building envelope. Measurement results at measurement points 1A to 5D as shown in table 1 below.

Table 1. Result of measurement level of illuminance

Time	Floor	Measurement Point				
		1A	2B	3C	4D	5E
		Level of illuminance (lux)				
Morning	1st Floor	3504	2458	1678	924	535
	2nd Floor	4042	2688	1743	999	583
	3rd Floor1	4568	3290	2182	1584	140
Noon	1st Floor	3641	1985	983	544	244
	2nd Floor	4125	2059	1094	555	282
	3rd Floor1	4493	2461	1573	1035	68
Afternoon	1st Floor	2671	1445	688	337	219
	2nd Floor	2938	1645	980	604	181
	3rd Floor1	3158	2017	1171	831	55

Table 1 shows that the farther away from the opening of the building envelope, the lower the illuminance level. For example: the illuminance level in the morning on the 1st floor is point 1A at 3504 lux and point 5E at 535 lux. This level indicates that the farther away from the building envelope, the illuminance rate decreases. In the zone C area, the highest illuminance level illuminance is during the day and the lowest in the afternoon. The results of the analysis show that the maximum illuminance level is during the daylight, this happens on floors 1, 2 and 3.

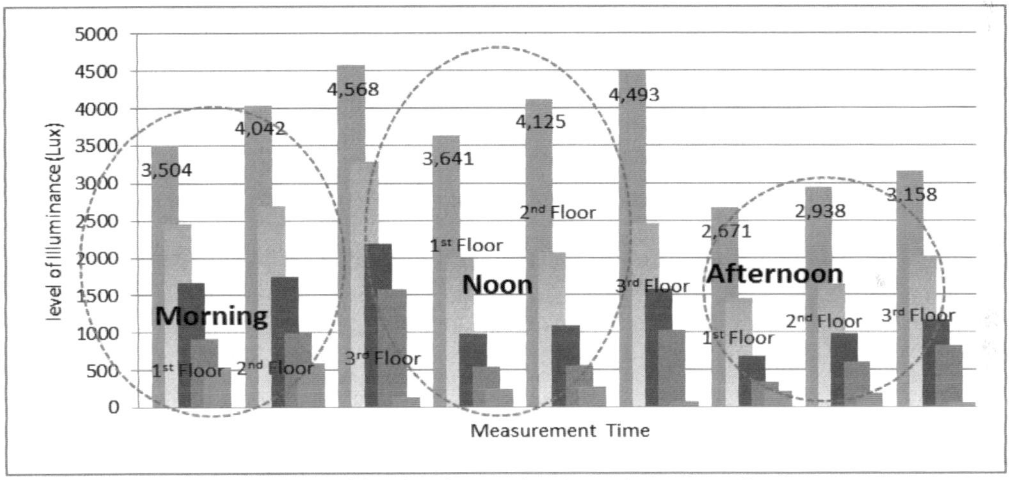

Figure 10. Iluminance level graph on zone C

The average level of illuminance in Zone C, is made in graphical form to facilitate analysis of the decrease in illuminance level into space (figure 11). Measurement results in the middle area of the meeting room (Zone C) are the measurement points A1, B2, C3, D4 and E5 as follows:

- Morning: 3504 lux (1st floor), 4042 lux (2nd floor) and 4568 lux (3rd floor)
- Daytime: 3641 lux (1st floor), 4125 lux (2nd floor) and 4493 lux (3rd floor)
- In the afternoon: 2671 lux (1st floor), second floor 2938 lux and 3158 lux (3rd floor

The mean level of measurement results in Zone C (points A1, B2 and C3) by the clear sky conditions as shown in Figure 12 below. The illuminance level on the 1st floor is 3504-2671 lux, the second floor is between 4042-2938 lux and the third floor is between 4568-3158 lux. This shows that the distribution of light in the morning, afternoon and evening will increase if the building gets higher.

Furthermore, analysis the measurement data in the C zone area is by making the mean levels on floors 1,2 and 3 in the morning, afternoon and evening, but by not entering the A1 measuring point, because the level fluctuates. This is done to confirm the occurrence of a decrease in illuminance level if the measuring point is further away from the building envelope opening.

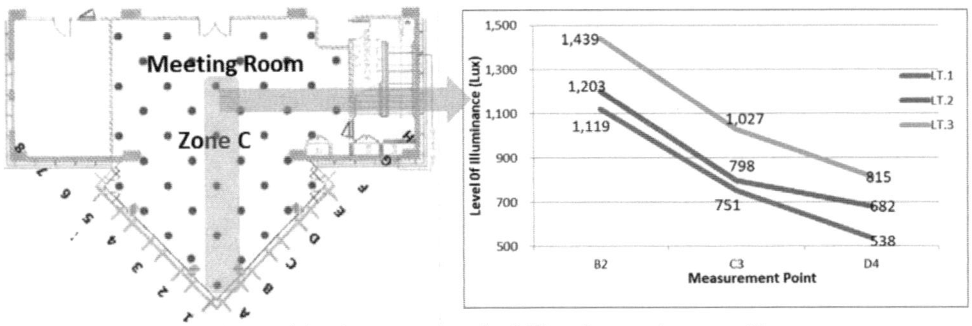

Figure 11. The mean level of illuminance in zone C

Figure 10 is a graph of mean levels in Zone C (points B2, C3 and D4). This graph shows that the illuminance level from point B2 to D4 decreases. That is, the farther from the opening of the building envelope the illuminance level is lower

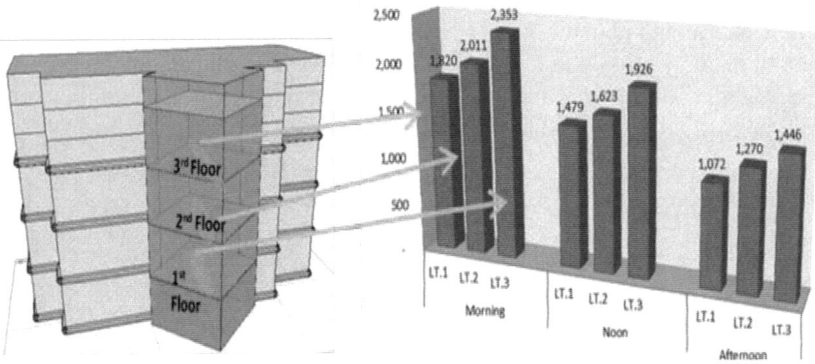

Figure 12. The mean value of illuminance level on 1st, 2st, and 3rd Floor

Figure 11 shows that the highest illuminance level on the 3rd floor is 2353 lux (morning), 1926 (afternoon) and 1446 lux (afternoon). While the lowest illuminance level is on the 1st floor, namely 1820 lux (morning), 1479 lux (noon) and 1072 lux (afternoon). This shows that the higher the floor of the building, the level of illuminance increases.

4.6 *Comparison of illuminance level average on zone A and B*

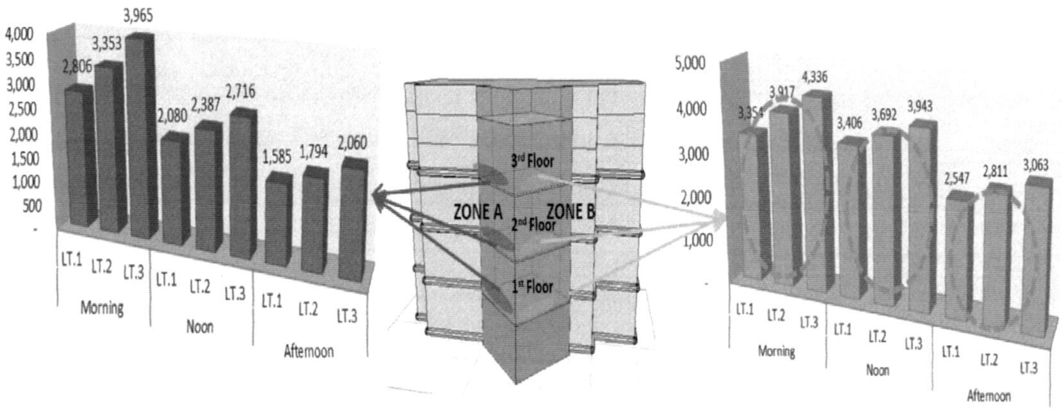

Figure 13. Illuminance levels of building envelope on zona A and B

Figure 12 shows that the highest illuminance level in zones A and B is in the morning. The higher the floor of the building, the level of illuminance will increase. For example illuminance level in the morning on the first floor of 2806 lux, the second floor is 3353 lux while on the third floor is 3965 lux. This shows that the higher the building, the higher the level of illumination. Likewise, the conditions in Zone B were 3354 lux (1st floor), 3917 lux (2nd and 4336th floor (3rd floor). To facilitate comparison of illuminance levels in zones A and B, the graphs were made as shown in Figure 13 below.

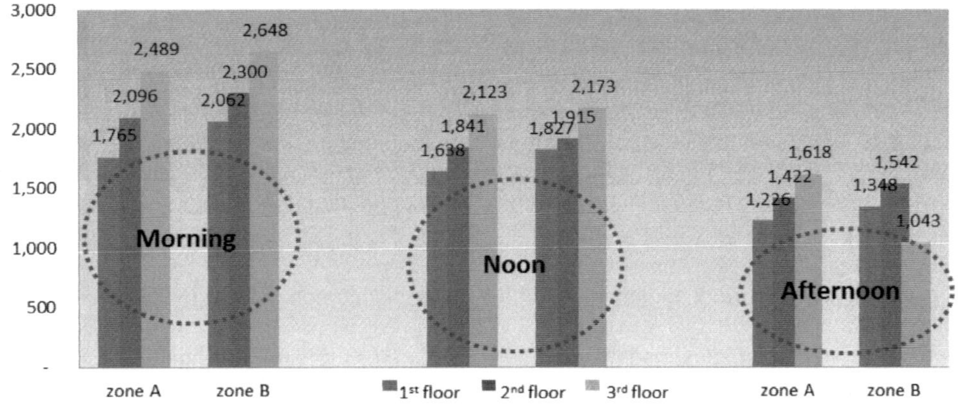

Figure 14. Comparison of illuminance level in zone A and B

This picture shows the similarities in zones A and B, when the increase of illuminance level from 1st floor to the second floor to 3rd floor. For example in the morning in zone A of 1745 lux (1st floor), 2096 lux (2nd floor) and 2489 lux (3rd floor). Whereas in zone B is 2062 lux (1st floor), 2300 lux (2nd floor) and 2648 lux (3rd floor). This graph also shows that the illuminance level in zone B is higher than in zone A.

5. Conclusion

Based on the measurement results, it can be seen that the illuminance level in the meeting room of the building COT exceeds the illuminance standard recommended by SNI 03-6575-2001. The average level of meeting rooms on floors 1-3 is 1459 lux (maximum) and 538 lux (minimum), while the standard illuminance recommendations are 300 lux. Based on this description, it can be seen that the illuminance level in the COT meeting room greatly exceeds the illuminance standard.

The results of the analysis show that the illuminance level on the 3rd floor is higher than on the 1st and 2nd floors. On the 3rd floor, the building envelope opening is wider than openings on floors 1 and 2 so it can be seen that the opening area in the building envelope affects the illuminance level in space.

In the morning, afternoon and evening, the average level on the 2nd floor is higher than the 1st floor, so it can be seen that in the same condition outside the building, there is an increase in illuminance level in the meeting room on the 2nd floor. So it can be concluded that the level of illuminance affects the height building floor. The distribution of light enters the space higher in the area near the building envelope, the farther away from the opening of the building envelope, the lower the level of illuminance into the space. Based on this analysis, it can be seen that the depth of space affects the distribution of natural light

References

[1] Sugianto 1998 *Building in Indonesia with a Tropical Climate Humidity from Aspect Building Physics* Jakarta (Original in language)

[2] Rahim R 2009 *Theory and Application of sky Luminance Distribution in Indonesia* Nasional Library (Original in Indonesia language)

[3] Satwiko P 2009 Building Physics. Andy Publisher (Original in language)

[4] Jamala N *The Effect of building Façade on Natural lighting (Case Study: Building Of Phinisi Tower UNM)* Proceeding by AIP Publishing 978-0-7354-1499-0 (American Institute of Physics) Vol.1831- 02006, 2017

[5] Indonesia National Standard 03-2396-2001 *The Design of The System of Natural Light in The Building* (Original in Indonesia language)
[6] Indonesia National Standard 03-6575-2001 *The Design of The System of Artificial Light in The Building* (Original in Indonesia language)
[7] Australian standard, 1990, *Interior lighting: General Principles and Recommendations*, Published by Standards Australian, North Sidney NSW, p.46
[8] Manurung 2012 *Natural Light in Architecture* Andy Publisher
[9] Rahim, R. et. al. 2004. *Classification of daylight and Radiation Data into Three Sky Condition by Cloud Ratio and Sunshine Duration*, Journal Energy and Building, Elsevier, Vol 36, 2004 pp.660-666
[10] Nakamura H et. al. 1985 *Luminance Distribution of Intermediate Sky* Journal of light and Vis. Environment 9 (1) p n6-13

Acknowledgments
Appreciation is extended to Annajma Nurul Wika on her work to conducted measured on illuminance level in building.

IOP Conf. Series: Materials Science and Engineering **619** (2019) 012045 doi:10.1088/1757-899X/619/1/012045

Control of a tethered Quadrotor using a quaternion feedback

Ryuta Takeuchi, Keigo Watanabe, and Isaku Nagai,

Graduate School of Natural Science and Technology, Okayama University,
3-1-1 Tsushima-naka, Kita-ku, Okayama 700-8350 Japan

Email: takeuchi-r@usmm.sys.okayama-u.ac.jp, {watanabe, in}@sys.okayama-u.ac.jp

Abstract. In Japan, superannuation of infrastructures poses a problem and the application of unmanned aerial vehicles (UAVs) such as a Quadrotor etc. is being actively studied as one of ways of inspection. Such a Quadrotor needs any position control independent of GPS, when being used for the inspection of tunnels and of interior of building. In previous research, Ouchi et al. proposed a method for controlling the position of a Quadrotor using a tether. In addition, they derived the dynamical model of the Quadrotor considering the influence of a tether, and conducted some experiments on the attitude control. The objective of this study is to realize angle and angular velocity control in the tethered Quadrotor by applying quaternion feedback control proposed by Fresk et al. In the quaternion feedback controller, the angular velocity obtained from the gyro sensor is directly available in the controller design, so that there is a possibility that more stable attitude control can be performed. After giving the overview of the tethered Quadrotor and describing Quaternion feedback control, this paper gives some experiments to verify that it is more stable with an actual machine, compared to the conventional method.

1. Introduction

The applications of Unmanned Aerial Vehicle (UAV) attracts attention as a means of gathering information at the time of disaster, photographing from the air, delivering home-delivery items, inspecting infrastructures such as tunnels and bridges, and so on[1][2][3].

In Japan, superannuation of infrastructures poses a problem, because proportion of road bridges spending 50 years after construction will be increased from the present 16\% to about 65\% in the next 20 years[4]. From this reason, research is being conducted to use a Quadrotor that is superior in various performance than other UAVs for inspection[5]. It needs to control the position of the Quadrotor, if it is used for inspection. In general, the position control of the Quadrotor is performed by manual operations using a joystick or by using position information obtained from the GPS[6][7]. Although it is necessary to bring the Quadrotor close to the inspection object if possible, expert skill is required in an operation with a joystick. On the other hand, position control by GPS has problems such as difficulty in receiving signals in a tunnel or inside a building. Therefore, Ouchi et al.[8] proposed a tethered Quadrotor, as shown in figure 1, attaching a tether to the Quadrotor as position control without using GPS. In this method, the position information is calculated from the length and inclination of the tether attached to the airframe. Their Quadrotor has the PD controller only for the attitude angle and position control, but it is desirable for a stable flight that the angular velocity is also controlled. Normally, the attitude angle of the Quadrotor is estimated by time integration of the angular velocity obtained from the gyro sensor.

When updating this attitude angle, there is a singular attitude in which the degrees of freedom degenerate.

Therefore, the differential term of a PD controller based on the Euler angle is not necessarily consistent with the actual angular velocity, and thus the angular velocity obtained from the gyro sensor cannot be directly used in such a controller. Thus, the angular velocity of the tethered Quadrotor obtained by a gyro sensor is directly controlled in this paper by applying quaternion feedback control proposed by Fresk et al.[9]. Since the time differentiation of the quaternions is expressed by the angular velocity vector, the value obtained from the gyro sensor can be directly used in a controller design.In applying this method, some simulations of using the controller and flying experiments are carried out with a real system. As a result, more stable attitude control is shown to be performed, compared with the conventional control method.

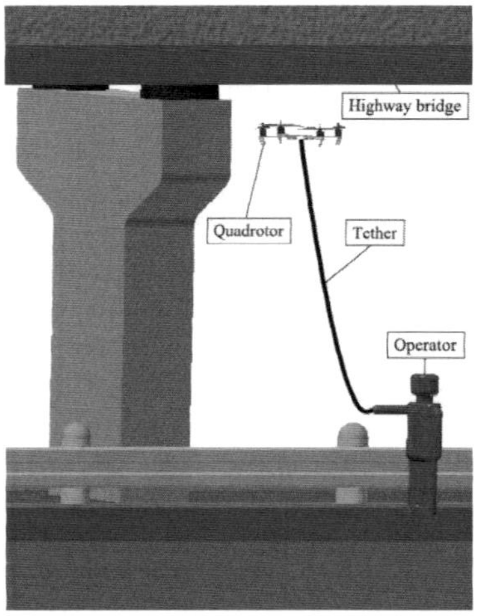

Figure 1. Schematic view of the inspection using the tethered Quadrotor.

2. Dynamical model of the tethered Quadrotor

A dynamical model of the Quadrotor is shown below. This model is partly changed from that derived by Ouchi et al. In the dynamical model derived by them, only the thrust from two-rotors can be used for controlling the roll-axis or pitch-axis motion. Therefore, when viewed from the upper side of the aircraft, let us rotate the original coordinate system, which was defined by them, 45 degrees counterclockwise, to obtain the present coordinate system as defined in this paper. Figure 2 shows the appearance of a Quadrotor and the coordinate systems, respectively. If a right-handed coordinate system is adopted in each coordinate system, then the body coordinate system of the Quadrotor is denoted by B, whereas the inertial one is represented by E.

Let the position of center of gravity of the airframe in the inertial coordinate system be(x_g, y_g, z_g), and the rotational angles around the X-, Y-, and Z-axis be ϕ, θ, and ψ, respectively. The body mass of the tethered Quadrotor is m; the gravitational acceleration is g; the length from the center of the body to the center of the rotor is l; I_x, I_y, and I_z are the moment inertia of around each axis; J_r is the rotation moment of the rotor; and the sum of the rotational speed of the rotor is Ω. The tension caused by the tether is T, the length of the tether is l_t, and the center of gravity of the airframe to the fixed point of the tether in the z direction is O_z.

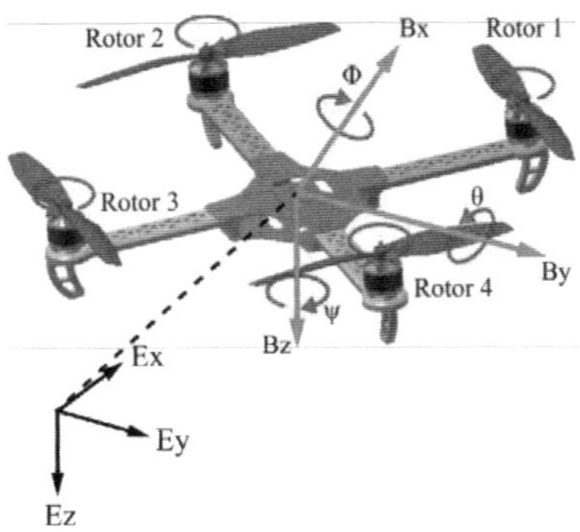

Figure 2. Definition of the coordinate systems for the Quadrotor.

Assume that the control input for the translational position in the Z-axis direction is given by U_1, and the control inputs for the roll angle, pitch angle and yaw angle are denoted by U_2, U_3, and U_4, respectively. The rotational part of the dynamical model is as follows:

$$\ddot{\phi} = \dot{\theta}\dot{\psi}\left(\frac{I_y - I_z}{I_x}\right) - \frac{J_r}{I_x}\dot{\theta}\Omega + \frac{l}{\sqrt{2}I_x}U_2 - \frac{O_z T_y}{I_x l_t} \quad (1)$$

$$\ddot{\theta} = \dot{\phi}\dot{\psi}\left(\frac{I_z - I_x}{I_y}\right) + \frac{J_r}{I_y}\dot{\phi}\Omega + \frac{l}{\sqrt{2}I_x}U_3 + \frac{O_z T_x}{I_y l_t} \quad (2)$$

$$\ddot{\psi} = \dot{\phi}\dot{\theta}\left(\frac{I_x - I_y}{I_z}\right) + \frac{1}{I_z}U_4 \quad (3)$$

Then, the translational part of the dynamical model is as follows:

$$\ddot{x} = -A_3\frac{U_1}{m} + (A_1 x + A_2 y + A_3 z)\frac{T}{m l_t} \quad (4)$$

$$\ddot{y} = -A_6\frac{U_1}{m} + (A_4 x + A_5 y + A_6 z)\frac{T}{m l_t} \quad (5)$$

$$\ddot{z} = -A_9\frac{U_1}{m} + (A_7 x + A_8 y + A_9 z)\frac{T}{m l_t} \quad (6)$$

Here, the A_1 to A_9, are given by:

$$A_1 = \cos\theta \sin\psi \quad (7)$$

$$A_2 = \sin\phi \sin\theta \cos\psi - \cos\phi \sin\psi \quad (8)$$

$$A_3 = \cos\phi \sin\theta \cos\psi + \sin\phi \sin\psi \quad (9)$$

$$A_4 = \cos\phi \sin\psi \quad (10)$$

$$A_5 = \sin\phi \sin\theta \sin\psi + \cos\phi \cos\psi \quad (11)$$

$$A_6 = \cos\phi \sin\theta \sin\psi - \sin\phi \cos\psi \quad (12)$$

$$A_7 = -\sin\theta \quad (13)$$

$$A_8 = \sin\phi \cos\theta \quad (14)$$

$$A_9 = \cos\phi \cos\theta \quad (15)$$

Defining that the thrust coefficient of a propeller is b, its drag coefficient is d, the number of each rotor is i, the rotational speed is ω_i, the generated thrust is f_i, and the generated torque is τ_i, Ω, U_1, U_2, U_3, and U_4 are reduced to:

$$\Omega = -\omega_1 + \omega_2 - \omega_3 + \omega_4 \quad (16)$$

$$U_1 = f_1 + f_2 + f_3 + f_4 = b(\omega_1^2 + \omega_2^2 + \omega_3^2 + \omega_4^2) \quad (17)$$

$$U_2 = -f_1 + f_2 + f_3 - f_4 = b(-\omega_1^2 + \omega_2^2 + \omega_3^2 - \omega_4^2) \quad (18)$$

$$U_3 = f_1 + f_2 - f_3 - f_4 = b(\omega_1^2 + \omega_2^2 - \omega_3^2 - \omega_4^2) \quad (19)$$

$$U_4 = \sum_{i=1}^{4}\tau_i = d(\omega_1^2 - \omega_2^2 + \omega_3^2 - \omega_4^2) \quad (20)$$

3. Controller of the tethered Quadrotor

3.1. Controller for the attitude angle

In this article, the attitude of the tethered Quadrotor is controlled using a quaternion feedback control method proposed by Fresk et al. This method consists of two P controllers, each of which is for the Quaternion and the angular velocity, respectively. Defining the target quaternion as $\tilde{\boldsymbol{q}}_r$, and the quaternion of the airframe as $\tilde{\boldsymbol{q}}_m$, the error of the quaternion is given by

$$\tilde{\boldsymbol{q}}_e = \tilde{\boldsymbol{q}}_r \otimes \tilde{\boldsymbol{q}}_m^* \tag{21}$$

Since the rotational direction is different depending on the sign of the scalar part in $\tilde{\boldsymbol{q}}_e$, the control input $\boldsymbol{u}_q$ for the quaternion is as follows.

$$\boldsymbol{u}_q = \begin{cases} -[q_{e1}\ q_{e2}\ q_{e3}]^T & (q_{e0} \geq 0) \\ [q_{e1}\ q_{e2}\ q_{e3}]^T & (q_{e0} < 0) \end{cases} \tag{22}$$

Defining the angular velocity of the airframe for each axis as ω_x, ω_y, and ω_z, the control input $\boldsymbol{u}_\omega$ for the angular velocity is as follows:

$$\boldsymbol{u}_\omega = \left[\omega_x\ \omega_y\ \omega_z\right]^T \tag{23}$$

The P gains of controller for the quaternion is K_{Pq}, The P gains of controller for the angular velocity is $K_{p\omega}$, the control input $\boldsymbol{\tau}$ for each rotational axis movement is as follows:

$$\boldsymbol{\tau} = [U_1\ U_2\ U_3]^T = -K_{Pq}\boldsymbol{u}_q - K_{p\omega}\boldsymbol{u}_\omega \tag{24}$$

3.2. Controller for the position

The position of the Quadrotor is controlled by changing the attitude of the airframe.
The Quadrotor can move in the X- and Y-axis directions by tilting the airframe ϕ and $-\theta$ in each direction, respectively. Therefore, in this paper, position control in the X- and Y-axis directions is performed by changing the value of the target quaternion in the attitude controller shown in Eq. (21).
The target quaternion corresponding to the target attitude angles $(\phi_d, \theta_d, \psi_d)$ is given by

$$\tilde{\boldsymbol{q}}_r = \begin{bmatrix} \cos(\phi_d/2)\cos(\theta_d/2)\cos(\psi_d/2) - \sin(\phi_d/2)\sin(\theta_d/2)\sin(\psi_d/2) \\ \sin(\phi_d/2)\cos(\theta_d/2)\cos(\psi_d/2) - \cos(\phi_d/2)\sin(\theta_d/2)\sin(\psi_d/2) \\ \cos(\phi_d/2)\sin(\theta_d/2)\cos(\psi_d/2) - \sin(\phi_d/2)\cos(\theta_d/2)\sin(\psi_d/2) \\ \cos(\phi_d/2)\cos(\theta_d/2)\sin(\psi_d/2) - \sin(\phi_d/2)\sin(\theta_d/2)\cos(\psi_d/2) \end{bmatrix} \tag{25}$$

That is, the target value of the attitude angle is generated from the error between the target position of the airframe and the current position, and it is converted into the target quaternion. Here, a PD controller is used for the position control. Defining the P gains of the controller as K_{Px} and K_{Py}, the D gains of the controller as K_{Dx} and K_{Dy}, and the target values of position of the airframe as x_d and y_d, the PD position controllers are given by:

$$\theta_d = -K_{Px}(x - x_d) - K_{Dx}\dot{x} \tag{26}$$
$$\phi_d = -K_{Py}(y - y_d) - K_{Dy}\dot{y} \tag{27}$$

The height of the airframe in the Z-axis direction is directly manipulated by an operator by pulling the tether, so that there is no controller for height.

3.3. Measurement of the angular velocity and attitude

In this paper, the internal gyro sensor of MPU-9250 manufactured by Invensense is used for measuring the angular velocity of the tethered Quadrotor.MPU-9250 is an IMU that has a tri-axis gyro sensor, a tri-axis acceleration sensor, and a tri-axis geomagnetic sensor. In addition, the attitude angle of the airframe is estimated by combining the values of the gyro sensor and the angular velocity sensor.

For the estimation, the sensor fusion algorithm proposed by Madgwick et al.[10] is used.

3.4. Measurement of the positon

In this study, the position of the airframe in the X- and Y- axis directions is determined from the inclination and length of the tether attached to the Quadrotor. Figure 3 shows the relationship between the inclination of the tether and the position of the airframe. Let the positions in Ex-, Ey-, and Ez-axes of the airframe be x_e, y_e, and z_e. The slopes of the tether against the perpendicular line given from the center of the airframe directed to Ex-axes and Ey-axis are defined by α and β. The airframe position in the Ex- and Ey-axes are given by

$$x_e = z_e \tan\alpha \tag{28}$$

$$y_e = z_e \tan\beta \tag{29}$$

$$z_e = \sqrt{l_t^2 \frac{(\cos\alpha)^2(\cos\beta)^2}{(\cos\alpha)^2+(\cos\beta)^2-(\cos\alpha)^2(\cos\beta)^2}} \tag{30}$$

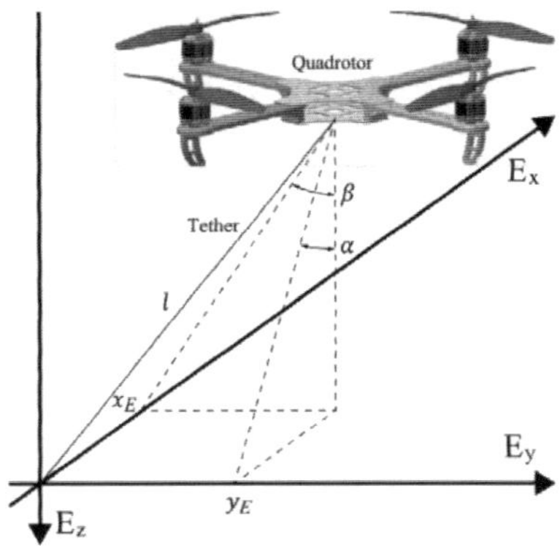

Figure 3. The relationship between the tether inclination and the airframe positions.

Figure 4 shows a module of measuring the inclination of the tether. This module consists of an aluminum frame and two potentiometers, whose structure is a gimbal, and the inclination of the tether is independently measured as a voltage by each potentiometer.

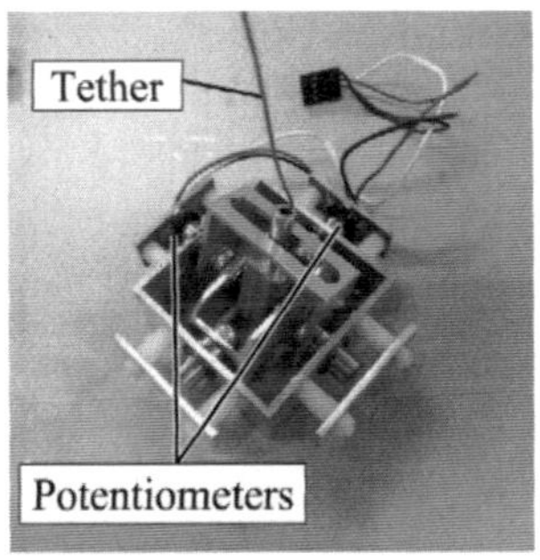

Figure 4. A module measuring the inclination of the tether.

4. Simulation with the quaternion feedback controller

Some simulations were conducted for applying the quaternion feedback controller to the tethered Quadrotor. It is verified through simulations whether the quaternion feedback controller shown in Section 3 can control the attitude of the tethered Quadrotor.

4.1. Simulation conditions

The dynamical model shown in Section 2 is used as the object to be simulated and it is analyzed by MATLAB. Let $\boldsymbol{x} = \left[x\ y\ z\ \phi\ \theta\ \psi\ \dot{\phi}\ \dot{\theta}\ \dot{\psi}\ \ddot{\phi}\ \ddot{\theta}\ \ddot{\psi}\right]^T = [0\ 0\ 0\ 0\ 0\ 0\ 0\ 0\ 0\ 0\ 0\ 0]^T$ be the initial value. Then, after one second, the target quaternion is changed stepwise to the quaternion representing $(\phi_d, \theta_d, \psi_d) = (\pi/6, \pi/6, \pi/6)$. In addition, the 4th-order Runge-Kutta method is used as a solver for differential equations. Table 1 shows the parameters used in the simulation.

Table 1. Simulation parameters.

Variable	Parameter
m	0.65 (kg)
I_x	0.00598 (kg· $\mathrm{m}^2 \cdot \mathrm{s}^2$)
I_y	0.00598 (kg· $\mathrm{m}^2 \cdot \mathrm{s}^2$)
I_z	0.0117 (kg· $\mathrm{m}^2 \cdot \mathrm{s}^2$)
J_r	0.0000065 (kg· $\mathrm{m}^2 \cdot \mathrm{s}^2$)
b	0.00000313
d	0.000000075
l	0.23 (m)
g	9.81 ($\mathrm{m/s}^2$)

4.2. Results and considerations

Figure 5, figure 7 and figure 9 shows the simulation results with $K_{Pq} = 10$ and $K_{p\omega} = 0.6$. Figure 6, figure 8 and figure 10 shows the simulation results with $K_{Pq} = 10$ and $K_{p\omega} = 0.05$. It is found, from figure 5, figure 7 and figure 9 that the airframe attitude follows the change in the target value and converges to it. It is found, from figure 6, figure 8 and figure 10 that the airframe attitude follows the change in the target value but oscillates largely. As a result, it was confirmed that the attitude of the airframe can be controlled by using the quaternion feedback controller. Moreover, since oscillations occur when the control input to the angular velocity is small, it is seen that the control of the angular velocity is useful for stabilizing the airframe.

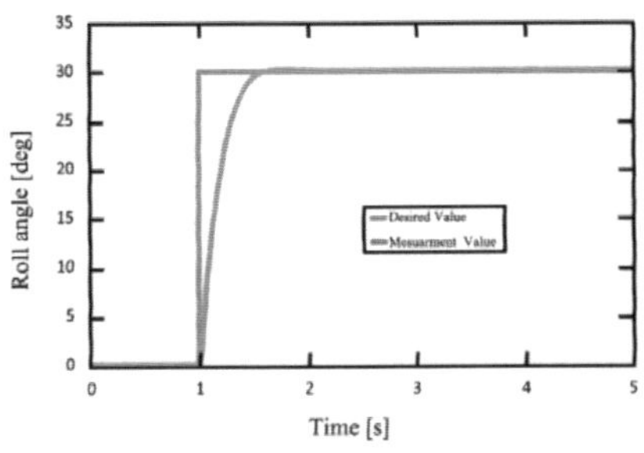

Figure 5. Roll angle with $K_{Pq} = 10$ and $K_{p\omega} = 0.6$.

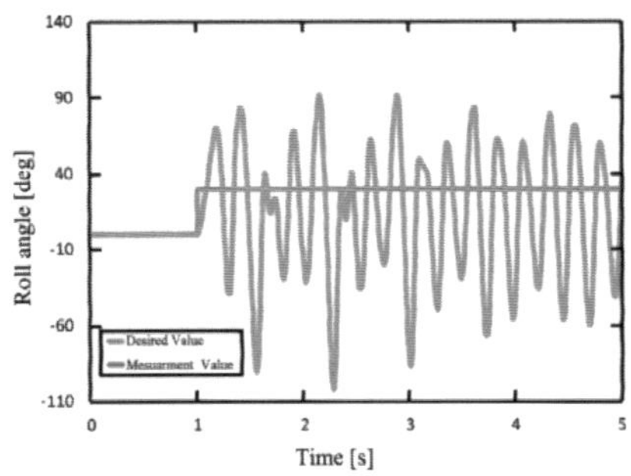

Figure 6. Roll angle with $K_{Pq} = 10$ and $K_{p\omega} = 0.05$.

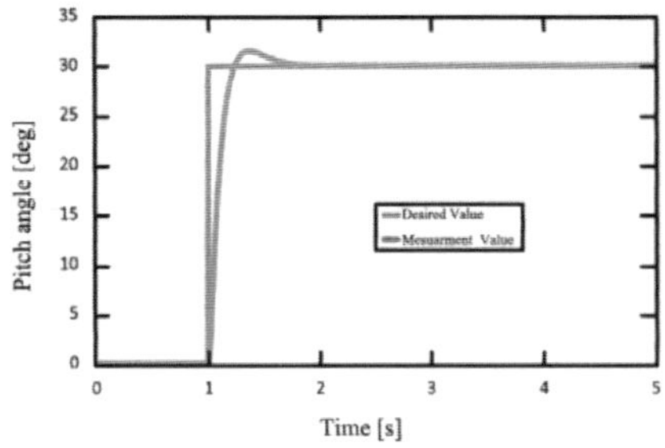

Figure 7. Pitch angle with $K_{Pq} = 10$ and $K_{p\omega} = 0.6$.

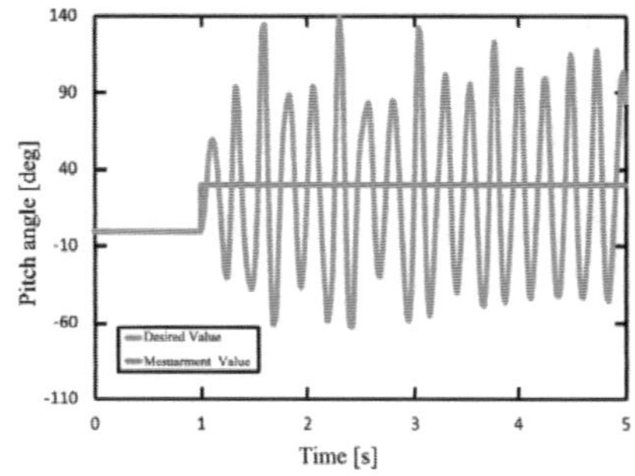

Figure 8. Pitch angle with $K_{Pq} = 10$ and $K_{p\omega} = 0.05$.

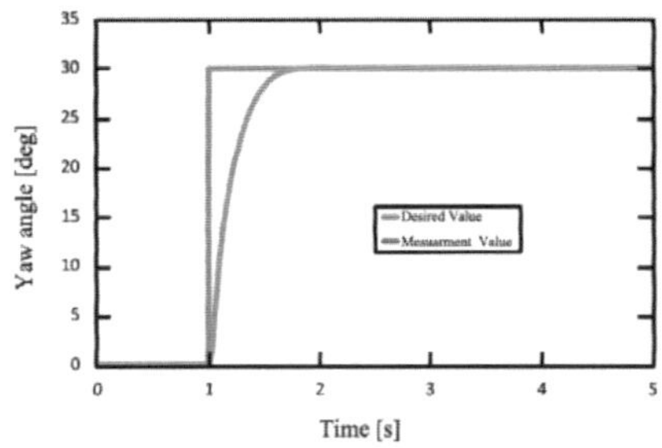

Figure 9. Yaw angle with $K_{Pq} = 10$ and $K_{p\omega} = 0.6$.

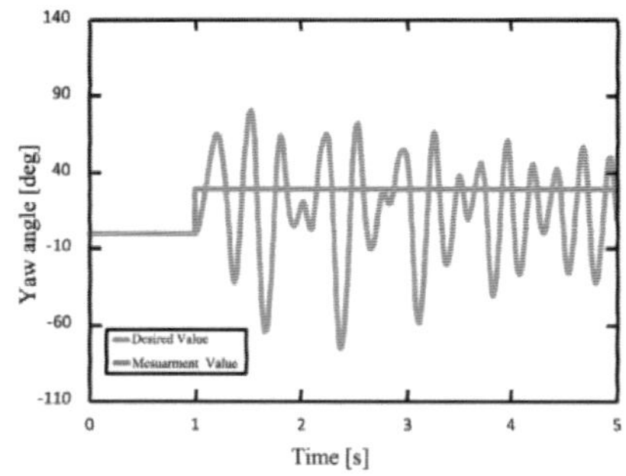

Figure 10. Yaw angle with $K_{Pq} = 10$ and $K_{p\omega} = 0.05$.

5. Flight experiments with a real system

5.1. Experimental Conditions

Figure 11 shows an experimental setup, where the MCU installed in the aircraft has a wireless LAN function and can realize wireless communication with a PC.

Figure 11. Experimental setup

Figure 12. Experimental scene.

The logging data of the airframe are obtained by using such a wireless LAN. The target quaternion in flight is set to the quaternion representing of $(\phi_d, \theta_d, \psi_d) = (0,0,0)$. Then, the roll and pitch angles of the airframe inflight are logged. The same experiment is carried out with a conventional PD controller. Then compare these results. Figure 12 shows the experimental scene. In this case, the flight experiment was carried out in an indoor space, whose size was 4.4 m in width, 4.4 m in depth, and 2 m in height, as shown in figure 12.

5.2. Results and considerations

Figure 13 shows the experimental results of roll angle. Figure 14 shows the experimental results of pitch angle. Table 2 and table 3 shows the maximum and average errors in each method. There is no significant difference in the average errors in either approach. The maximum errors in the quaternion feedback control are smaller in the roll and pitch angles, than those in the conventional method. This implies that the quaternion feedback controller can realize more stable flight in the tethered Quadrotor.

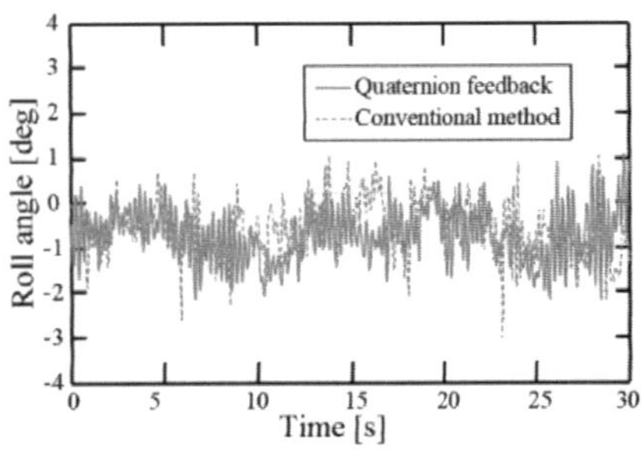

Figure 13. Experimental results of roll angle.

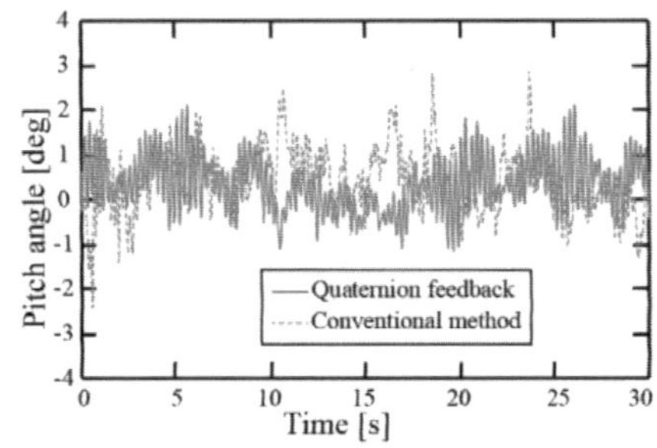

Figure 14. Experimental results of pitch angle.

Table 2. Experimental results with the quaternion feedback controller.

Axis	Maximum error (deg)	Average error (deg)
Roll angle	2.186035	−0.70809
Pitch angle	2.109319	0.344004

Table 3. Experimental results with the quaternion feedback controller.

Axis	Maximum error (deg)	Average error (deg)
Roll angle	3.00174	−0.56831
Pitch angle	2.843081	0.44037

6. Conclusion

This paper has given an overview about the tethered Quadrotor and introduced the quaternion feedback control to stabilize its attitude control system. Also, simulation experiments and actual airframe experiments were carried out by applying the quaternion feedback control to the tethered Quadrotor. As a result, the maximum error was shown to be improved over the conventional control method. In future, we are going to conduct some flight experiments, using the in actual airframe combined with a position controller, outdoors.

References

[1] Teodor T, Korbinian S, Philipp L, Andreas D, Michael K, Elmar M, Iris L G, Felix R, Micheal S, and Darius B 2012 Toward a fully autonomous UAV: Research platform for indoor and outdoor urban search and rescue *Trans. of Robotics and Automation Magazine* **19** 3 46-56.

[2] Koeda M, Matsumoto Y, and Ogasawara T 2005 AR-based assistance system to search disaster victims using teleoperated unmanned helicopter *Trans. of the Society of Instrument and Control Engineers.* **41** 12 1019-1025

[3] Pounds P, Bersak D, and Dollar A 2011 Grasping from the air: Hovering capture and load stability *Proc. of the 2011 Int. Conf. on Robotics and Automation* 2491-2498

[4] Yamada K, Adachi H, Itami M, and Kato S 2014 Research of flying robot of bridge inspection: Implementation of control to avoid contact with structure *Proc. of the 2014 JSME Conf. on Robotics and Mechatronics* **2A1**- B02(1)-**2A1**-B02(4)

[5] Yano H, Takeuchi E, Ohno K, and Tadokoro S 2014 Attempt of hammering test using multicopter *Proc. of the 2014 JSME Conf. on Robotics and Mechatronics* **2A1**-G01(1)-**2A1**-G01(4)

[6] Watanabe K and Izumi K 2006 On the controllability of rotor-type flying robots: Why is the drive of four rotors effective for Quadrotor? *Proc. of The Society of Instrument and Control Engineers System Integration (SICE SI 7th)* 820-821

[7] Machida S, Noda S, Masuta H, and LIM H 2012 Development of four rotor flying robot *Proc. of The 2012 JSME Conf. on Robotics and Mechatronics* **2A2**-L10(1)-**2A2**-L10(4)

[8] Ouchi Y, Watanabe K, Kinoshita K, and Nagai I 2014 Position Control of a Quadrotor Using a Tether *Proc. of the 1st Int. Symposium on Smart Material and Mechatronics* pp 46-49.

[9] Emil F and George N 2013 Full Quaternion Based Attitude Control for a Quadrotor *Proc. of 2013 European Control Conference* 3864-3869

[10] Sebastian O H M, Andrew J L H, and Ravi V 2011 Estimation of IMU and MARG orientation using a gradient descent algorithm *Proc. of 2011 IEEE International Conference on Rehabilitation Robotics* 179-185.

Thrust analysis of propulsion mechanisms with pectoral fins in a Manta robot

Tomoya Watanabe, Keigo Watanabe, and Isaku Nagai

Graduate School of Natural Science and Technology, Okayama University,
3-1-1 Tsushima-naka, Kita-ku, Okayama 700-8530, Japan

E-mail: watanabe-t@usmm.sys.okayama-u.ac.jp, watanabe@sys.okayama-u.ac.jp, in@sys.okayama-u.ac.jp

Abstract. The development of autonomous underwater robots is required for the ecological survey of aquatic life. Conventional underwater robots have a common mechanism for propulsion with screw propellers. Noises generated by screw propellers have a possibility of giving a bad effect on the biological behavior. Robots mimicking fish propulsion mechanisms are less likely to seriously affect living things. In our laboratory, a Manta robot that has a propulsion mechanism with pectoral fins has been developed. The operation range of the conventional Manta robot was limited by a cable for remote control. It needs an autonomous navigation system based on localization not so as to be restricted by cables. Constructing the model that takes into consideration the thrust of propulsion mechanisms with pectoral fins is expected to be used for localization. In this paper, we analyze the thrust of propulsion mechanisms with pectoral fins, using a Wei's model. In addition, we compare the analysis results with ones obtained by an actual Manta robot, and verify whether the analysis is correct.

1. Introduction

The development of fisheries resources and the ecological survey of aquatic life are carried out in the sea [1]. However, it is dangerous for humans to directly investigate underwater, and the activity time is restricted. Thus, it needs to develop autonomous underwater robots [2]. For general underwater exploration robots, screw propellers are used as propulsion mechanisms [3]. However, the propulsion mechanism using the screw propeller has a problem that underwater noises caused by cavitation adversely affect to organisms underwater [4], a problem that catches aquatic animals and plants, and so on.

Mikuriya et al. [5] developed a Manta robot that has a propulsion mechanism with pectoral fins, which are constructed by mimicking the pectoral fin of the manta ray. The Manta robot generates a propulsive force by moving the fins attached to the propulsion mechanism with pectoral fins, so that it can make forward and backward motion, turning motion, diving motion, and surfacing motion. Since the propulsion mechanism with pectoral fins is not using a screw propeller, it has little influence on aquatic life, and it is suitable for ecological survey of aquatic life. There is a problem that the operation range in the conventional Manta robot is limited, because such a robot communicates by a cable. Radio waves are attenuated in the water, so that it is difficult to use GPS and remote control by wireless communication. It needs an autonomous navigation system with localization for Manta robots to conduct an ecological survey of aquatic life not so as to be restricted by cables.

In this research, we aim to construct a model for localization so that it takes into consideration the thrust generated by the propulsion in Manta robots mechanism with pectoral fins. Since the localization method using a model does not use many sensors, it is easy to be applied to a Manta robot which is one of smallsized underwater robots. In addition, by integrating such a model with information provided by other localization methods, an improvement of accuracy of localization can be expected. In this paper, we analyze the thrust of the propulsion mechanisms with pectoral fins in Manta robots, by using a thrust model developed by Wei et al. [6]. In addition, comparing the analysis results with ones obtained by an actual Manta robot, it is verified whether the analysis is correct.

2. Outline of Our Manta Robot

Figure 1 shows a top view of our Manta robot.

Figure 1. Manta robot.

The Manta robot has a pair of the propulsion mechanism with pectoral fins on the left and right sides of the fuselage and has a control unit in the center. Each propulsion mechanism with a pectoral fin is composed of six fin-rays which is a skeleton of the fin. There are fins between fin-rays to push out water, and a stretchable silicon wrap is used for the material of the fin. For a total of 12 fin-rays, a servomotor is attached to each one independently. The Manta robot gains a thrust by moving up and down fin-rays and generating a traveling wave in the fin. The center part of the robot includes the control unit for controlling the propulsion mechanism with pectoral fins and some weights for setting the specific gravity of the Manta robot to about 1. The size of the Manta robot is 405 mm in length, 620 mm in width, 90 mm in height and 5.3 kg in weight when all the fin-rays are in a horizontal state.

3. Thrust model

The thrust generated by the propulsion mechanism with pectoral fins is expressed using a model. In this research, we used the thrust model developed by Wei et al. [6].

3.1. Force received from fluid

The force F [N] that an object in a flow receives from a fluid is expressed by the following equation:

$$F = \frac{1}{2}\rho C U^2 S \tag{1}$$

Here, ρ [kg/m^3] is the density of a fluid, C is the drag coefficient, U [m/s^2] is the flow velocity, and S [m^2] is the reference area. The C is determined by the shape of the object and by Reynolds number, and it is a dimensionless number obtained by experiments.

F [N] can be decomposed into forces parallel to the surface of the object and forces perpendicular to it, and in the case of smooth, thin, flat objects, the force parallel to the surface can be neglected. Therefore, the force that the Manta robot's fin receives from the fluid is a component perpendicular to the fin's surface in F [N].

3.2. Setting the coordinate system

In the motion analysis of the propulsion mechanism with pectoral fins, the coordinate system shown in figure 2 is defined in the propulsion mechanism with pectoral fins for the Manta robot.

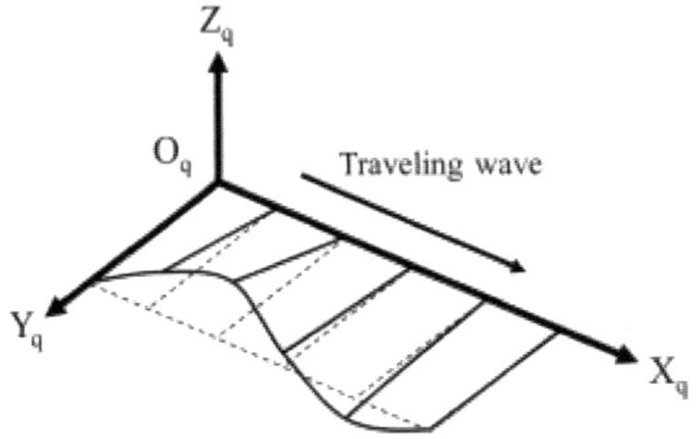

Figure 2. Coordinate system for the fin.

3.3. Waveforms formed by fin-rays

The propulsion mechanism with pectoral fins forms a traveling wave by moving fin-rays up and down, thus generates a propulsive force. When the traveling wave is a cosine wave, the angle $\theta(s,t)$ [rad] of the fin-ray at the point s [mm] on the X_q axis is expressed by the following equation:

$$\theta(s,t) = \theta_{\max}\cos\left(2\pi f t - \frac{2\pi}{\lambda}s + \phi_0\right) \quad (2)$$

Here, $\theta_{\max}$ [rad] is the maximum angle of the fin-ray, f [Hz] is the frequency of the fin-ray, t [s] is the time, λ [mm] is the wavelength, and ϕ_0 [rad] is the initial angle of the fin-ray.

3.4. Thrust by the propulsion mechanism with pectoral fins

An arbitrary point P on the fin can be expressed as follows:

$$\boldsymbol{P} = \left[\, x_q(s,h,t) \;\; y_q(s,h,t) \;\; z_q(s,h,t) \right]^T \quad (3)$$

with

$$x_q(s,h,t) = s, \quad y_q(s,h,t) = h\cos[\theta(s,t)], \quad z_q(s,h,t) = h\sin[\theta(s,t)]$$

if the X_q coordinate is s [mm] and the distance from the origin O_q is h [mm]. The point P is shown in figure 3.

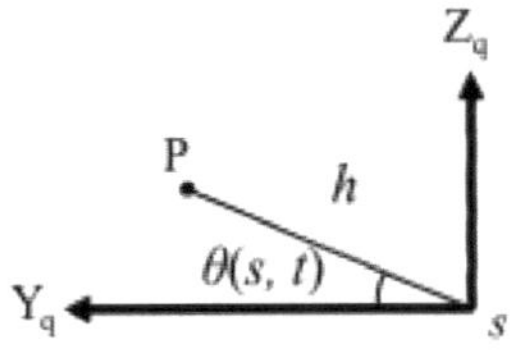

Figure 3. Point P on the fin.

The velocity vector $\boldsymbol{v}_p$ of the point P is expressed as follows:

$$\boldsymbol{v}_p = \begin{bmatrix} 0 \\ -h\theta_t' \sin[\theta(s,t)] \\ h\theta_t' \cos[\theta(s,t)] \end{bmatrix} \quad (4)$$

where θ'_t denotes the time derivative of $\theta(s,t)$. The normal vector $\boldsymbol{n}$ of the point P is expressed by the following equation:

$$\boldsymbol{n} = \frac{\partial \boldsymbol{P}(s,h,t)}{\partial s} \times \frac{\partial \boldsymbol{P}(s,h,t)}{\partial h} = \begin{bmatrix} -h\theta'_s \\ -\sin[\theta(s,t)] \\ \cos[\theta(s,t)] \end{bmatrix} \tag{5}$$

where θ'_s is the derivative of $\theta(s,t)$ with respect to s. Note here that × represents the cross product. Therefore, the normal unit vector $\boldsymbol{n}_0$ of the point P is as follows:

$$\boldsymbol{n}_0 = \frac{\boldsymbol{n}}{||\boldsymbol{n}||} = \frac{1}{[(h\theta'_s)^2+1]^{1/2}} \begin{bmatrix} -h\theta'_s \\ -\sin[\theta(s,t)] \\ \cos[\theta(s,t)] \end{bmatrix} \tag{6}$$

From equations (1), (4) and (6), the propulsive force $\boldsymbol{F}_n$ [N] by the propulsion mechanism with pectoral fins is expressed by the following equation:

$$\begin{aligned} \boldsymbol{F}_n &= -\frac{1}{2}\rho C \iint_D (\boldsymbol{v}_\mathrm{p} \cdot \boldsymbol{n}_0)^2 \boldsymbol{n}_0 dS \\ &= -\frac{1}{2}\rho C(\lambda f)^2 \iint_D \frac{(h\theta'_s)^2 \mathrm{sgn}(h\theta'_s)}{(h\theta'_s)^2+1} \begin{bmatrix} -h\theta'_s \\ -\sin[\theta(s,t)] \\ \cos[\theta(s,t)] \end{bmatrix} dsdh \end{aligned} \tag{7}$$

where · denotes the inner product.

4. Verification of the analysis and actual measured values in thrust

We analyze the thrust generated by the propulsion mechanism with pectoral fins of the Manta robot using MATLAB. In addition, we compare the measured thrust values by the previous research in Mikuriya et al. [5] with the present analysis result.

4.1. Rectangle method

To calculate the thrust generated by the propulsion mechanism with pectoral fins, a rectangle method was used as a numerical integration method. The rectangle method is a method of dividing the region of the model to be analyzed into square elements, and thus obtaining an approximate solution by summing the analysis results of each node.

4.2. Analysis conditions

Equation (7) is calculated using a difference method and the thrust by the propulsion mechanism with pectoral fins is analyzed. Three kinds of operational conditions were used the thrust measurement experiment of a Manta robot conducted by Mikuriya et al. [5]. Referring to these, we calculate the parameters so that the fins generate the same motion as that in [5], and name them the cases a, b and c, respectively. The measured swimming speed of the Manta robot in the cases a, b and c are 0.0873 m/s, 0.135 m/s, and 0.184 m/s, respectively. The fin shape is a rectangle as shown in figure 4. The fin in the propulsion mechanism with pectoral fins was divided into many squares whose one side is 1 mm and analyzed using a rectangle method.

Several parameters used for the analysis are shown in table 1. ρ [kg/m^3] was set to 998.2 kg/m^3 as the density of water at 20°C. C was given 1.12 which is the drag coefficient of a square thin plate [7]. The average thrusts per cycle are calculated on the X_q axis.

IOP Conf. Series: Materials Science and Engineering **619** (2019) 012046 doi:10.1088/1757-899X/619/1/012046

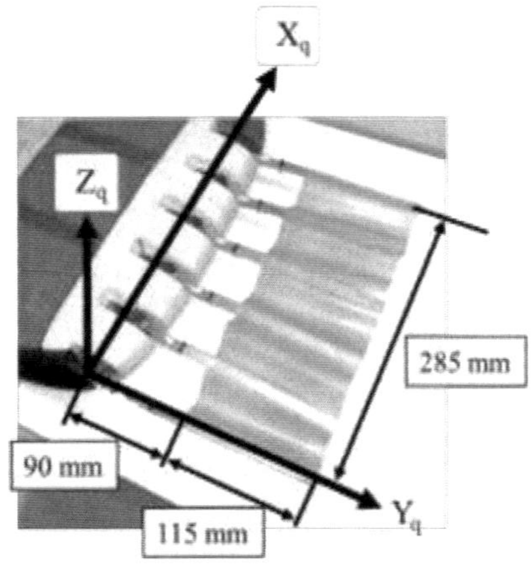

Figure 4. Fin shape.

Table 1. Parameters used for the analysis.

	Cases		
Parameters	a	b	c
λ [mm]	1026	821.4	684.8
θ_{max} [rad]	0.317	0.380	0.443
f[Hz]	0.8	0.9	1.0
ϕ_0 [rad]	0	0	0
ρ [kg/m^3]	998.2	998.2	998.2

4.3. Analysis results

From the previous study in Mikuriya et al. [5], the measured thrusts for the case a, b and c are 0.189 N, 0.689 N, and 1.15 N, respectively. The analysis results, when $C = 1.12$, show that the average thrusts per cycle in the cases of a, b and c are 0.265 N, 0.659 N, and 1.36 N, respectively. Figure 5 shows the measured thrusts and the analysis results for the cases of a, b and c when $C = 1.12$.

4.4. Setting of drag coefficient

The C is in general a dimensionless number obtained by experiments. Here, a more appropriate drag coefficient is determined using the analysis result. The drag coefficient at which the sum of errors between the three measured thrusts and three analysis results becomes a minimum is 0.95. The drag coefficient of fins of the Manta robot was set to 0.95 and analyzed again. It was found, from the analysis results, that the average thrusts per cycle for the cases of a, b and c were 0.225 N, 0.559 N, and 1.14 N, respectively, when $C = 0.95$. Figure 6 shows the measured thrusts and the analysis results for the cases of a, b and c when $C = 0.95$.

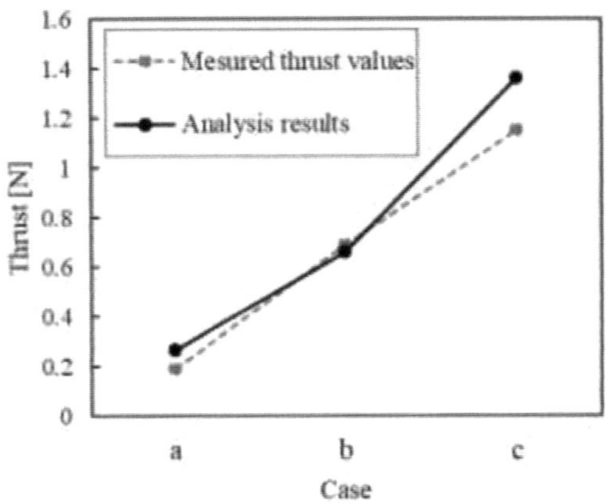

Figure 5. The analysis results when C = 1.12.

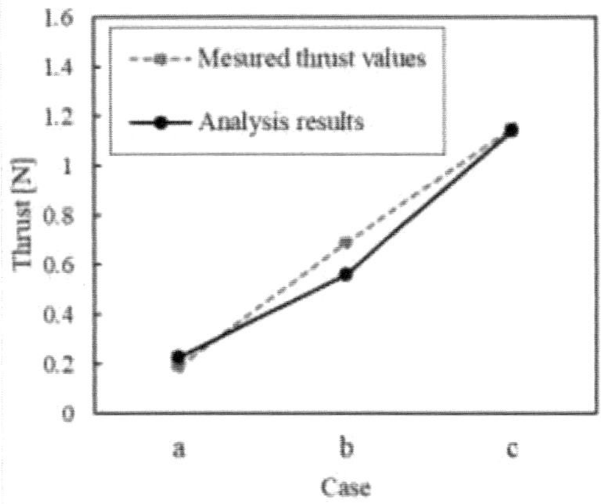

Figure 6. The analysis results when C = 0.95.

4.5. Considerations

The total error of between the measured thrusts and the analysis results is 0.316 N when C = 1.12, whereas, it is 0.176 N when C = 0.95. When C = 0.95, the total difference of between the measured thrusts and the analysis result is 44.3 % lower than that of when C = 1.12. Therefore, it is thought that 0.95 is appropriate for the drag coefficient of the Manta robot's fin. Figure 5 and figure 6 show that the measured thrusts increase as the swimming speed increases. The analysis results are also shown to be the same trend as the measured values. From this result, it is considered that a localization system can be constructed for the Manta robot based on the thrust model.

5. Conclusion

In this paper, we have conducted the thrust analysis of the Manta robot by using a thrust model by the propulsion mechanism with pectoral fins. In addition, we compared the analysis results with ones obtained by an actual the Manta robot, and verified whether the analysis is correct. As a result, it was found that the analysis result and the measured thrusts showed the same trend. In future, we will model the movement of Manta robot based on the analyzed thrust and will aim to construct a localization system using such a model.

References

[1] Nye J A, Link J S, Hare J A and Overholtz W J 2009 Changing spatial distribution of fish stocks in relation to climate and population size on the northeast United States continental shelf *Marine Ecology Progress Series* **44** 615-632

[2] Ura T 2004 Robots for underwater world *Journal of the Robotics Society of Japan* **22** 692-696

[3] Sitorus P E, Nazaruddin Y Y, Leksono E and Budiyono A 2009 Design and implementation of paired pectoral fins locomotion of labriform fish applied to a fish robot *Journal of Bionic Engineering* **6** 37-45

[4] Low K H 2009 Modelling and Parametric study of modular undulating fin rays for fish robots *Mechanism and Machine Theory* **393** 111-129

[5] Mikuriya K, Watanabe K, Hamano Y and Nagai I 2015 Development of a new manta robot considering the propulsive resistance *International Journal on Smart Material and Mechatronics* **3** 151-155

[6] Wei Q P, Wang S, Dong X, Shang L J and Tan M 2013 Design and kinetic analysis of a biomimetic underwater vehicle with two undulating long-fins *Acta Automatica Sinica* **39** 1330-1338

[7] Harada S 1985 *Industrial Fluid Mechanics* [in Japanese] (Tokyo: Maki Syoten) p187

IOP Conf. Series: Materials Science and Engineering **619** (2019) 012047 doi:10.1088/1757-899X/619/1/012047

Development of a foldable personal mobility vehicle with one drive wheel

Takanori Sakata, Isaku Nagai, and Keigo Watanabe,

Graduate School of Natural Science and Technology, Okayama University,
3-1-1 Tsushima-naka, Kita-ku, Okayama 700-8530, Japan

E-mail: sakata-t@usmm.sys.okayama-u.ac.jp, in@sys.okayama-u.ac.jp,
watanabe@sys.okayama-u.ac.jp

Abstract. As a form of new transportation, cooperation of Personal Mobility Vehicles (PMVs) and a public transportation facility attracts attention from a viewpoint of an environmental load. Various compact PMVs adopting the attitude control technology of an inverted pendulum have been already developed. However, it is reported that attitude control becomes unstable under the influence of road surface irregular and level difference in the vehicle using this attitude control. Thus, a foldable three-wheel style PMV is developed in this research. By folding up, carrying into a public transportation facility called a railroad or a bus becomes easy. Moreover, static stability is sustained to the body by being grounded with a total of three wheels, two of which are used in front part whereas one is used in rear part. This needs no introduction of attitude control, which is to make the attitude stable under the influence of a road surface, so that the risk of a fall of a driver can be reduced. Furthermore, the number of sensors and the number of motors can be reduced by setting not to use attitude control and a drive wheel to one. Thereby, the manufacture cost of the body can be held down. Therefore, it is economical, so that it is thought that general spread is easy. This paper reports the outline and design of hardware to be developed, the running speed control technique, and some running performance experiments to be conducted.

1. Introduction

Cars are used as a means of transportation in various scenes. This progress of motorization involves problems of air pollution and global warming due to the increased emission of CO2 and NOx [1]. Therefore, as a new form of transportation, cooperation of personal mobility vehicles (PMVs) and public transportation facilities attracts attention from a viewpoint of the environmental load [2]. Parallel two-wheel type PMVs including Segway, adopting the attitude control technology of inverted pendulums [3], have been already developed. However, the attitude control is unstable due to the influence of road surface irregularities and level differences [4][5]. As the result, this type of PMV has a risk that the attitude control becomes unstable under irregularities in our living environment and the rider finally falls over.

Therefore, we develop a foldable three-wheel style PMV in this research. By folding the vehicle up, we can carry it into a public transportation facility such as a railroad or a bus. Therefore, it can be used not only for moving to stations or bus stops, but also for moving after getting off the railway or the bus. Furthermore, due to the fact that we can carry it into public transportation facilities, it is considered that such users contribute to decreasing of illegal parking around stations or bus stops. In addition, since this

PMV is grounded by three wheels, two of which are used in front part whereas one is used in rear part, this vehicle has static stability. Therefore, even if the attitude control is not well functioned, the rider can balance himself and maintain his posture. As a result, it need not adopt attitude control, and a risk of falling over due to road irregularities decreases. In this paper, we describe the overview of our PMV, its mechanical design, a revolution speed control method, and the running performance of a foldable PMV with one drive wheel. In addition, we report the result of an experiment evaluating our prototype vehicle.

2. Overview of PMV under development

Figure 1 shows a whole appearance of PMV under development. This PMV is foldable as shown in figure 2.

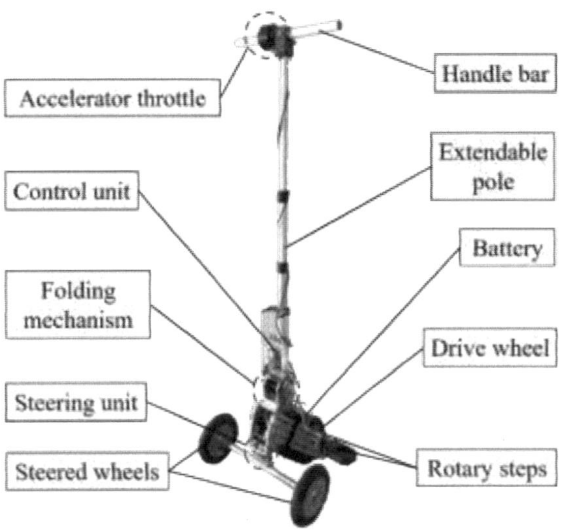

Figure 1. Whole appearance of the prototype.

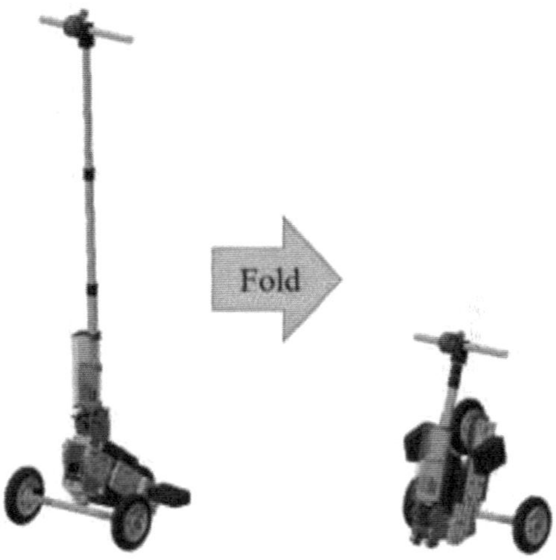

Figure 2. Running mode and folded mode.

In the running mode, the machine size is 40 cm in width, 50 cm in maximum depth and 139 cm in maximum height. In the folded mode, the size is 40 cm in width, 33 cm in depth and 63 cm in height. Therefore, the depth is 34 % smaller, and the height is 55 % smaller than the running mode. In addition, since this PMV is grounded by three wheels, two of which are used in front part whereas one is used in rear part, this vehicle has static stability. Therefore, even if the attitude control is not well functioned, the rider can balance himself and maintain his posture. As a result, adopting attitude control is not necessarily needed, and a risk of falling over due to the irregularities decreases. Furthermore, a T-shaped handle and rotary steps are adopted to improve the performance of overcoming a level difference in roads. These parts allow us easily to use techniques such as wheelie and unweighting which are used when bicycles overcome level differences. By lifting the front follower wheels, it is possible to approach the differences with only the drive wheel. Moreover, by using an unweighting technique, the vehicle can overcome the level differences even if the driving motor is not so powerful. Acceleration and deceleration of the running speed are achieved by operating a throttle mounted on a handle bar. The maximum running speed is 6 km/h which is equivalent to the electric wheelchair common in terms of the vehicle assumed to be used on walkways. In addition, 6-inch air tires are adopted for all wheels. Since a skateboard track is adopted as the steering mechanism, the front wheels are steered by shifting the rider's weight to the left or right. The weight of the vehicle is 7.8 kg, and the maximum weight of the rider is 90 kg.

3. Design of drive mechanism

In this section, after giving the overview of a drive system, we examine the maximum revolution speed and torque of the drive wheel, and the performance of the encoder used for measuring the revolution speed.

3.1. Overview of drive mechanism

Figure 3 shows the overview of the drive mechanism.

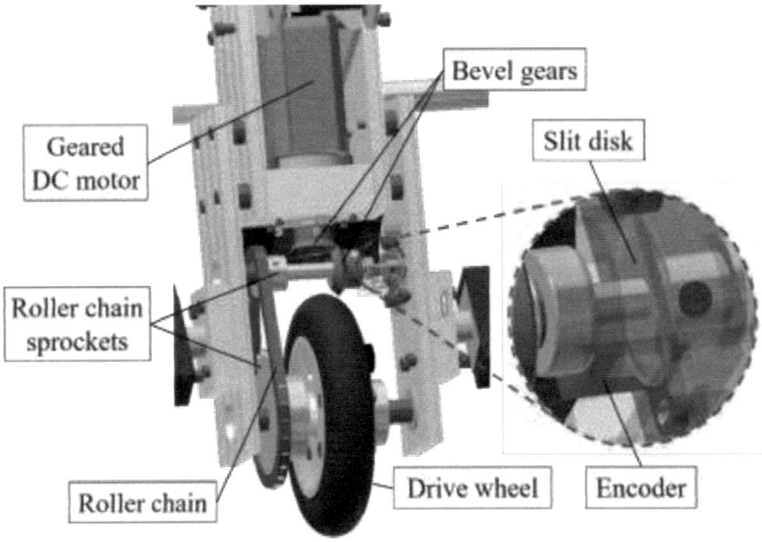

Figure 3. Overview of the drive part.

A geared DC motor taken out from KYOCERA Industrial Tools Corporation's electric driver drill "BDM-180" is used as a driving motor. The maximum revolution speed is 2000 rpm and the maximum torque is 20 N·m. The reduction ratio between bevel gears is 1, and the reduction ratio between roller chain sprockets is 3. Therefore, assuming that transmission efficiencies between the bevel gears and between the roller chain sprockets are 0.98 and 0.8 respectively, the maximum revolution speed of the drive wheel and the maximum torque become 666 rpm and 47 N·m respectively. As the rotary encoder and the slit disk used for feeding back the revolution speed of the driving wheel, KODENSHI Corporation's "KE-203" and "KE-203 DISK" are adopted respectively.

3.2. Examination about maximum revolution speed

When the vehicle runs at 6 km/h which is the maximum speed, the drive wheel rotates at 212 rpm. As mentioned in subsection 3.1, the drive wheel in the vehicle can rotate at up to 666 rpm. Considering these facts, the drive mechanism has a margin of 68 % in a generatable revolution speed of the drive wheel. Therefore, it is said that the drive mechanism performance is sufficient.

3.3. Examination about maximum output torque

When the vehicle runs on a walkway, there are some scenes such as plane running, slope running, and overcoming a level difference between a roadway and a walkway. Since it is considered that plane running is also possible if slope running is possible, only the slope running and the overcoming level differences are examined in this paper. The total mass of summing a rider to the vehicle is 97.8 kg and the gravitational acceleration is assumed to be 9.8 m/s^2. We ignore the moment of inertia of the wheels to ease the examinations in the following sections.

3.3.1. In the case of slope running. Figure 4 shows a schematic diagram when the vehicle is running on a slope.

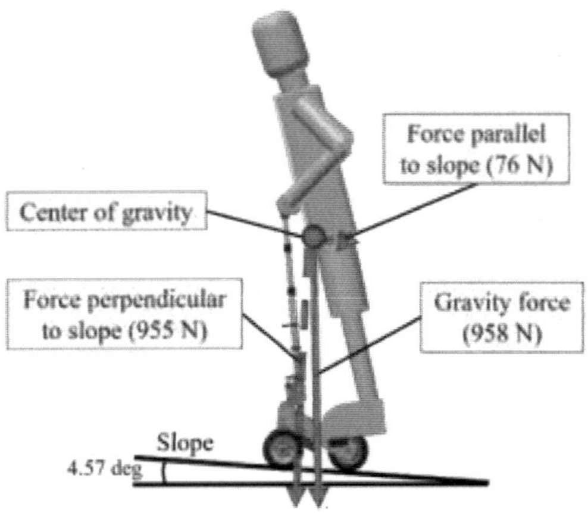

Figure 4. Case of running on slope.

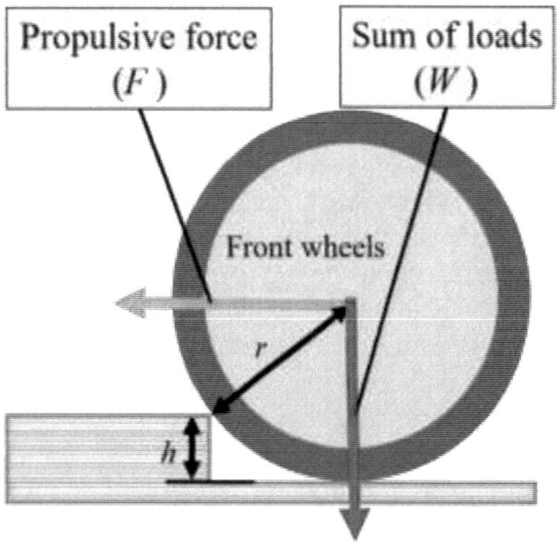

Figure 5. Case of overcoming level differences.

In Japan, the maximum slope of walkways is stipulated as 8 % (4.57 deg) at the maximum according to "Standards related to the general structure of walkways" issued by the Ministry of Land, Infrastructure and Transport. Therefore, dividing the gravity working to the system into the direction parallel to the slope and the direction perpendicular to the slope, the former is reduced to 76.4 N whereas the latter is 955.4 N. Therefore, letting the rolling friction coefficient between the tire and a walkway be 0.005, the rolling frictional force of the tire is calculated as 4.8 N. Thus, in order to climb the slope, the propulsive force is required larger than 81.2 N, which is the sum of the frictional force and the force parallel to the slope. On the other hand, since the radius of wheel is 0.075 m and the maximum torque of the drive wheel is 47 N·m, the maximum propulsive force without slippage is 627 N. Therefore, assuming that the ratio between the loads on the front and rear wheels is given as 15:9, which is a ratio calculated by 3D-CAD "Inventor 2016", the normal force from the slope in the rear wheel is 358.1 N. If the maximum static friction coefficient between the tire and the road surface is 0.8, the maximum static friction force is estimated as 286 N. Thus, the maximum propulsive force that can actually be output is 286 N and there is a margin of 72 % to climb a slope of the gradient 8 %.

3.3.2. In the case of overcoming level differences. Figure 5 shows a schematic view around the front wheels in the case of overcoming a level difference in a road. Let F be the propulsive force, W be the sum of loads acting on the two front wheels, h be the height of a level difference, and r be the radius of the wheels. The following equation is a requirement for that the front wheels, which are followers, overcome the level difference when the vehicle runs at a low speed.

$$|F| > \frac{\sqrt{h(2r-h)}}{r-h}|W| \tag{1}$$

Solving this equation for h, it yields that

$$h < r - \frac{|W|r}{\sqrt{|F|^2 + |W|^2}} \tag{2}$$

Like the case of slope running, assuming that a ratio of the loads on the front and rear wheels is given as 15:9, the front wheel receives a load of 599 N, whereas the rear wheel receives a load of 359 N. Letting the coefficient of static friction between wheels and a walkway surface be 0.8, the maximum static friction force is estimated to be 288 N. As described in the previous section, the maximum propulsion force of drive wheel is 627 N without the slippage, so that the maximum propulsion force that is actually able to output is 288 N. Therefore, assuming that $r = 0.075$ m, $W = 599$ N and $F = 288$ N, the height of a level difference h that the vehicle can overcome is calculated as less than 0.0074 m. Most of level differences supposed to exist in Japanese walkways are the difference between walkways and roadways, and protrusions of braille blocks. Following "Standards related to the general structure of walkways" issued by the Ministry of Land, Infrastructure and Transport, the standard height difference between the walkways and roadways is defined to be 0.02 m. On the other hand, the protrusion height of braille blocks is defined to be 0.005 m in Japan Industrial Standards. Therefore, it is considered for the drive mechanism that it has a sufficient performance to overcome protrusions of braille blocks, but not enough for level differences between the walkways and roadways. To overcome the level differences, the rider needs to use techniques such as unweighting and wheelie.

3.4. Examination about the performance of encoder

The photo interrupter KE-203 has a response frequency of 20 kHz. In the case of running at 6 km/h, which is the maximum speed of the vehicle under development, the revolution speed on the axis to be measured is about 11 rps. Therefore, since the number of slits on the KE-203 DISK is 100, the response frequency of 1.1 kHz is required for the encoder. As described above, the KE-203 satisfies the required performance, so that the revolution speed is satisfactorily measured.

4. Revolution speed control for drive wheel

In this section, we explain about the revolution speed control for the drive wheel, the microcomputer for calculation, the motor driver, the control method, and the accelerator throttle.

4.1. Microcomputer for calculation and motor driver

Espressif Systems' microcomputer "ESP-WROOM-32" is adopted for performing the revolution speed control. In addition, Dimension Engineering's motor driver with a regenerative brake function "SyRen25" is adopted. The ESP-WROOM-32 sends a PWM signal as the operation input to the SyRen25.

The speed control period is set to 20.0 ms (50 Hz), and the period of the PWM is set to 2.0 ms (500 Hz). In the SyRen25, the output voltage is determined by the time length in which the PWM signal is high. The output voltage becomes the maximum in forward direction when the high level time is 2.0 ms, and it is 0 when the time is 1.5 ms. Therefore, the motor rotates in the maximum speed when the PWM duty ratio is 100 %, and stops when the ratio is 75 %. In addition, the resolution for the PWM signal in ESP-WROOM-32 is 12-bit.

4.2. Method to control revolution speed

The control system has two controllers: a PI controller and an open-loop controller. The system finally controls a motor by switching them. Figure 6 shows the switching of the controllers. First, a current revolution speed of the drive wheel measured by an encoder is compared with a target revolution speed, which is determined by the amount of inputs in the accelerator throttle. When the current speed is lower than the target value, in other words, when acceleration is required, the PI controller operates. On the other hand, when the current speed exceeds the target value, in other words, when deceleration is required, the open-loop controller operates. The speed control is executed every 0.02 ms.

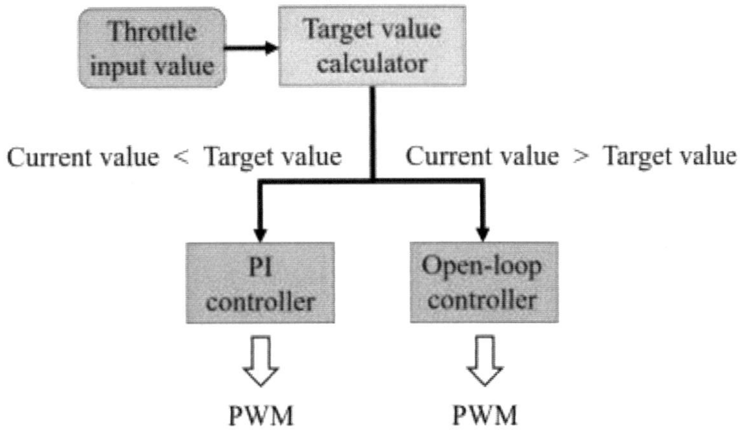

Figure 6. Switching of controllers in revolution speed control system of drive wheel.

4.2.1. PI controller. The PI control algorithm is shown in equation (3) and equation (4).

$$\Delta u(n) = K_p\{e(n) - e(n-1)\} + K_i \cdot e(n) \tag{3}$$
$$u(n) = u(n-1) + \Delta u(n) \tag{4}$$

Here, K_p, K_I and $e(n)$ denote a proportional gain, an integral gain, and an error deviation between the current measured speed value and the target value respectively. $u(n)$ is the current manipulated variable expressed in the PWM duty ratio.$\Delta u(n)$ represents the change amount of the manipulated variables. We empirically set the proportional gain and the integral gain to 0.30 and 0.075 respectively. Furthermore, in order to prevent a rider from breaking his balance due to sudden speed changes, $\Delta u(n)$ is limited to be between -0.0375 % and 0.0375 %. For measuring the drive wheel revolution speed used as a feedback value, the encoder described in subsection 3.4 is used. The encoder pulses every the control cycle are measured by a counter in 4-multiplied. The maximum target value in revolution speed control is designed to be 212.2 rpm. When the drive wheel rotates at this revolution speed, the vehicle runs at 6 km/h in a case of that there is no slippage between the drive wheel and the road surface.

4.2.2. Open-loop controller. The open-loop control algorithm is shown in equation 5.

$$u(n) = u(n-1) - C \tag{5}$$

In equation (5), C denotes a constant duty ratio, whose value is 0.015 %. In the mechanism described in section 3, there is a case that the measurement speed increased rapidly due to an influence of backlash in the power transmission unit during deceleration. Due to this rapid change, the manipulated variable becomes excessively large if the feedback control is used during deceleration. This causes too large regenerative electric power. As a result, a protection circuit works and shuts down the SyRen25. The shutdown makes the vehicle stop suddenly, and the rider finally falls over. Therefore, influences due to this problem are decreased by using the open-loop control during deceleration.

4.3. Accelerator throttle

An overview of the accelerator throttle is shown in figure 7 and figure 8. The throttle consists of ABS resin frames, two tension coil springs, a neodymium magnet with the magnetic flux density of 280 mT, and two Hall sensors "A1324LUA-T". The rider controls the running speed by adjusting the amount of revolution in the throttle lever. Due to the extension coil springs, a restoring force always occurs against the lever. The revolution range of the lever is 50 deg, and the neodymium magnet rotates around the same axis as the lever revolution. As a result, the magnetic flux density around the Hall sensors changes, and the output signals of the sensors change. Each Hall sensor is installed so that the relative angle is

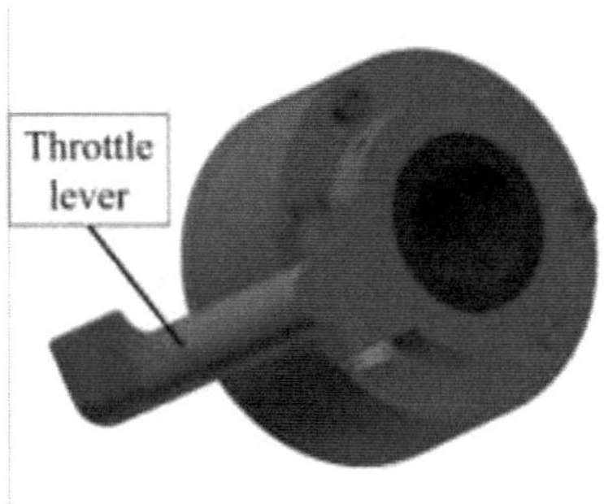

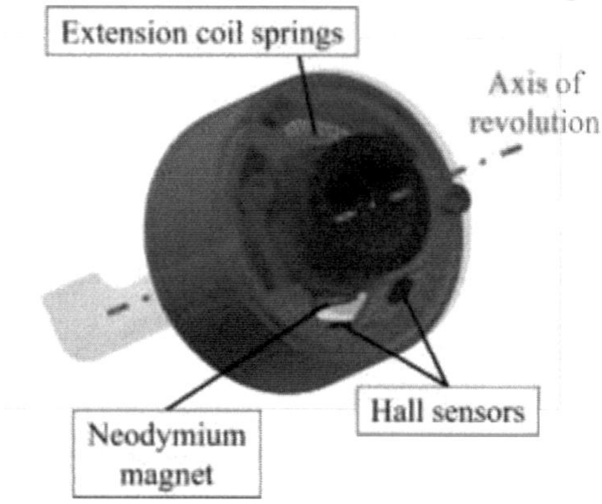

Figure 7. Overview of accelerator throttle. **Figure 8.** Internal structure of accelerator throttle.

50 deg around the same axis, and each output pin is connected to an I/O port of the ESP-WROOM-32. The voltage signal from each sensor is quantified by A-D conversion, and the amount of revolution in the throttle lever is measured based on the voltage. On the revolution of the lever, the voltage value from each sensor changes nonlinearly, whereas the difference in the signal values linearly changes. Therefore, a difference in A-D conversion values of the signals from each sensor is used to measure the amount of revolution. Then, as shown in figure 9, the target value of the revolution speed control is linearly determined from 0 to 212.2 rpm with respect to the amount of revolution in the throttle lever, where 0 to 5 deg is an idle range.

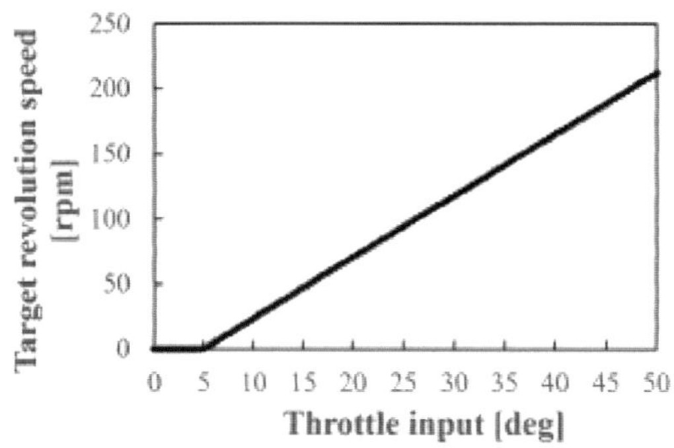

Figure 9. Relationship between amount of revolution in throttle lever and target speed.

5. Experiment for evaluating control performance of running speed

In this section, we describe some experiments for evaluating the followability of running speed to the target speed, under the condition of using the drive mechanism explained in section 3 and the revolution speed control mentioned in section 4.

5.1. Experimental method

We used the control algorithm and the parameters described in subsection 4.2. In these experiments, the running speed of the vehicle is converted from the revolution speed of the drive wheel. Moreover, the target value of the running speed is converted from the target value of the revolution speed control. We conducted experiments for two cases: one is that nobody was boarded on the vehicle and the other is that an adult rider with a weight of 72 kg was boarded. The target value of running speed was increased by 0.02 km/h every control cycle for 6 s from the start of the experiment. As a result, the target value reached 6 km/h at 6 s from the start. And then, the vehicle ran for 10 s with the same target value. The target value was changed to 4 km/h at the time when just 10 s has passed, and the vehicle finally ran at the same target speed for 15 s. After that, the target value was changed to 0 km/h for the vehicle to stop. The logging data was remotely recorded via a Bluetooth serial port.

5.2. Results and considerations of experiment

Figure 10 shows a transition of running speed in a case without loads, and figure 11 shows the transition in a case in which 72 kg weighted man was boarded.

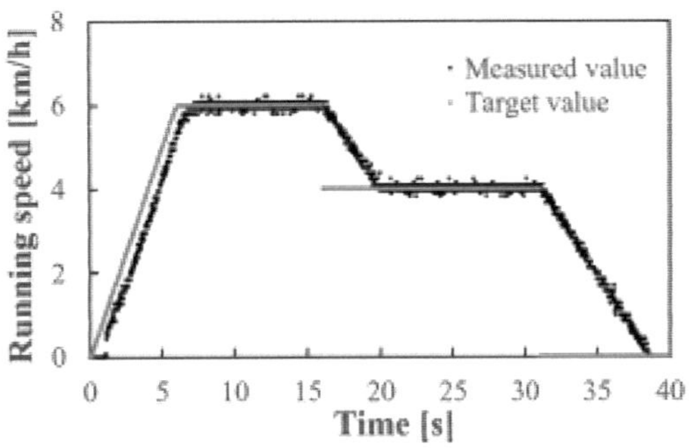

Figure 10. Running speed transition in no load case.

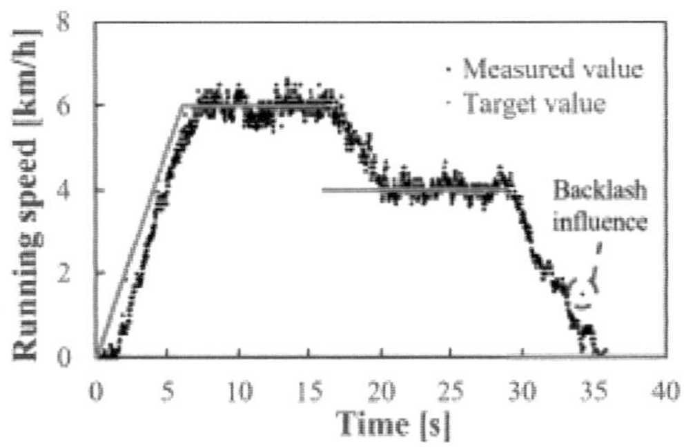

Figure 11. Running speed transition in case of 72 kg man boarding.

It is confirmed that there is no significant delay in tracking the target values of the running speed between the both cases. Therefore, we can say that the followability of the vehicle is high. In addition, it was confirmed that the vehicle was able to gently stop without being affected by the backlash. However, since experiments were conducted for only two cases in this experiment, it needs to evaluate the influence due to load differences by conducting the same experiment by various riders with different weights.

6. Conclusion and future work

In this paper, we have proposed a foldable personal mobility vehicle with one drive wheel. In particular, we described the overview, the mechanical design, and the revolution speed control method, and moreover gave some results and considerations about the running experiment using the prototype.

As future work, we will perform improving a mechanical backlash, conducting the control performance evaluation of the running speed with different weight riders, measuring the minimum turning radius, and evaluating the performance to overcome level differences in roads.

References

[1] Chapman L 2007 Transport and climate change: a review J. Transp. Geogr. 15 pp.354-367
[2] Suda Y, Hirayama Y, Yamaguchi D and Aki M 2012 Communication with Personal Mobility Vehicle and Driver *Monthly journal of the Institute of Industrial Science* **64** 193-196
[3] Mori S, Nishihara H and Furuta K 1976 Control of Unstable Mechanical System; Control of Pendulum *International Journal of Control* **23** 673-692
[4] Nakagawa C, Nakao K, Suda Y and Hayashi R 2009 Stability of the Dynamically Stabilized Two-Wheeled Vehicle Traveling on a Rough Road *JMTL* **2** 78-89
[5] Sugano T, Yokoyama K, Suzuki T and Takahashi M 2013 Modeling and Motion Analysis of a Mobile Inverted Pendulum Considering a Change of Slope Angle *Transactions of the Japan Society of Mechanical Engineers, Series (C)* **79** 216-227

IOP Conf. Series: Materials Science and Engineering **619** (2019) 012048 doi:10.1088/1757-899X/619/1/012048

Interacting Propeller Thrust for Navigating Omni-Directional Remotely Operated Vehicle

Rahimuddin[1], Haryanti Rivai, Hasnawiyah Hasan

Marine Engineering Department, Engineering Faculty of Hasanuddin University
E-mail: rahimnav@unhas.ac.id

Abstract. The role of underwater vehicles in surveying, observation, search and rescue activities are increasing along with the development of the underwater positioning technology, such as sonar technology, underwater modem, etc. These technology help to develop the navigation algorithms, such intelligent system are developed in operating the vehicle autonomously or controlled. The Remotely Operated Vehicle (ROV) is equipped four fixed horizontal propellers installed at 45-degree angle to axis and one vertical propeller. The propellers are controlled simultaneously to drive the vehicle in multi-direction. Mapping movement strategy of interacting propeller were studied to control the vehicle's movement and increase the of ROV performance.

1. Introduction

Underwater vehicle consists of two types of design based on how the vehicle operated; remotely operated and autonomous. The remotely operated vehicle (ROV) operated from surface ship through tether cable and it has a limited distance and restricted to the length of the tether cable. Autonomous underwater vehicle (AUV) is designed to operate autonomously and it has a limited operation time by the life time of the battery. Both vehicles are named Underwater Unmanned Vehicle (UUV), equipped an advanced control system regard to the objective task of the vehicle as a work class vehicle or survey class vehicle.

Performance of underwater vehicle is highly depending to the vehicle's design such as number of propellers and the control strategy to drive the propellers. Under operation condition, the vehicle could be influenced by various forces act internally or externally. Internal forces resulted by the weight distribution and the buoyancy forces, and interacting propeller forces, while external forces resulted by the dynamic of the water fluid around the vehicle. Interaction of the whole forces resulted in a resultant force that will determine the vehicle movements, speed, and its direction. Another side, the design of the vehicle is highly determined by the task of the vehicle. The vehicle with single task or various tasks. The specific design for a single task mostly applies a fixed device while the vehicle with various task designed to be able to install various equipment for different tasks [1]. The design with different task has a different shape but it should be designed properly to reduce the effect of stability and the vehicle's performance.

There are several scenarios how the vehicle or robot moves in water, among others; propulsion system resembles a fish fin model, robot that resembles snake motion in water. However, generally the survey and work-class type vehicles are driven using propellers. The number of propellers used varies according to the vehicle design and navigation scenario embedded in the vehicle. The vehicles for survey class generally use 1 up to 3 propellers. The work-class vehicles use more propellers. The number of propellers used is based on the moving vehicle scenarios and to give the vehicle an ability to withstand the force from the currents and keeping position ability [2][3][4]

The pressure force acting on the surface of the vehicle increases with the depth proportionally. The vehicle is designed to have a good strength to the water pressure. The increase of depth 1.0 meter the increase force 1.0N per area in cm^2. The vehicles that weight slightly lower than the displacement weight tends to float the vehicle. It requires a navigation mechanism that governs the vehicle to be at the desired depth using horizontal fins. Using fins is a way to keep the vehicle moves at certain depth and speed continuously. The vehicle moves continuously to sustain force maintaining vertical movement. The vehicle also can be equipped with ballast system to adjust the vehicle weight equal to the water displacement weight by pumping water in to or out to the ballast tank.

In 1950s and '60s, the early design of ROV developed to have ability for recovery operation and to perform seabed rescue. The operation depth up to 10,000 feet, could be equipped with a various mechanical tools and sensors, fitted with high resolution digital cameras, sonar and a fibre optic which can combine up to eight channels of video, sonar, USBL (ultra-short baseline) positioning data, serial data communications and navigation data on a single fibre optic.

Generally, underwater vehicle can be divided in two types; Autonomous Underwater Vehicle (AUV) and Remotely Operated Vehicle (ROV). An autonomous underwater vehicle (AUV) is a robot which travels underwater without requiring direct input from an operator. In 1990s, there are many designs of AUV were developed for scientific, bottom surveys, and pipeline survey operations. The vehicle has capabilities of depth ratings up to 6,000m [5].

In 2013, Benjamin developed a circular ellipse AUV equipped with three propellers using servo. the propeller direction can be rotated at an angle exceeding 180 degrees, with moving ability in three-dimension to all directions or omni-directional movement[6].

Another typical vehicle that can be driven from the operator at surface vehicles is a remotely operated underwater vehicle, ROV. The vehicle used as work-class vehicle in deep water industries such as offshore oil and gas. The vehicle is connected to a surface ship by a buoyant tether or with a load-carrying umbilical cable. The umbilical cable is an armoured cable that contains a group of electrical conductors and fibre optics that carry electric power, video, and data signals from the vehicle and vice versa. In high-power applications, most of the tools and equipment are powered by hydraulic system. The system used for propulsion and power equipment such as torque tools and manipulator arms. Most ROVs are equipped with at least one video camera and light, some sensors to expand the vehicle's capabilities such manipulator or cutting arm, water samplers, and instruments that measure water clarity, water temperature, water density, sound velocity, light penetration, and temperature.

After years, most ROV perform for offshore oil and gas industry with so-called "work-class" ROVs. The vehicle became essential tools in the 1970s and '80s when the operation of the vehicle exceeded the human divers. Today, the design of ROV developed to provide operation in the construction of sub-sea oil and gas in their subsequent inspection, repair and maintenance and very hugely in terms of depth rating, size, complexity and capabilities.

The typical design of ROV comprises a large of floatation pack above to provide the necessary buoyancy and high stability. Various electrical power systems are attached for operation such as torque tools, cutter and arm manipulators include the main propulsion system. Electrical power used to drive the hydraulic pump that is used for propulsion as an alternative to the more normal electrically power thruster [7].

2. Concept Design of ROV

The concept of the developed ROV has an omni-directional movement ability. With this ability, the vehicle can conduct survey and observation activities without tactical diameter during maneuver. The vehicle can move to sideways, backward, forward or move at certain angle direction. This vehicle designed to be equipped additional equipment at the bottom if necessary, such robotic arm. It is different to omni-directional vehicle designed by Benjamin in 2013 where the vehicle used for survey and monitoring with three rotatable propellers [6].

Two important parts considered in the design of the vehicle to be stable are position of the center of gravity and center of the buoyancy force. Both determine the vehicle's stability characteristic, where the vehicle's stability is an ability of the vehicle to return to the equilibrium position when there any external force or moment disturb. Good stability when the center of gravity is underneath the center of buoyancy force and the longer distance between center of gravity and center of buoyancy the higher moment stability. Therefore, the vehicle designed to place the buoyant structure at the top of the vehicle and weights at the bottom.

In the design of propeller, the vehicle uses ducted propeller where the ducted propeller has advantages than the conventional propeller. The ducted propeller can increase the efficiency of the thrust propeller, has lower diameter compare to non-ducted propeller, and the ducted propeller can protect the propeller from other objects. Ducted propeller installed on the main structure with fixed mounting angles. However, besides the fixed angle mounted on the structure, rotatable propeller also able to be installed that can decrease the number of propellers installed for application to multi-direction movement. In this paper, the fixed propeller angle has a simpler in design and installation.

In another side, the vehicle designed to have a capability to keep the vehicle's position in stable position even under force disturbance from current. The vehicle also designed has the same weight with the buoyancy force that can ease the vehicle move vertically.

Design of the ballast system also consider the total weight required and the centre of gravity. Therefore, the design of weight distribution and buoyancy for every part are calculated properly that it will not decrease the vehicle's stability. The higher stability can make the vehicle very stiff and decrease the ability of the vehicle to rotate in pitch and roll motion. The concept design of the ROV as shown in figure1.

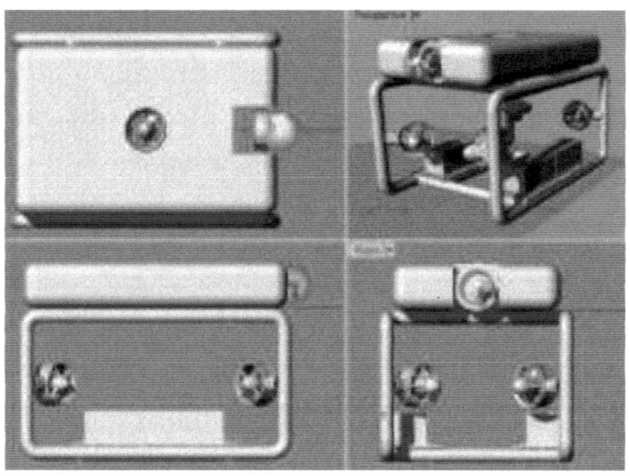

Figure 1. Developed Design of ROV

In this research, this early ROV designed to operate in 100m under the water surface. The dimension of the developed design has 58.60cm length, 42.21cm breadth and 32.60cm high. The model equipped five propellers with 7.0cm diameter, one buoyant tank at top, four dry batteries placed at bottom as additional source power and to increase the stability. Four propellers are installed at each corner with 45 degrees angle to the axis and one propeller placed vertically at the buoyant tank. The propellers used to control the vehicle motions in six degrees. The design speed of vehicle is 0.5m/s.

2.1. Propulsion System

Propulsion systems of the ROV are designed using five propellers. Four propellers are used to drive the vehicle in horizontal and longitudinal direction. One propeller is placed at the upper side in line with the centre of gravity to drive the vehicle vertically. Four propellers installed at each corner and crossing each other at 45 and 135 degrees to its directional axis to drive the robot move in multidirectional.

The brushless motors were used as the prime mover of the propellers. The vehicle's movement resulted by vectored forces of the propeller's thrust and it designed to have ability in omni-directional movement. The vehicle uses ducted propeller with three blades (Figure 2) with some advantages as follows; It has more efficient in producing thrust than a conventional propeller [8], at the same static thrust, a ducted propeller has a smaller diameter than conventional propeller, and Ducted propeller offers enhanced safety from contact to others object.

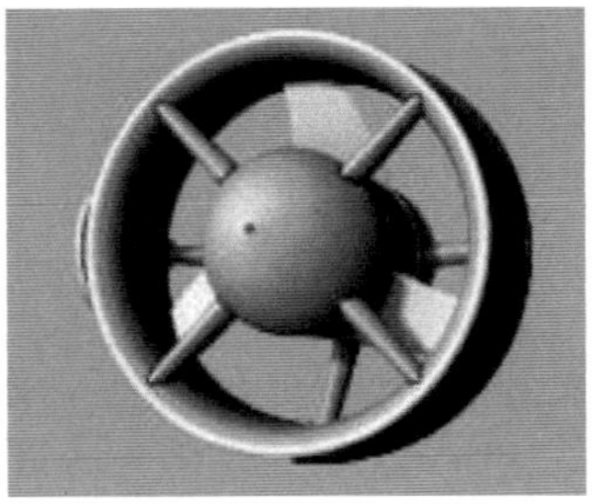

Figure 2. Ducted propeller with 3 blades

2.2. Navigation ROV

The navigation system of the vehicle uses joystick as the console for motion command signal. Four propellers used for transversal and longitudinal movement while one propeller for vertical movement. All propellers controlled simultaneously follow the signal from the joystick. Combinations of navigation movement and propeller thrust direction in navigation are shown in Figure 3 and 4.

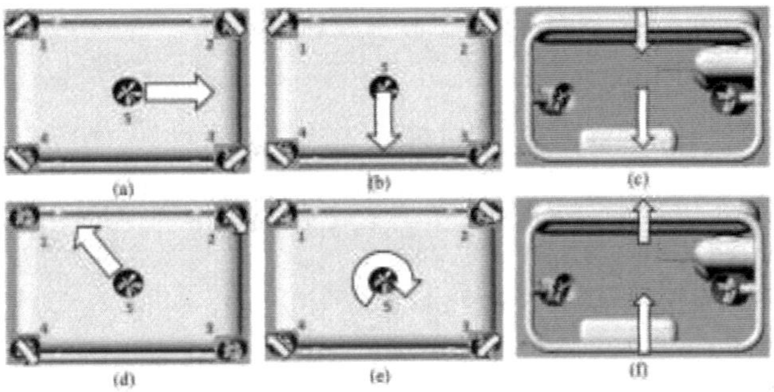

Figure 3. Navigation concept of ROV, yellow arrow is propeller thrust direction and white arrow is vehicle movement directions. (a) surge (b) sway, (c) Heave down, (d) combining surge and sway, (e) yaw and (f) heave up.

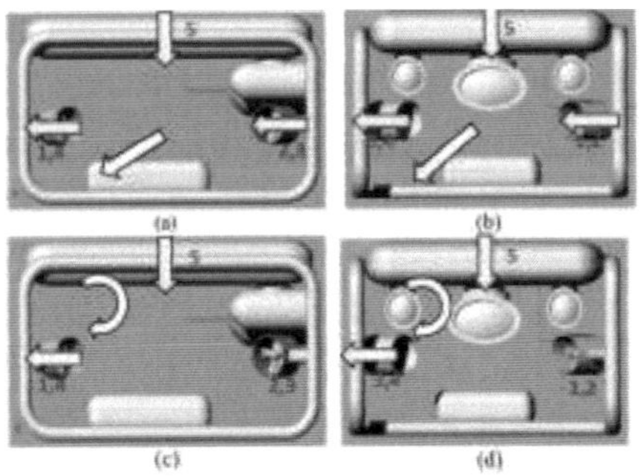

Figure 4. Navigation concept of ROV, yellow arrow is propeller thrust direction and white arrow is vehicle movement directions. (a) Combining surge and heave (b) combining sway and heave (c) asymmetric pitch, and (d) asymmetric roll.

In general, navigation concept of the vehicle designed to have ability moving in six degrees of freedom; surge, sway and heave (translation motion) and roll, pitch and yaw (rotation motion). Combining those motion, the vehicle can move in Omni-directional movement. However, roll and pitch motion are limited due to high static stability of the vehicle. Installation propeller at each corner of the vehicle consider the centre of vehicle's movement where propeller installed at proper position and angle that will not result a contra thrust between propellers. Positioning the propeller in such can affect the performance of propeller thrust and moment ineffectively.

3. Kinematic Model

An underwater vehicle with omni-directional capability was developed by Meyer at.al,[6]. The vehicle equipped three propellers with servo motor to rotate the propellers direction. The propeller thrust direction were designed can rotate automatically to provide the vehicle moves in multi direction. In this ROV design, the propellers attached at fixed angle; 45 and 135 degrees. The vehicle movement follows the Joystick command by rotating each propeller in a particular rotation. Propeller rotation is calculated based on a combination of the Joystick signal in x-signal and y-signal. The ratio between the signals is the angle of the direction of the vehicle direction and the resultant signal is a command to provide the propeller thrust. The combination signals are calculated to provide the number of rotations for each propeller. The calculation of propeller rotation is done using the kinematic model of the vehicle written as follows;

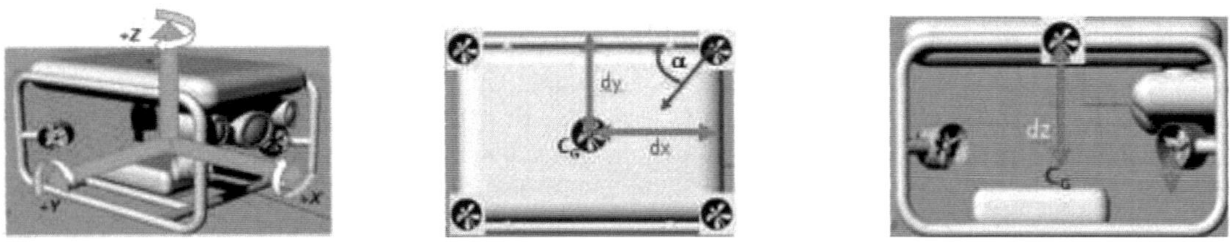

Figure 5. Design of movement axis of the vehicle

$$\begin{bmatrix} F_x \\ F_y \\ F_z \\ M_x \\ M_y \\ M_z \end{bmatrix} = \begin{bmatrix} c\alpha & c\beta & c\alpha & c\beta & 0 \\ -s\alpha & s\beta & -s\alpha & s\beta & 0 \\ 0 & 0 & 0 & 0 & 1 \\ -dx_1 s\alpha & dx_2\, s\beta & 0 & 0 & dz_5 \\ 0 & dy_2 c\beta & dy_3 c\alpha & 0 & dz_5 \\ dy_1 c\alpha & -dy_2 c\beta & -dy_3 c\alpha & -dy_4 c\beta & 0 \end{bmatrix} \begin{bmatrix} T_1 \\ T_2 \\ T_3 \\ T_4 \\ T_5 \end{bmatrix} \qquad [F]=[A][T], \quad [T]=[A^{-1}][F] \tag{1}$$

$$T_i = K_T \times D_i^4 \times n_i^2 \qquad n_i = \sqrt{\frac{T_i}{K_T \times D_i^4}} \tag{2}$$

In equation (1), transformation matric has 6 columns and 5 rows dimension. Regularly, there no inverse matric derived using a non-square matric. However, the inverse matric could be generated using a quasi-inverse matric for non-square dimension as shown by Ruben [9]. Then, the thrust [F] for all propeller transformed in to thrust [T] and the number of rotations for all propeller are obtained.

The strategy of navigating the vehicle analysed and the ratio of resultant force of interacting propellers obtained for all angles based on the angle of mapping motion required. Based on the kinematic model, the relation between angle of motion and the force ratio are showed as follows;

$$\tan\psi \cong \frac{F_y}{F_x} \qquad \tan\theta \cong \frac{F_z}{F_x} \qquad \tan\phi \cong \frac{F_z}{F_y} \tag{3}$$

Where *F, M, T* are force, moment and propeller thrust respectively, *c* is cosines, *s* is sinus, *d* is horizontal propeller arm to centre of rotation, *l* is vertical propeller arm to centre of rotation, *D* is propeller diameter, *n* is propeller rotation and K_T is propulsion coefficient of thrust. α, β are angles of propeller where α is angle for propeller 1 and 3, and β is angle for propeller 2 and 4. The direction of rotation motion symbols are ϕ, θ, ψ for roll, pitch, and yaw respectively.

The strategy of navigation the vehicle, drive the motor simultaneously to propel the vehicle at certain motion direction in such scenarios are;

a. Scenario-1 navigating the vehicle straight forward,
b. Scenario-2 navigating the vehicle sideward,
c. Scenario-3 navigating the vehicle downward,
d. Scenario-4 navigating the vehicle to 45 degrees sideward (drift),
e. Scenario-5 navigating the vehicle to 45 degrees side-downward,
f. Scenario-6 navigating the vehicle to 45 degrees back-downward,
g. Scenario-7 navigating the vehicle to roll,
h. Scenario-8 navigating the vehicle to pitch,
i. Scenario-9 navigating the vehicle to yaw at certain degree.

Simulation of the navigating scenarios through interacting thrust propeller only consider the thrust propellers. Forces resulted by the vehicle's motions were not considered. Thrust propeller linearized in range 0.0 and 1.0-unit force. The negative value indicating the propeller rotation is counter clockwise and positive value indicating the propeller rotation is clockwise. Then, all movement scenario found as follows.

Table 1. Interacting propeller thrust, force and moment in each axis

Scenario	P1	P2	P3	P4	P5	Fx	Fy	Fz	Mx	My	Mz
1	1	1	1	1	0	2,83	0,00	0,00	0,00	0,00	0,00
2	-1	1	-1	1	0	0,00	2,83	0,00	0,00	0,00	0,00
3	0	0	0	0	1	0,00	0,00	1,00	0,00	0,00	0,00
4	0	1	0	1	0	1,41	1,41	0,00	0,00	0,00	0,00
5	-1	1	-1	1	1	0,00	2,83	1,00	0,00	0,00	0,00
6	-1	-1	-1	-1	1	-2,83	0,00	1,00	0,00	0,00	0,00
7	0	0	-1	1	1	0,00	1,41	1,00	0,00	0,00	0,00
8	-1	0	0	-1	1	-1,41	0,00	1,00	0,00	0,00	0,00
9	-1	-1	1	1	0	0,00	0,00	0,00	0,00	0,00	0,85

In the Table-1 shows the maximum force moving forward and sideward are 2.83-unit force, drift is 1,41-unit force for each horizontal axis by four propellers. The maximum moment resulted at yaw motion with maximum 0.83-unit force. At the scenario-7 and 8 show no moment resulted as the propeller installed at the same high of the vehicle's centre gravitation.

Based on the navigation scenario, interacting propeller forces and moments were simulated with scenario from moving forward (scenario1) to Sideward (scenaio.2), Upward (scenario.3), Sideward-front (scenario.4) as shown in the following figures.

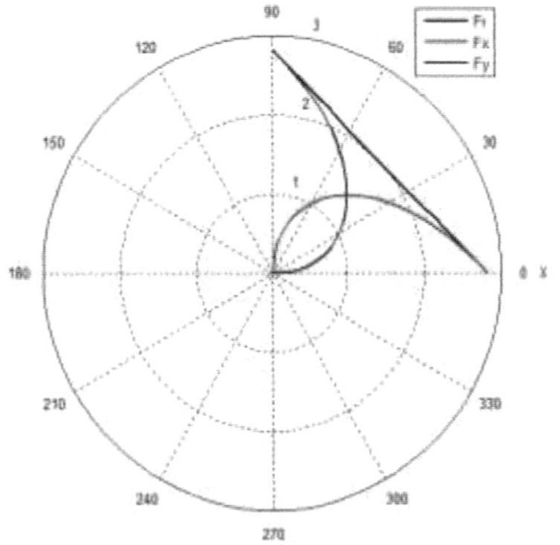

Figure 6. Forces in X, Y axis and the force resultant for the vehicle navigation from scenario 1 to scenario 2. The propeller 1 and 3 thrust simulated gradually change from +1.0 to -1.0 while propeller thrust 2 and 4 are +1.0.

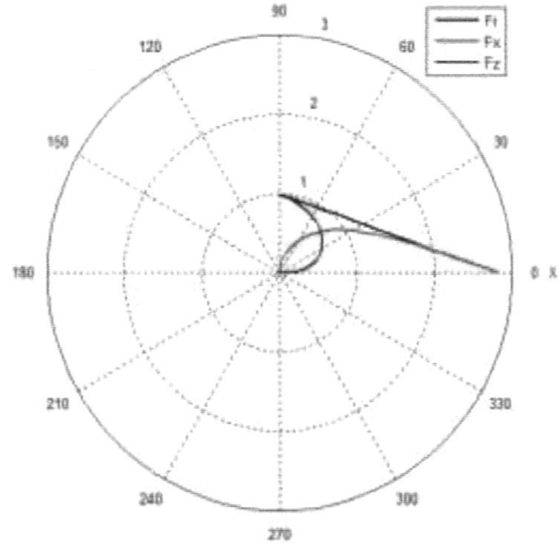

Figure 7. Forces in Fx, Fy, and Fr when the vehicle navigates from scenario 1 to scenario 3. The propeller 1, 2, 3, and 4 thrust were simulated change gradually from 1 to 0.0 while propeller thrust 5 changes gradually from 0.0 to 1.0.

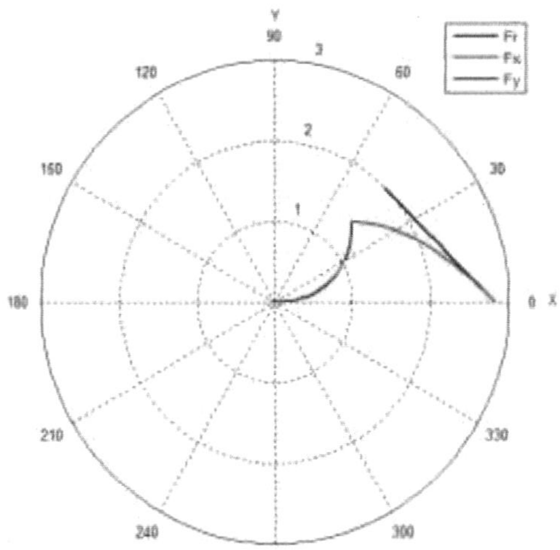

Figure 8. Forces in Fx, Fy, and Fr when the vehicle navigates from scenario 1 to scenario 4. The thrust of propeller 1 and 3 were simulated change gradually from 1 to 0.0 while propeller thrust 2 and 4 were constant at 1.0.

4. Experimental

The experiments were conducted in the small pool three metres diameter. This test were conducted to determine the ability of the propeller system drives the vehicle in accordance with the instructions given by JoyStick. The vehicle testing includes; manoeuvring to the right, manoeuvring to the left, drift to the right and sway. The tests were shown in the following figures.

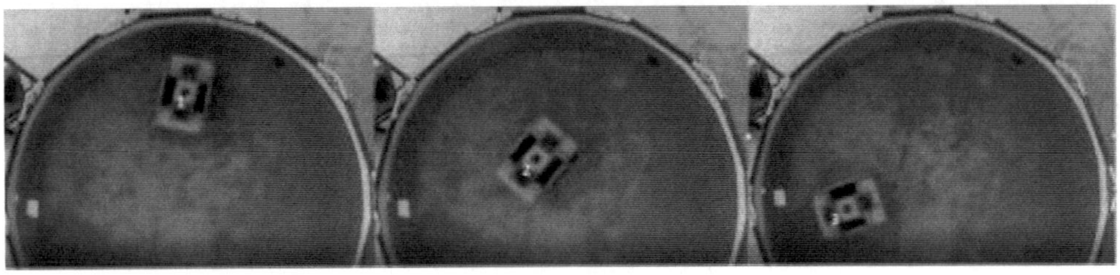

Figure 9. The vehicle moves from top to left. The vehicle manoeuvres with direction of the moving from 270 degrees (facing to –y axis) to 180 degrees (facing to –x axis).

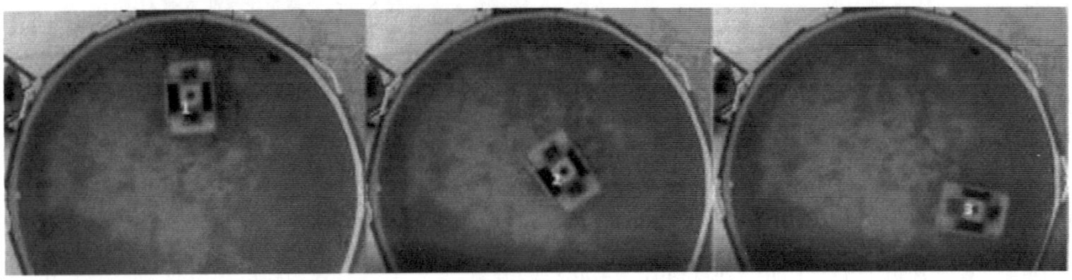

Figure 10. The vehicle moves from top to right. The vehicle manoeuvres with direction of the moving from 270 degrees (facing to –y axis) to 90 degrees (facing to +x axis).

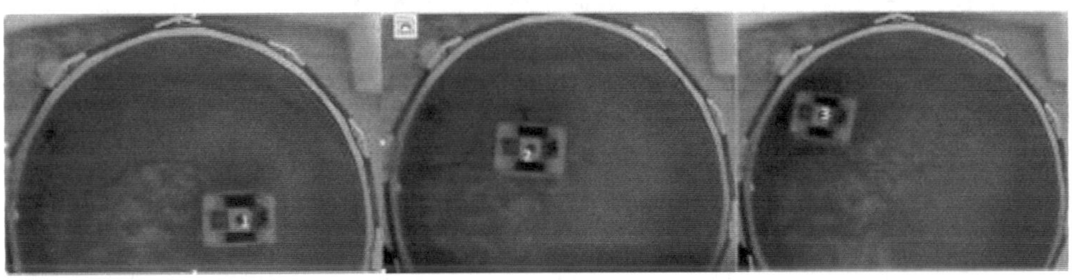

Figure 11. The vehicle drift from the centre position to the top left. The vehicle moves 135 degrees with the facing direction to the –x axis. The vehicle keeps the facing angle constantly.

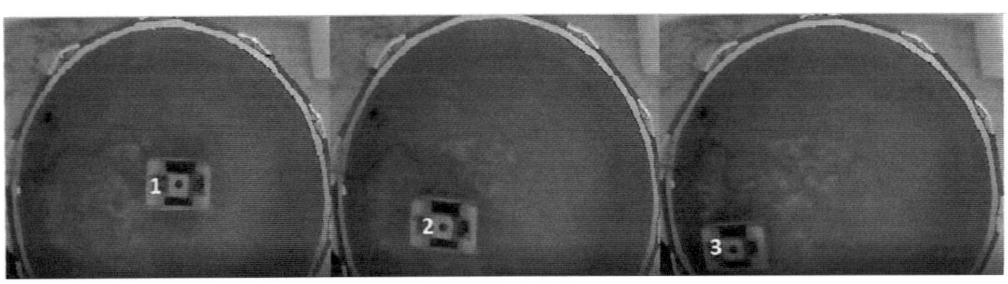

Figure 12. The vehicle drift from the centre position to the bottom left. The vehicle moves to 225 degrees with the facing direction to the –x axis. The vehicle keeps the facing angle constantly.

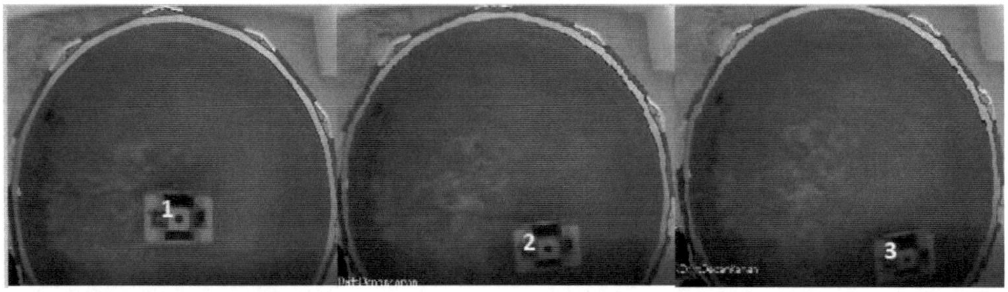

Figure 13. The vehicle drift from the centre position to the bottom right. The vehicle moves to 315 degrees with the facing direction to the +x axis. The vehicle keeps the facing angle constantly.

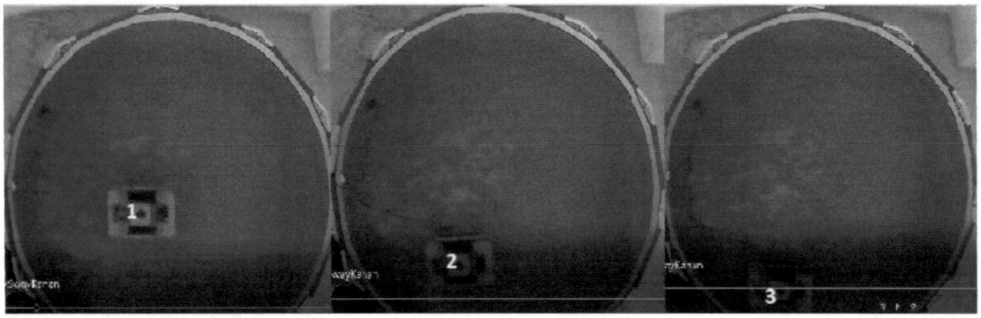

Figure 14. The vehicle drift from the centre position to the bottom. The vehicle moves to -y axis while the facing direction to the –x axis. The vehicle keeps the facing angle constantly.

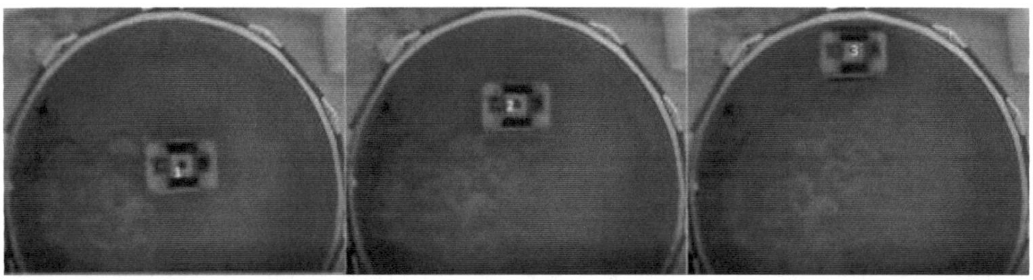

Figure 15. The vehicle drift from the centre position to the top. The vehicle moves to +y axis while the facing direction to the –x axis. The vehicle keeps the facing angle constantly.

5. Discussion

The navigation simulation that has been carried out showed the force or moment resulted in each scenario. When the vehicle moves forward, the resultant force resulted at a maximum force of 2.83 units force and 0,83-unit moment. The resultant force resulted was 70% of the total thrust given by all propellers. However, the vehicle has ability to move in omni-directions of movement except pitching and rolling were in the simulation of force and moment interaction shows there no moment for both motions. The test conducted in the small pool with 3,0 metres diameter. Based on the tests of the vehicle for manoeuvring, drift, and sway showed the vehicle can follow the scenario. The signal of propeller drivers generated by the Joystick's signal, the microcontroller calculated and feed the signal to the motor driver.

6. Conclusion

The simulation of interacting force and moment based on the navigation scenario and the experiment results showed the vehicle has ability to move in omni-directional motion while rolling and pitching motion are not available. The experiments showed the vehicle manoeuvred to left and right, sway and drift movement. The kinematic model and the generated inverse solution matrix showed a good response to the vehicle movement.

7. Future work

This research will continue on developing the robust control system for vehicle's keeping position under current disturbance. Numerical simulation and experimental will be performed to validate the simulation results at the end of this study.

References

[1] Gomes R M F, Sousa A, Fraga S L, Martins A, Sousa J B and Pereira F L 2005 A new ROV design: Issues on low drag and mechanical symmetry *Ocean. 2005 - Eur.* **2** 957–62
[2] Harsamizadeh Tehrani N, Heidari M, Zakeri Y and Ghaisari J 2010 Development, depth control and stability analysis of an underwater Remotely Operated Vehicle (ROV) *Ieee Icca 2010* 814–9
[3] Curtin T B, Crimmins D M, Curcio J, Benjamin M and Roper C 2005 Autonomous Underwater Vehicles: Trends and Transformations *Mar. Technol. Soc. J.* **39** 65–75
[4] Smallwood D, Bachmayer R and Whitcomb L 1999 A new remotely operated underwater vehicle for dynamics and control research *Proc. Int. Symp. Unmanned Untethered Submers. Technol.* 370–7
[5] J. Yuh 2000 Design and Control of Autonomous Underwater Robots: A Survey *Kluwer Acad. Publ.* **8** 7–24
[6] Meyer B, Ehlers K, Osterloh C and Maehle E 2013 Smart-E An Autonomous Omnidirectional Underwater Robot *Paladyn, J. Behav. Robot.* **4** 204–10
[7] Robert Boque 2015 Underwater robots: a review of technologies and applications *Int. J.* **42** 186–91
[8] Yilmaz S, Erdem D and Kavsaoglu M S 2015 Performance of a ducted propeller designed for UAV applications at zero angle of attack flight: An experimental study *Aerosp. Sci. Technol.* **45** 376–86
[9] Jr R D C and Manayon M B 2017 On the Quasi-inverse of a Non-square Matrix : An Infinite Solution **11** 1337–51

Acknowledgements

This research was funded by Kementerian Riset dan Pendidikan Tinggi of Indonesia. Thanks to all support team in this research, the students of marine engineering in the laboratory of control and electrical of marine engineering department of Hasanuddin University.

Clitoria ternatea flower as natural dyes for Dye-sensitized solar cells

Azwar Hayat[1*], A. Erwin E. Putra[1], Novriany Amaliyah[1], and Shyam. S. Pandey[2]

[1]Mechanical Engineering Department, Faculty of Engineering, Hasanuddin University, Poros Malino Km. 6 Gowa, South Sulawesi, Indonesia.
[2]LSSE, Kyushu Institute of Technology, Japan, 2-4 Hibikino, Wakamatsu-ku, Kitakyushu-shi, Fukuoka, Japan.

*azwar.hayat@unhas.ac.id

Abstract. *Clitoria ternatea* flowers have been investigated as a natural sensitizer in Dye-sensitized solar cell (DSSC). Dried *C. ternatea* flowers ware immersed in ethanol for 48 hours in dark room to obtain anthocyanin pigments. We compared the photovoltaic performance of the sensitizers with Titanium dioxide (TiO_2) only and TiO_2 with *scattering layers. C. ternatea* dyes have higher Power Conversion Efficiency (PCE) when applied in TiO_2 with *scattering layers* compared to standard TiO_2 semiconductors. *Scattering layer* (SL) has increased the light path inside the photo-anode film that doubled the short-circuit current (Jsc) and enhanced the open circuit voltage (Voc) compared to TiO_2 based cell without the SL. In conclusion, from the result, higher electron injections can be achieved by light management on photo-anode film with *scattering layers* as one of strategy to enhance the efficiency for further low cost and environment-friendly solar cell.

1. Introduction

Dye-sensitized solar cells (DSSCs) are gaining more attention due to their relatively lower cost of production [1]. DSSCs consist of two electrodes sandwich together filled with electrolyte. Both working and counter electrode are supported with conductive glass mainly made of Fluorine doped Tin Oxide (FTO). The Working electrode is made of FTO glass coated with thin layer of semiconductor materials such as Titanium dioxide (TiO_2), Tin dioxide (SnO_2), Zinc oxide (ZnO), etc [2-7]. Thin layer of semiconductor is covered with dye as electrons generator [8]. Counter electrode is made of FTO glass coated with thin layer of carbon, Graphene or platinum [9-11]. In order to maintain the electron regeneration, Electrolyte is employed between working and counter electrode. It is made of redox couple such as iodide/tri-iodide [12], Co(II)/Co(III)[13], etc. In DSSCs, one of the main components to increase the efficiency is dye sensitizer. Ruthenium-based sensitizer is the most widely used in DSSCs research area now. Ruthenium-based as well as metal free dyes have been reported having photon harvesting area mainly in the visible region of solar spectrum [14]. The difficulties of synthesize and higher price of ruthenium dyes made researcher effort to find out organic or inorganic dyes as co-sensitizer to fill the gap in ruthenium dye [15-18]. Ruthenium dyes are manufactured in sophisticated ways that raise the cost of the DSSC. Most of the sensitized dyes need anchoring groups in order to attach in nanoporous surface. In order to attach dyes on TiO_2 semiconductor surface, it could be accomplished by several types of bonding, for example, hydrophobic and covalent bonding [19]. Natural dye has been used for decades in industries. In DSSC, dyes from the plant as sensitizer are getting more attention due to lower cost production and easier in preparation [20-25]. In this works, the extracts of *Clitoria ternatea* flowers as dye sensitizer in DSSC. *C. ternatea* dye solution in ethanol was measured with UV-Vis Spectrophotometer. The DSSC was made using two configurations of TiO_2 semiconductors which are TiO_2 only and TiO_2 with scattering layers. The Incident photon-to-current Efficiency (IPCE) and current-voltage (I-V) characteristics were measured to define the photo-conversion efficiency (PCE).

2. Experimental

2.1. Plant material

Scientific classification of *Clitoria teratea is*
Kingdom: Plantae
Order: Fabales
Family: Fabaceae
Genus: *Cltoria*
Species: *C. ternatea*
Local name: Blue Pea, Bunga Telang, Butterfly Pea.

Clitoria ternatea is plant species belongings to Fabaceae Family. It is a native plant to Africa and introduced to lowlands of Asia, Central and South America. *C. ternatea* flowers are widely used as herb tea and coloring for food in Southeast Asia. Other than that, It is used as traditional Ayurvedic medicine in India[26] and traditional Chinese medicine [27]

2.2. Dye preparation

Clitoria ternatea flowers used in this study were cleaned with water and dried at room temperature in shaded places for more than three weeks. It were cut into small pieces and then crushed into powder with Food Grinder (FTC-Z100). Powdered flowers have 20-180 mesh fineness. Prior to usage, it was diluted in ethanol solvent with the composition of 10 gram of *C. ternatea* powder and 100 ml of ethanol. The mixture was sonicated for 60 minutes and then kept in dark place for 72 hours. In order to separate the remained powder, it was filtered with a vacuum pump and stored in a dark glass container. The dark container was used to protect the dye solution from direct light exposure.

2.3. Preparation of dye-sensitized solar cell

Fluorine doped Tin Oxide (FTO) glass was cut into 15 x 20 mm. It was cleaned in an ultrasonic cleaner with distilled water, Isopropanol, acetone sequentially 15 minutes for each process. After dried, the glass was coated with Titanium Dioxide (TiO_2) using blade coating method. The configuration of working electrode for this study were two types, which are TiO_2 only (smaller particle size, 30NRD) and TiO_2 with Scattering layers (Smaller and larger particle size, 30NRD and 400NR). Photoanode TiO_2 only was prepared by coated cleaned FTO glass with TiO_2 with small particle size for dye adsorption. TiO_2 film with scattering layer was prepared by coated FTO glass sequentially with two types of TiO_2 paste which are large (400nm) particle size serve as scattering layer and then sintered 450°C for 30 minutes. After that, the second layer was made by screen printing the smaller (30nm) particle size of TiO_2 on the top of the first layer of 400nm particle size TiO_2, and then sintered 450°C for 30 minutes. The average thickness after sintering process for TiO_2 only was 7 µM, TiO_2 with scattering layer was 10 µM (SURFCOM130A). The cell area was 0.25 cm^2. Baked working electrode then immersed in the dye solution for 24 hours until dye attached in semiconductor indicated from color changes. Counter electrode using Platinum (Pt) sputtered on FTO glass. Both working and counter electrode assembled in sandwich structure separated with 25 µm (Solaronix Spacer) thickness of the spacer used to create a gap for electrolyte injection. Electrolyte containing of 500 mM of Lithium Iodide (LiI), 50 mM Iodine (I_2), 600 mM Ethylmethylimidazolium dicyanoimide (MeEtIm-DCA), and 580 mM *tert*-butylpyridine (TBP) in Acetonitrile was used to fabricate the DSSCs.

Photovoltaic characterization the DSSCs were made using a solar simulator (Bunko-Keiki Co. Ltd., Model Solar Simulator CEP-2000SRR). The photovoltaic measurement used a black metal mask to avoid the effect of optical reflection from the light source to the cell from the side of the glass.

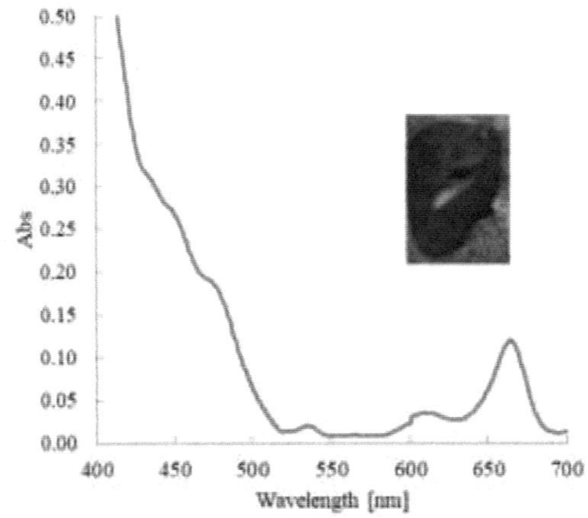

Figure 1. UV-Vis Absorption spectra of Natural dyes: *C. ternatea* in ethanol.

3. Results and Discussions

The Natural dyes from C. ternatea flowers extract have been utilized as sensitizers in this study. UV-Vis spectrophotometer (Jusco V-670) was used to characterize the light absorption spectra of dye. The absorption spectra of *C. ternatea* in ethanol solution are shown in figure 1. The maximum peak was recorded at 665 nm. It was found that *C. ternatea* flowers have lower solubility in ethanol that required immersing the powdered flowers for 72 hours before suitable usage in DSSC fabrication.

The Current-Voltage (I-V) characteristics of DSSC fabricated with TiO_2 only and TiO_2 with scattering layers were shown in figure 2. The main aspects adjusting the Photoconversion efficiency (PCE) are the short circuit density (Jsc) since open circuit voltage (Voc) and fill factors (FF). It can be seen that PCE of a device fabricated using TiO_2 with scattering layers has increased dramatically from 0.069% to 0.182%. This result was contributed to the enhancement of Jsc parameters. The dramatic boost in Jsc was increased more than twice from 0.203 mA/cm^2 to 0.481 mA/cm^2. This effect was donated by an increase of light path and better light management caused by scattering layers [28]. However, fill factors and Voc were not improved so much.

Table 1. I-V Characteristic of DSSCs with *C. ternatea* dye.

	TiO_2 Only	TiO_2 + SL
Efficiency [%]	0.069	0.182
FF	0.678	0.700
Voc [V]	0.501	0.539
Jsc [mA/cm^2]	0.203	0.481
Mask Area [cm^2]	0.214	0.214

Table 1 shows result of TiO_2 only have Jsc of 0.203 mA/cm^2, Voc 0.501 V, FF 0.678 and PCE 0.069% and TiO_2 with SL have Jsc of 0.481 mA/cm^2, Voc 0.539 V, FF 0.700 and PCE 0.182% for device with TiO_2 semiconductor. IPCE graph in figure 3 shown that photon to current density on TiO_2 with scattering layer was higher than TiO_2 only configuration. Higher electron injection was confirmed from wider IPCE in 300 to 500 nm wavelength.

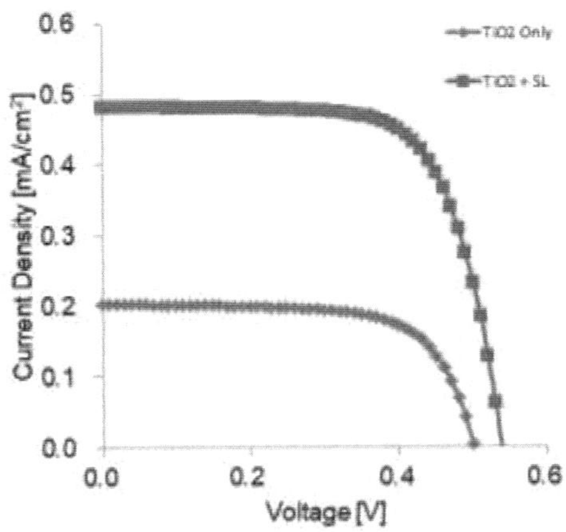

Figure 2. I-V Characteristic of DSSCs

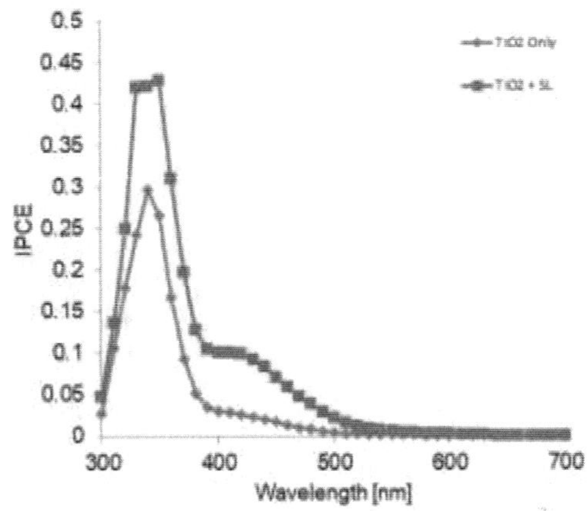

Figure 3. IPCE Spectra of DSSCs

4. Conclusions

DSSC with Natural dye, *Clitoria ternatea* flowers, was fabricated with two configurations of TiO_2 semiconductors which are TiO_2 only and TiO_2 with scattering layer (SL). Cell fabricated with TiO_2 + SL configuration produced higher PCE than a cell with TiO_2 only configuration. Longer light diffusion path on TiO_2 with SL lead to higher electron injection due to a greater probability of light interact with dye in TiO_2 film. IPCE graph confirmed that photon to current density on TiO_2 film with SL was higher. In conclusion, from the result, higher electron injections can be achieved by appropriate light management as one of strategy to enhance the efficiency for further low cost and environment-friendly solar cell.

References

[1] B. O'Reagan, M. Grätzel (1991), A low-cost, high-efficiency solar cell based on dye-sensitized colloidal TiO_2, *Nature*, 535, 737

[2] Ahmad, M. S., Pandey, A. K., & Rahim, N. A. (2017). Advancements in the development of TiO2 photoanodes and its fabrication methods for dye sensitized solar cell (DSSC) applications. A review. *Renewable and Sustainable Energy Reviews*, *77*, 89-108.

[3] Divya, C., Janarthanan, B., & Chandrasekaran, J. (2017). Review on Recent Advances in Titanium Dye Oxide Nanoparticles for Dye Sensitized Solar Cell Applications.

[4] Das, P. P., Roy, A., & Sujatha Devi, P. (2016). Zn2SnO4 as an Alternative Photoanode for Dye Sensitized Solar Cells: Current Status and Future Scope. *Transactions of the Indian Ceramic Society*, *75*(3), 147-154.

[5] Wang, D., Liu, S., Shao, M., Li, Q., Gu, Y., Zhao, J., ... & Fang, Y. (2018). Aqueous Solution-Processed Multifunctional SnO2 Aggregates for Highly Efficient Dye-Sensitized Solar Cells. *Industrial & Engineering Chemistry Research*, *57*(21), 7136-7145.

[6] Omar, A., & Abdullah, H. (2014). Electron transport analysis in zinc oxide-based dye-sensitized solar cells: a review. *Renewable and Sustainable Energy Reviews*, *31*, 149-157.

[7] Vittal, R., & Ho, K. C. (2017). Zinc oxide based dye-sensitized solar cells: A review. *Renewable and Sustainable energy reviews*, *70*, 920-935.

[8] Sugathan, V., John, E., & Sudhakar, K. (2015). Recent improvements in dye sensitized solar cells: A review. *Renewable and Sustainable Energy Reviews*, *52*, 54-64.

[9] Chen, M., & Shao, L. L. (2016). Review on the recent progress of carbon counter electrodes for dye-sensitized solar cells. *Chemical Engineering Journal*, *304*, 629-645.

[10] Singh, E., & Nalwa, H. S. (2015). Graphene-based dye-sensitized solar cells: a review. *Science of Advanced Materials*, *7*(10), 1863-1912.

[11] Lim, S. C., Hsiao, M. C., Lu, M. D., Tung, Y. L., & Tuan, H. Y. (2018). Synthesis of Germanium-Platinum Nanoparticles as High-Performance Catalysts for Spray-Deposited Large-Area Dye-sensitized Solar Cells (DSSC) and Hydrogen Evolution Reaction (HER). *Nanoscale*.

[12] Chang, P. J., Cheng, K. Y., Chou, S. W., Shyue, J. J., Yang, Y. Y., Hung, C. Y., ... & Chou, P. T. (2016). Tri-iodide reduction activity of shape-and composition-controlled PtFe nanostructures as counter electrodes in dye-sensitized solar cells. *Chemistry of Materials*, *28*(7), 2110-2119.

[13] Wu, K. L., Huckaba, A. J., Clifford, J. N., Yang, Y. W., Yella, A., Palomares, E., ... & Nazeeruddin, M. K. (2016). Molecularly engineered Ru (II) sensitizers compatible with cobalt (II/III) redox mediators for dye-sensitized solar cells. *Inorganic chemistry*, *55*(15), 7388-7395.

[14] A. Yella, H.-W. Lee, H. N. Tsao, C. Yi, A. K. Chandiran, Md. K. Nazeeruddin, E. W.-G. Diau, C.-Y. Yeh, S. M. Zakeeruddin and M. Grätzel (2011), "Porphyrin-sensitized solar cells with cobalt (II/III)–based redox electrolyte exceed 12 percent efficiency", *Science*, 334, 629-633

[15] Mehmood, U., Hussein, I. A., Harrabi, K., Tabet, N., & Berdiyorov, G. R. (2016). Enhanced photovoltaic performance with co-sensitization of a ruthenium (II) sensitizer and an organic dye in dye-sensitized solar cells. *RSC Advances*, *6*(10), 7897-7901.

[16] Wang, P., Yang, L., Wu, H., Cao, Y., Zhang, J., Xu, N., ... & Grätzel, M. (2018). Stable and Efficient Organic Dye-Sensitized Solar Cell Based on Ionic Liquid Electrolyte. *Joule*.

[17] Aghazada, S., Gao, P., Yella, A., Marotta, G., Moehl, T., Teuscher, J., ... & Nazeeruddin, M. K. (2016). Ligand engineering for the efficient dye-sensitized solar cells with ruthenium sensitizers and cobalt electrolytes. *Inorganic chemistry*, *55*(13), 6653-6659.

[18] Islam, A., Akhtaruzzaman, M., Chowdhury, T. H., Qin, C., Han, L., Bedja, I. M., ... & Reynolds, J. R. (2016). Enhanced photovoltaic performances of dye-sensitized solar cells by co-sensitization of benzothiadiazole and squaraine-based dyes. *ACS applied materials & interfaces*, *8*(7), 4616-4623.

[19] Zhang, H., He, B., Tang, Q., & Yu, L. (2015). Bifacial dye-sensitized solar cells from covalent-bonded polyaniline–multiwalled carbon nanotube complex counter electrodes. *Journal of Power Sources*, *275*, 489-497.

[20] Ludin, N. A., Mahmoud, A. A. A., Mohamad, A. B., Kadhum, A. A. H., Sopian, K., & Karim, N. S. A. (2014). Review on the development of natural dye photosensitizer for dye-sensitized solar cells. *Renewable and Sustainable Energy Reviews*, *31*, 386-396.

[21] Shalini, S., Prasanna, S., Mallick, T. K., & Senthilarasu, S. (2015). Review on natural dye sensitized solar cells: operation, materials and methods. *Renewable and Sustainable Energy Reviews*, *51*, 1306-1325.

[22] Goswami, D., Sinha, D., & De, D. (2017, March). Nanostructured ZnO and natural dye based DSSC for efficiency enhancement. In *Science Technology Engineering & Management (ICONSTEM), 2017 Third International Conference on* (pp. 556-560). IEEE.

[23] Yoon, S. J., Lim, I., Kim, J. H., Adhikari, S., Lee, W. Y., Lee, J. K., ... & Han, S. H. (2016). Deprotonated curcumin as a simple and quick available natural dye for dye sensitized solar

cells. *Energy Sources, Part A: Recovery, Utilization, and Environmental Effects*, *38*(2), 183-189.

[24] Teoli, F., Lucioli, S., Nota, P., Frattarelli, A., Matteocci, F., Di Carlo, A., ... & Forni, C. (2016). Role of pH and pigment concentration for natural dye-sensitized solar cells treated with anthocyanin extracts of common fruits. *Journal of Photochemistry and Photobiology A: Chemistry*, *316*, 24-30.

[25] Hayat, A., Putra, A. E. E., Amaliyah, N., Hayase, S., & Pandey, S. S. (2018, March). Piper Ornatum and Piper Betle as Organic Dyes for TiO2 and SnO2 Dye Sensitized Solar Cells. In *Journal of Physics: Conference Series* (Vol. 979, No. 1, p. 012062). IOP Publishing.

[26] Mukherjee PK, Kumar V, Kumar NS, Heinrich M (2008). The Ayurvedic medicine Clitoria ternatea-From traditional use to scientific assessment. *J Ethnopharmacol*. 120 (3): 291–301

[27] Fantz, Paul R. (1991). Ethnobotany of *Clitoria* (Leguminosae). *Economic Botany*. New York Botanical Garden Press. 45 (4): 511–20.

[28] Goa X.D, Li X.M, Gan X.Y (2013), Enhancing the light harvesting capacity of photo anode films in DSSC. *Chap 7, Solar cell research and application prospective*, INTECH 2013.

Acknowledgment.
The work was part of Grand “PDUPT 2018” from Indonesia Directorate General of Higher Education.

Finite element analysis on vibration of a flexible single-link manipulator moved translationally

D Dermawan[1,2,*], H Abbas[1], R Syam[1], Z Djafar[1], A K Muhammad[2]

[1]Department of Mechanical Engineering, Hasanuddin University, Indonesia 90245
[2]Department of Mechanical Engineering, State Polytechnic of Ujung Pandang, Indonesia 90245

*E-mail: dermawan@poliupg.ac.id

Abstract. The objectives of this study are to formulate the equation of motion of a flexible single-link manipulator system and to develop computational codes by a finite-element method in order to perform dynamics simulation on vibration of the manipulator system. The system used in this research consist of an aluminium beam as a flexible link, a clamp-part to hold the link, a DC motor and a lead screw set to move the link translationally. Computational codes on time history responses and FFT (Fast Fourier Transform) processing were developed to calculate dynamic behaviour of the link. The simulation result show the dynamic behaviour of the system on free vibration and forced vibration by excitation force due to the translational motion of the single-link manipulator system.

Keywords: Finite-element method, flexible manipulator, translational motion.

1. Introduction

In industrial applications and robotics system, a single-link flexible manipulator is expected to achieve maximum workability during operations such as high speed with safe operation, improved positioning accuracy, lighter weight, and lower energy consumption. Problems that often arise in flexible-link manipulator are vibrations caused by its flexibility which causes deformations that interfere with the performance of the system.

In the last decade, a number of researchers have investigated dynamic behaviour of flexible manipulator. Muhammad et al investigated dynamic of flexible manipulator with rotational motion using finite-element method including vibration control [1]–[8]. Muhammad et al used two node element in the investigation.

This study aims to formulate the equation of motion of the system and to develop computational codes with finite-element methods to find out time history responses and FFT (Fast Fourier Transform) processing of the link. The results of the plot time history response and FFT generated the natural frequency of the system.

Grounded in the previous work on computer simulation of a flexible single-link manipulator [8] and a flexible two-link manipulator [3] which failed to include the translational motion, this study put an emphasis on the translational motions in which these motions are often used in industrial applications and robotics. Another previous study also reported that the result of numeric simulation of a two-link flexible manipulator was used to verify the control scheme effectively [9]. In relation to the current research, some similarities were also identified such as the use of a kinetic energy, potential energy, and Lagrange equation to derive the equation of motion[10]. Beyond the similarities, Bien's work focuses on translational and rotational joint of two-link manipulator while this piece of work focuses on translational motion with a clamp-part to hold the link.

Prior to the application of active-force control on vibration of a single-link manipulator [2], it is necessary to have a clear picture of the characteristic of the link without vibration control. Obtaining the value of the free vibration is necessary to find out the natural frequency of the link. Furthermore, the natural frequency is pivotal to determine the type of effective control.

2. Formulation by finite-element method

Grounded in a finite-element method [11], each partition of the link is partially calculated to find out the matrix of mass *(M)* and stiffness *(K)* which generated a natural frequency of system. The link with a cantilever structure with a length of 33 cm is divided into 6 elements and move translationally by a DC motor to determine the vibration characteristics of the link. The degrees of freedom of the finite-element are divided into two types namely the lateral deformation $v(t)$ and the rotational angle $\Psi(t)$. The physical properties of the system consists of the length (L_i), the cross-sectional area (S_i), and the area moment of inertia (I_i). Each element of the mechanical properties (Young's modulus and mass density) are denoted as E_i and ρ_i.

2.1. Kinematics

Some text Figure 1 plots the directions of translational motion on the link into three types namely the displacement (Y), velocity ($\dot{Y}$), and acceleration ($\ddot{Y}$)

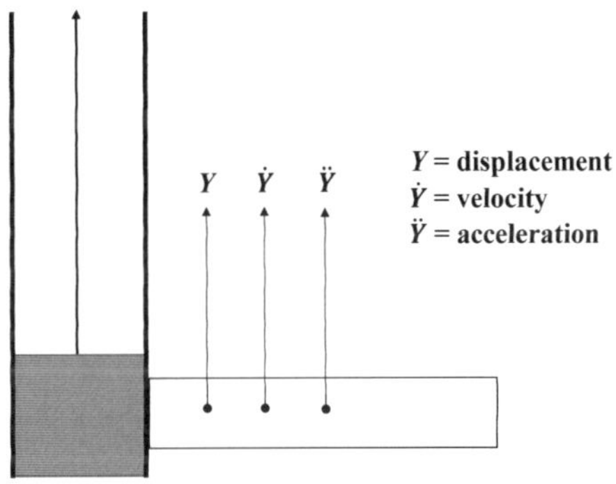

Figure 1. Directions of translational motion on the link

Figure 2 shows the position vector at point P on the link in the global coordinate frame and translating coordinate frame. $O - XY$ is the global coordinate frame and $O - yx$ is the translational coordinate frame. When the link move translationally, the displacement at X axis is x_p and the displacement at Y axis is Yp. The deflection due to the translational motion is denoted as v_p.

The vector position $\boldsymbol{r}(x, t)$ at point P on the link at time $t = t$, in the global coordinate frame $O - XY$ shown in the figure 2 is given by:

$$r(x,t) = X_p(x,t)\boldsymbol{I} + Y_p(x,t)\boldsymbol{J} \tag{1}$$

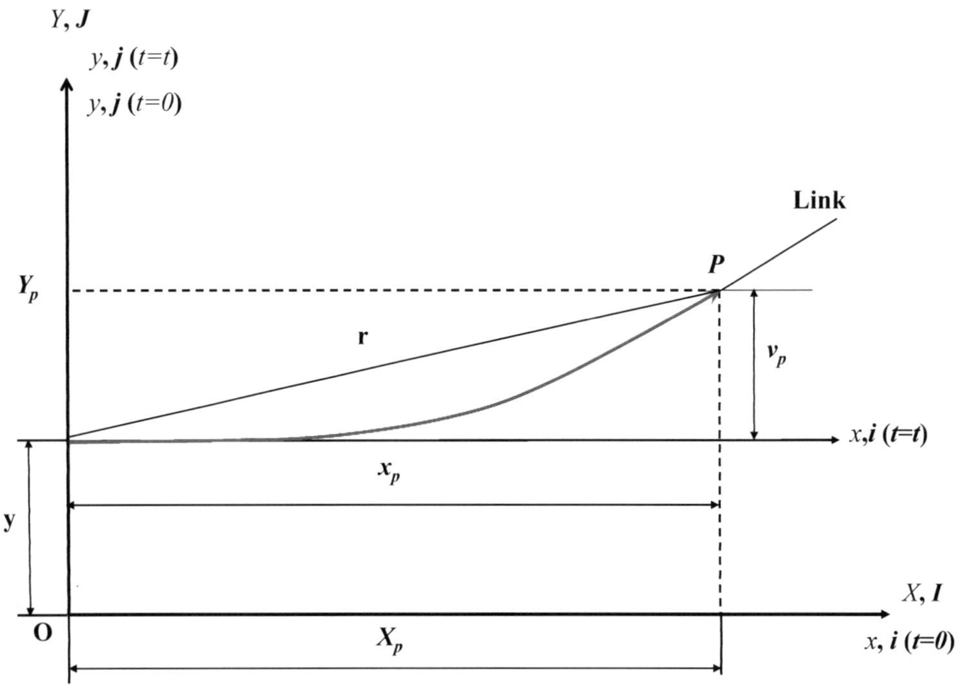

Figure 2. Position vector of an arbitrary point P of the link in the translational coordinate frames

The position vector of P in frame $O-XY$ is

$$X_p(x,t) = x_p \tag{2}$$
$$Y_p(x,t) = y + v_p \tag{3}$$

The velocity vector of P in frame $O-xy$, given by

$$\dot{\boldsymbol{r}}(x,t) = \dot{X}_p(x,t)\boldsymbol{I} + \dot{Y}_p(x,t)\boldsymbol{J} \tag{4}$$

2.2. Finite-element method

Figure 3. illustrates the partitions of the link which are divided into six elements. Each node has the lateral deformation $v(t)$ and the rotational angle $\Psi(t)$.

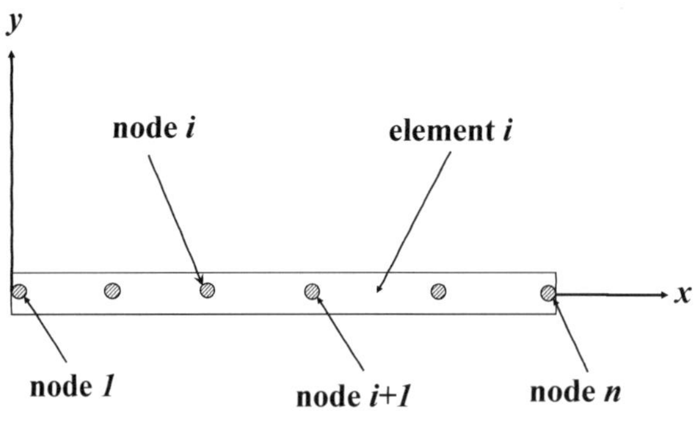

Figure 3. Translating coordinate frame of the link

Then, figure 4 shows the element coordinate frame of the *i-th* element.

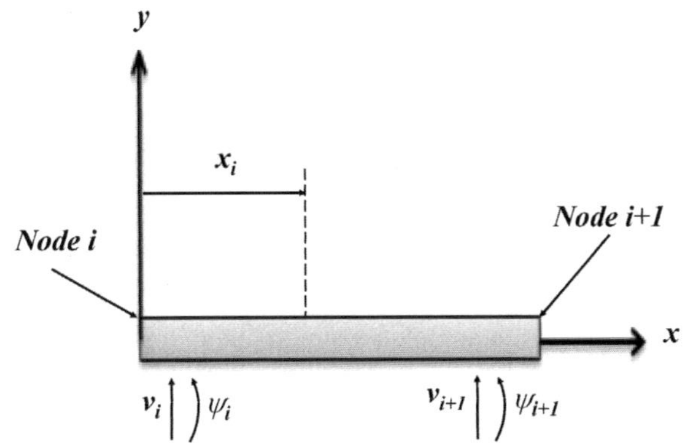

Figure 4. Element coordinate frame of the *i-th* element

The nodal displacement vector is

$$\delta_i = \{v_i \quad \varphi_i \quad v_{i+1} \quad \varphi_{i+1}\}^T \tag{5}$$

2.3. Equation of motion

The discretization generated from the previous section, the velocity square of *P* can be obtained

$$\dot{\boldsymbol{r}}^T.\dot{\boldsymbol{r}} = \dot{x}_p{}^2 + \dot{y}^2 + \dot{v}_p{}^2 + 2\dot{y}\dot{v}_p \tag{6}$$

The kinetic energy (*Ti*) of system can be given by

$$\boldsymbol{T_i} = \int_{v_i} \rho_i.\dot{\boldsymbol{r}}^T.\dot{\boldsymbol{r}}.dv_i \tag{7}$$

The kinetic energy (*Ti*) of system can be expressed as

$$\boldsymbol{T_i} = \frac{1}{2}m_i\dot{x}_p{}^2 + \frac{1}{2}m_i\dot{y}^2 + \frac{1}{2}\dot{\delta}_i{}^T\boldsymbol{M}_i\dot{\delta}_i + \dot{y}\boldsymbol{f}_{\boldsymbol{t}_i}{}^T\dot{\delta}_i \tag{8}$$

Where

$$\dot{y}\boldsymbol{f}_{\boldsymbol{t}_i}{}^T\dot{\delta}_i = \frac{m_i}{12}[6 \quad l_i \quad 6 \quad -l_i]$$

The potential energy of the system can be written as

$$\boldsymbol{U_i} = \frac{1}{2}\delta_i{}^T K_i\delta_i \tag{9}$$

The dynamic model is formulated using the Lagrange equation can be defined by

$$M_i\ddot{\delta}_i - \ddot{y}(t)f_{ti}{}^T + C_i\dot{\delta}_i + K_i\delta_i = 0 \tag{10}$$

The equation of motion of the *i-th* element is given by:

$$M_i\ddot{\delta}_i + C_i\dot{\delta}_i + K_i\delta_i = \ddot{y}(t)f_{ti}{}^T \tag{11}$$

Where mass matrix (M_i), damping matrix (C_i), stiffness matrix (K_i), and the excitation force of the translation motion ($\ddot{y}(t)f_{ti}{}^{T}$)
Matrix M_i, C_i, K_i and vector $\ddot{y}(t)\boldsymbol{f}_{ti}{}^{T}$ in Eq. (11) are represented as

$$\boldsymbol{M}_i = \frac{\rho_i S_i L_i}{420}\begin{bmatrix} 156 & 22L_i & 54 & -13L_i \\ 22L_i & 4{L_i}^2 & 13L_i & -3{L_i}^2 \\ 54 & 13L_i & 156 & -22L_i \\ -13L_i & -3{L_i}^2 & -22L_i & 4{L_i}^2 \end{bmatrix} \quad (12)$$

$$\boldsymbol{Ki} = \frac{E_i I_{zi}}{{L_i}^3}\begin{bmatrix} 12 & 6L_i & -12 & 6L_i \\ 6L_i & 4{L_i}^2 & -6L_i & 2{L_i}^2 \\ -12 & -6L_i & 12 & -6L \\ 6L_i & 2{L_i}^2 & -16L_i & 4{L_i}^2 \end{bmatrix} \quad (13)$$

$$\boldsymbol{C_i} = \alpha . K_i \quad (14)$$

$$\boldsymbol{f_{ti}^T} = -\frac{\rho_i S_i L_i}{12}\{6,\ l_i,\ 6, -l_i\} \quad (15)$$

The length of the i-th element, the length from element 1 to i, and the Rayleigh damping factor are denoted by l_i, $l_{1\text{-}i}$, and α [12] respectively
In the end, the equation of motion of the single-link with n element considering the boundary conditions is written by

$$\boldsymbol{M_n \ddot{\delta}_n + C_n \dot{\delta}_n + K_n \delta_n = \ddot{y}(t) f_{t_n}} \quad (16)$$

3. Dynamic behaviour of the system

3.1. Computational model

Figure 5 illustrates the system of a flexible single-link manipulator which has a track of the link, clamp as the holder of the link, a servo motor to rotate the link, and a DC motor to make a translation motion.

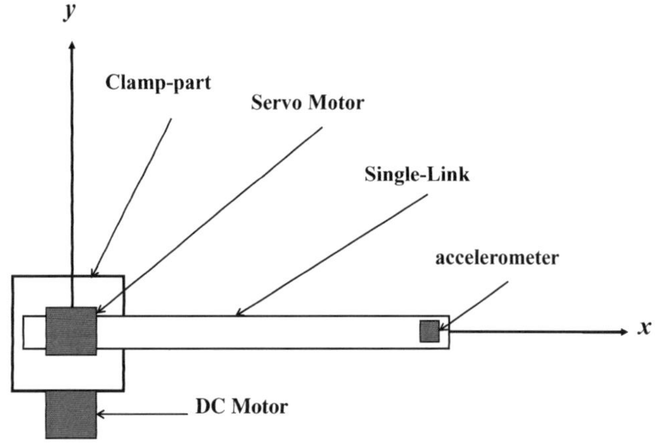

Figure 5. Model system of flexible Single-link manipulator

IOP Conf. Series: Materials Science and Engineering **619** (2019) 012050 doi:10.1088/1757-899X/619/1/012050

3.2. Physical parameters of dynamic models

Table 1. Physical parameters of the link

Property	**Symbol**	**value**
Total length (m)	L	3.30×10^{-1}
Length of the link (m)	l_i	3.00×10^{-1}
breadth of cross section (m)	b_i	2.50×10^{-2}
height of cross section (m)	h_i	1.00×10^{-3}
Cross section area of the link (m^2)	S_i	1.95×10^{-5}
Cross section area moment of inertia around i-axis of the link (m^4)	I_i	2.75×10^{-12}
Young's Modulus of the link (GPa)	E_i	7.00×10^{1}
Density of the link (kg/m^3)	ρ_i	2.70×10^{3}
Damping factor of the link	α	0.10×10^{-3}

3.3. Time history response on free vibration

Time history response on free vibration simulation on free vibration was conducted using an impulse force as an external one. Figure 6 shows the simulation time history response of lateral deformation v_p on the free vibration. Furthermore, the computational codes on time history response of single-link were developed. Figure 6 shows the calculated lateral deformation at Node 6 of the system under the impulse force.

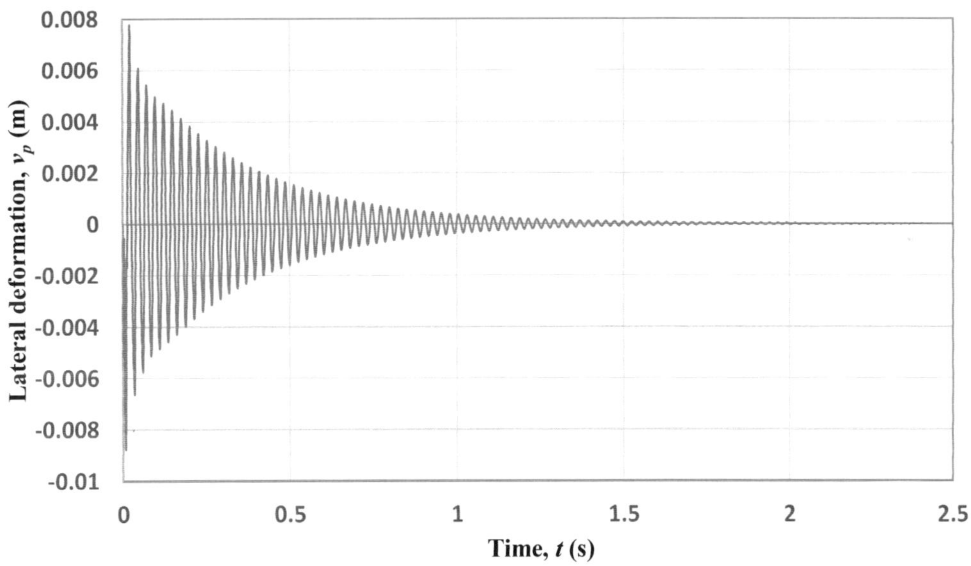

Figure 6. Time history response of the system

3.4. FFT (Fast Fourier Transform) Processing.

The calculated result from time history response to lateral deformation with free vibration on the system was transferred by FFT processing to find its frequencies. Figure 7 shows the calculated natural frequencies of the flexible link manipulator. The first, second, and third natural frequency are 38.57 [Hz], 109.37 [Hz], and 174.32 [Hz].

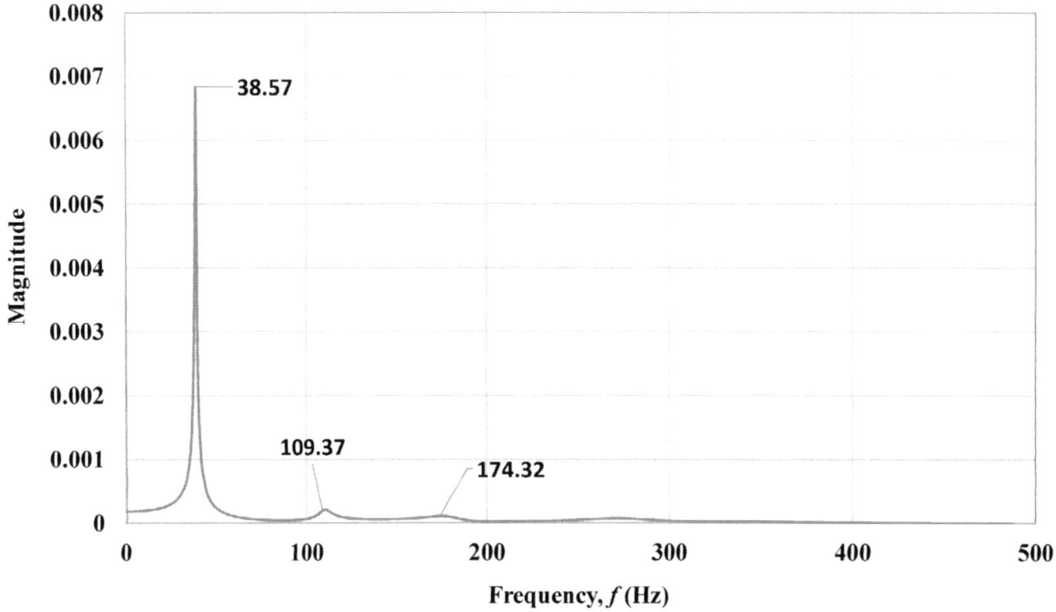

Figure 7. Calculated natural frequencies of the link

3.5. Time history response of the system with excitation force.

The calculated result from time history response to lateral deformation with excitation force on the system is shown in the figure 8. Showing at 0.5 (*s*) in the timeline, the lateral deformation becomes large due to the excitation force ($\boldsymbol{f_t}$ = 5).

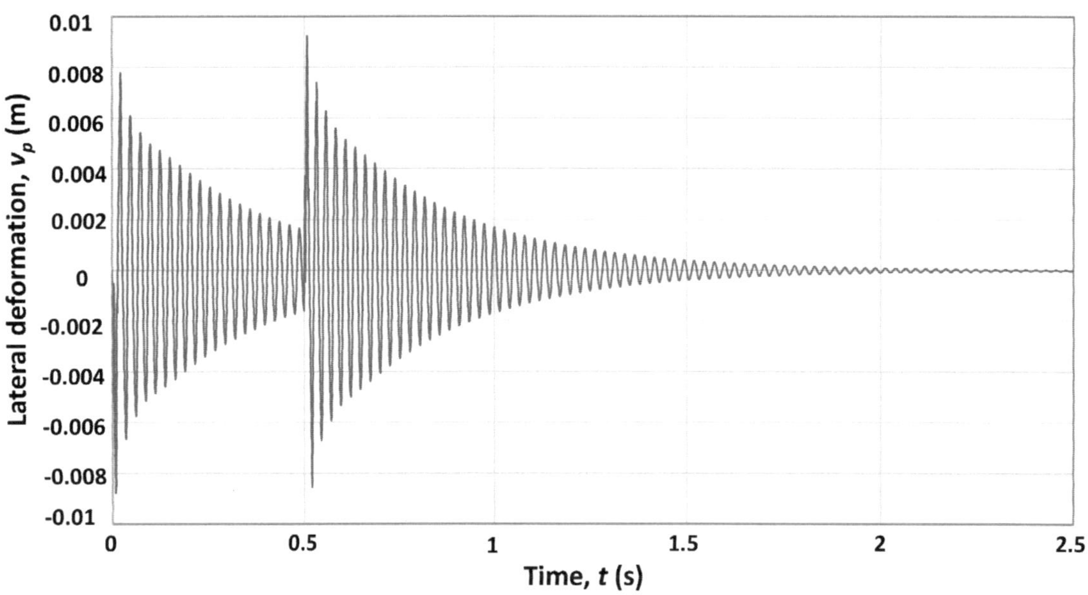

Figure 8. Time history response of the system with excitation force

IOP Conf. Series: Materials Science and Engineering **619** (2019) 012050 doi:10.1088/1757-899X/619/1/012050

4. Conclusions

The equation of motion on the single-link flexible manipulator moved translationally has been discretized using a finite-element method. The computational codes were developed to simulate the dynamic system using the Scilab application. The simulation and calculated result of time history response, natural frequencies, and vibration mode showed validated formulation, computational codes, and model system. Fast Fourier Transform (FFT) has also generated the first, second, and third natural frequencies values as (f1) 38.57 [Hz], (f2)109.37 [Hz], and (f3)174.32 [Hz].

Further research recommends to combine the translational and the rotational motion on the single-link flexible manipulator by finding the equation of translational and rotational motion. The further step will continue doing an experiment of the model.

References

[1] A. K. Muhammad, S. Okamoto, and J. H. Lee, "Comparison Between the One Piezoelectric Actuator and the Two Ones on Vibration Control of a Flexible Two-Link Manipulator Using Finite Element Method," *Int. J. Mech. Eng.*, vol. **5**, no. 1, pp. 25–42, 2016.

[2] A. K. Muhammad, S. Okamoto, and J. H. Lee, "Finite Element Analysis for Active-force Control on Vibration of a Flexible Single-link Manipulator," *Int. J. Smart Mater. Mechatronics*, vol. **2**, no. April, p. 18, 2016.

[3] A. K. Muhammad, S. Okamoto, and J. H. Lee, "Computer Simulations and Experiments on Vibration Control of a Flexible Link Manipulator Using a Piezoelectric Actuator Computer Simulations and Experiments on Vibration Control of a Flexible Link Manipulator Using a Piezoelectric Actuator," *Eng. Lett.*, vol. **3**, no. 23, 2015.

[4] A. K. Muhammad, S. Okamoto, and J. H. Lee, "Active-Force Control on Vibration of a Flexible Single-Link Manipulator Using a Piezoelectric Actuator," *Trans. Eng. Technol.*, pp. 1–15, 2015.

[5] A. K. Muhammad, S. Okamoto, and J. H. Lee, "Active-Force Control on Vibration of a Flexible Single-Link Manipulator Using a Piezoelectric Actuator," in *Transactions on Engineering Technologies*, springer, 2015, pp. 1–15.

[6] A. K. Muhammad, S. Okamoto, and J. H. Lee, "Comparison of proportional-derivative and active-force controls on vibration of a flexible single-link manipulator using finite-element method," *Artif. Life Robot.*, vol. **19**, no. 4, pp. 375–381, 2014.

[7] A. K. Muhammad, S. Okamoto, and J. H. Lee, "Comparisons of proportional and active-force controls on vibration of a flexible link manipulator using a piezoelectric actuator through calculatiions and experiments," *Eng. Lett.*, vol. **22**, no. 3, p. 8, 2014.

[8] A. K. Muhammad, S. Okamoto, and J. H. Lee, "Computer simulations on vibration control of a flexible single-link manipulator using finite- element method," in *The Nineteenth International Symposium on Artificial Life and Robotics 2014*, 2014.

[9] X. Yang and Z. Zhong, "Dynamics and Terminal Sliding Mode Control of Two-Link Flexible Manipulators with Noncollocated Feedback," *IFAC Proc. Vol.*, vol. **46**, no. 20, pp. 218–223, 2013.

[10] D. X. Bien, C. A. My, and P. B. Khoi, "Dynamic Modeling and Control of a Flexible Link Manipulators with Translational and Rotational Joints," *VNU J. Sci. Math. – Phys.*, vol. **34**, no. 1, pp. 52–66, 2018.

[11] S. S. Rao, *The finite element method in engineering*, Fifth Edit. Elsevier, 2011.

[12] M. Lalanne, P. Berthier, and J. Der Hagopian, *Mechanical Vibrations for Engineers*. <cel-00315803> HAL, 1984.

IOP Conf. Series: Materials Science and Engineering **619** (2019) 012051 doi:10.1088/1757-899X/619/1/012051

Computational Simulation on Facial Expressions Using Finite Element Analysis

Andi Amijoyo Mochtar, Rafiuddin Syam

Hasanuddin University, Departement of Mechanical Engineering, Jl. Poros Malino KM. 6 Gowa, 92171, Sulawesi Selatan, Indonesia
Email: andijoyo@unhas.ac.id

Corrresponding author email: rafiuddinsyam@gmail.com

Abstract. The purposes of this research are to identify mechanical properties of silicone rubber used as artificial skin and perform computational simulation on facial expressions using Finite Element Analysis (FEA). The tensile test was carried out to find mechanical properties of silicone rubber. Calculations on facial expressions were carried out using the mechanical properties of artificial skin of silicone rubber identified by the tensile test.

Keywords: facial expression, artificial skin, silicone rubber, tensile test, FEA.

1. Introduction

In recent years, the applications of robotics become important for human life in the worldwide. The robot on facial expressions has developed by following human facial expressions [1]. It is required to examine the positions and directions of forces loaded on a robot face in order to make the robot express the same facial expressions as human's ones [2]. The directions and the magnitudes of forces depend on the shapes of robot facial expressions [3].

Firstly, the tensile test was carried out on silicone rubber (RTV2 VP 7550) used as artificial skin of robot face. Secondly, the mechanical properties of artificial skin were identified using the experimental and FEA's results. Thirdly, the facial shape was measured by 3D scanning. In addition, the facial model was made by meshing the 3D scanning data. Finally, the calculated results of facial expressions using FEA were compared with real human facial expressions.

2. Experimental on mechanical properties of artificial skin

2.1. Experimental Setup

The schematic representations of tensile test set-up are shown in Figure 1. The tensile test set-up consists of clamps, weight cups, and bucket container that are mounted on the aluminum frame. The width and height of aluminum frame are 450 mm and 850 mm, respectively. The weight cup has 53 g. The elongation of test pieces were measured by using the aluminum scale.

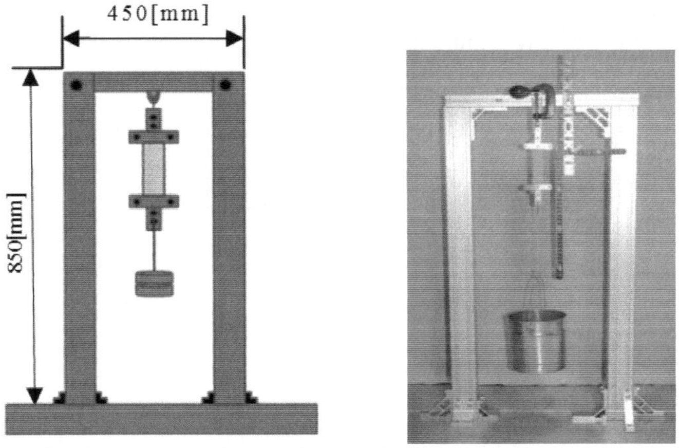

A.Design on experimental tensile strength

B. Photograph experimental tensile strength

Figure 1. Experimental setup for tensile strength

2.2. Experimental method

The tensile test was performed to examine the mechanical properties of the silicone rubber as the artificial skin. Seven test pieces have been used. The tensile test was carried out under the same conditions, for example, about the room temperature, loading time, and a kind of material of the test pieces.

Every test piece was clamped in top side as shown in Figure 1.B. The top side was fixed and the bottom side was pulled down with the load. The loading to test piece was continued until the test pieces broke. The elongations of test piece were measured after 10 sec for an every loading.

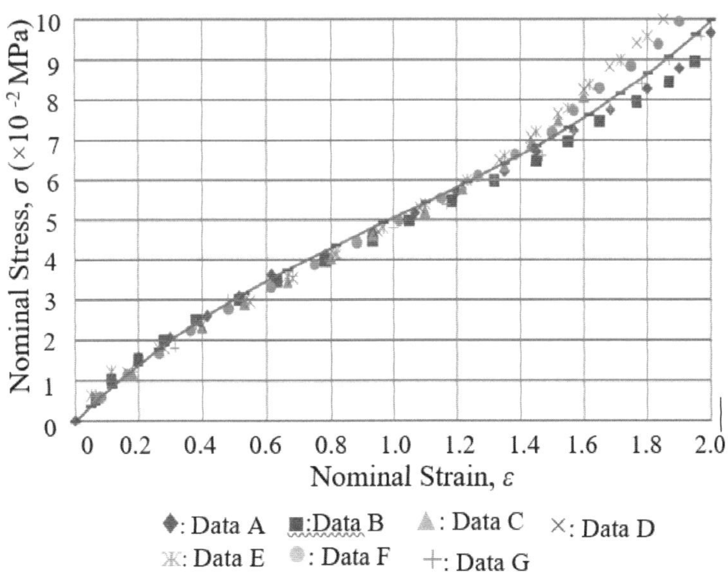

Figure 2. Relationship between nominal stress and strain of silicone rubber used as artificial skin

2.3. Finite Element Analysis (FEM) on mechanical properties of artificial skin

The FEA was performed in order to examine the mechanical properties of silicone rubber. Quadrilateral shell elements of 4-node were used as finite elements. The length, width, and thickness of test pieces were 60 mm, 30 mm, and 3 mm, respectively. Test piece model was divided to 200 elements. Six Degrees of Freedom (DOF) were clamped for an each node on the top tip of test piece. The loads were applied to the other tip of test piece. The magnitudes of forces were varied from 0.52 N to 15.60 N. Nonlinear analysis were carried out since the deformations are large.

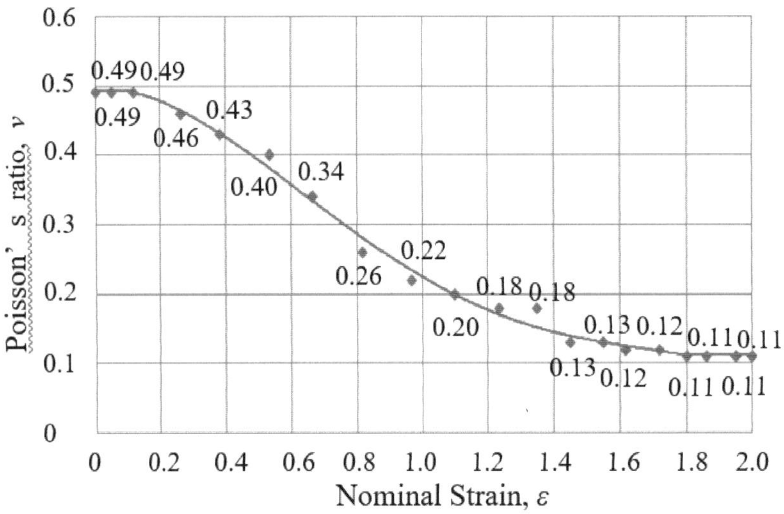

Figure 3. Poisson's ratio determined by comparing the calculated and experimental results

2.4. Identification of mechanical properties of artificial skin using experimental and calculated results

Figure 2 shows the relationship between nominal stress, σ and strain, ε. The maximum nominal stress and strain were 9.7×10^{-2} MPa and 2.0, respectively.

Figure 3 shows the relationship between Poisson's ratio and strain. The Poisson's ratios were determined by comparing the calculated and experimental results. The poisson's ratio of silicon rubber based on calculation and experiment is 0.49.

3. Measurement of facial shape

3.1. Measurement system

The schematic representations of scanning system are shown in Figure 4. The experimental devices consist of the 3D red line laser, the Charge Coupled Devices (CCD) video camera, the PC, and the calibration panel with scale points to adjust the size of an scanned object. The position of scanned object is in front of the calibration panel with the distance of 1,000 [mm]. The positions of video camera and red line laser are parallel to each other and in front of the scanned object.

3.2. 3D Scanning method and results

Scanning process consists of three steps. In the first step, the object is calibrated in front of calibration panel. In the second step, photo image of the object is captured by using the CCD video camera. In the third step, the object is scanned by using the red line laser.

The photos and scanned images consist of three parts: front face, 45° right side, and 45° left side. The photos and scanned images were merged. The merged of three parts has bring out the precision of image result.

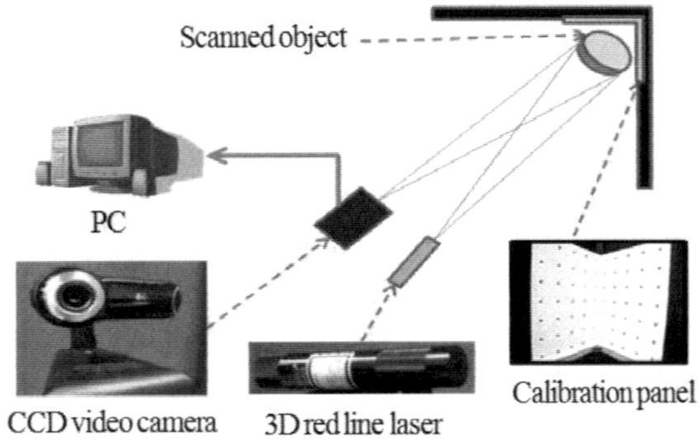

A. Arrangement of scanning system

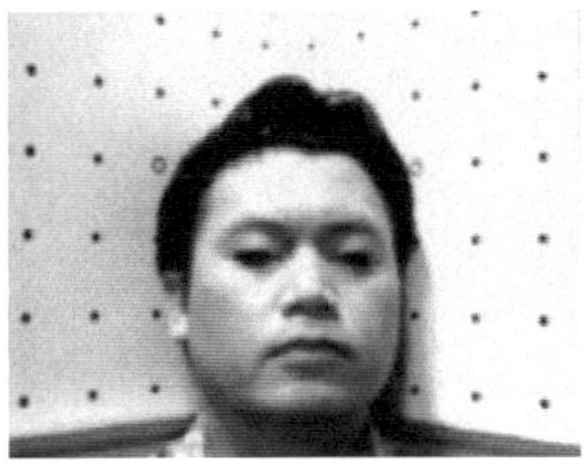

B.Scanned object and calibration panel

Figure 4. Experimental setup for facial scanning

4. Making meshes of facial model

4.1. Meshing by scanned 3D facial data

Figure 5 shows the comparison of facial model between initial condition and final model.

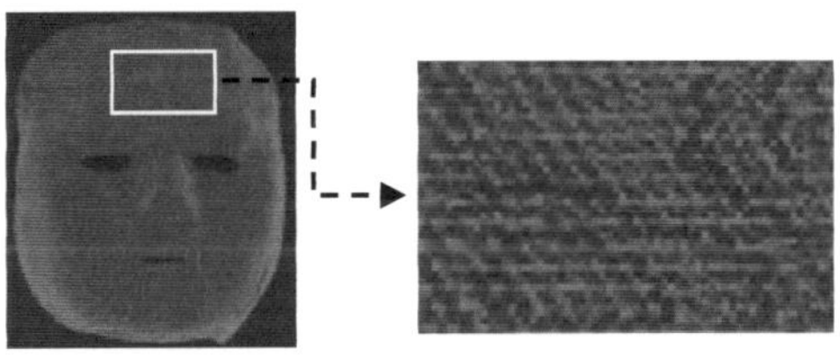

A. Before meshing

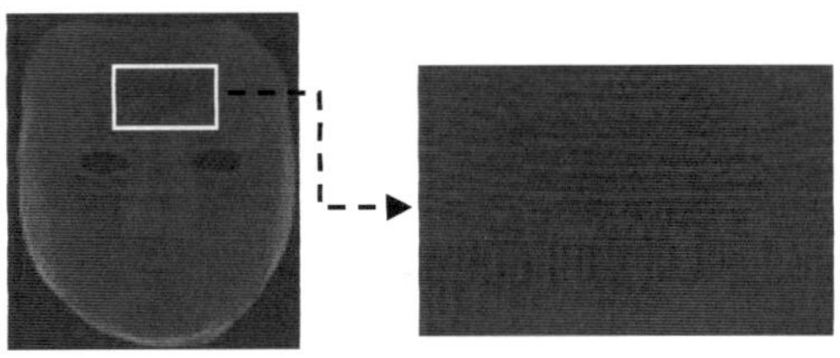

B. After meshing

Figure 5. Comparison of facial model between initial condition and final model

The meshing was included analyzing the geometry model and optimizing the meshing in all the surface of elements. Meshing the facial model is determined by the shape object surface. The quadrilateral shell element was used in meshing face because of the best quality surface.

4.2. Optimization of meshes

IOP Conf. Series: Materials Science and Engineering **619** (2019) 012051 doi:10.1088/1757-899X/619/1/012051

The optimization of meshes was carried out to repair defective elements on the surface. The defective elements were caused by overlapping or disconnected elements. Therefore, the overlapping elements should be removed and the disconnected elements should be linked.

5. Calculated results of facial expressions using finite element analysis

Figure 7 shows the comparison between the calculated and real human facial expressions. Six facial expressions have been carried out in this study: happiness, surprise, sadness, anger, disgust, and fear.

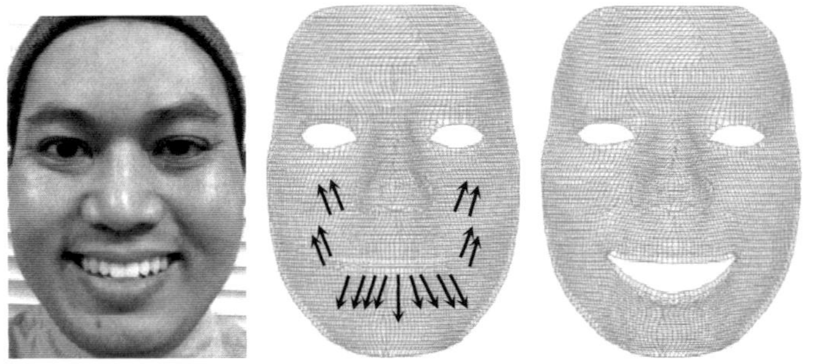

A1. Human facial expressions B1. Direction of forces calculation C1. Calculated result

1. Happiness

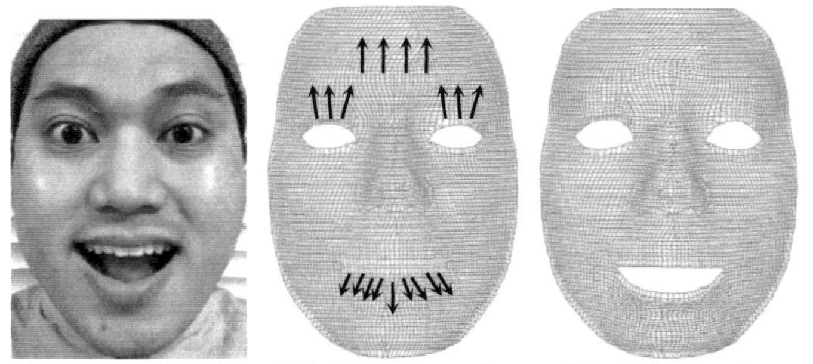

A2. Human facial expressions B2. Direction of forces calculation C2. Calculated result

2. Surprise

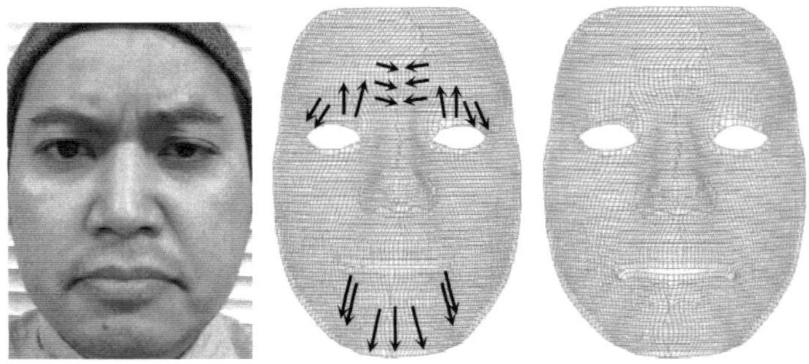

A3. Human facial expressions B3. Direction of forces calculation C3. Calculated result

3. Sadness

Figure 6. Expressions for Happiness, Surprise and Sadness between real human facial expression and calculated results by using the action units

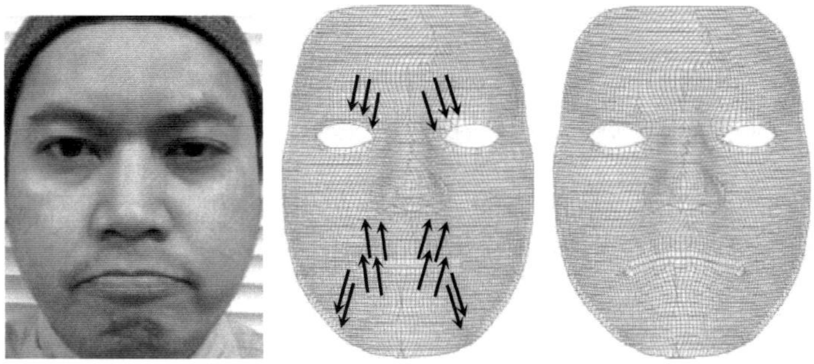

A4. Human facial expressions B4. Direction of forces calculation C4. Calculated result

4. Anger

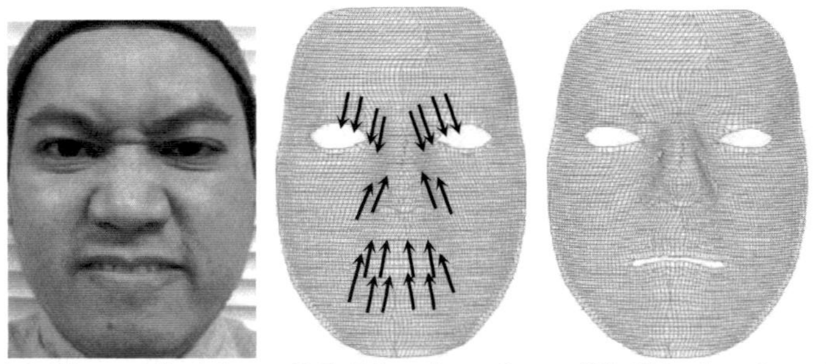

A5. Human facial expressions B5. Direction of forces calculation C5. Calculated result

5. Disgust

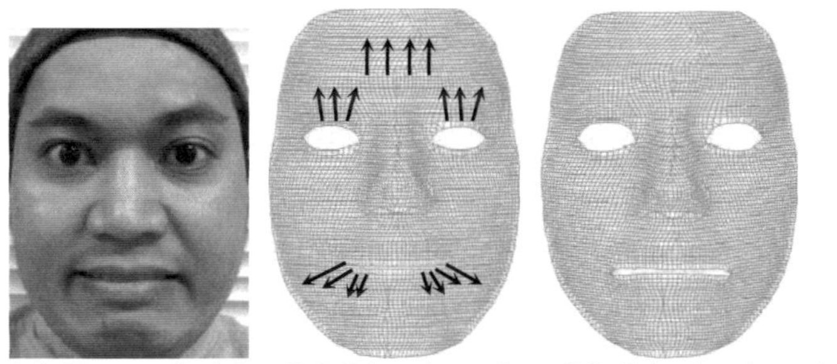

A6. Human facial expressions B6. Direction of forces calculation C6. Calculated result

6. Fear

Figure 7. Expressions between for Anger, Disgust and Fear between real human facial expression and calculated results by using the action units

The directions and magnitudes of forces in calculated are the most important to follow the human facial expressions. The calculated results effectively showed the facial expressions like human.

6. Conclusion

Mechanical properties of silicone rubber have been identified by tensile test and FEA. The properties have been used in computational simulation for facial expressions using FEA. The calculated results effectively show the facial expressions like human.

References

[1] M. I. Khan, "Facial Expression Recognition for Human-Robot Interface," **vol. 9**, no. 4, pp. 300–306, 2009.

[2] Ekman, P. and Friesen, W.V., “The Facial Action Coding System”, Consulting Psychologist Press, 1978.
[3] Andi Amijoyo Mochtar, “Computational Simulation on Facial Expression and Experimental Tensile Strength for Silicone Rubber as Artificial Skin,” J. Phys. Conf. Ser. **962** 012025, 2018.

IOP Conf. Series: Materials Science and Engineering **619** (2019) 012052 doi:10.1088/1757-899X/619/1/012052

Performance of a 90°-Bent Bladed Vertical Axis Wind Turbine Model with Various Numbers of Blades

Luther Sule, Andi Amijoyo Mochtar and Jalaluddin

Hasanuddin University, Departement of Mechanical Engineering, Jl. Poros Malino KM. 6 Gowa, 92171, Sulawesi Selatan, Indonesia

Email Corrresponding author: andijoyo@unhas.ac.id

Abstract. This study aims to design a vertical axis wind turbine that can effectively harness the power of wind. This was performed by designing and testing a vertical axis wind turbine with 2, 3 and 4 blades bent at 90° angles in wind speeds between 1.6 m/s and 4.2 m/s for every type of wind turbine. The results show that 2-blade vertical axis wind turbines coefficient of power (C_P) = 0.28 is most capable of converting wind power into electricity with wind speeds of 3.4 m/s and 3-blade wind turbines C_P = 0.40 in wind speeds of 2.2 m/s, and 4 blade wind turbines C_P = 0.25 performed best in wind speeds of 2.4 m/s. The C_P of all three wind turbines decreased as wind speeds increased to 4.2 m/s. The examination and calculation analysis results indicate that 3-blade wind turbines work more effectively than 2-blade and 4-blade wind turbines due to the asymmetric positioning creating a relatively small drag, and the distance between each blade and the shaft of the wind turbine creates a rift allowing for wind to flow which results in the blades hitting other. This increases the blades momentum leading to less turbulence for 3-blade wind turbines.
Keywords: wind turbines, wind speeds, blades bent, blades momentum, number of blades, coefficient of power

1. Introduction

Humans have been harnessing wind energy for 5,500 years[1]. As technology develops, so do the potential uses of wind energy[2]. Humans have been using wind energy to support their activity for long time ago. Currently, wind energy potential has attracted the attention of researchers to invent a form of energy utilization in the form of turbines and wind turbines during this era of global warming.

Based on data by National Meteorological Agency of Indonesia (BMKG), the average of annual wind speeds in Indonesia is 4-5 m/s[3]. Therefore, an appropriate wind energy conversion system (SKEA) capable of harnessing wind energy in Indonesia is needed.

Generally, wind turbines were used to generate electricty by harnessing wind energy. In the beginning, wind turbines were made to accommodate the needs of farmers in digging paddies and irrigation among others. The basic work principle of electricity-producing wind turbines is converting the mechanical energy of wind into rotation energy by the turbines[4,5]. This rotation is then used to run an electricity-producing generator.

One type of wind turbine is a vertical axis with a shaft or a perpendicular main rotary axis[6–8]. The main advantage of this type of build is that the turbine does not need to be aimed towards wind direction to produce electricity[5,9]. This is useful in places where wind direction varies greatly. Vertical axis wind turbines are able to harness wind from different directions, as well as having a good self-starting mechanism which allows for the rotary to rotate with little wind[6]. In addition, the resulting torque from this wind turbine is relatively high.

Figure 1. Vertical Axis Wind Turbine Model with 2, 3, and 4 of 90° Bent Blades

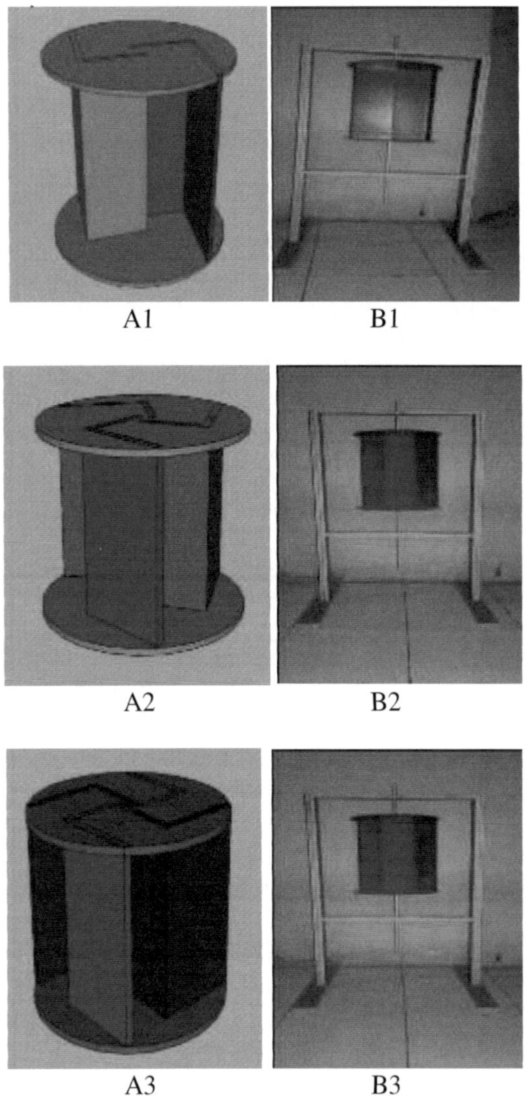

Figure 2 Scheme of axis turbine A1-A3 and vertical axis wind turbine with 2,3, and 4 of 90° bent blades B1-B3

This study focuses on vertical axis wind turbines due to their ability to better utilize lower and fluctuating wind speeds in Indonesia. Although this type of wind turbine is less efficient than other builds, this turbine has other advantages such as its simplistic build, ability to utilize wind from all directions, and independence from over speed rotation protection components. In addition, it also has good starting torque and a self- starting mechanism that can perform well in producing electricity in small-scaled generators. In order to be efficient, some modifications have been made on the wind turbines. Additional fins, geometric alterations and blade shape, additional stages, and use of valves are some changes made. This study employs the use of 90°-bent blade vertical axis wind turbines.

2. Research methodology

The type of research used in this study is experimental research which involves observing and measuring predetermined variables to assess 90°-bent blade vertical axis wind turbines with varying number of fans (2, 3 and 4). A stand fan is placed in a vary distance from the front of blade due to define a speed of wind.

2.1. Research variable

Two types of research variables used in this study: independent and dependent variables. Firstly, independent variables give a variation of treatment to the wind turbines in this study. In this study, the variables are the varying wind speeds starting from 1.6 to 4.2 m/s, as well as the use of multiple fan blades (2, 3 and 4). Secondly, dependent variables are the results of the study. In this research, the dependent variables has some circumstances i.e., shaft power of 90°-bent blades vertical axis wind turbines, torque resulting from the 90°-bent blades vertical axis wind turbines, and efficiency level.

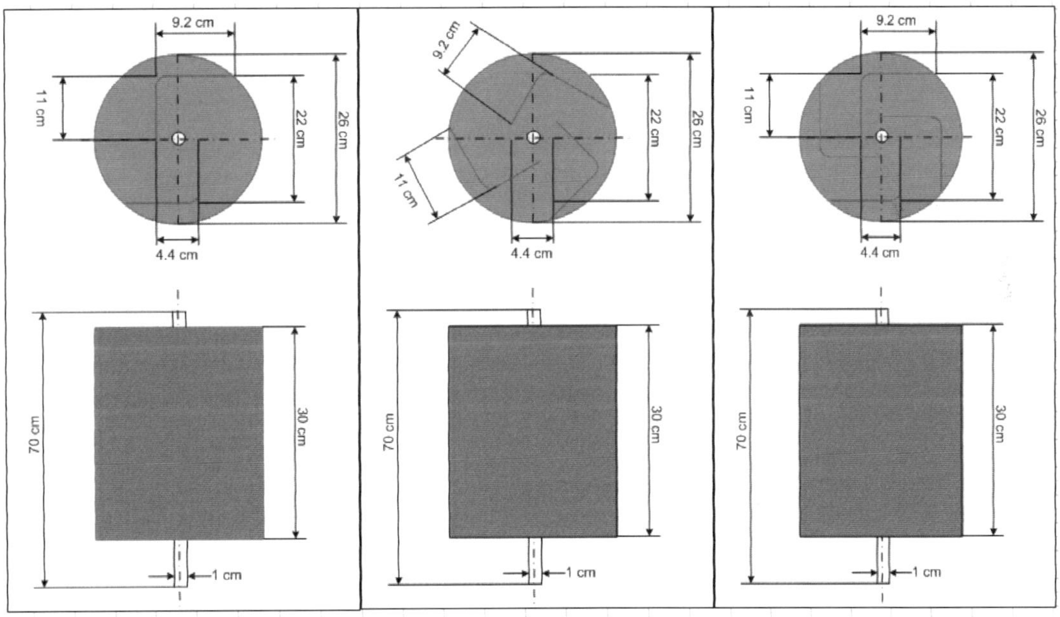

Figure 3. Vertical axis wind turbine with 2, 3 and 4 of 90° bent blades dimensions

2.2. Experimental setup

Prepare the study tools and materials. Set/make a series of savonius wind turbine bent blades.
Prepare various bent blades in sets of 2, 3 and 4. Make a series of savonius wind turbines with 2, 3 and 4 blades as shown in Figure 1. Design savonius wind turbines with 2, 3 and 4 bent blades is shown in Figure 2.

Figure 3 and Figure 4 shows the vertical axis wind turbine with dimension and position experimental device. The blower fan is turned on due to gain wind energy. Then, the wind speed of the fan is adjust in order to arrange the wind power into the axis turbine. Measure and record the rotation of the wind turbine axis with a digital tachometer, tachometer sensor attached to the wind turbine axis below the pulley. The wind speed of air flow into the turbine blades is measured by using an anemometer that installed between the fan and the turbine blades. Measure voltage and current outflow of the DC generator with a multimeter. Procedure step is repeated for each different blade (2, 3 and 4).

2.3. Collected data

Collected data was conducted by measuring wind speeds using an anemometer. The wind speeds gauged by the blowers for each experiment ranged from 1.6 to 4.2 m/s. A tachometer was used to measure the shaft's rotations per minute (rpm) for each experiment with varying wind speeds. A multimeter was used to compute generator output by assessing DC current and voltage.

3. Result and discussions

Data processing was done by calculating the results with certain formulas and presenting them through tables and graphics. Figure 5 shows the relationship between wind speed and coefficient power where higher lead to a greater increase of break horse power (BHP). This increase is because the coefficient power is the ratio between BHP and wind force. This can be seen from the formula

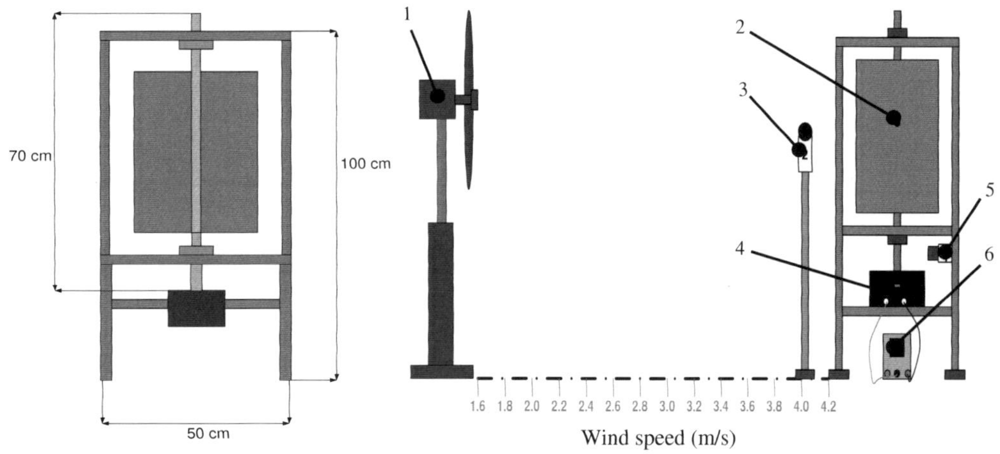

Description:

1. Fan
2. 90° of bent wind turbine
3. Anemometer
4. DC generator
5. Tacometer
6. Multimeter

Figure 4. Scheme of frame and shaft savonius wind turbine using 90° of bent blades

It can be seen from Figure 5 that wind turbines with 3 blades has a greater coefficient power than turbines with 2 and 4 blades. Wind turbines with 2 blades have an optimum coefficient power of 0.28 for wind speeds of 3.4 m/s which decreases as wind speeds increase to a maximum of 4.2 m/s.

The 3-blade turbines have an optimum coefficient power of 0.40 in wind speeds of 2.2 m/s which decreases as wind speeds increase to 4.2 m/s. 4-blade wind turbines have an optimum coefficient power of 0.25 when wind speed reaches 2.4 m/s which decreases as wind speed increases to 4.2 m/s. This decrease in coefficient power is caused by an imbalance between increase in axis force and wind increase in wind speeds. Higher wind speeds lead to greater losses which cause the coefficient power value to decline.

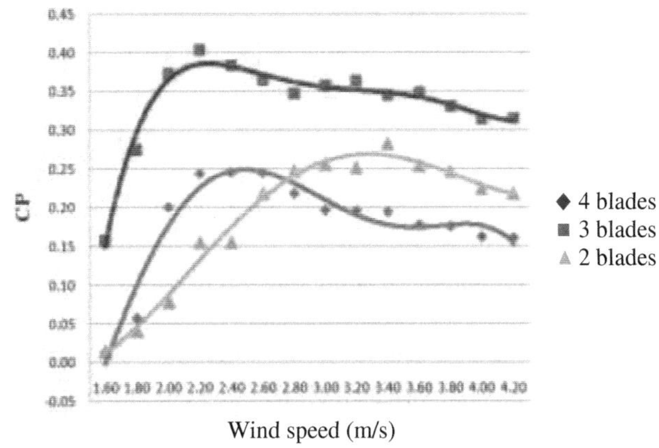

Figure 5. Curve of wind speed and coefficient power (CP)

$$C_p = \frac{BHP}{P_{wind}} \tag{1}$$

Figure 6 shows that 3 of 90° bent blade wind turbines have a greater rpm number compared to 2- and 4-blade turbines. This is because an odd number of fan blades experiences the least amount of wind draft which results in a smaller negative torque. This allows optimum turbine rotation. 2-blade turbines have a greater rpm number than turbines with 4 blades. This is due to the greater mass belonging to 4-blade turbines compared to that of 2-blade turbines which allows for faster rpm in the latter.

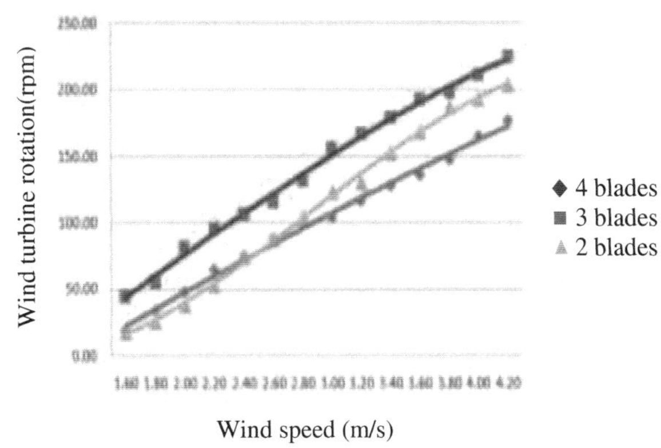

Figure 6. Curve of wind speed and wind turbine rotation

Figure 7 shows that 3 of 90° bent blades are optimal in turning wind power into generator power than tubines with 2 and 4 of 90° bent blade. This is due to the higher rpm for 3-blade turbines with 90° bends in wind speeds of 4.2 m/s (wind power of 2.99 watts) than 2- and 4-blade turbines. The 2 of 90°

bent blades have a generator power of 0.537 watts, the 3 of 90° bent blades have 0.774 watt generator output value, while 4-blade 90° bends only generate 0.395 watts.

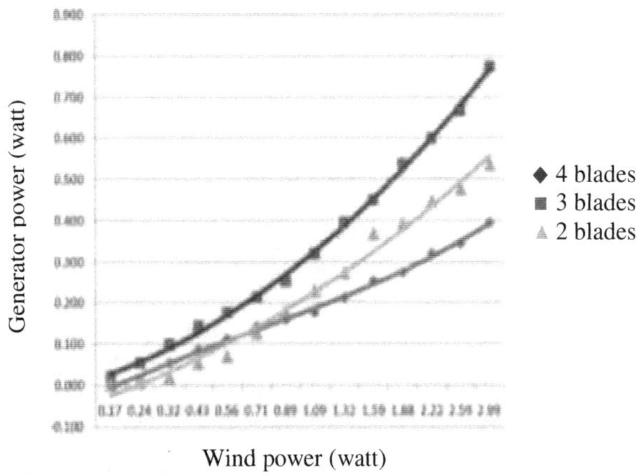

Figure 7 Curve of wind power as input power and generator power

Figure 8 shows 3-blade wind turbines have a far greater coefficient power value than that of 2- and 4-blade turbines. This is due to the 3-blade turbine's ability to capture wind power due to less wind friction on returning blades which leads to less negative torque allowing for the turbine to spin at a maximum rate. These result in a greater output compared to 2- and 4-blade turbines.
The highest coefficient power reached by the 2-blade wind turbine is 0.28 in wind speeds of 3.4 m/s. 3-blade wind turbines reached a coefficient power of 0.4 in wind speeds of 2.2 m/s. The 4-blade wind turbines reached a coefficient power of 0.25 in wind speeds of 2.4 m/s.

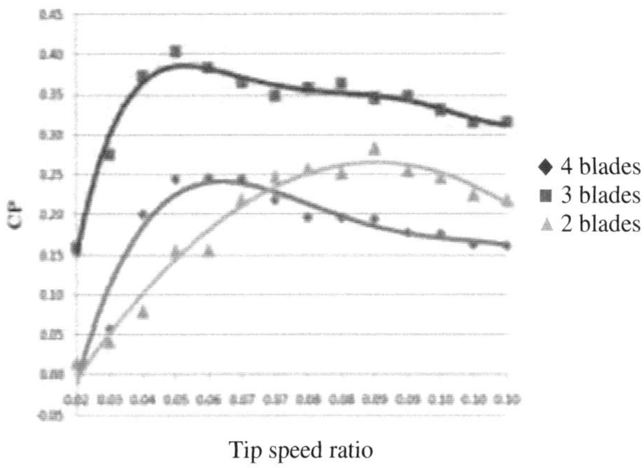

Figure 8 Curve of tip speed ratio and coefficient power

Figure 9 shows that tip speed ratio (TSR) has a trend similar to that of rpm. In this case, TSR shows the comparison between output speed (rotor) and input speed (wind). In other words, TSR was parallel. The peak of graph for TSR is caused by the savonius characteristic that is unable to convert in high wind speeds. The graph and data shows that the TSR value for 3-blade wind turbines are far better than for 2 and 4-blade turbines. 2-blade turbines reached a TSR 0.1, 3-blade turbines reached 0.11 and 4-blade turbines reached a TSR of 0.8.

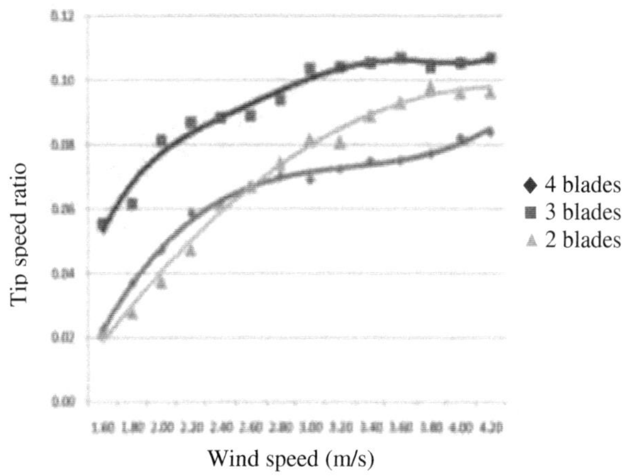

Figure 9 Curve of wind speed and tip speed ratio

From Figure 10 and Figure 11, it can be concluded that the coefficient power value for savonius wind turbines is far better in comparison to 2 90° bent blade vertical axis wind turbines and America multiblade wind turbines.. 2 90° bent blade vertical axis wind turbines have a greater coefficient power when compared to American multiblade wind turbines.. However, the TSR value for the 2-blade wind turbine is lowest compared to that of the savonius and American multiblade turbines.

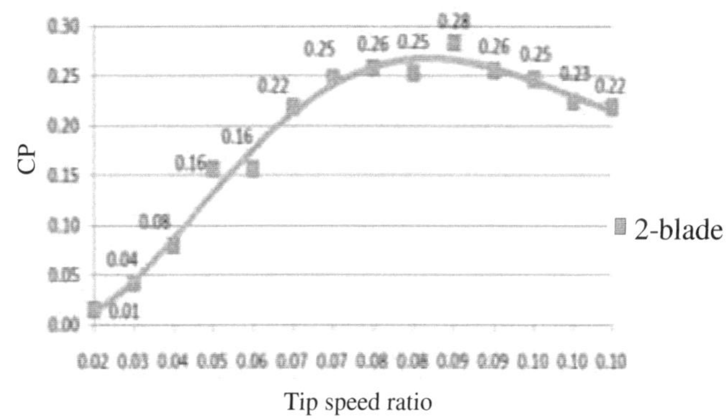

Figure 10. Relation between Tip Speed Ratio and Coefficient Power for 2 of 90° Bent Blade Vertical Axis Wind Turbines

IOP Conf. Series: Materials Science and Engineering **619** (2019) 012052 doi:10.1088/1757-899X/619/1/012052

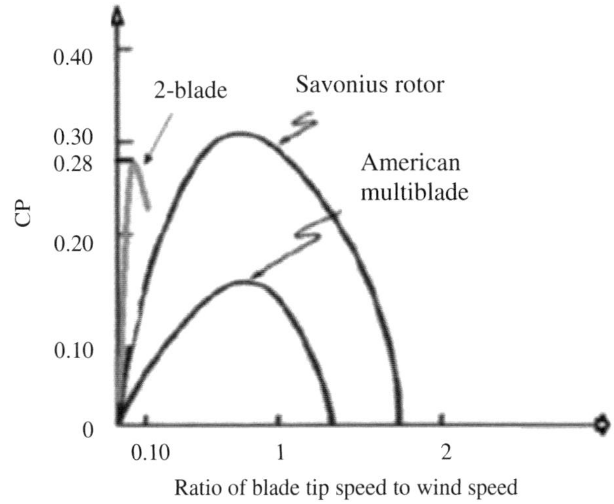

Figure 11. comparison between tip speed ratio and coefficient power for 2 of 90° bend blade vertical axis wind turbine, savonius wind turbine, and American multiblade wind turbine

Figure 12 and Figure 13 show that the coefficient power of the 3 of 90° bent blade vertical axis wind turbine is far greater in comparison to that of the savonius and American multiblade wind turbines. However, the TSR value for the 3 of 90° bent blade vertical axis wind turbine is lowest in comparison to that of the savonius and American multiblade turbines.

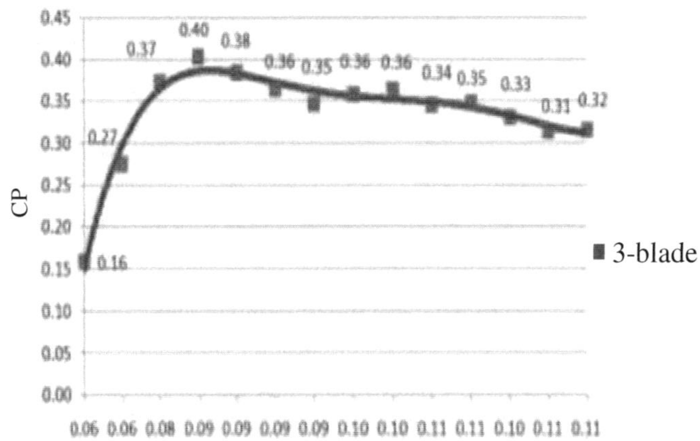

Figure 12. Relation between Tip Speed Ratio and Coefficient Power for 3of 90° Bent Blade Vertical Axis Wind Turbines

IOP Conf. Series: Materials Science and Engineering **619** (2019) 012052 doi:10.1088/1757-899X/619/1/012052

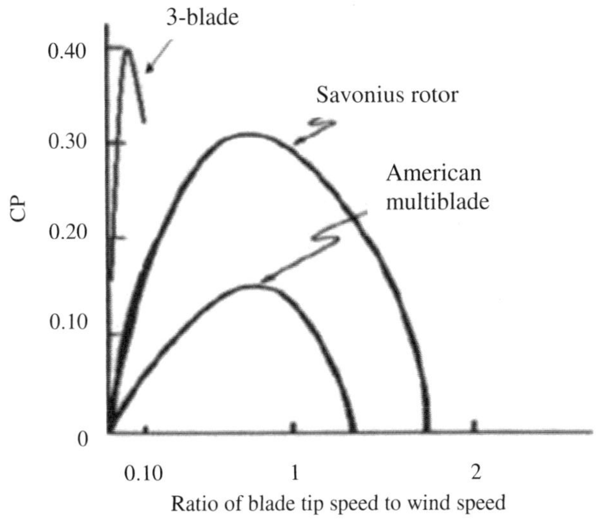

Figure 13. Comparison between tip speed ratio and coefficient power for 3of 90° bent blade vertical axis wind turbines, savonius wind turbines, and American multiblade wind turbines

Figure 14 and Figure 15 indicate that the coefficient power of the savonius wind turbine is greater than that of the 4 90° bent blade vertical axis wind turbine and the American multiblade wind turbine. The 4 of 90° bent blade wind turbine has a better coefficient power than that of the American multiblade, but the TSR value of this wind turbine is lowest in comparison to the savonius and American multiblade turbines.

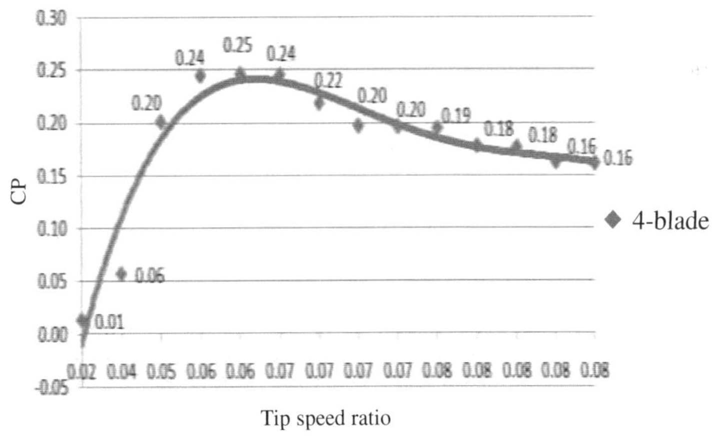

Figure 14. Relation between tip speed ratio and coefficient power for 4 90° bent blade vertical axis wind turbines

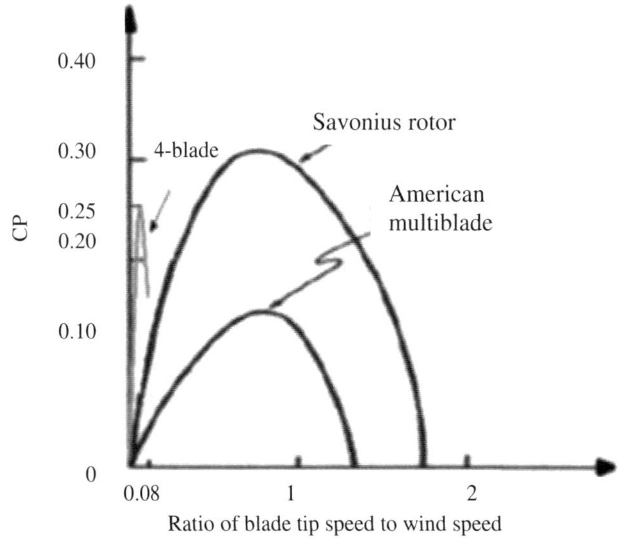

Figure 15. Comparison between Tip Speed Ratio and Coefficient Power for 4 90° Bent Blade Vertical Axis Wind Turbines, Savonius Wind Turbines, and American Multiblade Wind Turbines

Conclusions

The three 90°bent blade of vertical axis wind turbine is the best design compared to the Savonius and American multiblade wind turbines. It can be utilized in wind speeds starting from 1 m/s.

The design of the 90° bent blade of vertical axis wind turbine has an aspect ratio (h / R) of 2.8, making it able to reach higher coefficient powers when compared to Savonius and American multiblade turbines. The three 90° bent blade of vertical axis wind turbine is able to reach coefficient powers of approximately 0.4 greater than the Savonius design which can reach a coefficient power of 0.3.

The 90° bent blade of vertical axis wind turbines are able to convert wind energy in electricity. 2-blade models can reach a maximum capacity of 28 percent in winds of 3.4 m/s, 3-blade models has a maximum capacity of 40 percent in winds of 2.2 m/s, and 4-blade turbines have a maximum capacity of 25 percent in wind speeds of 2.4 m/s. The 3-blade vertical axis wind turbine model is the most effective model among the three designs made for this research.

References

[1] M. M. Aslam Bhutta, N. Hayat, A. U. Farooq, Z. Ali, S. R. Jamil, and Z. Hussain, "Vertical axis wind turbine - A review of various configurations and design techniques," *Renew. Sustain. Energy Rev.*, vol. 16, no. 4, pp. 1926–1939, 2012.

[2] M. J. Khan, G. Bhuyan, M. T. Iqbal, and J. E. Quaicoe, "Hydrokinetic energy conversion systems and assessment of horizontal and vertical axis turbines for river and tidal applications: A technology status review," *Appl. Energy*, vol. 86, no. 10, pp. 1823–1835, 2009.

[3] N. M. Soeripno Martosaputro, "Blowing the Wind Energy in Indonesia," *Energy Procedia*, vol. 47, pp. 273–282, 2014.

[4] Q. Li, T. Maeda, Y. Kamada, J. Murata, K. Furukawa, and M. Yamamoto, "Effect of number of blades on aerodynamic forces on a straight-bladed Vertical Axis Wind Turbine," *Energy*, vol. 90, pp. 784–795, 2015.

[5] S. Giorgetti, G. Pellegrini, and S. Zanforlin, "CFD investigation on the aerodynamic interferences between medium-solidity Darrieus Vertical Axis Wind Turbines," *Energy Procedia*, vol. 81, pp. 227–239, 2015.
[6] R. Howell, N. Qin, J. Edwards, and N. Durrani, "Wind tunnel and numerical study of a small vertical axis wind turbine," *Renew. Energy*, vol. 35, no. 2, pp. 412–422, 2010.
[7] G. Müller, M. F. Jentsch, and E. Stoddart, "Vertical axis resistance type wind turbines for use in buildings," *Renew. Energy*, vol. 34, no. 5, pp. 1407–1412, 2009.
[8] M. Elkhoury, T. Kiwata, and E. Aoun, "Experimental and numerical investigation of a three-dimensional vertical-axis wind turbine with variable-pitch," *J. Wind Eng. Ind. Aerodyn.*, vol. 139, pp. 111–123, 2015.
[9] B. D. Altan and M. Atilgan, "An experimental and numerical study on the improvement of the performance of Savonius wind rotor," *Energy Convers. Manag.*, vol. 49, no. 12, pp. 3425–3432, 2008.

IOP Conf. Series: Materials Science and Engineering **619** (2019) 012053 doi:10.1088/1757-899X/619/1/012053

Experimental Analysis of Smart Green House with Rotary Garden

Onny S. Sutresman, Rafiuddin Syam, Abdul Hay Muchsin, Zulkifli Djafar, Syahrier Arief, Andi Amijoyo Mochtar, Azwar Hayat, Onny S Sutresman, Amiruddin, Abdul Mufti Radja*

Mechanical Engineering Department, Engineering Faculty, Hasanuddin University
*Architecture Engineering Department, Engineering Faculty, Hasanuddin University

the corresponding author's e-mail: rafiuddin@unhas.ac.id

Abstract. To overcome climate change that affects agriculture in all over the world, shows the estate and farming for the raising of the extreme weather (e.g., floods, droughts, *etc*). Therefore, the smart greenhouse is essential in providing sustainable food for humans. The application of Smart Greenhouse, The ultimate and most common monitored variables are temperature, moisture, the intensity of sunlight and environmental conditions to support the using of sunlight a source of electricity. This is necessary for the Smart Greenhouse to manipulates the environment in appropriate to its plant's needs. In this case, the temperature and humidity in the greenhouse are important to keeps the productivity of the plants. So, it's using automatic system to keep suitable environmental conditions as explained before for the plants. In this research , the smart Greenhouses is shaped Ferris wheel to make sprinkling and radiation of sunlight for plants is equally distributed for all plants and its parts, Based on the results of this experiments with rotary mechanical system for Smart Greenhouse as well shown that Smart Greenhouse can preserve suitable temperature and moisture for growing the plants.

Keywords. the temperature, moisture, the light of the sun, and rotary motion for sprinkling

1. Introduction

Currently, the problem faced in the world is the adequacy of agricultural food evenly throughout the world. Uneven distribution of agricultural food occurs all over the world including production, distribution and affordability. Then, there is often disruption of agricultural food production and has greatly detrimental to people's health as well as directly initiating a major confidence crisis to government and agricultural food companies.[1]

Other researcher explains an indoor intelligent agricultural IOT system is designed and implemented by the author in order to attack this conundrum [2]. Jun et.al also make the system directs an new trend for agricultural development the capability of parallel extension, and the system can connect to large-scale indoor farms gradually thus to make these farms combining with each other organically.

The researcher [3] introduce the climate-smart agriculture (CSA) concept. That is gaining considerable traction at international and national levels to meet the challenges of addressing agricultural planning under climate change. He [3] also describe CSA is a concept that calls for integration of the need for adaptation and the possibility of mitigation in agricultural growth strategies to support food security.

Climate change that affects agricultural around the world, shows plantation and plants to the extreme weather condition such as flood, drought, etc. Therefore, of the smart green house in environmental is important to provide food for humans, while they are preparing the place of ideal cultivation for the environment. But to have a common idea about suitable greenhouse for nowadays needs It necessary to have that detailed information, at least, of the most relevant variables that affect the environment. Smart Greenhouse is a controlled building of agricultural conditions to keep the plants from outside extreme conditions. Smart Greenhouse technology gives farmers the freedom to choose plants at any time of the year. The quality of and productivity of plants in the smart greenhouse very much dependent on the quality of management and the scheme good management defined by the quality of the amount of information collected out of the smart greenhouse of the living environment. Therefore, continuous environmental monitoring variables such as temperature, humidity, and soil moisture provide information to farmers, this information needed to understand how each factor influences growth and how to manage maximum plant productivity[4]. The authors already done a preliminary research on smart greenhouse can be found at [5].

In this research, the writers used rotary smart greenhouse as productive cultivation and that sprinkling and radiation of light for plants is equally distributed for all plants that evenly using the sun as a source of electric power, rotary smart greenhouse is also able to minimize the use of land with a number of plants more than other smart greenhouses.

2. Smart Green House Concept

Greenhouse or nowaday known as the Smart Greenhouse is not a new item in the agribusiness sector, especially horticultural agribusinesses such as vegetables and decorative plants. However, it does not guarantee that all Indonesian farmers understand and know about this greenhouse. Based on these considerations, in this study will be reviewed the general description of what exactly and the benefits of a greenhouse as a supporter of our agribusiness.

Figure 1. Smart greenhouse (Poly-Tex, 2017)

The advantages of Greenhouse are settings schedule of productions i.e., to increase production output, improving production quality, to minimize pesticides and to increase the perfomance of the system. The Greenhouse is also widely used as a collection room for various types of high-value plants. Inside the greenhouse visitors can see a variety of interesting plants, even rare, so it can be an attraction. There is a special collection of cactus, orchids or various types of plants with the atmosphere made in the wild. In Indonesia greenhouse like this can be found in many botanical gardens and places of agro-tourism.

3. Automatic Control System of Smart Green House

The process of controlling the rotary smart greenhouse is a process that is carried out continuously, over a long period of time and requires a high accuracy of control. Therefore automatic control is needs, in this case, to reduce the problems such as on the manual control, among others: fatigue, uniformity and human inaccurate. In this automatic control, the control stages such as measuring, comparing, counting and correcting are performed by the instrument repeatedly. With automatic control can be achieved the purpose of the smooth operation, control of security and product quality.

3.1. Temperature and humidity sensors

DHT11 is a digital sensor that can measure the temperature and humidity of the surrounding air. This sensor is very easy to use along with Arduino. It has excellent stability levels as well as highly accurate calibration features

Figure 2. Temperature and Humidity Sensors DHT11

3.2. Relay Devices

The list of authors should be indented 25 mm to match the abstract. The style for the names is initials then surname, with a comma after all but the last two names, which are separated by 'and'. Initials should not have full stops—for example **A J Smith** and *not* **A. J. Smith**. First names in full may be used if desired. If an author has additional information to appear as a footnote, such as a permanent address or to indicate that they are the corresponding author, the footnote should be entered after the surname.

A relay is a device for disconnecting or connecting a circuit, almost the same as a switch only the relay works automatically and can be used as a remote control device, see figure 2.7. The relay consists of an iron-core coil and a switch. If this iron-core coil is electrified then the relay will work.

Figure 3. Relay devices

3.3. Arduino Uno Microcontroller

IOP Conf. Series: Materials Science and Engineering **619** (2019) 012053 doi:10.1088/1757-899X/619/1/012053

Arduino is developed by a team of people from different parts of the world. The core members of this team are Massimo Banzi Milano (Italy), David Cuartielles Malmoe (Sweden), Tom Igoe (New York, US), Gianluca Martino Torino (Italy) dan David A. Mellis Boston (MA, USA)

The main component of the Arduino board is an 8-bit microcontroller with ATmega brand made by Atmel Corporation. Arduino boards use different types of ATmega depending on the specifications, for example, Arduino Uno uses ATmega328 while the more sophisticated Arduino Mega 2560 uses ATmega2560.

Figure 4. Board of Arduino Uno

3.4. Solar Cell System

Solar Cell is one type of photovoltaic light sensor, a sensor that can change the intensity of light into electrical voltage. Solar Cells are made up of solar cells, also known as photovoltaic or "PV" cells. These cells are made of silicon pieces that have opposite electrical charges and cause electrons to flow when exposed to sunlight. At the most basic level, the solar panel when connected to the right material along with the inverter and battery will collect energy from the sun and release the usable knobs as shown in figure 5.

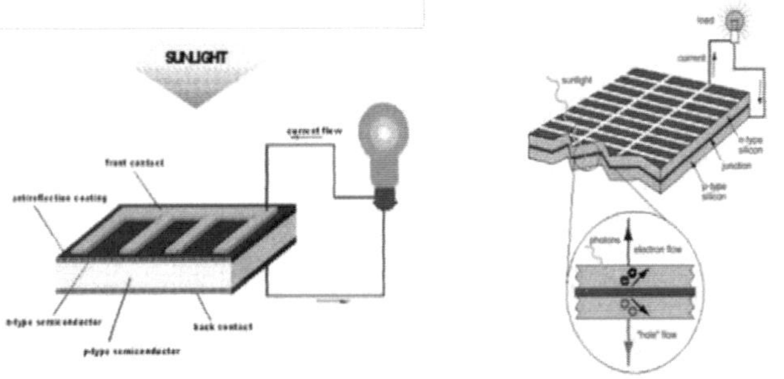

Figure 5. Solar Cell work diagram

In solar cells, there is a junction between two thin layers made of semiconductor material, each known as a "P" (positive) semiconductor and an "N" (negative) semiconductor type. N-type semiconductors are made from crystalline silicon and there are also a number of other materials (generally phosphorus) in the range that the material can provide an excess of free electrons.

4. Research Method

In the Figure 6 shows the model of the rotary design of Smart Greenhouse for producing productive plants. The system could be rotate with slow speed.

Solar Cell is one type of photovoltaic light sensor, a sensor that can change the intensity of light into electrical voltage. Solar Cells are made up of solar cells, also known as photovoltaic or "PV" cells. These cells are made of silicon pieces that have opposite electrical charges and cause electrons to flow when exposed to the light.

At the most basic level, the solar panel when connected to the right material along with the inverter and battery will collect energy from the sun and release the usable knobs as shown in figure 5.

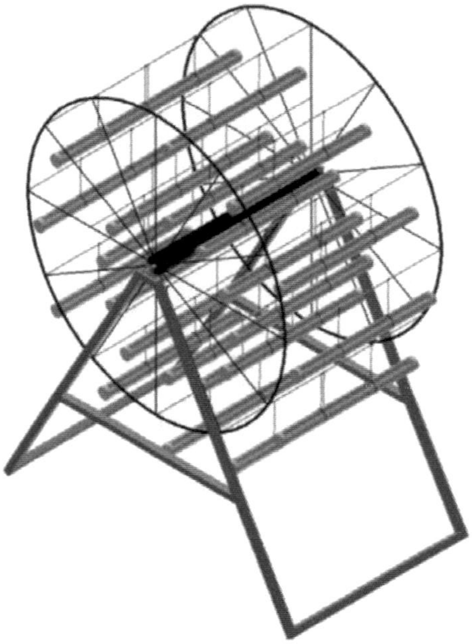

Figure 6. Smart rotary greenhouse design for productive plants

Stages of Design Rotary Smart Greenhouse for productive plants

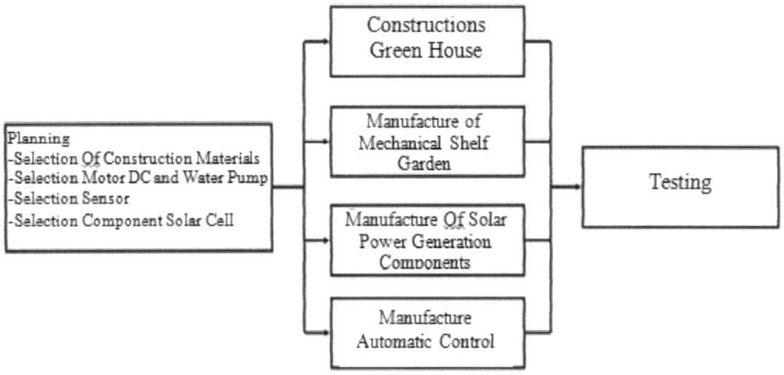

Figure 7. Stages of the smart rotary design of a green house for productive plants

4.1. The Automatic control system on smart greenhouse rotary

In the greenhouse there are several automatic conditioning control systems that are used:

a. Plant watering control system.

b. Plant positioning control system.

This program is to determine how the rotary smart greenhouse can work automatically, starting from the system of temperature and humidity readings, adjusting the movement of DC motors in accordance with the movement of water pumps. Software used is Arduino as shown in figure 8.

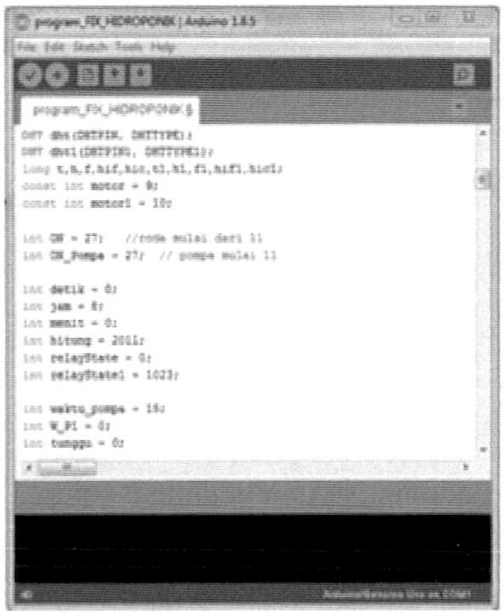

Figure 8. Arduino Software Windows Version

After the program from Arduino software for windows is uploaded on the Arduino UNO board, the next step is to combine DC motor and relay with Arduino UNO. In this step that needs to be prepared a relay, Arduino UNO board, power supply and cable. Connect the input port on the relay to the arduino uno board digital port. The greenhouse tool control circuit scheme is as shown below.

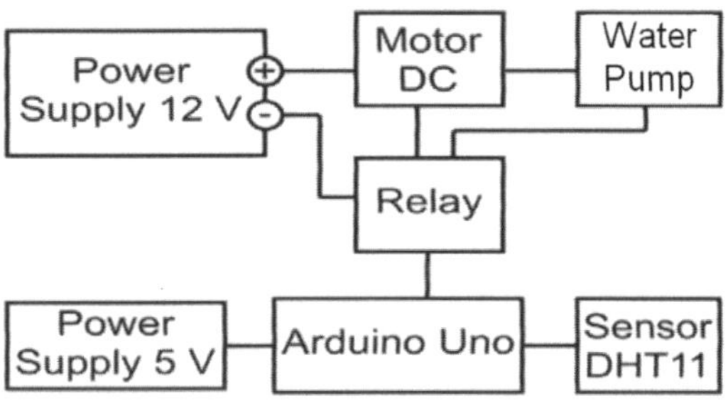

Figure 9. Scheme of rotary smart greenhouse control unit

4.2. Temperature and humidity testing in the rotary smart greenhouse.

In conducting the test or data retrieval, the temperature and humidity sensor (DHT 11) is installed in two different places within the rotary smart greenhouse and outside the rotary smart greenhouse with the intention to compare temperature and humidity between inside and outside the smart greenhouse.

Temperature data's collection is continuous for 12 months starting in March 2017 until February 2018. Temperature data's collection is carried out to determine the amount of temperature and humidity in the greenhouse are regularly doused in accordance with the program and temperature and humidity outside the smart greenhouse without watering or with existing weather condition.

From the results of the test then inserted in the form of graphs as seen in the following figure:

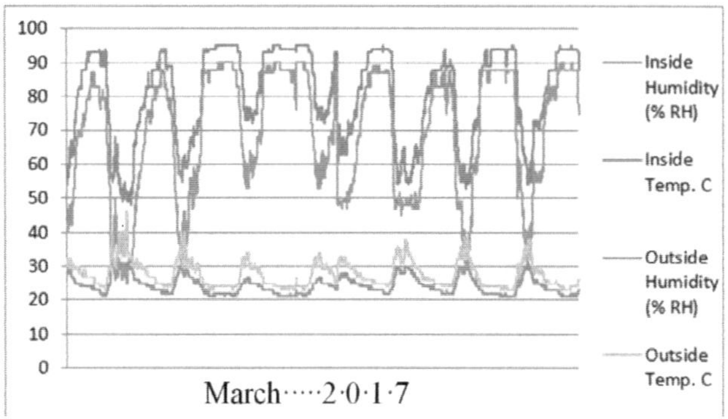

Figure 10. Data on March 2017

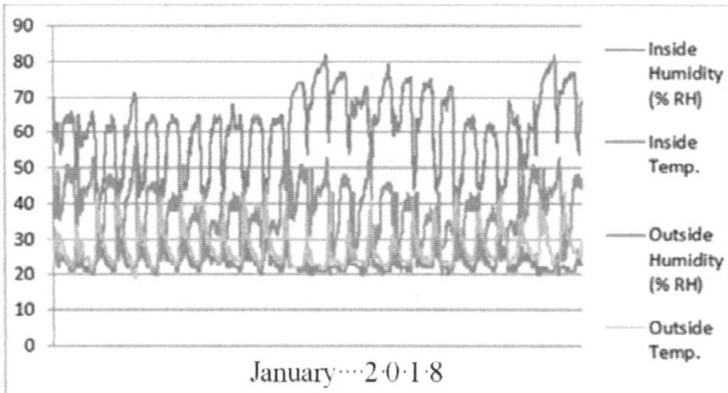

Figure 11. Temperature and Humidity on January 2018

5. Conclusions

From the test results of rotary smart greenhouse system design for productive crops that are done continuously for 12 months can be summarized as follows:

1. Rotary smart greenhouse with the productive crop condition is considered capable of optimizing temperature and humidity between 23oC- 30oC and 40% - 80%.
2. Taking temperature and humidity data using DHT 11 sensors can show the comparison between temperature and humidity inside and outside the rotary environment of smart greenhouse for productive crops.
3. With a watering system designed in a rotary smart greenhouse, this can be watered to all parts of the plant equally.
4. The smart green house with rotary systems works effectively with a decided rotation so that all plants get good sunlight during the day and can be well exposed to water.
5. Rotary smart greenhouse is effective for narrow land and with the number of plants that quite a lot and can be adjusted with to existing dimensions.

6. References

[1] Li juan, Huang Pan Liu, 2013, Key Technologies and Alogrithms' Application in Agricultural Food Supply Chain Tracking System in E-commerce, CCTA 2013: Computer and Computing Technologies in Agriculture VII pp 269-281, (Part of the IFIP Advances in Information and Communication Technology book series, IFIPAICT, vol. 420)

[2] Jun Yang, Mengchen Liu, Jiayi Lu, Yiming Miao, M. Anwar Hossain, Mohammed F. Alhamid, Botanical Internet of Things: Toward Smart Indoor Farming by Connecting People, Plant, Data and Clouds, Mobile Netw Appl (2018) **23**:188–202, https://doi.org/10.1007/s11036-017-0930-x

[3] Solomon Asfaw, Giacomo Branca, Introduction and Interview Chapter, Climate Smart Agriculture, Building Resilience to Climate Change, Springer International Publishing, 2018,

isbn 978-3-319-61194-5, doi 10.1007/978-3-319-61194-5.
[4] Shaker, M. & Imran, A., 2013. Greenhouse Micro Climate Monitoring Based On WSN with Smart Irrigation Technique. World Academy of Science, Engineering and Technolog, **7**(1), p. 12.
[5] Rafiuddin, S., Wahyu, H. P. & Budi, J., 2015. Controlling Smart Green House Using Fuzzy Logic Method. International Journal on Smart Material and Mechatronics, 2(2), pp. 116-120.

Acknowledgments
Authors wishing to acknowledge assistance from Hasanuddin University and Directorate General High Education, Ministry of Technology Research and High Education of Indonesia Republic.

Kinematic Motion Control for Robot Mobile Manipulators as Fire Fighters

Rafiuddin Syam, Onny S. Sutresman, Fahruddin[1], Adi Tonggiroh[2], Amiruddin, Abdul Mufti Radja[3]

Mechanical Engineering Dept., Engineering Faculty, Hasanuddin University
[1] Mechanical Engineering Dept., Engineering Faculty, Muslim Indonesia University,
[2] Geology Engineering Dept., Engineering Faculty, Hasanuddin University
[3] Architecture Engineering Dept., Engineering Faculty, Hasanuddin University,

the corresponding author's e-mail: rafiuddin@unhas.ac.id

Abstract. The purpose of this research is to make mobile manipulator robot as firefighting, to derivate kinematical formulation for mobile robot manipulator system, beside that is to make control system of such robot. The design and experiments were conducted in a mobile manipulator study as a firefighter. The result of this unmanned robot unmanned robot design can still carry the appropriate the standard sprinkle of firefighting system. The design of mobile manipulutor mechanism is done by using 3 SolidWorks 2010 design software design. The process of making robot begins with mobile robot as base frame and then installed the robot manipulator. Direct control method was done in this research by using arduino mega as interface system. The conclusion of this research is that robot can move with direct control and autonavigation for tracking trajectory case as firefighting system.
Keywords: Firefighting, direct control, Arduino, Control System

1. Introduction

A team of firefighters with a fast response system both natural and artificial system is increasingly needed [1]. Infact that, in the fire disaster there are two things those are dangerous fire and smoke hazard [2]. Furthermore, a fire disaster is a fire that is not controlled, not desired because it can cause harm to both property and fatalities.

Furthermore, fire can be happened if there is a balance of three elements consisting of fuel, oxygen, and heat. The relationship of these three components is usually called a fire triangle, so that if one of the elements is removed then the fire will be extinguished. Based on the theory of fire triangle then the principle of extinguishing technique is to damage the balance of mixing the three elements, for stopping the combustion process by breaking the chain reaction.

This principle can be done for firefighters with the following techniques [3]:

1. Cooling Method; fires can be extinguished by cooling the surface of a burning material using a water spray until the temperature below the flash point. But, in the case for fuels with low flash point such as gasoline, for cooling by water is not effective. Usually cooling is used to extinguishing fires involving fuel with high flash point.
2. Smothering method; a fire is limited by disconnecting the fuel with oxygen or air required for the combustion process. Covering a fire with CO2 or foam will stop the air supply for fire.
3. Starvation method; this method of extinguishing the fire by separating the combustible material by closing the flow of fuel into the fire or stop the fuel supply.

4. Breaking the chemical chain reaction; the combustion process combines by three elements to produce other gases such as H2S, NH3, HCN (according to a burning object). An important reaction is a free atom of O and H known as the radical atoms that make up OH and rupture into H2 and O. The radical O atom can form a larger flame. So that the way this outage is to break the chain of combustion reaction with fire extinguishers that work chemically.

Firefighting is a very dangerous task but it is still often done by human operators [3]. The need for rescue robot action planning was identified over 25 years ago by Professor Satoshi Tadokoro's group, he also proposed the Rescue Robo Cup5 competition framework. Then, According to Robin R. Murphy, 2001, rescue robots can be grouped into four major types as follows[4]:

a. Unmanned Ground Vehicles (UGVs) - UGV is working on the ground and can help rescue teams find and interact with trapped victims, in areas that are dangerous and difficult for rescue workers to enter the area.
b. Unmanned Surface Vehicles (USVs)- USVs floats on the surface of the water, and as searching robot then can help rescue teams to the right locate for preparing equipment to the victims.
c. Unmanned Underwater Vehicles (UUVs)- UUVs has the ability to search through water and identify fatalities, subjects or hazardous materials.
d. The UAV works without contact with the ground surface and can assist in transporting medical assistance to the victim and to present a rough scenario of the crash site

The science of robotics is very often used in hazardous industries to undertake dangerous technological processes [5].

Figure 1. Mobile manipulator robot as firefighting robot

2. Design of Firefighting Rescue Robot

2.1. Design of rescue robot

The initial stage of planning is to calculate the load will be handled by mobile robot manipulator as a fire extinguisher system. After that, determine the type of motor that will be used. So that obtained the selected motor data. The authors already done an example manipulator robot applied to another system at [6]. Power obtained from torque and motor rotation. Power P (kw) must be transmitted and rotation spindle (rpm) is given. In this case it is necessary to check the power P is. If P is the required average power it shall be divided by the mechanical efficiency η of the transmission system to obtain the necessary initial driving force. Great power may be needed at startup. Thus it is often necessary to correct the required average power by using correction factors in the planning.

2.2. Gears transmission

2.2.1. Gears straight. The gears are defined as the power transmission wheel, the wheels that transmit motion from the drive to the actuated. The use of gears as a transmission is commonly in industry, for example in small and meticulous measuring instruments such as watches, to reducing gears on large turbines that are powered by tens of mega-watts, machine tools, motor vehicles, lifting machines, transportation machinery etc . An example of straight gears can be found at Figure 2.

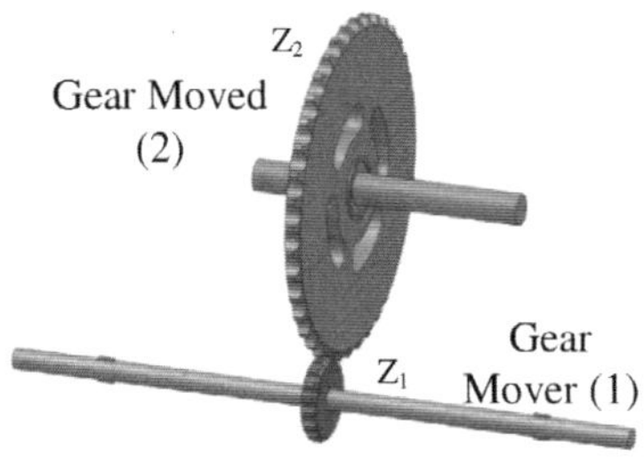

Figure 2. Direct Gear Transmission

If Z_1 is the number of gears on the motor shaft gear, Z_2 is the number of gears on the output shaft T_1 is the torque on the motor shaft, and T_2 is the torque on the output shaft. The output torque can be calculated by the formula,

$$T_2 = \frac{Z_2}{Z_1} T_1 \tag{1}$$

The direction of rotation of the shaft in the direct connection gear transmission is always opposite for each serial connection. T_o get the same rotation direction as on the motor shaft the gear must be arranged with an odd number.

Torque that occurs on the thread:

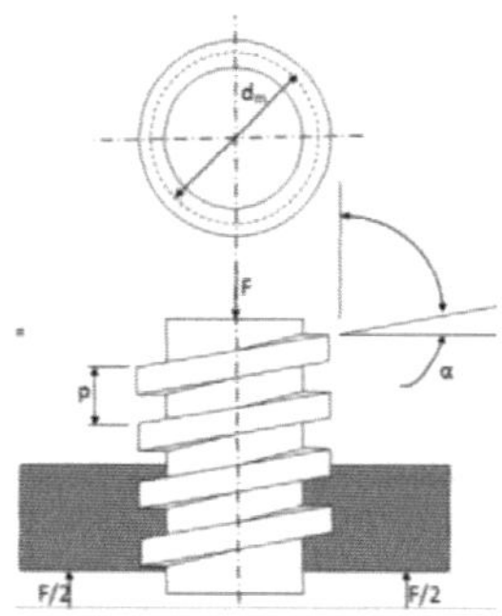

Figure 3. The direction of force on the pair of thread and nut

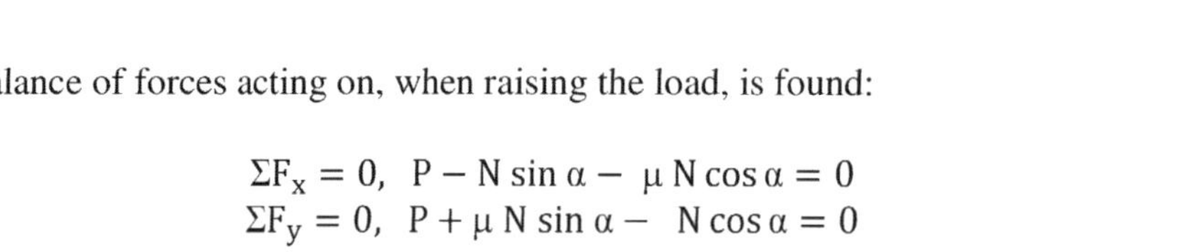

Figure 4. The force equilibrium on the thread (a. up, b. Down)

The force balance of forces acting on, when raising the load, is found:

$$\Sigma F_x = 0, \quad P - N \sin\alpha - \mu N \cos\alpha = 0 \tag{2}$$
$$\Sigma F_y = 0, \quad P + \mu N \sin\alpha - N \cos\alpha = 0 \tag{3}$$

While the balance of force of forces that work, when lowering the load, is obtained:

$$\Sigma F_x = 0, \quad -P - N\sin\alpha + \mu N\cos\alpha = 0 \tag{4}$$
$$\Sigma F_y = 0, \quad F - \mu N\sin\alpha - N\cos\alpha = 0 \tag{5}$$

From the equation (1) and (2), it is found:

$$P = \frac{F(\sin\alpha + \mu\cos\alpha)}{\cos\alpha - \mu\sin\alpha} \tag{6}$$

Then, with the same path, equation (3) and (4), can be define:

$$P = \frac{F(\mu\cos\alpha + \sin\alpha)}{\cos\alpha - \mu\sin\alpha} \tag{7}$$

Also, from the equation (5) and (6), multiply by 1/cos α and tan α = l/πdm, so that, force P to raise and lower the load:

$$P_n = \frac{F[\left(\frac{l}{\pi dm}\right) + \mu]}{1 - [\mu\left(\frac{l}{\pi dm}\right)]} \tag{8}$$
$$P_t = \frac{F[\mu - \left(\frac{l}{\pi dm}\right)]}{1 + [\mu\left(\frac{l}{\pi dm}\right)]} \tag{9}$$

So that the torque acts on the driving thread:

$$Mt = P\frac{d_m}{2} \tag{10}$$

When equation (7) and (8), substitute to equation (9), it was found:

$$Mt_n = \frac{Fdm\,(l + \pi\,\mu\,dm)}{2\,(\pi\,dm - \mu\,l)} \tag{11}$$
$$Mt_t = \frac{Fdm\,(\pi\,\mu\,dm - l)}{2\,(\pi\,dm + \mu\,l)} \tag{12}$$

2.2.2. Belt Transmission. This transmission is a power transmission / torque system from one shaft to the other through a circular belt on a pulley attached to the axes. The belt friction character and the pulley surface greatly affect the transmission capability. So the amount of tension in the belt determines the amount of torque that can be transmitted.

The belt can transfer the power elastically, so no elastic coupling is required. And it is not noisy and can accept and reduce shock loads, also easy and cheap in the manufacture and requires little maintenance

3. Research Method

The research method used is design and experiment. Details of the method used are shown below.

3.1. Mechanics System Design

The design of the mechanical system we created in the making of robot consists of 2 (two) units, namely: mechanical system of mobile axle driving wheel, mechanical system manipulator as sprinkle mechanism, as follow:

*3.1.1. Design of Mechanical System of Robot Drive Wheel.*The design of mechanical systems of the robot drive wheel is based on the environmental conditions that will be passed by the robot.

Tracking trajectory that will be traversed by the robot is a trajectory that is essentially uneven and made of soil used for wheel berbentu chain. Nonholnomic mobile robot is design in this robot. The number of chain wheels used are two (2) pieces each driven by a motor.

3.1.2. Design sprinkle mechanism as manipulator robot. Sprinkler system is considered as a robot manipulator. Sprinkle is one component of the robot that serves to direct the liquid fluid during the blackout process takes place. The sprinkler is directed according to the position of the heat source that needs spraying. At the time of the blackout process, sprinkler direction control requires a large force due to the pressure of the liquid fluid through the sprinkler hole and the momentum occurring at the sprinkler ends due to fluid backflow that slid out of the sprinkler. The driving source of the sprinkler motion control mechanism uses a DC motor. To ensure that the sprinkler moves well in the desired direction, a mechanical system design of the sprinkler drive is shown in FIG. 4.

Figure 5. Design of the Sprinkler Motion Control mechanism.

4. Result and Discussion
The result of the research is design of robot control which able to overcome the blackout process with various position of height of heat source.

Figure 6. Model Fire Fighting Robot

4.1. Sprinkler Motion Control System
Sprinkler is one part of the firefighting robot. The drive mechanism of the sprinkler can be seen in Figure. 11.

Figure 7. Sprinkle system as manipulator robot

Using a DC motor connected by several components forming a finite kinematic circuit, including: motor shaft, gear drive, chain, gear actuated, threaded shaft, driving nut, drive shaft, and sprinkler holder. When the motor is rotating, the driving gear moves another gear through a chain.

The specification of DC motor data and sprinkler mechanical systems are used:

Table 1. Specification of DC Motor

Torque (T_m)	100 kg.cm
Rotation (n)	500 rpm
Threaded Pitch (p)	1,6 mm
Volt (V)	12 Volt
Current (I)	4 ampere
Shaft Diameter (d)	1,5 cm
Length of motor (L)	12,5 cm
Diameter of motor (dm)	5 cm
Mass (m)	930 gram
Threaded angle (α)	4°

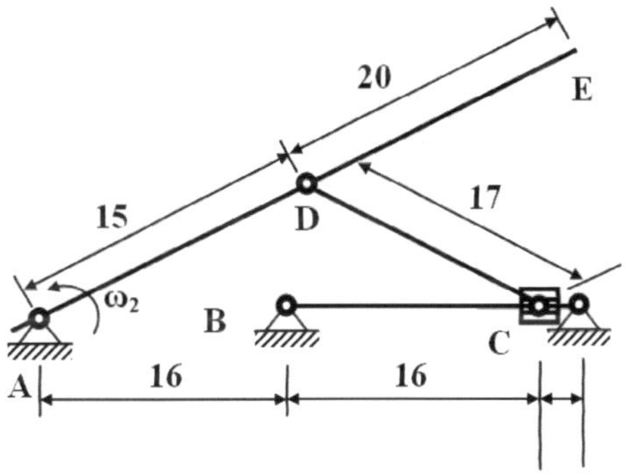

Figure 8. Free-body diagram of mechanical system of sprinkler motion controller

Polygon of Speed Movement System Mechanical Control of Sprinkler Motion

Table 2 Polygon Speed of Sprinkle Mechanism

Point	Distance
A-D	15 cm
D-E	20 cm
D-C	17 cm
A-B	16 cm
B-C	16 cm

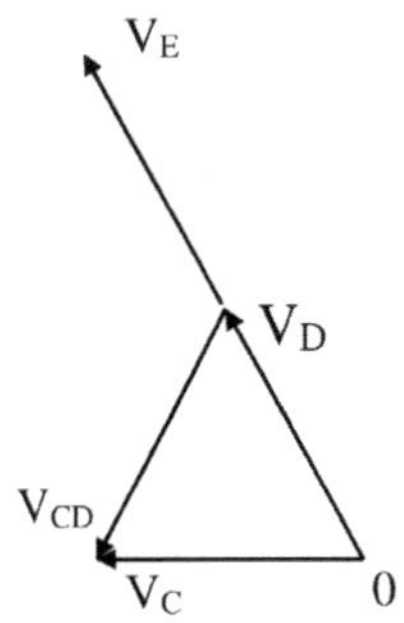

Figure 9. Polygon velocity of sprinkler system as a robot manipulator.

4.2. Force on Sprinkler system

Using the equation (1) The force that occurs in the driving gear (F_1) and (F_2) is 50 kg.

Figure 10. Transmission of gear 1 to gear 2

The force transfer system used from the gear 2 (F_2) to the screw nut of the steering wheel (F_3) is by using the principle of screw pairs of bolts and nuts. The magnitude of the screw angle is inversely proportional to the magnitude of the forwarded force. For more details, the principle can be seen in the following figure:

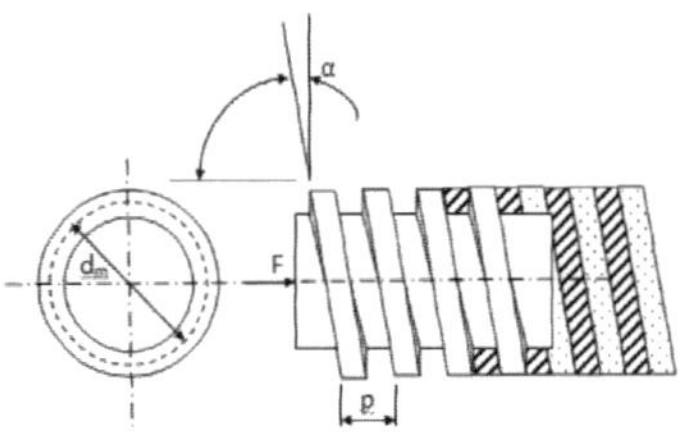

Figure 11. Transmission System to a shaft (Syam, 2018)

Using the equation (10) The force occurring in the threaded stem F_3 = 100kg. Force that occur during thread (F_4):

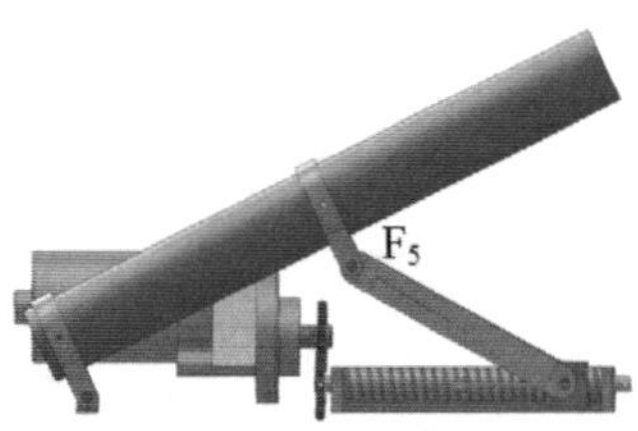

Figure 12. The force on the nut screw (F_4) and the connecting rod (F_5) as a nozzle

By using the equilibrium equation of force then it obtained F4 = 299 kg. Given the force (F_4) on the threaded stem it will produce a force (F_4) reaction on the nut screw in the opposite direction.

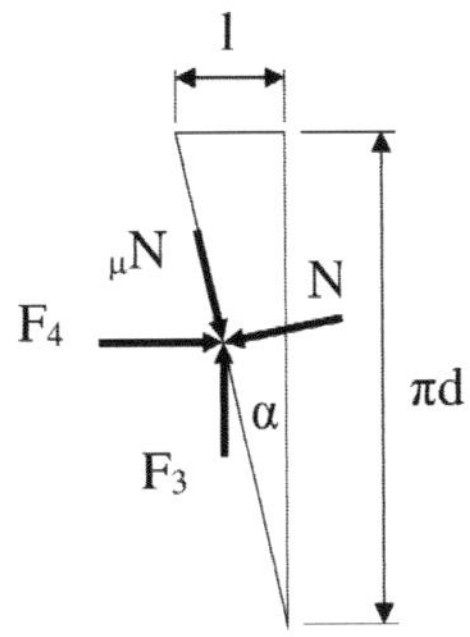

Figure 13. Freebody diagram (F_3)

The force on the connecting rod to the sprinkler holder of magnitude changes according to the angular change between the threaded axis of the rod and the sprinkler connecting rod. If the angle formed between the threaded rod and with the sprinkle connecting rod is 30°, then the lift force on the sprinkler holder can be calculated with the result $F_5 = F_4 . \cos 30°$.

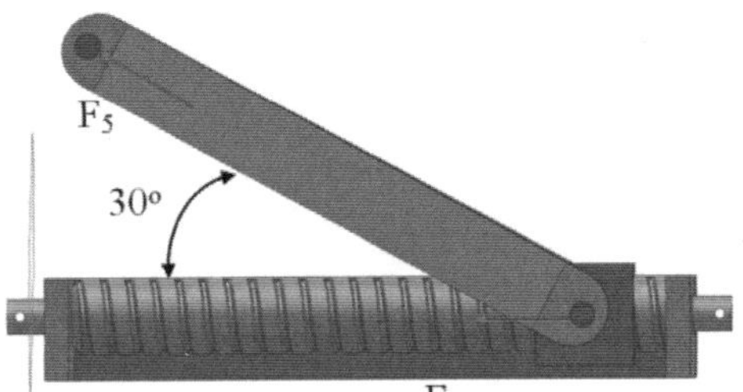

Figure 14. The angle between the screw axis with the connecting rod

The result of sprinkler vertical motion is used to control the altitude of the blackout direction during the process. As it happens that the blackout targets constantly change according to the position of the heat source. From the test results with at some position of the heat source, it is seen that sprinkler movement control able to steer well the blackout path after the previous position of the robot to take the right position.

4.3. Experiment of tracking trajectory problem of firefighting robot

GPS tracking method is done in this experiment. This experiment used GPS tracking trajectory method for mobile manipulator robot as firefighting system. The trajectory of the robot can be found at Fig. 15. The mobile robot follows the trajectory then the effectiveness shown on Figure152. On this path there are 17 waypoints starting at -5.2306 LS, 119.5024 LB to -5.2307 LS, 119.5024 LB, then -5.2306 LS, 119.5024 LB and end-point -5.2307 LS, 199.5024 LB.. The result of this experiment can be found at Figure 40. The mean waypoint distance is 25 cm with the total length of 4 meters.

In Figure 15. Visible cross passing by mobile robot. There are 3 stages: Stage A, B and C. Turns 90 degrees from A to B. The amount of mobile robot deviation as a fire extinguisher during the trial can be seen in Figure 47.

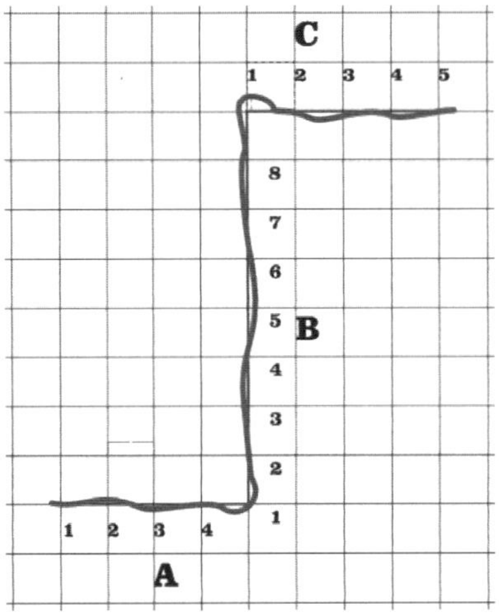

Figure 15. Error history of tracking Trajectory Mobile robot rescue as a fire fighting

Sometime after the mobile robot goes there are some deviations. Trajectory tracking experiments during the first trajectory occur several deviations on the axes X_a, Y_b, and X_c but the farthest extent of the waypoint is on the X_c axis at waypoint 1 with a distance of 7.43 cm apart and deviate at the coordinate point -5.2307 South Latitude, 199.5024 Longitude.

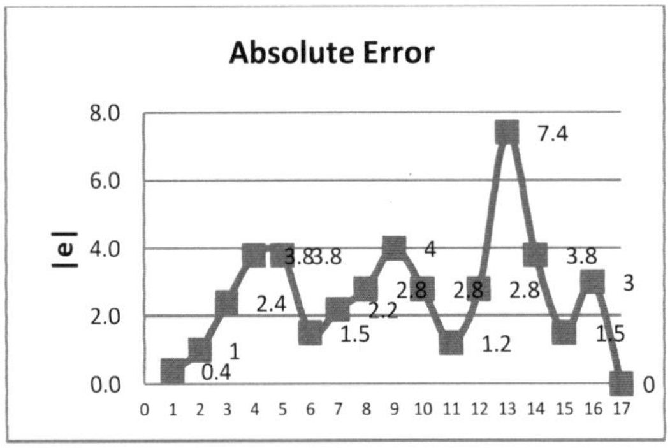

Figure 16. Error history during tracking trajectory

Root mean square error, RMS Error method is used to calculate tracking trajectory erro in this experiment. The Error total of all trajectories is 0.1757 m, as shown in figure 16.

5. Conclusions

It was found some conclusions:

1. Experiment of tracking trajectory with direct control method succeeded by using RMS error method.
2. The rotary motion of the driving DC motor is transformed into linear motion by the actuation mechanism and the threaded rod.
3. Torque of DC motor is 100 kg.cm can be converted into lift force on the sprinkler holder by a connecting rod of 149 kg.

6. References

[1] Marzena Półkaa , Szymon Ptaka* , Łukasz Kuzioraa, 2017, *The use of UAV's for search and rescue operations*, a The Main School of Fire Service, Slowackiego 52/54 St., Warsaw 01-629, Poland
[2] Zhan Yang, Ai-huaYi, Jian-yongLiu, Xia Zhao, 2016, *Study of Fire Hazard of Flooring Materials on Data of Cone Calorimeter*, Guangdong Province Enterprise Key Laboratory of Materials and

Elements Fire Testing Technology, Guangzhou, China, bGuangzhou Building Material Institute Limited Company, Guangzhou, China

[3] The Nam, Patrick Sebastian, Abu Bakar Sayuti Saman, 2012*, Autonomous Fire Fighting Mobile Platform,* Universiti Teknologi, Petronas, Bandar Seri Iskandar, 31750 Trono, Perak, Malaysia, International Symposium on Robotics and Intelligent Sensors 2012)

[4] Robin R. Murphy, Disaster Robotics, ISBN:9780262027359 pp. 240, B&W Illus.February 2014.

[5] Craig, J. John, 2005, Introduction to Robotics, Pearson Prentice Hall, USA

[6] S.Y. Agayeva, 2016, *The Kinematics of Final Point of The Holder Robot Manipulators,* Azerbaijan State Oil and Industry University, Baku, Azerbaijan

[7] Syam, Rafiuddin.,et.al., 2018, Kinematics Analysis of End Effector for Carrier Robot of Feeding Broiler Chicken System, Journal of Physics: Conference Series, Volume 962, available at http://iopscience.iop.org/article/10.1088/1742-6596/962/1/012013.

Acknowledgments

Authors appreciate to Hasanuddin University and Directorate General High Education, Ministry of Technology Research and High Education of Indonesia Republic.

Synthesis and Characterization of Ni-doped ZnO Thin Films Grown by Sol-Gel Spin Coating

P.L. Gareso[1*], Monika Pakabu, N. Rauf, E. Juarlin, and Didik Aryanto[2]

[1]Department of Physics, Faculty of Mathematics and Natural Sciences, Hasanuddin University, Jl. Perintis Kemerdekaan KM 10 Tamalanrea-Makassar 90245, Indonesia.
[2]Indonesian of Science Institution (LIPI) Serpong, Indonesia

Email: *pgareso@fmipa.unhas.ac.id, pgareso@gmail.com

Abstract. ZnO thin film doped with nickel has been grown by sol-gel spin coating technique. The thin films were prepared using zinc acetate dehydrate, ISO propanol, and ethanolamine as a precursor, solvent, and stabilizer respectively. For dopant, Nickel (II) acetate tetrahydrate with different concentration of Ni was added to the mixture. The sol-gel was deposited on a corning glass substrate at room temperature with a speed rate of 3000 rpm during 60 seconds. The Ni-doped ZnO thin films were put into a tube furnace for heat treatment at 500°C for 60 minutes. The characterization of the film was studied using X-ray diffraction (X-RD), and the optical transmittance UV-Vis measurements. The X-RD results show that a good crystallite was clearly observed in a hexagonal wurtzite structure with preferential orientation of (101) plane. The optical UV-Vis measurements show that the transmittance value increased with increasing the concentration of a dopant percentage. The optical band gap of Ni-doped ZnO was in the range of 3.20 – 3.23 eV.

Keywords: Sol-gel, spin coating, ZnO thin films

1. Introduction

In recent years, ZnO thin film has received considerable attention due to its optical properties. ZnO material is categorized as a wide band gap that has the band gap energy of 3.3 eV and has a large exciton binding energy of 60 meV [1,2]. The wide range application in optoelectronic and electronic devices of ZnO materials have been applied such as light emitting diodes (LED) [3], piezoelectric transduction [4], and gas sensors [5]. Also, ZnO has high transmittance value in the visible region which is good for solar cell [6]. The band gap energy of the ZnO thin films can be modified by adding with a small concentration of atom into the film. In this case, nickel atom was used as a dopant to change the structural and optical properties of the ZnO thin films.

Varying different techniques have been applied to grow ZnO thin film doped with nickel such as sputtering [7], spray pyrolysis [8], and sol-gel process [9]. Among of these techniques, sol-gel method approach several advantages such as low cost to study the structural and optical characterization, and easy to control the chemical compound.

This study reports the synthesis and characterization of the structural and optical properties of ZnO thin films doped with nickel by various concentration of nickel. The films were fabricated by sol-gel spin coating on corning glass substrates after heat treatment at 500°C during 60 minutes. The samples, then, were characterized using x-ray diffraction (X-RD) to study the grain size and the strain of the

IOP Conf. Series: Materials Science and Engineering **619** (2019) 012055 doi:10.1088/1757-899X/619/1/012055

films. UV-Vis spectrophotometer measurement was performed to investigate the transmittance value and the optical band gap of Ni-doped ZnO

2. Experimental Set-Up

ZnO thin films were fabricated using-sol-gel method by dissolving zinc acetate dehydrate as precursor into 20 ml ISO propanol as a solvent. This solution was added with 0.6 mL ethanolamine as stabilizer. Then, Nickel acetate tetrahydrate with different concentration was added to the mixture. This Mixture was stirred on the magnetic stirred hotplate at 60^0C during 15 minutes with the speed rate of 1000 rpm.

Before the sol-gel was deposited on a corning glass, the glass was cleaned with ethanol, and acetone into each solution during 10 minutes to remove away the unwanted particles; Then, the sol-gel of Ni-doped ZnO thin films were deposited on the corning glass with a speed rate of 3000 rpm during 30 seconds. Next, the samples were heated at 300^0C for 10 minutes at rapid thermal processor (RTP) to remove the solvent from the glass. Finally, the Ni-ZnO thin films were inserted into the furnace for annealing in the temperature range of 300^0C until 500^0C for 60 minutes. Figure.1 shows the flowchart for preparing TiO_2 multilayers thin film.

The structural properties of Ni-doped ZnO thin films were characterized using X-ray diffraction measurement with a single scan diffractometer Cu Kα radiation (λ = 1.5406 Å. The current and the voltage of X-RD were kept a constant at value of 30 mA and 40 kV, with the scan speed of diffractometer was 2°/min. The grain size and the strain of Ni-doped ZnO thin films were determined using the Debye-Scherrer method and William-Hall (UDM) formulation. The Optical transmittance measurements were carried out to characterize the optical properties of the films using a single beam UV-Vis spectrophotometer with a wavelength range of 250 – 800 nm.

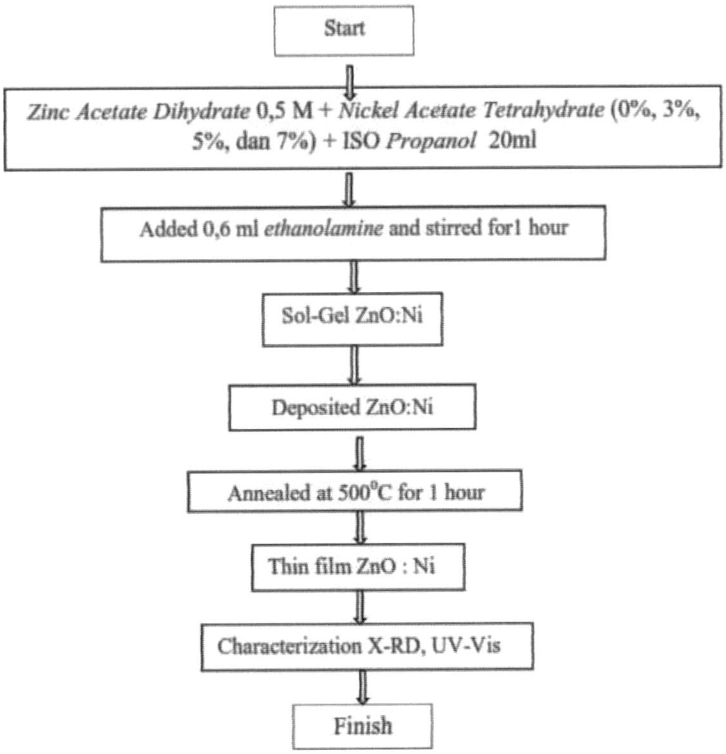

Figure.1. The flowchart for of the ZnO:Ni with various concentration of nickel.

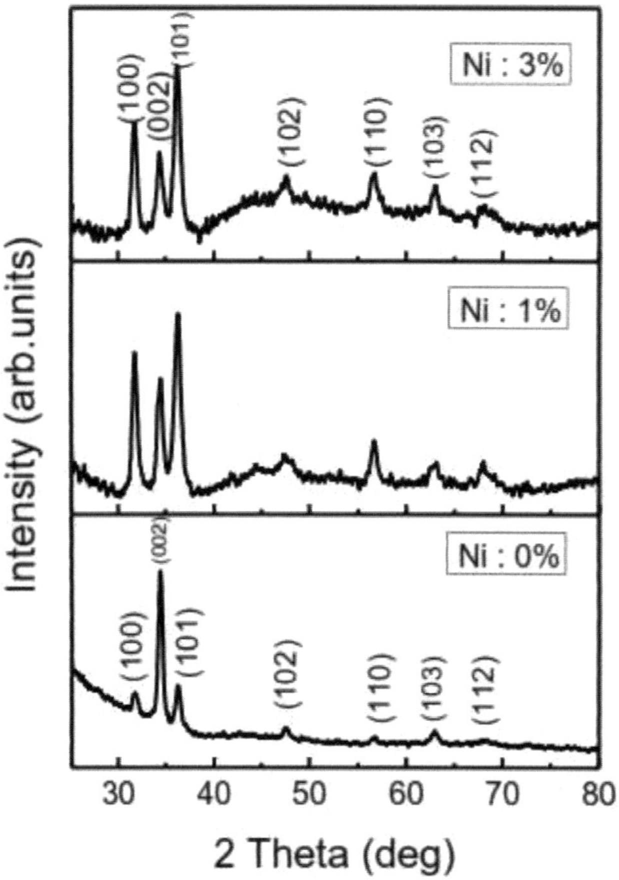

Figure.2.The X-ray diffraction (X-RD) patterns of Ni-ZnO thin films with various concentrations of nickel 0%, 1%, and 3%, respectively annealed at 500^{0}C for 60 minutes.

3. Results and Discussion

3.1. Structural characterization

Figure.2 displays the X-ray diffraction patterns of ZnO:Ni after heat treatment at 500^{0}C for 60 minutes with various concentration of nickel 0%, 1%, and 3%, respectively. It was clearly observed from figure.2 that the X-RD spectra without nickel appearing at diffraction angle of 31.84^0, 34.49^0, 36.24^0, 47.47^0, 56.72^0, 62.89^0, and 67.73^0, respectively. These angles are related to the plane of (100), (002), (101), (102), (110), (103), and (112) respectively (JCPDS: 04-03-6608). Also, it is shown in Figure.2, the intensity of (002) plane was higher than that other plane. The grain size of this plane is larger than that (002) plane at nickel concentration of 1% and 3% indicating that the preferential plane orientation of ZnO thin film without nickel was taken place in (002) plane. However, this intensity reduced significantly when the concentration of nickel was increased to 3%. In contrast to (002) plane, the intensity of (101) plane is higher in the sample where nickel has concentration of 3%. This indicated that ZnO thin films with nickel concentration of 3% have plane orientation in (101) plane. The change of preferential of plane orientation is probably due to the incorporation of nickel atom in the ZnO films. This can be looked in Table.1 where the grain size of the (101) plane with 3% nickel concentration is larger than that 1% nickel. Moreover, there was a shift of diffraction angle to 31.76^0, 34.33^0, 36.16^0, 47.50^0, 56.58^0, 62.93^0, and 68.26^0 respectively in ZnO:Ni the thin film (Ni=3%) with the same plane of the samples without nickel.

Tabel.1. The structure parameters of Ni-doped ZnO thin films after annealing at 500^{0}C for 60 minutes with various concentrations of nickel 1% and 3%, respectively.

Ni Concentration (%)	2θ	(hkl)	Scherrer Method [8]			William Hall Plot-Method [9] Uniform Deformation Method (UDM)	
			Size (nm)	Average Size (nm)	Strain	Size (nm)	Strain (x10^{-3})
1	31.78	(100)	14.39	12.74	0.0092	21.91	4.24
	34.44	(002)	13.38		0.0092		
	36.24	(101)	11.21		0.0104		
	47.80	(102)	18.17		0.0049		
	56.69	(110)	13.88		0.0055		
	63.03	(103)	10.05		0.0069		
	67.91	(112)	8.14		0.0080		
3	31.76	(100)	13.93	10.01	0.0095	18.5	13.55
	34.33	(002)	13.17		0.0093		
	36.16	(101)	12.67		0.0092		
	47.50	(102)	8.18		0.0011		
	56.58	(110)	8.74		0.0088		
	62.93	(103)	8.12		0.0086		
	68.26	(112)	5.28		0.0012		

Results from X-ray diffraction (X-RD) measurements were used to calculate the grain size (D) of the Ni-doped ZnO thin films with various different concentration of nickel using the Debye-Scherrer formulation [10],

$$D = \frac{K\,\lambda}{\beta\cos\theta} \tag{1}$$

where k is a constant number, λ is the wavelength of x-ray diffraction (Cu Kα, λ = 1.5406 Å), β is the full width at half-maximum (FWHM) of the peak, and θ is the Bragg angle, respectively. Based on Figure (1), the structures parameters of Ni-doped ZnO are tabulated in Table.1. As can be shown in Table.1, the average grain size of Ni-doped ZnO was calculated using the Debye-Scherrer formulation. Based on this formulation, the average grain size of Ni-doped ZnO with 1% of nickel concentration is 12.74 nm. This size reduced slightly to 10.01 nm when the concentration of nickel increased to 3%. Also, we used the William-Hall Method (UDM) to determine the grain size and the strain of the thin films using the equation (2) [11], to confirm the grain size results of Ni-doped ZnO thin film that was calculated from the Debye-Scherrer formulation.

$$\beta_{hkl}cos\theta = \frac{k\alpha}{D} + (4\varepsilon\ sin\theta) \tag{2}$$

The grain size and the strain of Ni-doped ZnO thin film were calculated using the UDM methods by assuming that the strain is uniform in all crystallographic direction. From equation (2), the term of $\beta_{hkl}cos\theta$ in the left hand side of equation (2) is potted to $(4sin\theta)$ of the second term in the right hand side. The results would give the slope and y-intercept of the fitted line where the slope indicates the strain and the y-intercept represents of the crystalline size of Ni-doped ZnO thin film, respectively. Figure 3 displays the UDM figure of the fitted line in the equation (2) for Ni-doped ZnO thin film after annealing at 500°C for 60 minutes with various concentration of nickel. As shown in Figure 3, the slope of the fitted line in Figure.3a (concentration of nickel is 1%) is smaller than that Figure.3b (concentration of nickel is 3%) indicating that the strain of Ni-doped ZnO with nickel 1% is lower than nickel 3%. Meanwhile, the grain size of the film was 21.91 nm for nickel 1%, and 18.5 nm for nickel 3%.

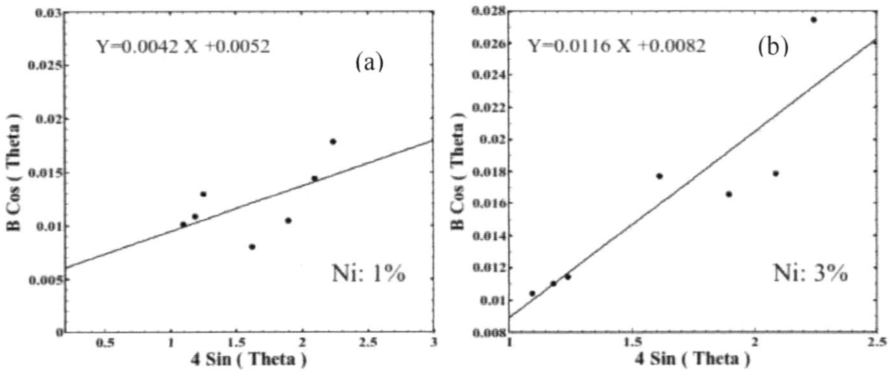

Figure.3. The William-Hall (UDM) analysis of Ni-doped ZnO thin film with various concentrations of nickel 1%, and 3%, respectively annealed at 500^{0}C for 60 minutes. The fitting-data results shows that y-intercept represents the grain size and the slope represents the strain of the films.

3.2. Optical characterization

Figure.4 depicts the optical transmittance UV-Vis of ZnO thin films doped after annealing at 500°C for 60 minutes with various nickel concentration of 0%, 1%, and 3%, respectively. With nickel concentration of 0%, the transmittance value was around 15% at wavelength of 400 nm. This value increased at was around 25% when the nickel concentration of 1%. When the concentration of nickel was further increased to 3%, the transmittance value reduced again to about 15%. This number is lower than that previous study where the transmittance value was around 80% [12,13]. In this study, they used aluminum atom as a dopant. From the transmittance spectra, the band gap energy of Ni-doped ZnO thin films can be obtained with use the equation [14] :

$$(\alpha h\upsilon)^2 = A\left(h\upsilon - E_g\right) \tag{3}$$

where α is the absorption coefficient, $h\upsilon$ is the photon energy, E_g is the optical band gap and A is a constant.

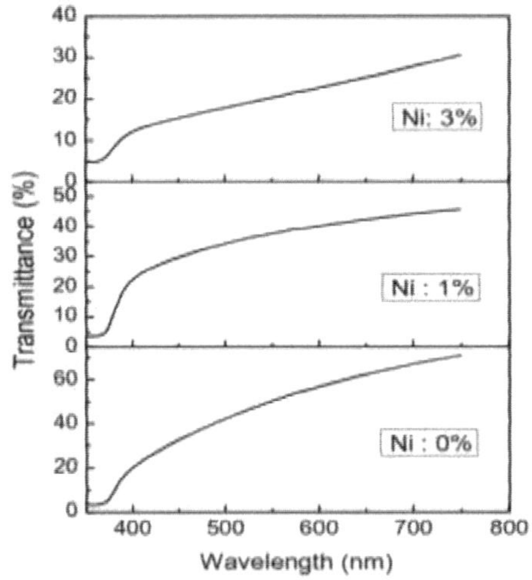

Figure.4. The optical transmittance of Ni-doped ZnO thin films with various concentrations of nickel 0%, 1% and 3%, respectively.

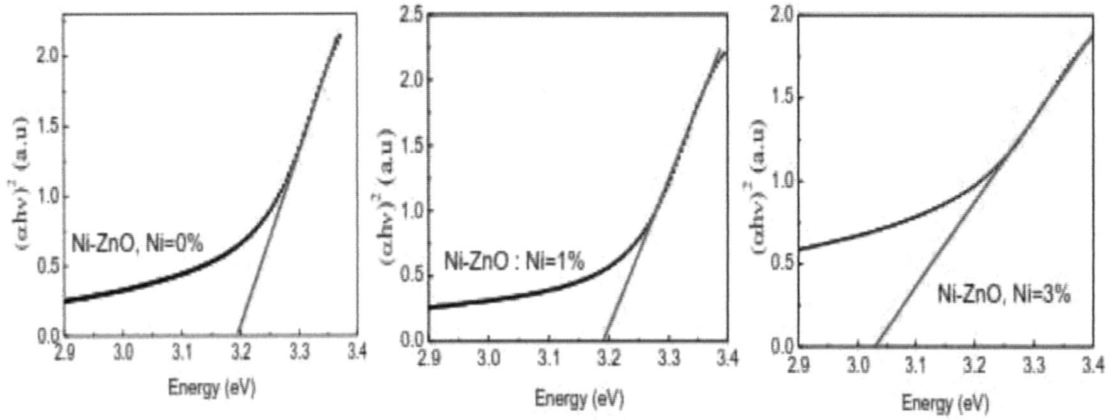

Figure.5. Plot of $(\alpha h\upsilon)^2$ in respect to hυ Ni-doped ZnO thin films with different concentration of nickel after annealing at 500°C for 60 minutes.

Figure. 5 displays the linier extrapolation of $(\alpha h\upsilon)^2$ in regard to energy (hv). The intercept of linier line toward the x-abscissa would give the optical band gap energy of Ni-doped ZnO thin films. As shown in Figure.5, the band gap energy was 3.195 eV without nickel atom in the samples. This value is well confirmed with the previous study where the optical band gap energy in this study was 3.2 eV. The band gap energy reduced to 3.189 eV when the concentration of nickel was 1%. Again, this optical band gap was further reduced to 3.028 eV at the concentration of nickel was 3%. The reduction of the band gap energy in Ni-doped ZnO is probably due to the improvement of crystalline of the film after the nickel concentration was increased.

4. Conclusion

We have studied the structural and optical characterization of Ni-doped ZnO thin films with varying the nickel concentration after annealing at 500°C for 60 minutes. The X-RD results show that a good crystallite was observed in a hexagonal wurtzite structure with preferential orientation of (002) plane without nickel. The plane orientation was change to the plane of (101) when the nickel concentration increased to 3%. The optical UV-Vis measurements show that the transmittance value increased with increasing the concentration of a dopant percentage. The optical band gap of Ni-doped ZnO thin films was in the range of 3.20 – 3.03 eV.

References

[1] Lin B, Fu Z, and Jia Y, 2001, *Appl. Phys. Lett*, **Vol 79**, 943
[2] Aksoy S, and Caglar Y, 2012, *Superlattices and Microstructures*, **Vol 51**, 613
[3] Jeong W J, Kim S K, and Park G C, 2006, *Thin Solid Films*, **Vol 506**, 180
[4] Tomar M, Gupta V, Sreenivas K, and Mansing A, 2005, *IEEE Trans on Devices and Materials Reliability*, **Vol 5**, 494
[5] Rau U, and Schmidth M, 2001, *Thin Solid Films*, **Vol 387**, 141
[6] Hupkes J, Rech B, Kluth O, Repmann T, Zwaygardt B, Muller J, Drese R, and Wutting M, 2006, *Solar Energy Materials & Solar Cells*, **Vol 90**, 3054
[7] Cztermastek H, *Optoelectronics Rev*, 2004, **Vol 12**, 49
[8] Romero R, Lopez M C, Leinen D, Martin F, and Ramos-Barrado J R, 2004, *Mater.Sci. Eng. B*, **Vol 110**, 87
[9] Nagarani N, and Vasu V, 2013, *J. on Photonics and Spintronics*, **Vol 2**, 19
[10] Fallah H R, Ghasemi M, Hassanzadeh A, and Steki H, 2006. *Physica B*, **Vol 373**, 274
[11] Pandiyarajan T, and Karthikeyan B, 2012, *J. Nanopart. Res*, **Vol 14**, 647

[12] Gareso P L, Rauf N, Juarlin E, Sugianto, and Maddu A, 2013, *3rd Int. Conf. On Theoretical and Applied Physics (AIP Conf. Proc)*, **Vol 1617** (Malang: Jawa Timur –Indonesia) 60
[13] Kuo S Y, Chen W C, Lai F I, Cheng C P, Kuo H C, Wang S C, and Hsieh W F, 2006, *J. Crystal Growth*, **Vol 287**, 78
[14] Tauc J, Grigorovichi R, and Vancu A, 1966, *Phys. Status Solidi*, **Vol 15**, 627

Acknowledgement
P.L. Gareso acknowledges the financial support from the higher education of Indonesia (RISTEKDIKTI) through the research scheme of Competence Based Research (PBK) under contract number of 1718/UN4.21/PL.00.00/2018.

IOP Conf. Series: Materials Science and Engineering **619** (2019) 012056 doi:10.1088/1757-899X/619/1/012056

Optimization of rotation speed parameters and number of grinding wheels on the quality and production capacity of chicken feed pellets

Syaharuddin Rasyid[1], Muas Muchtar[1], and Tri Agus Susanto[1]

[1]Department of Mechanical Engineering, Politeknik Negeri Ujung Pandang, Makassar, 90245, Indonesia.

E-mail: syaharuddinrasyid@poliupg.ac.id

Abstract. This paper presents an experimental design approach for the optimization process parameters for the chicken feed pellets. To achieve this goal, the speed parameters and number of grinding wheels are selected and two levels of this parameter are considered. Design of expert (DOE) of tests was used for experimental design and analysis of results. The chicken feed ingredients mixture is formed into pellets using a grinding wheel pellet machine with a variation of 150, 200 and 250 rpm rotation and variations in the number of grinding wheels 4, 6, and 8 pieces. The highest pellet production capacity (26.2 Kg/hour) occurs at 200 rpm rotating speed and 4 pieces of grinding wheels. The highest engine efficiency (87.6%) occurs at 200 rpm rotating speed and 4 pieces of grinding wheels. The highest pellet durability (91.6%) occurs at 200 rpm rotating speed and 4 pieces of grinding wheels. Optimal machining parameters are recommended to produce a pellet production capacity response of 25.6 Kg/hour, 85.53% efficiency machine, and 91.473% durability pellets are 150-225 rpm rotation speed range and 4-5 pieces grinding wheels.

1. Introduction

The process of making animal feed involves the use of various raw materials to produce compound feeds. Compound feeds must be in accordance with certain specifications regarding nutritional composition based on the description specified for nutritional quality, hygienic and physical. These specifications require knowledge of the properties of different ingredients to optimize processing while controlling nutritional quality for a given form of feed. Therefore, cooperation is needed from various disciplines such as nutrition science, science, and feed technology for further progress in producing animal feed [1].

Feed processing provides an opportunity to increase broiler production. According to Nolan, the biggest cost in producing broilers from the total production costs is feed (60-70%) [2,3]. There are several strategies to improve feed processing techniques; however, the cost of each strategy must be carefully considered for performance improvements that can be achieved and negative effects on target animals [4]. For this reason, several strategies are needed to improve feed processing techniques; however, the cost of each strategy must be carefully considered for performance improvements that can be achieved and negative effects on target animals [4].

Pelleting is the most common thermal processing method in the production of poultry feed. The main aim of pelleting is to agglomerate smaller feed particles by the use of mechanical pressure, moisture and heat. A major step in the pelleting process is the conditioning of mash prior to pelleting, which is generally accomplished by adding steam to the mash feed [3,5].

Opening a new animal feed factory in the area is an effort to optimize the potential of large local raw materials, the number of educated workers in the field of animal husbandry, and can reduce unemployment. This effort can reduce the dependence of animal feed from large factories. So that the production of pellet animal feed with adaptive technology using local equipment and raw materials must continue to be improved [6, 7].

Making chicken feed in the form of pellets in large capacity requires a machine. Various kinds of chicken feed pellet machines have been circulating, but the price is still expensive, which makes it very difficult for chicken farmers to buy it. Thus, a new breakthrough is needed by making pellet machines using cheap and lightweight materials, small driving forces to produce pellet machines that can be provided by most farmers' communities [7, 8].

Feed processing refers to any treatment to which animal feed undergoes prior to ingestion [9]. Feed processing technology has witnessed substantial improvements, from a hand scoop shovel as the basic mixing tool [10] to various processing operations which are currently performed utilizing modern feed technology [11]. The widely used processing operations in feed manufacturing plants are receiving the raw materials, grinding or particle size reduction, proportioning or batching, mixing, heating or thermal treatment (or pellet shaping), packaging, warehousing, and loading. Each of these operations can influence feed quality and poultry performance. Feed processing refers to the treatment provided fodder for consumption [9].

Feed ingredients, especially cereal grains, are cleaned and ground before being mixed into the diet. Grinding or particle size reduction modifies the physical characteristics of the material to increase the surface area of the larger nutrient digestion, improve blending ability and homogeneity of mixed feeds [4, 12].

Hammer and roller mills are the most commonly used equipment to reduce the particle size of feed ingredients. At the factory hammer, particle size reduction is carried out by affecting the slow-moving material with a set of hammers moving high speed. Hammer mills generally produce spherical particles with polished surfaces [13]. Particle size distributions produced at hammer mills vary widely around the geometric mean, with most and many small-sized particles [13, 14]. In the roller mill, size reduction is carried out through the compression strength between pairs of rotating rolls, resulting in a more uniform particle size distribution with a lower proportion of fine material [13].

Two basic methods, namely, cycle (batch) can be used to reach proportions. In a cycle or batching system, each material is weighed into batches, while continuous systems involve concurrent and continuous material addition [15]. Proportional materials and proper mixing are needed to achieve a homogeneous mixture.

DOE has been used by several researchers to optimize various types of manufacturing processes. In this study, a model has been developed to predict the optimization of engine speed parameters and the number of grinding wheels in the process of making chicken feed pellets using a tapered form wheel pellet machine.

In this work using a comparison of the mixture of chicken feed ingredients constructed by the author, several experiments have been carried out on the effects of machining parameters such as rotational speed and a number of grinding wheels.

2. Experimental Procedures

2.1. Machine Specifications

Pellet machine wheels grinding system (Figure 1a) which is used to measure the effect of rotation speed and number of grinding wheels on the capacity of pellet, machine efficiency, and durability of the pellet. The specification of pellet machine wheels grinding system is: machine dimensions (600 x 470 x 1150 mm), a cylindrical tube (280 x 310 mm), disc printing (diameter 265 x 10 mm), wheel rollers (diameter 60 x 110 mm), dynamo motor (1HP, 1450 rpm), the speed rotation (100-400 rpm). The shape and the number of the grinding wheels is shown in Figure 2. The workings of this tool are a disc mold rotated by the motor and wheel roller rotates. The feed mixture was inserted into the hole-mold after crushed by the roller wheels. The pellets were formed after passing through the hole-mold.

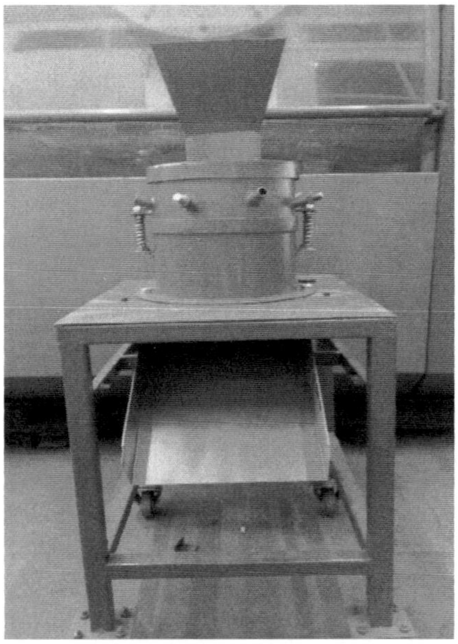

Figure 1. The pellet machine with grinding wheels system.

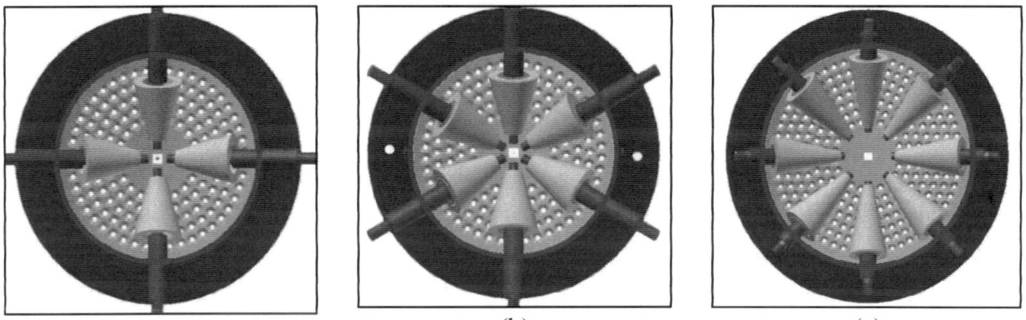

(a) (b) (c)

Figure 2. The grinding wheels models, (a) four pieces, (b) sixth pieces, and (c) eight pieces.

2.2. *Stage of Testing Method*

Testing was conducted to obtain the composition of the feed mixture that is optimal. The first stage is to prepare 500 grams of feed materials are added to the adhesive (starch) 50 grams and then mixed with water 400 ml. The next stage, the feed pellets mixture are moulded on the machine roller wheel system. Testing parameters are rotational speed of 150 rpm, 200 rpm, and 250 rpm at a hole-mold diameter of 6 mm, and a number of grinding wheels 4 pieces, 6 pieces, and 8 pieces. Production capacity and pellet durability are calculated by equations (1) and (2).

$$\text{Production Capacity} = \frac{weight\ of\ pellet\ (Kg)}{Times\ (Hour)} \tag{1}$$

$$\text{Durability Pellet} = \frac{Weight\ of\ durablitiy\ pellet\ (Kg)}{Weight\ of\ broken\ pellet\ (Kg)}\ \text{x}\ 100\% \tag{2}$$

2.3. Experimental design and statistical analysis.

To explore the effect of the operational factors on the response in the region of investigation, a DOE at two levels was performed. Rotation speed (rpm, A) and a number of grinding wheels (pcs., B) were selected as independent factors. The range of values and coded levels of the factors are given in Table 1. A polynomial equation (Eq. 3) was used to predict the response as a function of independent factors and their interactions. An interaction is the failure of the one factor to produce the same effect on the response at different levels of another factor [16]. In this work, there were four independent factors; therefore, the response for the quadratic polynomials becomes:

Table 1. Independent factors and their levels for DOE of chicken feed pellets process.

Independent Factors	Unit	Level		
		-1	0	1
Rotation speed (A)	(°C)	150	200	250
grinding wheels (B)	(pcs.)	4	6	8

$$Y = \beta_0 + \sum \beta_i x_i + \sum \beta_i x_i^2 + \sum\sum \beta_{ii} x_i x_j \quad (3)$$

where b_0, b_i, b_{ii}, b_{ij} are the constant, linear, square and interaction regression coefficient terms, respectively, and xi and xj are the independent factors (A and B).

Design-Expert 6 software was used for multiple regression analysis, analysis of variance (ANOVA), and analysis of ridge maximum of data in the response surface regression (RSREG) procedure. The goodness of the model was evaluated by the coefficient of determination R^2 and its statistical significance was checked by the F-test.

3. Result and Discussion

This study demonstrates the effect of rotation speed and number of grinding wheels for the optimization of the pellet feed chicken production. The design is used to obtain 11 design points within the whole range of two factors for experiments. The designs and the response are given in Table 2. Following the experiments, the response surface is approximated by DOE.

Table 2. Design layout and experimental results.

Std	Rotation Speed (rpm)	Number of Grinding Wheels (pcs.)	Coded A	Coded B	Production Capacity (Kg/hr.)	Machine Efficiency (%)	Durability Pellet (%)
1	150	4	-1	-1	26.0	87.3	91.6
2	200	4	0	-1	26.2	87.6	91.6
3	250	4	1	-1	25.8	86.6	91.4
4	150	6	-1	0	25.5	85.3	91.4
5	200	6	0	0	25.5	86.3	91.5
6	200	6	0	0	25.6	86.4	91.4
7	200	6	0	0	25.7	86.5	91.3
8	250	6	1	0	25.2	85.3	91.1
9	150	8	-1	1	24.5	83.9	90.9
10	200	8	0	1	24.8	84.7	90.8
11	250	8	1	1	24.7	84.0	90.4

IOP Conf. Series: Materials Science and Engineering **619** (2019) 012056 doi:10.1088/1757-899X/619/1/012056

3.1. Production Capacity

Results for production capacity at rotation speed and number of grinding wheels show that it fits the quadratic model. The ANOVA for the production capacity data is given in Table 3. Having its Prob>F of much less than 0.01, the quadratic model is valid. As for the coefficients, the rotation speed and number of grinding wheels was considered a significant factor.

Table 3. ANOVA with CI = 95% for model and factors of the production capacity.

Source	Sum of Squares	DF	Mean Square	F Value	Prob > F	
Model	2.96	5	2.96	55.42	0.0002	significant
A	0.015	1	0.015	1.41	0.2890	
B	2.67	1	2.67	250.00	<0.0001	
A^2	0.016	1	0.016	14.84	0.0120	
B^2	0.025	1	0.025	2.37	0.1839	
AB	0.040	1	0.040	3.75	0.1106	
Residual	0.053	5	0.011			
Cor Total	3.01	10				

The obtained empirical equation of the production capacity in the form of an actual factor is as stated in equation (1),

$$\text{production capacity} = 25.60 - 0.050A - 0.67B - 0.25A^2 - 0.10B^2 + 0.100AB \tag{4}$$

Where A is rotation speed (rpm) and B is a number of grinding wheels (pcs.).

For convenience, the equation can be displayed as response surface contour as well as three-dimensional surfaces, as shown in Figure 3.

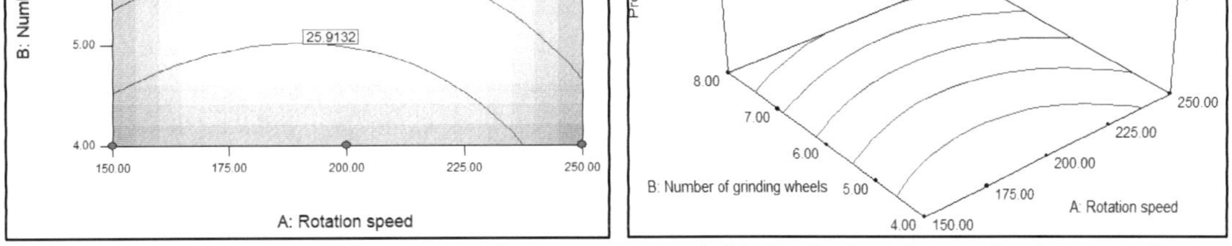

Figure 3. Response surface graph of (a) contours and (b) 3D Surface for production capacity.

Figure 3 shows the pellet production capacity of the test results of pellet machines with rotational speed variations (150, 200, and 250 rpm) and the number of grinding wheels (4, 6, and 8 pieces). The higher the rotational speed, the greater the production capacity, but at a rotating speed from 200 rpm to

IOP Conf. Series: Materials Science and Engineering **619** (2019) 012056 doi:10.1088/1757-899X/619/1/012056

250 rpm there is a decrease in production capacity. Figure 3 also shows the addition of the number of grinding wheels causes the production capacity to decrease. The average production capacity in this study was 25.4 Kg/hour. This pellet production capacity is better than the results of previous tests [7, 17]

3.2. Machine Efficiency

Results for machine efficiency at various rotation speed and number of grinding wheels show that it fits the linear model. The ANOVA for the machine efficiency data is given in Table 4. Having its Prob>F of much less than 0.01, the linear model is valid. As for the coefficients, both of the rotation speed and number of grinding wheels was considered as a significant factor. Machine efficiency was insensitive to the change in rotation speed and number of grinding wheels.

Table 4. ANOVA with CI = 95% for model and factors of the machine efficiency.

Source	Sum of Squares	DF	Mean Square	F Value	Prob > F	
Model	10.20	2	5.10	13.49	0.0027	significant
A	0.060	1	0.060	0.16	0.7008	
B	10.14	1	10.14	10.14	0.0008	
Residual	3.03	8	0.38			
Cor Total	834.85	10				

The obtained empirical equation of machine efficiency in the form of an actual factor is as stated in equation (2),

$$Machine\ Efficiency = 85.54 - 0.10A - 1.30B \quad (5)$$

Where A is rotation speed (rpm) and B is a number of grinding wheels (pcs.). For convenience, the equation can be displayed as response surface contour as well as three-dimensional surfaces, as shown in Figure 4.

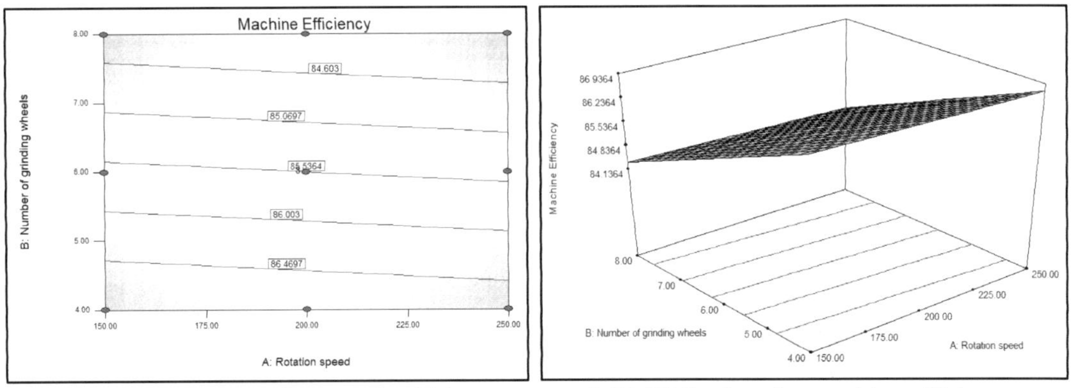

Figure 4. Response surface graph of (a) contours and (b) 3D Surface for machine efficiency.

Figure 4 shows the pellet production capacity of the test results of pellet machines with rotational speed variations (150, 200, and 250 rpm) and the number of grinding wheels (4, 6, and 8 pieces). The higher the rotational speed and the more grinding wheels the smaller the engine efficiency capacity.

3.3. Durability Pellet

Results for durability pellet at various rotation speed and number of grinding wheels show that it fits the quadratic model. The ANOVA for the durability pellet data is given in Table 5. Having its Prob>F of much less than 0.01, the quadratic model is valid. As for the coefficients, both of the rotation speed and number of grinding wheels was considered as a significant factor. Durability pellet was insensitive to the change in rotation speed and number of grinding wheels.

Table 5. ANOVA with CI = 95% for model and factors of durability pellet.

Source	Sum of Squares	DF	Mean Square	F Value	Prob > F	
Model	1.54	5	0.31	16.78	0.0038	significant
A	0.17	1	0.17	9.10	0.0295	
B	1.04	1	1.04	56.90	0.0006	
A^2	0.086	1	0.086	4.70	0.0825	
B^2	0.14	1	0.14	7.59	0.0401	
AB	0.022	1	0.022	1.23	0.3181	
Residual	0.092	5	0.018			
Cor Total	1.63	10				

The obtained empirical equation of durability pellet in the form of an actual factor is as stated in equation (3),

$$\text{Durability pellet} = 91.47 - 0.17A - 0.42B - 0.18A^2 - 0.23B^2 - 0.75AB \qquad (6)$$

Where A is rotation speed (rpm) and B is a number of grinding wheels (pcs.).

For convenience, the equation can be displayed as response surface contour as well as three-dimensional surfaces, as shown in Figure 5.

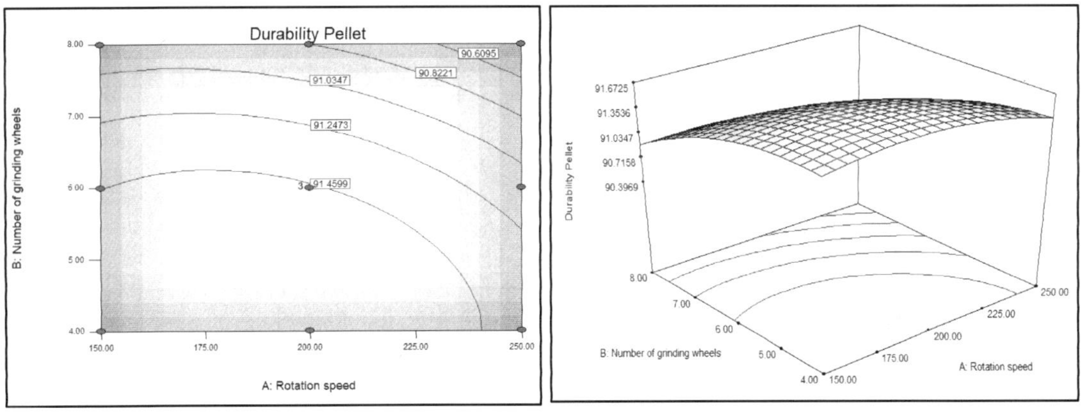

Figure 5. Response surface graph of (a) contours and (b) 3D Surface for durability pellet.

Figure 5 shows the pellet durability of the test results of pellet machines with rotational speed variations (150, 200, and 250 rpm) and the number of grinding wheels (4, 6, and 8 pieces). The higher the rotational speed, the greater the percentage of pellet durability, but at a rotational speed from 200 rpm to 250 rpm there is a decrease in the percentage of pellet durability. Figure 3 also shows the addition of the number of grinding wheels causes the percentage of pellet durability to decrease. The average pellet durability in this study was 91.2%. The durability of this pellet is better than the results of previous tests [7, 17]

3.4. Optimization

Now that the empirical model for all casting responses as a function of stirring speed and stirring time has been obtained, the selection of the optimal casting parameter setting can be performed. One can adjust the expected range of each casting response and the range of stirrer speed and stirring time in line with expectations for all foundry responses can be determined. For example, that in order to obtain optimal mechanical properties, minimum grain size, and optimum shape factor, the stirring parameters should be carried out at 150-225 rpm rotation speed range and 4-5 pieces grinding wheels. To achieve this criterion, the rotation speed range and number of grinding wheels must be within the yellow plot of the overlay (Figure 6) of all production pellets responses.

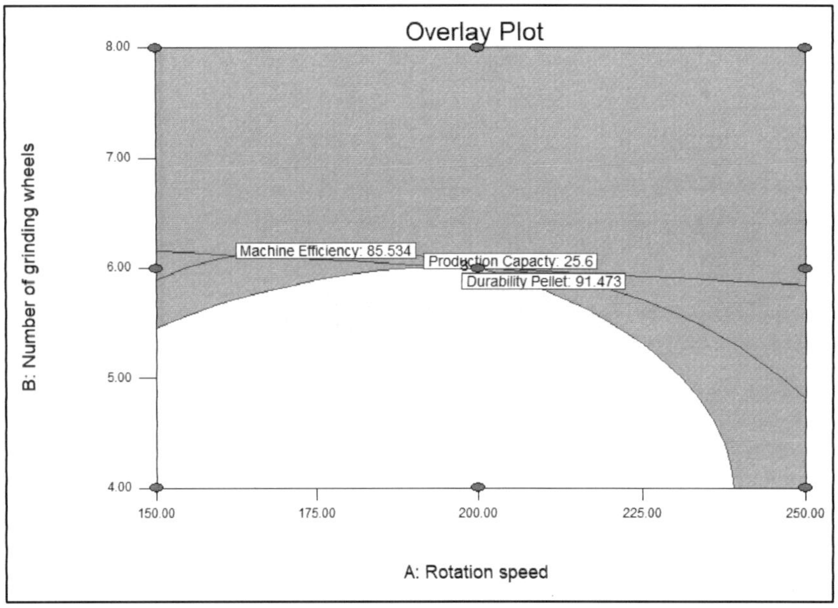

Figure 6. Overlay plot of the input factors for the predetermined response criteria of pellet production of 25.6 Kg/hour, 85.53 % efficiency machine, and 91.473 % durability pellets.

4. Conclusion

Research on making chicken feed pellets using a roller wheel pellet machine with rotational speed variations and the number of grinding wheels has been studied. The results obtained can be synergized as follows: The highest pellet production capacity (26.2 Kg/hour) occurs at 200 rpm rotating speed and 4 pieces of grinding wheels. The highest engine efficiency (87.6%) occurs at 200 rpm rotating speed and 4 pieces of grinding wheels. The highest pellet durability (91.6%) occurs at 200 rpm rotating speed and 4 pieces of grinding wheels. Optimal machining parameters are recommended to produce a pellet production capacity response of 25.6 Kg/hour, 85.53% efficiency machine, and 91.473% durability pellets are 150-225 rpm rotation speed range and 4-5 pieces grinding wheels.

References

[1] Thomas M, Van D P 1996 J. Ani. Feed Sci. Tech. 61 89-1I2.
[2] Nolan A, McDonnell K, Devlin G J, Carroll J P, Finnan J 2010. En. J. 3, 1–11.
[3] Abdollahi M.R, Ravindran v, Svihus b 2013 Ani. Feed Sci. and Tech. 179 1-23.
[4] Behnke K C 1996 Ani. Feed Sci. and Tech. 62 (1) 49–57.
[5] Schoeff, R.W., Fairchild, F.J., Bursiek, B., Castaldo, D., 2005. Feed Man. Tech. pp. 1–13.
[6] Glover D and Kusterer K 2016 Cont. Farm. Rur. Dev. Spr.
[7] Rasyid S, Muchtar M, and Susanto T A 2018 AIP Con. Proc. 1977, 020019, 020019-(1-5).

IOP Conf. Series: Materials Science and Engineering **619** (2019) 012056 doi:10.1088/1757-899X/619/1/012056

[8] Tanaka M 1993 U.S. Patent 5,186,959.
[9] Maier, D.E., Bakker A F W 1992 J. Agric. Eng. Res. 53, 305–319.
[10] Schoeff, R.W, Fairchild, F.J, Bursiek, B., Castaldo, D., 2005. Feed Man. Tech. V. pp. 1–13.
[11] Deyoe, C.W., 1976. Feed Man. Tech. pp. 19–20.
[12] Behnke, K.C., 1996. Ani. Feed Sci. Tech. 62, 49–57.
[13] Koch, K., 1996. Feed Man. pp. 1–4.
[14] Svihus, B., Kløvstad, K.H., Perez, V., Zimonja, O., Sahlstrom, S., Schuller, R.B., Jeksrud, W.K., Prestløkken, E., 2004.Anim. Feed Sci. Tech. 117, 281–293.
[15] Fairchild, F.C., Moorehead, D.F., 2005. Feed Man. Tech. V, pp. 127–136.
[16] Montgomery D C 1997 Des. and Ana. of the Exp.
[17] Rasyid S, Tri A Susanto T A, and Nur R 2017 Mat. Sci. and Eng. **180** (2017) 012026

Acknowledgments

This paper was composed based on work supported by RISTEKDIKTI grant. I would like to thank The Ministry of Research, Technology and Higher Education of the Republic of Indonesia and Politeknik Negeri Ujung Pandang for funding and support.

IOP Conf. Series: Materials Science and Engineering **619** (2019) 012057 doi:10.1088/1757-899X/619/1/012057

Codes Development of Industrial Mechatronic System

Sri Hasnawati Ulandari[1)], Irdyah Inayah[1)], Zulfhis Shara Firstiawaty[1)], Lewi[1)], Simon Ka'ka[1)], Rafiuddin Syam[2)]

Centre for Mechatronics and Control System, Mechanical Engineering, Polytechnic of Ujung Pandang[1)]
Departement of Mechanical Engineering, University of Hasanuddin[2)]

the corresponding author's e-mail: rafiuddinsyam@gmail.com

Abstract. Industrial Mechatronic System (IMS) is a system designed focuses on the simulation of industrial production activities. This study aims to re-operate Industrial Mechatronic System and robot arm as a whole system by means of treating and repairing some of the components and develop the universal and partial codes. The system consists of six stations and a robot arm with different functions. Some of the stations and the robot arm cannot be operated properly and cannot move per-station by using the factory default codes. To operate each station required the development of coding using Programmable Logic Controller (PLC) additional reserve to avoid deletion of codes that exist in the main PLC with according to it's work principle. The results showed that the Industrial Mechatronic System can be used to simulate the production system on a PLC-based industry that can be operated universal and partial.

1. Introduction

Developments in science and technology are so rapid has brought a considerable impact on human society to learn and develop science. In every industry fast and precise work required to face the demand. To face these needs, it is not enough just to rely on human power itself, but also the technology and more advanced controllers. Today many industrial equipment that is equipped with a fully automated equipment.

Along with the advance of the mindset of human resources so as to exclude ideas and creative mind to create a wide variety of human needs aiming device to facilitate human life. One of them in the development of mechatronic products. Mechatronics is a discipline that combines or synergy of engineering, electronic engineering, informatics and engineering techniques arrangement (or engineering control) to design, create or produce, operate and maintain a system to achieve the desired goal. Currently it has many universities open mechatronic majors, and not a few students who are interested to cultivate this science.

Due to this mechatronic system has been widely used in the industrial world, we must first recognize this system in college before plunging into the world of industry. On campus we have been there a system that includes already using mechatronic systems. This tool is a means of supporting the Mechatronics Laboratory practicum at the Polytechnic of Ujung Pandang. This tool is called Industrial Mechatronic System, which this tool serves to provide information or a shadow to the students of mechatronic systems in the industry. This tool control system using PLC (Programmable Logic Control).

However, today these tools cannot be used as a whole, because there are some damaged components and problems on the configuration. As for the condition of Industrial Mechatronic System in this time that the conveyor does not run smoothly, several stations and a robot arm that cannot be operated properly and this tool cannot move a per-station for using the factory default program.

To address the shortcomings outlined above, we as students of Mechatronics feel motivated to make improvements and refinements of existing tools, and create a new control system that uses a system of Programmable Logic Controller (PLC). So that the tool is no longer using the default control system of the plant and also can be controlled independently, which means that the tool can be moved per-station.

2. Literature Review

2.1. Definition of Industrial Mechatronic System (IMS)

Industrial Mechatronic System (IMS) provides a simulation system that is similar to the existing system in the industry with varying degrees of difficulty. Industrial Mechatronic System (IMS) is a modular system that can be run freely and module design is also easily expanded. This means it can add to the knowledge of students before knowing and also expand knowledge after knowing and studying these systems [1][11].

As for the parts of the Industrial Mechatronic System that is:

- Selection station serves to place the workpiece on the workpiece carrier.
- Assembly station serves to combine or unite the workpiece.
- Processing station serves to lock or process the workpiece.
- Testing station serves to check the workpiece. storage station responsible for maintaining or developing the workpiece on each shelf / place based on the color and material of the workpiece.
- Robot arm used to move the workpiece from the storage station to disassembly station.

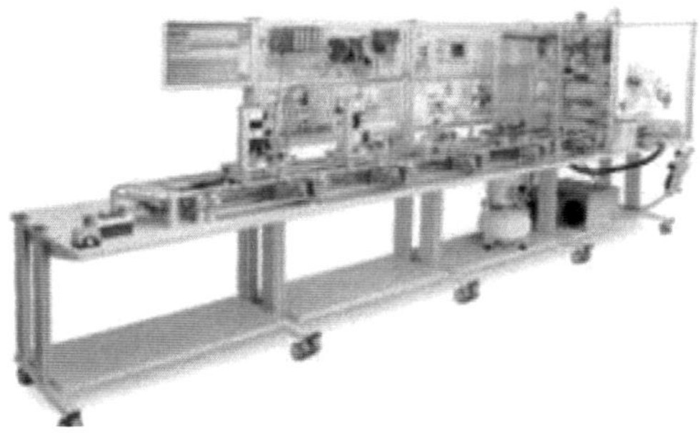

Figure 1. Industrial Mechatronic System

2.2. Composer Components Industrial Mechatronic System (IMS)

2.2.1. Industrial Mechatronic System (IMS). Industrial Mechatronic System consist of: Sensor is a device for detecting / measuring something that is used to change a variety of mechanical, magnetic, heat, light and chemistry into a voltage and electric current. The sensor itself consists of a transducer with or without an amplifier / signal processor which is formed in the sensing system. In an environment control systems and robotics, sensors give equality resembling the eyes, ears, nose, tongue then be processed by the controller as the brain [2].

As for the various sensors used in Industrial Mechatronic System are:

- Magnetic sensors serve to monitor / monitoring the movement of the cylinder.
- Sensors monitor the mechanical function of the level of the workpiece.
- Inductive sensors detect metal function.
- The optical sensor function identifies white superstructure.
- The capacitive sensor serves to detect the presence of the workpiece and raise the stopper.

IOP Conf. Series: Materials Science and Engineering **619** (2019) 012057 doi:10.1088/1757-899X/619/1/012057

An electric motor is an electric machine that converts electrical energy into mechanical energy. Construction of motor and generator is basically the same. DC motor develops a big moment and allows setting the number of rounds without stages. The number of motor rotation can exceed the rotational field [3].

In industrial mechatronic system (IMS) only uses two types of cylinders are single acting cylinders and double-acting cylinder. The explanation of the second cylinder in Table 1. as follows:

Table 1. Various Kinds of Linear Actuators

NAME ACTUATOR	INFORMATION	SYMBOL
Single work cylinder	• Cylinders with pressure only work in one direction (step forward) • Step back with the external force	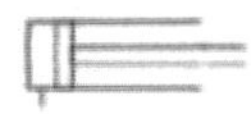
Single work cylinder	• Step back by spring	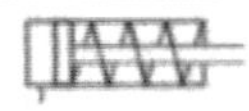
Double work cylinder	• Cylinders with pressure can work in both directions (step forward and backward): With a single piston rod	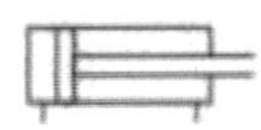
	• With double piston rod	

The vacuum generator is used to produce air vacuum or air suction. Used in conjunction with a suction cup to move a variety of workpieces. This tool works on the principle of venturi meters (vacuum) [4].

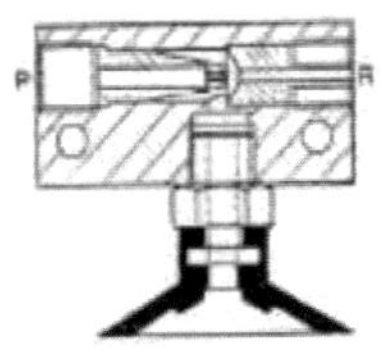

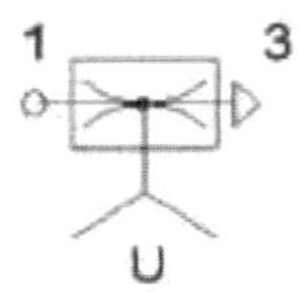

Figure 2. Vacuum Generator

2.2.2. Robot Arm (Manipulator). Industrial Mechatronic System consist of: The robot arm includes part of the tool Industrial Mechatronic System. There are many types of robot arm RS03N one type robot with specification shown in Table 2.

Table 2. Specifications Robot Arm Type RS03N and Controller E76

Type Robot	RS03N
degrees of Freedom	6 DOF (Degree of Freedom)
Weight	20 kg
Motorcycle	Brushless AC Servo Motor
Battery Type	Maxell Lithium Battery 3.6V AA ER6C
Controller Type	E-Controller E76
Memory capacity	8 MB
interface checker	USB, Ethernet, RS232C

IOP Conf. Series: Materials Science and Engineering **619** (2019) 012057 doi:10.1088/1757-899X/619/1/012057

Figure 3. Robot Type RS03N

Type of controller used for robot arms are E-Controller E76, where in the controller is very compact and is designed specifically for small arm robot as RS03N, 05N, 05L, 06L and 10N. although compact in design, these controllers offer the performance and the ability to upgrade high and high-performance CPU provides accurate control, high-speed program execution as well as the loading and storage of files very quickly. [5]

Figure 4. E-Controller E76

Figure 5. Teach Pendant.

2.3. Configuration Control on Industrial Mechatronic System

2.3.1. Programmable Logic Controller (PLC). Industrial Mechatronic System consist of: Programmable Logic Controller (PLC) is a specialized type of microprocessor-based controller that utilizes a programmable memory to store instructions and to implement functions such as logic, timing, counting and arithmetic to control machines and processes [6]. PLC concept is as follows:

a. Programmable; demonstrated that can easily be customized program created and ability in terms of program memory that has been made.
b. Logic demonstrate the ability to process input arithmetic and logic, which perform operations compare, add, multiply, divide, subtract, and negation.
c. Controller; demonstrate the ability to control and manage the process so as to produce the desired output.

2.3.2. PLC programming structure. Programmable Logic Controller (PLC) provide various types of user program blocks which relate to data that can be stored. Depending on the purpose of the process, the program can be arranged in a different block.[7]

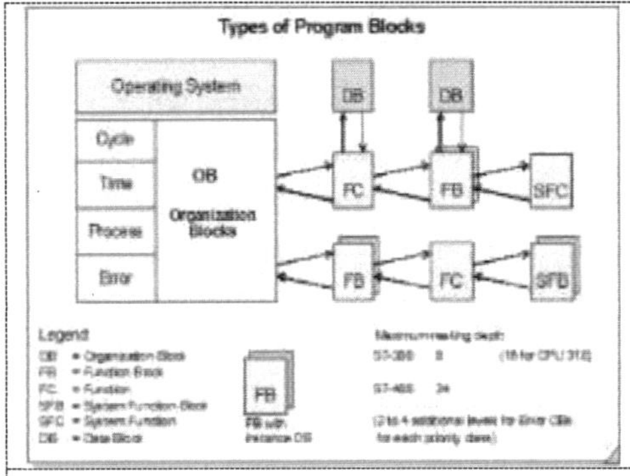	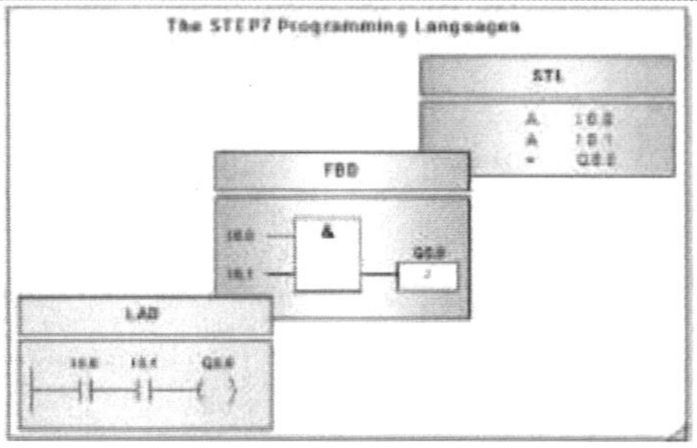
Figure 6. Type of Program Block	**Figure 7.** PLC Programming Language

There are several options that can be used programming languages in STEP 7 as follows:

a. Ladder Diagram (LAD), a programming language that is similar to the circuit diagram. This programming language is often an attraction for programming that has a background as an electrical drafting and, because it uses symbols such as coil, contact, etc.
b. Statement List (STL), which are sets of instructions statement STEP 7. This programming language is preferred by programmers familiar with using various programming languages.
c. Function Block Diagram (FBD), a programming language that uses a box-box functionality. FBD benefit can be used by "non-programmer" for every box-box has indicated a specific function such as operation logic functions. [8]

2.3.3. Communication Profibus. Some important aspects required in a PLC system as the interconnection network PLC. PROFIBUS is a good solution for network interconnection between the PLC, and the PLC with other interfaces. Profibus is an interconnection in the PLC network covering both types of cables, connectors, and interfaces such as the Repeater, Power Rail Booster and many others. PROFIBUS standardized in accordance with the International Electrotechnical Commission (IEC) 61 158 on the standardization of field bus specifications. Profibus is available in many forms for various applications [9].

According to the Profibus functions are divided into three types of PROFIBUS, the PROFIBUS DP, PROFIBUS PA, PROFIBUS FMS. PROFIBUS Distributed Input / Output (PROFIBUS DP) This connection is used for the distribution of field devices such as SIMATIC ET 200, or drivers with a very fast response time. PROFIBUS DP has a standard - compliant components from a wide variety of tools and producer's addition Siemens also can connect. PROFIBUS Process Automation (PROFIBUS PA) is a development of PROFIBUS DP with Intrinsically Safe Transmission technology in its reference to the standard IEC 61158-2. PROFIBUS Field bus Message Specification (PROFIBUS FMS) is a PROFIBUS that is designed specifically for field bus communication process. [10]

2.3.4. Maintenance Tool. Maintenance tool consist of:
Treatments are activities undertaken to improve, maintain, and restore the equipment in good condition and ready to use. In relation to the care of laboratory equipment, maintenance is intended as a preventive effort that equipment is not damaged or maintained in good condition, ready to operate. Besides, the treatment is also intended as an attempt to set or restore laboratory equipment is already damaged or less decent, so ready to be used for practicum students.

The Goals of Treatment are to Maintenance of laboratory equipment include:

- In order for the laboratory equipment is always primed, ready to be used optimally
- Extend lifespan
- Ensure smooth learning activities

- Ensure safety and comfort for the users
- Knowing early damage or injury symptoms
- Avoid sudden failure

Maintenance is the ways or methods to conduct laboratory equipment maintenance work that can be done among others by:

- Prevention, for example by giving a warning through drawing or writing, regulations, codes of conduct for users of laboratory / workshop, gave preservatives.
- Store, for example storing laboratory equipment in order to avoid damage.
- Cleaning, so that the laboratory equipment is always clean from dirt which could damage, such as dust and moisture that can cause corrosion.
- Maintain, for example, with oiling of mechanical equipment.
- Inspect or check the condition of laboratory equipment to determine the presence of symptoms.
- Realign or tune-up, calibration of equipment for the facility or the equipment under normal conditions or standards.
- Repair minor damage that occurs in the laboratory equipment limit certain damages that may still be repaired, so ready for use in practicum students.
- Replace components of laboratory equipment that is damaged.

3. Research Methods

The design diagram as a guide in the research process, development codes, repair and maintenance tool Industrial Mechatronic System as follows:

Figure 8. Flowchart System Design

IOP Conf. Series: Materials Science and Engineering **619** (2019) 012057 doi:10.1088/1757-899X/619/1/012057

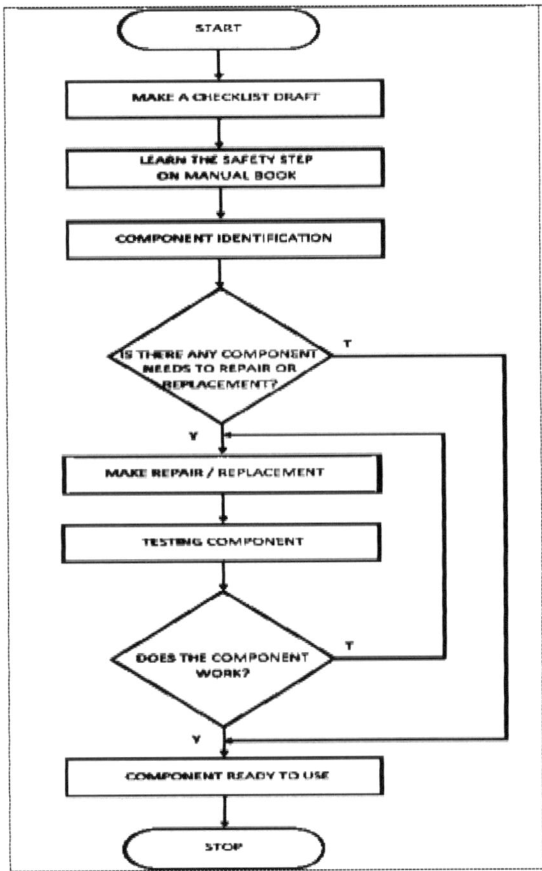

Figure 9. Flowchart Maintenance and Repair

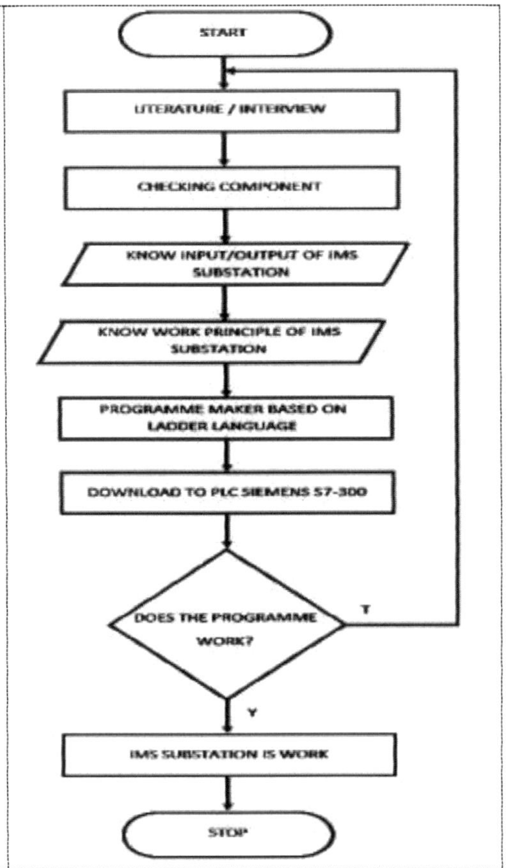

Figure 10. Programming Flowchart

3.1. Literature/Interview

In a research tool, the first step to do is to find as much data as well as information via CD-ROM Lab Soft, various print and electronic, weblogs, videos, as well as a general encyclopaedia.

3.2. Simulation

Trying to simulate sub-station by using the system UniTrain-1 as an introduction and to check the function of each component of IMS sub-station. UniTrain-1 system is a simple application with didactic structured introduction to the control of each sub-system and a preparation for integration and process control of production lines with industrial standard equipment using PLC.

In addition to using the tool to simulate the system UniTrain-1, can also be simulated directly using the factory default program so it can be observed working principle and function of components.

Once all components are functioning properly, then the next simulation is done is communicate between sub-station. Communication used for interconnection of the sub-station Industrial Mechatronic System is a Profibus DP interface. Cable connections to Profibus is the length of 1,5meter and 2x connection plug, Profibus line with two terminals for the Master and Slave unit.

IOP Conf. Series: Materials Science and Engineering **619** (2019) 012057 doi:10.1088/1757-899X/619/1/012057

Profibus DP is commonly used in industry and is a realistic application in automation engineering. The basics of this system is presented in graphic and practice-oriented using automation techniques Uni-Train-1.

	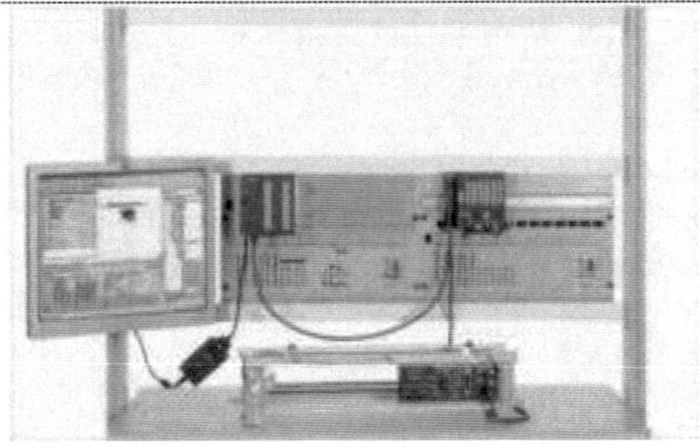
Figure 11. System UniTrain-1	**Figure 12.** System UniTrain-1 with Profibus Communication

3.3. Manufacture Program

After doing a simulation and already know all of the code, byte and Profibus adress, then the tool is set for the program.

4. Result and Discussion

4.1. Result

4.1.1. Maintenance tool. The maintenance carried out by the authors, namely, Cleaning components of Industrial Mechatronic System. The author's purpose cleaning components Industrial Mechatronic System is that the components Industrial Mechatronic System always clean from dirt which could damage, such as dust and moisture that can cause corrosion.

Figure 13. Components that have been cleared

Then, The author's intent installing the parts correctly in order for the tool Industrial Mechatronic System in normal and safe to operate.

The components are installed properly as follows:

- Equitable pedestal subsystem
  Levelling pedestal subsystem is very important. If the pedestal subsystem uneven, will lead the process in the subsystem is not running smoothly. Therefore, the authors smoothing pedestal subsystem so that the process runs smoothly and safely. So as to maintain the smooth operation of the learning.
- Improvements position sensor

The position sensor that can affect whether or not the tool runs smoothly. Therefore, in the final project writer, fix all the incorrect position of the sensor, so that the tool can function properly and there are no unwanted errors.

- Repair vacuum position
  The author's purpose in order to improve the position of vacuum is a vacuum can suck the workpiece exactly so that when the workpiece moves up, the workpiece remains stable.

Making the list this checklist aims to examine or check the existence of early equipment failure or the presence of symptoms. While the manufacture of SOP is intended as a precaution. Because the SOP contains warnings through pictures or writing, regulations, codes of conduct for laboratory users. So, before doing practicum student must pay attention to the SOP and fill in the checklist tool if the tool is ready to use or not.

4.1.2. Repair tool. As for some of the components that have been made by the author of the repair process, namely:

- Repair limit sensor processing station
  Improvements made to the author of the limit sensor processing station that broke caused concussion of the carrier during the process of practicum connect sensors with sensor stand.

Figure 14. Limit Sensor Repaired

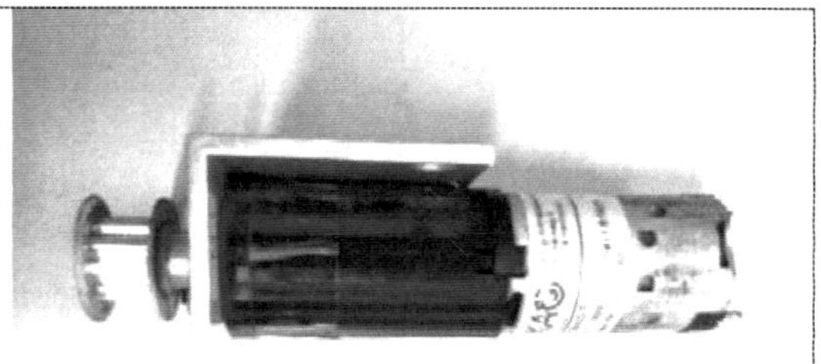

Figure 15. DC motor which has been tensioned between Pulley with Shaft

- Improvements to the Z-axis DC motor of storage station
  Improvements to the DC motor as the Z axis on storage station only on mechanical improvements. Repairs on the author is tighten bolts with shaft pulley. As the lead screw is loose when the practicum, students forcibly pull belt to reach the starting position the Z axis storage station (B1). The cause of the Z axis does not reach its initial position when the storage station to operate sometimes water flow sensor is inactive due to several things like air pressure compressor which does not reach a minimum pressure of 6 bar and no workpiece on the carrier. This causes the Z-axis DC motor continues to rotate until the emergency sensor is active, so the storage station ceased to operate and cannot return to the starting position despite the reset button has been pressed.

- Battery replacement robot arm
  Weak battery voltage causes the robot arm cannot be operated either manually or automatically, the authors replace the battery with a new battery that has the same voltage is 3.6 V.

Figure 16. New Battery had Mounted on a Robot Arm

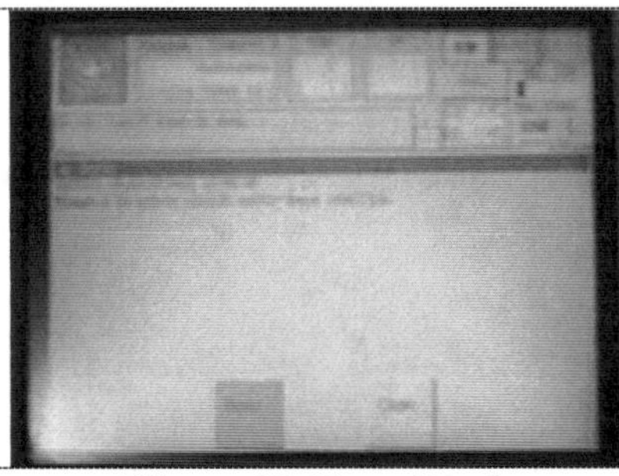

Figure 17. Error Display on Teach Pendant Shows Safety Fence is Open

- Enabling automatic mode of robot arm
  Robots have to operate but still a manual mode, the authors seek mode to activate automatic configuration of robots. The cause of the automatic mode is not active because of safety fence jumper on PLC panel is not connected. After the author connects the safety fence jumper on PLC panel, error safety fence is open on the teach pendant can be reset and the robot has to operate automatically.

Figure 18. Jumper Safety Fence Panel PLC

- Enabling people safety sensor robot arm
  When the robot arm operates automatically, then the movement according to the steps of the program without human control and therefore needed a sensor that serves as a safety of human beings or objects that can interfere with the movement of the robot. The sensor is also not active simultaneously with a robot arm that is inactive due to a weak battery voltage. After the author looking for a configuration to activate the sensor, such as finding the angle right between the receivers, mirrors and transmitter, installation of jumpers safety fence on the panel PLC correctly that only one jumper that connects and activates the OSSD (Output Signal Switching Device) by pressing release on the panel PLC then safety sensors have been active people.

IOP Conf. Series: Materials Science and Engineering **619** (2019) 012057 doi:10.1088/1757-899X/619/1/012057

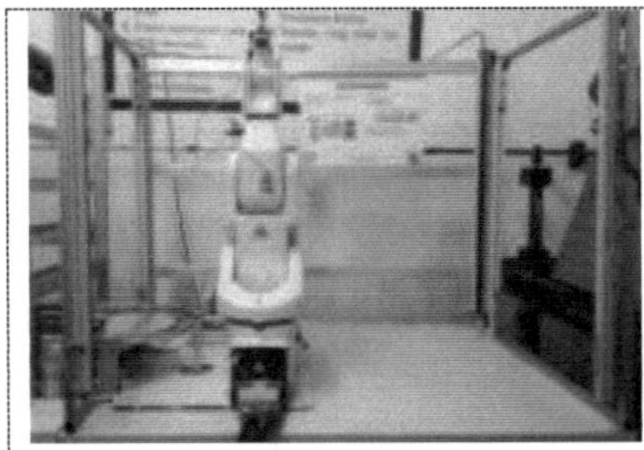

Figure 19. Position Receiver, Mirror and Transmitter

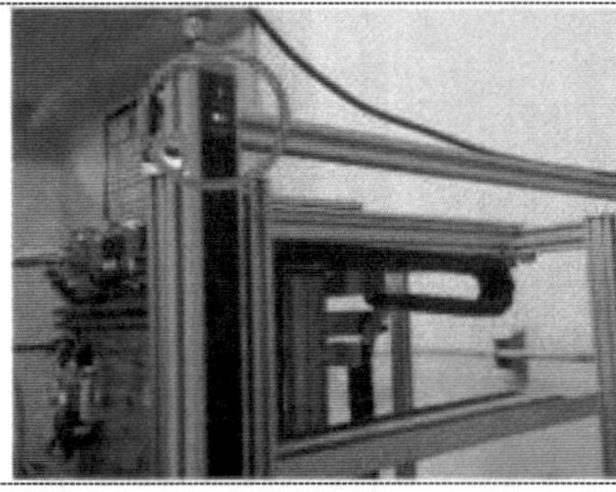

Figure 20. OSSD (Output Signal Switching Device) in the OFF Condition

Figure 21. Release Button and Safety Fence Jumper

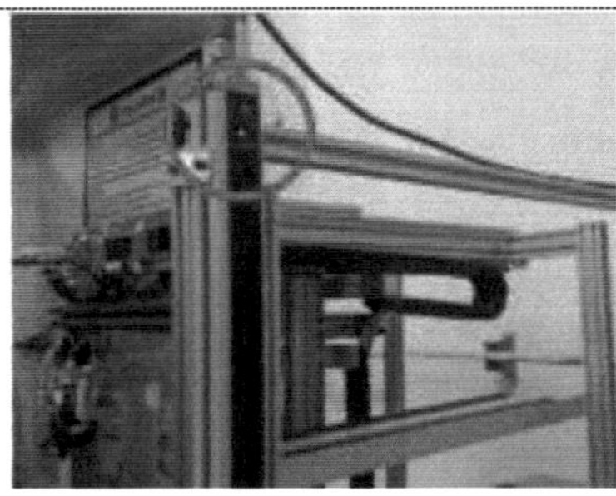

Figure 22. OSSD (Output Signal Switching Device) in the ON condition

4.1.3. Programming development.

- Programming development of station 1 to station 5

  In this final project, the author has developed the programming on Industrial Mechatronic System without removing the factory default program that is to create a program partial and universal. That is the program partial namely, Industrial Mechatronic System has been able to operate on a per-station or in whole. The byte address and Profibus address to create manual and automatic program are:

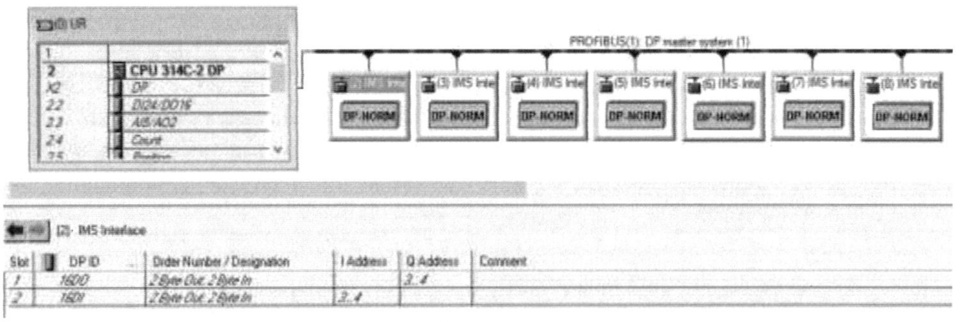

Figure 23. Byte Address and PROFIBUS Address

IOP Conf. Series: Materials Science and Engineering **619** (2019) 012057 doi:10.1088/1757-899X/619/1/012057

Below is a table Byte and PROFIBUS address.

Table 3. Byte and PROFIBUS Address

No	*Sub-Station*	*Profibus Address*	*Byte Address*	Nilai X
1	*Selection Station*	2	3...4	3
2	*Assembly Station*	6	11...12	11
3	*Processing Station*	4	7...8	7
4	*Testing Station*	3	5...6	5
5	*Storage Station*	8	15...16	15
6	> *Storage Station*	5	9...10	9
7	< *Selection Station*	7	13...14	13

- Home position robot arm (manipulator)
  Due to the long arm robot is not operated, this led to the starting position the robot arm is not in the home position in the teach pendant and there is no program to return to the home position. Meanwhile, in order to operate properly robot arm should be in a state home position. Therefore, in the final project, the authors conducted a programming development program was the production of home position. Writer inputting programs into the home position teach pendant, this so young students can restore the robot arm into the home position.
- Editing program arm robot (manipulator)
  The author did a little modification to the default program robot arm, the program at the time of the robot arm will be put one by one part of the workpiece.

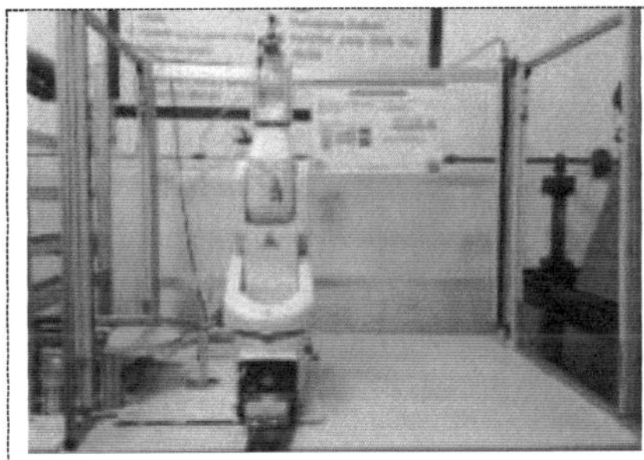

Figure 24. Home Position Robot

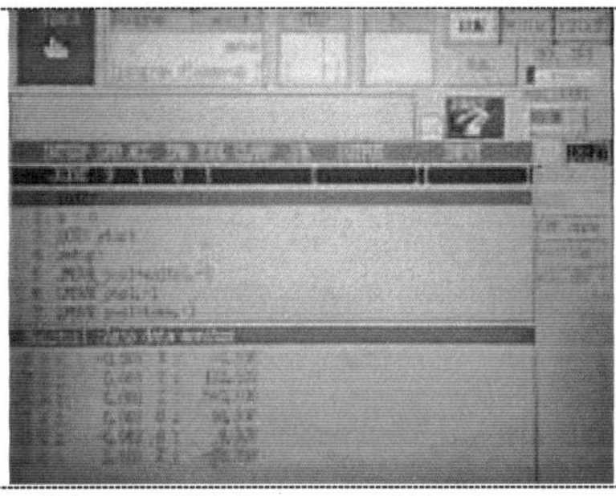

Figure 25. Program Display the Home Position Robot Teach Pendant

4.1.4. Making the learning module and job sheet. Job sheet learning modules and is designed to help students achieve the purpose for doing practical work. Job sheet learning module and this is a learning tool that provides materials, methods, SOPs and check lists listing tool.

4.2. Discussion

In initial tests of the tool, the authors do not use the system simulation UniTrain-1 because there are some CDs to install Lab soft has been damaged, so the authors studied the working principle and configuration tools Industrial Mechatronic System of built-in program.

PLC is part of the Industrial Mechatronic System that serves as the controller, while PROFIBUS is used to communicate between stations, thereby reducing the use of wires and use the input output (I / O) PLC.

Use of PROFIBUS on Industrial Mechatronic System that communicates between the PLC with other interfaces, using the type of PROFIBUS-DP (Distributed Input / Output). Where the PROFIBUS types, matched with the PLC used in the final project the author.

Since all the stations on Industrial Mechatronic System has the same device, thereby using PROFIBUS addressing to differentiate the devices from one another.

In the Industrial Mechatronic System also has the disassembly station that serves to separate parts of the workpiece. The process to move a workpiece from the storage station to disassembly station using the help of a robot arm.

The robot arm using a separate controller that is E-Controller E76 connected to the connector on the panel PLC robot Industrial Mechatronic System uses DB-9 cable to communicate with all the station robotic arm on Industrial Mechatronic System.

Robot arm people equipped with safety sensors to prevent accidents caused by human error or other factors. People Safety sensor works automatically, a system that has been programmed with a coherent and in some cases are not in the program cannot be avoided so that people safety sensor serves as a feedback tool when something happens that is not in accordance with the procedure.

5. Conclusion

Form the research conducted, it can be concluded that:

1. Has developed a system of partial control on Industrial Mechatronic System and robot arm which can simulate the working system in the industry with some components of the pneumatic, DC motors and sensors.
2. Has developed a system of automatic control on the Industrial Mechatronic System and robot arm which can simulate the working system in the industry with some components of the pneumatic, DC motors and sensors.
3. Job sheet been prepared containing the overall experimental procedure stations, operational standards implementation, and check the list of learning modules PLC to support the process of student practicum.

References

[1] Lucas Nuelle. 2018. *Individual Components for Mechatronics.* Available at. https://www.lucas-nuelle.com
[2] Lilik, Gunarta. 2011. *Get to know sensors and actuators* (Original in Bahasa: *Mengenal Sensor dan Aktuator*) available at: http://skp.unair.ac.id/repository/Guru-Indonesia/MengenalSensordan_lilikgunarta_12437.pdf.
[3] P. Handayani, T.Y. Putro. 2008. Maintenance of Motors and Electric Generators, Electronic System Maintenance and Repair Techniques (Original in Bahasa: Pemeliharaan Motor dan Generator Listrik, Teknik Pemeliharaan dan Perbaikan Sistem Elektronika) Jakarta: Directorate Tutorial of Vocational High School.
[4] A. Setiawan, 2004. Plane Pneumatic Robot Arm Design Using Programmable Logic Controller (Original in Bahasa: Rancangan Lengan Robot Pneumatik Pemindah Plat Menggunakan Programmable Logic Controller). Diponegoro University, Indonesia.
[5] Kawasaki. 2018. *Product E-Controller.* Available at http://www.robotics.kawasaki.com.
[6] Budiyanto. 2017. Pneumatic System Control Using Siemens S7 PC-300 PLC (Original in Bahasa: Pengendalian Sistem Pneumatik Menggunakan *PLC Siemens S7 PC-300).* Jakarta. Tutorial of Vocational High School.
[7] Lewi. 2015. Learning Materials Characteristics of PLC(Original in Bahasa: *Bahan Ajar Karakteristik PLC).* Makassar: State Polytechnics Ujung Pandang, Makassar, Indonesia.

IOP Conf. Series: Materials Science and Engineering **619** (2019) 012057 doi:10.1088/1757-899X/619/1/012057

[8] Centre, Tanoto Information. 2010. *Programmable Logic Controller.* Available at http://tanotocentre.wordpress.com, diakses 10 Januari 2018.
[9] Pati, Teguh. 2013. What is Profibus. (Original in Bahasa: Apa itu Profibus). Available at http://teguhpati.blogspot.co.id.
[10] Anwar, Chairil. 2016. Profibus paper (Original in Bahasa: Makalah Profibus). University of Elektro Katolik Widya Mandala Surabaya, Indonesia.
[11] R. Syam, 2015, Fuzzy Logic Control for Pneumatic Excavator Model, International Journal of Applied Engineering Research ISSN 0973-4562 Volume 10, Number 9 pp. 21647-21657

IOP Conf. Series: Materials Science and Engineering **619** (2019) 012058 doi:10.1088/1757-899X/619/1/012058

Numerical solutions to BVPs governed by the anisotropic modified Helmholtz equation for trigonometrically graded media

N. La Nafie[1], M. I. Azis[2,*], Fahruddin[3]

[1]Department of Chemistry, Hasanuddin University, Makassar, Indonesia
[2]Department of Mathematics, Hasanuddin University, Makassar, Indonesia
[3]Department of Biology, Hasanuddin University, Makassar, Indonesia

E-mail: mohivanazis@yahoo.co.id (*Corresponding author)

Abstract. A Boundary Element Method (BEM) is used for obtaining solutions to anisotropic trigonometrically graded media (FGM) boundary value problems (BVPs) governed by the modified Helmholtz type equation. A technique of transforming the variable coefficient governing equation to a constant coefficient equation is utilized for deriving a boundary integral equation. Some particular problems are considered to illustrate the application of the BEM. The results show the convergence, consistency, and accuracy of the BEM solutions.

1. Introduction

The BEM has been successfully used for solving many types of problems of homogeneous media. Some works on homogeneous media problems have been recently done by Azis et. al [1, 2] and Haddade et. al [3] in which the authors considered pollutant transport problems governed by 2D diffusion-convection, and also Azis [4] in which numerical solutions for the Helmholtz boundary value problems were obtained.

However, this is generally not the case for FGM. There are two techniques usually used to deal with problems of FGM. The first one uses a technique of deriving a relevant Green function or fundamental solution to the FGM problem. Cheng [5] had applied this technique. The second technique is by transforming the variable coefficient governing equation to a constant coefficient equation. Some progress in using the second technique has been made. For examples, Clements and Azis [6] considered the case for isotropic FGM. For anisotropic FGM some works have been studied by Azis and Clements [7, 8, 9] and Azis et. al [10]. In addition to this, recently Salam et. al [11, 12] and Azis [13] have been working respectively on a class of elliptic problems and diffusion convection reaction equation for anisotropic FGM.

This paper discusses derivation of a boundary integral equation for numerically solving 2D boundary value problems governed by the dimensionless modified Helmholtz type equation of the form

$$\frac{\partial}{\partial x_i}\left[\kappa_{ij}\left(x_1,x_2\right)\frac{\partial \nu\left(x_1,x_2\right)}{\partial x_j}\right]-\beta^2\left(x_1,x_2\right)\nu\left(x_1,x_2\right)=0 \qquad (1)$$

where the coefficients κ_{ij} and β^2 depend on x_1 and x_2 and the repeated summation convention (summing from 1 to 2) is employed.

IOP Conf. Series: Materials Science and Engineering **619** (2019) 012058 doi:10.1088/1757-899X/619/1/012058

For (1) to be a second order elliptic partial differential equation, we assume that the matrix of coefficients $[\kappa_{ij}]$ is a real symmetric positive definite matrix. Explicitly equation (1) may be written as

$$\frac{\partial}{\partial x_1}\left(\kappa_{11}\frac{\partial \nu}{\partial x_1}\right)+2\frac{\partial}{\partial x_1}\left(\kappa_{12}\frac{\partial \nu}{\partial x_2}\right)+\frac{\partial}{\partial x_2}\left(\kappa_{22}\frac{\partial \nu}{\partial x_2}\right)-\beta^2\nu=0$$

Further, the coefficients κ_{ij} and β are required to be twice differentiable functions of the two independent variables x_1 and x_2. The analysis is also relevant for isotropic media as a special case of anisotropic media which occurs when $\kappa_{11}=\kappa_{22}$ and $\kappa_{12}=0$.

A variety of problems of both isotropic and anisotropic inhomogeneous media are usually modelled with equation (1). Infiltration problems (when $\beta^2>0$), and antiplane strain in elastostatics and plane thermostatic problems (when $\beta^2=0$) are the areas for which the governing equation is of the type (1). Recently some works on the modified Helmholtz equation for isotropic and/or homogeneous media have been done. For example, Clements and Lobo in [14] and Solekhudin and Ang in [15] considered the modified Helmholtz equation as the governing equation for isotropic and homogeneous media which is a special case of the equation (1).

2. The boundary value problem

Referred to a Cartesian frame Ox_1x_2 a solution to (1) is sought which is valid in a region Ω in R^2 with boundary $\partial\Omega$ which consists of a finite number of piecewise smooth closed curves. On $\partial\Omega_1$ the dependent variable $\nu(\mathbf{x})$ $(\mathbf{x}=(x_1,x_2))$ is specified and on $\partial\Omega_2$

$$P(\mathbf{x})=\kappa_{ij}\left(\partial\nu/\partial x_j\right)n_i \tag{2}$$

is specified where $\partial\Omega=\partial\Omega_1\cup\partial\Omega_2$ and $\mathbf{n}=(n_1,n_2)$ denotes the outward pointing normal to $\partial\Omega$.

3. The boundary integral equation

The boundary integral equation is derived by transforming the variable coefficient equation (1) to a constant coefficient equation. The coefficients κ_{ij} and β are required to take the form

$$\kappa_{ij}(\mathbf{x}) = \overline{\kappa}_{ij}g(\mathbf{x}) \tag{3}$$

$$\beta^2(\mathbf{x}) = \overline{\beta}^2 g(\mathbf{x}) \tag{4}$$

where the $\overline{\kappa}_{ij}$ and $\overline{\beta}$ are constants and g is an exponential function of $\mathbf{x}$. Use of (3) and (4) and in (1) yields

$$\overline{\kappa}_{ij}\frac{\partial}{\partial x_i}\left(g\frac{\partial \nu}{\partial x_j}\right)-\overline{\beta}^2 g\nu=0 \tag{5}$$

Let

$$\nu(\mathbf{x})=g^{-1/2}(\mathbf{x})\,\psi(\mathbf{x}) \tag{6}$$

so that (5) may be written in the form

$$\overline{\kappa}_{ij}\frac{\partial}{\partial x_i}\left[g\frac{\partial\left(g^{-1/2}\psi\right)}{\partial x_j}\right]-\overline{\beta}^2 g^{1/2}\psi=0$$

That is

$$\overline{\kappa}_{ij}\left[\left(\frac{1}{4}g^{-3/2}\frac{\partial g}{\partial x_i}\frac{\partial g}{\partial x_j}-\frac{1}{2}g^{-1/2}\frac{\partial^2 g}{\partial x_i\partial x_j}\right)\psi+g^{1/2}\frac{\partial^2\psi}{\partial x_i\partial x_j}\right]-\overline{\beta}^2 g^{1/2}\psi=0 \tag{7}$$

IOP Conf. Series: Materials Science and Engineering **619** (2019) 012058 doi:10.1088/1757-899X/619/1/012058

Use of the identity

$$\frac{\partial^2 g^{1/2}}{\partial x_i \partial x_j} = -\frac{1}{4} g^{-3/2} \frac{\partial g}{\partial x_i} \frac{\partial g}{\partial x_j} + \frac{1}{2} g^{-1/2} \frac{\partial^2 g}{\partial x_i \partial x_j}$$

permits (7) to be written in the form

$$g^{1/2} \overline{\kappa}_{ij} \frac{\partial^2 \psi}{\partial x_i \partial x_j} - \psi \overline{\kappa}_{ij} \frac{\partial^2 g^{1/2}}{\partial x_i \partial x_j} - \overline{\beta}^2 g^{1/2} \psi = 0 \tag{8}$$

If we restrict the function $g(\mathbf{x})$ to take the trigonometrical form

$$g(\mathbf{x}) = \{A [\cos(\alpha_m x_m) + \sin(\alpha_m x_m)]\}^2 \quad \overline{\kappa}_{ij} \alpha_i \alpha_j = -\overline{\beta}^2 \tag{9}$$

where α_m are constant, then

$$\overline{\kappa}_{ij} \frac{\partial^2 g^{1/2}}{\partial x_i \partial x_j} + \overline{\beta}^2 g^{1/2} = 0 \tag{10}$$

Subtitution (10) into (8) gives

$$\overline{\kappa}_{ij} \frac{\partial^2 \psi}{\partial x_i \partial x_j} = 0 \tag{11}$$

Also, substitution of (3) and (6) into (2) gives

$$P = -P_g \psi + P_\psi g^{1/2} \tag{12}$$

where

$$P_g(\mathbf{x}) = \overline{\kappa}_{ij} \frac{\partial g^{1/2}}{\partial x_j} n_i \qquad P_\psi(\mathbf{x}) = \overline{\kappa}_{ij} \frac{\partial \psi}{\partial x_j} n_i$$

A boundary integral equation for the solution of (11) is given in the form

$$\sigma(\mathbf{x}_0) \psi(\mathbf{x}_0) = \int_{\partial\Omega} [\Theta(\mathbf{x}, \mathbf{x}_0) \psi(\mathbf{x}) - \Psi(\mathbf{x}, \mathbf{x}_0) P_\psi(\mathbf{x})] \, ds(\mathbf{x}) \tag{13}$$

where $\mathbf{x}_0 = (a, b)$, $\sigma = 0$ if $(a, b) \notin \Omega \cup \partial\Omega$, $\sigma = 1$ if $(a, b) \in \Omega$, $\sigma = \frac{1}{2}$ if $(a, b) \in \partial\Omega$ and $\partial\Omega$ has a continuously turning tangent at (a, b).

The so called fundamental solution Ψ in (13) is any solution of the equation

$$\overline{\kappa}_{ij} \frac{\partial^2 \Psi}{\partial x_i \partial x_j} = \delta(\mathbf{x} - \mathbf{x}_0)$$

and the Θ is given by

$$\Theta(\mathbf{x}, \mathbf{x}_0) = \overline{\kappa}_{ij} \frac{\partial \Psi(\mathbf{x}, \mathbf{x}_0)}{\partial x_j} n_i$$

where δ is the Dirac delta function. Following Azis [16], for two-dimensional problems Ψ and Θ are given by

$$\begin{aligned} \Psi(\mathbf{x}, \mathbf{x}_0) &= \tfrac{K}{2\pi} \ln R \\ \Theta(\mathbf{x}, \mathbf{x}_0) &= \tfrac{K}{2\pi} \tfrac{1}{R} \overline{\kappa}_{ij} \tfrac{\partial R}{\partial x_j} n_i \end{aligned} \tag{14}$$

IOP Conf. Series: Materials Science and Engineering **619** (2019) 012058 doi:10.1088/1757-899X/619/1/012058

where

$$
\begin{aligned}
K &= \ddot{\tau}/\zeta \\
\zeta &= \left[\overline{\kappa}_{11} + 2\overline{\kappa}_{12}\dot{\tau} + \overline{\kappa}_{22}\left(\dot{\tau}^2 + \ddot{\tau}^2\right)\right]/2 \\
R &= \sqrt{(\dot{x}_1 - \dot{a})^2 + (\dot{x}_2 - \dot{b})^2} \\
\dot{x}_1 &= x_1 + \dot{\tau}x_2 \\
\dot{a} &= a + \dot{\tau}b \\
\dot{x}_2 &= \ddot{\tau}x_2 \\
\dot{b} &= \ddot{\tau}b
\end{aligned}
$$

where $\dot{\tau}$ and $\ddot{\tau}$ are respectively the real and the positive imaginary parts of the complex root τ of the quadratic

$$\overline{\kappa}_{11} + 2\overline{\kappa}_{12}\tau + \overline{\kappa}_{22}\tau^2 = 0$$

The derivatives $\partial R/\partial x_j$ needed for the calculation of the Θ in (14) are given by

$$
\begin{aligned}
\frac{\partial R}{\partial x_1} &= \frac{1}{R}(\dot{x}_1 - \dot{a}) \\
\frac{\partial R}{\partial x_2} &= \dot{\tau}\left[\frac{1}{R}(\dot{x}_1 - \dot{a})\right] + \ddot{\tau}\left[\frac{1}{R}\left(\dot{x}_2 - \dot{b}\right)\right]
\end{aligned}
$$

Use of (6) and (12) in (13) yields

$$
\begin{aligned}
\sigma(\mathbf{x}_0)\, g^{1/2}(\mathbf{x}_0)\, \nu(\mathbf{x}_0) = &\int_{\partial\Omega} \left\{ \left[g^{1/2}(\mathbf{x})\, \Theta(\mathbf{x}, \mathbf{x}_0) - P_g(\mathbf{x})\, \Psi(\mathbf{x}, \mathbf{x}_0) \right] \nu(\mathbf{x}) \right. \\
&\left. - \left[g^{-1/2}(\mathbf{x})\, \Psi(\mathbf{x}, \mathbf{x}_0) \right] P(\mathbf{x}) \right\} ds(\mathbf{x}) \qquad (15)
\end{aligned}
$$

This equation provides a boundary integral equation for determining ν and P at all points of Ω.

4. Numerical examples

Some particular boundary value problems will be solved numerically by employing the integral equation (15). The main aim is to show the validity of the analysis used above for deriving the boundary integral equation (15) and the appropriateness of the BEM in solving the problems through the derived boundary integral equation (15). Standard boundary element method is employed to obtain numerical results. The integrals in equation (15) are evaluated numerically using the Bode's quadrature (see Abramowitz and Stegun [17]).

4.1. Problem 1

We will consider a problem with analytical solution. The aim is to verify the convergence and accuracy of the BEM. We choose an anisotropic medium with

$$\overline{\kappa}_{11} = 1, \overline{\kappa}_{12} = 0.5, \overline{\kappa}_{22} = 1$$

and we take an inhomogeneity function $g(\mathbf{x})$ satisfying (9) of the form

$$g(\mathbf{x}) = \left\{2\left[\cos(0.5x_1 + 0.75x_2) + \sin(0.5x_1 + 0.75x_2)\right]\right\}^2$$

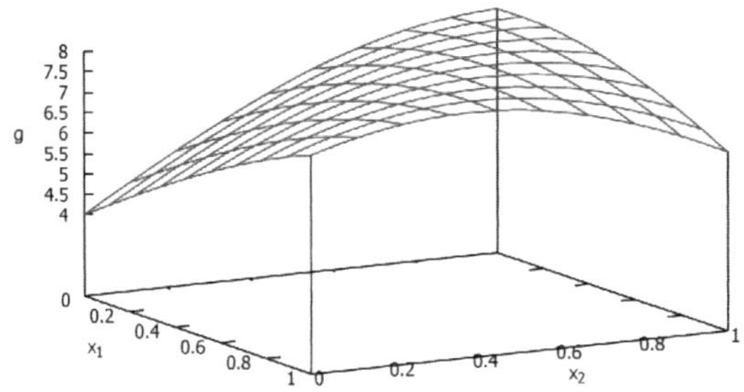

Figure 1. An trigonometrical inhomogeneity function $g(\mathbf{x}) = \{2\left[\cos\left(0.5x_1 + 0.75x_2\right) + \sin\left(0.5x_1 + 0.75x_2\right)\right]\}^2$

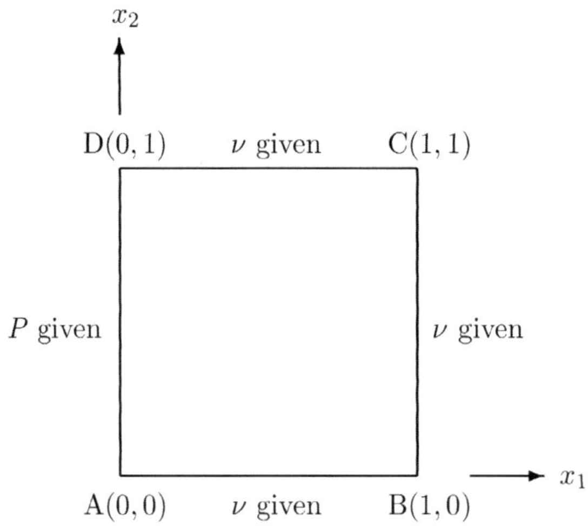

Figure 2. The geometry of Problem 1

Figure 1 shows the plot of $g(\mathbf{x})$. Also, for $g(\mathbf{x})$ to satisfy equation (9) we take $\overline{\beta}^2 = 1.1875$. The analytical solution is

$$\nu(\mathbf{x}) = \frac{3.5\left(1 + 1.5x_1 + 2.5x_2\right)}{2\left[\cos\left(0.5x_1 + 0.75x_2\right) + \sin\left(0.5x_1 + 0.75x_2\right)\right]}$$

Figure 2 shows the domain Ω and the boundary conditions.
As shown in Table 1 the BEM solution converges to the analytical solution as the number of segments increases. The accuracy of the BEM solutions occurs at the fourth figure of the decimal.

4.2. Problem 2
Now, a problem for a homogeneous isotropic medium with $g(\mathbf{x}) = 9$, $\overline{\kappa}_{11} = \overline{\kappa}_{22} = 1, \overline{\kappa}_{12} = 0$ and $\overline{\beta} = 0$ is under consideration. Figure 3 shows the geometry Ω and the boundary conditions.

Table 1. BEM and analytical solutions for Problem 1

(x_1, x_2)	ν	$\partial\nu/\partial x_1$	$\partial\nu/\partial x_2$	ν	$\partial\nu/\partial x_1$	$\partial\nu/\partial x_2$
	BEM 160 segments			BEM 320 segments		
(.1,.5)	3.1749	1.3824	2.4149	3.1743	1.3841	2.4118
(.3,.5)	3.4583	1.4590	2.5161	3.4580	1.4596	2.5133
(.5,.5)	3.7612	1.5765	2.6823	3.7609	1.5762	2.6799
(.7,.5)	4.0919	1.7381	2.9176	4.0914	1.7370	2.9155
(.9,.5)	4.4598	1.9503	3.2311	4.4591	1.9487	3.2293
(.5,.1)	2.7605	1.3882	2.4356	2.7615	1.3876	2.4302
(.5,.3)	3.2484	1.4294	2.4735	3.2486	1.4289	2.4704
(.5,.7)	4.3327	1.8336	3.0640	4.3320	1.8336	3.0620
(.5,.9)	5.0006	2.2256	3.6539	4.9995	2.2255	3.6527
	BEM 640 segments			Analytical		
(.1,.5)	3.1740	1.3848	2.4104	3.1737	1.3856	2.4089
(.3,.5)	3.4578	1.4599	2.5119	3.4576	1.4603	2.5106
(.5,.5)	3.7607	1.5762	2.6787	3.7606	1.5761	2.6775
(.7,.5)	4.0912	1.7364	2.9146	4.0910	1.7359	2.9137
(.9,.5)	4.4587	1.9472	3.2289	4.4583	1.9459	3.2285
(.5,.1)	2.7620	1.3872	2.4279	2.7625	1.3869	2.4256
(.5,.3)	3.2487	1.4285	2.4688	3.2488	1.4282	2.4673
(.5,.7)	4.3316	1.8337	3.0610	4.3313	1.8337	3.0600
(.5,.9)	4.9989	2.2255	3.6517	4.9984	2.2256	3.6507

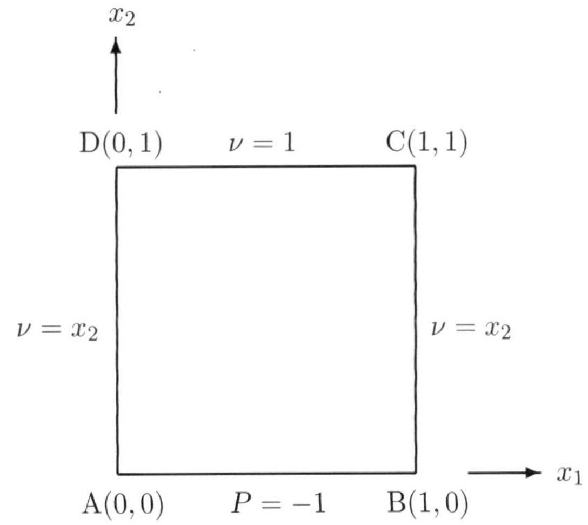

Figure 3. The geometry of Problem 2 and Problem 3

Table 2 shows the results of the BEM solution using 80, 160, 320 and 640 segments of equal length. As expected, the results converge as the number of segments increases and also they are symmetrical about the axes $x_1 = 0.5$ as the boundary conditions are also symmetrical about the axes $x_1 = 0.5$.

Table 2. BEM solution for Problem 2

(x_1, x_2)	ν	$\partial\nu/\partial x_1$	$\partial\nu/\partial x_2$	ν	$\partial\nu/\partial x_1$	$\partial\nu/\partial x_2$
	BEM 80 segments			BEM 160 segments		
(.1,.5)	.5221	.2153	.9230	.5222	.2155	.9222
(.3,.5)	.5576	.1288	.8031	.5578	.1288	.8020
(.5,.5)	.5708	-.0000	.7607	.5710	.0000	.7595
(.7,.5)	.5576	-.1288	.8031	.5578	-.1288	.8020
(.9,.5)	.5221	-.2154	.9230	.5222	-.2155	.9222
(.5,.1)	.3472	-.0000	.2866	.3479	.0000	.2855
(.5,.3)	.4353	-.0000	.5773	.4357	.0000	.5761
(.5,.7)	.7339	.0000	.8588	.7339	.0000	.8575
(.5,.9)	.9106	.0000	.9005	.9103	-.0000	.8990
	BEM 320 segments			BEM 640 segments		
(.1,.5)	.5223	.2156	.9218	.5224	.2157	.9217
(.3,.5)	.5579	.1289	.8014	.5579	.1289	.8011
(.5,.5)	.5711	-.0000	.7589	.5711	-.0000	.7586
(.7,.5)	.5579	-.1289	.8014	.5579	-.1289	.8011
(.9,.5)	.5223	-.2156	.9218	.5224	-.2157	.9217
(.5,.1)	.3482	-.0000	.2851	.3484	.0000	.2849
(.5,.3)	.4360	-.0000	.5754	.4361	.0000	.5751
(.5,.7)	.7338	-.0000	.8568	.7338	-.0000	.8565
(.5,.9)	.9101	-.0000	.8983	.9100	-.0000	.8979

4.3. Problem 3

For this problem, the geometry of the medium Ω and the boundary conditions are again as depicted in Figure 3. And the medium is assumed to be a trigonometrically graded isotropic material with $g(\mathbf{x}) = \{2\left[\cos(0.5x_1 + 0.75x_2) + \sin(0.5x_1 + 0.75x_2)\right]\}^2$, $\overline{\kappa}_{11} = \overline{\kappa}_{22} = 1, \overline{\kappa}_{12} = 0$ and $\overline{\beta}^2 = 0.8125$. As expected, the results in Table 3 shows convergence of the BEM solution and they are not symmetrical anymore as the material is not homogeneous.

5. Conclusion

The boundary integral equation (15) is derived from this governing equation (1) and then from (15) a BEM is constructed for calculation of numerical solutions to the problems for anisotropic trigonometrically graded media. The results in Section 4 show that the BEM solution gives a convergence, consistency, and accuracy. Therefore the results also prove that the analysis used for deriving the boundary integral equation (15) is valid. Together with its ease in implementation, it may be concluded that BEM is a useful numerical method for solving such kind of problems.

Acknowledgements

The authors would like to thank The Hasanuddin University and The Ministry of Higher Education of Indonesia for the support provided.

Table 3. BEM solution for Problem 3

(x_1, x_2)	ν	$\partial\nu/\partial x_1$	$\partial\nu/\partial x_2$	ν	$\partial\nu/\partial x_1$	$\partial\nu/\partial x_2$
	BEM 80 segments			BEM 160 segments		
(.1,.5)	.5176	.1605	.8972	.5178	.1602	.8965
(.3,.5)	.5406	.0670	.7685	.5408	.0668	.7674
(.5,.5)	.5443	-.0285	.7314	.5444	-.0286	.7304
(.7,.5)	.5309	-.0991	.7755	.5310	-.0991	.7745
(.9,.5)	.5089	-.1053	.9042	.5090	-.1059	.9035
(.5,.1)	.3233	-.0304	.3227	.3238	-.0305	.3218
(.5,.3)	.4136	-.0322	.5657	.4139	-.0323	.5646
(.5,.7)	.7042	-.0241	.8675	.7041	-.0241	.8663
(.5,.9)	.8930	-.0135	1.0292	.8926	-.0132	1.0276
	BEM 320 segments			BEM 640 segments		
(.1,.5)	.5179	.1601	.8960	.5179	.1600	.8958
(.3,.5)	.5408	.0667	.7669	.5408	.0667	.7667
(.5,.5)	.5444	-.0286	.7299	.5445	-.0286	.7296
(.7,.5)	.5310	-.0991	.7741	.5310	-.0991	.7738
(.9,.5)	.5090	-.1060	.9032	.5090	-.1061	.9030
(.5,.1)	.3240	-.0305	.3214	.3241	-.0305	.3213
(.5,.3)	.4141	-.0323	.5641	.4141	-.0323	.5639
(.5,.7)	.7040	-.0241	.8657	.7040	-.0241	.8654
(.5,.9)	.8924	-.0131	1.0270	.8923	-.0131	1.0266

References

[1] Azis M I, Kasbawati, Haddade A and Thamrin S A 2018 On some examples of pollutant transport problems solved numerically using the boundary element method *Journal of Physics: Conference Series* **979**(1)

[2] Azis M I, Asrul L, Khaeruddin and Paharuddin 2018 BEM solutions for unsteady transport problems in anisotropic media *JP Journal of Heat and Mass Transfer* **15**(4) 915

[3] Haddade A, Salam N, Khaeruddin and Azis M I 2017 A boundary element method for 2D diffusion-convection problems in anisotropic media *Far East Journal of Mathematical Sciences* **102**(8) 1593

[4] Azis M I 2019 Numerical solutions for the Helmholtz boundary value problems of anisotropic media *Journal of Computational Physics* **381** 42

[5] Cheng A H-D 1984 Darcy's Flow with Variable Permeability: A Boundary Integral Solution *Water Resources Research* **20** 980

[6] Clements D L and Azis M I 2000 A Note on a Boundary Element Method for the Numerical Solution of Boundary Value Problems in Isotropic Inhomogeneous Elasticity *Journal of the Chinese Institute of Engineers* **23**(3) 261

[7] Azis M I and Clements D L 2014 On some problems concerning deformations of functionally graded anisotropic elastic materials *Far East Journal of Mathematical Sciences* **87**(2) 173

[8] Azis M I and Clements D L 2014 A Boundary Element Method for Transient Heat Conduction Problem of Nonhomogeneous Anisotropic Materials *Far East Journal of Mathematical Sciences*, **89**(1) 51

[9] Azis M I and Clements D L 2008 Nonlinear transient heat conduction problems for a class of inhomogeneous anisotropic materials by BEM *Engineering Analysis with Boundary Elements* **32**(12) 1054

[10] Azis M I, Toaha S, Bahri M and Ilyas N 2018 A boundary element method with analytical integration for deformation of inhomogeneous elastic materials *Journal of Physics: Conference Series* **979**(1) 012072

[11] Salam N, Haddade A, Clements D L and Azis M I 2017 A boundary element method for a class of elliptic boundary value problems of functionally graded media *Engineering Analysis with Boundary Elements* **84**(3) 186

[12] Azis M I 2019 Numerical solutions to a class of scalar elliptic BVPs for anisotropic exponentially graded media *Journal of Physics: Conference Series* **1218** 012001

[13] Azis M I 2019 Standard-BEM solutions to two types of anisotropic-diffusion convection reaction equations with variable coefficients *Engineering Analysis with Boundary Elements* **105** 87

[14] Clements D L and Lobo M 2010 A BEM for time dependent infiltration from an irrigation channel *Engineering*

Analysis with Boundary Elements **34** 1100

[15] Solekhudin I and Ang K-C 2012 A DRBEM with a predictor-corrector scheme for steady infiltration from periodic channels with root-water uptake *Engineering Analysis with Boundary Elements* **36** 1199

[16] Azis M I 2017 Fundamental solutions to two types of 2D boundary value problems of anisotropic materials *Far East Journal of Mathematical Sciences* **101**(11) 2405

[17] Abramowitz M and Stegun I A 1972 *Handbook of mathematical functions: with formulas, graphs and mathematical tables*, Dover Publications, Washington

BEM solutions to a class of elliptic BVPs for anisotropic trigonometrically graded media

St. N. Jabir[1], M. I. Azis[2,*], Z. Djafar[3], B. Nurwahyu[2]
[1]Department of Electrical Engineering, ATI Polytechnic, Makassar, Indonesia
[2]Department of Mathematics, Hasanuddin University, Makassar, Indonesia
[3]Department of Mechanical Engineering, Hasanuddin University, Makassar, Indonesia

E-mail: mohivanazis@yahoo.co.id (*Corresponding author)

Abstract. Boundary value problems (BVPs) governed by a class of elliptic equations for anisotropic trigonometrically graded media are solved using Boundary Element Method (BEM). The variable coefficient governing equation is transformed to a constant coefficient equation which is then transformed to a boundary integral equation. The results show the convergence, consistency, and accuracy of the BEM solutions.

1. Introduction

Several types of constant coefficient equations have been solved using BEM (see for examples [1, 2, 3, 4]. But in general this is not the case for variable coefficient equation. There is some progress in using BEM to solve several types of variable coefficient governing equations (see for examples [5, 6, 7, 8, 9, 10, 11, 12, 13])

The governing equation considered by Salam et. al in [11] takes the form

$$\frac{\partial}{\partial x_i}\left[\lambda_{ij}\left(x_1,x_2\right)\frac{\partial\varsigma\left(x_1,x_2\right)}{\partial x_j}\right]=0 \tag{1}$$

This paper is intended to extend the work by Salam et. al [11] for problems with governing equation (1) to for 2D boundary value problems governed by another type of (dimensionless) elliptic equation of the form

$$\frac{\partial}{\partial x_i}\left[\lambda_{ij}\left(x_1,x_2\right)\frac{\partial\varsigma\left(x_1,x_2\right)}{\partial x_j}\right]+\beta\left(x_1,x_2\right)\varsigma\left(x_1,x_2\right)=0 \tag{2}$$

where the coefficients λ_{ij} depend on x_1 and x_2 and the repeated summation convention (summing from 1 to 2) is employed.

The matrix of coefficients $[\lambda_{ij}]$ is a real symmetric positive definite matrix so that equation (2) is a second order elliptic partial differential equation and may be written explicitly as

$$\frac{\partial}{\partial x_1}\left(\lambda_{11}\frac{\partial\varsigma}{\partial x_1}\right)+2\frac{\partial}{\partial x_1}\left(\lambda_{12}\frac{\partial\varsigma}{\partial x_2}\right)+\frac{\partial}{\partial x_2}\left(\lambda_{22}\frac{\partial\varsigma}{\partial x_2}\right)+\beta\varsigma=0$$

Further, the coefficients λ_{ij} and β are required to be twice differentiable functions of the two independent variables x_1 and x_2. The analysis here is specially relevant to an anisotropic medium but it equally applies to isotropic media. For isotropy, the coefficients in (2) take the form $\lambda_{11} = \lambda_{22}$ and $\lambda_{12} = 0$ and use of these equations in the following analysis immediately yields the corresponding results for an isotropic medium.

Steady infiltration problems (when $\beta < 0$, see for examples [14, 15]), acoustic problems (when $\beta > 0$, see for examples [16, 17]), and antiplane strain in elastostatics and plane thermostatic problems (when $\beta = 0$) are the areas for which the governing equation is of the type (2).

The technique of transforming (2) to a constant coefficient equation will again be used for obtaining a boundary integral equation for the solution of (2).

2. The boundary value problem

Referred to a Cartesian frame Ox_1x_2 a solution to (2) is sought which is valid in a region Ω in R^2 with boundary $\partial\Omega$ which consists of a finite number of piecewise smooth closed curves. On $\partial\Omega_1$ the dependent variable $\varsigma(\mathbf{x})$ ($\mathbf{x} = (x_1, x_2)$) is specified and on $\partial\Omega_2$

$$P(\mathbf{x}) = \lambda_{ij} \left(\partial\varsigma/\partial x_j\right) n_i \tag{3}$$

is specified where $\partial\Omega = \partial\Omega_1 \cup \partial\Omega_2$ and $\mathbf{n} = (n_1, n_2)$ denotes the outward pointing normal to $\partial\Omega$.

3. The boundary integral equation

The boundary integral equation is derived by transforming the variable coefficient equation (2) to a constant coefficient equation. The coefficients λ_{ij} and β are required to take the form

$$\lambda_{ij}(\mathbf{x}) = \overline{\lambda}_{ij} g(\mathbf{x}) \tag{4}$$
$$\beta(\mathbf{x}) = \overline{\beta} g(\mathbf{x}) \tag{5}$$

where the $\overline{\lambda}_{ij}$ and $\overline{\beta}$ are constants and g is a differentiable function of $\mathbf{x}$. Use of (4) and (5) and in (2) yields

$$\overline{\lambda}_{ij} \frac{\partial}{\partial x_i}\left(g \frac{\partial\varsigma}{\partial x_j}\right) + \overline{\beta} g\varsigma = 0 \tag{6}$$

Let

$$\varsigma(\mathbf{x}) = g^{-1/2}(\mathbf{x})\,\psi(\mathbf{x}) \tag{7}$$

so that (6) may be written in the form

$$\overline{\lambda}_{ij} \frac{\partial}{\partial x_i}\left[g \frac{\partial\left(g^{-1/2}\psi\right)}{\partial x_j}\right] + \overline{\beta} g^{1/2}\psi = 0$$

That is

$$\overline{\lambda}_{ij}\left[\left(\frac{1}{4} g^{-3/2} \frac{\partial g}{\partial x_i}\frac{\partial g}{\partial x_j} - \frac{1}{2} g^{-1/2} \frac{\partial^2 g}{\partial x_i \partial x_j}\right)\psi + g^{1/2} \frac{\partial^2 \psi}{\partial x_i \partial x_j}\right] + \overline{\beta} g^{1/2}\psi = 0 \tag{8}$$

Use of the identity

$$\frac{\partial^2 g^{1/2}}{\partial x_i \partial x_j} = -\frac{1}{4} g^{-3/2} \frac{\partial g}{\partial x_i}\frac{\partial g}{\partial x_j} + \frac{1}{2} g^{-1/2} \frac{\partial^2 g}{\partial x_i \partial x_j}$$

permits (8) to be written in the form

$$g^{1/2}\overline{\lambda}_{ij} \frac{\partial^2 \psi}{\partial x_i \partial x_j} - \psi \overline{\lambda}_{ij} \frac{\partial^2 g^{1/2}}{\partial x_i \partial x_j} + \overline{\beta} g^{1/2}\psi = 0 \tag{9}$$

If we further restrict the function $g(\mathbf{x})$ to take the exponential form

$$g(\mathbf{x}) = \{A[\cos(\alpha_m x_m) + \sin(\alpha_m x_m)]\}^2 \qquad \overline{\beta} = -\overline{\lambda}_{ij}\alpha_i\alpha_j < 0 \tag{10}$$

where α_m are constant, then

$$\overline{\lambda}_{ij}\frac{\partial^2 g^{1/2}}{\partial x_i \partial x_j} - \overline{\beta} g^{1/2} = 0 \tag{11}$$

Subtitution (11) into (9) implies a constant coefficients equation

$$\overline{\lambda}_{ij}\frac{\partial^2 \psi}{\partial x_i \partial x_j} = 0 \tag{12}$$

Also, substitution of (4) and (7) into (3) gives

$$P = -P_g \psi + P_\psi g^{1/2} \tag{13}$$

where

$$P_g(\mathbf{x}) = \overline{\lambda}_{ij}\frac{\partial g^{1/2}}{\partial x_j}n_i \qquad P_\psi(\mathbf{x}) = \overline{\lambda}_{ij}\frac{\partial \psi}{\partial x_j}n_i$$

A boundary integral equation for the solution of (12) is given in the form

$$\kappa(\mathbf{x}_0)\psi(\mathbf{x}_0) = \int_{\partial\Omega}[\Gamma(\mathbf{x},\mathbf{x}_0)\psi(\mathbf{x}) - \Phi(\mathbf{x},\mathbf{x}_0)P_\psi(\mathbf{x})]\,ds(\mathbf{x}) \tag{14}$$

where $\mathbf{x}_0 = (a,b)$, $\kappa = 0$ if $(a,b) \notin \Omega \cup \partial\Omega$, $\kappa = 1$ if $(a,b) \in \Omega$, $\kappa = \frac{1}{2}$ if $(a,b) \in \partial\Omega$ and $\partial\Omega$ has a continuously turning tangent at (a,b).

The so called fundamental solution Φ in (14) is any solution of the equation

$$\overline{\lambda}_{ij}\frac{\partial^2 \Phi}{\partial x_i \partial x_j} = \delta(\mathbf{x} - \mathbf{x}_0)$$

and the Γ is given by

$$\Gamma(\mathbf{x},\mathbf{x}_0) = \overline{\lambda}_{ij}\frac{\partial \Phi(\mathbf{x},\mathbf{x}_0)}{\partial x_j}n_i$$

where δ is the Dirac delta function. Following Azis in [18], for two-dimensional problems Φ and Γ are given by

$$\begin{aligned}\Phi(\mathbf{x},\mathbf{x}_0) &= \tfrac{K}{2\pi}\ln R \\ \Gamma(\mathbf{x},\mathbf{x}_0) &= \tfrac{K}{2\pi}\tfrac{1}{R}\overline{\lambda}_{ij}\tfrac{\partial R}{\partial x_j}n_i\end{aligned} \tag{15}$$

where

$$\begin{aligned}K &= \ddot{\tau}/\zeta \\ \zeta &= \left[\overline{\lambda}_{11} + 2\overline{\lambda}_{12}\dot{\tau} + \overline{\lambda}_{22}\left(\dot{\tau}^2 + \ddot{\tau}^2\right)\right]/2 \\ R &= \sqrt{(\dot{x}_1 - \dot{a})^2 + (\dot{x}_2 - \dot{b})^2} \\ \dot{x}_1 &= x_1 + \dot{\tau}x_2 \\ \dot{a} &= a + \dot{\tau}b \\ \dot{x}_2 &= \ddot{\tau}x_2 \\ \dot{b} &= \ddot{\tau}b\end{aligned}$$

where $\dot{\tau}$ and $\ddot{\tau}$ are respectively the real and the positive imaginary parts of the complex root τ of the quadratic

$$\overline{\lambda}_{11} + 2\overline{\lambda}_{12}\tau + \overline{\lambda}_{22}\tau^2 = 0$$

The derivatives $\partial R/\partial x_j$ needed for the calculation of the Γ in (15) are given by

$$\begin{aligned}\frac{\partial R}{\partial x_1} &= \frac{1}{R}\left(\dot{x}_1 - \dot{a}\right)\\ \frac{\partial R}{\partial x_2} &= \dot{\tau}\left[\frac{1}{R}\left(\dot{x}_1 - \dot{a}\right)\right] + \ddot{\tau}\left[\frac{1}{R}\left(\dot{x}_2 - \dot{b}\right)\right]\end{aligned}$$

Use of (7) and (13) in (14) yields

$$\begin{aligned}\kappa\left(\mathbf{x}_0\right) g^{1/2}\left(\mathbf{x}_0\right) \varsigma\left(\mathbf{x}_0\right) &= \int_{\partial\Omega} \left\{\left[g^{1/2}\left(\mathbf{x}\right) \Gamma\left(\mathbf{x}, \mathbf{x}_0\right) - P_g\left(\mathbf{x}\right) \Phi\left(\mathbf{x}, \mathbf{x}_0\right)\right] \varsigma\left(\mathbf{x}\right)\right.\\ &\quad \left. - \left[g^{-1/2}\left(\mathbf{x}\right) \Phi\left(\mathbf{x}, \mathbf{x}_0\right)\right] P\left(\mathbf{x}\right)\right\} ds\left(\mathbf{x}\right) \qquad (16)\end{aligned}$$

This equation provides a boundary integral equation for determining ς and P at all points of Ω.

4. Numerical examples

In order to show the appropriateness of the BEM and the validity of the analysis used above for deriving the boundary integral equation (16), some particular boundary value problems will be solved. The integrals in equation (16) are evaluated numerically using the Bode's quadrature (see Abramowitz and Stegun [19]).

4.1. Examples with analytical solutions

In order to see the convergence and accuracy of the BEM we will consider some examples of problems with analytical solutions. The parameters for the trigonometrical inhomogeneity function $g(\mathbf{x})$ are $A = 3, \alpha_1 = 0.15, \alpha_2 = 0.55$. Plot of $g(\mathbf{x})$ is shown in Figure 1.

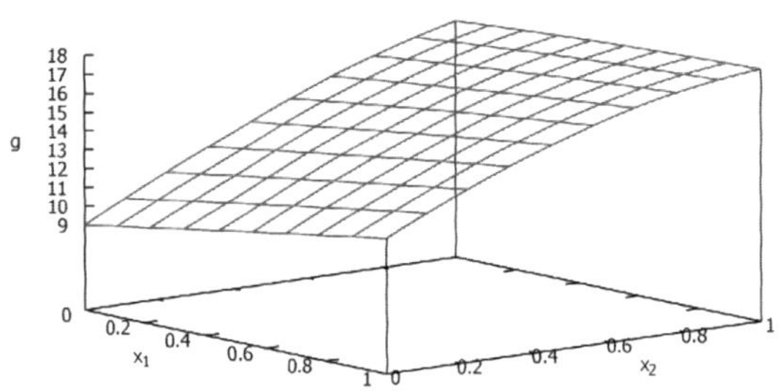

Figure 1. A trigonometrical inhomogeneity function $g(\mathbf{x}) = \{3\left[\cos\left(0.15x_1 + 0.55x_2\right) + \sin\left(0.15x_1 + 0.55x_2\right)\right]\}^2$

The geometry of the region Ω and the boundary conditions are as depicted in Figure 2.

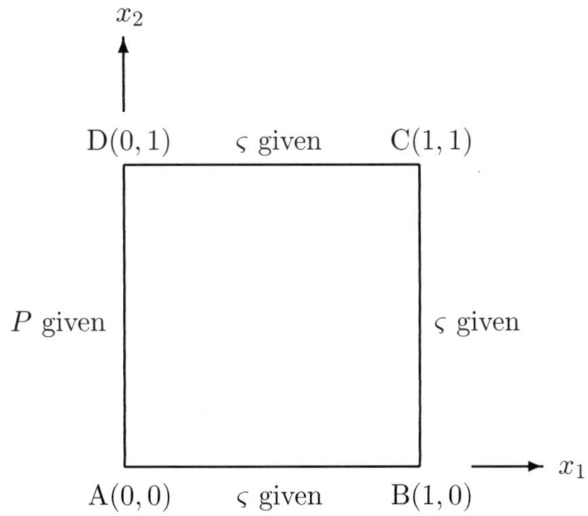

Figure 2. The geometry of all problems in Section 4.1

The values of the constant coefficients $\overline{\lambda}_{ij}$ for the governing equation (2) are

$$\overline{\lambda}_{11} = 0.25, \overline{\lambda}_{12} = 0.25, \overline{\lambda}_{22} = 1$$

Therefore from (9)

$$\overline{\beta} = -\overline{\lambda}_{ij}\alpha_i\alpha_j = -0.349375$$

The function $g(\mathbf{x})$ satisfies (9). Therefore equation (12) has to be the corresponding constant coefficient equation for $\psi(\mathbf{x})$. For test problems, we will take $\psi(\mathbf{x})$ as a linear function satisfying (12)

$$\psi(\mathbf{x}) = B\left(\gamma_0 + \gamma_1 x_1 + \gamma_2 x_2\right)$$

4.1.1. Problem 4.1.1 We choose analytical solution with parameters for $\psi(\mathbf{x})$

$$B = 2, \gamma_0 = 1, \gamma_1 = 0.85, \gamma_2 = 0.25$$

so that from (7)

$$\varsigma\left(\mathbf{x}\right) = 2\left(1 + 0.85x_1 + 0.25x_2\right) / \left\{3\left[\cos\left(0.15x_1 + 0.55x_2\right) + \sin\left(0.15x_1 + 0.55x_2\right)\right]\right\}$$

Table 1 shows the results of the analytical and BEM solutions with 20, 40 and 80 elements of equal length. The BEM solution converges to the analytical solution as the number of elements increases.

IOP Conf. Series: Materials Science and Engineering **619** (2019) 012059 doi:10.1088/1757-899X/619/1/012059

Table 1. BEM and analytical solutions for Problem 4.1.1

(x_1, x_2)	ς	$\partial\varsigma/\partial x_1$	$\partial\varsigma/\partial x_2$	ς	$\partial\varsigma/\partial x_1$	$\partial\varsigma/\partial x_2$
	BEM 20 elements			BEM 40 elements		
(.1,.5)	.6501	.3993	-.0567	.6489	.4012	-.0569
(.3,.5)	.7286	.3896	-.0636	.7282	.3921	-.0666
(.5,.5)	.8060	.3850	-.0707	.8059	.3854	-.0736
(.7,.5)	.8823	.3797	-.0753	.8824	.3795	-.0777
(.9,.5)	.9582	.3835	-.0795	.9580	.3773	-.0798
(.5,.1)	.8655	-.0026	-.2967	.8607	.3935	-.1999
(.5,.3)	.8262	.3973	-.1351	.8266	.3933	-.1353
(.5,.7)	.7970	.3861	-.0221	.7965	.3822	-.0211
(.5,.9)	.7999	.6441	.1031	.7976	.3931	.0378
	BEM 80 elements			Analytical		
(.1,.5)	.6486	.4020	-.0576	.6483	.4029	-.0587
(.3,.5)	.7280	.3931	-.0679	.7280	.3935	-.0692
(.5,.5)	.8059	.3856	-.0749	.8059	.3857	-.0762
(.7,.5)	.8824	.3794	-.0789	.8823	.3792	-.0800
(.9,.5)	.9578	.3756	-.0803	.9576	.3741	-.0809
(.5,.1)	.8616	.4044	-.2137	.8622	.4060	-.2159
(.5,.3)	.8269	.3934	-.1367	.8271	.3934	-.1382
(.5,.7)	.7963	.3825	-.0222	.7960	.3826	-.0233
(.5,.9)	.7967	.3823	.0262	.7962	.3840	.0248

4.1.2. Problem 4.1.2 We choose analytical solution with parameters

$$B = 4, \gamma_0 = 1, \gamma_1 = 0.85, \gamma_2 = 0.25$$

so that

$$\varsigma(\mathbf{x}) = 4(1 + 0.85x_1 + 0.25x_2) / \{3[\cos(0.15x_1 + 0.55x_2) + \sin(0.15x_1 + 0.55x_2)]\}$$

Table 2 shows the results of the analytical and BEM solutions with 20, 40 and 80 elements of equal length. The BEM solution converges to the analytical solution as the number of elements increases.

Table 2. BEM and analytical solutions for Problem 4.1.2

(x_1, x_2)	ς	$\partial\varsigma/\partial x_1$	$\partial\varsigma/\partial x_2$	ς	$\partial\varsigma/\partial x_1$	$\partial\varsigma/\partial x_2$
	BEM 20 elements			BEM 40 elements		
(.1,.5)	1.3003	.7986	-.1134	1.2978	.8024	-.1138
(.3,.5)	1.4572	.7793	-.1272	1.4563	.7843	-.1331
(.5,.5)	1.6119	.7701	-.1415	1.6118	.7708	-.1473
(.7,.5)	1.7646	.7595	-.1505	1.7648	.7591	-.1555
(.9,.5)	1.9164	.7671	-.1591	1.9159	.7546	-.1597
(.5,.1)	1.7309	-.0053	-.5933	1.7213	.7870	-.3998
(.5,.3)	1.6524	.7946	-.2702	1.6532	.7866	-.2705
(.5,.7)	1.5939	.7722	-.0442	1.5931	.7645	-.0423
(.5,.9)	1.5997	1.2882	.2062	1.5952	.7861	.0756
	BEM 80 elements			Analytical		
(.1,.5)	1.2971	.8040	-.1152	1.2967	.8058	-.1174
(.3,.5)	1.4561	.7861	-.1358	1.4559	.7871	-.1384
(.5,.5)	1.6118	.7712	-.1499	1.6117	.7714	-.1524
(.7,.5)	1.7647	.7589	-.1578	1.7647	.7585	-.1600
(.9,.5)	1.9156	.7511	-.1607	1.9153	.7482	-.1618
(.5,.1)	1.7232	.8087	-.4273	1.7243	.8120	-.4318
(.5,.3)	1.6537	.7868	-.2735	1.6542	.7868	-.2764
(.5,.7)	1.5926	.7650	-.0444	1.5920	.7652	-.0467
(.5,.9)	1.5934	.7646	.0525	1.5924	.7681	.0497

4.2. Examples without analytical solutions

In this section we will consider some examples of problems without simple analytical solutions. We setup some problems for a homogeneous isotropic material by taking $\overline{\lambda}_{11} = \overline{\lambda}_{22} = 1, \overline{\lambda}_{12} = 0$ and with symmetrical boundary conditions. The aim is to see the consistency of the BEM of whether it produces symmetrical solutions.

4.2.1. Problem 4.2.1 For this problem we take $g(\mathbf{x}) = \{A[\cos(\alpha_i x_i) + \sin(\alpha_i x_i)]\}^2 = 1$ with $A = 1, \alpha_1 = 0, \alpha_2 = 0$ so that $\overline{\beta} = \overline{\lambda}_{ij}\alpha_i\alpha_j = 0$. And the symmetrical boundary conditions are as shown in Figure 3.

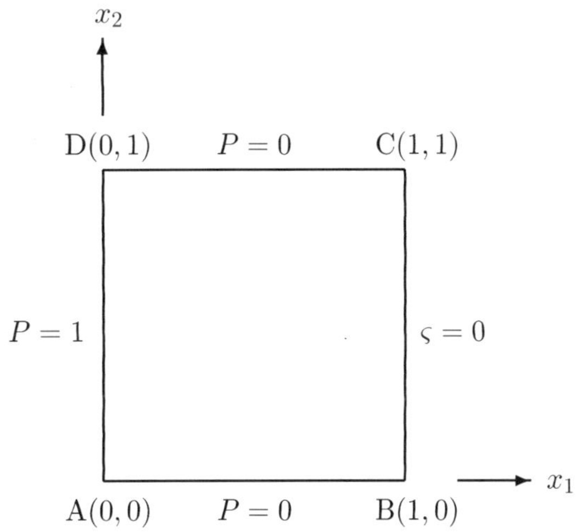

Figure 3. The geometry of Problem 4.2.1

Table 3 shows the results of the BEM solution using 40, 80, 160 and 320 elements of equal length. As expected, the results converge as the number of elements increases and also they are symmetrical about the axes $x_2 = 0.5$.

Table 3. BEM solution for Problem 4.2.1

(x_1, x_2)	ς	$\partial\varsigma/\partial x_1$	$\partial\varsigma/\partial x_2$	ς	$\partial\varsigma/\partial x_1$	$\partial\varsigma/\partial x_2$
	BEM 40 elements			BEM 80 elements		
(.1,.5)	.8924	-.9973	.0000	.8967	-.9987	-.0000
(.3,.5)	.6932	-.9952	.0000	.6971	-.9980	-.0000
(.5,.5)	.4942	-.9943	.0000	.4975	-.9977	-.0000
(.7,.5)	.2955	-.9932	.0000	.2980	-.9973	-.0000
(.9,.5)	.0970	-.9919	.0000	.0986	-.9968	-.0000
(.5,.1)	.4940	-.9975	.0005	.4974	-.9982	.0002
(.5,.3)	.4942	-.9945	.0007	.4975	-.9978	.0002
(.5,.7)	.4942	-.9945	-.0007	.4975	-.9978	-.0002
(.5,.9)	.4940	-.9975	-.0005	.4974	-.9982	-.0002
	BEM 160 elements			BEM 320 elements		
(.1,.5)	.8985	-.9994	-.0000	.8993	-.9997	.0000
(.3,.5)	.6987	-.9991	-.0000	.6994	-.9996	.0000
(.5,.5)	.4989	-.9990	-.0000	.4995	-.9996	.0000
(.7,.5)	.2991	-.9989	-.0000	.2996	-.9995	.0000
(.9,.5)	.0993	-.9987	-.0000	.0997	-.9994	.0000
(.5,.1)	.4988	-.9992	.0001	.4995	-.9997	.0000
(.5,.3)	.4989	-.9991	.0001	.4995	-.9996	.0000
(.5,.7)	.4989	-.9991	-.0001	.4995	-.9996	-.0000
(.5,.9)	.4988	-.9992	-.0001	.4995	-.9997	-.0000

4.2.2. Problem 4.2.2 Now we consider a problem with $g(\mathbf{x}) = 1$ again and the symmetrical boundary conditions are as shown in Figure 4. Table 4 shows the results of the BEM solution using 40, 80, 160 and 320 elements of equal length. The results converge as the number of elements increases and also they are symmetrical about the axes $x_1 = 0.5$.

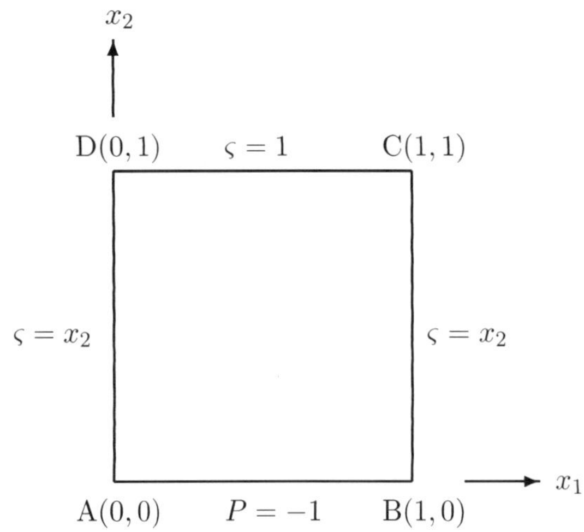

Figure 4. The geometry of Problem 4.2.2

Table 4. BEM solution for Problem 4.2.2

(x_1, x_2)	ς	$\partial\varsigma/\partial x_1$	$\partial\varsigma/\partial x_2$	ς	$\partial\varsigma/\partial x_1$	$\partial\varsigma/\partial x_2$
	BEM 40 elements			BEM 80 elements		
(.1,.5)	.5003	.0031	1.0020	.5001	.0014	1.0001
(.3,.5)	.5008	.0016	1.0004	.5004	.0007	1.0003
(.5,.5)	.5010	-.0000	1.0005	.5004	.0000	1.0004
(.7,.5)	.5008	-.0016	1.0004	.5004	-.0007	1.0003
(.9,.5)	.5003	-.0031	1.0020	.5001	-.0014	1.0001
(.5,.1)	.1012	-.0000	.9987	.1005	-.0000	.9993
(.5,.3)	.3010	-.0000	.9992	.3004	-.0000	.9998
(.5,.7)	.7012	-.0000	1.0022	.7006	.0000	1.0012
(.5,.9)	.9019	.0000	1.0037	.9009	-.0000	1.0019
	BEM 160 elements			BEM 320 elements		
(.1,.5)	.5001	.0006	1.0001	.5000	.0003	1.0000
(.3,.5)	.5002	.0004	1.0002	.5001	.0002	1.0001
(.5,.5)	.5002	.0000	1.0002	.5001	-.0000	1.0001
(.7,.5)	.5002	-.0004	1.0002	.5001	-.0002	1.0001
(.9,.5)	.5001	-.0006	1.0001	.5000	-.0003	1.0000
(.5,.1)	.1002	.0000	.9996	.1001	-.0000	.9998
(.5,.3)	.3002	.0000	.9999	.3001	-.0000	.9999
(.5,.7)	.7003	.0000	1.0006	.7001	-.0000	1.0003
(.5,.9)	.9005	-.0000	1.0010	.9002	-.0000	1.0005

5. Conclusion

The scalar elliptic governing equation (2) is used for modeling physical problems such as steady infiltration problems (when $\beta < 0$), acoustic problems (when $\beta > 0$), and antiplane strain in elastostatics and plane thermostatic problems (when $\beta = 0$). The boundary integral equation (16) was derived from this governing equation (2) and straight from (16) a BEM was then constructed for calculation of numerical solutions to the problems for anisotropic trigonometrically graded media. The results show the convergence, consistency, and accuracy of the BEM solutions. Together with its ease in implementation, it may be concluded that BEM is a good numerical method for solving such kind of problems.

Acknowledgements

The author would like to thank The Hasanuddin University and The Ministry of Research, Technology and Higher Education of the Republic of Indonesia for the provided support.

References

[1] Azis M I, Kasbawati, Haddade A and Thamrin S A 2018 On some examples of pollutant transport problems solved numerically using the boundary element method *Journal of Physics: Conference Series* **979**(1)

[2] Azis M I, Asrul L, Khaeruddin and Paharuddin 2018 BEM solutions for unsteady transport problems in anisotropic media *JP Journal of Heat and Mass Transfer* **15**(4) 915

[3] Haddade A, Salam N, Khaeruddin and Azis M I 2017 A boundary element method for 2D diffusion-convection problems in anisotropic media *Far East Journal of Mathematical Sciences* **102**(8) 1593

[4] Azis M I 2019 Numerical solutions for the Helmholtz boundary value problems of anisotropic media *Journal of Computational Physics* **381** 42

[5] Cheng A H-D 1984 Darcy's Flow with Variable Permeability: A Boundary Integral Solution *Water Resources Research* **20** 980

[6] Clements D L and Azis M I 2000 A Note on a Boundary Element Method for the Numerical Solution of Boundary Value Problems in Isotropic Inhomogeneous Elasticity *Journal of the Chinese Institute of Engineers* **23**(3) 261

IOP Conf. Series: Materials Science and Engineering **619** (2019) 012059 doi:10.1088/1757-899X/619/1/012059

[7] Azis M I and Clements D L 2014 On some problems concerning deformations of functionally graded anisotropic elastic materials *Far East Journal of Mathematical Sciences* **87**(2) 173

[8] Azis M I, Toaha S, Bahri M and Ilyas N 2018 A boundary element method with analytical integration for deformation of inhomogeneous elastic materials *Journal of Physics: Conference Series* **979**(1) 012072

[9] Azis M I and Clements D L 2014 A Boundary Element Method for Transient Heat Conduction Problem of Nonhomogeneous Anisotropic Materials *Far East Journal of Mathematical Sciences*, **89**(1) 51

[10] Azis M I and Clements D L 2008 Nonlinear transient heat conduction problems for a class of inhomogeneous anisotropic materials by BEM *Engineering Analysis with Boundary Elements* **32**(12) 1054

[11] Salam N, Haddade A, Clements D L and Azis M I 2017 A boundary element method for a class of elliptic boundary value problems of functionally graded media *Engineering Analysis with Boundary Elements* **84**(3) 186

[12] Azis M I 2019 Numerical solutions to a class of scalar elliptic BVPs for anisotropic exponentially graded media *Journal of Physics: Conference Series* **1218** 012001

[13] Azis M I 2019 Standard-BEM solutions to two types of anisotropic-diffusion convection reaction equations with variable coefficients *Engineering Analysis with Boundary Elements* **105** 87

[14] Clements D L and Lobo M 2010 A BEM for time dependent infiltration from an irrigation channel *Engineering Analysis with Boundary Elements* **34** 1100

[15] Solekhudin I and Ang K-C 2012 A DRBEM with a predictor-corrector scheme for steady infiltration from periodic channels with root-water uptake *Engineering Analysis with Boundary Elements* **36** 1199

[16] Barucq H, Bendali A, Fares M, Mattesi V and Tordeux S 2017 A symmetric Trefftz-DG formulation based on a local boundary element method for the solution of the Helmholtz equation *Journal of Computational Physics* **330** 1069

[17] Loeffler C F, Mansur W J, Barcelos H D M and Bulcão A 2015 Solving Helmholtz problems with the boundary element method using direct radial basis function interpolation *Engineering Analysis with Boundary Elements* **61** 218

[18] Azis M I 2017 Fundamental solutions to two types of 2D boundary value problems of anisotropic materials *Far East Journal of Mathematical Sciences* **101**(11) 2405

[19] Abramowitz M and Stegun I A 1972 *Handbook of mathematical functions: with formulas, graphs and mathematical tables*, Dover Publications, Washington

IOP Conf. Series: Materials Science and Engineering **619** (2019) 012060 doi:10.1088/1757-899X/619/1/012060

Numerical solutions to anisotropic BVPs for quadratically graded media governed by a Helmholtz equation

S. Hamzah[1], **M. I. Azis**[2,*], **A. K. Amir**[2]

[1]Department of Civil Engineering, Hasanuddin University, Makassar, Indonesia
[2]Department of Mathematics, Hasanuddin University, Makassar, Indonesia

E-mail: `mohivanazis@yahoo.co.id`

Abstract. A Boundary Element Method (BEM) is used for obtaining solutions to anisotropic quadratically graded media boundary value problems (BVPs) governed by a Helmholtz type equation. A technique of transforming the variable coefficient governing equation to a constant coefficient equation is utilized for deriving a boundary integral equation. Some particular problems are considered to illustrate the application of the BEM. The results show the convergence, consistency, and accuracy of the BEM solutions.

1. Introduction

The BEM has been successfully used for solving many types of problems of homogeneous media. Some works on homogeneous media problems have been recently done by Azis et. al [1, 2] and Haddade et. al [3] in which the authors considered pollutant transport problems governed by 2D diffusion-convection, and also Azis [4] in which numerical solutions for the Helmholtz boundary value problems were obtained.

However, this is generally not the case for FGM. There are two techniques usually used to deal with problems of FGM. The first one uses a technique of deriving a relevant Green function or fundamental solution to the FGM problem. Cheng [5] had applied this technique. The second technique is by transforming the variable coefficient governing equation to a constant coefficient equation. Some progress on using the second technique has been made. For examples, Clements and Azis [6] considered the case for isotropic FGM. For anisotropic FGM some works have been studied by Azis and Clements [7, 8, 9] and Azis et. al [10]. In addition to this, recently Salam et. al [11], Azis [12] and Azis [13] have been working respectively on a class of elliptic problems and diffusion convection reaction equation for anisotropic FGM.

This paper discusses derivation of a boundary integral equation for numerically solving 2D boundary value problems governed by the (dimensionless) Helmholtz type equation of the form

$$\frac{\partial}{\partial x_i}\left[\lambda_{ij}(x_1,x_2)\frac{\partial\mu(x_1,x_2)}{\partial x_j}\right]+\beta^2(x_1,x_2)\,\mu(x_1,x_2)=0 \tag{1}$$

where the coefficients λ_{ij} and β^2 depend on x_1 and x_2 and the repeated summation convention (summing from 1 to 2) is employed.

For (1) to be a second order elliptic partial differential equation, we assume that the matrix of coefficients $[\lambda_{ij}]$ is a real symmetric positive definite matrix. Explicitly equation (1) may be written as

$$\frac{\partial}{\partial x_1}\left(\lambda_{11}\frac{\partial \mu}{\partial x_1}\right)+2\frac{\partial}{\partial x_1}\left(\lambda_{12}\frac{\partial \mu}{\partial x_2}\right)+\frac{\partial}{\partial x_2}\left(\lambda_{22}\frac{\partial \mu}{\partial x_2}\right)+\beta^2\mu=0$$

Further, the coefficients λ_{ij} and β are required to be twice differentiable functions of the two independent variables x_1 and x_2. The analysis is also relevant for isotropic media as a special case of anisotropic media which occurs when $\lambda_{11}=\lambda_{22}$ and $\lambda_{12}=0$.

A variety of problems of both isotropic and anisotropic inhomogeneous media are usually modeled with equation (1). Acoustic problems (when $\beta^2>0$), and antiplane strain in elastostatics and plane thermostatic problems (when $\beta^2=0$) are the areas for which the governing equation is of the type (1). Recently some works on the Helmholtz equation for isotropic and/or homogeneous media have been done. For example, Barucq et. al. in [14], Loeffler et. al. in [15] considered Helmholtz equation as the governing equation for isotropic and/or homogeneous media which is a special case of the equation (1).

2. The boundary value problem

Referred to a Cartesian frame Ox_1x_2 a solution to (1) is sought which is valid in a region Ω in R^2 with boundary $\partial\Omega$ which consists of a finite number of piecewise smooth closed curves. On $\partial\Omega_1$ the dependent variable $\mu(\mathbf{x})$ ($\mathbf{x}=(x_1,x_2)$) is specified and on $\partial\Omega_2$

$$P(\mathbf{x})=\lambda_{ij}\left(\partial\mu/\partial x_j\right)n_i \tag{2}$$

is specified where $\partial\Omega=\partial\Omega_1\cup\partial\Omega_2$ and $\mathbf{n}=(n_1,n_2)$ denotes the outward pointing normal to $\partial\Omega$.

3. The boundary integral equation

The boundary integral equation is derived by transforming the variable coefficient equation (1) to a constant coefficient equation. The coefficients λ_{ij} and β are required to take the form

$$\lambda_{ij}(\mathbf{x}) = \overline{\lambda}_{ij}g(\mathbf{x}) \tag{3}$$

$$\beta^2(\mathbf{x}) = \overline{\beta}^2 g(\mathbf{x}) \tag{4}$$

where the $\overline{\lambda}_{ij}$ and $\overline{\beta}$ are constants and g is an exponential function of $\mathbf{x}$. Use of (3) and (4) and in (1) yields

$$\overline{\lambda}_{ij}\frac{\partial}{\partial x_i}\left(g\frac{\partial \mu}{\partial x_j}\right)+\overline{\beta}^2 g\mu=0 \tag{5}$$

Let

$$\mu(\mathbf{x})=g^{-1/2}(\mathbf{x})\,\psi(\mathbf{x}) \tag{6}$$

so that (5) may be written in the form

$$\overline{\lambda}_{ij}\frac{\partial}{\partial x_i}\left[g\frac{\partial\left(g^{-1/2}\psi\right)}{\partial x_j}\right]+\overline{\beta}^2 g^{1/2}\psi=0$$

That is

$$\overline{\lambda}_{ij}\left[\left(\frac{1}{4}g^{-3/2}\frac{\partial g}{\partial x_i}\frac{\partial g}{\partial x_j}-\frac{1}{2}g^{-1/2}\frac{\partial^2 g}{\partial x_i\partial x_j}\right)\psi+g^{1/2}\frac{\partial^2\psi}{\partial x_i\partial x_j}\right]+\overline{\beta}^2 g^{1/2}\psi=0 \tag{7}$$

Use of the identity

$$\frac{\partial^2 g^{1/2}}{\partial x_i \partial x_j} = -\frac{1}{4} g^{-3/2} \frac{\partial g}{\partial x_i} \frac{\partial g}{\partial x_j} + \frac{1}{2} g^{-1/2} \frac{\partial^2 g}{\partial x_i \partial x_j}$$

permits (7) to be written in the form

$$g^{1/2} \overline{\lambda}_{ij} \frac{\partial^2 \psi}{\partial x_i \partial x_j} - \psi \overline{\lambda}_{ij} \frac{\partial^2 g^{1/2}}{\partial x_i \partial x_j} + \overline{\beta}^2 g^{1/2} \psi = 0 \tag{8}$$

If we further restrict the function $g(\mathbf{x})$ to take the exponential form

$$g(\mathbf{x}) = [A(\alpha_0 + \alpha_1 x_1 + \alpha_2 x_2)]^2 \tag{9}$$

where α_m are constant, then

$$\overline{\lambda}_{ij} \frac{\partial^2 g^{1/2}}{\partial x_i \partial x_j} = 0 \tag{10}$$

Subtitution (10) into (8) implies a constant coefficients equation

$$\overline{\lambda}_{ij} \frac{\partial^2 \psi}{\partial x_i \partial x_j} + \overline{\beta}^2 \psi = 0 \tag{11}$$

Also, substitution of (3) and (6) into (2) gives

$$P = -P_g \psi + P_\psi g^{1/2} \tag{12}$$

where

$$P_g(\mathbf{x}) = \overline{\lambda}_{ij} \frac{\partial g^{1/2}}{\partial x_j} n_i \qquad P_\psi(\mathbf{x}) = \overline{\lambda}_{ij} \frac{\partial \psi}{\partial x_j} n_i$$

A boundary integral equation for the solution of (11) is given in the form

$$\eta(\mathbf{x}_0) \psi(\mathbf{x}_0) = \int_{\partial\Omega} [\Gamma(\mathbf{x}, \mathbf{x}_0) \psi(\mathbf{x}) - \Phi(\mathbf{x}, \mathbf{x}_0) P_\psi(\mathbf{x})] \, ds(\mathbf{x}) \tag{13}$$

where $\mathbf{x}_0 = (a, b)$, $\eta = 0$ if $(a, b) \notin \Omega \cup \partial\Omega$, $\eta = 1$ if $(a, b) \in \Omega$, $\eta = \frac{1}{2}$ if $(a, b) \in \partial\Omega$ and $\partial\Omega$ has a continuously turning tangent at (a, b).

The so called fundamental solution Φ in (13) is any solution of the equation

$$\overline{\lambda}_{ij} \frac{\partial^2 \Phi}{\partial x_i \partial x_j} + \overline{\beta}^2 \Phi = \delta(\mathbf{x} - \mathbf{x}_0)$$

and the Γ is given by

$$\Gamma(\mathbf{x}, \mathbf{x}_0) = \overline{\lambda}_{ij} \frac{\partial \Phi(\mathbf{x}, \mathbf{x}_0)}{\partial x_j} n_i$$

where δ is the Dirac delta function. Following Azis [16], for two-dimensional problems Φ and Γ are given by

$$\begin{aligned} \Phi(\mathbf{x}, \mathbf{x}_0) &= \begin{cases} \frac{K}{2\pi} \ln R & \text{if } \overline{\beta}^2 = 0 \\ \frac{\imath K}{4} H_0^{(2)}(\omega R) & \text{if } \overline{\beta}^2 > 0 \end{cases} \\ \Gamma(\mathbf{x}, \mathbf{x}_0) &= \begin{cases} \frac{K}{2\pi} \frac{1}{R} \overline{\lambda}_{ij} \frac{\partial R}{\partial x_j} n_i & \text{if } \overline{\beta}^2 = 0 \\ \frac{-\imath K \omega}{4} H_1^{(2)}(\omega R) \overline{\lambda}_{ij} \frac{\partial R}{\partial x_j} n_i & \text{if } \overline{\beta}^2 > 0 \end{cases} \end{aligned} \tag{14}$$

where

$$\begin{aligned}
K &= \ddot{\tau}/\zeta \\
\omega &= \sqrt{|\overline{\beta}^2/\zeta|} \\
\zeta &= \left[\overline{\lambda}_{11} + 2\overline{\lambda}_{12}\dot{\tau} + \overline{\lambda}_{22}\left(\dot{\tau}^2 + \ddot{\tau}^2\right)\right]/2 \\
R &= \sqrt{(\dot{x}_1 - \dot{a})^2 + (\dot{x}_2 - \dot{b})^2} \\
\dot{x}_1 &= x_1 + \dot{\tau}x_2 \\
\dot{a} &= a + \dot{\tau}b \\
\dot{x}_2 &= \ddot{\tau}x_2 \\
\dot{b} &= \ddot{\tau}b
\end{aligned}$$

where $\dot{\tau}$ and $\ddot{\tau}$ are respectively the real and the positive imaginary parts of the complex root τ of the quadratic

$$\overline{\lambda}_{11} + 2\overline{\lambda}_{12}\tau + \overline{\lambda}_{22}\tau^2 = 0$$

and $H_0^{(2)}$, $H_1^{(2)}$ denote the Hankel function of second kind and order zero and order one respectively and $\imath$ represents the square root of minus one. The derivatives $\partial R/\partial x_j$ needed for the calculation of the Γ in (14) are given by

$$\begin{aligned}
\frac{\partial R}{\partial x_1} &= \frac{1}{R}(\dot{x}_1 - \dot{a}) \\
\frac{\partial R}{\partial x_2} &= \dot{\tau}\left[\frac{1}{R}(\dot{x}_1 - \dot{a})\right] + \ddot{\tau}\left[\frac{1}{R}\left(\dot{x}_2 - \dot{b}\right)\right]
\end{aligned}$$

Use of (6) and (12) in (13) yields

$$\begin{aligned}
\eta(\mathbf{x}_0)\, g^{1/2}(\mathbf{x}_0)\, \mu(\mathbf{x}_0) &= \int_{\partial\Omega} \left\{ \left[g^{1/2}(\mathbf{x})\, \Gamma(\mathbf{x}, \mathbf{x}_0) - P_g(\mathbf{x})\, \Phi(\mathbf{x}, \mathbf{x}_0) \right] \mu(\mathbf{x}) \right. \\
&\quad \left. - \left[g^{-1/2}(\mathbf{x})\, \Phi(\mathbf{x}, \mathbf{x}_0) \right] P(\mathbf{x}) \right\} ds(\mathbf{x}) \qquad (15)
\end{aligned}$$

This equation provides a boundary integral equation for determining μ and P at all points of Ω.

4. Numerical examples

Some particular boundary value problems will be solved numerically by employing the integral equation (15). The main aim is to show the validity of the analysis used above for deriving the boundary integral equation (15) and the appropriateness of the BEM in solving the problems through the derived boundary integral equation (15). Standard boundary element method is employed to obtain numerical results. The integrals in equation (15) are evaluated numerically using the Bode's quadrature (see Abramowitz and Stegun [17]).

4.1. Problem 1

In order to see the convergence and accuracy of the BEM we will consider a problem with analytical solution. We choose a medium which anisotropic with

$$\overline{\lambda}_{11} = 1, \overline{\lambda}_{12} = 0.5, \overline{\lambda}_{22} = 1$$

and we take an inhomogeneity function $g(\mathbf{x})$ satisfying (9) of the form

$$g(\mathbf{x}) = \left[2\left(1 + 0.5x_1 + 0.75x_2\right)\right]^2$$

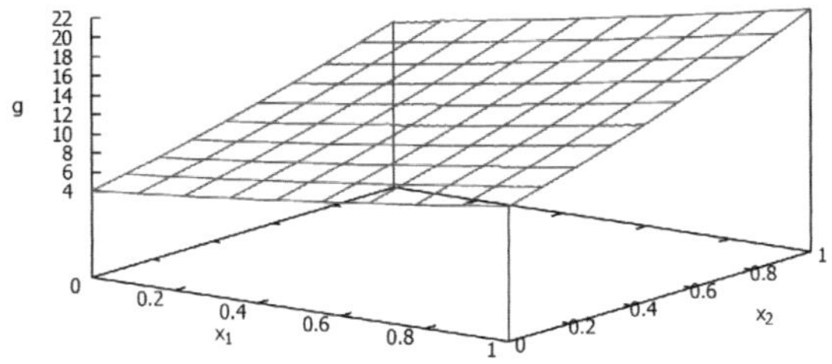

Figure 1. A quadratical inhomogeneity function $g(\mathbf{x}) = [2(1 + 0.5x_1 + 0.75x_2)]^2$

Plot of $g(\mathbf{x})$ is shown in Figure 1. For the analytical solution μ we will take

$$\psi(\mathbf{x}) = 3.5\left[\cos(0.75x_1 + 0.25x_2) + \sin(0.75x_1 + 0.25x_2)\right]$$

so that from (11) $\overline{\beta}^2 = 0.8125$. Therefore from (6) the analytical solution

$$\mu(\mathbf{x}) = \frac{3.5\left[\cos(0.75x_1 + 0.25x_2) + \sin(0.75x_1 + 0.25x_2)\right]}{2(1 + 0.5x_1 + 0.75x_2)}$$

The geometry of the region Ω and the boundary conditions are as depicted in Figure 2.

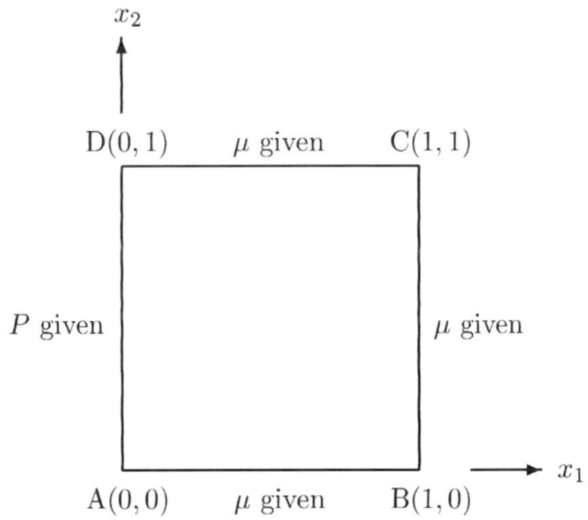

Figure 2. The geometry of Problem 1

The results are shown in Table 1. The BEM solution converges to the analytical solution as the number of elements increases. The accuracy of the BEM solutions lie on the fourth figure of the decimals.

4.2. Problem 2

We consider a problem for a homogeneous isotropic medium with $g(\mathbf{x}) = 25$, $\overline{\lambda}_{11} = \overline{\lambda}_{22} = 1, \overline{\lambda}_{12} = 0$ and $\overline{\beta}^2 = 1$. The geometry of the region Ω and the boundary conditions are as

Table 1. BEM and analytical solutions for Problem 1

(x_1, x_2)	μ	$\partial\mu/\partial x_1$	$\partial\mu/\partial x_2$	μ	$\partial\mu/\partial x_1$	$\partial\mu/\partial x_2$
	BEM 160 elements			BEM 320 elements		
(.1,.5)	1.4477	.2111	-.5216	1.4476	.2115	-.5218
(.3,.5)	1.4715	.0304	-.5521	1.4715	.0307	-.5523
(.5,.5)	1.4613	-.1284	-.5669	1.4614	-.1282	-.5671
(.7,.5)	1.4215	-.2668	-.5695	1.4215	-.2668	-.5696
(.9,.5)	1.3558	-.3864	-.5619	1.3559	-.3863	-.5623
(.5,.1)	1.7305	-0.y1263	-.8036	1.7307	-.1263	-.8040
(.5,.3)	1.5842	-.1229	-.6670	1.5843	-.1229	-.6673
(.5,.7)	1.3558	-.1384	-.4913	1.3558	-.1382	-.4914
(.5,.9)	1.2637	-.1506	-.4329	1.2636	-.1507	-.4329
	BEM 640 elements			Analytical		
(.1,.5)	1.4476	.2116	-.5219	1.4476	.2118	-.5220
(.3,.5)	1.4715	.0308	-.5524	1.4715	.0309	-.5525
(.5,.5)	1.4614	-.1281	-.5672	1.4614	-.1281	-.5673
(.7,.5)	1.4216	-.2668	-.5696	1.4216	-.2668	-.5697
(.9,.5)	1.3559	-.3863	-.5622	1.3560	-.3863	-.5622
(.5,.1)	1.7308	-.1264	-.8041	1.7308	-.1265	-.8042
(.5,.3)	1.5843	-.1229	-.6674	1.5844	-.1229	-.6676
(.5,.7)	1.3558	-.1381	-.4915	1.3558	-.1380	-.4916
(.5,.9)	1.2636	-.1506	-.4330	1.2636	-.1505	-.4331

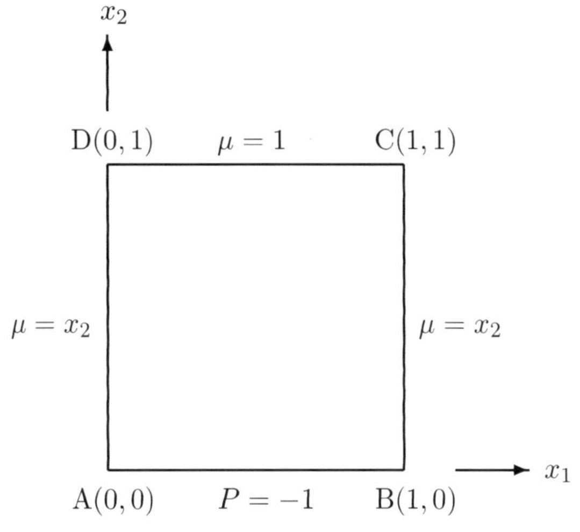

Figure 3. The geometry of Problem 2 and Problem 3

depicted in Figure 3. As shown in Figure 3 the boundary conditions are symmetrical about the axes $x_1 = 0.5$. Table 2 shows the results of the BEM solution using 80, 160, 320 and 640 elements of equal length. As expected, the results converge as the number of elements increases and also they are symmetrical about the axes $x_1 = 0.5$.

4.3. Problem 3

Now we consider a problem for an exponentially graded isotropic material with $g(\mathbf{x}) = [2(1 + 0.5x_1 + 0.75x_2)]^2$, $\overline{\lambda}_{11} = \overline{\lambda}_{22} = 1, \overline{\lambda}_{12} = 0$ and $\overline{\beta}^2 = 1$. Again, the geometry of the

Table 2. BEM solution for Problem 2

(x_1, x_2)	μ	$\partial\mu/\partial x_1$	$\partial\mu/\partial x_2$	μ	$\partial\mu/\partial x_1$	$\partial\mu/\partial x_2$
	BEM 80 elements			BEM 160 elements		
(.1,.5)	.5442	.4144	.9197	.5445	.4148	.9189
(.3,.5)	.6096	.2279	.7830	.6099	.2280	.7817
(.5,.5)	.6327	.0000	.7321	.6330	-.0000	.7307
(.7,.5)	.6096	-.2279	.7830	.6099	-.2280	.7817
(.9,.5)	.5442	-.4144	.9197	.5445	-.4148	.9189
(.5,.1)	.4204	.0000	.2363	.4213	-.0000	.2352
(.5,.3)	.5016	.0000	.5548	.5023	-.0000	.5533
(.5,.7)	.7856	-.0000	.7759	.7856	-.0000	.7745
(.5,.9)	.9349	-.0000	.6975	.9347	-.0000	.6961
	BEM 320 elements			BEM 640 elements		
(.1,.5)	.5447	.4150	.9184	.5448	.4150	.9182
(.3,.5)	.6101	.2281	.7810	.6102	.2281	.7807
(.5,.5)	.6332	.0000	.7300	.6333	.0000	.7297
(.7,.5)	.6101	-.2281	0.y7810	.6102	-.2281	.7807
(.9,.5)	.5447	-.4150	.9184	.5448	-.4150	.9182
(.5,.1)	.4217	.0000	.2347	.4219	-.0000	.2345
(.5,.3)	.5026	.0000	0.y5526	.5027	-.0000	.5523
(.5,.7)	.7857	.0000	.7738	.7857	.0000	.7734
(.5,.9)	.9346	-.0000	.6954	.9346	-.0000	.6950

Table 3. BEM solution for Problem 3

(x_1, x_2)	μ	$\partial\mu/\partial x_1$	$\partial\mu/\partial x_2$	μ	$\partial\mu/\partial x_1$	$\partial\mu/\partial x_2$
	BEM 80 elements			BEM 160 elements		
(.1,.5)	.5745	.6503	.8986	.5749	.6497	.8970
(.3,.5)	.6657	.2731	.7528	.6661	.2726	.7512
(.5,.5)	.6876	-.0461	.7146	.6879	-.0464	.7133
(.7,.5)	.6501	-.3230	.7813	.6503	-.3231	.7803
(.9,.5)	.5609	-.5619	.9235	.5611	-.5623	.9224
(.5,.1)	.4535	-.0657	.3500	.4543	-.0663	.3490
(.5,.3)	.5525	-.0614	.6121	.5530	-.0619	.6108
(.5,.7)	.8295	-.0286	.6839	.8295	-.0286	.6825
(.5,.9)	.9532	-.0101	.5336	.9529	-.0099	.5320
	BEM 320 elements			BEM 640 elements		
(.1,.5)	.5752	.6494	.8968	.5753	.6491	.8965
(.3,.5)	.6663	.2724	.7506	.6664	.2722	.7502
(.5,.5)	.6880	-.0465	.7126	.6881	-.0465	.7123
(.7,.5)	.6505	-.3231	.7796	.6505	-.3231	.7793
(.9,.5)	.5612	-.5621	.9219	.5613	-.5620	.9217
(.5,.1)	.4547	-.0664	.3485	.4549	-.0664	.3484
(.5,.3)	.5533	-.0620	.6101	.5534	-.0620	.6098
(.5,.7)	.8295	-.0287	.6818	.8295	-.0288	.6814
(.5,.9)	.9528	-.0098	.5312	.9527	-.0098	.5309

region Ω and the boundary conditions are as depicted in Figure 3. Table 3 shows the results of the BEM solution using 80, 160, 320 and 640 elements of equal length. As expected, the results converge as the number of elements increases but they are not symmetrical anymore as the material is not homogeneous.

5. Conclusion

The Helmholtz type governing equation (1) is sometimes used for modeling physical problems such as acoustic problems (when $\beta^2 > 0$), and antiplane strain in elastostatics and plane thermostatic problems (when $\beta^2 = 0$). The boundary integral equation (15) is derived from this governing equation (1) and then from (15) a BEM is constructed for calculation of numerical solutions to the problems for anisotropic quadratically graded media. The results show that the BEM solution gives a convergence, consistency, and accuracy. Therefore the results also prove that the analysis used for deriving the boundary integral equation (15) is valid. Together with its ease in implementation, it may be concluded that BEM is a useful numerical method for solving such kind of problems.

Acknowledgements

The author would like to thank The Hasanuddin University and The Ministry of Research, Technology and Higher Education of the Republic of Indonesia for the provided support.

References

[1] Azis M I, Kasbawati, Haddade A and Thamrin S A 2018 On some examples of pollutant transport problems solved numerically using the boundary element method *Journal of Physics: Conference Series* **979**(1)
[2] Azis M I, Asrul L, Khaeruddin and Paharuddin 2018 BEM solutions for unsteady transport problems in anisotropic media *JP Journal of Heat and Mass Transfer* **15**(4) 915
[3] Haddade A, Salam N, Khaeruddin and Azis M I 2017 A boundary element method for 2D diffusion-convection problems in anisotropic media *Far East Journal of Mathematical Sciences* **102**(8) 1593
[4] Azis M I 2019 Numerical solutions for the Helmholtz boundary value problems of anisotropic media *Journal of Computational Physics* **381** 42
[5] Cheng A H-D 1984 Darcy's Flow with Variable Permeability: A Boundary Integral Solution *Water Resources Research* **20** 980
[6] Clements D L and Azis M I 2000 A Note on a Boundary Element Method for the Numerical Solution of Boundary Value Problems in Isotropic Inhomogeneous Elasticity *Journal of the Chinese Institute of Engineers* **23**(3) 261
[7] Azis M I and Clements D L 2014 On some problems concerning deformations of functionally graded anisotropic elastic materials *Far East Journal of Mathematical Sciences* **87**(2) 173
[8] Azis M I and Clements D L 2014 A Boundary Element Method for Transient Heat Conduction Problem of Nonhomogeneous Anisotropic Materials *Far East Journal of Mathematical Sciences*, **89**(1) 51
[9] Azis M I and Clements D L 2008 Nonlinear transient heat conduction problems for a class of inhomogeneous anisotropic materials by BEM *Engineering Analysis with Boundary Elements* **32**(12) 1054
[10] Azis M I, Toaha S, Bahri M and Ilyas N 2018 A boundary element method with analytical integration for deformation of inhomogeneous elastic materials *Journal of Physics: Conference Series* **979**(1) 012072
[11] Salam N, Haddade A, Clements D L and Azis M I 2017 A boundary element method for a class of elliptic boundary value problems of functionally graded media *Engineering Analysis with Boundary Elements* **84**(3) 186
[12] Azis M I 2019 Numerical solutions to a class of scalar elliptic BVPs for anisotropic exponentially graded media *Journal of Physics: Conference Series* **1218** 012001
[13] Azis M I 2019 Standard-BEM solutions to two types of anisotropic-diffusion convection reaction equations with variable coefficients *Engineering Analysis with Boundary Elements* **105** 87
[14] Barucq H, Bendali A, Fares M, Mattesi V and Tordeux S 2017 A symmetric Trefftz-DG formulation based on a local boundary element method for the solution of the Helmholtz equation *Journal of Computational Physics* **330** 1069
[15] Loeffler C F, Mansur W J, Barcelos H D M and Bulcão A 2015 Solving Helmholtz problems with the boundary element method using direct radial basis function interpolation *Engineering Analysis with Boundary Elements* **61** 218

[16] Azis M I 2017 Fundamental solutions to two types of 2D boundary value problems of anisotropic materials *Far East Journal of Mathematical Sciences* **101**(11) 2405
[17] Abramowitz M and Stegun I A 1972 *Handbook of mathematical functions: with formulas, graphs and mathematical tables*, Dover Publications, Washington

IOP Conf. Series: Materials Science and Engineering **619** (2019) 012061 doi:10.1088/1757-899X/619/1/012061

Numerical solutions to anisotropic FGM BVPs governed by the modified Helmholtz type equation

R. Syam[1], Fahruddin[3], M. I. Azis[2,*], A. Hayat[1]

[1]Department of Mechanical Engineering, Hasanuddin University, Makassar, Indonesia
[2]Department of Mathematics, Hasanuddin University, Makassar, Indonesia
[3]Department of Biology, Hasanuddin University, Makassar, Indonesia

E-mail: mohivanazis@yahoo.co.id (*Corresponding author)

Abstract. The Boundary Element Method (BEM) is used for obtaining solutions to anisotropic functionally graded media (FGM) boundary value problems (BVPs) governed by the modified Helmholtz type equation. A technique of transforming the variable coefficient governing equation to a constant coefficient equation is utilized for deriving a boundary integral equation. Some particular problems are considered to illustrate the application of the BEM. The results show the convergence, consistency, and accuracy of the BEM solutions.

1. Introduction

The BEM has been successfully used for solving many types of problems of homogeneous media. Some works on homogeneous media problems have been recently done by Azis et. al [1, 2] and Haddade et. al [3] in which the authors considered pollutant transport problems governed by 2D diffusion-convection, and also Azis [4] in which numerical solutions for the Helmholtz boundary value problems were obtained.

However, this is generally not the case for inhomogeneous media. There are two techniques usually used to deal with inhomogeneous media. The first one uses a relevant Green function or fundamental solution to the inhomogeneous problem. Cheng [5] had applied this technique. The second technique is by transforming the variable coefficient governing equation to a constant coefficient equation. Some progress on using the second technique has been made. For examples, Clements and Azis [6] considered the case for isotropic inhomogeneous media. For anisotropic inhomogeneous some works have been studied by Azis and Clements [7, 8, 9] and Azis et. al [10]. In addition to this, recently Salam et. al [11], Azis [12] and Azis [13] have been working respectively on a class of elliptic problems and diffusion convection reaction equation for anisotropic inhomogeneous media. In these works a boundary integral equation was derived after a transformation of the variable governing equation to a constant coefficient equation. The governing equation considered by Salam et. al [11] takes the form

$$\frac{\partial}{\partial x_i}\left[\lambda_{ij}\left(x_1,x_2\right)\frac{\partial \nu\left(x_1,x_2\right)}{\partial x_j}\right]=0$$

where the coefficients λ_{ij} depend on x_1 and x_2 and the repeated summation convention (summing from 1 to 2) is employed.

This paper discusses derivation of a boundary integral equation for numerically solving 2D boundary value problems governed by the dimensionless modified Helmholtz type equation of the form

$$\frac{\partial}{\partial x_i}\left[\lambda_{ij}(x_1,x_2)\frac{\partial \nu(x_1,x_2)}{\partial x_j}\right]-\beta^2(x_1,x_2)\,\nu(x_1,x_2)=0 \tag{1}$$

A variety of problems of both isotropic and anisotropic inhomogeneous media are usually modeled with equation (1). Steady infiltration problems (when $\beta^2>0$, see for example [14, 15]), and antiplane strain in elastostatics and plane thermostatic problems (when $\beta^2=0$) are the areas for which the governing equation is of the type (1).

The technique of transforming (1) to constant coefficient equations will again be used for obtaining a boundary integral equation for the solution of (1). Some constraint on the class of coefficients λ_{ij} and β will be placed for which the solution obtained is valid. The analysis of this paper is purely formal; the main aim being to construct effective BEM for class of equations which falls within the type (1).

2. The boundary value problem

Referred to a Cartesian frame Ox_1x_2 a solution to (1) is sought which is valid in a region Ω in R^2 with boundary $\partial\Omega$ which consists of a finite number of piecewise smooth closed curves. On $\partial\Omega_1$ the dependent variable $\nu(\mathbf{x})$ ($\mathbf{x}=(x_1,x_2)$) is specified and on $\partial\Omega_2$

$$P(\mathbf{x})=\lambda_{ij}\left(\partial\nu/\partial x_j\right)n_i \tag{2}$$

is specified where $\partial\Omega=\partial\Omega_1\cup\partial\Omega_2$ and $\mathbf{n}=(n_1,n_2)$ denotes the outward pointing normal to $\partial\Omega$.

For all points in Ω the matrix of coefficients $[\lambda_{ij}]$ is a real symmetric positive definite matrix so that throughout Ω equation (1) is a second order elliptic partial differential equation. Therefore equation (1) may be written explicitly as

$$\frac{\partial}{\partial x_1}\left(\lambda_{11}\frac{\partial\nu}{\partial x_1}\right)+2\frac{\partial}{\partial x_1}\left(\lambda_{12}\frac{\partial\nu}{\partial x_2}\right)+\frac{\partial}{\partial x_2}\left(\lambda_{22}\frac{\partial\nu}{\partial x_2}\right)-\beta^2\nu=0$$

Further, the coefficients λ_{ij} and β are required to be twice differentiable functions of the two independent variables x_1 and x_2.

The method of solution will be to obtain boundary integral equations from which numerical values of the dependent variables ν and P may be obtained for all points in Ω. The analysis here is specially relevant to an anisotropic medium but it equally applies to isotropic media. For isotropy, the coefficients in (1) take the form $\lambda_{11}=\lambda_{22}$ and $\lambda_{12}=0$ and use of these equations in the following analysis immediately yields the corresponding results for an isotropic medium.

3. The boundary integral equation

The boundary integral equation is derived by transforming the variable coefficient equation (1) to a constant coefficient equation. The coefficients λ_{ij} and β are required to take the form

$$\lambda_{ij}(\mathbf{x}) = \overline{\lambda}_{ij}g(\mathbf{x}) \tag{3}$$
$$\beta^2(\mathbf{x}) = \overline{\beta}^2 g(\mathbf{x}) \tag{4}$$

where the $\overline{\lambda}_{ij}$ and $\overline{\beta}$ are constants and g is a differentiable function of $\mathbf{x}$. Use of (3) and (4) and in (1) yields

$$\overline{\lambda}_{ij}\frac{\partial}{\partial x_i}\left(g\frac{\partial\nu}{\partial x_j}\right)-\overline{\beta}^2 g\nu=0 \tag{5}$$

Let

$$\nu(\mathbf{x}) = g^{-1/2}(\mathbf{x})\,\psi(\mathbf{x}) \tag{6}$$

so that (5) may be written in the form

$$\overline{\lambda}_{ij}\frac{\partial}{\partial x_i}\left[g\frac{\partial\left(g^{-1/2}\psi\right)}{\partial x_j}\right] - \overline{\beta}^2 g^{1/2}\psi = 0$$

That is

$$\overline{\lambda}_{ij}\left[\left(\frac{1}{4}g^{-3/2}\frac{\partial g}{\partial x_i}\frac{\partial g}{\partial x_j} - \frac{1}{2}g^{-1/2}\frac{\partial^2 g}{\partial x_i\partial x_j}\right)\psi + g^{1/2}\frac{\partial^2\psi}{\partial x_i\partial x_j}\right] - \overline{\beta}^2 g^{1/2}\psi = 0 \tag{7}$$

Use of the identity

$$\frac{\partial^2 g^{1/2}}{\partial x_i\partial x_j} = -\frac{1}{4}g^{-3/2}\frac{\partial g}{\partial x_i}\frac{\partial g}{\partial x_j} + \frac{1}{2}g^{-1/2}\frac{\partial^2 g}{\partial x_i\partial x_j}$$

permits (7) to be written in the form

$$g^{1/2}\overline{\lambda}_{ij}\frac{\partial^2\psi}{\partial x_i\partial x_j} - \psi\overline{\lambda}_{ij}\frac{\partial^2 g^{1/2}}{\partial x_i\partial x_j} - \overline{\beta}^2 g^{1/2}\psi = 0$$

It follows that if g is such that

$$\overline{\lambda}_{ij}\frac{\partial^2 g^{1/2}}{\partial x_i\partial x_j} = 0 \tag{8}$$

then the transformation (6) carries the variable coefficients equation (5) to the constant coefficients equation

$$\overline{\lambda}_{ij}\frac{\partial^2\psi}{\partial x_i\partial x_j} - \overline{\beta}^2\psi = 0 \tag{9}$$

And if g is such that

$$\overline{\lambda}_{ij}\frac{\partial^2 g^{1/2}}{\partial x_i\partial x_j} + \overline{\beta}^2 g^{1/2} = 0 \tag{10}$$

then

$$\overline{\lambda}_{ij}\frac{\partial^2\psi}{\partial x_i\partial x_j} = 0 \tag{11}$$

Also, substitution of (3) and (6) into (2) gives

$$P = -P_g\psi + P_\psi g^{1/2} \tag{12}$$

where

$$P_g(\mathbf{x}) = \overline{\lambda}_{ij}\frac{\partial g^{1/2}}{\partial x_j}n_i \qquad P_\psi(\mathbf{x}) = \overline{\lambda}_{ij}\frac{\partial\psi}{\partial x_j}n_i$$

A boundary integral equation for the solution of (9) and (11) is given in the form

$$\kappa(\mathbf{x}_0)\,\psi(\mathbf{x}_0) = \int_{\partial\Omega}\left[\Gamma(\mathbf{x},\mathbf{x}_0)\,\psi(\mathbf{x}) - \Phi(\mathbf{x},\mathbf{x}_0)\,P_\psi(\mathbf{x})\right]ds(\mathbf{x}) \tag{13}$$

where $\mathbf{x}_0 = (a,b)$, $\kappa = 0$ if $(a,b) \notin \Omega\cup\partial\Omega$, $\kappa = 1$ if $(a,b)\in\Omega$, $\kappa = \frac{1}{2}$ if $(a,b)\in\partial\Omega$ and $\partial\Omega$ has a continuously turning tangent at (a,b).

IOP Conf. Series: Materials Science and Engineering **619** (2019) 012061 doi:10.1088/1757-899X/619/1/012061

The so called fundamental solution Φ in (13) is any solution of the equation

$$\overline{\lambda}_{ij}\frac{\partial^2\Phi}{\partial x_i \partial x_j} - \overline{\beta}^2\Phi = \delta\left(\mathbf{x}-\mathbf{x}_0\right)$$

and the Γ is given by

$$\Gamma\left(\mathbf{x},\mathbf{x}_0\right) = \overline{\lambda}_{ij}\frac{\partial\Phi\left(\mathbf{x},\mathbf{x}_0\right)}{\partial x_j}n_i$$

where δ is the Dirac delta function. Following Azis [16], for two-dimensional problems Φ and Γ are given by

$$\begin{aligned}
\Phi\left(\mathbf{x},\mathbf{x}_0\right) &= \begin{cases} \frac{K}{2\pi}\ln R & \text{if } \overline{\beta}^2 = 0 \\ \frac{-K}{2\pi}K_0\left(\omega R\right) & \text{if } \overline{\beta}^2 > 0 \end{cases} \\
\Gamma\left(\mathbf{x},\mathbf{x}_0\right) &= \begin{cases} \frac{K}{2\pi}\frac{1}{R}\overline{\lambda}_{ij}\frac{\partial R}{\partial x_j}n_i & \text{if } \overline{\beta}^2 = 0 \\ \frac{K\omega}{2\pi}K_1\left(\omega R\right)\overline{\lambda}_{ij}\frac{\partial R}{\partial x_j}n_i & \text{if } \overline{\beta}^2 > 0 \end{cases}
\end{aligned} \tag{14}$$

where

$$\begin{aligned}
K &= \ddot{\tau}/\zeta \\
\omega &= \sqrt{|\overline{\beta}^2/\zeta|} \\
\zeta &= \left[\overline{\lambda}_{11} + 2\overline{\lambda}_{12}\dot{\tau} + \overline{\lambda}_{22}\left(\dot{\tau}^2 + \ddot{\tau}^2\right)\right]/2 \\
R &= \sqrt{(\dot{x}_1 - \dot{a})^2 + (\dot{x}_2 - \dot{b})^2} \\
\dot{x}_1 &= x_1 + \dot{\tau}x_2 \\
\dot{a} &= a + \dot{\tau}b \\
\dot{x}_2 &= \ddot{\tau}x_2 \\
\dot{b} &= \ddot{\tau}b
\end{aligned}$$

where $\dot{\tau}$ and $\ddot{\tau}$ are respectively the real and the positive imaginary parts of the complex root τ of the quadratic

$$\overline{\lambda}_{11} + 2\overline{\lambda}_{12}\tau + \overline{\lambda}_{22}\tau^2 = 0$$

and K_0, K_1 denote the modified Bessel function of order zero and order one respectively and $\imath$ represents the square root of minus one. The derivatives $\partial R/\partial x_j$ needed for the calculation of the Γ in (14) are given by

$$\begin{aligned}
\frac{\partial R}{\partial x_1} &= \frac{1}{R}\left(\dot{x}_1 - \dot{a}\right) \\
\frac{\partial R}{\partial x_2} &= \dot{\tau}\left[\frac{1}{R}\left(\dot{x}_1 - \dot{a}\right)\right] + \ddot{\tau}\left[\frac{1}{R}\left(\dot{x}_2 - \dot{b}\right)\right]
\end{aligned}$$

Use of (6) and (12) in (13) yields

$$\begin{aligned}
\kappa\left(\mathbf{x}_0\right)g^{1/2}\left(\mathbf{x}_0\right)\nu\left(\mathbf{x}_0\right) &= \int_{\partial\Omega}\left\{\left[g^{1/2}\left(\mathbf{x}\right)\Gamma\left(\mathbf{x},\mathbf{x}_0\right) - P_g\left(\mathbf{x}\right)\Phi\left(\mathbf{x},\mathbf{x}_0\right)\right]\nu\left(\mathbf{x}\right)\right. \\
&\quad \left. - \left[g^{-1/2}\left(\mathbf{x}\right)\Phi\left(\mathbf{x},\mathbf{x}_0\right)\right]P\left(\mathbf{x}\right)\right\}ds\left(\mathbf{x}\right)
\end{aligned} \tag{15}$$

This equation provides a boundary integral equation for determining ν and P at all points of Ω.

4. Numerical examples

Some particular boundary value problems will be solved numerically by employing the integral equation (15). The main aim is to show the validity of the analysis for deriving the boundary integral equation (15) and the appropriateness of the BEM in solving the problems through the derived boundary integral equation (15). Standard boundary element method is employed to obtain numerical results. The integrals in equation (15) are evaluated numerically using the Bode's quadrature (see Abramowitz and Stegun [17]).

4.1. Examples with analytical solutions

In order to see the convergence and accuracy of the BEM we will consider some examples of problems with analytical solutions. Two possible multiparameter forms of the inhomogeneity function $g(\mathbf{x})$ satisfying (8) and (10) are quadratical and trigonometrical functions respectively. For all problems considered in this section we take quadratical and trigonometrical functions $g(\mathbf{x})$ respectively as

$$g(\mathbf{x}) = \left[A\left(\alpha_0 + \alpha_1 x_1 + \alpha_2 x_2\right)\right]^2 \quad (16)$$

$$g(\mathbf{x}) = \left\{A\left[\cos\left(\alpha_1 x_1 + \alpha_2 x_2\right) + \sin\left(\alpha_1 x_1 + \alpha_2 x_2\right)\right]\right\}^2 \quad (17)$$

$$\text{where } \overline{\beta}^2 = \overline{\lambda}_{11}\alpha_1^2 + 2\overline{\lambda}_{12}\alpha_1\alpha_2 + \overline{\lambda}_{22}\alpha_2^2$$

with $A = 1.5, \alpha_0 = 1, \alpha_1 = 1, \alpha_2 = 1$. Plots of $g(\mathbf{x})$ are shown in Figures 1–2. The geometry of the region Ω and the boundary conditions are as depicted in Figure 3. The values of the constant coefficients $\overline{\lambda}_{ij}$ for the governing equation (1) are

$$\overline{\lambda}_{11} = 1, \overline{\lambda}_{12} = 1, \overline{\lambda}_{22} = 2$$

For the case when $g(\mathbf{x})$ satisfies equation (8), $\psi(\mathbf{x})$ must satisfy (9). For the analytical solution ν we will take

$$\psi(\mathbf{x}) = B\exp\left(\alpha_1 x_1 + \alpha_2 x_2\right) \quad \overline{\beta}^2 = \overline{\lambda}_{11}\gamma_1^2 + 2\overline{\lambda}_{12}\gamma_1\gamma_2 + \overline{\lambda}_{22}\gamma_2^2$$

$$\psi(\mathbf{x}) = B\left(\gamma_0 + \gamma_1 x_1 + \gamma_2 x_2\right) \quad \overline{\beta}^2 = 0$$

Whereas if $g(\mathbf{x})$ satisfies equation (10) then $\psi(\mathbf{x})$ satisfies (11). And we will take

$$\psi(\mathbf{x}) = B\left(\gamma_0 + \gamma_1 x_1 + \gamma_2 x_2\right)$$

Then the analytical solution ν can be obtained from equation (6). The parameters for ψ are taken to be

$$B = 1.5, \gamma_0 = 1, \gamma_1 = 1, \gamma_2 = 1$$

4.1.1. Quadratically graded media: $\mathbf{g(x)}$ *is of the form* (16)

Problem 4.1.1.1: $\overline{\boldsymbol{\beta}}^2 > \mathbf{0}$ *in equation* (9) We take analytical solution

$$\nu\left(\mathbf{x}\right) = \frac{B\exp\left(\gamma_1 x_1 + \gamma_2 x_2\right)}{A\left(\alpha_0 + \alpha_1 x_1 + \alpha_2 x_2\right)}$$

so that $\overline{\beta}^2 = \overline{\lambda}_{ij}\gamma_i\gamma_j = 5$. The results are shown in Table 1. The BEM solution converges to the analytical solution as the number of segments increases.

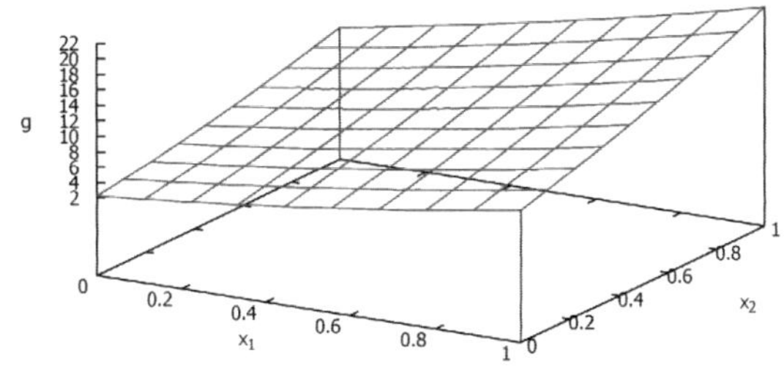

Figure 1. A quadratic inhomogeneity function $g(\mathbf{x}) = [1.5\,(1 + x_1 + x_2)]^2$

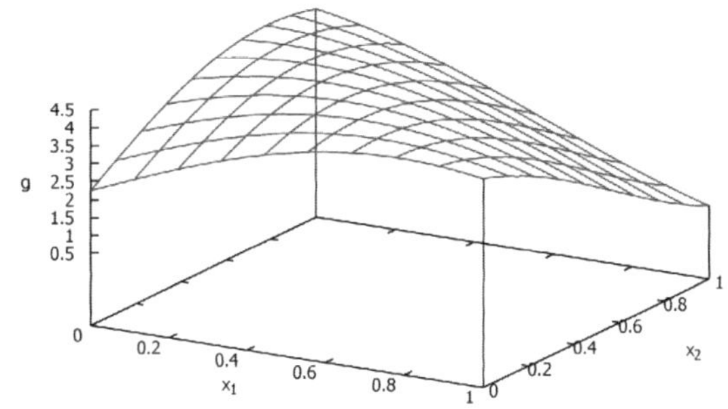

Figure 2. A trigonometrical inhomogeneity function $g(\mathbf{x}) = \{1.5\,[\cos(x_1 + x_2) + \sin(x_1 + x_2)]\}^2$

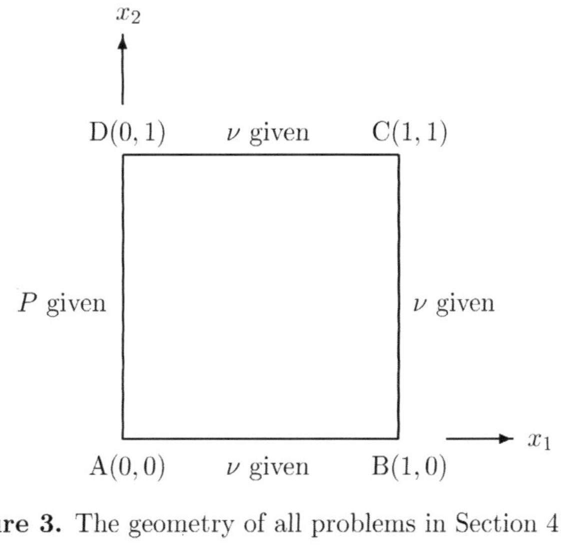

Figure 3. The geometry of all problems in Section 4.1

Table 1. BEM and analytical solutions for Problem 4.1.1.1

(x_1, x_2)	ν	$\partial\nu/\partial x_1$	$\partial\nu/\partial x_2$	ν	$\partial\nu/\partial x_1$	$\partial\nu/\partial x_2$
	BEM 160 segments			BEM 320 segments		
(.1,.5)	1.1406	.4244	.4301	1.1397	.4259	.4285
(.3,.5)	1.2379	.5488	.5533	1.2372	.5491	.5514
(.5,.5)	1.3605	.6794	.6839	1.3598	.6794	.6817
(.7,.5)	1.5105	.8231	.8275	1.5098	.8231	.8253
(.9,.5)	1.6913	.9886	.9882	1.6905	.9871	.9868
(.5,.1)	1.1379	.4282	.4345	1.1384	.4269	.4311
(.5,.3)	1.2368	.5491	.5552	1.2366	.5493	.5523
(.5,.7)	1.5113	.8238	.8265	1.5102	.8234	.8248
(.5,.9)	1.6924	.9878	.9871	1.6910	.9859	.9871
	BEM 640 segments			Analytical		
(.1,.5)	1.1393	.4264	.4278	1.1388	.4271	.4271
(.3,.5)	1.2368	.5493	.5505	1.2364	.5495	.5495
(.5,.5)	1.3595	.6796	.6806	1.3591	.6796	.6796
(.7,.5)	1.5095	.8232	.8242	1.5091	.8232	.8232
(.9,.5)	1.6901	.9863	.9863	1.6897	.9856	.9856
(.5,.1)	1.1386	.4270	.4291	1.1388	.4271	.4271
(.5,.3)	1.2365	.5495	.5509	1.2364	.5495	.5495
(.5,.7)	1.5097	.8233	.8240	1.5091	.8232	.8232
(.5,.9)	1.6904	.9860	.9863	1.6897	.9856	.9856

Problem 4.1.1.2: $\overline{\boldsymbol{\beta}}^{\mathbf{2}} = \mathbf{0}$ *in equation* (9) Now we choose analytical solution

$$\nu(\mathbf{x}) = \frac{B(\gamma_0 + \gamma_1 x_1 + \gamma_2 x_2)}{A(\alpha_0 + \alpha_1 x_1 + \alpha_2 x_2)}$$

The results are shown in Table 2. Again, the BEM solution converges to the analytical solution as the number of segments increases.

4.1.2. Trigonometrically graded media: $\boldsymbol{g}(\mathbf{x})$ *is of the form* (17)

Problem 4.1.2 The analytical solution is

$$\nu(\mathbf{x}) = \frac{B(\gamma_0 + \gamma_1 x_1 + \gamma_2 x_2)}{A[\cos(\alpha_1 x_1 + \alpha_2 x_2) + \sin(\alpha_1 x_1 + \alpha_2 x_2)]}$$

Table 3 shows the results of the analytical and BEM solutions with 160, 320 and 640 segments of equal length. The BEM solution converges to the analytical solution as the number of segments increases.

4.2. Examples without analytical solutions

In this section we will consider some examples of problems without simple analytical solutions. We setup some problems for a homogeneous isotropic material by taking $g(\mathbf{x}) = 1.5, \overline{\lambda}_{11} = \overline{\lambda}_{22} = 1, \overline{\lambda}_{12} = 0$ and with symmetrical boundary conditions. This function $g(\mathbf{x})$ satisfies equation (8) thus we will take $\psi(\mathbf{x})$ that satisfies (9) to be put into the integral equation (13). The aim is to see the consistency of the BEM of whether it produces symmetrical solutions.

Table 2. BEM and analytical solutions for Problem 4.1.1.2

(x_1, x_2)	ν	$\partial\nu/\partial x_1$	$\partial\nu/\partial x_2$	ν	$\partial\nu/\partial x_1$	$\partial\nu/\partial x_2$
	BEM 160 segments			BEM 320 segments		
(.1,.5)	1.0003	-.0012	.0019	1.0002	-.0006	.0009
(.3,.5)	1.0001	-.0006	.0020	1.0001	-.0003	.0010
(.5,.5)	1.0000	-.0001	.0017	1.0000	-.0000	.0008
(.7,.5)	1.0001	.0004	.0012	1.0000	.0002	.0006
(.9,.5)	1.0002	.0015	.0004	1.0001	.0008	.0002
(.5,.1)	.9992	.0007	.0026	.9996	.0003	.0014
(.5,.3)	.9997	.0004	.0020	.9998	.0002	.0010
(.5,.7)	1.0003	-.0003	.0013	1.0002	-.0002	.0007
(.5,.9)	1.0006	-.0003	.0010	1.0003	-.0002	.0005
	BEM 640 segments			Analytical		
(.1,.5)	1.0001	-.0003	.0005	1.0000	.0000	.0000
(.3,.5)	1.0000	-.0002	.0005	1.0000	.0000	.0000
(.5,.5)	1.0000	-.0000	.0004	1.0000	.0000	.0000
(.7,.5)	1.0000	.0001	.0003	1.0000	.0000	.0000
(.9,.5)	1.0001	.0004	.0001	1.0000	.0000	.0000
(.5,.1)	.9998	.0002	.0007	1.0000	.0000	.0000
(.5,.3)	.9999	.0001	.0005	1.0000	.0000	.0000
(.5,.7)	1.0001	-.0001	.0003	1.0000	.0000	.0000
(.5,.9)	1.0001	-.0001	.0002	1.0000	.0000	.0000

Table 3. BEM and analytical solutions for Problem 4.1.2

(x_1, x_2)	ν	$\partial\nu/\partial x_1$	$\partial\nu/\partial x_2$	ν	$\partial\nu/\partial x_1$	$\partial\nu/\partial x_2$
	BEM 160 segments			BEM 320 segments		
(.1,.5)	1.1516	.5016	.5062	1.1513	.5025	.5048
(.3,.5)	1.2731	.7250	.7285	1.2730	.7253	.7271
(.5,.5)	1.4475	1.0391	1.0418	1.4475	1.0391	1.0404
(.7,.5)	1.6998	1.5214	1.5228	1.6997	1.5209	1.5217
(.9,.5)	2.0778	2.3355	2.3330	2.0774	2.3333	2.3324
(.5,.1)	1.1502	.5055	.5062	1.1506	.5040	.5051
(.5,.3)	1.2725	.7258	.7283	1.2727	.7258	.7270
(.5,.7)	1.7003	1.5207	1.5236	1.6999	1.5206	1.5220
(.5,.9)	2.0785	2.3328	2.3346	2.0778	2.3314	2.3335
	BEM 640 segments			Analytical		
(.1,.5)	1.1512	.5031	.5042	1.1511	.5035	.5035
(.3,.5)	1.2730	.7255	.7264	1.2729	.7258	.7258
(.5,.5)	1.4474	1.0392	1.0398	1.4474	1.0392	1.0392
(.7,.5)	1.6997	1.5208	1.5212	1.6996	1.5206	1.5206
(.9,.5)	2.0773	2.3325	2.3320	2.0772	2.3315	2.3315
(.5,.1)	1.1509	.5037	.5042	1.1511	.5035	.5035
(.5,.3)	1.2728	.7258	.7264	1.2729	.7258	.7258
(.5,.7)	1.6998	1.5206	1.5213	1.6996	1.5206	1.5206
(.5,.9)	2.0775	2.3318	2.3324	2.0772	2.3315	2.3315

IOP Conf. Series: Materials Science and Engineering **619** (2019) 012061 doi:10.1088/1757-899X/619/1/012061

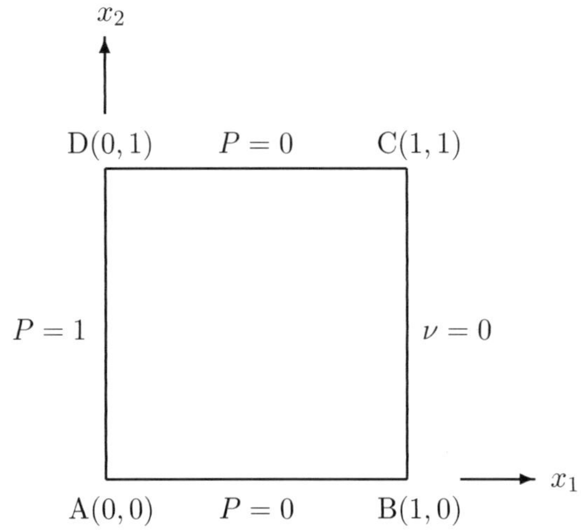

Figure 4. The geometry of Problem 4.2.1

Table 4. BEM solution for Problem 4.2.1

(x_1, x_2)	ν	$\partial\nu/\partial x_1$	$\partial\nu/\partial x_2$	ν	$\partial\nu/\partial x_1$	$\partial\nu/\partial x_2$
	BEM 80 segments			BEM 160 segments		
(.1,.5)	.2946	-.4120	-.0000	.2952	-.4124	-.0000
(.3,.5)	.2176	-.3606	-.0000	.2181	-.3611	-.0000
(.5,.5)	.1494	-.3240	-.0000	.1498	-.3244	-.0000
(.7,.5)	.0871	-.3003	-.0000	.0874	-.3007	-.0000
(.9,.5)	.0284	-.2886	-.0000	.0287	-.2891	-.0000
(.5,.1)	.1493	-.3241	.0000	.1498	-.3245	.0000
(.5,.3)	.1494	-.3240	.0001	.1498	-.3244	.0000
(.5,.7)	.1494	-.3240	-.0001	.1498	-.3244	-.0000
(.5,.9)	.1493	-.3241	-.0000	.1498	-.3245	-.0000
	BEM 320 segments			BEM 640 segments		
(.1,.5)	.2954	-.4126	.0000	.2956	-.4127	.0000
(.3,.5)	.2183	-.3613	.0000	.2184	-.3614	.0000
(.5,.5)	.1499	-.3246	.0000	.1500	-.3247	.0000
(.7,.5)	.0876	-.3009	.0000	.0876	-.3010	.0000
(.9,.5)	.0288	-.2893	.0000	.0288	-.2894	.0000
(.5,.1)	.1499	-.3246	-.0000	.1500	-.3247	-.0000
(.5,.3)	.1499	-.3246	.0000	.1500	-.3247	.0000
(.5,.7)	.1499	-.3246	-.0000	.1500	-.3247	-.0000
(.5,.9)	.1499	-.3246	.0000	.1500	-.3247	.0000

4.2.1. Problem 4.2.1: $\overline{\boldsymbol{\beta}}^2 > \mathbf{0}$ *in equation* (9) We take $\overline{\beta}^2 = 1$ and boundary conditions are as shown in Figure 4. Table 4 shows the results of the BEM solution using 80, 160, 320 and 640 segments of equal length. As expected, the results converge as the number of segments increases and also they are symmetrical about the axes $x_2 = 0.5$.

4.2.2. Problem 4.2.2: $\overline{\boldsymbol{\beta}}^2 = \mathbf{0}$ *in equation* (9) We consider a problem with $\overline{\beta}^2 = 0$ and the boundary conditions are as shown in Figure 5. Table 5 shows the results of the BEM solution

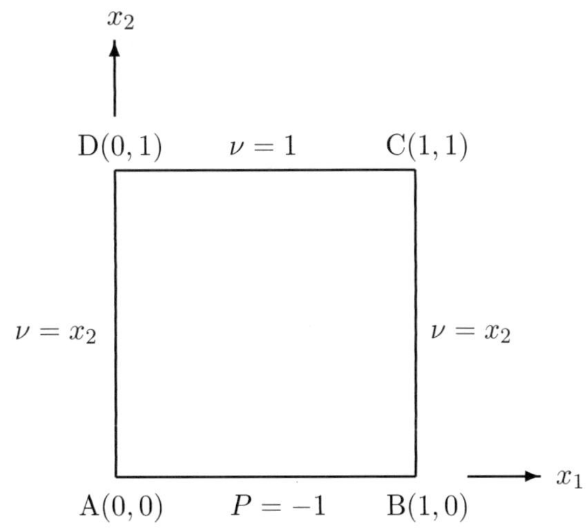

Figure 5. The geometry of Problem 4.2.2

Table 5. BEM solution for Problem 4.2.2

(x_1, x_2)	ν	$\partial\nu/\partial x_1$	$\partial\nu/\partial x_2$	ν	$\partial\nu/\partial x_1$	$\partial\nu/\partial x_2$
	BEM 80 segments			BEM 160 segments		
(.1,.5)	.5139	.1351	.9519	.5139	.1349	.9514
(.3,.5)	.5362	.0808	.8771	.5362	.0807	.8763
(.5,.5)	.5444	.0000	.8506	.5445	.0000	.8498
(.7,.5)	.5362	-.0808	.8771	.5362	-.0807	.8763
(.9,.5)	.5139	-.1351	.9519	.5139	-.1349	.9514
(.5,.1)	.2547	-.0000	.5539	.2550	-.0000	.5533
(.5,.3)	.3847	-.0000	.7357	.3849	-.0000	.7350
(.5,.7)	.7214	-.0000	.9122	.7213	.0000	.9112
(.5,.9)	.9070	-.0000	.9385	.9066	-.0000	.9373
	BEM 320 segments			BEM 640 segments		
(.1,.5)	.5140	.1349	.9512	.5140	.1348	.9510
(.3,.5)	.5362	.0806	.8759	.5362	.0806	.8757
(.5,.5)	.5445	-.0000	.8494	.5445	-.0000	.8491
(.7,.5)	.5362	-.0806	.8759	.5362	-.0806	.8757
(.9,.5)	.5140	-.1349	.9512	.5140	-.1348	.9510
(.5,.1)	.2552	-.0000	.5531	.2553	-.0000	.5530
(.5,.3)	.3850	-.0000	.7346	.3851	-.0000	.7345
(.5,.7)	.7212	-.0000	.9106	.7211	-.0000	.9104
(.5,.9)	.9064	-.0000	.9366	.9063	-.0000	.9363

using 80, 160, 320 and 640 segments of equal length. The results converge as the number of segments increases and also they are symmetrical about the axes $x_1 = 0.5$.

5. Conclusion

The modified Helmholtz type governing equation (1) is sometimes used for modeling physical problems such as steady infiltration problems (when $\beta^2 > 0$), and antiplane strain in elastostatics and plane thermostatic problems (when $\beta^2 = 0$). The boundary integral equation (15) is derived from this governing equation (1) and then from (15) a BEM is constructed for calculation of numerical solutions to the problems for anisotropic functionally graded media including quadratically and trigonometrically graded media. The results show that the BEM solution gives a convergence, consistency, and accuracy. Therefore the results also prove that the analysis used for deriving the boundary integral equation (15) is valid. Together with its ease in implementation, it may be concluded that BEM is a useful numerical method for solving such kind of problems.

Acknowledgements

The author would like to thank The Hasanuddin University and The Ministry of Research, Technology and Higher Education of the Republic of Indonesia for the provided support.

References

[1] Azis M I, Kasbawati, Haddade A and Thamrin S A 2018 On some examples of pollutant transport problems solved numerically using the boundary element method *Journal of Physics: Conference Series* **979**(1)
[2] Azis M I, Asrul L, Khaeruddin and Paharuddin 2018 BEM solutions for unsteady transport problems in anisotropic media *JP Journal of Heat and Mass Transfer* **15**(4)
[3] Haddade A, Salam N, Khaeruddin and Azis M I 2017 A Boundary Element Method for 2D Diffusion-Convection Problems in Anisotropic Media *Far East Journal of Mathematical Sciences* **102**(8)
[4] Azis M I 2019 Numerical solutions for the Helmholtz boundary value problems of anisotropic media *Journal of Computational Physics* **381** 42
[5] Cheng A H-D 1984 Darcy's Flow with Variable Permeability: A Boundary Integral Solution *Water Resources Research* **20** 980
[6] Clements D L and Azis M I 2000 A Note on a Boundary Element Method for the Numerical Solution of Boundary Value Problems in Isotropic Inhomogeneous Elasticity *Journal of the Chinese Institute of Engineers* **23**(3) 261
[7] Azis M I and Clements D L 2014 On some problems concerning deformations of functionally graded anisotropic elastic materials *Far East Journal of Mathematical Sciences* **87**(2) 173
[8] Azis M I and Clements D L 2014 A Boundary Element Method for Transient Heat Conduction Problem of Nonhomogeneous Anisotropic Materials *Far East Journal of Mathematical Sciences*, **89**(1) 51
[9] Azis M I and Clements D L 2008 Nonlinear transient heat conduction problems for a class of inhomogeneous anisotropic materials by BEM *Engineering Analysis with Boundary Elements* **32**(12) 1054
[10] Azis M I, Toaha S, Bahri M and Ilyas N 2018 A boundary element method with analytical integration for deformation of inhomogeneous elastic materials *Journal of Physics: Conference Series* **979**(1) 012072
[11] Salam N, Haddade A, Clements D L and Azis M I 2017 A boundary element method for a class of elliptic boundary value problems of functionally graded media *Engineering Analysis with Boundary Elements* **84**(3) 186
[12] Azis M I 2019 Numerical solutions to a class of scalar elliptic BVPs for anisotropic exponentially graded media *Journal of Physics: Conference Series* **1218** 012001
[13] Azis M I 2019 Standard-BEM solutions to two types of anisotropic-diffusion convection reaction equations with variable coefficients *Engineering Analysis with Boundary Elements* **105** 87
[14] Clements D L and Lobo M 2010 A BEM for time dependent infiltration from an irrigation channel *Engineering Analysis with Boundary Elements* **34** 1100
[15] Solekhudin I and Ang K-C 2012 A DRBEM with a predictor-corrector scheme for steady infiltration from periodic channels with root-water uptake *Engineering Analysis with Boundary Elements* **36** 1199
[16] Azis M I 2017 Fundamental solutions to two types of 2D boundary value problems of anisotropic materials *Far East Journal of Mathematical Sciences* **101**(11) 2405
[17] Abramowitz M and Stegun I A 1972 *Handbook of mathematical functions: with formulas, graphs and mathematical tables*, Dover Publications, Washington

AUTHOR INDEX

Abadi, Sukma 248
Abbas, H 406
Abdullah, Dharmawan 255
Agustinus, M 261
Alfaroq 142
Alie, Muhammad Zubair Muis 51, 63
Alimuddin, I 105
Amaliyah, Novriany 400
Amir, A. K. 208, 499
Amiruddin 432, 440
Amrin, Waode Ainun Apriyani 150
Anis, IR Muhammad 76
Annisa, Nurul 335
Anugrah, Muh Rifai 57
Apollo 10
Areni, Intan Sari 164, 176
Arief, Syahrier 432
Arif, Effendy 94, 239
Arman 298
Arsyad, Hairul 239, 335
Arsyad, Muhammad 69
Artini, Sitti Wahyuni 1
Aryani, D 187
Aryanto, Didik 450
Asniawaty 221
Asriani, Sherly 350
Azis, M. I. 129, 208, 480, 489, 499, 508
Azis, Nasruddin 94
Azwar, Misa Amira 1
Baeda, A.Y. 33
Baeda, Achmad Yasir 255, 261
Dermawan, D. 406
Dewiani 1
Djafar, Z. 406, 489
Djafar, Zulkifli 432
Fahruddin 292, 440, 480, 508
Farida, M 115
Firman 150
Firstiawaty, Zulfhis Shara 466
Gaffar, A. 269
Gareso, P.L. 450
Hakim, Lukmanul 335
Hamzah, S. 129, 499
Hasan, Hasnawiyah 390
Hayat, A. 508
Hayat, Azwar 400, 432
Hidayat, Alif 261
Hidayatullah, N A 187
Himran, Syukri 16
Ilyas, N. 208
Inayah, Irdyah 466
Indrabayu 164, 176
Irfan, U R 105
Ishak, Muhammad Taufik 221
Jabir, St. N. 489
Jalaluddin 76, 94, 421
Jamal, Jamal 248, 307
Jamala, Nurul 196, 350, 358
Jaya, A. 115
Jeky 57
Jorge, A. Belen 342
Juarlin, E. 450
Juswan 63
Kaikatui, Rapha Nichita 176
Ka'ka, S 234, 319
Ka'ka, Simon 16, 466
Kamil, Kusno 350
Kasbawati 150
Kasim, H. 342
Kusno, Asniawaty 350, 358
La Nafie, N. 208, 480
Lamba, Y. 342
Lewi 319, 466
Lewi, L 234
Mazmur, Anwar 313
Miyara, Akio 76
Mochtar, Andi Amijoyo 76, 414, 421, 432
Muas, M. 10
Muchsin, Abdul Hay 432
Muchtar, Muas 457
Mufti, Siti Khumaerah 164
Muhammad, A K 234, 319, 406
Muhammad, Abdul Kadir 57, 142
Mulyadi, Musrady 278
Mulyadi, Rosady 196
Muslimin, Zaenab 176
Mustafa, M.W. 269
Nagai, Isaku 367, 376, 382
Nomura, Shinfuku 284
Nugraha, Muhammad Ifan 24
Nur, Rusdi 10
Nuratisah 342
Nurwahyu, B. 489
Pakabu, Monika 450
Pala'biran, Merandy 176
Palantei, Elyas 1, 176
Pandey, Shyam. S. 400
Paotonan, Chairul 33, 63, 261
Pao'tonan, Chairul 255
Pasalli, I B 105
Patabang, Daud 94
Putera, Andi Cahya Adi 142
Putra, A. Erwin E. 400
Rachman, Taufiqur 63, 261
Radja, Abdul Mufti 432, 440
Rahim, Ramli 196
Rahimuddin 390
Rahman, A 234, 319
Rahman, Abdul 142
Rahman, Sabaruddin 33, 261

AUTHOR INDEX

Ramadhan, Muhammad Iqra 51
Rasyid, Syaharuddin..239, 457
Rauf, N. 450
Raya, I. 342
Renreng, Ilyas...16, 239
Risal, Syah 10
Rivai, Haryanti 390
Rizkiani, Tsumiratin...51
Saini, M. 269
Sakata, Takanori...382
Sampelawang, Petrus..39
Sato, T 115
Sirande, Normalia...39
Sofyan, Bondan T. 330
Sufriadin 122
Sule, Luther 421
Sulistyowati, Dwi Retno...221
Sultan, A.R. 269
Suluh, Sallolo 39
Susanto, Tri Agus..457
Sutresman, Onny 16
Sutresman, Onny S 432
Sutresman, Onny S. 432, 440
Suwasti, Sri 248
Suyuti, Ansar 176
Syahid, Muhammad...239, 330
Syam, R. 406, 508
Syam, Rafiuddin....... 24, 57, 82, 142, 414, 432, 440, 466
Syamsuddin, E. 129
Syarif, Syafruddin 176
Syarifuddin 313
Takebe, Hiromichi 298
Takeuchi, Ryuta 367
Tandioga, R 234, 319
Tandioga, Remigius 24
Tarakka, Rustan 76
Taufik, A 234, 319
Taufik, Akhmad 24, 57
Tonggiroh, Adi 82, 440
Trianto, M 122
Ulandari, Sri Hasnawati 466
Umar, Hasdinar 33, 261
Utomo, Ardiansyah Tri 24
Wang, L 187
Watanabe, Keigo 367, 376, 382
Watanabe, Tomoya 376
Widodo, S. 122
Wika, Annajma Nurul 358
Yunus, A. M. Shiddiq 43, 278
Zakir, M. 342